HANDBOOK OF MARINE MICROALGAE

HANDBOOK OF MARINE MICROALGAE

BIOTECHNOLOGY ADVANCES

Edited by

SE-KWON KIM
Pukyong National University, Busan, South Korea

AMSTERDAM • BOSTON • HEIDELBERG • LONDON
NEW YORK • OXFORD • PARIS • SAN DIEGO
SAN FRANCISCO • SINGAPORE • SYDNEY • TOKYO
Academic Press is an imprint of Elsevier

Academic Press is an imprint of Elsevier
125 London Wall, London EC2Y 5AS, UK
525 B Street, Suite 1800, San Diego, CA 92101-4495, USA
225 Wyman Street, Waltham, MA 02451, USA
The Boulevard, Langford Lane, Kidlington, Oxford OX5 1GB, UK

ISBN: 978-0-12-800776-1

British Library Cataloguing in Publication Data
A catalogue record for this book is available from the British Library

Library of Congress Catalog Number
A catalog record for this book is available from the Library of Congress

For information on all Academic Press publications
visit our website at http://store.elsevier.com/

Working together
to grow libraries in
developing countries

www.elsevier.com • www.bookaid.org

Publisher: Janice Audet
Acquisition Editor: Kristi A.S. Gomez
Editorial Project Manager: Pat Gonzalez
Production Project Manager: Caroline Johnson
Designer: Matthew Limbert

Typeset by TNQ Books and Journals
www.tnq.co.in

Contents

13. Medicinal Effects of Microalgae-Derived Fatty Acids

LUÍSA BARREIRA, HUGO PEREIRA, KATKAM N. GANGADHAR, LUÍSA CUSTÓDIO, AND JOÃO VARELA

14. Innovative Microalgae Pigments as Functional Ingredients in Nutrition

EFTERPI CHRISTAKI, ELEFTHERIOS BONOS, AND PANAGIOTA FLOROU-PANERI

15. Application of Diatom Biosilica in Drug Delivery

JAYACHANDRAN VENKATESAN, BABOUCARR LOWE, AND SE-KWON KIM

16. Microalgal Nutraceuticals

J. PANIAGUA-MICHEL

17. Microalgae as a Novel Source of Antioxidants for Nutritional Applications

KOEN GOIRIS, KOENRAAD MUYLAERT, AND LUC DE COOMAN

18. Production of Biopharmaceuticals in Microalgae

BERNARDO BAÑUELOS-HERNÁNDEZ, JOSUÉ I. BELTRÁN-LÓPEZ, AND SERGIO ROSALES-MENDOZA

19. Nutritional and Pharmaceutical Properties of Microalgal *Spirulina*

THANH-SANG VO, DAI-HUNG NGO, AND SE-KWON KIM

Contributors

Cynthia Alcántara Valladolid University, Department of Chemical Engineering and Environmental Technology, Valladolid, Spain

Sarmidi Amin Agency for Assessment and Application Technology (BPPT), Indonesia

Cristiano V.M. Araújo Institute of Marine Sciences of Andalusia (CSIC), Research group of Ecotoxicology, Ecophysiology and Biodiversity of Aquatic Systems, Excellence International Campus of the Seas (CEIMAR), Puerto Real, Cádiz, Spain; Universidad Laica Eloy Alfaro de Manabí (ULEAM), Central Department of Research (DCI), Manta, Ecuador

Ali Bahadar Department of Chemical and Materials Engineering, King Abdulaziz University, Rabigh, Saudi Arabia; School of Chemical and Materials Engineering (SCME), National University of Sciences and Technology (NUST), Islamabad, Pakistan

Lieselot Balduyck KU Leuven Kulak, Research Unit Food and Lipids, Kortrijk, Belgium; KU Leuven, Leuven Food Science and Nutrition Research Centre (LFoRCe), Leuven, Belgium

Bernardo Bañuelos-Hernández Universidad Autónoma de San Luis Potosí, Laboratorio de Biofarmacéuticos Recombinantes, Facultad de Ciencias Químicas, San Luis Potosí, Mexico

Luísa Barreira University of Algarve, CCMAR, Centre of Marine Sciences, Faro, Portugal

Josué I. Beltrán-López Universidad Autónoma de San Luis Potosí, Laboratorio de Biofarmacéuticos Recombinantes, Facultad de Ciencias Químicas, San Luis Potosí, Mexico

Michael J. Betenbaugh Johns Hopkins University, Department of Chemical & Biomolecular Engineering, Baltimore, MD, USA

Arghya Bhattacharya Applied Microbiology Laboratory, Centre for Rural Development and Technology, Indian Institute of Technology (IIT) Delhi, Hauz Khas, New Delhi, India

Eleftherios Bonos Laboratory of Nutrition, School of Veterinary Medicine, Faculty of Health Sciences, Aristotle University of Thessaloniki, Thessaloniki, Greece

Charlotte Bruneel KU Leuven Kulak, Research Unit Food and Lipids, Kortrijk, Belgium; KU Leuven, Leuven Food Science and Nutrition Research Centre (LFoRCe), Leuven, Belgium

Laura Bulgariu "Gheorghe Asachi" Technical University of Iaşi, Faculty of Chemical Engineering and Environmental Protection, Department of Environmental Engineering and Management, Iaşi, Romania

Faizal Bux Institute for Water and Wastewater Technology, Durban University of Technology, Durban, South Africa

Poonam Choudhary Applied Microbiology Laboratory, Centre for Rural Development and Technology, Indian Institute of Technology (IIT) Delhi, Hauz Khas, New Delhi, India

Efterpi Christaki Laboratory of Nutrition, School of Veterinary Medicine, Faculty of Health Sciences, Aristotle University of Thessaloniki, Thessaloniki, Greece

Samuel Cirés James Cook University, College of Marine and Environmental Sciences, Townsville, QLD, Australia; James Cook University, Centre for Sustainable Fisheries and Aquaculture, Townsville, QLD, Australia; James Cook University, Comparative Genomics Centre, Townsville, QLD, Australia

Luísa Custódio University of Algarve, CCMAR, Centre of Marine Sciences, Faro, Portugal

Cecilia Faraloni CNR, Istituto per lo Studio degli Ecosistemi Sezione di Firenze, Firenze, ITALY

Panagiota Florou-Paneri Laboratory of Nutrition, School of Veterinary Medicine, Faculty of Health Sciences, Aristotle University of Thessaloniki, Thessaloniki, Greece

Imogen Foubert KU Leuven Kulak, Research Unit Food and Lipids, Kortrijk, Belgium; KU Leuven, Leuven Food Science and Nutrition Research Centre (LFoRCe), Leuven, Belgium

Katkam N. Gangadhar University of Algarve, CCMAR, Centre of Marine Sciences, Faro, Portugal; New University of Lisbon, Institute of Chemical and Biological Technology, Lisbon, Portugal

Maria Gavrilescu "Gheorghe Asachi" Technical University of Iaşi, Faculty of Chemical Engineering and Environmental Protection, Department of Environmental Engineering and Management, Iaşi, Romania; Academy of Romanian Scientists, Bucharest, Romania

Koen Goiris KU Leuven Technology Campus Ghent, Laboratory of Enzyme, Fermentation and Brewing Technology, Ghent, Belgium

A. Catarina Guedes Interdisciplinary Centre of Marine and Environmental Research (CIIMAR/CIMAR), University of Porto, Porto, Portugal

Benoit Guieysse Massey University, School of Engineering and Advanced Technology, Palmerston North, New Zealand

Abhishek Guldhe Institute for Water and Wastewater Technology, Durban University of Technology, Durban, South Africa

Sanjay Kumar Gupta Institute for Water and Wastewater Technology, Durban University of Technology, Durban, South Africa

Bernardo J. Guzman Johns Hopkins University, Department of Chemical & Biomolecular Engineering, Baltimore, MD, USA

Kirsten Heimann James Cook University, College of Marine and Environmental Sciences, Townsville, QLD, Australia; James Cook University, Centre for Sustainable Fisheries and Aquaculture, Townsville, QLD, Australia; James Cook University, Comparative Genomics Centre, Townsville, QLD, Australia

S.W.A. Himaya The Institute for Molecular Bioscience, The University of Queensland, St Lucia, QLD, Australia

Seong-Joo Hong Marine Bioenergy Research Center, Department of Biological Engineering, Inha University, Korea

Roger Huerlimann James Cook University, College of Marine and Environmental Sciences, Townsville, QLD, Australia; James Cook University, Centre for Sustainable Fisheries and Aquaculture, Townsville, QLD, Australia

Pavan P. Jutur DBT-ICGEB Centre for Advanced Bioenergy Research, New Delhi, India

Kyong-Hwa Kang Department of Marine-Bio. Convergence Science and Marine Bioprocess Research Center, Pukyong National University, Busan, Republic of Korea

Prachi Kaushik Applied Microbiology Laboratory, Centre for Rural Development and Technology, Indian Institute of Technology (IIT) Delhi, Hauz Khas, New Delhi, India

M. Bilal Khan Center for Advanced Studies in Energy (CAE), National University of Sciences and Technology (NUST), Islamabad, Pakistan

M.A. Asim K. Jalwana School of Electrical Engineering and Computer Sciences (SEECS), National University of Sciences and Technology (NUST), Islamabad, Pakistan

Se-Kwon Kim Marine Bioprocess Research Center, Pukyong National University, Busan, Republic of Korea; Specialized Graduate School Science and Technology Convergence, Pukyong National University, Department of Marine-Bio Convergence Science, Busan, Republic of Korea

Ayse Kose Ege University, Engineering Faculty, Department of Bioengineering, Izmir, Turkey

Man Kee Lam Chemical Engineering Department, Universiti Teknologi PETRONAS, Bandar Seri Iskandar, Perak, Malaysia

Choul-Gyun Lee Marine Bioenergy Research Center, Department of Biological Engineering, Inha University, Korea

Keat Teong Lee Low Carbon Economy (LCE) Research Group, School of Chemical Engineering, Universiti Sains Malaysia, Nibong Tebal, Pulau Pinang, Malaysia

Yuan Kun Lee National University of Singapore, Yong Loo Lin School of Medicine, Department of Microbiology, Singapore

Baboucarr Lowe Pukyong National University, Department of Marine-Bio Convergence Science, Busan, South Korea

F. Xavier Malcata University of Porto, Department of Chemical Engineering, Porto, Portugal; LEPABE—Laboratory of Process Engineering, Environment, Biotechnology and Energy, Porto, Portugal

Anushree Malik Applied Microbiology Laboratory, Centre for Rural Development and Technology, Indian Institute of Technology (IIT) Delhi, Hauz Khas, New Delhi, India

Panchanathan Manivasagan Pukyong National University, Department of Marine-Bio Convergence Science and Marine Bioprocess Research Center, Busan, South Korea

Ignacio Moreno-Garrido Institute of Marine Sciences of Andalusia (CSIC), Research group of Ecotoxicology, Ecophysiology and Biodiversity of Aquatic Systems, Excellence International Campus of the Seas (CEIMAR), Puerto Real, Cádiz, Spain

Raúl Muñoz Valladolid University, Department of Chemical Engineering and Environmental Technology, Valladolid, Spain

Koenraad Muylaert KU Leuven Kulak, Research Unit Aquatic Biology, Kortrijk, Belgium

Asha A. Nesamma DBT-ICGEB Centre for Advanced Bioenergy Research, New Delhi, India

Daphne H.P. Ng National University of Singapore, Yong Loo Lin School of Medicine, Department of Microbiology, Singapore

Yi Kai Ng National University of Singapore, Yong Loo Lin School of Medicine, Department of Microbiology, Singapore

Dai-Hung Ngo Marine Bioprocess Research Center, Pukyong National University, Busan, Republic of Korea

Victor H. Oh Johns Hopkins University, Department of Chemical & Biomolecular Engineering, Baltimore, MD, USA

Jorge Olmos Soto Molecular Microbiology Laboratory, Department of Marine Biotechnology, Centro de Investigación Científica y de Educación Superior de Ensenada (CICESE), Ensenada, B.C., México

Suphi S. Oncel Ege University, Engineering Faculty, Department of Bioengineering, Izmir, Turkey

George A. Oyler Johns Hopkins University, Department of Chemical & Biomolecular Engineering, Baltimore, MD, USA; Synaptic Research, Baltimore, MD, USA

J. Paniagua-Michel Laboratory for Bioactive Compounds and Bioremediation, Department of Marine Biotechnology, Centro de Investigación Científica y de Educación Superior de Ensenada (CICESE), Ensenada, BC, México

Hugo Pereira University of Algarve, CCMAR, Centre of Marine Sciences, Faro, Portugal

José C.M. Pires Universidade do Porto, LEPABE, Departamento de Engenharia Química, Faculdade de Engenharia, Porto, Portugal

Esther Posadas Valladolid University, Department of Chemical Engineering and Environmental Technology, Valladolid, Spain

Kurniadhi Prabandono Ministry for Marine affairs and Fisheries—Republic of Indonesia

Sanjeev K. Prajapati Applied Microbiology Laboratory, Centre for Rural Development and Technology, Indian Institute of Technology (IIT) Delhi, Hauz Khas, New Delhi, India

P.T. Pratheesh School of Biosciences, Mahatma Gandhi University, Kottayam, Kerala, India

Zhong-Ji Qian College of Food Science and Technology, Guangdong Ocean University, Zhanjiang, PR China

Ismail Rawat Institute for Water and Wastewater Technology, Durban University of Technology, Durban, South Africa

Sergio Rosales-Mendoza Universidad Autónoma de San Luis Potosí, Laboratorio de Biofarmacéuticos Recombinantes, Facultad de Ciencias Químicas, San Luis Potosí, Mexico

Julian N. Rosenberg Johns Hopkins University, Department of Chemical & Biomolecular Engineering, Baltimore, MD, USA; Synaptic Research, Baltimore, MD, USA

BoMi Ryu School of Pharmacy, The University of Queensland, Brisbane, QLD, Australia

Rakesh Chandra Saxena Indian Institute of Petroleum, Dehradun, India

Kashif M. Shaikh DBT-ICGEB Centre for Advanced Bioenergy Research, New Delhi, India

Hui Shen National University of Singapore, Yong Loo Lin School of Medicine, Department of Microbiology, Singapore

Jasvinder Singh Indian Institute of Petroleum, Dehradun, India

Poonam Singh Institute for Water and Wastewater Technology, Durban University of Technology, Durban, South Africa

Ganapathy Sivakumar Arkansas State University, Arkansas Biosciences Institute and College of Agriculture and Technology, Jonesboro, AR, USA

Isabel Sousa-Pinto Interdisciplinary Centre of Marine and Environmental Research (CIIMAR/CIMAR), University of Porto, Porto, Portugal; University of Porto, Department of Biology, Faculty of Sciences, Porto, Portugal

Leanne Sparrow James Cook University, College of Marine and Environmental Sciences, Townsville, QLD, Australia; James Cook University, Centre for Sustainable Fisheries and Aquaculture, Townsville, QLD, Australia

Keat H. Teoh Arkansas State University, Arkansas Biosciences Institute and College of Agriculture and Technology, Jonesboro, AR, USA

Giuseppe Torzillo CNR-Istituto per lo Studio degli Ecosistemi Sezione di Firenze, Firenze, Italy

Fazilet Vardar Ege University, Engineering Faculty, Department of Bioengineering, Izmir, Turkey

João Varela University of Algarve, CCMAR, Centre of Marine Sciences, Faro, Portugal

Jayachandran Venkatesan Pukyong National University, Department of Marine-Bio Convergence Science and Marine Bioprocess Research Center, Busan, South Korea

M. Vineetha Department of Microbiology, Government Arts and Science College, Kozhinjampara, Palakkad, Kerala, India

Thanh-Sang Vo Marine Bioprocess Research Center, Pukyong National University, Busan, Republic of Korea

Geng Yu Johns Hopkins University, Department of Chemical & Biomolecular Engineering, Baltimore, MD, USA

Luc De Cooman KU Leuven Technology Campus Ghent, Laboratory of Enzyme, Fermentation and Brewing Technology, Ghent, Belgium

Preface

The marine system has a huge amount of unexplored animals, plants, and microorganisms of great interest to researchers. Among these, the marine algae have been extensively studied for various biological and biomedical applications. Marine microalgae are typically found in the marine system and are explored for industrial applications. Microalgae have a capacity to produce polymers, toxins, fatty acids, enzymes, which can be useful for pharmaceutical, nutraceutical, and cosmeceutical developments.

The present book *Handbook of Marine Microalgae: Biotechnology Advances* describes the characteristic features of marine microalgae substances, their source, isolation, cultivation, production, and applications (biological, biomedical, and industrial applications).

Chapter 1 of this book provides the general introduction to the topics covered in the present book and deals with marine microalgae technology in all aspects.

Chapters 2−6 describe the diversity, significance, classification, isolation, and mass production of microalgae.

Chapters 7−12 explain bioenergy, the historical background use of microalgae, and biofuel (biohydrogen and bioethanol) production from microalgae.

Chapters 13−19 deal with microalgae usage in cosmeceuticals, nutraceuticals, and pharmaceuticals in detail.

Chapters 22−27 explain the use of genetic engineering in marine microalgae to optimize bioenergy production.

Chapters 28−31 investigate the usage of marine microalgae in water treatment and bioremediation.

Chapters 34−36 deal with toxins in microalgae and their usage.

I express my sincere thanks to all the authors who have provided the state-of-the-art contributions included in this book. I am also thankful to Academic Press publishers for their continual support, which is essential for the successful completion of the present work.

I hope that the fundamental as well as applied contributions in this book might serve as potential research and development leads for the benefit of humankind. Microalgae biotechnology will be an important field in the future aimed at the enrichment of targeted algal species, which further establishes a sustainable oceanic environment. The current book is intended to be a handbook for emerging students and experts in the field of biotechnology.

Prof. Se-Kwon Kim
Pukyong National University
Busan, South Korea

Acknowledgments

I would like to thank the Publisher, Elsevier, for their continuous encouragement and suggestions to get this wonderful compilation published. I would also like to extend my sincere gratitude to all the contributors for providing their help, support, and advice to accomplish this task. Further, I would like to thank Dr. Jayachandran Venkatesan, Dr. Panchanathan Manivasagan, and Dr. S.W.A. Himaya, who worked with me throughout the course of this book project. I strongly recommend this book for marine algae researchers/students/industrialists, and hope that it helps to enhance their understanding in this field.

Prof Se-Kwon Kim
Pukyong National University
Busan, South Korea

Marine Microalgae Biotechnology: Present Trends and Future Advances

Jayachandran Venkatesan, Panchanathan Manivasagan, Se-Kwon Kim

Pukyong National University, Department of Marine-Bio Convergence Science and Marine Bioprocess Research Center, Busan, South Korea

1. INTRODUCTION

Microalgae are unicellular species, commonly found in marine and freshwater with the size ranging from a few micrometers to a few hundreds of micrometers. It has been estimated that 2×10^5 to 8×10^5 species exist. Microalgae are a promising source for several bioactive compounds (Cardozo et al., 2007; Norton et al., 1996). Polymers, peptides, fatty acids, carotenoids, toxins, and sterols are important bioactive products produced by microalgae. Microalgae do not contain stems and roots as do higher plants. The three most important classes of microalgae in terms of abundance are the diatoms (*Bacillariophyceae*), the green algae (*Chlorophyceae*), and the golden algae (*Chrysophyceae*). The cyanobacteria or blue-green algae (*Cyanophyceae*) are also referred to as microalgae, i.e., *Spirulina* (*Arthrospira platensis* and *Arthrospira maxima*).

2. ISOLATION AND CULTURE

Isolation, production, and culture of microalgae are important steps to producing a commercial product in sufficient amounts. Microalgae growths are dependent on light, water temperature, nutrient concentration, salinity, and pH (Brennan and Owende, 2010; Mutanda et al., 2011). Collection, sampling, and preservation techniques must be optimized to grow a high-quality microalgae sample. Three important isolation techniques are commonly used for microalgae collection:

1. Streaking
2. Serial dilution
3. Single-cell isolation (He et al., 2012).

Different kinds of culture techniques are available to culture microalgae; culture media and culture conditions are the main aspects to be considered in microalgae cultivation.

Several important nutrients (nitrate, urea, ammonium, vitamins, phosphorous, nitrogen, iron, manganese, selenium, cobalt, nickel, and zinc) are required for the production of any microalgae species (Andersen, 2005; Harrison and Berges, 2005).

Microalgae can be cultured mainly in two different kinds of systems:

1. Open cultivation
2. Photobioreactor.

The traditional open cultivation method (lakes and ponds) has been used since 1950. The photobioreactor system has more advantages than open cultivation, avoiding several issues such as contamination and environmental disturbance. There are three distinct algae production mechanisms: photoautotrophic, heterotrophic, and mixotrophic.

3. APPLICATIONS OF MARINE MICROALGAE

Microalgae have been widely used for various applications including human and animal nutrition, cosmetics, pharmaceuticals, CO_2 capture, bioenergy production, and nutrient removal from wastewater. Biological properties of algae and their components are well studied in the following areas of research: antioxidants, antimicrobials, anticancer agents, anti-inflammatory and cardiovascular health, anti-obesity, and anti-diabetic activity (Dominguez, 2013).

3.1 Animal Feed

Microalgae have been investigated as human and animal foods for over six decades (Liu et al., 2014). Although several hundreds of microalgae species have been investigated for food applications, only a few have been used in aquaculture. The most common species are *Chlorella*, *Tetraselmis*, *Isochrysis*, *Pavlova*, *Phaeodactylum*, *Chaetoceros*, *Nannochloropsis*, *Skeletonema*, and *Thalassiosira* genera (Brown et al., 1996; Spolaore et al., 2006). Marine algae have been used as food additives, for bivalve mollusks (oysters, scallops, clams, and mussels) and fishmeal. There is a promising work in the area of using microalgae as a food additive to increase weight, oil content, and protein deposition in muscle. The nutritional composition of microalgae is made up of carbohydrates, proteins, vitamins, lipids, antioxidants, and other trace elements. Kang et al. (2012) reported the nutritional composition of microalgae (*Navicular incerta*): crude lipids—8.76%, crude proteins—50.38%, and carbohydrates—10.84%, respectively. A general overview of microalgae in food usage is explained in Figure 1.

3.2 Fatty Acids

Microalgae are traditionally considered good sources of fatty acids (Benemann, 1989; Borowitzka, 2013). The accumulation of fatty acids by microalgae is well developed and presented elsewhere (Griffiths and Harrison, 2009; Rodolfi et al., 2009). The presence of eicosapentaenoic acid (C20:5) and docosahexaenoic acid (C22:6) is of interest for their health benefits (Harris et al., 2008; Mozaffarian and Rimm, 2006). Cyanobacterium *Spirulina* is rich in linolenic acid and a good source for polyunsaturated fatty acid (PUFA) (Mahajan and Kamat, 1995).

3.3 Nutraceuticals

Microalgae are particularly of interest as a source of nutraceuticals because algae can produce a number of biomolecules, viz, beta-carotene, lutein, astaxanthin, chlorophyll, phycobiliprotein, and PUFAs, which are useful for human and animal health and development. Marine microalgae pigments are carotenoids, chlorophylls, and phycobiliproteins, which have

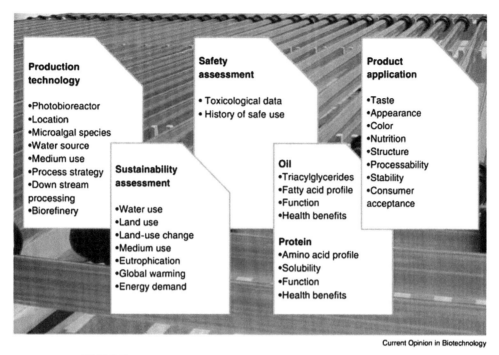

Production technology

•Photobioreactor
•Location
•Microalgal species
•Water source
•Medium use
•Process strategy
•Down stream processing
•Biorefinery

Sustainability assessment

•Water use
•Land use
•Land-use change
•Medium use
•Eutrophication
•Global warming
•Energy demand

Safety assessment

• Toxicological data
• History of safe use

Oil
•Triacylglycerides
•Fatty acid profile
•Function
•Health benefits

Protein
•Amino acid profile
•Solubility
•Function
•Health benefits

Product application

•Taste
•Appearance
•Color
•Nutrition
•Structure
•Processability
•Stability
•Consumer acceptance

Current Opinion in Biotechnology

FIGURE 1   Food commodities from microalgae (Draaisma et al., 2013).

health-promoting properties such as vitamin precursors, antioxidants, immune enhancers, and anti-inflammatory agents. Accordingly, microalgae pigments can find commercial applications as innovative functional ingredients in the food and feed industries, as well as in pharmaceuticals and in cosmetics (Borowitzka, 2013). Microalgae that have become more prevalent in food supplements and nutraceuticals are *Nostoc, Botryococcus, Anabaena, Chlamydomonas, Scenedesmus, Synechococcus, Perietochloris,* and *Porphyridium* because they contain vitamins and essential elements such as potassium, zinc, iodine, selenium, iron, manganese, copper, phosphorus, sodium, nitrogen, magnesium, cobalt, molybdenum, sulfur, and calcium. Algae are also high producers of essential amino acids and omega 6 (arachidonic acid) and omega 3 (docosahexaenoic acid, eicosapentaenoic acid) fatty acids (Simoons, 1990; West and Zubeck, 2012). Average nutritional compositions of the microalgae expressed as g per 100 g dry weight are presented in Table 1.

3.4 Cosmeceuticals

Microalgae-derived skin care products have been developed in the form of anti-aging creams, refreshing or regenerating care products, emollients and anti-irritants in peelers, sunscreen creams, and hair care products. These cosmetic products have been developed with marine microalgae extracts or bioactive components. By the early 2000s, numerous cosmetic companies in Europe and the United States started to launch cosmetics that used extractions of microalgae such as *Spirulina, Chlorella, Arthrospira, Anacystis, Halymenia, Nannochloro,* and *Dunaliella* which act on the epidermis to erase vascular imperfections by boosting collagen synthesis and thus possibly prevent wrinkle formation (Stolz and

TABLE 1   Average Nutritional Composition of the Microalgae Expressed as Grams per 100 g Dry Weight (West and Zubeck, 2012)

Component	Spirulina	Dunaliella	Haematococcus	Chlorella	Aphanizomenon
Protein	63	7.4	23.6	64.5	1
Fat	4.3	7	13.8	10	3
Carbohydrates	17.8	29.7	38	15	23
Chlorophyll	1.15	2.2	0.4 (red), 1.1 (green)	5	1.8
Magnesium	0.319	4.59	1.14	0.264	0.2
Beta-carotene	0.12	1.6	0.054	0.086	0.42
Vitamin B1 (thiamin)	0.001	0.0009	0.00047	0.0023	0.004
Vitamin B2 (riboflavin)	0.0045	0.0009	0.0017	0.005	0.0006
Vitamin B3 (niacin)	0.0149	0.001	0.0066	0.025	0.013
Vitamin B5 (pantothenic acid)	0.0013	0.0005	0.0014	0.0019	0.0008
Vitamin B6 (pyridoxine)	0.00096	0.0004	0.00036	0.0025	0.0013
Vitamin B9 (folic acid)	0.000027	0.00004	0.00029	0.0006	0.0001
Vitamin B12 (cobalamine)	0.00016	0.000004	0.00012	0.000008	0.0006

Obermayer, 2005). Mycosporine-like amino acids from *Spirulina*, *Chlorella*, and *Dunaliella* are known to act as sunscreens to reduce ultraviolet (UV)-induced damage (Atkin et al., 2006; Balskus and Walsh, 2010; Dionisio-Se Se, 2010; Garciapichel et al., 1993; Priyadarshani and Rath, 2012). Carotenoids such as astaxanthin, lutein, zeaxanthin, and canthaxanthin found in *Haematococcus* and *Dunaliella* protect against sun damage and also are reported to have antioxidant activity (Gierhart and Fox, 2013; Guerin et al., 2003; Tominaga et al., 2012; Walker et al., 2005). Beta-carotene and astaxanthin are important microalgal products and are commonly used in cosmetic applications (as protection against oxidation of essential PUFAs, protection against UV light effects) (Lorenz and Cysewski, 2000).

3.5 Pharmaceuticals

Marine microalgae are rich in biologically active compounds, which can be used for pharmaceutical and nutraceutical development. Microalgae have a capacity to produce toxins that can be used for pharmaceutical applications. Cyanobacteria are known to produce extracellular and intracellular metabolites, which possess antifungal, antibacterial, and antiviral properties (Baquero et al., 2008; Noaman et al., 2004; Thillairajasekar et al., 2009). Figure 2 shows the important compounds of blue green alga Lyngbya majuscula which can be used for drug development.

1. Cytotoxic activity is important in anticancer drugs (Sirenko and Kirpenko, 1999; Simmons et al., 2005)

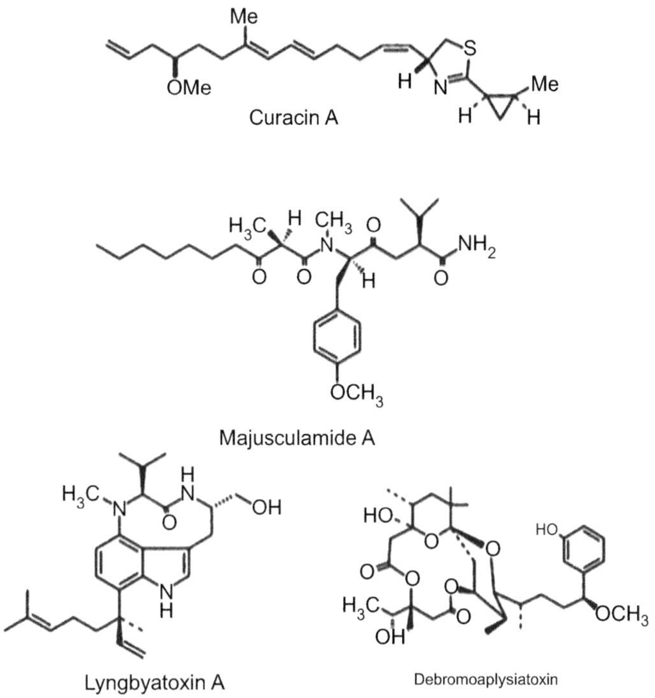

FIGURE 2 Variant structural types found in an identical species of the blue-green alga *Lyngbya majuscula*.

2. Antiviral activities are found mainly in cyanobacteria but also in apochlorotic diatoms and the conjugaphyte *Spirogyra*, where certain sulfolipids are active, for example, against the herpes simplex virus (Muller-Feuga et al., 2003)

3. Antimicrobial activity is under investigation to find new antibiotics, although currently the success rate is about 1% (Muller-Feuga et al., 2003)

4. Antifungal activity is found in different extracts of cyanobacteria (Nagai et al., 1992)

5. Neuroprotective agents from *Spirulina* (Chamorro et al., 2006; McCarty, 2007; Nuhu, 2013)

6. Antioxidants (Miranda et al., 1998)

7. Anti-inflammatory activity (Jin et al., 2006)

8. Cardiovascular health effects (Doughman et al., 2007)

9. Gastric and hepatic protective effects (Abdel-Wahhab, Ahmed & Hagazi, 2006)

10. Antidiabetic and antiobesity properties (Mayer and Hamann, 2005)

11. Antiviral activity (Hayashi et al., 2008; Hernández-Corona et al., 2002; Shimizu, 1996)

12. Asthma (Senevirathne and Kim, 2011).

Diatoms are a kind of microalgae and consist of biosilica; they are used for drug delivery due to their pore size and drug-holding capacity. Isolation, culture and characterization methods of diatom are important to use in commercial applications (drug delivery applications). Diatom biosilica have been used to load several anti-inflammatory drugs and release them at a sustainable rate (Dolatabadi and de la Guardia, 2011; Gordon et al., 2009; Losic et al., 2010; Nassif and Livage, 2011).

3.6 Biofuels

Algae have been widely used for fuel production because of their high photosynthetic efficiency, high biomass production, and fast growth (Miao et al., 2004). Microalgae contain proteins, carbohydrates, and lipids; the lipids can be converted into biodiesel, carbohydrates into ethanol and H_2, and proteins into the raw material of biofertilizer (Raja et al., 2013). Biofuel from microalgae can be processed by using thermochemical and biochemical conversion. The thermochemical process can be divided in to gasification, liquefaction, pyrolysis, and direct combustion; meanwhile, the biochemical process can be divided into anaerobic digestion, fermentation, and photobiological activity. By using a gasification process, the biomass produces CH_2, H_2, CO_2, and ammonia.

3.6.1 Bioethanol

Bioethanol is an alternative biofuel to gasoline (Dale, 2007; Demirbaş, 2000; Naik et al., 2010), which can be produced through yeast fermentation of carbohydrates, specifically from sugary and starchy feedstock, such as sugar cane, sugar beets, corn, and wheat (Balat, 2009; Naik et al., 2010). Certain microalgae species are able to accumulate large quantities of carbohydrate within their cells. The carbohydrate is usually stored at the outer layer of cell wall (Chen et al., 2013). Through hydrolysis reaction, the carbohydrate can be hydrolyzed into fermentable sugar (e.g., glucose) for subsequent bioethanol production via fermentation process (Harun et al., 2010).

3.6.2 Biodiesel

Microalgae have a strong capacity to produce lipids, which can be easily converted to biodiesel. Transesterification using homogeneous and heterogeneous catalysts and in situ transesterification are possible methods to produce biofuel from microalgae lipids (Lam and Lee, 2012). Several advantages have been described for the

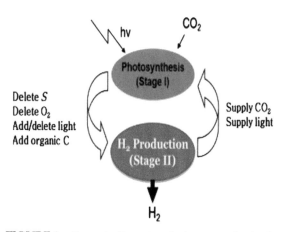

FIGURE 3   Concept of two-stage hydrogen production by microalgae (Rashid et al., 2013).

production of biodiesel from microalgae (Chisti, 2007; Li et al., 2008; Schenk et al., 2008).

3.6.3 Biohydrogen

Biohydrogen from microalgae is an alternative source of energy. A two-stage method of production has been used: carbon fixation and anaerobic digestion (Rashid et al., 2013). This concept of a two-stage hydrogen production by microalgae is shown in Figure 3.

3.7 Wastewater Treatment

Microalgae-based wastewater treatment relies on the ability of phototrophic microorganisms to supply oxygen to aerobic organic pollutant degraders and enhance the removal of nutrients and pathogens. Conventional methods for the removal of heavy metals from wastewater include chemical precipitation, coagulation, ion exchange, membrane processing, electrochemical techniques, adsorption on activated carbon, etc. (Dabrowski et al., 2004; Gautam et al., 2014; Llanos et al., 2010; Wan Ngah and Hanafiah, 2008).

4. CONCLUSIONS

Microalgae are microscopic plants that contain potential bioactive materials in the form of proteins, lipids, glycerols, carotenes, and vitamins (Avagyan, 2008; Priyadarshani and Rath, 2012). Potential bioactive metabolites from microalgae can play a vital role in human health and nutrition. The designing of new functional foods, nutraceuticals, and pharmaceuticals from marine microalgae makes them one of the most valuable marine sources. The use of transgenic microalgae for commercial applications has not yet been reported but holds significant promise (Spolaore et al., 2006). Drug development is the most promising aspect of microalgal biotechnology, although screening of possibilities remains limited (Tramper et al., 2003). Finally, marine microalgae promise to be an important alternative as a future bioenergy source.

Acknowledgments

This research was supported by a grant from the Marine Bioprocess Research Center of the Marine Biotechnology Program, funded by the Ministry of Oceans and Fisheries, Republic of Korea.

References

Abdel-Wahhab, M.A., Ahmed, H., Hagazi, M.M., 2006. Prevention of aflatoxin B1-initiated hepatotoxicity in rat by marine algae extracts. J. Appl. Toxicol. 26 (3), 229–238.

Andersen, R.A., 2005. Algal Culturing Techniques. Academic Press.

Atkin, S.L., Backett, S.T., Mackenzie, G., 2006. Topical Formulations Containing Sporopollenin. Google Patents.

Avagyan, A., 2008. Microalgae: big feed potential in a small package. Feed In. 29 (2), 16–18.

Balat, M., 2009. Bioethanol as a vehicular fuel: a critical review. Energy Sources, Part A 31 (14), 1242–1255.

Balskus, E.P., Walsh, C.T., 2010. The genetic and molecular basis for sunscreen biosynthesis in cyanobacteria. Science 329 (5999), 1653–1656.

Baquero, F., Martínez, J.-L., Cantón, R., 2008. Antibiotics and antibiotic resistance in water environments. Curr. Opin. Biotechnol. 19 (3), 260–265.

Benemann, J., 1989. The Future of Microalgal Biotechnology. Algal and Cyanobacterial Biotechnology. Longman Scientific & Technical, Harlow, UK, 317–337.

Borowitzka, M.A., 2013. High-value products from microalgae—their development and commercialisation. J. Appl. Phycol. 25 (3), 743–756.

Brennan, L., Owende, P., 2010. Biofuels from microalgae—a review of technologies for production, processing, and extractions of biofuels and co-products. Renewable Sustainable Energy Rev. 14 (2), 557–577.

Brown, M.R., Barrett, S.M., Volkman, J.K., Nearhos, S.P., Nell, J.A., Allan, G.L., 1996. Biochemical composition of new yeasts and bacteria evaluated as food for bivalve aquaculture. Aquaculture 143 (3), 341–360.

Cardozo, K.H.M., Guaratini, T., Barros, M.P., Falcão, V.R., Tonon, A.P., Lopes, N.P., Campos, S., Torres, M.A., Souza, A.O., Colepicolo, P., Pinto, E., 2007. Metabolites from algae with economical impact. Comp. Biochem. Physiol., Part C: Toxicol. Pharmacol. 146 (1–2), 60–78.

Chamorro, G., Pérez-Albiter, M., Serrano-García, N., Mares-Sámano, J.J., Rojas, P., 2006. Spirulina maxima pretreatment partially protects against 1-methyl-4-phenyl-1,2,3,6-tetrahydropyridine neurotoxicity. Nutr. Neurosci. 9 (5–6), 207–212.

Chen, C.-Y., Zhao, X.-Q., Yen, H.-W., Ho, S.-H., Cheng, C.-L., Lee, D.-J., Bai, F.-W., Chang, J.-S., 2013. Microalgae-based carbohydrates for biofuel production. Biochem. Eng. J. 78, 1–10.

Chisti, Y., 2007. Biodiesel from microalgae. Biotechnol. Adv. 25 (3), 294–306.

Dabrowski, A., Hubicki, Z., Podkościelny, P., Robens, E., 2004. Selective removal of the heavy metal ions from waters and industrial wastewaters by ion-exchange method. Chemosphere 56 (2), 91–106.

Dale, B.E., 2007. Thinking clearly about biofuels: ending the irrelevant 'net energy' debate and developing better performance metrics for alternative fuels. Biofuels Bioprod. Biorefin. 1 (1), 14–17.

Demirbaş, A., 2000. Conversion of biomass using glycerin to liquid fuel for blending gasoline as alternative engine fuel. Energy Convers. Manage. 41 (16), 1741–1748.

Dionisio-Se Se, M.L., 2010. Aquatic microalgae as potential sources of UV-screening compounds. Philippine J. Sci. 139, 5–19.

Dolatabadi, J.E.N., de la Guardia, M., 2011. Applications of diatoms and silica nanotechnology in biosensing, drug and gene delivery, and formation of complex metal nanostructures. TrAC Trends Anal. Chem. 30 (9), 1538–1548.

Dominguez, H., 2013. Functional Ingredients from Algae for Foods and Nutraceuticals. Elsevier.

Doughman, S.D., Krupanidhi, S., Sanjeevi, C.B., 2007. Omega-3 fatty acids for nutrition and medicine: considering microalgae oil as a vegetarian source of EPA and DHA. Curr. Diabetes Rev. 3 (3), 198–203.

Draaisma, R.B., Wijffels, R.H., Slegers, P.M., Brentner, L.B., Roy, A., Barbosa, M.J., 2013. Food commodities from microalgae. Curr. Opin. Biotechnol. 24 (2), 169–177.

Garciapichel, F., Wingard, C.E., Castenholz, R.W., 1993. Evidence regarding the UV sunscreen role of a mycosporine-like compound in the Cyanobacterium Gloeocapsa sp. Appl. Environ. Microbiol. 59 (1), 170–176.

Gautam, R.K., Mudhoo, A., Lofrano, G., Chattopadhyaya, M.C., 2014. Biomass-derived biosorbents for metal ions sequestration: adsorbent modification and activation methods and adsorbent regeneration. J. Environ. Chem. Eng. 2 (1), 239–259.

Gierhart, D.L., Fox, J.A., 2013. Protection against Sunburn and Skin Problems with Orally-ingested High-dosage Zeaxanthin. Google Patents.

Gordon, R., Losic, D., Tiffany, M.A., Nagy, S.S., Sterrenburg, F.A., 2009. The glass menagerie: diatoms for novel applications in nanotechnology. Trends Biotechnol. 27 (2), 116–127.

Griffiths, M.J., Harrison, S.T., 2009. Lipid productivity as a key characteristic for choosing algal species for biodiesel production. J. Appl. Phycol. 21 (5), 493–507.

Guerin, M., Huntley, M.E., Olaizola, M., 2003. Haematococcus astaxanthin: applications for human health and nutrition. Trends Biotechnol. 21 (5), 210–216.

Harris, W.S., Kris-Etherton, P.M., Harris, K.A., 2008. Intakes of long-chain omega-3 fatty acid associated with reduced risk for death from coronary heart disease in healthy adults. Curr. Atheroscler. Rep. 10 (6), 503–509.

Harrison, P.J., Berges, J.A., 2005. Marine culture media. Algal Culturing Tech. 21–34.

Harun, R., Singh, M., Forde, G.M., Danquah, M.K., 2010. Bioprocess engineering of microalgae to produce a variety of consumer products. Renewable Sustainable Energy Rev. 14 (3), 1037–1047.

Hayashi, K., Nakano, T., Hashimoto, M., Kanekiyo, K., Hayashi, T., 2008. Defensive effects of a fucoidan from brown alga Undaria pinnatifida against herpes simplex virus infection. Int. Immunopharmacol. 8 (1), 109–116.

He, M., Li, L., Liu, J., 2012. Isolation of wild microalgae from natural water bodies for high hydrogen producing strains. Int. J. Hydrogen Energy 37 (5), 4046–4056.

Hernández-Corona, A., Nieves, I., Meckes, M., Chamorro, G., Barron, B.L., 2002. Antiviral activity of Spirulina maxima against herpes simplex virus type 2. Antiviral Res. 56 (3), 279–285.

Jin, D.-Q., Lim, C.S., Sung, J.-Y., Choi, H.G., Ha, I., Han, J.-S., 2006. Ulva conglobata, a marine algae, has neuroprotective and anti-inflammatory effects in murine hippocampal and microglial cells. Neurosci. Lett. 402 (1), 154–158.

Kang, K.-H., Qian, Z.-J., Ryu, B., Kim, D., Kim, S.-K., 2012. Protective effects of protein hydrolysate from marine microalgae Navicula incerta on ethanol-induced toxicity in HepG2/CYP2E1 cells. Food Chem. 132 (2), 677–685.

Lam, M.K., Lee, K.T., 2012. Microalgae biofuels: a critical review of issues, problems and the way forward. Biotechnol. Adv. 30 (3), 673–690.

Li, Y., Horsman, M., Wu, N., Lan, C.Q., Dubois-Calero, N., 2008. Biofuels from microalgae. Biotechnol. Prog. 24 (4), 815–820.

Liu, J., Sun, Z., Gerken, H., 2014. Recent Advances in Microalgal Biotechnology. OMICS Group International.

Llanos, J., Williams, P., Cheng, S., Rogers, D., Wright, C., Perez, A., Canizares, P., 2010. Characterization of a ceramic ultrafiltration membrane in different operational states after its use in a heavy-metal ion removal process. Water Res. 44 (11), 3522–3530.

Lorenz, R.T., Cysewski, G.R., 2000. Commercial potential for Haematococcus microalgae as a natural source of astaxanthin. Trends Biotechnol. 18 (4), 160–167.

Losic, D., Yu, Y., Aw, M.S., Simovic, S., Thierry, B., Addai-Mensah, J., 2010. Surface functionalisation of diatoms with dopamine modified iron-oxide nanoparticles: toward magnetically guided drug microcarriers with biologically derived morphologies. Chem. Commun. 46 (34), 6323–6325.

Mahajan, G., Kamat, M., 1995. γ-Linolenic acid production from Spirulina platensis. Appl. Microbiol. Biotechnol. 43 (3), 466–469.

Mayer, A., Hamann, M.T., 2005. Marine pharmacology in 2001–2002: marine compounds with anthelmintic, antibacterial, anticoagulant, antidiabetic, antifungal, anti-inflammatory, antimalarial, antiplatelet, antiprotozoal, antituberculosis, and antiviral activities; affecting the cardiovascular, immune and nervous systems and other miscellaneous mechanisms of action. Comp. Biochem. Physiol., Part C: Toxicol. Pharmacol. 140 (3), 265–286.

McCarty, M.F., 2007. Clinical potential of Spirulina as a source of phycocyanobilin. J. Med. Food 10 (4), 566–570.

Miao, X., Wu, Q., Yang, C., 2004. Fast pyrolysis of microalgae to produce renewable fuels. J. Anal. Appl. Pyrolysis 71 (2), 855–863.

Miranda, M., Cintra, R., Barros, S., Mancini-Filho, J., 1998. Antioxidant activity of the microalga Spirulina maxima. Braz. J. Med. Biol. Res. 31 (8), 1075–1079.

Mozaffarian, D., Rimm, E.B., 2006. Fish intake, contaminants, and human health: evaluating the risks and the benefits. JAMA 296 (15), 1885–1899.

Muller-Feuga, A., Moal, J., Kaas, R., 2003. The Microalgae of Aquaculture. Live Feeds in Marine Aquaculture. Oxford, 206–252.

Mutanda, T., Ramesh, D., Karthikeyan, S., Kumari, S., Anandraj, A., Bux, F., 2011. Bioprospecting for hyper-lipid producing microalgal strains for sustainable biofuel production. Bioresour. Technol. 102 (1), 57–70.

Nagai, H., Torigoe, K., Satake, M., Murata, M., Yasumoto, T., Hirota, H., 1992. Gambieric acids: unprecedented potent antifungal substances isolated from cultures of a marine dinoflagellate Gambierdiscus toxicus. J. Am. Chem. Soc. 114 (3), 1102–1103.

Naik, S.N., Goud, V.V., Rout, P.K., Dalai, A.K., 2010. Production of first and second generation biofuels: a comprehensive review. Renewable Sustainable Energy Rev. 14 (2), 578—597.

Nassif, N., Livage, J., 2011. From diatoms to silica-based biohybrids. Chem. Soc. Rev. 40 (2), 849—859.

Noaman, N.H., Fattah, A., Khaleafa, M., Zaky, S.H., 2004. Factors affecting antimicrobial activity of *Synechococcus leopoliensis*. Microbiol. Res. 159 (4), 395—402.

Norton, T.A., Melkonian, M., Andersen, R.A., 1996. Algal biodiversity. Phycologia 35 (4), 308—326.

Nuhu, A.A., 2013. *Spirulina* (*Arthrospira*): an important source of nutritional and medicinal compounds. J. Mar. Biol. 2013.

Priyadarshani, I., Rath, B., 2012. Commercial and industrial applications of micro algae—a review. J. Algal Biomass Utln. 3 (4), 89—100.

Raja, A., Vipin, C., Aiyappan, A., 2013. Biological importance of marine algae—an overview. Int. J. Curr. Microbiol. App. Sci. 2 (5), 222—227.

Rashid, N., Rehman, M.S.U., Memon, S., Ur Rahman, Z., Lee, K., Han, J.-I., 2013. Current status, barriers and developments in biohydrogen production by microalgae. Renewable Sustainable Energy Rev. 22, 571—579.

Rodolfi, L., Chini Zittelli, G., Bassi, N., Padovani, G., Biondi, N., Bonini, G., Tredici, M.R., 2009. Microalgae for oil: strain selection, induction of lipid synthesis and outdoor mass cultivation in a low-cost photobioreactor. Biotechnol. Bioeng. 102 (1), 100—112.

Schenk, P.M., Thomas-Hall, S.R., Stephens, E., Marx, U.C., Mussgnug, J.H., Posten, C., Kruse, O., Hankamer, B., 2008. Second generation biofuels: high-efficiency microalgae for biodiesel production. BioEnergy Res. 1 (1), 20—43.

Senevirathne, M., Kim, S.-K., 2011. Marine macro- and microalgae as potential agents for the prevention of asthma: hyperresponsiveness and inflammatory subjects. Adv. Food Nutr. Res. 64, 277—286.

Shimizu, Y., 1996. Microalgal metabolites: a new perspective. Ann. Rev. Microbiol. 50 (1), 431—465.

Simmons, T.L., Andrianasolo, E., McPhail, K., Flatt, P., Gerwick, W.H., 2005. Marine natural products as anticancer drugs. Mol. Cancer Ther. 4 (2), 333—342.

Simoons, F.J., 1990. Food in China: A Cultural and Historical Inquiry. CRC Press.

Sirenko, L., Kirpenko, Y.A., 1999. Influence of metabolites of certain algae on human and animal cell cultures. Int. J. Algae 1 (1).

Spolaore, P., Joannis-Cassan, C., Duran, E., Isambert, A., 2006. Commercial applications of microalgae. J. Biosci. Bioeng. 101 (2), 87—96.

Stolz, P., Obermayer, B., 2005. Manufacturing microalgae for skin care. Cosmet. Toiletries 120 (3), 99—106.

Thillairajasekar, K., Duraipandiyan, V., Perumal, P., Ignacimuthu, S., 2009. Antimicrobial activity of *Trichodesmium erythraeum* (Ehr)(microalga) from South East coast of Tamil Nadu, India. Int. J. Integr. Biol. 5 (3), 167—170.

Tominaga, K., Hongo, N., Karato, M., Yamashita, E., 2012. Cosmetic benefits of astaxanthin on humans subjects. Acta Biochim. Pol. 59 (1), 43—47.

Tramper, J., Battershill, C., Brandenburg, W., Burgess, G., Hill, R., Luiten, E., Müller, W., Osinga, R., Rorrer, G., Tredici, M., 2003. What to do in marine biotechnology? Biomol. Eng. 20 (4), 467—471.

Walker, T.L., Purton, S., Becker, D.K., Collet, C., 2005. Microalgae as bioreactors. Plant Cell Rep. 24 (11), 629—641.

Wan Ngah, W., Hanafiah, M., 2008. Removal of heavy metal ions from wastewater by chemically modified plant wastes as adsorbents: a review. Bioresour. Technol. 99 (10), 3935—3948.

West, M., Zubeck, M., 2012. Evaluation of microalgae for use as nutraceuticals and nutritional supplements. J. Nutr. Food Sci. 2, 147. http://dx.doi.org/10.4172/2155-9600.1000147.

An Introduction to Microalgae: Diversity and Significance

Jasvinder Singh, Rakesh Chandra Saxena

Indian Institute of Petroleum, Dehradun, India

1. INTRODUCTION

Microalgae are largely a diverse group of microorganisms comprising eukaryotic photoautotrophic protists and prokaryotic cyanobacteria (sometimes called blue-green algae). These microbes contribute to half of global photosynthetic activity and are virtually found in euphotic niches. (Anderson, 1996). Moreover, these microalgae form the source of the food chain for more than 70% of the world's biomass (Wiessner et al., 1995). Microalgae can be cultivated photosynthetically using solar energy and carbon dioxide as a carbon source in shallow lagoons or raceway ponds on marginal land or closed ponds. Usage of plastic tubes in ponds can achieve up to seven times the production efficiency when compared to open ponds (Singh and Gu, 2010).

As per the literature on molecular biology, algae are understood to be a group of microorganisms that have independently acquired chloroplasts, that is, intracellular structures with their own photosynthesis mechanism (Gibbs, 1992). Nevertheless, some disagreement still prevails in the taxonomic classification of algae (Bold and Wynne, 1985). Algae are classified into more than a dozen major groups, predominantly based on pigment composition, storage profile of products, and diversity of ultrastructural features. Recently, families of the various algal groups and their relationships to other taxonomic groups have been determined using various techniques of molecular biology (Wainright et al., 1993; Hecht, 1993). Several interesting and surprising results of these studies are presented in a line map in Figure 1. It seems that various algal groups are scattered all over this line map. The blue-green algae, usually known as cyanobacteria, are prokaryotes, which are linked to many common bacteria. These algae are also considered to be the progenitors of the chloroplasts of some higher algae and plants (Gibbs, 1992). At the other end of the spectrum, green algae are closely related to higher plants. Mesokaryotes (dinoflagellates and euglenoids) possess certain characteristics that are intermediate between prokaryotes and eukaryotes (Lee, 1989). These mesokaryotes may be more closely associated with the red algae and the slime molds than with other algal groups. While molecular methods of phylogenetic analysis are innovative and controversial, the available information so far indicates that algae are heterogeneous at the molecular level. The extensive phylogenetic diversity of algae

11

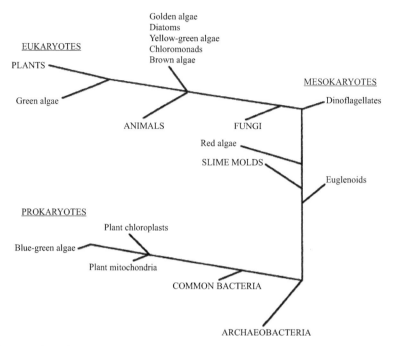

FIGURE 1   Phylogenic scheme based on the analysis of ribosomal RNS sequences. *Reproduced from BioScience vol. 46, No. 4., with permission.*

shown in Figure 1 reflects many aspects of their existence.

2. MICROALGAE DIVERSITY

The diversity of microalgae is vast and represents an intact resource. The scientific literature indicates the existence of 200,000 to several million species of microalgae when compared to about 250,000 species of higher plants (Norton et al., 1996). The genetic and phenotypic diversity of microalgae is obvious in their nearly ubiquitous distribution in the biosphere. Green microalgae usually grow in fresh water and seawater, whereas several other species of microalgae grow in extremely saline environments, such as the Great Salt Lake in UT, USA, and the Dead Sea in Israel. Within these aqueous habitats, some algae grow inside a few hundred micrometers of the water layer, others populate the subsurface water column, and a few grow at the limits of the photic zone, that is, 200–300 m below the water surface. Microalgae also grow in rich humus soil, desert sands, rocks, snowfields, and in more unusual sites, like the fur of sloths and polar bears. A number of species of microalgae occur virtually in every type of terrestrial environment, including the most harsh, such as walls of urban buildings (Rindi et al., 2007), biotic crusts in hot deserts (Lewis and Flechtner, 2002; Flechtner, 2007), Antarctic snow (Broady, 1996), and air at 2000-m height (Sharma et al., 2007). Often these are small in size (mostly 5–50 μm) and characterized by a simple morphology, usually unicellular. Accordingly, most species are not observable as individual specimens and become visually noticeable only when producing large populations, typically in the form of black, green, red, or brown patches. It can be stated that microalgae can either be free living or exist in association with other organisms, as in the case of lichens (Radmer, 1996).

2.1 Geographical Diversity

Territories with a coastline on the Mediterranean Sea between 45° and 30°N are suitable regions for algae farming, in particular in those territories at the south of the Mediterranean that experience warmer climates and whose temperature do not go too much below 15 °C throughout the year (Singh and Gu, 2010). This type of warmer climate of the Mediterranean region can facilitate the algal growth in the open or closed pond system. These suitable conditions can be the most efficient, economic, and favorable way to grow the algal biomass. Innovative technologies in algae harvesting have also succeeded in open pond farms located in slightly cooler regions by covering them with special material and making them behave in a similar way as a greenhouse. This can certainly facilitate algae farming at increased latitude.

The environmental parameters favorable for potential algae cultivation are being explored in a number of countries in the Mediterranean basin. Israel, for example, has for a long time cultivated algae and used them in medicines as well as nutrient production. Recently, these Mediterranean countries have also begun producing several strains for fuel production. The southern Mediterranean Sea belt countries, for example, Morocco, Algeria, Tunisia, and Egypt, have great potential for algae farming because of higher environmental temperatures and enormous unused desert land. At the same time, countries like Libya, Cyprus, and Turkey also have plenty of marginal land to harvest algae. For these countries a limited water resource is not a constraint as it is well established now that algae do not require freshwater but rather can grow with recycled brackish or saline water. With the high temperatures in the Mediterranean region, the open or closed pond system would probably be the most efficient and suitable to grow algae.

2.2 Diversity in Terrestrial Green Microalgae

Green microalgae are among the most prevalent microorganisms occurring in terrestrial environments (Rindi et al., 2009). A majority of these organisms performing oxygenic photosynthesis in terrestrial environments represent a very heterogeneous and evolutionarily diverse collection. Usually a general impression on the diversity and biogeography of these organisms is entirely based on the concepts of morphological species. However, ultrastructural and molecular data generated in the last three decades have uncovered a scenario in substantial contrast with morphological classifications. It has become obvious that these organisms have been affected by an extreme morphological convergence, which has restricted their morphology to a narrow range, not indicative of their great genetic diversity. Their pattern is very simple and uniform, usually referable to a few types (unicellular, uniseriate filamentous, sarcinoid colony) and offering very few characteristics useful for taxonomic and systematic purposes.

Terrestrial microalgae belong primarily to three diverse evolutionary pedigrees: the blue-green algae (or Cyanobacteria), the green algae (Chlorophyta and Streptophyta), and the diatoms (Bacillariophyceae, Ochrophyta). By and large, it is demonstrated that blue-green algae represent the main component of the terrestrial microalgal vegetation in tropical regions, whereas green algae represent the dominant element in temperate regions (Fritsch, 1907; John, 1988). From a statistical point of view, the green algae and the blue-green algae include the majority of the species described. Nevertheless, the understanding of the patterns of geographical distribution in terrestrial algae is inadequate, mainly due to poor understanding of the diversity of these organisms.

Green algae are photosynthetic eukaryotes bearing double-membrane-bound plastids containing *chlorophyll a* and *b*, accessory pigments found in embryophytes, for example, β carotene and xanthophylls, and a unique stellate structure linking nine pairs of microtubules in the flagellar base (Lewis and McCourt, 2004). These are part of the most diverse groups of eukaryotes and include morphological forms ranging from flagellated unicells, coccoids, branched or unbranched filaments to multinucleated macrophytes and taxa with parenchymatous tissues (Pröschold and Leliaert, 2007). According to Yoon et al. (2004) molecular dating data reveal that the most recent common ancestor of all green algae may have existed 1100−1200 million years ago.

From a systematic point of view, the green algae have been traditionally a difficult group. In the past the classification of these organisms has undergone several major rearrangements, mostly due to the fact that different criteria, based on different types of evidence (morphological, ultrastructural, molecular), have been adopted at different stages (Lewis and McCourt, 2004; Pröschold and Leliaert, 2007). Significant advancements have been made in the last 30 years, in which new types of data have complemented the bulk of morphological information produced in the two previous centuries. The development of electron microscopy in the 1960s revealed many important ultrastructural characteristics of the green algae, which have proved to be key features for classification.

3. MICROALGAE CELL STRUCTURE—A CLOSE LOOK

Microalgae are free-living microorganisms of the kingdom Protista which can be found in a variety of aquatic habitats. Similar to plant cells, these life-sustaining microorganisms are photoautotrophic and have chloroplasts. Although microalgae do not contain roots, stems, or leaves as do higher plants, they do exhibit certain characteristics similar to cellular organelles. Like all eukaryotes, but unlike the bacteria and archaea

domains, algae cells contain membrane-bound organelles, including nuclei containing their genetic information. However, each cell is only one of its kind of life form and can sustain life on its own, having a rigid cell wall, layered structure that surrounds the cell, and consisting of polysaccharides unlike animal cells. The cell wall membrane allows materials to pass through, but its biochemical composition functions as a selective barrier. The cell wall is a more porous, rigid structure outside the membrane. There are passageways, called plasmodesmata, connecting cell to cell through the wall and membrane. Most algae are unicellular organisms; nevertheless, there are a few multicellular groups such as seaweeds and colonial species such as filamentous "string" algae.

Microalgae also contain lipid bodies, which are a ready source of stored energy generated from photosynthesis. These characteristics combined with their photosynthetic nature distinguish them from bacteria and other unicellular microorganisms. Algae, like so-called higher vegetation, are vital in their capacity to produce their own food through the route of photosynthesis. The algal cell component that makes this possible is the chloroplast. The chloroplast has a complex internal arrangement with a chlorophyll molecule at the core. Through a complex series of biochemical reactions, chlorophyll uses carbon dioxide from the atmosphere and light energy to manufacture the sugar glucose.

The nucleus is the largest cell organelle, and every cell possesses one. The nucleus in a microalgal cell is the control center, which contains deoxyribonucleic acid (DNA), the genetic material of the cell. DNA contains the biological source code that coordinates many complex cell functions. The endoplasmic reticulum (ER) of the cell is a complex inner membrane structure like a crumpled plastic bag with all of its folds and wrinkles and the small space of the inside of the bag all compressed. This organelle's function is constant synthesis and transportation of complex proteins and other important cell-building blocks containing lipids for usage within the cell and other parts of the organism. Ribosomes are small organelles attached to some of the ER and are active in protein synthesis.

All cells possess a Golgi apparatus, which acts as a kind of "cell gland," providing material for the building and maintenance of the cell and cell wall membrane. The Golgi apparatus repackages and effectively ships proteins and other materials manufactured in the ER out to other parts of the cell, functioning like the shipping department of the cell. For example, in the pancreas, the Golgi apparatus is responsible for the production of digestive enzymes and protection of the cell from harmful substances (by isolating them).

Mitochondria are the powerhouses of the cell. Mitochondria in the microalgae cells burn substances to process respiration, using oxygen and generating the compound adenosine triphosphate (ATP). ATP is the metabolic carrier of the cell, providing the chemical energy necessary to perform vital cellular processes. As per their requirements, cells may possess up to 2000 of these power-generating plants.

The vacuole is a bladder-resembling organelle that can occupy considerable space in the cell and disposes waste. Vacuoles are often such a large presence within the cell that they exert pressure and help the cell maintain its structure and shape. Algae do not have vascular structures, which are tubes that are found within plants to transport nutrients. Unlike the reproductive structures of plants, most algae reproduce asexually or by cell separation; since they do not need to generate elaborate support and reproductive structures, they can devote more of their energy into trapping and converting light energy and CO_2 into biomass.

3.1 Photosynthetic Mechanism of Microalgae

Though most microalgae use photosynthesis for their food, and like plants some even have roots and leaves, they are not considered plants (Chaudhary et al., 2014). Some algae have roots but these usually support algal structures and are not proper roots. Characteristically, algae photosynthesize, generating energy directly from the Sun's radiation. This characteristic designates most algae as autotrophs, in contrast to heterotrophs, which consume other organisms. On the other hand, some species of algae are mixotrophs, with both characteristics, and some are primarily heterotrophic. The algal photosynthesis is primarily based on the Calvin cycle wherein ribulose-1,5-bisphosphate reacts with CO_2 to synthesize 3-phosphoglyceric acid (3-PGA), which is consumed during production of glucose and other metabolites (John et al., 2011). An ethanogenic recombinant of *Rhodobacter* sp. was developed for carbon redirection from the Calvin cycle (Benemann, 1979).

Microalgae have better photon conversion effectiveness and can produce and build up large quantities of carbohydrate biomass (Melis and Happe, 2001; Harun et al., 2010a). Aquatic algal cells are buoyant, and do not contain structural biopolymers such as hemicellulose and lignin that are vital for better plant growth in earthy environment. Microalgae can transform approximately 6% of the total incident radiation energy into fresh biomass (Odum, 1971). Comparatively, it can be observed that terrestrial crops usually have less photosynthetic conversion efficiencies.

4. MICROALGAE AS A POTENTIAL SOURCE FOR FUELS AND CHEMICALS

A wide range of products can be synthesized from algae (Harun et al., 2010a; Brennan and Owende, 2010). These products vary from human nutrients for livestock feed to organic chemicals for pharmaceuticals, pigments, and various other industry and energy applications, like biobutanol, acetone, biodiesel, bioethanol, and biomethane. The protein and carbohydrate contents in various strains of microalgae are high, up to 50% of dry weight (Singh and Gu, 2010). The maximum lipid contents in microalgae are also around 40% on wt. basis, which is reasonably good. All these factors make microalgae a potential source for bio-oil production. Various products obtainable from microalgae are briefly discussed in the following subsections.

4.1 Dietary and Pharmaceutical Source

Microalgae are a potential source of various food supplements and biomaterials used in the pharmaceutical industry. Some of these are omega-3 fatty acids, eicosapentaenoic acid (EPA), docosahexaenoic acid (DHA), and chlorophyll. Omega-3 fatty acids are usually obtained from fish oil. But poor supplies of fish oil in recent years, together with problems due to its unpleasant taste and poor oxidative stability, have left this route less promising (Luiten et al., 2003). The literature highlights (Harun et al., 2010a) that microalgae naturally contain omega-3 fatty acids that can be purified to provide a high-value food supplement. The practical sources of omega-3 in microalgae are normally EPA and DHA. In comparison to fish, microalgae are self-producing omega-3, thus making the process straightforward as well as economical (Belarbi et al., 2000).

4.2 Livestock Feed

Livestock feed is another useful product that can be obtained from algae. A number of researchers have examined the biochemical composition of various algae for their suitability as a substitute or primary livestock feed. It has also been reported that microalgae play a key role in high-grade animal nutrition food, from aquaculture to farm animals. Systematic nutritional and toxicological evaluations have demonstrated the suitability of algae biomass as a valuable feed supplement or substitute for conventional animal feed sources (Dhargalkar and Verlecar, 2009). They have investigated that certain edible seaweeds can be used as food due to a lower number of calories, low fat content, and high concentration of minerals, vitamins, and proteins. Many studies have reported the use of algae as aquaculture feed. Microalgae species *Hypnea cervicornis* and *Cryptonemia crenulata*, which are particularly rich in protein, were tested in shrimp diets (da Silva and Barbosa, 2008). The studies carried out by da Silva and Barbosa reveal that the amount of algae in fish meal results in a significant increase of shrimp growth rates. A better growth weight and protein efficiencies ratio was observed by Azaza et al. (2008) in the case of tilapia fish farming with algae as a nutritional food source in feed. In addition, *Phorphidium valderianum*, marine cyanobacteria, were successfully used as feed for aquaculture based on their nutritional and nontoxic performance (Thajuddin and Subramaniyan, 2005). In addition to their importance in aquaculture, Spolaore et al. (2006) have reported that up to 5–10% of conventional protein sources in poultry feed can effectively be replaced by algae. Ginzberg et al. (2000) studied the role of the algae, *Porphyridium* sp., as feed supplement in the metabolism of chickens. Their investigations show that cholesterol of egg yolk was reduced by about 10% and the color of egg yolk became darker, indicating a higher content of carotenoids. Belay et al. (1996) reviewed the potential of *Arthrospira* (*Spirulina*) in animal feed. Although *Arthrospira* is widely used as food additive and can replace up to 50% of protein diets in existing feeds, it was concluded that protein sources from soya and fish meal were more cost-effective and thus preferred to *Arthrospira* (Humphrey, 2004). Furthermore, a study on the addition of *Laminaria digitata* suggested that algae-supplemented feed increased pig weight up to 10% on a daily basis (He et al., 2002).

4.3 Biofuels Products from Algae: Biodiesel, Bioethanol, and Biomethane

4.3.1 Biodiesel

At this time, biodiesel has come to mean a very specific chemical modification of natural oils. Oilseed crops such as rapeseed and soybean oil have been extensively evaluated as sources of biodiesel by various researchers (Meher et al., 2006; Van Gerpan, 2005; Mata et al., 2010; Khan et al., 2009; Song et al., 2008). The viability of microalgae for biodiesel production has also been studied (see the comprehensive review by Mata et al. (2010)). Khan et al. (2009) have critically evaluated the prospects of biodiesel production from microalgae. They have emphasized the need to explore the possibilities of producing biodiesel from microalgae, as it will not compete with the land and cereal crops. Microalgae are a potential raw material for biodiesel production, as they meet all of the following requirements. They possess a high growth rate and provide lipid fraction for biodiesel production (Song et al., 2008). Microalgal lipids are mostly neutral lipids with a lower degree of unsaturation. This makes microalgal lipids a potential replacement for fossil fuels.

4.3.2 Bioethanol

Recent attempts at producing ethanol are focusing on microalgae as a feedstock for the fermentation process. Microalgae are rich in carbohydrates and proteins that can be used as carbon sources for fermentation. Table 1 lists the amount of carbohydrates and proteins measured from various algal species. Bacteria, yeast, or fungi are microorganisms used to ferment carbohydrates to produce ethanol under anaerobic conditions. In addition to ethanol as a main product, CO_2 and water are also formed as by-products. According to the following simplified reaction, stoichiometric yields are 0.51 kg ethanol and 0.49 kg CO_2, per kg of carbon sugar, that is, glucose:

$$C_6H_{12}O_6 \rightarrow 2CH_3CH_2OH + 2CO_2:$$

TABLE 1 Proteins and Carbohydrates Contents from Various Species of Microalgae

S.N.	Algae strain	Protein, % dwt	Carbohydrate, % dwt
1	*Anabaena cylindrical*	43 − 56	25 − 30
2	*Chlamydomonas rheinhardii*	48	17
3	*Chlorella pyrenoidosa*	57	26
4	*Chlorella vulgaris*	51 − 58	12 − 17
5	*Dunaliella bioculata*	49	4
6	*Dunaliella salina*	57	32
7	*Euglena gracilis*	39 − 61	14 − 18
8	*Porphyridium cruentum*	28 − 45	40 − 57
9	*Prymnesium parvum*	28 − 45	25 − 33
10	*Scenedesmus dimorphus*	8 − 18	21 − 52
11	*Scenedesmus obliquus*	50 − 56	10 − 17
12	*Scenedesmus quadricauda*	47	-
13	*Spirogyra* sp.	6 − 20	33 − 64
14	*Spirulina maxima*	28 − 39	13 − 16
15	*Spirulina platensis*	52	8 − 14
16	*Synechoccus* sp.	46 − 63	15
17	*Tetraselmis maculate*	52	15

The most recent work on bioethanol production by fermentation has been reported by Harun et al. (2010b) They have carried out experiments for studying the suitability of microalgae (*Chlorococum* sp.) as a substrate, using yeast for fermentation. A productivity level of around 38 wt% has been reported, which supports the suitability of microalgae as a promising substrate for bioethanol production. In his doctoral work, Moen (1997) demonstrated that brown seaweed produces higher bioethanol compared to other algae species. Another study by Hirayama et al. (1998) proposed a self-fermentation of algae to obtain ethanol. The reported advantages of this technique over conventional fermentation are a comparatively simple process with shorter fermentation time. Ueda et al. (1996) have patented a two-stage process for microalgae fermentation. In the first stage, microalgae undergo fermentation in an anaerobic and dark environment and ethanol is produced, which can be purified to be used as fuel. The CO_2 produced in the fermentation process was recycled to algae cultivation ponds as a nutrient to grow microalgae.

Although limited reports on algae fermentation are available, a number of advantages have been reported in the production of bioethanol from algae. The fermentation process involves less intake of energy and the process is much simpler in comparison to the biodiesel production system. In addition, CO_2 produced as a by-product of the fermentation process can be recycled as carbon sources for microalgae cultivation, thus reducing the greenhouse gas emissions as well. However, the technology for the commercial production of bioethanol from microalgae is still in development and under investigation.

4.3.3 Biomethane

The remaining algae biomass slurry left after fermentation may be used in the anaerobic digestion process while keeping the pH in the range of 6–9. This process produced methane, which can further be converted to produce electricity (Ueda et al., 1996) Thus, it is imperative that the bioethanol production by fermentation is also useful for simultaneous production of biogas. This has resulted in considerable attention toward the application of methane fermentation technology to algae to produce valuable by-products such as biogas. Biogas produced from anaerobic microorganisms by anaerobic digestion mainly consists of a mixture of methane (55–75%) and CO_2 (25–45%).

Residual biomass from anaerobic digestion can be further reprocessed to make fertilizers. In addition to being renewable and sustainable, this would encourage sustainable agricultural practices in providing greater efficiencies and reduce

algae production costs. Due to the absence of lignin and lower cellulose, microalgae exhibit good process stability and high conversion efficiency for anaerobic digestion (Vergara-Fernandez et al., 2008).

5. PRODUCTS STORAGE MECHANISM AND GENETIC MODIFICATION POTENTIAL OF MICROALGAE

Major advances in microalgal genomics have been accomplished during the last decade. With the latest biotechnological advances, efforts at genomic alteration of microalgal cells are rapidly increasing. However, the full potential of genetic engineering can be fully realized only if usual breeding methods become firmly established, thus permitting useful mutations to be easily combined (Radkovits et al., 2010). Historically, green algae, namely *Chlamydomonas reinhardtii*, have been the focus of molecular and genetic phycological research. Therefore, most of the tools developed for the expression of transgenes and gene knock-over are specific for this kind of species. Current genetic engineering research is focusing on microalgae that are of greater interest in industrial applications and environmental conservation (Radkovits et al., 2010). Numerous approaches have been developed to improve microalgae biomass or lipid production and CO_2 capturing efficiency (Zeng et al., 2011). A brief account of these is discussed in the following subsections.

The basic energy-storing biochemical of plants including microalgae is starch; and the rate-limiting molecules of starch synthesis are adenosine diphosphate-glucose pyrophosphorylase (AGPase) and 3-PGA (Radakovitis et al., 2010; Stark et al., 1992; Zabawinski et al., 2001). Numerous studies are reported on the catalytic and allosteric properties of AGPases in crop plants to increase starch production (Smith, 2008;

Tjaden et al., 1998; Tetlow et al., 2004). A large amount of cellular AGPases are far from the pyrenoid and this could result in 3-PGA inactivation. Nevertheless, some AGPases, such as Mos(1−198)/SH2 AGPase, have activity even without an activator (Boehlin et al., 2009). In some microalgal species, starch production is improved when Mos(1−198)/SH2 AGPase or other AGPases are overexpressed. Although the exact mechanism of starch catabolism in most microalgae is rarely known, some studies have reported decreasing starch degradation in microalgae cells (Ball and Starch, 2009; Beer et al., 2009). Catabolism of starch provides stock and intermediate of lipid and protein synthesis, and this is sometimes the key rate-limiting step of lipid and protein synthesis (Yu et al., 2005).

5.1 Genetic Engineering of Microalgae to Enhance Carbohydrate and Protein Storage

One of the main biochemicals in microalgae is protein. Protein synthesis is the most important as well as complex in all cells. The main steps of the protein synthesis mechanism in a cell are amino acid synthesis, peptide chain condensation reaction, and modification of the primary protein. The rate of amino acid synthesis can be adjusted by changes in the expression levels of the enzymes involved. Enzyme synthesis rate can be altered by changing the activities of the related gene (Merrick, 1992). For example, when a synthetic amino acid production rate is too high, the pathway of the enzyme gene can be suppressed. However, the encoding gene cannot be inhibited in case the synthetic amino acid concentration declines, thus the expression of enzyme required for amino acid synthesis increases. Few studies have been reported on knocking out relevant gene segments involved, and it is indicated that this not only results in altering protein storage but also leads to toxic effects on cellular

growth and production (Maheshwaran et al., 1991; Wang et al., 2003).

5.2 Genetic Engineering of Microalgae to Enhance Lipid Storage

Several microalgae do not produce large amounts of lipids at logarithmic growth. However, when they come across environmental stress, such as lack of nitrogen, they slow down their rate of production and start producing energy storage products, such as lipids and starch (Hu et al., 2008). Interestingly, it is observed that overexpression of genes controls the lipid synthesis pathway and thus affects microalgae abundance. This is attributed to the increasing rate of lipid synthesis that could result in cell division reduction. In this case, overexpression of lipid synthesis genes may still be beneficial if they can be controlled by an inducible promoter that can be activated once the microalgae cells have reached high densities and have entered into the stationary phase. For example, in *C. reinhardtii*, copper-responsive elements are inducible promoters (Kropat et al., 2005; Quinn and Merchant, 1995) and in diatoms nitrate-responsive species are inducible promoters (Poulsen and Kroger, 2005).

One more approach to increase lipid accumulation is to decrease lipid catabolism. In the process of lipid catabolism, acyl-CoA oxidase, acyl CoA synthase, carnitine acyltransferase I, and fatty acyl CoA dehydrogenase are the vital enzymes of β-oxidation of fatty acids. Studies have been reported (Derelle et al., 2006; Molnar et al., 2009) on strategic knocking out of some of these enzyme genes to increase lipid storage. Microalgae initiate TAG storage during light cycles and use it during dark cycles to provide cellular energy for cell growth and proliferation. Consequently, inhibition of β-oxidation would prevent the loss of TAG during the dark cycles, but also possibly at the cost of reduced growth rate. Several publications have shown that knocking out genes involved in *Saccharomyces cerevisiae* β-oxidation not only leads to increased amounts of intracellular free fatty acids but also leads to extracellular fatty acid secretion (Michinaka et al., 2003; Scharnewski et al., 2008). However, due to the fact that cells rely on β-oxidation of fatty acids for cellular energy under certain physiological conditions, this also might have deleterious effects on cellular growth and proliferation.

5.3 Genetic Engineering of Microalgae to Improve Photosynthetic Efficiency

Light plays a vital role in all plant growth as the source of energy. Microalgae are believed to be a great model of photosynthetic efficiency. Optimal light delivery affects the photosynthetic efficiency of microalgae (Scott et al., 2010). Irrespective of the method used to cultivate the microalgae biomass, the photosynthetic apparatus must consist of photosystems, where light energy is used for photochemical reactions, surrounded by antennae in the chloroplast complexes that harvest the light energy and transport to the photosystems. Microalgae have evolved to absorb more light than needed for their photosynthetic requirements. The excess light energy is dissipated as heat and fluorescence (Wang et al., 2003):

$$CO_2 + H_2O + 9.5 h\nu \rightarrow \frac{1}{6} C_6 H_{12} O_6 + O_2$$

According to the above equation, light conversion efficiency as well as the ability of CO_2 fixation increases with an increase in photon utilization in microalgae (Melis, 2009). Several attempts have been made to improve the photosynthetic efficiency and reduce the effects of photoinhibition on microalgal growth. Most of these are focused on reducing the size of the chlorophyll antenna (Radakovits et al., 2010; Lee et al., 2002) and that can be achieved in following ways. On a macro-scale, microalgae growing at high-light intensity are kept for a long period to induce mutagenesis in which the

antenna size decreases. However, this change is readily reversible when the cells are subsequently transferred to light with low intensity (Melis et al., 1998). Also, at high-light intensity levels there is less efficient use of absorbed light energy and also biochemical damage to photosynthetic machinery (photoinhibition), making light energy utilization even less efficient. Thus the highest photosynthetic efficiency is realized at a low-light intensity (Scott et al., 2010). This approach may seem nonimpulsive, but it has two positive effects; first, it permits a higher light penetration in high-density cultures, and second, it allows a higher maximum rate of photosynthesis due to the fact that cells are less likely to be subjected to photoinhibition since their light-harvesting complexes absorb less light (Melis, 2009). Adjustment of antenna size through nutrient levels management has received less attention. However, conditions of nutrient denial leading to increased productivity of TAGs may cause a reduction in chlorophyll levels per cell (Melis, 1996), suggesting that nutrient deprivation may have multiple beneficial consequences for feedstock production. On a microscale, genetic exploitation could be employed to modify the gene sequences involved. Earlier reported studies endorse random mutagenesis strategies to generate mutants with fewer or smaller chlorophyll antennas, but recent publications efficiently used an RNAi-based strategy to knock down both LHCI and LHCII in *C. reinhardtii* (Mussgnug et al., 2007; Wobbe et al., 2009). This strategy can be widely applied to many different microalgae more easily than a random mutagenesis approach. It is quite obvious that manipulation of light-harvesting complexes can lead to increased biomass productivity under high light under controlled laboratory conditions. However, it remains to be seen how well these mutants will perform in larger-scale cultures with more varied conditions and perhaps with competition from wild invasive microalgae or bacteria species.

6. TECHNO-ECONOMIC IMPACT AND COMMERCIALIZATION POTENTIAL

The techno-economics of microalgae have become considerably more important in recent decades. Applications range from simple biomass production for food and feed to valuable products for ecological applications and, more importantly, energy needs as microalgae can be cultivated year round. Algae farming can also address socioeconomic issues by generating local job opportunities and transferring the technologies to developing countries in particular.

Economically, the quantity of oil production from microalgae exceeds the yields of the best oil seed crops (Singh and Gu, 2010). The quick growth prospective and several species of microalgae with oil content in the range of 20–50% (dry weight of biomass) is another advantage for its choice as a potential biomass. The exponential growth rates of microalgae can double their biomass in periods as short as 3.5 h (Chisti, 2007; Spolaore et al., 2006). In addition, in spite of growth in aqueous media, algae need smaller quantities of water than regular land crops, reducing load on freshwater resources (Dismukes et al., 2008). Subsequently, microalgae can also be cultivated in brackish water on non-arable land without affecting land use, environmental parameters (Searchinger et al., 2008), and food–fodder profile (Chisti, 2007).

There is a twofold potential for treatment of organic effluent from the agro-food industry for algae cultivation (Cantrel et al., 2008). Apart from providing growth medium, the nutrients for its cultivation, for example, nitrogen and phosphorus, can also be obtained from wastewater. A significant advantage to the environment is that algae cultivation does not require herbicide or pesticide application (Rodolfi et al., 2008). In addition, these can also produce valuable coproducts such as proteins and residual biomass after oil extraction, which may be used as feed or fertilizer

(Spolaore et al., 2006), or fermented to produce bioethanol or biomethane (Hirano et al., 1997). The biochemical composition of the algal biomass can be transmuted by varying growth environments, and thus considerably enhance the oil yield (Qin, 2005).

Nevertheless, commercialization of algae biofuels is also reliant on the economics of the process. Ease of implementation is an additional important factor for success of a new technology or process. In view of this scenario, "cheaper" and "easier" processes are important strategies for promoting commercial utilization of microalgae. Since algae production systems are a complex composite of several subsets of systems (i.e., production, harvesting, extraction, drying systems), reducing the number of steps in algae biofuels production is essential to providing easier, better, and lower-cost systems.

Another crucial economic challenge for algae producers is to discover low-cost oil extraction and harvesting methods. With the advent of cheaper photobioreactors, this cost is likely to reduce significantly in the next few years. Currently, reducing these costs is critical to algae biofuel companies for successful commercial implementation. Extraction systems with estimates up to $15 per gallon of oil produced depending on the extraction method can be less than cost-effective. Another example is to employ a method that uses algae cells as miniprocessors and refineries in a process referred to as "milking the algae" that will consume CO_2 and excrete hydrocarbon fuels directly.

7. CONCLUSIONS

Microalgae have been identified as a potential resource for a number of value-added products as well as feedstock for future green fuels. Their wide diversity makes them an easily cultivable resource that does not interfere with the food, fodder, or other products from terrestrial crops. They require much fewer resources as compared to other crops. They are easily modifiable by simple genetic engineering methods to enhance the yields of desired products. However, significant issues are yet to be resolved before microalgae to biofuel production becomes cost-effective and makes an impact on the world's supply of transportation fuels. The key issues that need to be addressed are minimizing the capital and operational costs, cost of drying and extraction, and research work to increase productivity by developing more efficient harvesting systems. It is estimated that full commercialization of algae oil will begin to take place in the United States in roughly 4–5 years.

References

Andersen, R.A., 1996. In: Hunter-Cevera, J.C., Belt, A. (Eds.), Maintaining Cultures for Biotechnology and Industry. Academic Press, San Diego, pp. 29–64.

Azaza, M.S., Mensi, F., Ksouri, J., Dhraief, M.N., Brini, B., Abdelmouleh, A., 2008. Growth of Nile tilapia (Oreochromis niloticus L.) fed with diets containing graded levels of green algae ulva meal (Ulva rigida) reared in geothermal waters of southern Tunisia. J. Appl. Ichthyol. 24, 202–207.

Ball, S.G., Starch, P.D., 2009. Metabolism. In: Harris, E.H., Stern, D.B. (Eds.), The Chlamydomonas Sourcebook Second Edition: Organellar Metabolic Processes, second ed. Academic Press, Oxford, pp. 1–40.

Beer, L.L., Boyd, E.S., Peters, J.W., Posewitz, M.C., 2009. Engineering algae for biohydrogen and biofuel production. Curr. Opin. Biotech. 20, 264–271.

Belarbi, E.H., Molina, E., Chisti, Y., 2000. A process for high yield and scaleable recovery of high purity eicosapentaenoic acid esters from microalgae and fish oil. Process. Biochem. 35, 951–969.

Belay, A., Kato, T., Ota, Y., 1996. Spirulina (Arthrospira): potential application as an animal feed supplement. J. Appl. Phycol. 8, 303–311.

Benemann, J.R., 1979. Production of nitrogen fertilizer with nitrogen-fixing blue-green algae. Enzym. Microb. Technol. 1 (2), 83–90.

Boehlein, S.K., Shaw, J.R., Stewart, J.D., Hannah, L.C., 2009. Characterization of an autonomously activated plant ADP-glucose pyrophosphorylase. Plant Physiol. 149, 318–326.

Bold, H.C., Wynne, M.J., 1985. Introduction to the Algae. Prentice-Hall, Englewood, NJ.

Brennan, L., Owende, P., 2010. Biofuels from microalgae—a review of technologies for production, processing, and extractions of biofuels and co-products. Renewable Sustainable Energy Rev. 14, 557–577.

Broady, P.A., 1996. Diversity, distribution and dispersal of Antarctic terrestrial algae. Biodiversity Conserv. 5, 1307–1335.

Cantrell, K.B., Ducey, T., Ro, K.S., Hunt, P.G., 2008. Livestock waste-to-bioenergy generation opportunities. Bioresour. Technol. 99 (17), 7941–7953.

Chaudhary, L., Pradhan, P., Soni, N., Singh, P., Tiwari, A., April–June 2014. Algae as a feedstock for bioethanol production: new entrance in biofuel world. Int. J. ChemTech Res. 6 (2), 1381–1389.

Chisti, Y., 2007. Biodiesel from microalgae. Biotechnol. Adv. 25 (3), 294–306.

Derelle, E., Ferraz, C., Rombauts, S., Rouze, P., Worden, A.Z., Robbens, S., 2006. Genome analysis of the smallest free-living eukaryote Ostreococcus tauri unveils many unique features. Proc. Natl Acad. Sci. U.S.A. 103, 11647–11652.

Dhargalkar, V.K., Verlecar, X.N., 2009. Southern Ocean seaweeds: a resource for exploration in food and drugs. Aquaculture 287, 229–242.

Dismukes, G.C., Carrieri, D., Bennette, N., Ananyev, G.M., Posewitz, M.C., 2008. Aquatic phototrophs: efficient alternatives to land-based crops for biofuels. Curr. Opin. Biotechnol. 19 (3), 235–240.

Flechtner, V.R., 2007. North American microbiotic soil crust communities: diversity despite challenge. In: Seckbach, J. (Ed.), Algae and Cyanobacteria in Extreme Environments. Springer, Dordrecht, The Netherlands, pp. 539–551.

Fritsch, F., 1907. The subaerial and freshwater algal flora of the tropics. A phyto geographical and ecological study. Ann. Bot. 21, 235–275.

Gibbs, S.P., 1992. The evolution of algal chloroplasts. In: Lewin, R.A. (Ed.), Origins of Plastids. Chapman & Hall, New York, pp. 107–117.

Ginzberg, A., Cohen, M., Sod-Moriah, U.A., Shany, S., Rosenshtrauch, A., Arad, S., 2000. Chickens fed with biomass of the red microalga Porphyridium sp. have reduced blood cholesterol level and modified fatty acid composition in egg yolk. J. Appl. Phycol. 12, 325–330.

Van Gerpen, J., 2005. Biodiesel processing and production. Fuel Process. Technol. 86, 1097–1107.

Harun, R., Singh, M., Forde, G.M., Danquah, M.K., 2010a. Bioprocess engineering of microalgae to produce a variety of consumer products. Renewable Sustainable Energy Rev. 14 (3), 1037–1047.

Harun, R., Danquah, M.K., Forde Gareth, M., 2010b. Microalgal biomass as a fermentation feedstock for bioethanol production. J. Chem. Technol. Biotechnol. 85, 199–203.

He, M.L., Hollwich, W., Rambeck, W.A., 2002. Supplementation of algae to the diet of pigs: a new possibility to improve the iodine content in the meat. J. Anim. Physiol. Anim. Nutr. 86, 97–104.

Hecht, J., 1993. Animals and fungi are closer than anyone expected. New Sci. (1877), 16.

Hirano, A., Ueda, R., Hirayama, S., Ogushi, Y., 1997. CO_2 fixation and ethanol production with microalgal photosynthesis and intracellular anaerobic fermentation. Energy 22 (2–3), 137–142.

Hirayama, S., Ueda, R., Ogushi, Y., Hirano, A., Samejima, Y., Hon-Nami, K., 1998. Ethanol production from carbon dioxide by fermentative microalgae. Stud. Surf. Sci. Catal. 114, 657–660.

Hu, Q., Sommerfeld, M., Jarvis, E., Ghirardi, M., Posewitz, M., Seibert, M., 2008. Microalgal triacylglycerols as feedstocks for biofuel production: perspectives and advances. Plant J. 54, 621–639.

Humphrey, A.M., 2004. Chlorophyll as a colour and functional ingredient. J. Food Sci. 69, 422–425.

John, D.M., 1988. Algal growths on buildings: a general review and methods of treatment. Biodeterioration Abstr. 2, 81–102.

John, R.P., Anisha, G., Nampoothiri, K.M., Pandey, A., 2011. Micro and macroalgal biomass: a renewable source for bioethanol. Bioresour. Technol. 102 (1), 186–193.

Khan, S.A., Rashmi, Hussain, Mir Z., Prasad, S., Banerjee, U.C., 2009. Prospects of biodiesel production from microalgae in India. Renewable Sustainable Energy Rev. 13, 2361–2372.

Kropat, J., Tottey, S., Birkenbihl, R.P., Depege, N., Huijser, P., Merchant, S.A., 2005. Regulator of nutritional copper signaling in Chlamydomonas is an SBP domain protein that recognizes the GTAC core of copper response element. Proc. Natl. Acad. Sci. U.S.A. 102, 18730–18735.

Lee, R.E., 1989. Phycology. Cambridge University Press, Cambridge, UK.

Lewis, L.A., Flechtner, V.R., 2002. Green algae (Chlorophyta) of desert microbiotic crusts: diversity of North American taxa. Taxon 51, 443–451.

Lewis, L.A., McCourt, R.M., 2004. Green algae and the origin of land plants. Am. J. Bot. 91, 1535–1556.

Luiten, E.E.M., Akkerman, I., Koulman, A., Kamermans, P., Reith, H., Barbosa, M.J., Sipkema, D., 2003. Realizing the promises of marine biotechnology. Biomol. Eng. 20, 429–439.

Lee, J.W., Mets, L., Greenbaum, E., 2002. Improvement of photosynthetic CO_2 fixation at high light intensity through reduction of chlorophyll antenna size. Appl. Biochem. Biotech. 98, 37–48.

Maheswaran, S., Mccormack, J.E., Sonenshein, G.E., 1991. Changes in phosphorylation of Myconcogene and Rb antioncogene protein products during growth arrest of the murine lymphoma Wehi-231 cell-line. Oncogene 6, 1965–1971.

Mata, T.M., Martins Antonio, A., Caetano Nidia, S., 2010. Microalgae for biodiesel production and other applications: a review. Renewable-Sustainable Energy Rev. 14, 217–232.

Merrick, W.C., 1992. Mechanism and regulation of eukaryotic protein-synthesis. Microbiol. Rev. 56, 291–315.

Meher, L.C., Vidya, S.D., Naik, S.N., 2006. Technical aspects of biodiesel production by transesterification—a review. Renewable-Sustainable Energy Rev. 10, 248–268.

Melis, A., Neidhardt, J., Benemann, J.R., 1998. Dunaliella salina (Chlorophyta) with small chlorophyll antenna sizes exhibit higher photosynthetic productivities and photon use efficiencies than normally pigmented cells. J. Appl. Phycol. 10, 515–525.

Melis, A., 1996. Extcitation energy transfer: functional and dynamic aspects of LHC (cab) proteins. In: Ort, D.R., Yocum, C.F. (Eds.), Oxygenic Photosynthesis: The Light Reactions. Kluwer Academic Publishers, Dordrecht, pp. 523–533.

Melis, A., Happe, T., 2001. Hydrogen production. Green algae as a source of energy. Plant physiol. 127 (3), 740–748.

Melis, A., 2009. Solar energy conversion efficiencies in photosynthesis: minimizing the chlorophyll antennae to maximize efficiency. Plant Sci. 177, 272–280.

Moen, E., 1997. Biological Degradation of Brown Seaweeds. Doctoral thesis submitted to Norwegian University of Science and technology. Referred by Kelly, M.S. and Dworjanyn, S., in "The Potential of Marine Biomass for Anaerobic Biogas Production" Argyll, Scotland: Scottish Association for Marine Science Oban; 2008.

Molnar, A., Bassett, A., Thuenemann, E., Schwach, F., Karkare, S., Ossowski, S., 2009. Highly specific gene silencing by artificial microRNAs in the unicellular alga Chlamydomonas reinhardtii. Plant J. 58, 165–174.

Michinaka, Y., Shimauchi, T., Aki, T., Nakajima, T., Kawamoto, S., Shigeta, S., 2003. Extracellular secretion of free fatty acids by disruption of a fatty acyl-CoA synthetase gene in Saccharomyces cerevisiae. J. Biosci. Bioeng. 95, 435–440.

Mussgnug, J.H., Thomas-Hall, S., Rupprecht, J., Foo, A., Klassen, V., McDowall, A., 2007. Engineering photosynthetic light capture: impacts on improved solar energy to biomass conversion. Plant Biotechnol. J. 5, 802–814.

Norton, T.A., Melkonian, M., Andersen, R.A., 1996. Algal biodiversity. Phycologia 35, 308–326.

Odum, H.T., 1971. Environment, Power and Society. Wiley-Interscience, New York, USA.

Pröschold, T., Leliaert, F., 2007. Systematics of the green algae: conflict of classic and modern approaches. In: Brodie, J., Lewis, J. (Eds.), Unravelling the Algae: The Past, Present and Future of Algal Systematics, The Systematics Association Special Volume Series, vol. 75. CRC Press, Boca Raton, London and New York, pp. 123–153.

Poulsen, N., Kroger, N., 2005. A new molecular tool for transgenic diatoms—control of mRNA and protein biosynthesis by an inducible promoter-terminator cassette. Febs J. 272, 3413–3423.

Qin, J., 2005. Bio-Hydrocarbons from Algae—Impacts of Temperature, Light and Salinity on Algae Growth. Rural Industries Research and Development Corporation, Barton, Australia.

Quinn, J.M., Merchant, S., 1995. 2 copper-responsive elements associated with the *Chlamydomonas* Cyc6 gene-function as targets for transcriptional activators. Plant Cell 7, 623—638.

Radakovits, R., Jinkerson, R.E., Darzins, A., Posewitz, M.C., 2010. Genetic engineering of algae for enhanced biofuel production. Eukaryot. Cell 9, 486—501.

Radmer, Richard J., April, 1996. Algal diversity and commercial algal products. BioScience 46 (4), 263—270. Marine Biotechnology.

Rindi, F., McIvor, L., Sherwood, A.R., Friedl, T., Guiry, M.D., Sheath, R.G., 2007. Molecular phylogeny of the green algal order Prasiolales (Trebouxiophyceae, Chlorophyta). J. Phycol. 43, 811—822.

Rindi, et al., 2009. An overview of the biodiversity and biogeography of terrestrial green algae. In: Rescigno, Vittore, et al. (Eds.), Biodiversity Hotspots. Nova Science Publishers, Inc.

Rodolfi, L., Zittelli, G.C., Bassi, N., Padovani, G., Biondi, N., Bonini, G., et al., 2008. Microalgae for oil: strain selection, induction of lipid synthesis and outdoor mass cultivation in a low-cost photobioreactor. Biotechnol. Bioeng. 102 (1), 100—112.

da Silva, R.L., Barbosa, J.M., 2008. Seaweed meal as a protein source for the white shrimp *Litopenaeus vannamei*. J. Appl. Phycol. 1—5.

Scharnewski, M., Pongdontri, P., Mora, G., Hoppert, M., Fulda, M., 2008. Mutants of *Saccharomyces cerevisiae* deficient in acyl-CoA synthetases secrete fatty acids due to interrupted fatty acid recycling. FEBS J. 275, 2765—2778.

Scott, S.A., Davey, M.P., Dennis, J.S., Horst, I., Howe, C.J., Lea-Smith, D.J., 2010. Biodiesel from algae: challenges and prospects. Curr. Opin. Biotech. 21, 277—286.

Searchinger, T., Heimlich, R., Houghton, R.A., Dong, F., Elobeid, A., Fabiosa, J., 2008. Use of US croplands for biofuels increases greenhouse gases through emissions from land-use change. Science 319 (5867), 1238—1240.

Sharma, N.K., Rai, A.K., Singh, S., Brown, R.M., 2007. Airborne algae: their present status and relevance. J. Phycol. 43, 615—627.

Singh, J., Gu, S., 2010. Commercialization potential of microalgae for production of biofuels. Renewable Sustainable Energy Rev. 14, 2596—2610.

Smith, A.M., 2008. Prospects for increasing starch and sucrose yields for bioethanol production. Plant J. 54, 546—558.

Song, D., Fu, J., Shi, D., 2008. Exploitation of oil-bearing microalgae for biodiesel. Chin J. Biotechnol. 24, 341—348.

Spolaore, P., Joannis-Cassan, C., Duran, E., Isambert, A., 2006. Commercial applications of microalgae. J. Biosci. Bioeng. 101, 87—96.

Stark, D.M., Timmerman, K.P., Barry, G.F., Preiss, J., Kishore, G.M., 1992. Regulation of the amount of starch in plant-tissues by ADP glucose pyrophosphorylase. Science 258, 287—292.

Thajuddin, N., Subramanian, G., 2005. Cyanobacterial biodiversity and potential applications in biotechnology. Curr. Sci. 89, 47—57.

Tetlow, I.J., Morell, M.K., Emes, M.J., 2004. Recent developments in understanding the regulation of starch metabolism in higher plants. J. Exp. Bot. 55, 2131—2145.

Tjaden, J., Mohlmann, T., Kampfenkel, K., Henrichs, G., Neuhaus, H.E., 1998. Altered plastidic ATP/ADP-transporter activity influences potato (Solanum tuberosum L.) tuber morphology, yield and composition of tuber starch. Plant J. 16, 531—540.

Ueda, R., Hirayama, S., Sugata, K., Nakayama, H., 1996. Process for the Production of Ethanol from Microalgae. U.S. Patent 5,578,472.

Vergara-Fernandez, A., Vargas, G., Alarcon, N., Velasco, A., 2008. Evaluation of marine algae as a source of biogas in a two-stage anaerobic reactor system. Biomass Bioenergy 32, 338—344.

Wang, J.Y., Zhu, S.G., Xu, C.F., 2003. Biosynthesis of lipids. In: Wang, J.Y., Zhu, S.G., Xu, C.F. (Eds.), Biochemistry, third ed. Higher Education Express, Beijing, pp. 257—298.

Wainright, P.O., Hinkle, G., Sogin, M.L., Stickel, S.K., 1993. Monophyletic origins of the metazoa: an evolutionary link with the fungi. Science 260, 340—342.

Wiessner, W., Schnepf, E., Starr, R.C., 1995. Algae, Environment and Human Affairs. Biopress Ltd., Bristol.

Wobbe, L., Blifernez, O., Schwarz, C., Mussgnug, J.H., Nickelsen, J., Kruse, O., 2009. Cysteine modification of a specific repressor protein controls the translational status of nucleus-encoded LHCII mRNAs in *Chlamydomonas*. Proc. Natl. Acad. Sci. U.S.A 106, 13290—13295.

Yoon, H.S., Hackett, J.D., Ciniglia, C., Pinto, G., Bhattacharya, D., 2004. A molecular timeline for the origin of photosynthetic eukaryotes. Mol. Biol. Evol. 21, 809—818.

Yu, T.S., Zeeman, S.C., Thorneycroft, D., Fulton, D.C., Dunstan, H., Lue, W.L., 2005. alpha -Amylase is not required for breakdown of transitory starch in *Arabidopsis* leaves. J. Biol. Chem. 280, 9773—9779.

Zabawinski, C., Van den Koornhuyse, N., D'Hulst, C., Schlichting, R., Giersch, C., Delrue, B., et al., 2001. Starch-less mutants of Chlamydomonas reinhardtii lack the small subunit of a heterotetrameric ADP-glucose pyrophosphorylase. J. Bacteriol. 183, 1069 1077.

Zeng, X., Danquah, M.K., Chen, X.D., Lu, Y., 2011. Microalgae bioengineering: from CO_2 fixation to biofuel production. Renewable Sustainable Energy Rev. 15, 3252—3260.

Microalgal Classification: Major Classes and Genera of Commercial Microalgal Species

Kirsten Heimann[1,2], *Roger Huerlimann*[1,2]

[1]James Cook University, College of Marine and Environmental Sciences, Townsville, QLD, Australia;
[2]James Cook University, Centre for Sustainable Fisheries and Aquaculture, Townsville, QLD, Australia

1. INTRODUCTION TO MICROALGAL CLASSIFICATION

The hierarchical organization of life into kingdoms, phyla (divisions), classes, etc. predates the idea of evolution (existing forms of life arose from preexisting ancestors). As human social organization is hierarchical, hierarchical classification is deeply seated in human nature (Ragan, 1998). This causes problems with classification with regard to monophyly, particularly of the "lower" or less-complex unicellular and multicellular forms, the so-called protists to which the eukaryotic microalgae and the multicellular brown seaweeds belong. Development of powerful light microscopy and of the idea of evolution (Darwin and Wallace), as well as observations that some life forms did not fit neatly into either the animal or plant kingdom, that is, in terms of motility, trophic status, etc. led several taxonomists to suggest a third kingdom (as summarized in (Rothschild, 1989)), the kingdom Protozoa (Owen, 1860), the Primigenum containing the

Protoctista (Hogg, 1860), and the Protista (Haeckel, 1866), the latter being based on a phylogenetic framework. As such, the original kingdom Protista included amoebae, bacteria (initially until removed by Copeland in 1938), myxomycetes, euglenoids, diatoms, some green algae (e.g., *Volvox*), radiolaria, foraminifera, rhizopods, sponges, and some dinoflagellates, while the seaweeds (red, brown, and macroscopic green algae) as well as filamentous cyanobacteria, desmids, fungi, lichens, and charophytes were grouped with the kingdom Plantae and ciliates to the kingdom Animalia (reviewed in Ragan, 1998).

Later, Copeland (1956, as reviewed in Ragan, 1998) renamed the kingdom Protista to Protoctista to include the brown and red algae, the pyrrophytes (including cryptomonads, dinoflagellates, and euglenoids), chytrids, fungi, myxomycetes, rhizopods, heliozoans, myxozoans, sacrodinids, gregarines, sporozoans, and ciliates, while the green algae remained with the Plantae. The kingdom Protista thus represents a collection of unrelated organisms that could not easily be classified elsewhere (Maneveldt and Keats, 2004). With regard to phylogenetic relationships, retaining the kingdom Protista is not defendable from a cladistic point of view due to its polyphyletic nature, but classification aside, this group of organisms can be referred to as protists, for the sake of convenience. The three-kingdom classification was later successively expanded to eight kingdoms: the Eubacteria (containing the cyanobacteria) and the Archae (formerly in one kingdom Monera), Animalia, Plantae, Fungi, Archaezoa, Protozoa, and the new kingdom, Chromista, the latter containing the brown algae (phaeophytes), the diatoms (bacillariophytes), the golden-brown algae (chrysophytes), the oomycetes, and the silicoflagellates (Cavalier-Smith, 1993); this was later reduced back to six kingdoms, the Bacteria, the Protozoa being basal to the Animalia, Plantae, and Chromista (Cavalier-Smith, 1998), using ultrastructural and reproductive traits. Inclusion of molecular data and information derived from multigene trees led to the reclassification of the Chromista and Protozoa (Cavalier-Smith, 2010) (Table 1).

The new kingdom Chromista is a good example for classification flux. Originally defined by ultrastructural characteristics of plastid (photosynthetic forms have plastids with chlorophyll a and c, which are bordered by a periplastidial membrane and situated additionally in the rough endoplasmic reticulum (ER), which is continuous with the nuclear envelope) and the arrangements of flagellar appendages (presence of stiff bipartite or tripartite hairs on either one or both flagella), which was then expanded to include organisms as per the chromalveolate theory based on the argument that the complex protein import machinery required for plastids surrounded by more than two envelopes was unlikely to evolve more than once (Sym and Maneveldt, 2011). Genetic data have added more complexity to the classification of the algae, as the plastids are of endosymbiotic origin (see below); thus algae/plants have one genome (the plastidial one) in addition to the two common eukaryotic ones (mitochondrial—mitochondria also arose from a single endosymbiotic event with an aerobic bacterium—and nuclear genomes). This makes it much harder to trace the origin of genes (acquired endosymbiotically (e.g., vertically) versus lateral and horizontal gene transfers). In addition, taxon sampling efforts can also skew conclusions regarding phylogeny (Sym and Maneveldt, 2011). For example, analyses of the nuclear eukaryotic elongation factor 2 and the plastid-targeted gene product glyceraldehyde-3-phosphate dehydrogenase support the close relationship of the alveolates with the stramenopiles but not with the haptophytes and cryptophytes (as reviewed in Sym and Maneveldt, 2011). The argument whether the kingdom Chromista is mono- or polyphyletic is raging (compare Cavalier-Smith, 2010 with Baurain et al., 2010) and is based on whether one accepts multiple secondary endosymbiotic events for plastid acquisition or not.

In essence, many scientists have attempted to reclassify the algae/protists and no attempt has been satisfactory, and, given that algae are the research basis of many different fields in biology (not just systematics), many different classifications are currently in use. The constant renaming/reclassification and moving of organisms from one hierarchical category to another without following the clear aim to better represent likely phylogenies is of no help in resolving this issue and certainly does not assist in a unifying, even if not monophyletic, representation of the genera and species within the current scientific literature. Since, as elaborated using the kingdoms Protista and Chromista, it is presently impossible to arrive at a classification that unambiguously groups monophyletic lineages, it is perhaps best to dissuade further hierarchically inspired classification attempts and to adopt groupings that can be more readily refined, as more information (support or otherwise) becomes available (see Figure 1; Adl et al., 2005).

TABLE 1 Algae Phyla Distribution as per Six-Kingdom Classification Scheme by Cavalier-Smith (1998, 2010) with Genera of Commercial Importance Included Where Possible

Kingdom	Subkingdom	Infrakingdom	Phylum/Subphylum	Examples
Plantae	Biliphyta[a]	Glaucophyta[a]	Glaucophyta	
		Rhodophyta[a]	/Rhodellophytina[a]	
			/Macrorhodophytina[a]	
	Viridiplantae	Chlorophyta	/Chlorophytina[a]	Prasinophytae[a]
				Tetraphytae[a] (*Ulva*, *Chlamydomonas*, *Chlorella*, *Tetraselmis*)
			/Phragmophytina[a]	Charophytae
				Rudophytae[a]
Chromista	Harosa[a]	Heterokonta	Ochrophyta	Diatoms, pedinellids, silicoflagellates, pelagophytes, brown algae, xanthophytes, chrysophytes, eustigmatophytes (e.g., *Nannochloropsis*), raphidophytes
			Bigyra	Labyrinthulae, e.g., Traustochytrids
		Alveolata	Myozoa	Dinoflagellates (e.g., *Crypthecodinium*), Perkinsis, Apicomplexa (incl. *Chromera*)
			Ciliophora	Ciliates
		Rhizaria	Cercozoa	Chlorachniophyta, e.g., *Bigelowiella*
			Retaria	Foraminifera, Radiozoa
	Hacrobia[a]		Haptophyta	*Pavlova*, *Prymnesium*, *Isochrysis*, *Chrysochromulina*
			Cryptista	e.g., *Rhodomonas*, *Cryptomonas*
			Heliozoa	Centrohelia
Protozoa	Eozoa	Euglenozoa	Euglenozoa	Euglenoidea, e.g., *Euglena*, Diplonemea, Postgaardea[a], Kinetoplastea

[a] *New classifications.*

2. ORIGIN OF MICROALGAL DIVERSITY

The extant vast algal diversity arose from a complicated evolutionary history, giving rise to widespread acquisition of photosynthesis over different taxonomic groups. Eukaryotic life is currently divided into six major supergroups, including Opisthokonta, Amoebozoa, Archaeplastida, Rhizaria, Chromalveolata, and Excavata (Adl et al., 2005; Keeling et al., 2005), of

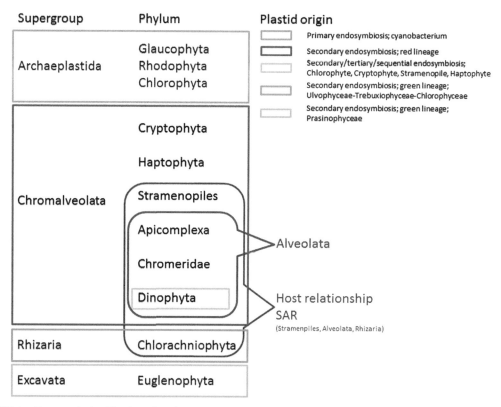

FIGURE 1   Taxonomic classification, plastid origins, and host phylogenetic relationships of protistan/algal phyla containing plastid-bearing members.

which the last four contain photosynthetic members (see Figure 1).

The endosymbiosis of a cyanobacterium with a heterotrophic, eukaryotic host is called primary endosymbiosis and leads to a primary endosymbiotic plastid. There is strong evidence that the plastids of the Viridiplanta (land plants and green algae), the Rhodophyta (red algae), and the Glaucophyta (a small group of freshwater microalgae), all containing plastids surrounded by two envelope membranes, evolved from a single primary endosymbiotic event involving a cyanobacterium, forming the supergroup Archaeaplastida (Archibald, 2012; Ball et al., 2011) (see Figure 1).

The endosymbiosis of a photosynthetic eukaryotic organism containing a primary plastid with a heterotrophic, eukaryotic host is called secondary endosymbiosis and leads to a secondary endosymbiotic plastid. While the Archaeaplastida only contain organisms with primary plastids, secondary and tertiary plastids are restricted to members of the Rhizaria (Chlorarachniophyta), Chromalveolata (Cryptophyta, Stramenopiles, Haptophyta, Apicomplexa, Chromerida, and certain Dinoflagellata), and Excavata (Eugleonophyta) (see Figure 1).

The involvement of two organisms, a host and an endosymbiont, each with different phylogenetic relationships of their own, impedes the resolution of the complex relationships of these algae. This is even more confounded by the presence of plastid-lacking organisms, which either lost their plastids or never acquired one. Looking at the phylogenetic relationships on a host level, recent phylogenetic analyses group the Stramenopiles together with the Alveolata (comprised of Apicomplexa, Chromerida, and Dinoflagellata) and the green plastid-containing Chlorarachniophyta (Rhizaria), which are commonly abbreviated as SAR (see Figure 1). The Haptophyta form a sister group to the SAR, the Cryptophyta being more closely related to the Viridiplantae (Burki et al., 2012), while the Euglenophyta (Excavata) are only distantly related to all the other groups (Figure 1).

On a plastid level, secondary endosymbiotic events can be divided based on the origin of the plastid, either from a Chlorophyta (green lineage) or a Rhodophyta (red lineage), while no endosymbiotic events from Glaucophyta are reported. In the green lineage, two independent endosymbiotic events from an ancestral core Chlorophyta (Ulvophyceae—Trebuxiophyceae—Chlorophyceae) and a Prasinophycea gave rise to the Chlorarachniophyta (Rhizaria) and Euglophyta (Excavata), respectively (Rogers et al., 2007; Turmel et al., 2009) (see Figure 1). This is supported by the former host organisms of Rhizaria and Excavata, not being closely related, while their individual chloroplasts are genetically closely related (Archibald, 2009; Keeling, 2010). By far the most diversity can be found in the red lineage, which includes the Cryptophyta, Haptophyta, Stramenopiles, Apicomplexa, and Chromerida (see Figure 1). There is strong genetic and phenotypic evidence that all plastids were derived from the same ancestral red alga (Bodył et al., 2009; Keeling, 2010); however, the exact circumstances of this acquisition have yet to be resolved.

The phylum Dinoflagellata, which includes plastids from at least four different sources, forms a special case. The plastids of dinoflagellates are either derived from tertiary endosymbiosis, where a heterotrophic eukaryote took up a secondary plastid-containing alga, or from sequential secondary endosymbiosis, where a plastid was lost and replaced by endosymbiosis of a primary plastid-containing alga. Therefore, the phylum Dinoflagellata contains plastids derived from Chlorophyta, Cryptophyta, Stramenopiles (heterokontophyta), and Haptophyta.

3. MAJOR PHYLA/CLASS CHARACTERISTICS OF COMMERCIAL MICROALGAL GENERA

3.1 Chlorophyta

The members of the phylum Chlorophyta can be found in freshwater, marine, or even terrestrial environments and include unicellular and multicellular members possessing chlorophylls *a* and *b* in a single chloroplast surrounded by two envelope membranes. *Chlorella vulgaris*, *Dunaliella salina*, and *Haematococcus pluvialis* are unicellular representatives of the phylum Chlorophyta and used today in commercial productions, while *Parietochloris incisa* and *Botryococcus braunii* show potential for lipid and hydrocarbon production, respectively (see below). The UTS (Ulvophyceae, Trebouxiophyceae (*Chlorella*, *Perietochloris*, *Botryococcus*)), chlorophyceae (*Dunaliella*, *Haemotococcus*) classes are characterized by closed mitosis, a cruciate flagellar microtubular root system, absence of a multilayered structure, and phycoplast formation-assisted cytokinesis (Leliaert et al., 2012). At the light microscopical level, the freshwater alga *C. vulgaris* is a fairly nondescript round nonmotile reproductive cell ranging in size from 2 to 10 μm with a single chloroplast, containing amylose and amylopectin-based starch granules, a single pyrenoid containing high amounts of ribulose-1,5-bisphosphate carboxylase/oxygenase (RuBisCO) for CO_2 fixation (Safi et al., 2014). It reproduces vegetatively, producing four daughter cells that are liberated from the mother autospore cell well by rupture. The cell wall develops in complexity with cell age being tougher in older cells, with glucosamine incorporation providing cell wall rigidity, but sporopollenin incorporation was also reported when grown heterotrophically (Martinez et al., 1991; Safi et al., 2014). Unlike *Chlorella*, vegetative cells of the marine *D. salina* are photoautotrophic biflagellate (flagella are of equal length), naked (without cell walls but surrounded by a glycocalix) unicells, which divide vegetatively by binary fission and sexually by producing 32 daughter cells that are liberated by rupture of the zygote wall, which contains sporopollenin (Borowitzka and Borowitzka, 1988). Cell shape and size vary, but shape is typically oval (4–15 μm wide and 6–25 μm long) under good growth conditions, becoming round under unfavorable environmental or growth conditions (Tran et al., 2013). Like *Chlorella*, cells of *D. salina* contain a single cup-shaped chloroplast with a central pyrenoid surrounded by starch granules (Cavalier-Smith, 1998). *H. pluvialis* is a freshwater unicell with biflagellate and palmelloid forms. Motile stages are pear-shaped (8–50 μm in diameter) biflagellate, with two flagella of equal length emerging from the anterior papilla and cells are surrounded by a cellulose wall, while a thickened gelatinous layer traversed by simple or branched plasma strands is observed in palmelloid stages (Boussiba, 2000). The cup-shaped chloroplast contains several scattered pyrenoids, and several contractile vacuoles are present in the cytosol. Unfavorable conditions induce encystment, when the cell produces astaxanthin and surrounds itself with a thick cellulose wall impregnated with a sporopollenin-like substance (Boussiba, 2000). The Mt. Tateyama (Toyema Prefecture, Japan) isolate of *P. incisa* was previously classified as *Myrmecia incisa* but has since been renamed due to the presence of an inconspicuous pyrenoid surrounded sparsely by thin starch granules within the deeply incised parietal chloroplast (Watanabe et al., 1996). Cells are round to oval (10–15 μm in diameter) covered typically by a thick cell wall, which is uniform and even in round cells, but may be thinner in the elongated parts of ovoid cells. *B. braunii* is a floating colonial trebouxiophyte where 50–100 oval cells are embedded in a complex hydrocarbon extracellular matrix, containing *n*-hexane-extractable liquid hydrocarbons and fibrous polymerized hydrocarbons, which are cross-linked by race-specific hydrocarbons (see below) and not extractable with *n*-hexane (Weiss et al., 2012). Colonies are surrounded by a carbohydrate-based retaining wall composed of arabinose and galactose, while cell walls surrounding the cells are the typical α-1,4- and β-1,3-linked polysaccharides (Weiss et al., 2012).

3.2 Rhodophyta

The members of the phylum Rhodophyta include mainly marine multicellular species, while freshwater or unicellular species are rare. Commercialization of the unicellular *Porphyridium cruentum* is the research and development phase in Israel and France (Table 2). Cells are spherical with an eccentric nucleus and contain a large single chloroplast surrounded by two envelope membranes with a single central pyrenoid (Gantt and Conti, 1965). Phycobilisomes, containing phycocyanin and phycoerythrin, line the stromal side of thylakoids. Cells are surrounded by fibrillar and diffuse mucilage with thickness increasing with increasing age and composed of carbohydrates with xylose being one of the main sugars (Gantt and Conti, 1965). *Rhodella reticulata* is another member of the Porphyridales with promising characteristics for phycobilin production. The brown-to-olive-colored unicells are 8–32 μm in diameter coccoid, nonmotile with a single pyrenoid-containing chloroplast, and the mucilaginous sheath may be unilaterally thickened (Deason et al., 1983). The plastid is deeply incised toward the central pyrenoid with the resulting chloroplast lobes radiating out to the cell surface. The pyrenoid is traversed by thylakoids, a distinguishing feature of *R. reticulata* from other species of the genus (Deason et al., 1983).

TABLE 2 Morphological and Biochemical Characteristics of the Labyrinthulomycetes

Morphological feature	*Schizochytrium*[a]	*Auranthiochytrium*[a]	*Thraustochytrium*[b]	*Ulkenia*[d]
Thallus	Thin walled Globose Pale yellow	Thin walled Globose Orange	Amoeboid colonies Light orange Spherical Pale cream	Thin walled Globose-pear-shaped thallus naked and amoeboid at maturity exiting the sporangium through a single opening
Colony	Large	Small	—	Small
	Binary fission	Continuous binary fission	No binary fission protoplasts divides into tetrads and octads for direct release of zoospores	
Zoospores	Biflagellate heterokonts Rentiform-ovoid	Biflagellate heterokonts Rentiform-ovoid	Biflagellate heterokonts Release from thallus	Biflagellate heterokonts Release from zoosporangium which originates from the thallus
Pigments	β-carotene only	Astaxanthin Phenicoxanthin Canthaxanthin β-carotene	Astaxanthin[c]	Astaxathin Phenicoxanthin Echininone β-carotene
Dominant FAs	Arachidonic acid (AA—20% of TFA)	Docosahexaenoic acid (DHA) (∼23% of TFA[b])	Oleic acid (18:1) (∼71% of TFA)	DHA (50−70% of TFA) depending on strain

AA, arachidonic acid; DHA, docosahexaenoic acid; FAs, fatty acids; TFA, total fatty acids.

[a] *Yokohama, R., Honda, D., 2007. Taxonomic rearrangement of the genus* Schizochytrium *sensu lato based on morphology, chemotaxonomic characteristics, and 18S rRNA gene phylogeny (Thraustochytriacea, Labyrinthulomycetes): emandation for Schizochytrium and erection of* Auranthichytrium *and* Oblongichytrium *gen. nov. Myoscience 48, 199−211.*

[b] *Arafiles, K.H.V., Alcantara, J.C.O., Cordero, P.R.F., Batoon, J.A.L., Galura, F.S., Leaño, E.M., Dedeles, G.R., 2011. Cultural optimization of thraustochytrids for biomass and fatty acid production. Mycosphere 2(5), 521−531.*

[c] *Yamaoka, Y., Carmona, M.L., Oota, S., 2004. Growth and carotenoid production of* Thraustochytrium *sp. CHN-1 cultured under superbright red and blue light-emitting diodes. Biosci. Biotechnol. Biochem. 68 (7), 1594−1597. http://dx.doi.org/10.1271/bbb.68.1594.*

[d] *Yokohama, R., Salleh, B., Honda, D., 2007. Taxonomic rearrangement of the genus* Ulkenia *sensu lato based on morphology, chemotaxonomical characteristics, and 18S rRNA gene phylogeny (Thraustochytriaceaea, Labyrinthulomycetes): emendation for Ulkenia and erection of Botryochytrium, Parietichytrium and Sicyoidochytrium gen. nov. Mycoscience 48, 329−341.*

3.3 Haptophyta

The algae belonging to the phylum Haptophyta are mainly marine and unicellular or colonial, although some freshwater species are known. *Isochrysis* aff. *galbana* (T-ISO) and *Pavlova salina* are the two best-known examples of Haptophyta used as feed microalgae in aquaculture. All haptophytes contain one or two pyrenoid-containing chloroplast(s) and an antapical nucleus with the nuclear envelope being continuous with the chloroplast ER and a peripheral ER underneath the plasma membrane

(Edvardsen et al., 2000). A fan-shaped Golgi apparatus is located near the flagellar basal bodies and the base of the haptonema. Based on molecular and morphological differences, the authors removed *P. salina* from the genus and erected a new genus, *Rebecca*, with the following morphological characteristics: yellowish-green biflagellate cells, containing a single chloroplast, are elongate and slightly compressed bearing two flagella of unequal length (the longer is covered with fine nontubular hairs with a pit or canal at its base), a short haptonema and a stigma (eyespot is absent). T-ISO was also recently reexamined morphologically and genetically and is considered distinctly different in both aspects from *Isochrysis galbana*, and was consequently reclassified as *Tisochrysis lutea*, with the first letter of the genus referring to the Tahitian origin and the species name describing the orange color of older cultures (Bendif et al., 2013). Morphologically, *T. lutea* is almost indistinguishable from *I. galbana*, but the color of old cultures of the latter is brown. Cells are typically covered with several layers of organic scales, are round to oblong, 3–7.5 μm in diameter, biflagellate with apically inserting flagella of equal length (7 μm) and a short, scaly haptonema of 100-nm length possessing a central swelling. The single plastid is parietal containing a single pyrenoid. Nonmotile cells are embedded in thin layers of mucilage.

3.4 Stramenopiles

3.4.1 Eustigmatophyceae

The members of the class Eustigmatophyceae characterized by containing only chlorophyll *a*, not *c* as the other stramenopiles, are all unicellular and can be found in marine, freshwater, and terrestrial environments. *Nannochloropsis oculata* is used frequently as a live aquaculture feed (see below) and will be described here. The genus *Nannochloropsis* contains five species, of which only *N. limnetica* occurs in freshwater, with few differentiating features (Suda et al., 2002). Unicells are oval to round, not exceeding 5 μm in diameter, with a characteristic papilla, nonmotile containing a single mitochondrion, Golgi body, and plastid, the latter with an eccentric pyrenoid (Suda et al., 2002).

3.4.2 Bacillariophyceae

The members of the class Bacillariophyceae can be found in freshwater and marine environments, and are mainly unicellular, although some are colonial. A key feature of this class is the production of a siliceous frustule covering the cell, and many are used as aquaculture feed (see below). The centric diatoms *Skeletonema costatum*, *Chaetoceros muelleri*, and *Thalassiosira pseudonana* are cultivated at commercial scales (Table 2); distinctive diagnostic features will be briefly described here. The chain-forming diatom *S. costatum* is characterized by cylindrical cells containing two chloroplasts where the face of the valve margin bears long processes (costae), which connect with processes originating from the adjacent daughter cell valve face (Sarno and Kooistra, 2005). The valve face diameter for the genus is 2–38 μm. Detailed light and scanning electron microscopy, complemented with molecular work by Sarno and Kooistra (2005) and Zingone et al. (2005), however, shows that the species *S. costatum* represents a species complex, and the reader will need to refer to these publications in order to differentiate between the species in the *S. costatum* species complex. However, in the context of industrial use, it is relevant to note that *S. costatum* cultures CCAP 1077/3 (Genbank accession number X85395), CCMP 780, UBC18/C (Genbank accession number M54988.1), and CCAP1077/6 (CS76 see reference in (Sarno and Kooistra, 2005)) have been misidentified, with CCAP1077/6 from Australian waters being *S. pseudocostatum*, while the others are *S. grethae*. As a model for physiological studies, *T. pseudonana* was the first marine phytoplankton for which the genome was completely sequenced (Armbrust et al., 2004). The genus *Thalassiosira* contains >100 species, with T. *pseudonana* being a marine small (4−6 μm), centric, unicellular diatom belonging to the Coscinodiscophyceae and quite indistinctive by light microscopy. The detailed morphology of the siliceous frustule (cell wall) varies with the availability of silica; however, a ring of 6−14 strutted marginal processes is present, with one labiate process also present along with one subcentral strutted process (Hasle and Syvertsen, 1997). The centric diatom C. *muelleri* is frequently used in aquaculture, as the cells are weakly silicified and the cells are solitary. The valves of vegetative cells are elliptical (4.5−20 μm in diameter) in valve view with four long setae (one on each corner of the epi- and hypovalves, respectively), while auxospores are spherical (6 μm in diameter) and more heavily silicified, but nonetheless unornamented by light and scanning electron microscopy (Reinke, 1984). In contrast, the valve diameter of resting spores is similar to the vegetative cells, but setae are absent (Reinke, 1984).

3.4.3 *Labyrinthulomycetes*

The Labyrinthulomycetes are heterotrophic, filamentous protists that are classified in the Stramenopiles, comprised mainly of marine species. Most are saprotrophic decomposers, with some acting as parasites (Tsui et al., 2009). *Auranthiochytrium*, *Schizotrichium*, *Thraustochytrium*, and *Ulkenia* are of commercial interest for pigment and fatty acid production, which together with morphological criteria are being used to distinguish between the genera (see Table 2 for a summary of morphological and chemotaxonomic details).

3.5 Dinophyta

The phylum Dinophyta contains members that are unicellular and mainly marine, with some freshwater species. Only approximately half of the Dinophyta are photosynthetic, while the other 50% are heterotrophs without chloroplasts. *Crypthecodinium cohnii* is of commercial interest for the heterotrophic production of

docosahexaenoic acid (DHA). It is a marine, heterotrophic, colorless dinoflagellate characterized by dinokont flagellation, where the transverse flagellum lies in a medial encircling cingulum, which is displaced and descends downward on the left, when viewed from ventral (Prabowo et al., 2013). The sulcus, which houses the longitudinal flagellum, is broad. Cells can be motile or nonmotile, are oval or round (sometimes in ventral or dorsal view), and two cell sizes were observed in culture: 5−20 and 8−30 μm, which might represent different life cycle stages, i.e., the larger, nonmotile cells could be cysts, which are characterized by thin cellulosic plates. These cysts give rise to multiple daughter cells produced by multiple fission of the cyst protoplast. Planozygotes resulting from sexual reproduction within the same clonal culture are characterized by bearing two transverse and two longitudinal flagella (Prabowo et al., 2013).

4. BIOTECHNOLOGICAL APPLICATIONS: MICROALGAE

4.1 Bioremediation of Waste Waters

The remediation of wastewaters does not typically rely on the cultivation of a single species, as the end product of the process is the provision of clean(er) water as opposed to a biomass-derived product. In most large-scale operations, the native algal flora in the to-be-remediated wastewater is encouraged to grow fast (bloom) and through this fast growth absorb the inorganic nutrients it requires into the biomass. A classic example is the use of an algal turf scrubber (ATS) system for the cultivation of periphyton (biofilms) for the remediation of secondary effluent in Patterson, California. The system was used for phosphorous removal and was 152.4 m long and 6.5 m wide, handling between 436 and 1226 m^3 of effluent per day. Yearly phosphorous removal ($0.73 \, \mathrm{g \, m^{-2} \, d^{-1}}$) could be optimized by regulating hydraulic load rates and pH of the water, and the system produced biosolids with an average mean productivity of 35 g m^{-2} d^{-1} (Craggs et al., 1996). Another memorable commercial-scale microalgae-based wastewater remediation study was conducted in New Zealand, but this time, the resulting biomass was considered for value-adding coproduct production (biofuel), and the native microalgal flora was encouraged to bloom in suspension-based high rate algal pond (HIRAP) cultivation systems. Suspension systems have the disadvantage over biofilm cultivation in that dewatering of the biomass is required and HIRAPs very rarely contain more than 1 g dry weight (DW) L^{-1}. As such, dewatering is a major cost and energy issue. In the New Zealand HIRAP study, the systems were enriched with CO$_2$ and a native flora of green freshwater algae established, which could be separated from the large volume of water by flocculation (Park et al., 2011).

Microalgae have also been used for the remediation of wastewaters containing high metal loads, such as nickel from mining and metallurgy, stainless steel, electroplating, battery and accumulator manufacturing wastewaters. *C. vulgaris* was trialed in that regard and a biosorption of 60 mg Ni g^{-1} DW was achieved from a water source containing 250 mg L^{-1} (Aksu, 2002). Other experiments have recently been performed with native microalgae consortia in synthetic mine drainage to evaluate remediation and commercial potential (Orandi and Lewis, 2013). In this study, rotating biofilm cultivation was used to study the uptake, removal efficiency, and metal release by the biomass, which was enriched with typically present green freshwater microalgae, such as *Chlorella*, *Scenedesmus*, *Ankistrodesmus*, *Franceia*, *Mesotaenium*, and cyanobacteria (Orandi and Lewis, 2013, and references therein), which were also essentially present in the HIRAP wastewater treatment study, which suggests that these native microalgal consortia are stable and competent for nutrient- and metal-rich wastewater remediation applications (Figure 2).

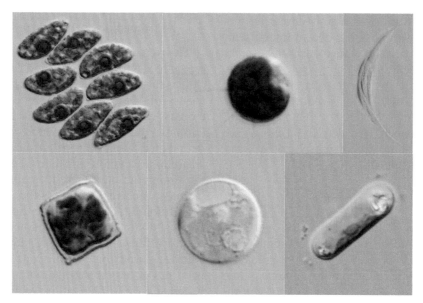

FIGURE 2   Micrographs of green freshwater microalgae growing in wastewater for bioremediation. From left to right, top: *Scenedesmus obliquus*, *Franceia* sp., *Ankistrodesmus* sp.; bottom: *Tetraedron* sp., *Chlorella* sp., *Mesotaenium* sp.

4.2 Bioproducts and Bioenergy

The commercial use of green microalgae primarily focuses on dietary supplementation through either whole algae products, or algal extracts. The freshwater microalgae *C. vulgaris* is the main alga used as dietary supplements for humans (Molnár et al., 2013) and animals (Kang et al., 2013). A summary of microalgal products and how these relate to the organism's taxonomy is provided in Table 3.

Green microalgae are also high in antioxidants, of which β-carotene and astaxanthin are of special interest. The halophile microalgae *D. salina* is commercially grown to produce β-carotene due to its high content, but lacks astaxanthin (Guedes et al., 2011). In contrast, the freshwater alga *H. pluvialis* is used to produce astaxanthin. The highest amounts of astaxanthin are found in the resting stages of *H. pluvialis*, which are formed under unfavorable environmental conditions like nutrient deprivation, increased salinity, high irradiance, and temperature (Collins et al., 2011). Astaxanthin, a strong antioxidant, is used as a nutritional supplement and anticancer agent, and exhibits preventive properties for diabetes, cardiovascular diseases, and neurodegenerative disorders, and also stimulates immunization (Ambati et al., 2014; Guedes et al., 2011). Other potential producers of astaxanthin are the freshwater algae *Chlorella zofingiensis* (Mulders et al., 2014) and *Chlorococcum* sp. (Ambati et al., 2014).

The production of lipids other than pigments in general, and fatty acids in particular, is usually not associated with green algae on a commercial scale. However, there are two exceptions, the production of arachidonic acid in *P. incisa* and long-chain hydrocarbons in *B. braunii*. The freshwater alga *P. incisa* (Chlorophyceae[C]) is considered for the commercial production of arachidonic acid, which is important in the production of infant formula (Tababa et al., 2012). *B. braunii* is a colonial microalga that can be found in freshwater, brackish lakes, reservoirs, and ponds, and is noted for its high lipid content, especially hydrocarbons (Metzger and Largeau, 2005). Currently there are three races of *B. braunii*, which are classified based on the type of hydrocarbons they accumulate. Race A mainly produces *n*-alkadiene and triene hydrocarbons (Metzger et al., 1985), race B produces triterpenoid hydrocarbons (Metzger et al., 1985) and methylated squalenes (Achitouv et al., 2004; Huang and Dale Poulter, 1989) and race L produce lycopadiene, a tetraterpenoid hydrocarbon (Metzger et al., 1990; Metzger and Casadevall, 1987).

Furthermore, there are three physiological states that have been characterized for *B. braunii*. In actively growing green colonies, nearly 17% of the dry weight is made up of a complex mixture of hydrocarbons, while the brown resting stages accumulate hydrocarbons up to 86% of their dry weight (Brown et al., 1969). The third state consists of large green cells with low hydrocarbon content, into which the resting stage develops once they are inoculated into fresh medium (Brown et al., 1969).

The majority of commercial Rhodophyta genera are multicellular and are mainly used for human consumption; however, certain genera of unicellular Rhodophyta are being used for commercial production. The genus *Porphyridium* (Porphyridiophyceae[C]) has great potential in the production of polysaccharides (Arad and Levy-Ontman, 2010), arachidonic acid (Ahern et al., 1983), or phycobilins (Kathiresan et al., 2007). Microalgal polysaccharides, which have a variety of biotechnological uses in medicine, nutrition, and cosmetics, could serve as replacement of the depleted stocks of naturally harvested red and brown macroalgae (Arad and Levy-Ontman, 2010). Arachidonic acid from red algae can be used in the same way as from green algae. Lastly, Phycobilins are water-soluble pigments found in the chloroplasts of Rhodophyta, which can be used for a wide range of applications including food and cosmetic

TABLE 3   Microalgae Bioproducts and Companies

Microalgal species	Bioproduct	Major producing countries (companies)
GREEN MICOALGAE		
Chlorella vulgaris	Dietary supplement	Taiwan, Germany, Japan[a] (Sun *Chlorella*, Japan with headquarters in California, Indonesia, England, Germany, Brazil, China), Yaeyama Shokusan (Japan), Wudi Xinhui (China)
Dunliella salina	β-carotene	Australia[a] (Cognis), Israel[a] (NBT), USA, China
Haematococcus pluvialis	Astaxanthin	USA, India, Israel[a]
RED MICROALGAE		
Porphyridium cruentum[e]	Polysaccharides, phycobilins	Israel, France[b]
HAPTOPHYTES		
Pavlova salina	EPA, DHA aquaculture feed	Global[b]
Tisochrysis lutea (formerly Isochrysis aff. galbana, T-ISO)	DHA aquaculture feed	Global[b]
STRAMENOPILES		
Chaetoceros muelleri	Aquaculture feed	Global[b]
Nannochloropsis spp.	Aquaculture feed	Global[b]
Skeletonema spp.	Aquaculture feed	Global[b]
Thalassiosira pseudonana	Aquaculture feed	Global[b]
Schizochytrium sp.	DHA, EPA enrichment of nutritional products (e.g., cereals, etc.)	DSM Nutritional Products LLC[c] Omega Tech[d]
DINOPHYTA		
Crypthecodinium cohnii	DHA enrichment of infant formulae etc.	Martek[a]

[a] Spolaore, P., Jaonnis-Cassan, C., Duran, E., Isambert, A., 2006. Commercial applications of microalgae. J. Biosci. Biotech. 101, 87–96.
[b] Paul, N., Tseng, C.K., Borowitzka, M., 2012. Seaweed and Microalgae. In: Lucas, J.S., Southgate, P.C. (Eds.), Aquaculture: Farming Aquatic Animals and Plants. second ed. John Wiley & Sons Ltd., Chichester, p. 268.
[c] Jones, C., 2012. DHA and EPA rich oil from the microalga Schizochytrium. NFU 786. Foods Standards Agency, London.
[d] ANZFA, 2002. DHA-rich dried marine micro algae (Schizochytrium sp.) and DHA-rich oil derived from Schizochytrium sp. as novel food ingredients. Australia New Zealand Food Authority.
[e] Research-and-development phase.

colorants (Borowitzka, 2013), biotechnological uses like fluorescent tags for flow cytometry and immunology (Glazer, 1994), or medical uses as photosensitizers in cancer treatment (Hu et al., 2008; Pan et al., 2013). *Rhodella* (Rhodellophyceae[C]) is another genus with commercial potential

for the production of phycobilins (Yen et al., 2013).

The main commercially used genera of Haptophyta are *Isochrysis* (Prymnesiophyceae[C]) and *Pavlova* (Pavlovophyceae[C]), which are important aquaculture feeds due to their high content of

DHA and generally high nutritional value (Guedes and Malcata, 2012). Due to the importance of DHA in human health, these two genera are also of interest as a replacement for fish oil, although the high production cost is still prohibitive for large-scale commercial production (Gladyshev et al., 2013). *Isochrysis* aff. *galbana* (T-ISO strain) has also been noted for its high content of fucoxanthin (Kim et al., 2012), a pigment that is being investigated for its medical properties in treating cancer, obesity, inflammation, and diabetes (Muthuirulappan and Francis, 2013; Peng et al., 2011). Lastly, the ubiquitous marine cocolithophore *Emiliana huxleyi* (Prymnesiophyceae[C]) plays an important role in earth's carbon cycle due to forming vast blooms. This characteristic of fast growth shows great potential for the renewable production of hydrocarbons either as gasses (Wu et al., 1999a) or as long-chain alkenes (Wu et al., 1999b).

The main genus with commercial application is *Nannochloropsis*, which generally has a high fatty acid content, mainly in the form of triacylglycerides. Furthermore, the fatty acids profile of *Nannochloropsis* contains a large proportion of polyunsaturated fatty acids (PUFAs), including the essential eicosapentaenoic acid (EPA) (Huerlimann et al., 2010). The main application of *Nannochloropsis* sp. is as aquaculture feed (Guedes and Malcata, 2012); however, they are also considered for the production of EPA (Winwood, 2013) and biofuel (Passell et al., 2013). Similarly, the freshwater Eustigmatophyceae *Monodus subterraneus* can be used for the commercial production of EPA (Lu et al., 2001).

Bacillariophyceae are known to accumulate fatty acids, especially PUFAs. The main commercial use of Bacillariophyceae is as feedstock for larval cultures in aquaculture, with the following genera being most widely used: *Chaetoceros*, *Nitzschia*, *Phaeodactylum*, *Skeletonema*, and *Thalassiosira*. Due to their high lipid and PUFA content, the Bacillariophyceae are also under consideration for the production of biofuels and essential fatty acids for human

consumption. *Phaeodactylum tricornutum* contains a large amount of EPA and fatty acids in general; therefore, it is considered for both applications (Silva Benavides et al., 2013). Another interesting candidate is *Odontella aurita*, which is being considered for the production of fucoxanthin (Xia et al., 2013) and aquaculture feed (Guedes and Malcata, 2012) and EPA (Haimeur et al., 2012).

The main commercial applications of Labyrinthulomycetes focus on the order Thraustochytrids. Even though Thraustochytrids are heterotrophic and not photosynthetic, their saprotrophic nature has many advantages, including being able to be grown on alternative carbon sources such as the readily available glycerol (Gupta et al., 2013) or food waste (Thyagarajan et al., 2014), and tolerating a wide range of salinities (Shabala et al., 2013). The main genera of Thraustochytrids with commercial potential are *Aurantiochytrium*, *Schizotrichium*, *Thraustrochytrium*, and *Ulkenia*, which are noted for their high lipid content in general and high DHA content in particular (Lee Chang et al., 2014). This leads to potential applications as fish feed (Conceição et al., 2010), the production of DHA for human consumption (Gupta et al., 2012), or the production of biofuels (Lee Chang et al., 2013).

Even though Dinophyta are mainly associated with harmful algal blooms, they are also known for their generally high fatty acid content, especially the valuable omega-3 PUFAs. The marine, chloroplast lacking, heterotrophic *C. cohnii* is the only commercially grown dinoflagellate species to date and exhibits a high DHA content used for the production of infant formulas (Atalah et al., 2007). Up to 20% of the dry weight of *C. cohnii* can consist of fatty acids, of which up to 30% can be DHA (Ward and Singh, 2005). As a heterotrophic algae, *C. cohnii* can be grown on organic carbon sources like glucose, ethanol, or acetic acid, with yeast extract as a nitrogen source (Mendes et al., 2009).

References

Achitouv, E., Metzger, P., Rager, M.-N., Largeau, C., 2004. C31–C34 methylated squalenes from a Bolivian strain of *Botryococcus braunii*. Phytochemistry 65, 3159–3165.

Adl, S.M., Simpson, A.G.B., Farmer, M.A., Andersen, R.A., Anderson, O.R., Barta, J.R., Bowser, S.S., Brugerolle, G.U.Y., Fensome, R.A., Fredericq, S., James, T.Y., Karpov, S., Kugrens, P., Krug, J., Lane, C.E., Lewis, L.A., Lodge, J., Lynn, D.H., Mann, D.G., McCourt, R.M., Mendoza, L., Moestrup, Ø., Mozley-Standridge, S.E., Nerad, T.A., Shearer, C.A., Smirnov, A.V., Spiegel, F.W., Taylor, M.F.J.R., 2005. The new higher level classification of eukaryotes with emphasis on the taxonomy of protists. J. Eukaryot. Microbiol. 52, 399–451.

Ahern, T.J., Katoh, S., Sada, E., 1983. Arachidonic acid production by the red alga *Porphyridium cruentum*. Biotechnol. Bioeng. 25, 1057–1070.

Aksu, Z., 2002. Determination of the equilibrium, kinetic and thermodynamic parameters of the batch biosorption of nickel(II) ions onto *Chlorella vulgaris*. Process. Biochem. 38, 89–99.

Ambati, R., Phang, S.-M., Ravi, S., Aswathanarayana, R., 2014. Astaxanthin: sources, extraction, stability, biological activities and its commercial applications—a review. Mar. Drugs 12, 128–152.

Arad, S., Levy-Ontman, O., 2010. Red microalgal cell-wall polysaccharides: biotechnological aspects. Curr. Opin. Biotechnol. 21, 358–364.

Archibald, J., 2012. The evolution of algae by secondary and tertiary endosymbiosis. Adv. Bot. Res. 64, 87–118.

Archibald, J.M., 2009. The puzzle of plastid evolution. Curr. Biol. 19, R81–R88.

Armbrust, E.V., Berges, J.A., Bowler, C., Green, B.R., Martinez, D., Putnam, N.H., Zhou, S., Allen, A.E., Apt, K.E., Bechner, M., Brzezinski, M.A., Chaal, B.K., Chiovitti, A., Davis, A.K., Demarest, M.S., Detter, J.C., Glavina, T., Goodstein, D., Hadi, M.Z., Hellsten, U., Hildebrand, M., Jenkins, B.D., Jurka, J., Kapitonov, V.V., Kröger, N., Lau, W.W.Y., Lane, T.W., Larimer, F.W., Lippmeier, J.C., Lucas, S., Medina, M., Montsant, A., Obornik, M., Schnitzler Parker, M., Palenik, B., Pazour, G.J., Richardson, P.M., Rynearson, T.A., Saito, M.A., Schwartz, D.C., Thamatrakoln, K., Valentin, K., Vardi, A., Wilkerson, F.P., Rokhsar, D.S., 2004. The genome of the diatom *Thalassiosira pseudonana*: ecology, evolution, and metabolism. Science 306, 79–86.

Atalah, E., Cruz, C.M.H., Izquierdo, M.S., Rosenlund, G., Caballero, M.J., Valencia, A., Robaina, L., 2007. Two microalgae *Crypthecodinium cohnii* and *Phaeodactylum tricornutum* as alternative source of essential fatty acids in starter feeds for seabream (*Sparus aurata*). Aquaculture 270, 178–185.

Ball, S., Colleoni, C., Cenci, U., Raj, J.N., Tirtiaux, C., 2011. The evolution of glycogen and starch metabolism in eukaryotes gives molecular clues to understand the establishment of plastid endosymbiosis. J. Exp. Bot. 62, 1775–1801.

Baurain, D., Brinkmann, H., Petersen, J., Rodríguez-Ezpeleta, N., Stechmann, A., Demoulin, V., Roger, A.J., Burger, G., Lang, B.F., Philippe, H., 2010. Phylogenomic evidence for separate acquisition of plastids in cryptophytes, haptophytes and stramenopiles. Mol. Biol. Evol. 27, 1698–1709.

Bendif, E.M., Probert, I., Schroeder, D.C., de Vargas, C., 2013. On the description of *Tisochrysis lutea* gen. nov. sp. nov. and *Isochrysis nuda* sp. nov. in the Isochrysidales, and the transfer of *Dicrateria* to the Prymnesiales (Haptophyta). J. Appl. Phycol. 25, 1763–1776.

Bodyl, A., Stiller, J.W., Mackiewicz, P., 2009. Chromalveolate plastids: direct descent or multiple endosymbioses. Trends Ecol. Evol. 24, 119–121.

Borowitzka, L.J., Borowitzka, M.A., 1988. β-Carotene (provitamin A) production with algae. In: Vandamme, E.J. (Ed.), Biotechnology of Vitamins, Pigments and Growth Factors. Elsevier Applied Science, London, pp. 15–26.

Borowitzka, M.A., 2013. High-value products from microalgae—their development and commercialisation. J. Appl. Phycol. 25, 743–756.

Boussiba, S., 2000. Carotenogenesis in the green alga *Haematococcus pluvialis*: cellular physiology and stress response. Physiol. Plant. 108.

Brown, A.C., Knights, B.A., Conway, E., 1969. Hydrocarbon content and its relationship to physiological state in the green alga *Botryococcus braunii*. Phytochemistry 8, 543–547.

Burki, F., Okamoto, N., Pombert, J.-F., Keeling, P.J., 2012. The evolutionary history of haptophytes and cryptophytes: phylogenomic evidence for separate origins. Proceed. Roy. Soc. B Biol. Sci. 279, 2246–2254.

Cavalier-Smith, T., 1993. Kingdom Protozoa and its 18 phyla. Microbiol. Rev. 57, 953–994.

Cavalier-Smith, T., 1998. A revised six-kingdom system of life. Biol. Rev. 73, 203–266.

Cavalier-Smith, T., 2010. Kingdoms Protozoa and Chromista and the eozoan root of the eukaryotic tree. Biol. Lett. 6, 342–345.

Collins, A.M., Jones, H.D.T., Han, D., Hu, Q., Beechem, T.E., Timlin, J.A., 2011. Carotenoid distribution in living cells of *Haematococcus pluvialis* (Chlorophyceae). PLoS One 6, e24302.

Conceição, L.E.C., Yúfera, M., Makridis, P., Morais, S., Dinis, M.T., 2010. Live feeds for early stages of fish rearing. Aquacult. Res. 41, 613–640.

Craggs, R.J., Adey, W.H., Jenson, K.R., St John, M.S., Green, F.B., Oswald, W.J., 1996. Phosphorus removal from wastewater using an algal turf scrubber. Water Sci. Technol. 33, 191–198.

Deason, T.R., Butler, G.L., Rhyne, C., 1983. *Rhodella reticulata* sp. nov., a new coccoid rhodophytan alga (Porphyridiales). J. Phycol. 19, 104–111.

Edvardsen, B., Eikerem, W., Green, J.C., Andersen, R.A., Moon-vand der Staay, S.Y., Medlin, L.K., 2000. Phylogenetic reconstructions of the Haptophyta inferred from 18s ribosomal DNA sequences and available morphological data. Phycologia 39, 19–35.

Gantt, E., Conti, S.F., 1965. The ultrastructure of *Porphyridium cruentum*. J. Cell Biol. 26, 365–381.

Gladyshev, M.I., Sushchik, N.N., Makhutova, O.N., 2013. Production of EPA and DHA in aquatic ecosystems and their transfer to the land. Prostaglandins Other Lipid Mediat. 107, 117–126.

Glazer, A.N., 1994. Phycobiliproteins — a family of valuable, widely used fluorophores. J. Appl. Phycol. 6, 105–112.

Guedes, A.C., Amaro, H.M., Malcata, F.X., 2011. Microalgae as sources of carotenoids. Mar. Drugs 9, 625–644.

Guedes, A.C., Malcata, F.X., 2012. Nutritional value and uses of microalgae in aquaculture. Aquaculture 10, 59–78.

Gupta, A., Barrow, C.J., Puri, M., 2012. Omega-3 biotechnology: thraustochytrids as a novel source of omega-3 oils. Biotechnol. Adv. 30, 1733–1745.

Gupta, A., Singh, D., Barrow, C.J., Puri, M., 2013. Exploring potential use of Australian thraustochytrids for the bioconversion of glycerol to omega-3 and carotenoids production. Biochem. Eng. J. 78, 11–17.

Haimeur, A., Ulmann, L., Mimouni, V., Guéno, F., Pineau-Vincent, F., Meskini, N., Tremblin, G., 2012. The role of *Odontella aurita*, a marine diatom rich in EPA, as a dietary supplement in dyslipidemia, platelet function and oxidative stress in high-fat fed rats. Lipids Health Dis. 11, 147.

Hasle, G.R., Syvertsen, E.E., 1997. Marine diatoms. In: Tomas, C.R. (Ed.), Identifying Marine Phytoplankton. Academic Press, San Diego, pp. 5–386.

Hu, L., Huang, B., Zuo, M-m., Guo, R-y., Wei, H., 2008. Preparation of the phycoerythrin subunit liposome in a photodynamic experiment on liver cancer cells1. Acta Pharmacol. Sin. 29, 1539–1546.

Huang, Z., Dale Poulter, C., 1989. Tetramethylsqualene, a triterpene from *Botryococcus braunii* var. *showa*. Phytochemistry 28, 1467–1470.

Huerlimann, R., de Nys, R., Heimann, K., 2010. Growth, lipid content, productivity, and fatty acid composition of tropical microalgae for scale-up production. Biotechnol. Bioeng. 107, 245–257.

Kang, H.K., Salim, H.M., Akter, N., Kim, D.W., Kim, J.H., Bang, H.T., Kim, M.J., Na, J.C., Hwangbo, J., Choi, H.C., Suh, O.S., 2013. Effect of various forms of dietary *Chlorella* supplementation on growth performance, immune characteristics, and intestinal microflora population of broiler chickens. J. Appl. Poult. Res. 22, 100–108.

Kathiresan, S., Sarada, R., Bhattacharya, S., Ravishankar, G.A., 2007. Culture media optimization for growth and phycoerythrin production from *Porphyridium purpureum*. Biotechnol. Bioeng. 96, 456–463.

Keeling, P.J., 2010. The endosymbiotic origin, diversification and fate of plastids. Philosoph. Transact. Roy. Soc. B Biol. Sci. 365, 729–748.

Keeling, P.J., Burger, G., Durnford, D.G., Lang, B.F., Lee, R.W., Pearlman, R.E., Roger, A.J., Gray, M.W., 2005. The tree of eukaryotes. Trends Ecol. Evol. 20, 670–676.

Kim, S., Kang, S.-W., Kwon, O.N., Chung, D., Pan, C.-H., 2012. Fucoxanthin as a major carotenoid in *Isochrysis* aff. *galbana*: characterization of extraction for commercial application. J. Korean Soc. Appl. Biol. Chem. 55, 477–483.

Lee Chang, K., Dumsday, G., Nichols, P., Dunstan, G., Blackburn, S., Koutoulis, A., 2013. High cell density cultivation of a novel *Aurantiochytrium* sp. strain TC 20 in a fed-batch system using glycerol to produce feedstock for biodiesel and omega-3 oils. Appl. Microbiol. Biotechnol. 97, 6907–6918.

Lee Chang, K., Nichols, C., Blackburn, S., Dunstan, G., Koutoulis, A., Nichols, P., 2014. Comparison of thraustochytrids *Aurantiochytrium* sp., *Schizochytrium* sp., *Thraustochytrium* sp., and *Ulkenia* sp. for production of biodiesel, long-chain omega-3 oils, and exopolysaccharide. Mar. Biotechnol. 16, 396–411.

Leliaert, F., Smith, D.R., Moreau, H., Herron, M.D., Verbruggen, H., Delwiche, C.F., de Clerk, O., 2012. Phylogeny and molecular evolution of the green algae. Crit. Rev. Plant Sci. 31, 1–46.

Lu, C., Rao, K., Hall, D., Vonshak, A., 2001. Production of eicosapentaenoic acid (EPA) in *Monodus subterraneus* grown in a helical tubular photobioreactor as affected by cell density and light intensity. J. Appl. Phycol. 13, 517–522.

Maneveldt, G.W., Keats, D.W., 2004. Chromista. eLS. John Wiley & Sons Ltd, p. 7.

Martinez, F., Ascaso, C., Orús, M.I., 1991. Morphometric and stereologic analysis of *Chlorella vulgaris* under heterotrophic growth conditions. Ann. Bot. 67, 239–245.

Mendes, A., Reis, A., Vasconcelos, R., Guerra, P., Lopes da Silva, T., 2009. *Crypthecodinium cohnii* with emphasis on DHA production: a review. J. Appl. Phycol. 21, 199–214.

Metzger, P., Allard, B., Casadevall, E., Berkaloff, C., Couté, A., 1990. Structure and chemistry of a new chemical race of *Botryococcus braunii* (Chlorophyceae) that produces lycopadiene, a tetraterpenoid hydrocarbon. J. Phycol. 26, 258–266.

Metzger, P., Berkaloff, C., Casadevall, E., Coute, A., 1985. Alkadiene- and botryococcene-producing races of wild strains of *Botryococcus braunii*. Phytochemistry 24, 2305–2312.

Metzger, P., Casadevall, E., 1987. Lycopadiene, a tetraterpenoid hydrocarbon from new strains of the green alga *Botryococcus braunii*. Tetrahedron Lett. 28, 3931—3934.

Metzger, P., Largeau, C., 2005. *Botryococcus braunii*: a rich source for hydrocarbons and related ether lipids. Appl. Microbiol. Biotechnol. 66, 486—496.

Molnár, S., Kiss, A., Virág, D., Forgó, P., 2013. Comparative studies on accumulation of selected microelements by *Spirulina platensis* and *Chlorella vulgaris* with the prospects of functional food development. J. Chem. Eng. Process. Technol. 4.

Mulders, K.J., Weesepoel, Y., Bodenes, P., Lamers, P.P., Vincken, J.-P., Martens, D.E., Gruppen, H., Wijffels, R.H., 2014. Nitrogen-depleted *Chlorella zofingiensis* produces astaxanthin, ketolutein and their fatty acid esters: a carotenoid metabolism study. J. Appl. Phycol. 1—16.

Muthuirulappan, S., Francis, S.P., 2013. Anti-cancer mechanism and possibility of nano-suspension formulation for a marine algae product fucoxanthin. Asian Pac. J. Cancer Prev. 14, 2213—2216.

Orandi, S., Lewis, D.M., 2013. Synthesising acid mine drainage to maintain and exploit indigenous mining micro-algae and microbial assemblies for biotreatment investigations. Environ. Sci. Pollut. Res. Int. 20, 950—956.

Pan, Q., Chen, M., Li, J., Wu, Y., Zhen, C., Liang, B., 2013. Antitumor function and mechanism of phycoerythrin from *Porphyra haitanensis*. Biol. Res. 46, 87—95.

Park, J.B.K., Craggs, R.J., Shilton, A.N., 2011. Wastewater treatment high rate algal ponds for biofuel production. Bioresour. Technol. 102, 35—42.

Passell, H., Dhaliwal, H., Reno, M., Wu, B., Ben Amotz, A., Ivry, E., Gay, M., Czartoski, T., Laurin, L., Ayer, N., 2013. Algae biodiesel life cycle assessment using current commercial data. J. Environ. Manage. 129, 103—111.

Peng, J., Yuan, J.-P., Wu, C.-F., Wang, J.-H., 2011. Fucoxanthin, a marine carotenoid present in brown seaweeds and diatoms: metabolism and bioactivities relevant to human health. Mar. Drugs 9, 1806—1828.

Prabowo, D.A., Hiraishi, O., Suda, S., 2013. Diverisity of *crypthecodinium* spp. (Dinophyceae) from Okinawa prefecture. Jpn. J. Mar. Sci. Technol. 21, 181—191.

Ragan, M., 1998. On the delineation and higher-level classification of the algae. Eur. J. Phycol. 33, 1—15.

Reinke, D.C., 1984. Ultrastructure of *Chaetoceros muelleri* (Bacillariophyceae): auxospore, resting spore and vegetative cell morphology. J. Phycol. 20, 153—155.

Rogers, M.B., Gilson, P.R., Su, V., McFadden, G.I., Keeling, P.J., 2007. The complete chloroplast genome of the chlorarachniophyte *Bigelowiella natans*: evidence for independent origins of chlorarachniophyte and euglenid secondary endosymbionts. Mol. Biol. Evol. 24, 54—62.

Rothschild, L.J., 1989. Protozoa, protista, protoctista: what's in a name? J. Hist. Biol. 22, 277—305.

Safi, C., Zebib, B., Merah, O., Pontalier, P.-Y., Vaca-Garcia, C., 2014. Morphology, composition, production, processing and applications of *Chlorella vulgaris*: a review. Ren. Sustain. Energy Rev. 35, 265—278.

Sarno, D., Kooistra, W.H.C.F., 2005. Diversity in the genus *Skeletonema* (Bacillariophyceae). II. An assessment of the taxonomy of *S. costatum*-like species with the description of four new species. J. Phycol. 41, 151—176.

Shabala, L., McMeekin, T., Shabala, S., 2013. Thraustochytrids can be grown in low-salt media without affecting PUFA production. Mar. Biotechnol. 15, 437—444.

Silva Benavides, A.M., Torzillo, G., Kopecký, J., Masojídek, J., 2013. Productivity and biochemical composition of *Phaeodactylum tricornutum* (Bacillariophyceae) cultures grown outdoors in tubular photobioreactors and open ponds. Biomass Bioenergy 54, 115—122.

Suda, S., Atsumi, M., Miyashita, H., 2002. Taxonomic characterization of a marine *Nannochloropsis* species, *N. oceanica* sp. nov. (Eustigmatophyceae). Phycologia 41, 273—279.

Sym, S.D., Maneveldt, G.W., 2011. Chromista. eLS. John Whiley & Sons Ltd., Chichester.

Tababa, H.G., Hirabayashi, S., Inubushi, K., 2012. Media optimization of *Parietochloris incisa* for arachidonic acid accumulation in an outdoor vertical tubular photobioreactor. J. Appl. Phycol. 24, 887—895.

Thyagarajan, T., Puri, M., Vongsvivut, J., Barrow, C.J., 2014. Evaluation of bread brumbs as a potential carbon source for the growth of thraustochytrid species for oil and omega-3 production. Nutrients 6, 2104—2114.

Tran, D., Vo, T., Portilla, S., Louime, C., Duan, N., Mai, T., Tran, D., Ho, T., 2013. Phylogenetic study of some strains of *Dunaliella*. Am. J. Environ. Sci. 9, 317—321.

Tsui, C.K.M., Marshall, W., Yokoyama, R., Honda, D., Lippmeier, J.C., Craven, K.D., Peterson, P.D., Berbee, M.L., 2009. Labyrinthulomycetes phylogeny and its implications for the evolutionary loss of chloroplasts and gain of ectoplasmic gliding. Mol. Phylogen. Evol. 50, 129—140.

Turmel, M., Gagnon, M.-C., O'Kelly, C.J., Otis, C., Lemieux, C., 2009. The chloroplast genomes of the green algae *Pyramimonas, Monomastix*, and *Pycnococcus* shed new light on the evolutionary history of prasinophytes and the origin of the secondary chloroplasts of euglenids. Mol. Biol. Evol. 26, 631—648.

Ward, O.P., Singh, A., 2005. Omega-3/6 fatty acids: alternative sources of production. Process. Biochem. 40, 3627—3652.

Watanabe, S., Hirabayashi, S.I., Cohen, Z., Vonshak, A., Richmond, A., 1996. *Parietochloris incisa* comb. nov. (Trebouxiophyceae, Chlorophyta). Phycol. Res. 44, 107—108.

Weiss, T.L., Roth, R., Goodson, C., Vitha, S.,I.,B., Azadi, P., Rusch, J., Holzenburg, A., Devarenne, T.P., Goodenough, U., 2012. Colony organization in the green alga *Botryococcus braunii* (Race B) is specified by a complex extracellular matrix. Eukaryot. Cell 11, 1424–1440.

Winwood, R.J., 2013. Recent developments in the commercial production of DHA and EPA rich oils from micro-algae. OCL 20, D604.

Wu, Q., Dai, J., Shiraiwa, Y., Sheng, G., Fu, J., 1999a. A renewable energy source—hydrocarbon gases resulting from pyrolysis of the marine nanoplanktonic alga *Emiliania huxleyi*. J. Appl. Phycol. 11, 137–142.

Wu, Q., Shiraiwa, Y., Takeda, H., Sheng, G., Fu, J., 1999b. Liquid-saturated hydrocarbons resulting from pyrolysis of the marine coccolithophores *Emiliania huxleyi* and *Gephyrocapsa oceanica*. Mar. Biotechnol. 1, 346–352.

Xia, S., Wang, K., Wan, L., Li, A., Hu, Q., Zhang, C., 2013. Production, characterization, and antioxidant activity of fucoxanthin from the marine diatom *Odontella aurita*. Mar. Drugs 11, 2667–2681.

Yen, H.-W., Hu, I.C., Chen, C.-Y., Ho, S.-H., Lee, D.-J., Chang, J.-S., 2013. Microalgae-based biorefinery — from biofuels to natural products. Bioresour. Technol. 135, 166–174.

Zingone, A., Percopo, I., Sims, P.A., Sarno, D., 2005. Diversity in the genus *Skeletonema* (Bacillariophyceae). I. A reexamination of the type material of *S. costatum* with the description of *S. grevillei* sp. nov. J. Phycol. 41, 95–104.

Microalgae Isolation and Basic Culturing Techniques

Poonam Singh, Sanjay Kumar Gupta, Abhishek Guldhe, Ismail Rawat, Faizal Bux

Institute for Water and Wastewater Technology,
Durban University of Technology, Durban, South Africa

1. INTRODUCTION

Microalgae are ubiquitous organisms that are the primary producers for life on the earth. Rapid growth rates and the simple structure of microalgae make them interesting organisms for biotechnological applications. Cellular components of microalgae such as carbohydrates and proteins can be utilized for various purposes including biofuels and products of nutritional and pharmaceutical value (Chisti, 2007; Harun et al., 2010; Singh et al., 2014). Researchers, industries, and governments are now focusing on the development of green processes. This has extended to the exploitation of microalgae, especially for biofuels and pharmaceutical products.

Microalgae are a diverse group of organisms and found in various natural habitats. They exhibit variation in their nutritional requirements as well as metabolite production (Rawat et al., 2013). Isolation and culture development of microalgal strains of interest are essential primary steps toward laboratory investigations and successful commercial application. Bioprospecting of microalgae for a particular product or application is a multistep process comprised of sampling of environments, purification and isolation of monocultures, determining nutrient requirements and cultivation parameters, screening of microalgal species, measuring growth, and developing suitable cultivation techniques (Mutanda et al., 2011).

This chapter describes the commonly adopted techniques in isolation, screening, growth measurement, determining cultivation conditions and nutrient requirements, and cultivation techniques. Choice of suitable technique depends upon the microalgal strain and its application.

2. MICROALGAL ISOLATION AND PURIFICATION TECHNIQUES

Sampling and isolation of microalgae from natural habitats is a well-established procedure. Depending on their different habitats, algal strains vary in their ease of cultivation under laboratory conditions. A key step in the isolation of microalgae is to provide culture conditions that mimic their natural habitats. Microalgae are

43

sensitive toward physiological conditions such as temperature, pH, and salinity (Brennan and Owende, 2010). Some microalgal species require specific nutrients, for example, diatoms need silica-supplemented media, and some microalgae only grow in the presence of bacteria or other algae. Purification of cultures to obtain algal monocultures is another challenging step. Isolation of a pure culture from its natural environment, cultivation under laboratory conditions, and maintenance of isolated cultures involve various techniques.

2.1 Sampling

The collection of microalgal samples is a crucial step for isolation of microalgae from their natural environment. Microalgae are found in different environmental conditions and habitats such as ice, hot water springs, fresh water, brackish water, rivers, oceans, dams, wastewater, rocks, saline bodies, coastal areas, soil, etc. Proper sampling technique, sampling season, habitat assessment, and preservation of samples are essential factors for collection of microalgae. Sampling techniques include syringe sampling, scraping, brushing, inverted petri dish method, etc. It is necessary to record abiotic factors such as light, water temperature, dissolved O_2 and CO_2, nutrient concentration, pH, and salinity; it is also important to record biotic factors such as pathogens and any competitors at the sampling site in order to mimic these conditions at the laboratory (Mutanda et al., 2011). Global positioning system coordinates must be recorded for reference and resampling. Microalgae are often present in consortium with complex population dynamics in natural habitats, thus it becomes inevitable to isolate the strain of interest from the collected samples.

2.2 Isolation of Single Cells

Isolation of a single cell of microalga is the technique whereby a cell is picked from the sample using a micropipette or glass capillary under microscopic observation. These single cells are transferred to sterile droplets of water or suitable media. This technique requires expertise and precision. Shear stress caused by micropipette or capillary tips can damage the cell. Damage is evident from a number of factors such as flagellates stop swimming, a shift in refraction of light by diatoms due to damaged frustules, and protoplasm leakage. Caution is necessary for successful implementation of this purification technique. Ultrapure droplets are required, especially for marine samples, to distinguish between microalgal cell and other particulates.

2.3 Serial Dilution

Serial dilution is the most common and established method of microalgal isolation from collected samples. Serial dilution is the stepwise dilution of a substance in solution. The dilution factor is usually constant at each step, resulting in a geometric progression of the concentration in a logarithmic fashion. A set of sterilized test tubes or flasks or wells with nutrient media can be used for dilution. Dilution sets are determined on the basis of known number of cells in the enriched culture. Depending upon the habitat and specific requirements of algae of interest, the medium can be supplemented to provide selective pressure toward a culture of interest. The success of the technique is highly dependent on the accuracy of a measured amount of cell culture during transfer from one medium to another. This method is useful for production of algal monocultures; however, these are generally not axenic cultures.

2.4 Streak Plate

Streaking is a common technique for isolation of a pure strain from a mixture of microorganisms for investigation and identification. This method requires the preparation of aseptic agar plates followed by introduction of a sterile inoculation loop into the liquid sample or pinching a morphologically distinct colony from the later membrane surface with a sterile pin tool. This technique is successful for most algal strains, especially coccoids, diatoms, soil microalgae, filamentous algae, and small cyanobacteria. This is the best method to isolate axenic cultures without requiring any manipulation or modification of technique. Agar concentration does not play a major role in the growth. Any concentration from 0.8% to 2% is sufficient for growth of microalgae on the agar surfaces. Isolation is accomplished by streaking the natural sample across the agar surface, identical to the technique used for isolating bacteria. A loop containing a small amount of sample is spread across the agar, with the use of one of several techniques or patterns. After streaking agar plates are incubated, providing optimum physiological conditions for the species/strain of interest. The incubation period varies from a few days for soil and freshwater algae up to several months for oceanic species. The single-cell algal strain can be further subcultured onto plates or into broth.

2.5 Density Centrifugation

Centrifugation applies gravity settling to separate organisms based on the cell size. This technique is usually applied to separate larger organisms from the microalgal cells. Density gradient (e.g., Silica sol, Percoll) centrifugal technique separates the different species of microalgae into different bands. This technique is primarily used to concentrate the number of cells of a desired microalgal strain. From concentrated cells the isolation can be achieved by coupling this with other isolation techniques. Centrifugation speed and time vary depending upon the target microalgal species. Centrifugation can also damage the delicate cells via shear stress.

2.6 Enrichment Media

Microalgal nutritional requirements vary depending on their natural habitat and cellular physiology. Enrichment cultures using selection pressure can be employed to obtain single-species isolation. Commonly used enrichment substances include specific nutrient mediums, soil extracts, nutrients like nitrate and phosphate, and trace metals (Mutanda et al., 2011). Adjustment of pH is a commonly adopted strategy to obtain bacteria-free cultures. Organic substances such as yeast extract, casein from various fruits and vegetable juices may also be added to the medium (Andersen, 2005). Natural habitats may be deficient in one or more nutrients required for microalgae growth. In nature, nutrients are recycled or provided by the physiological action of other organisms. Sampling reduces the recycling of nutrients and thus can cause death of microalgae. Enrichment substances are usually added in minimal quantities at sequential stages. Conventional methods of microalgal isolation have several limitations, thus there is the need to develop novel isolation techniques.

3. STRAIN SELECTION AND SCREENING CRITERIA OF MICROALGAE

Microalgal strain selection is important for efficient and economically viable production of desired products. Screening criteria depend upon the desired use or product for which microalgae are being exploited. For microalgal strain selection and screening, a number of factors such as growth physiology, tolerance toward biotic and abiotic factors, metabolite production, nutrient requirements, etc. are taken into consideration. Screening for metabolite production includes determination of cellular components such as proteins, lipids, and carbohydrates of microalgae. For CO_2 sequestration, tolerance toward high CO_2 concentrations is an important screening criterion. For mass culture robustness, stability and susceptibility to predators are considered (Mutanda et al., 2011). Fluorescent dyes such as Nile red (*9-diethylamino-5H-benzo[α] phenoxazine-5-one*) and BODIPY 505/515 (*4,4-difluoro-1,3,5,7-tetramethyl-4-bora- 3a,4a diaza-s-indacene*) are employed to screen microalgae for lipid accumulation, which can be used for biofuel production (Govender et al., 2012). Various chromatographic methods such as thin-layer chromatography, liquid and gas chromatography, and spectrophotometric methods such as near infrared, Fourier transform infrared, and nuclear magnetic resonance spectroscopy are applied to screen microalgae for metabolite production, which has significant applications in nutritional and pharmaceutical industries (Mutanda et al., 2011). Microalgae are a diverse group of organisms, thus rapid and efficient screening techniques need to be developed to screen a number of phenotypic characters at the same time.

4. MEASUREMENT OF ALGAL GROWTH

Accurate measurement of microalgal growth is a difficult task because of small cell size. However, several techniques are available to determine microalgal growth kinetics. Microalgal growth is usually expressed as biomass, cell number, or by determining pigments and proteins over the specific time period. Some of these methods are easy, while others require sophisticated instruments. Thus, choice of a growth measurement technique depends upon the scale and available laboratory facilities. Techniques commonly used for measurement of microalgal growth include spectrophotometry, gravimetric biomass determination, cell enumeration using counting chambers, and flow cytometry (Mutanda et al., 2011). Spectrophotometric determination of microalgal growth depends upon the culture density and chlorophyll content. This is a rapid and easy method, although suspended and dissolved solid particles may affect its accuracy. The wavelength (400–700 μm) used to determine the microalgal

growth is in the visible range of the electromagnetic spectrum. For measurement that is more accurate, this technique is coupled with gravimetric or counting methods to draw calibration curves. Gravimetric assessment of microalgal growth is based on the determination of the weight of dry or wet biomass of cells. Centrifugation is applied to concentrate the cell biomass, and the weight of the microalgal biomass pellet is determined gravimetrically. In the case of dry biomass measurement, separated pellets are dried prior to weight measurements. Other suspended solids and media components could hamper the accuracy of growth measurement, thus the cells are washed prior to gravimetric analysis. Microscopic counting of microalgal cells mounted on specialized slides with chambers is another enumeration technique. Determining cells and other contaminants resembling microalgal cells can cause inaccurate counting, however. Flow cytometry is an automated technique that measures fluorescence and scattering of light caused by microalgal cells. This highly sophisticated technique minimizes the human counting errors associated with the counting chamber method. A drawback associated with flow cytometry, however, is the need for costly and sophisticated instrumentation.

5. CULTURE TECHNIQUES FOR MICROALGAE

Several techniques are available for microalgae cultivation. The choice of technique is mainly based on the selected microalgal strain as well as its application. Choice of culture media, cultivation system, and maintenance of the abiotic and biotic cultivation parameters are the main aspects of microalgae cultivation.

5.1 Nutritional Requirements of Microalgae

Algae require a wide range of inorganic nutrients such as nitrogen, phosphorus, and vitamins and trace elements such as iron, manganese, selenium, zinc, cobalt, and nickel. The growth and the biomass productivity of any species depend on the type of nutrient sources, its quantity, and the relative proportions of C, N, and P of the culture medium. Nutrient sources, especially the nitrogen, such as nitrate, urea, and ammonium, and carbon sources, which are maintained by sparging of air + CO_2, are directly associated with the pH change especially in poorly buffered cultures, and therefore have a high impact on the algal growth. Sodium bicarbonate has been found to be better for carbon feedstock and less expensive than CO_2 sparging (Lam et al., 2012). Optimization of appropriate relative proportions of nutrient concentrations and sources, which is species specific, is imperative for the maximum growth yields and economics. In general, other than carbon sources, nitrogen and phosphorus in 16:1 ratio are believed to be the best composition for the proper growth of most of the algal species (Richmond, 2008).

5.1.1 Culture Media for Microalgae

The production of microalgal biomass is greatly influenced by the growth medium, nutrient concentration, and nutrient type (Andersen, 2005). Algal growth media can be grouped as freshwater and marine algal media. The suitability of growth media is monitored using photosynthetic health, chlorophyll content, and biomass yields of algae. The major components of each media are nitrogen, phosphorus, inorganic carbon, and essential elements such as Fe, Mn, Zn, Cu, etc. Common examples of media used for freshwater algae are BG 11, COMBO Medium, DY-V Medium, DY-III Medium, and MES Volvox Medium, and for marine algae, Black Sea medium, ES Medium, ASP Medium, Aquil medium, Allen's Cyanidium Medium, CCAP Artificial Seawater, Chry medium, ESAW Medium, ESM Medium, f/2 Medium, K Medium, L1 Medium, MNK Medium, Pro99 Medium, SN Medium, etc.

5.2 Cultivation System for Microalgae

Cultivation systems for microalgae are classified into two main categories: open cultivation and closed photobioreactors (PBRs). Cultivation systems are depicted in Figure 1.

5.2.1 Open Ponds and Raceways

The cultivation of microalgae in natural ponds or lagoons is the oldest cultivation technique and has been used since 1950; this was later modified into artificial raceways, but operated in a way to mimic the natural systems (Oswald, 1992). These microalgal cultivation systems may broadly be categorized as natural open ponds (lakes, lagoons, ponds) and artificial ponds (raceway ponds). The pond cultivation system popularly known as "raceway ponds" can be open type or closed type (Hase et al., 2000). Raceway ponds are basically shallow concrete trenches lined with thick plastic sheets with varying length and depth (usually 1–100 cm). Paddle wheels are provided in the raceway system for the circulation of algal culture and even distribution of nutrients as well as sunlight. The major advantages of cultivation of microalgae in open ponds or raceway ponds are its scalability, simple design, high production capacities, and lower operating and maintenance costs (Ugwu et al., 2008). However, microbial contamination such as bacteria, phycophases, parasites, zooplanktons, and growth of unwanted algal species, evaporative losses, low diffusion of atmospheric CO_2, requirement of large land area, poor productivity, limitation to a few strains, as well as extreme weather conditions (rain, temperature, and light intensities) are the major disadvantages of open pond systems (Ugwu et al., 2008). Thermal stratification or uneven distribution of nutrients is also one of the challenges in case of unavailability of electricity for long durations or failure/breakage of paddle wheels. Closed raceways are less vulnerable to

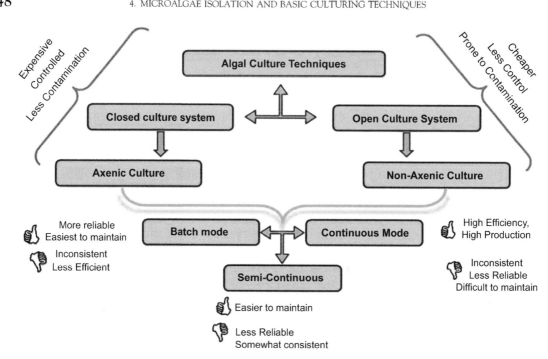

FIGURE 1   Generalized microalgal cultivation systems.

contamination and environmental disturbances in comparison to open raceway ponds.

5.2.2 Photobioreactors

Closed PBRs or Lumostats systems have been developed in order to overcome issues associated with open pond systems. PBRs are made up of glass/Plexiglas/transparent PVC material with internal or external illumination and provided with a controlled gas exchange and mixing/circulation of media. The size and type of these PBRs may vary with strain type and end use of biomass and scalability. The major algal growth limiting factors such as light type, intensity, duration, temperature, mixing and gas exchange, CO_2 and evaporation loss, etc. can be easily controlled and managed in such reactors. The closed PBRs are less prone to bacteria and other contamination and thus facilitate axenic algal cultivation of monocultures.

The design of PBRs is versatile because it can be used indoors or outdoors and is customizable according to the specific requirements and growth conditions. The major limitations with these systems are the economics and huge capital cost involved in installation and maintenance. Closed bioreactors include various types such as flat panel (flat plate), vertical/inclined tubular, helical, airlift, horizontal/serpentine tubular airlift, bubble column, membrane or hybrid type PBRs (Qiang and Richmond, 1996; Rubio et al., 1999; Ugwu et al., 2008).

5.3 Factors Limiting Microalgal Growth

A number of cultivation parameters such as light intensity, photoperiod, temperature, salinity, pH, mixing, etc. influence the growth of microalgae. Thus, it becomes imperative to optimize and maintain these parameters during the

TABLE 1   A generalized set of conditions for culturing microalgae

Parameters	Range	Optima
Temperature (°C)	16−27	18−24
Salinity (g l^{-1})	12−40	20−24
Light intensity (lux)[a]	1000−10,000	2500−5000
Photoperiod (light: dark)	−	16:8 (minimum)
	−	24:0 (maximum)
pH	7−9	8.2−8.7

[a] *Depends on volume and density.*
Adapted from FAO (1991).

cultivation period. Generalized parameters and their optimum ranges are depicted in Table 1.

5.3.1 Light Intensity and Quality

In all photoautotrophic organisms, growth, nutrient uptake, and all metabolic activities depend directly on the corresponding light quality, its intensity, and photoperiod, and can be explained by a light response curve or photosynthesis—irradiance (P—I) relationship/curve (Curtis and Megard, 1987; Meseck et al., 2005). The availability of light to algae cultures is crucial; the photosynthetic activity of the algal cells increases with the increase of light intensity up to the light saturation/threshold point (Mandalam and Palsson, 1998; Suh and Lee, 2003). Higher intensities damage the light receptors of the chloroplasts, which results in photoinhibition and thus reduced biomass production (Lee, 1999). Exposure to prolonged periods of high irradiance increases the biomass production and carotenoid contents in algal cells but decreases the light-harvesting pigments such as chlorophyll (Melis et al., 1998; Molina et al., 2001; Qiang and Richmond, 1994; Ugwu et al., 2008).

Photosynthetic and metabolic responses vary and are species specific with respect to the white, blue, green, and red wavelengths (Sing, 2010). The quality of the light is crucial, as the effect of each spectra of light on the physiology of microalgal cells is yet to be understood completely.

The energy of each solar spectrum varies and hence could affect the photosynthesis and algal cell growth as the penetration and distribution of each spectral light vary, hence it is unequal within the water column (Qiang et al., 1998). The white spectrum is believed to produce low-to-intermediate photosynthetic responses and growth. In various algal spp., comparatively higher growth, an increase in protein and carbohydrate ratio, and enhanced photosynthetic responses can be observed in the red and blue spectrum (Aidar et al., 1994). Reduced photosynthesis and growth, DNA damage, and altered lipid profile in UV-A and UV-B radiation have also been reported in various studies (Franklin et al., 2003).

5.3.2 Photoperiod

The growth and the circadian rhythm of photosynthesis in algal cells depend on the length of photoperiod and vary from species to species. Diurnal light and dark cycles have great influence on the algal photosynthesis, nutrient metabolism, synthesis of organic compounds, etc., therefore the productivity varies with the change in light and dark (L/D) cycles and seasons (Thompson, 1999). Increased photoperiod generally corresponds to increase in the growth of algae (Meseck et al., 2005). Photosynthetic productivity of the euphotic zone of algal cultures influenced by the L/D frequencies and higher growth rate biomass production in various species can be achieved in the range of 1−10 Hz (Kok, 1953). Low L/D frequencies (less than 1 Hz) may lower growth productivity, and very high L/D frequencies (more than 20 Hz) may damage the cells (Grobbelaar, 2009; Sing, 2010).

5.3.3 Temperature

Temperature plays an important role and directly influences the growth of microalgae through changes in metabolic activities, enzyme kinetics, conformation of vital structures, etc. and even the species dominance (Davison, 1991;

Fogg, 2001; Goldman and Ryther, 1976; Goldman and Mann, 1980). The temperature-dependent growth of algae may be exponential or linear depending on other environmental variables (Goldman and Carpenter, 1974; Thompson et al., 1992). Algal cells are adaptable to a wide range of temperatures, characterized by various physiological as well as biochemical responses, but this is species specific (Davison, 1991). Based on the temperature tolerance, the microalgae can be broadly divided into two groups: the eurythermal that can tolerate broad temperature fluctuations and stenothermal that have low-temperature shift tolerance (Marre, 1962). In general, increased temperature is associated with enhanced photosynthetic rates and high nutrient assimilation.

5.3.4 Media pH

The pH of the algal culture media plays a very important role in regulating the uptake of essential nutrients including nitrate and phosphate. The speciation of inorganic carbon sources such as $CO_2/HCO_3^-/CO_3^{2-}$, ionization of biochemical metabolites, precipitation of the phosphates, solubility and availability of trace elements, etc. are pH dependent (Andersen, 2005; Azov, 1982; Meseck, 2007). Moreover, various algal species are pH sensitive and have preferences over the various inorganic carbon sources, thus pH also regulates the species dominance within the mixed populations (Goldman et al., 1982a,b; Hansen, 2002).

5.3.5 Mixing/Turbulence

Mutual shading has been observed as one of the major limiting factors, especially in high cell density cultures. The cells that present in the euphotic zone receive maximum light, whereas cells that present below a few centimeters receive the least amount of light. Mixing of the cultures is very important with regard to efficient gas exchange, equal distribution of the nutrients and metabolites as well as light energies, and to prevent gravitational sedimentation of the algal cells (Richmond, 2008). In outdoor cultures, mixing is imperative to avoid thermal stratification. Despite mixing playing a very important role in achieving high growth rates and high biomass, rapid mixing is also associated with physical cell damage, which depends on the type of algal spp., its cell size, and type of cell wall (Kieran et al., 2000; Thomas and Gibson, 1990). Thus, optimization of the mixing of the culture and liquid depth is necessary to achieve even distribution of light and nutrients, and to prevent sedimentation. Various kinds of mixing such as mechanical stirring, aeration with air pumps, paddle wheels, and jet pumps, etc. are used depending on the type and scale of the culture system. Based on their shear tolerance, major microalgal groups can be arranged as green algae < blue-green algae < diatoms < dinoflagellates (Sing, 2010).

5.3.6 Salinity

The salinity tolerance in microalgal and cyanobacterial species varies substantially, and they can be grouped as hypotonic (usually fresh water species, which can sustain only low salt concentrations, and hypertonic marine species, which sustain high salt concentrations). Microalgae can be sustained in a wide range of salinity, through their inherent defensive osmoregulation abilities such as rigid cell walls, regulation of salt uptake, synthesis of uncharged low-molecular-weight compounds, and by excretion of water and salts (Kirst, 1990). However, a substantial increase of the salt concentration in growth medium beyond the threshold of any species leads to inhibition of photosynthesis and growth; in extreme conditions, plasmolysis or cell bursting can occur (Fogg, 2001; Kirst, 1990). Salinities in the range of $20-24 \, g \, L^{-1}$ have been found to be optimal for most of the microalgae (FAO, 1991).

5.3.7 Gas Exchange (CO_2 and O_2)

Algal cells are made up of approximately 45−50% carbon and require a continual uptake of carbon for growth (Becker, 1994). Carbon is provided by sparging of air or air $+ CO_2$. During gas exchange, dissolved CO_2 is converted into carbonic acid and utilized by algal cells during the photosynthesis. The high oxygen content leads to irreversible photo-oxidative damage to the photosynthetic apparatus due to an increased oxygen buildup, reactive oxygen species, and photorespiration (Molina et al., 2001; Richmond, 2008). Moreover, high oxygen concentrations promote the activity of oxygenase enzymes, causing preferential uptake of O_2 rather than CO_2, resulting in a significant loss of fixed carbon, and thus reduced growth and biomass productivity (Hartig et al., 1988). CO_2 also acts as a buffer, and proper gas exchange maintains CO_2/HCO_3^- balance against pH changes (Gross, 2013). A significant amount of CO_2 is unexploited during sparging due to very slow mass transfer of CO_2 and bubble size (macrobubbles, size $1-2$ mm diameter) whereas sparging of CO_2 microbubbles (size $10-50$ μm) offers unique benefits such as faster dissolution, slow rising, and high surface-to-volume ratio (Zimmerman et al., 2011). Therefore, it is imperative to maintain a proper O_2/CO_2 ratio in the sparging air to achieve maximum growth of algae.

5.3.8 Contamination

Contamination of microalgal cultures is one of the major limiting factors. Open pond systems are more prone to both chemical and biological contamination. Algae have a tendency to bioaccumulate various chemical contaminants from the growth medium. Studies have revealed that some exudates and metabolites such as 15-hydroxyeicosapentaenoic acid, palmitoleic acid, palmitic acid, and monounsaturated oleic acid, etc., secreted by the algae, are the potential growth inhibitors and chemical contaminants (Harris, 1970). The chemically or biologically contaminated biomass may not be suitable for human and/or animal consumption, thus can only be used for energy production purposes.

Fungi, bacteria, viruses, protozoa, zooplankton, and even unwanted algae are considered as potent biological contaminants, which results in unstable growth.

6. ADVANCED ISOLATION TECHNIQUES

6.1 Micromanipulation

Micromanipulation is a powerful tool for microalgal isolation, which allows the movement and culturing of a single cell. Conventional techniques for isolation of microalgae do not account for clumping and thus colony formation from a number of cells rather than a single cell (Frohlich and Konig, 2000). Traditionally, micromanipulation utilized capillary tubes to target microscopically identified cell/s and transfer these to sterile water or media. This is a manual task that is laborious and requires skilled operators (Jakob, 2013). Modern micromanipulation utilizes a micromanipulator and stereomicroscope employing microcapillary tubes or optical tweezers for cell separation with a high level of accuracy, making it an ideal tool for screening and isolation (Frohlich and Konig, 2000; Mutanda et al., 2011). Microcapillaries are usually made of glass with an outer diameter of 1 mm. These are melted under controlled conditions and stretched to give inner diameters of $2-10$ μm (Ishoy et al., 2006). Optical tweezers use a focused laser to capture and move a cell of interest to a compartment from where it can be transferred to sterile media (Wright et al., 2007). Due to the level of skill required and time taken, the technique is still in its infancy in most laboratories (Frohlich and Konig, 2000; Ishoy et al., 2006).

6.2 Automated Techniques

Flow cytometry coupled with fluorescence-activated cell sorting (FACS) is a rapid method of microalgal isolation and purification from environmental samples (Jakob, 2013). FACS is based on light scatter and fluorescence, resulting from the passing of a thin fluid stream through one or more laser beams. Cells scatter and/or absorb the laser beam and emit fluorescence, which provides information on cell size, integrity, and photosynthetic characteristics, which are closely related to morphological and photosynthetic characteristics that are utilized in conventional identification (Figueroa et al., 2010; La et al., 2012). One of the main advantages of FACS is the ability to obtain axenic cultures due to the removal of bacteria (Pereira et al., 2011). The method has gained popularity due to the efficiency and high throughput whereby cells in 5000–10,000 cells per second can be characterized and sorted (Doan and Obbard, 2012). Efficiency of cell sorting is based on the original sample, specifically with reference to the abundance, cell size, shape, and hardiness of the algae being sorted (Jakob, 2013). Greater efficiency is possible by enrichment of the culture/s prior to sorting. The technique is limited in terms of sorting of algae occurring as aggregates. This limitation may be overcome by sonication of the samples in order to disrupt cell-to-cell, cell-to-particle, and particle-to-particle connections within microalgae communities (La et al., 2012). The use of FACS in combination with fluorescent dyes such as Bodipy or Nile red for rapid screening of organisms overproduces metabolites of interest (Hyka et al., 2013; Pereira et al., 2011). Doan and Obbard (2012) utilized the technique for the selection of desirable mutants from a mixed population that would have otherwise been detected by antibiotic resistance or changes in morphology.

References

Aidar, E., Gianesella-Galvão, S., Sigaud, T., et al., 1994. Effects of light quality on growth, biochemical composition and photosynthetic production in *Cyclotella caspia* Grunow and *Tetraselmis gracilis* (Kylin) Butcher. J. Exp. Mar. Biol. Ecol. 180, 175–187.

Andersen, R.A., 2005. Algal Culturing Techniques. Elsevier/Academic Press.

Azov, Y., 1982. Effect of pH on inorganic carbon uptake in algal cultures. Appl. Environ. Microbiol. 43, 1300–1306.

Becker, E.W., 1994. Microalgae: Biotechnology and microbiology. Cambridge University press, Cambridge, Great Britain, p. 293.

Brennan, L., Owende, P., 2010. Biofuels from microalgae—a review of technologies for production, processing, and extractions of biofuels and co-products. Renewable Sustainable Energy Rev. 14, 557–577.

Chisti, Y., 2007. Biodiesel from microalgae. Biotechnol. Adv. 25, 294–306.

Curtis, P.J., Megard, R.O., 1987. Interactions among irradiance, oxygen evolution and NITRITE uptake by Chlamydomonas (Chlorophyceae) 1, 2. J. Phycol. 23, 608–613.

Davison, I.R., 1991. Environmental effects on algal photosynthesis: temperature. J. Phycol. 27, 2–8.

Doan, T.T.Y., Obbard, J.P., 2012. Enhanced intracellular lipid in Nannochloropsis sp. via random mutagenesis and flow cytometric cell sorting. Algal Res. 1, 17–21.

FAO, 1991. Manual on the Production and Use of Live Food for Aquaculture. Fisheries and Aquaculture Department, USA.

Figueroa, R.I., Garcés, E., Bravo, I., 2010. The use of flow cytometry for species identification and life-cycle studies in dinoflagellates. Deep Sea Res., Part II 57, 301–307.

Fogg, G., 2001. Algal adaptation to stress—some general remarks. In: Algal Adaptation to Environmental Stresses. Springer, pp. 1–19.

Franklin, L.A., Osmond, C.B., Larkum, A.W., 2003. Photoinhibition, UV-B and algal photosynthesis. In: Photosynthesis in Algae. Springer, pp. 351–384.

Frohlich, J., Konig, H., 2000. New techniques for isolation of single prokaryotic cells. Fems Microbiol. Rev. 20, 567–572.

Goldman, J.C., Carpenter, E.J., 1974. A kinetic approach to the effect of temperature on algal growth. Limnol. Oceanogr. 19, 756–766.

Goldman, J.C., Ryther, J.H., 1976. Temperature-influenced species competition in mass cultures of marine phytoplankton. Biotechnol. Bioeng. 18, 1125–1144.

Goldman, J.C., Mann, R., 1980. Temperature-influenced variations in speciation and chemical composition of marine phytoplankton in outdoor mass cultures. J. Exp. Mar. Biol. Ecol. 46, 29–39.

Goldman, J.C., Azov, Y., Riley, C.B., Dennett, M.R., 1982a. The effect of pH in intensive microalgal cultures. I. Biomass regulation. J. Exp. Mar. Biol. Ecol. 57, 1–13.

Goldman, J.C., Riley, C.B., Dennett, M.R., 1982b. The effect of pH in intensive microalgal cultures. II. Species competition. J. Exp. Mar. Biol. Ecol. 57, 15–24.

Govender, T., Ramanna, L., Rawat, I., Bux, F., 2012. Bodipy staining, an alternative to the Nile Red fluorescence method for the evaluation of intracellular lipids in microalgae. Bioresour. Technol. 114, 507–511.

Grobbelaar, J.U., 2009. Factors governing algal growth in photobioreactors: the "open" versus "closed" debate. J. Appl. Phycol. 21, 489–492.

Gross, M., 2013. Development and optimization of algal cultivation systems. In: Food Science and Technology. Iowa State University, Ames, Iowa.

Hansen, P.J., 2002. Effect of high pH on the growth and survival of marine phytoplankton: implications for species succession. Aquat. Microb. Ecol. 28, 279–288.

Harris, D., 1970. Growth inhibitors produced by the green algae (Volvocaceae). Archiv für Mikrobiologie 76, 47–50.

Hartig, P., Grobbelaar, J., Soeder, C., Groeneweg, J., 1988. On the mass culture of microalgae: areal density as an important factor for achieving maximal productivity. Biomass 15, 211–221.

Harun, R., Singh, M., Forde, G.M., Danquah, M.K., 2010. Bioprocess engineering of microalgae to produce a variety of consumer products. Renewable Sustainable Energy Rev. 14, 1037–1047.

Hase, R., Oikawa, H., Sasao, C., Morita, M., Watanabe, Y., 2000. Photosynthetic production of microalgal biomass in a raceway system under greenhouse conditions in Sendai city. J. Biosci. Bioeng. 89, 157–163.

Hyka, P., Lickova, S., Pribyl, P., Melzoch, K., Kovar, K., 2013. Flow cytometry for the development of biotechnological processes with microalgae. Biotechnol. Adv. 31, 2–16.

Ishoy, T., Kvist, T., Westermann, P., Ahring, B.K., 2006. An improved method for single cell isolation of prokaryotes from meso-, thermo- and hyperthermophilic environments using micromanipulation. Appl. Microbiol. Biotechnol. 69, 510–514.

Jakob, G., 2013. Surveying a diverse pool of microalgae as a bioresource for future biotechnological applications. J. Pet. Environ. Biotechnol. 04.

Kieran, P.M., Malone, D.M., MacLoughlin, P.F., 2000. Effects of hydrodynamic and interfacial forces on plant cell suspension systems. In: Influence of Stress on Cell Growth and Product Formation. Springer, pp. 139–177.

Kirst, G., 1990. Salinity tolerance of eukaryotic marine algae. Annu. Rev. Plant Biol. 41, 21–53.

Kok, B., 1953. Experiments on photosynthesis by Chlorella in flashing light. Algal Culture: From Laboratory to Pilot Plant 600, 63–75.

La, H.J., Lee, J.Y., Kim, S.G., Choi, G.G., Ahn, C.Y., Oh, H.M., 2012. Effective screening of Scenedesmus sp. from environmental microalgae communities using optimal sonication conditions predicted by statistical parameters of fluorescence-activated cell sorting. Bioresour. Technol. 114, 478–483.

Lam, M.K., Lee, K.T., Mohamed, A.R., 2012. Current status and challenges on microalgae-based carbon capture. Int. J. Greenhouse Gas Control 10, 456–469.

Lee, C.-G., 1999. Calculation of light penetration depth in photobioreactors. Biotechnol. Bioprocess Eng. 4, 78–81.

Mandalam, R.K., Palsson, B.Ø, 1998. Elemental balancing of biomass and medium composition enhances growth capacity in high-density Chlorella vulgaris cultures. Biotechnol. Bioeng. 59, 605–611.

Marre, E., 1962. Temperature. In: Lewin, A. (Ed.), Physiology and biochemistry of algae. R. Acad. Press, New York, London, pp. 541–550.

Melis, A., Neidhardt, J., Baroli, I., Benemann, J.R., 1998. Maximizing photosynthetic productivity and light utilization in microalgae by minimizing the light-harvesting chlorophyll antenna size of the photosystems. In: BioHydrogen. Springer, pp. 41–52.

Meseck, S.L., Alix, J.H., Wikfors, G.H., 2005. Photoperiod and light intensity effects on growth and utilization of nutrients by the aquaculture feed microalga, Tetraselmis chui (PLY429). Aquaculture 246, 393–404.

Meseck, S.L., 2007. Controlling the growth of a cyanobacterial contaminant, Synechoccus sp., in a culture of Tetraselmis chui (PLY429) by varying pH: implications for outdoor aquaculture production. Aquaculture 273, 566–572.

Molina, E., Fernández, J., Acién, F., Chisti, Y., 2001. Tubular photobioreactor design for algal cultures. J. Biotechnol. 92, 113–131.

Mutanda, T., Ramesh, D., Karthikeyan, S., Kumari, S., Anandraj, A., Bux, F., 2011. Bioprospecting for hyperlipid producing microalgal strains for sustainable biofuel production. Bioresour. Technol. 102, 57–70.

Oswald, W.J., 1992. Micro-algae and waste-water treatment. In: Borowitzka, M.A., Borowitzka, L.J. (Eds.), Microalgal Biotechnology. Cambridge University Press, Cambridge, pp. 305–328.

Pereira, H., Barreira, L., Mozes, A., Florindo, C., Polo, C., Duarte, C.V., Custodio, L., Varela, J., 2011. Microplate-based high throughput screening procedure for the isolation of lipid-rich marine microalgae. Biotechnol. Biofuels 4, 61.

Qiang, H., Richmond, A., 1994. Optimizing the population density in Isochrysis galbana grown outdoors in a glass column photobioreactor. J. Appl. Phycol. 6, 391–396.

Qiang, H., Richmond, A., 1996. Productivity and photosynthetic efficiency of Spirulina platensis as affected by light intensity, algal density and rate of mixing in a flat plate photobioreactor. J. Appl. Phycol. 8, 139–145.

Qiang, H., Zarmi, Y., Richmond, A., 1998. Combined effects of light intensity, light-path and culture density on output rate of Spirulina platensis (Cyanobacteria). Eur. J. Phycol. 33, 165–171.

Rawat, I., Ranjith Kumar, R., Mutanda, T., Bux, F., 2013. Biodiesel from microalgae: a critical evaluation from laboratory to large scale production. Appl. Energy 103, 444–467.

Richmond, A., 2008. Handbook of Microalgal Culture: Biotechnology and Applied Phycology. John Wiley & Sons.

Rubio, F.C., Fernandez, F., Perez, J., Camacho, F.G., Grima, E.M., 1999. Prediction of dissolved oxygen and carbon dioxide concentration profiles in tubular photobioreactors for microalgal culture. Biotechnol. Bioeng. 62, 71—86.

Sing, M.S.D.F., 2010. Strain Selection and Outdoor Cultivation of Halophilic Microalgae with Potential for Large Scale Biodiesel Production. Murdoch University.

Singh, B., Guldhe, A., Rawat, I., Bux, F., 2014. Towards a sustainable approach for development of biodiesel from plant and microalgae. Renewable Sustainable Energy Rev. 29, 216—245.

Suh, I.S., Lee, C.-G., 2003. Photobioreactor engineering: design and performance. Biotechnol. Bioprocess Eng. 8, 313—321.

Thomas, W.H., Gibson, C.H., 1990. Effects of small-scale turbulence on microalgae. J. Appl. Phycol. 2, 71—77.

Thompson, P., 1999. The response of growth and biochemical composition to variations in daylength, temperature, and irradiance in the marine diatom *Thalassiosira pseudonana* (Bacillariophyceae). J. Phycol. 35, 1215—1223.

Thompson, P.A., Guo, M.X., Harrison, P.J., 1992. Effects of variation in temperature. I. On the biochemical composition of eight species of marine phytoplankton. J. Phycol. 28, 481—488.

Ugwu, C.U., Aoyagi, H., Uchiyama, H., 2008. Photobioreactors for mass cultivation of algae. Bioresour. Technol. 99, 4021—4028.

Wright, G.D., Arlt, J., Poon, W.C., Read, N.D., 2007. Optical tweezer micromanipulation of filamentous fungi. Fungal genetics and biology. Fungal Genet. Biol. 44, 1—13.

Zimmerman, W.B., Tesař, V., Bandulasena, H., 2011. Towards energy efficient nanobubble generation with fluidic oscillation. Curr. Opin. Colloid. Interface Sci. 16, 350—356.

5

Mass Production of Microalgae

José C.M. Pires

Universidade do Porto, LEPABE, Departamento de Engenharia Química,
Faculdade de Engenharia, Porto, Portugal

1. MICROALGAE SCREENING AND ISOLATION TECHNIQUES

Isolation is a required procedure to achieve pure cultures and represents the first step in the selection of the species to any microalgal application. There are three main isolation techniques: (1) streaking; (2) serial dilution; and (3) single-cell isolations (Andersen and Kawachi, 2005; Black, 2008; He et al., 2012). Streaking is a rapid and simple process. A sample containing microorganisms is streaked in sterile petri dishes. Single microalgal species may be easily distinguished based on morphological characteristics: size, shape, or color. The observation of petri dish under the microscope leads to posterior selection of the suitable colony. The serial dilution protocol is a procedure that reduces the microorganisms' concentration in a liquid sample. It repeats the mixing of a certain amount of source culture with sterilized liquid (medium). Thus, an axenic culture may grow in one of the higher dilution tubes. Single-cell isolation based on traditional methods is time-consuming and requires sterilized cultivation media and equipment. An automated single-cell isolation method that has been successfully applied for microalgal cell sorting from water is flow cytometry (Davey and Kell, 1996; Montero et al., 2011; Sensen et al., 1993; Sieracki et al.,

2005). Despite being a fast and efficient method to isolate specific types of microorganisms from complex cell mixtures, the equipment is expensive and needs a trained and dedicated technician to run it properly. Other isolation methods that can be applied are micromanipulation, density centrifugation, UV radiation, and addition of antibiotics (Lee and Shen, 2004).

1.1 Species and Strain Selection

Microalgae are important in the food chain in marine environments. The main requirements for their growth are water, light, and soluble salts (Chisti, 2007; Oncel, 2013), and the optimal composition of the medium is specific for each microorganism. An important step in achieving microalgal culture on a large scale is selection of the best-suited species for growth. Local microalgal species should be collected because it can be expected that they have a competitive advantage under the local geographical, climatic, and ecological conditions. This selection is performed based on the considered application, such as food production, chemicals, drugs, biofuels, or wastewater treatment. For instance, concerning biofuel production, the evaluated characteristics of microalgae are (1) energy

Handbook of Marine Microalgae
http://dx.doi.org/10.1016/B978-0-12-800776-1.00005-4

yields (photoconversion efficiency); (2) valuable coproducts (hydrocarbons, proteins, lipids, etc.); (3) environmental tolerance (temperature, salinity, and pH); (4) mass culture performance (resistant to predators in open ponds and robust toward shear stresses in photobioreactors); (5) media requirements (vitamins, trace metals, etc.); and (6) data available on related strains (Chisti, 2007; Dillschneider et al., 2013). These characteristics are associated with screening and cultivation steps. The behavior of microalgal cells is also important in downstream processes. In the harvesting step, cell size and cell wall properties should promote autoflocculation (cheap harvesting process), and high sinking speed is required. Concerning biodiesel production, cell wall properties are important in oil extraction efficiency (Goncalves et al., 2013).

2. BASIC MICROALGAL CULTURING TECHNIQUES

2.1 Algal Nutrition

Being aquatic microorganisms, microalgae require access to some nutrients for their metabolic activities. The production of lipids, proteins, and carbohydrates is intrinsically linked to environmental conditions (light, temperature, and pH), nutrient availability, and presence of other microorganisms (Abdel-Raouf et al., 2012; Munoz and Guieysse, 2006). Carbon, hydrogen, and oxygen are nonmineral nutrients that are needed for microalgal growth. Hydrogen and oxygen are abundant in the media; thus, they never represent a limitation for microalgal growth. Carbon is one of the major nutrients that should be supplied to cultures. Depending on the type of carbon source, microalgae can be divided into two large groups: autotrophs and heterotrophs. Autotrophic organisms are able to use solar energy to convert inorganic forms of carbon (CO_2, carbonate, or bicarbonate). In contrast, heterotrophic organisms use the

chemical energy of organic forms of carbon (acetate or glucose) for their metabolic activities. Carbon is essential in algal growth and reproduction (represents about 50% of the cell dry weight) (Chisti, 2007; Milne et al., 2012). The fixed carbon is then used for respiration as an energy source and as raw material for cell division. Other macronutrients are nitrogen, phosphorus, sulfur, potassium, and magnesium. Nitrogen is an important element for the production of proteins and nucleic acids (Yen et al., 2014). It accounts for 7–20% of cell dry weight. This element can be available in the form of nitrate, nitrite, ammonia, or urea. Microalgae assimilate these nutrients from the medium, recycling them to meet changing physiological needs. The absence of this nutrient promotes the accumulation of lipids and reduction of protein content (Cakmak et al., 2012; Li et al., 2008; Rodolfi et al., 2009). Phosphorus is an important element in essential molecules, such as adenosine triphosphate (ATP; the energy carrier in cells), DNA, and RNA, and a key component of phospholipids (a major component of total lipid content) (Sydney et al., 2014). It represents about 1% of cell dry weight. In the medium, this element is presented in the form of orthophosphates. Nitrogen and phosphorus are usually limiting nutrients in natural environments (Bergstrom et al., 2008; Davey et al., 2008; Persic et al., 2009). Moreover, the limitation of phosphorus implies the reduction of synthesis and regeneration of substrates in the Calvin–Benson cycle, reducing the photosynthetic efficiency (Barsanti and Gualtieri, 2006). Sulfur is a structural component of some amino acids and vitamins and is important in the production of chloroplasts (Barsanti and Gualtieri, 2006). Potassium is important for several enzymes and is involved in protein synthesis and osmotic regulation (Sydney et al., 2014). Magnesium is an important element present in chlorophyll, a critical pigment for photosynthesis reaction. Micronutrients, such as iron, manganese, cobalt, zinc, copper, and molybdenum are required in small amounts. Iron is one of the most important trace elements for microalgal growth, due to its relevance in photosynthetic and respiration processes. It acts as a redox catalyst in photosynthesis and nitrogen assimilation. Its limitation decreases photosynthetic electron transfer, reducing NADPH (key element for the sequential biochemical reduction of carbon dioxide to carbohydrates in the dark reaction of photosynthesis) formation (Masojídek et al., 2004; Sydney et al., 2014).

2.2 Microalgae Culture Media

Andersen et al. (2005) presented several culture media (freshwater and seawater) compositions commonly used to produce microalgae. For instance, BG-11 medium has often been used to culture freshwater green algae and cyanobacteria (Grobbelaar, 2004). It is particularly rich in nitrate ($NaNO_3$ concentration of $1.5\,g\,L^{-1}$, equivalent to $1.09\,L^{-1}\,NO_3-N$) and exhibits an N:P ratio of 60:1, which is higher than the molar ratio of these elements in biomass (Redfield ratio 16:1) (Ernst et al., 2005; Geider and La Roche, 2002; Klausmeier et al., 2004). The Bold Basal Medium (BBM) is also used for freshwater algae and cyanobacteria. Dayananda et al. (2007) compared several culture media (including BG-11 and BBM) for the culture of *Botryococcus braunii* and concluded that BG-11 was the best medium regarding biomass and hydrocarbon production.

2.3 Algal Growth Kinetics and Measurement of Algal Growth

The growth of microalgal cultures is characterized by five phases (see Figure 1): (1) lag or induction phase; (2) exponential phase; (3) declining growth phase; (4) stationary phase; and (5) death phase (Grobbelaar, 2004). The lag phase corresponds to the period of physiological adaptation of cell metabolism to new nutrient or culture conditions. For instance, growth lag is observed when shade-adapted cells are exposed

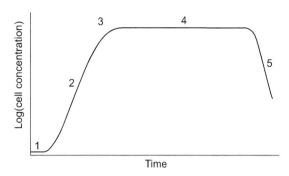

FIGURE 1   Growth phases of microalgal cultures: (1) lag or induction phase; (2) exponential phase; (3) declining growth phase; (4) stationary phase; and (5) death phase.

to higher light intensities. During this phase, a little increase of cell density may occur. To reduce the time of adaptation, the cultures can be inoculated with exponentially growing algae. During exponential phase, microalgae start to grow as a function of time according to an exponential function (growth rate follows a simple first-order rate law). Declining growth phase is characterized by the reduction of the cell division rate due to a physical or chemical limiting factor (nutrients, light, pH, carbon dioxide, and others). In the fourth stage, the limiting factor and the growth rate are balanced, maintaining a constant cell density. During the last phase, the culture conditions (depletion of a nutrient, overheating, pH disturbance, or contamination) do not favor sustainable growth, and the cell density starts to decrease.

To achieve high microalgal production rates, the cultures should be maintained in the exponential phase of growth. In continuous cultures, fresh medium is continuously added to microalgae, which allows permanent exponential growth cultivation. For instance, in a well-stirred photobioreactor, temporal variation of cell concentration is represented by the following mass balance equation:

$$\frac{dX}{dt} = \mu X - \frac{F}{V} X \qquad (1)$$

where μ is the specific growth rate, X the cell (or biomass) concentration, F the medium flow rate, and V the reactor effective volume. At steady state, the specific growth rate is equal to dilution rate (D).

$$\mu = \frac{F}{V} = D \qquad (2)$$

In batch cultures, during the exponential phase, the rate of increase of cells per unit of time is proportional to the cell concentration at the given time:

$$\frac{dX}{dt} = rX \qquad (3)$$

which is equivalent to:

$$X_t = X_0 \cdot e^{r \cdot t} \qquad (4)$$

where X_t and X_0 are cell concentrations at time t and at the beginning, respectively, and r is called the instantaneous rate of increase defined by the difference between the specific growth rate and the mortality rate. Another parameter to evaluate the growth kinetic of microalgae is the doubling time (T_2). It is defined by:

$$T_2 = \ln(2)/r \qquad (5)$$

Yield represents the production of mass per unit of volume, expressed in g L^{-1}. Biomass productivity is the yield per unit of time, being expressed in g L^{-1} h^{-1} or g L^{-1} d^{-1} (Pires et al., 2014).

2.4 Technological Aspects of Algae Culturing

Microalgae can be cultivated in indoor and outdoor environments. In indoor cultures, key parameters for microalgal growth (illumination, temperature, pH, nutrient levels, and contamination with predators) can be closely controlled. On the other hand, outdoor cultures presented high variability in these variables, since it is difficult to achieve higher production rates during extended periods (Lopez-Elias et al., 2005; Masojidek et al., 2003; Molina et al., 2001). Contamination with predators could be avoided if closed systems are chosen instead of open systems. The three operational modes for microalgal cultures are batch, continuous, and semicontinuous (Brown et al., 1993; Ruiz-Marin et al., 2010).

Table 1 compares open and closed systems for microalgal culture (Grobbelaar, 2009; Harun et al., 2010; Pires et al., 2012). Open bioreactors are the most used systems in microalgae cultivation due to their low investment and maintenance costs, which results in lower production costs (Pires et al., 2012; Stephenson et al., 2010; Tredici, 2004). They include shallow ponds, raceway ponds, tanks, and circular ponds (see Figure 2). They can have an area of 1–200 ha and can be constructed on degraded and nonagricultural lands, not competing with arable land used for the food industry. The depth should be 20–35 cm to ensure adequate exposure to light. Paddlewheels provide mechanical energy to the culture, keeping the cells is suspension. However, this type of bioreactor has some disadvantages. It presents poor light diffusion to cultures, which is an important drawback for autotrophic cultures. As an open bioreactor, there are losses of water (by evaporation) and CO_2 to the atmosphere, increasing the environmental impact of cultures. Due to the lack of control of these process parameters, microalgal production in open systems depends on the local climate (biomass yield ranges between 10 and 25 g m^{-2} d^{-1}, lower than the values achieved in closed systems: 25–50 g m^{-2} d^{-1}). Moreover, contamination with predators limits the commercial applications of the produced biomass. Thus, high biomass productivities are only achieved with microalgal strains resistant to severe environmental conditions, such as high salinity (*Dunaliella*), alkalinity (*Spirulina*), and nutrition (*Chlorella* spp.) (Harun et al., 2010; Lee, 2001). Besides its technological simplicity, production in open systems can be costly due to the downstream processing costs.

TABLE 1   Comparison Between Open and Closed Systems for Microalgal Production

	Parameter	Open systems	Closed systems
Environmental impact	Land footprint	High	Low
	Water footprint	High	Low
	CO_2 losses	High	Low
Biological issues	Algal species	Restricted	Flexible
	Contamination	High risk	Low risk
	Biomass productivity	Low	High
	Biomass composition	Variable	Reproducible
Process issues	Temperature control	No	Yes
	Weather dependence	High	Low
	Energy requirement	Low	High
	Process control	Difficult	Easy
	Use of wastewater	Yes	Yes
	Reactor cleaning	Not required	Required
Costs	Investment costs	Low	High
	Operational costs	Low	High
	Harvesting costs	Low	High

Adapted from Abdel-Raouf et al. (2012), Grobbelaar (2009), Harun et al. (2010), and Pires et al. (2012).

FIGURE 2   Open systems for microalgal culture: (a) raceway ponds. *(From http://algae-energy.co.uk.)* and (b) circular ponds. *(From http://www.antenna.ch/.)*

On the other hand, closed cultivation systems, usually designed as photobioreactors (PBRs), promote more controlled environmental conditions than open systems, in terms of CO_2 and water supply, optimal temperature and pH, distribution of light, and mixing regime (Grobbelaar, 2009; Harun et al., 2010). Consequently, high biomass productivities can be achieved. PBRs require less space, enhancing microalgal areal productivity. Figure 3 shows the main

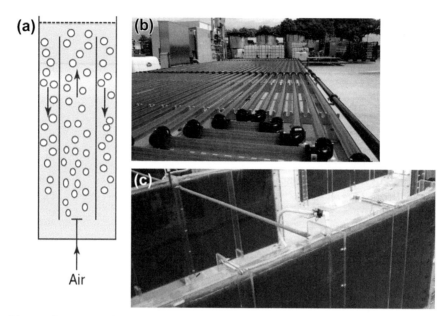

FIGURE 3   Main configurations of photobioreactors: (a) air-lift bioreactor (van Benthum et al., 1999); (b) tubular bioreactor *(from http://www.omega3.company)*; and (c) flat-plate bioreactor. *(From http://www.nanovoltaics.com.)*

configurations of PBRs. Air-lift reactors may play an important role in microalgae production, due to low level and homogeneous distribution of hydrodynamic shear that is an important disadvantage of closed production systems. In these reactors, mixing circulation and aeration is performed by gas injection, and the medium circulates in a cyclic pattern through channels built for this purpose. In addition, it promotes the fast exchange of O_2 and CO_2 between culture and gaseous stream. On the other hand, tubular PBRs are designed for outdoor cultures, having a large illumination surface. They present high CO_2 uptake efficiency, as it increases the contact time between gaseous and liquid phases (Stewart and Hessami, 2005). However, due to this long contact time, O_2 can accumulate in the culture, which is a stress factor to cells. Moreover, this typology of bioreactors presents high pumping costs. In all PBR typologies, flat-plate bioreactors are the ones that can achieve higher cell densities (in more than an order of

magnitude), presenting additional advantages: (1) lower power consumption; (2) high mass transfer rates between gaseous and liquid phases; (3) absence of dark volumes; and (4) high photosynthetic efficiency. The main drawbacks of PBRs are the investment and operational costs, reasons why the most current microalgal biomass production is performed in open ponds (Posten, 2009).

3. MASS PRODUCTION OF MICROALGAE

3.1 Principles of Algal Mass Production

The technology to produce microalgal biomass depends on the selected application and commercial value of the compounds that can be extracted. In the pharmaceutical and cosmetic industries, closed systems are preferred for microalgal production to obtain the desired products. The costs of microalgae production are supported with the high value of the extracted compounds. However, for environmental and energy applications, the production costs should be lower. Besides the selection of low-cost operating bioreactors, the integration of processes will enhance the economical viability of microalgal culture. Figure 4 shows a scheme of integration of biofuel production and CO_2 capture and nutrient removal from wastewater (Douskova et al., 2010; McGinn et al., 2011; Min et al., 2011). The use of wastewater (from determined sources) as microalgal culture medium reduces the requirement of nutrients (their price has increased in the last decades) and fresh water. Simultaneously, microalgal culture promotes the treatment of this effluent (secondary or tertiary). The high photosynthetic efficiency of microalgae can be used to capture CO_2 from flue gases. This air pollutant is considered one of the most important greenhouse gases, which are associated to the global warming. Microalgal culture is considered one environmentally sustainable option for CO_2 sequestration. The produced biomass is reached in lipids, which after extraction can be used to produce biodiesel. The residual biomass can be used to produce animal feed or biomethane (through anaerobic digestion). The net balance of carbon shows that microalgae biofuels can be zero or negative emission fuels (Lam et al., 2012).

3.2 Algal Production Systems

Carbon sources are important for microalgal growth. Photoautotrophic cultivation means that inorganic forms of carbon (CO_2 or bicarbonates) are supplied to the cultures. Light energy is transformed into chemical energy through photosynthesis (Ren et al., 2010). Other microalgal strains can use organic carbon (heterotrophic cultivation); however, this option is only used to produce high-value compounds. Mixotrophic nutrition mode is the combination of autotrophic and heterotrophic mechanisms.

3.2.1 Autotrophic Production Systems

Autotrophic production of microalgae can be performed in large ponds or in PBRs. CO_2 can be supplied using flue gases from power plants, obtained from combustion processes (Lam et al., 2012; Pires et al., 2012, 2014). This process can be

FIGURE 4 Integration of biofuel production, CO_2 capture, and nutrient removal from wastewaters.

used to recycle this greenhouse gas, reducing its emissions to the atmosphere and producing low carbon biofuels. Light supply is an important variable in autotrophic culture. Depending on its intensity, three regimes can be defined: (1) light-limitation phase; (2) light-saturation phase; and (3) light-inhibition phase (Lee et al., 2014; Ogbonna and Tanaka, 2000; Wu et al., 2014). To maximize the biomass production, light intensity should be supplied in saturation levels to the entire cultivation system. Increasing the cell concentration, self-shading may occur, reducing the light supply to lower layers of algae in PBRs. Therefore, autotrophic cultures do not usually achieve high cell densities (being one order of magnitude lower than the values achieved with heterotrophic cultures), a reason to require high culture volumes. Moreover, low cell densities lead to unsatisfactory biomass productivities and high harvesting costs. However, microalgae present higher photosynthetic efficiency than plants, making them the most important carbon-fixation group and oxygen producer on the planet. Wastewater treatment can be an important application for autotrophic microalgae (Perez-Garcia et al., 2011). These photosynthetic microorganisms absorb minerals oxidized by bacteria and, simultaneously, enrich water with oxygen to promote an aerobic environment. The produced oxygen is then used by bacteria for degradation of organic matter. The consortium between microalgae and bacteria has been extensively studied for wastewater treatment (Cruz et al., 2013; de-Bashan and Bashan, 2010; de-Bashan et al., 2004, 2011).

The type of bioreactor used in autotrophic culture has a strong effect on the biomass production cost, considering the investment and operational costs. The viability of this process could eventually be achieved by designing bioreactors to achieve high efficiencies in light conversion into biomass. The consequent increase of cell densities of cultures will then reduce the downstream processing costs.

3.2.2 Heterotrophic Production Systems

Heterotrophic microalgae are cultivated in fermenters using glucose or acetate as carbon and energy source. This production pathway has the main advantage that it uses currently available and developed fermentation technology. Other advantages are elimination of the light requirement, high cell density, and high cellular lipid content (Chen, 1996; Garci et al., 2000; Miao and Wu, 2004; Perez-Garcia et al., 2011). On the other hand, organic carbon sources represent the major cost of culture medium, increasing the cost of heterotrophic production of algal oils and making them less economically attractive than those from autotrophic cultures (Mitra et al., 2012). The composition of the medium increases the probability of contamination with bacteria or fungi. Moreover, the majority of algae are photosynthetic, and only few species present heterotrophic growth.

Besides organic carbon, other nutrients, such as nitrogen, are also required. Nitrogen availability has a strong impact on the profiles of lipids and fatty acids. A low concentration of this nutrient favors the accumulation of intracellular lipids (Chen and Walker, 2011; Li et al., 2013; Morales-Sanchez et al., 2013). In this nutrition mode, oxygen plays an important role in microalgae metabolism. The limitation in its supply reduces the observed specific growth rates and thus the biomass productivity (Wu and Shi, 2007).

3.2.3 Mixotrophic Production Systems

Mixotrophic microalgae are able to combine the two nutrition modes described above. They consume atmospheric CO_2 (in the presence of light through photosynthesis) as well as organic molecules and micronutrients from the growing environment. Some microalgal species are not truly mixotrophs but have the ability to switch between autotrophic and heterotrophic metabolisms, depending on the environmental conditions (Perez-Garcia et al., 2011). In autotrophic cultures, the accumulation of oxygen produced by photosynthesis can achieve levels that can inhibit microalgal growth, causing photoxidative damage (Sforza et al., 2012). Oxygen reutilization by mixotrophic cultures can prevent this phenomenon. In addition, availability of light energy is not a limitation for biomass production. Mixotrophic cultures usually present higher growth rates than autotrophic and heterotrophic cultures (Chojnacka and Noworyta, 2004; Mitra et al., 2012). Due to the presence of organic compounds in the medium, mixotrophic culture may be easily contaminated by heterotrophic microorganisms.

The main disadvantage of mixotrophic cultures is the significant cost of medium that naturally increases the biomass production cost (similar to heterotrophic cultures). To overcome this drawback, cheaper carbon sources from wastewaters of industrial and agricultural sectors can be used in microalgal culture. Another issue that should be taken into account is the design of PBR (similar to autotrophic cultures), to have a better control over culture conditions, increasing the biomass productivities.

3.3 Harvesting and Processing

3.3.1 Harvesting Techniques

Microalgal harvesting is the removal of biomass from the culture medium. This process can involve one or more steps (chemical, physical, or biological methods) and represents one of the most important challenges for biomass production on a commercial scale. Microalgae grow suspended in large amounts of water; they have similar specific gravity to water; they keep a stable, dispersed state due to their negatively charged membrane; thus, dewatering is energy and capital intensive. The harvesting process can contribute 20–30% of the total costs of microalgal production (Goncalves et al., 2013; Pires et al., 2012). The selection of adequate harvesting methods depends on the characteristics of the target microorganism and also the value

of the products that could be commercialized from the biomass.

Conventional methods are centrifugation, flocculation, and filtration, which can be applied individually or in combination. They present some economic or technological disadvantages, such as energy requirement (centrifugation), biomass contamination (chemical flocculation), or nonfeasible scale-up (Oh et al., 2001; Rossignol et al., 1999).

Screening is a preprocessing method applied to microalgal cultures. Its efficiency depends on the screen openings and the cells sizes (Show and Lee, 2014). Despite its simplicity and low investment, its efficiency in biomass recovery is low. Moreover, biofilms with microalgae and bacteria can be formed, which requires constant maintenance.

Chemical coagulation/flocculation is a low-cost harvesting method. This process can concentrate the cultures 20–100 times (Vandamme et al., 2013). It increases the particle size, reducing the energy requirement in dewatering process. Coagulation/flocculation is usually followed by gravity sedimentation. While coagulation involves pH adjustment or electrolyte addition, flocculation is based on the addition of cationic polymers (Banerjee et al., 2013; Show and Lee, 2014). The success of these processes is mainly influenced by cell concentration, microalgal surface properties, coagulant/flocculant concentration, pH value, and ionic strength of the broth (Oh et al., 2001). Multivalent metal salts, such as $FeCl_3$, $Al_2(SO_4)_3$, and $Fe_2(SO_4)_3$, have been tested. Their dissociation lowers the electrostatic repulsion between microalgal cells, enabling formation of cell aggregates. The referred flocculants are toxic when consumed at high concentrations. Ideally, the used flocculants should be inexpensive, nontoxic, and effective in low concentrations. Additionally, they should not limit the applications for the harvested biomass. Flocculation can also occur without addition of chemicals. Autoflocculation may occur naturally in microalgal cultures exposed to sunlight with limited CO_2 supply (Gonzalez-Fernandez and

Ballesteros, 2013; Salim et al., 2013; Vandamme et al., 2012); through photosynthesis, microalgae uptake CO_2 from medium, increasing its pH value. To study this phenomenon, researchers used calcium, magnesium, and sodium hydroxides to induce the pH increase. Harvesting based on high pH value has the advantage of creating conditions that negatively influence the growth of pathogenic microorganisms (Vandamme et al., 2012). However, this process may cause changes in cell composition. The formation of cell aggregates can also be possible due to biopolymers secreted by microalgae—bioflocculation (Christenson and Sims, 2011). Thus, chemical flocculants are not required, reducing the cost and toxicity of the process. However, the mixed culture of microalgae with bacteria or fungi results in microbiological contamination, which limits the biomass applications.

Flotation is usually defined as "inverted" sedimentation in which gas bubbles fed into the culture provide a lifting force required for particle transport and separation. It is also commonly applied after coagulation/flocculation process. For microalgal harvesting, flotation is more efficient than gravity sedimentation. Main advantages are the low space requirement, short operation time, high flexibility, and low initial equipment costs (Hanotu et al., 2012; Kwon et al., 2014; Milledge and Heaven, 2013).

Filtration is a physical separation process that requires the maintenance of a pressure drop across the system to force fluid flow through a membrane. Microalgae can deposit on the membrane, increasing the resistance to flow, reducing the filtration flux (with constant pressure drop). This phenomenon is called fouling and represents the main drawback of the process, which increases the operational cost (Christenson and Sims, 2011). Tangential flow filtration is considered more appropriate for the harvesting of smaller suspended algae, as it presents minor fouling problems. The culture medium flows tangentially across the membrane, keeping the cells in suspension. The major costs of filtration are related to membrane replacement and pumping. Therefore,

it is more effective only for small volumes. For volumes greater than $20 \, m^3 \, d^{-1}$, centrifugation may be more economical (Grima et al., 2003; Rossignol et al., 1999).

Centrifugation is the fastest harvesting method and can be applied to the great majority of microalgae. However, it is also the most expensive due to high energy requirements. Thus, this process is only applied for high-valued products (Grima et al., 2003; Lee et al., 2009; Sim et al., 1988). Another disadvantage is related to exposure of microalgal cells to high gravitational and shear forces that can damage cell structure.

3.3.2 Drying Techniques

Drying is the last harvesting step and is a required to remove moisture content (to 12% or less) to avoid interference with solvents used in downstream processes. It represents a significant fraction of the total production costs. It can be performed using dryers (spray drying, freeze drying—lyophilization, and fluidized bed drying) or by exposing the biomass to solar radiation (Brink and Marx, 2013). Sun-drying corresponds to lower production costs as well as power consumption. However, this process may not be efficient due to the high water content in biomass. Spray drying is the most commonly applied method. It is fast, which maintains the product quality; however, it is not economically feasible for low-value products, such as biofuels.

4. CONCLUSIONS

This chapter presents an overview of the technology associated with microalgal production. The current market for algae corresponds to the production of high-valued products (health foods, functional foods, antioxidants, and others). For environmental and energy applications (low-valued products), research efforts should be made to increase the economic viability of the cultures. The integration of processes could be the key to reduce production costs and associated environmental impacts. For these purposes, autotrophic and mixotrophic cultures can play an important role. In this context, the design of photobioreactors should be carefully studied to optimize the photosynthetic light energy utilization by microalgae. This limitation is associated with the major drawbacks of autotrophic production systems. Additionally, wastewater from adequate sources could be used as culture medium for microalgal culture, to critically assess the suitability, viability, and feasibility of the process. The consortia between bacteria and microalgae could be tested to (1) improve biomass production; (2) reduce wastewater treatment systems' operational costs and (3) enhance microalgae harvesting processes (by bioflocculation).

Acknowledgments

This work was supported by project "RL1 — Chemical Engineering: Process Optimization and Energy Conversion" of integrated program "LEPAE/CEFT — reinforcing R&D engineering competences in Energy, Environment & Health" (reference NORTE-07-0124-FEDER-000026), financed by *Fundo Europeu de Desenvolvimento Regional* (FEDER), through *Programa Operacional do Norte* (ON2) of *Quadro de Referência Estratégica Nacional* (QREN), and FCT/MEC (PIDDAC). J.C.M. Pires is grateful to his fellowship FEUP-ON2-26-FM-Engª Sistemas.

References

Abdel-Raouf, N., Al-Homaidan, A.A., Ibraheem, I.B.M., 2012. Microalgae and wastewater treatment. Saudi J. Biol. Sci. 19.

Andersen, R.A., Berges, J.A., Harrison, P.J., Watanabe, M.M., 2005. Appendix A—recipes for freshwater and seawater media. In: Andersen, R.A. (Ed.), Algal Culturing Techniques. Elsevier, Burlington, USA, pp. 429–538.

Andersen, R.A., Kawachi, M., 2005. Traditional microalgae isolation techniques. In: Andersen, R.A. (Ed.), Algal Culturing Techniques. Elsevier, Amsterdam, pp. 83–100.

Banerjee, C., Ghosh, S., Sen, G., Mishra, S., Shukla, P., Bandopadhyay, R., 2013. Study of algal biomass harvesting using cationic guar gum from the natural plant source as flocculant. Carbohyd. Polym. 92, 675–681.

Barsanti, L., Gualtieri, P., 2006. Algae: Anatomy, Biochemistry, and Biotechnology. CRC Press, Boca Raton, FL, USA.

de-Bashan, L.E., Bashan, Y., 2010. Immobilized microalgae for removing pollutants: review of practical aspects. Bioresour. Technol. 101, 1611–1627.

de-Bashan, L.E., Hernandez, J.P., Morey, T., Bashan, Y., 2004. Microalgae growth-promoting bacteria as "helpers" for microalgae: a novel approach for removing ammonium and phosphorus from municipal wastewater. Water Res. 38, 466–474.

de-Bashan, L.E., Schmid, M., Rothballer, M., Hartmann, A., Bashan, Y., 2011. Cell-cell interaction in the eukaryote-prokaryote model of the microalgae *Chlorella vulgaris* and the bacterium azospirillum brasilense immobilized in polymer beads. J. Phycol. 47, 1350–1359.

Bergstrom, A.K., Jonsson, A., Jansson, M., 2008. Phytoplankton responses to nitrogen and phosphorus enrichment in unproductive Swedish lakes along a gradient of atmospheric nitrogen deposition. Aquat. Biol. 4, 55–64.

Black, J.G., 2008. Microbiology: Principles and Explorations, eighth ed. John Wiley & Sons, Inc., United States of America.

Brink, J., Marx, S., 2013. Harvesting of Hartbeespoort Dam micro-algal biomass through sand filtration and solar drying. Fuel 106, 67–71.

Brown, M.R., Garland, C.D., Jeffrey, S.W., Jameson, I.D., Leroi, J.M., 1993. The gross and amino-acid compositions of batch and semicontinuous cultures of Isochrysis sp (Clone Tiso), Pavlova-Lutheri and Nannochloropsis-Oculata. J. Appl. Phycol. 5, 285–296.

Cakmak, T., Angun, P., Demiray, Y.E., Ozkan, A.D., Elibol, Z., Tekinay, T., 2012. Differential effects of nitrogen and sulfur deprivation on growth and biodiesel feedstock production of *Chlamydomonas reinhardtii*. Biotechnol. Bioeng. 109, 1947–1957.

Chen, F., 1996. High cell density culture of microalgae in heterotrophic growth. Trends Biotechnol. 14, 421–426.

Chen, Y.H., Walker, T.H., 2011. Biomass and lipid production of heterotrophic microalgae *Chlorella protothecoides* by using biodiesel-derived crude glycerol. Biotechnol. Lett. 33, 1973–1983.

Chisti, Y., 2007. Biodiesel from microalgae. Biotechnol. Adv. 25, 294–306.

Chojnacka, K., Noworyta, A., 2004. Evaluation of Spirulina sp. growth in photoautotrophic, heterotrophic and mixotrophic cultures. Enzym. Microb. Tech. 34, 461–465.

Christenson, L., Sims, R., 2011. Production and harvesting of microalgae for wastewater treatment, biofuels, and bioproducts. Biotechnol. Adv. 29, 686–702.

Cruz, I., Bashan, Y., Hernandez-Carmona, G., de-Bashan, L.E., 2013. Biological deterioration of alginate beads containing immobilized microalgae and bacteria during tertiary wastewater treatment. Appl. Microbiol. Biot. 97, 9847–9858.

Davey, H.M., Kell, D.B., 1996. Flow cytometry and cell sorting of heterogeneous microbial populations: the importance of single-cell analyses. Microbiol. Rev. 60, 641.

Davey, M., Tarran, G.A., Mills, M.M., Ridame, C., Geider, R.J., LaRoche, J., 2008. Nutrient limitation of pico-phytoplankton photosynthesis and growth in the tropical North Atlantic. Limnol. Oceanogr. 53, 1722–1733.

Dayananda, C., Sarada, R., Rani, M.U., Shamala, T.R., Ravishankar, G.A., 2007. Autotrophic cultivation of *Botryococcus braunii* for the production of hydrocarbons and exopolysaccharides in various media. Biomass Bioenergy 31, 87–93.

Dillschneider, R., Steinweg, C., Rosello-Sastre, R., Posten, C., 2013. Biofuels from microalgae: photoconversion efficiency during lipid accumulation. Bioresour. Technol. 142, 647–654.

Douskova, I., Kastanek, F., Maleterova, Y., Kastanek, P., Doucha, J., Zachleder, V., 2010. Utilization of distillery stillage for energy generation and concurrent production of valuable microalgal biomass in the sequence: biogas-cogeneration-microalgae-products. Energy Convers. Manage. 51, 606–611.

Ernst, A., Deicher, M., Herman, P.M.J., Wollenzien, U.I.A., 2005. Nitrate and phosphate affect cultivability of cyanobacteria from environments with low nutrient levels. Appl. Environ. Microbiol. 71, 3379–3383.

Garci, M.C.C., Sevilla, J.M.F., Fernandez, F.G.A., Grima, E.M., Camacho, F.G., 2000. Mixotrophic growth of *Phaeodactylum tricornutum* on glycerol: growth rate and fatty acid profile. J. Appl. Phycol. 12, 239–248.

Geider, R.J., La Roche, J., 2002. Redfield revisited: variability of C:N:P in marine microalgae and its biochemical basis. Eur. J. Phycol. 37, 1–17.

Goncalves, A.L., Pires, J.C.M., Simoes, M., 2013. Green fuel production: processes applied to microalgae. Environ. Chem. Lett. 11, 315–324.

Gonzalez-Fernandez, C., Ballesteros, M., 2013. Microalgae autoflocculation: an alternative to high-energy consuming harvesting methods. J. Appl. Phycol. 25, 991–999.

Grima, E.M., Belarbi, E.H., Fernandez, F.G.A., Medina, A.R., Chisti, Y., 2003. Recovery of microalgal biomass and metabolites: process options and economics. Biotechnol. Adv. 20, 491–515.

Grobbelaar, J.U., 2004. Algal nutrition. In: Richmond, A. (Ed.), Handbook of Microalgal Culture: Biotechnology and Applied Phycology. Blackwell Publishing Ltd, IA, pp. 97–105.

Grobbelaar, J.U., 2009. Factors governing algal growth in photobioreactors: the "open" versus "closed" debate. J. Appl. Phycol. 21, 489–492.

Hanotu, J., Bandulasena, H.C.H., Zimmerman, W.B., 2012. Microflotation performance for algal separation. Biotechnol. Bioeng. 109, 1663–1673.

Harun, R., Singh, M., Forde, G.M., Danquah, M.K., 2010. Bioprocess engineering of microalgae to produce a variety of consumer products. Renewable Sustainable. Energy Rev. 14, 1037–1047.

He, M.L., Li, L., Liu, J.G., 2012. Isolation of wild microalgae from natural water bodies for high hydrogen producing strains. Int. J. Hydrogen Energ. 37, 4046–4056.

Klausmeier, C.A., Litchman, E., Daufresne, T., Levin, S.A., 2004. Optimal nitrogen-to-phosphorus stoichiometry of phytoplankton. Nature 429, 171–174.

Kwon, H., Lu, M., Lee, E.Y., Lee, J., 2014. Harvesting of microalgae using flocculation combined with dissolved air flotation. Biotechnol. Bioproc. E 19, 143–149.

Lam, M.K., Lee, K.T., Mohamed, A.R., 2012. Current status and challenges on microalgae-based carbon capture. Int. J. Greenh. Gas Con. 10, 456–469.

Lee, A.K., Lewis, D.M., Ashman, P.J., 2009. Microbial flocculation, a potentially low-cost harvesting technique for marine microalgae for the production of biodiesel. J. Appl. Phycol. 21, 559–567.

Lee, E., Pruvost, J., He, X., Munipalli, R., Pilon, L., 2014. Design tool and guidelines for outdoor photobioreactors. Chem. Eng. Sci. 106, 18–29.

Lee, Y.-K., Shen, H., 2004. Basic culturing techniques. In: Richmond, A. (Ed.), Handbook of Microalgal Culture: Biotechnology and Applied Phycology. Blackwell Publishing, Australia, pp. 40–56.

Lee, Y.K., 2001. Microalgal mass culture systems and methods: their limitation and potential. J. Appl. Phycol. 13, 307–315.

Li, Y.Q., Horsman, M., Wang, B., Wu, N., Lan, C.Q., 2008. Effects of nitrogen sources on cell growth and lipid accumulation of green alga *Neochloris oleoabundans*. Appl. Microbiol. Biot. 81, 629–636.

Li, Y.Q., Mu, J.X., Chen, D., Han, F.X., Xu, H., Kong, F., Xie, F., Feng, B., 2013. Production of biomass and lipid by the microalgae *Chlorella protothecoides* with heterotrophic-Cu(II) stressed (HCuS) coupling cultivation. Bioresour. Technol. 148, 283–292.

Lopez-Elias, J.A., Voltolina, D., Enriquez-Ocana, F., Gallegos-Simental, G., 2005. Indoor and outdoor mass production of the diatom *Chaetoceros muelleri* in a Mexican commercial hatchery. Aquacult. Eng. 33, 181–191.

Masojídek, J., Koblížek, M., Torzillo, G., 2004. Photosynthesis in microalgae. In: Richmond, A. (Ed.), Handbook of Microalgal Culture: Biotechnology and Applied Phycology. Blackwell Publishing Ltd, Iowa, USA, pp. 20–39.

Masojidek, J., Papacek, S., Sergejevova, M., Jirka, V., Cerveny, J., Kunc, J., Korecko, J., Verbovikova, O., Kopecky, J., Stys, D., Torzillo, G., 2003. A closed solar photobioreactor for cultivation of microalgae under supra-high irradiance: basic design and performance. J. Appl. Phycol. 15, 239–248.

McGinn, P.J., Dickinson, K.E., Bhatti, S., Frigon, J.C., Guiot, S.R., O'Leary, S.J.B., 2011. Integration of microalgae cultivation with industrial waste remediation for biofuel and bioenergy production: opportunities and limitations. Photosynth. Res. 109, 231–247.

Miao, X.L., Wu, Q.Y., 2004. High yield bio-oil production from fast pyrolysis by metabolic controlling of *Chlorella prototothecoides*. J. Biotechnol. 110, 85–93.

Milledge, J., Heaven, S., 2013. A review of the harvesting of micro-algae for biofuel production. Rev. Environ. Sci. Bio. 12, 165–178.

Milne, J.L., Cameron, J.C., Page, L.E., Benson, S.M., Pakrasi, H.B., 2012. Algal technologies for biological capture and utilization of CO_2 require breakthroughs in basic research. ACS Sym. Ser. 1116, 107–141.

Min, M., Wang, L., Li, Y.C., Mohr, M.J., Hu, B., Zhou, W.G., Chen, P., Ruan, R., 2011. Cultivating *Chlorella* sp in a pilot-scale photobioreactor using centrate wastewater for microalgae biomass production and wastewater nutrient removal. Appl. Biochem. Biotech. 165, 123–137.

Mitra, D., van Leeuwen, J., Lamsal, B., 2012. Heterotrophic/mixotrophic cultivation of oleaginous *Chlorella vulgaris* on industrial co-products. Algal Res. 1, 40–48.

Molina, E., Fernandez, J., Acien, F.G., Chisti, Y., 2001. Tubular photobioreactor design for algal cultures. J. Biotechnol. 92, 113–131.

Montero, M.F., Aristizabal, M., Reina, G.G., 2011. Isolation of high-lipid content strains of the marine microalga *Tetraselmis suecica* for biodiesel production by flow cytometry and single-cell sorting. J. Appl. Phycol. 23, 1053–1057.

Morales-Sanchez, D., Tinoco-Valencia, R., Kyndt, J., Martinez, A., 2013. Heterotrophic growth of *Neochloris oleoabundans* using glucose as a carbon source. Biotechnol. Biofuels 6.

Munoz, R., Guieysse, B., 2006. Algal-bacterial processes for the treatment of hazardous contaminants: a review. Water Res. 40, 2799–2815.

Ogbonna, J.C., Tanaka, H., 2000. Light requirement and photosynthetic cell cultivation—development of processes for efficient light utilization in photobioreactors. J. Appl. Phycol. 12, 207–218.

Oh, H.M., Lee, S.J., Park, M.H., Kim, H.S., Kim, H.C., Yoon, J.H., Kwon, G.S., Yoon, B.D., 2001. Harvesting of *Chlorella vulgaris* using a bioflocculant from *Paenibacillus* sp. AM49. Biotechnol. Lett. 23, 1229–1234.

Oncel, S.S., 2013. Microalgae for a macroenergy world. Renewable Sustainable Energy Rev. 26, 241–264.

Perez-Garcia, O., Escalante, F.M.E., de-Bashan, L.E., Bashan, Y., 2011. Heterotrophic cultures of microalgae: metabolism and potential products. Water Res. 45, 11–36.

Persic, V., Horvatic, J., Has-Schon, E., Bogut, I., 2009. Changes in N and P limitation induced by water level fluctuations in Nature Park Kopacki Rit (Croatia): nutrient enrichment bioassay. Aquat. Ecol. 43, 27–36.

Pires, J.C.M., Alvim-Ferraz, M.C.M., Martins, F.G., Simões, M., 2012. Carbon dioxide capture from flue gases using microalgae: engineering aspects and biorefinery concept. Renewable Sustainable. Energy Rev. 16, 3043–3053.

Pires, J.C.M., Gonçalves, A.L., Martins, F.G., Alvim-Ferraz, M.C.M., Simões, M., 2014. Effect of light supply on CO_2 capture from atmosphere by *Chlorella vulgaris* and *Pseudokirchneriella subcapitata*. Mitigation Adaptation Strateg. Glob. Change 19 (7), 1109–1117.

Posten, C., 2009. Design principles of photo-bioreactors for cultivation of microalgae. Eng. Life Sci. 9, 165–177.

Ren, L.J., Ji, X.J., Huang, H., Qu, L.A., Feng, Y., Tong, Q.Q., Ouyang, P.K., 2010. Development of a stepwise aeration control strategy for efficient docosahexaenoic acid production by Schizochytrium sp. Appl. Microbiol. Biot. 87, 1649–1656.

Rodolfi, L., Zittelli, G.C., Bassi, N., Padovani, G., Biondi, N., Bonini, G., Tredici, M.R., 2009. Microalgae for oil: strain selection, induction of lipid synthesis and outdoor mass cultivation in a low-cost photobioreactor. Biotechnol. Bioeng. 102, 100–112.

Rossignol, N., Vandanjon, L., Jaouen, P., Quemeneur, F., 1999. Membrane technology for the continuous separation microalgae/culture medium: compared performances of cross-flow microfiltration and ultra-filtration. Aquacult. Eng. 20, 191–208.

Ruiz-Marin, A., Mendoza-Espinosa, L.G., Stephenson, T., 2010. Growth and nutrient removal in free and immobilized green algae in batch and semi-continuous cultures treating real wastewater. Bioresour. Technol. 101, 58–64.

Salim, S., Shi, Z., Vermue, M.H., Wijffels, R.H., 2013. Effect of growth phase on harvesting characteristics, autoflocculation and lipid content of Ettlia texensis for microalgal biodiesel production. Bioresour. Technol. 138, 214–221.

Sensen, C.W., Heimann, K., Melkonian, M., 1993. The production of clonal and axenic cultures of microalgae using fluorescence-activated cell sorting. Eur. J. Phycol. 28, 93–97.

Sforza, E., Simionato, D., Giacometti, G.M., Bertucco, A., Morosinotto, T., 2012. Adjusted light and dark cycles can optimize photosynthetic efficiency in algae growing in photobioreactors. Plos One 7.

Show, K.-Y., Lee, D.-J., 2014. Algal biomass harvesting. In: Pandey, A., Lee, D.-J., Chisti, Y., Soccol, C.R. (Eds.), Biofuels from Algae. Elsevier, USA, pp. 85–110.

Sieracki, M., Poulton, N., Crosbie, N., 2005. Automated isolation techniques for microalgae. In: Andersen, R.A. (Ed.), Algal Culturing Techniques. Elsevier, Amsterdam, pp. 101–116.

Sim, T.S., Goh, A., Becker, E.W., 1988. Comparison of centrifugation, dissolved air flotation and drum filtration techniques for harvesting sewage-grown algae. Biomass 16, 51–62.

Stephenson, A.L., Kazamia, E., Dennis, J.S., Howe, C.J., Scott, S.A., Smith, A.G., 2010. Life-cycle assessment of potential algal biodiesel production in the United Kingdom: a comparison of raceways and air-lift tubular bioreactors. Energy Fuel 24, 4062–4077.

Stewart, C., Hessami, M.A., 2005. A study of methods of carbon dioxide capture and sequestration—the sustainability of a photosynthetic bioreactor approach. Energy Convers. Manage. 46, 403–420.

Sydney, E.B., Novak, A.C., Carvalho, J.C., Soccol, C.R., 2014. Respirometric balance and carbon fixation of industrially important algae. In: Pandey, A., Lee, D.-J., Chisti, Y., Soccol, C.R. (Eds.), Biofuels from Algae. Elsevier, USA, pp. 67–84.

Tredici, M.R., 2004. Mass production of microalgae: photobioreactors. In: Richmond, A. (Ed.), Handbook of Microalgal Culture: Biotechnology and Applied Phycology. Blackwell Publishing, Oxford, UK, pp. 178–214.

van Benthum, W.A.J., van der Lans, R.G.J.M., van Loosdrecht, M.C.M., Heijnen, J.J., 1999. Bubble recirculation regimes in an internal-loop airlift reactor. Chem. Eng. Sci. 54 (18), 3995–4006.

Vandamme, D., Foubert, I., Fraeye, I., Meesschaert, B., Muylaert, K., 2012. Flocculation of *Chlorella vulgaris* induced by high pH: role of magnesium and calcium and practical implications. Bioresour. Technol. 105, 114–119.

Vandamme, D., Foubert, I., Muylaert, K., 2013. Flocculation as a low-cost method for harvesting microalgae for bulk biomass production. Trends Biotechnol. 31, 233–239.

Wu, Y.C., Wang, Z.J., Zheng, Y., Xiao, Y., Yang, Z.H., Zhao, F., 2014. Light intensity affects the performance of photo microbial fuel cells with *Desmodesmus* sp. A8 as cathodic microorganism. Appl. Energy 116, 86–90.

Wu, Z.Y., Shi, X.M., 2007. Optimization for high-density cultivation of heterotrophic Chlorella based on a hybrid neural network model. Lett. Appl. Microbiol. 44, 13–18.

Yen, H.-W., Hu, I.-C., Chen, C.-Y., Chang, J.-S., 2014. Design of photobioreactors for algal cultivation. In: Pandey, A., Lee, D.-J., Chisti, Y., Soccol, C.R. (Eds.), Biofuels from Algae. Elsevier, USA, pp. 23–45.

Microalgal Biotechnology: The Way Forward

Daphne H.P. Ng, Yi Kai Ng, Hui Shen, Yuan Kun Lee

National University of Singapore, Yong Loo Lin School of Medicine, Department
of Microbiology, Singapore

1. INTRODUCTION

Microalgae are a diverse group of eukaryotic or prokaryotic photosynthetic microorganisms that can grow rapidly and live in harsh conditions due to their unicellular or simple multicellular structure. They are ubiquitous and can be found in a variety of aquatic and terrestrial environments (Richmond, 2004). Green algae (Chlorophyta) and diatoms (Bacillariophyta) are examples of eukaryotic microalgae, while cyanobacteria are prokaryotic microalgae (Li et al., 2008a,b). It has been estimated that more than 50,000 species of microalgae exist, but only a limited number of approximately 30,000 have been studied and analyzed (Richmond, 2004).

Microalgae such as *Nostoc* and *Spirulina* have been used by indigenous populations as food since ancient times (Jensen et al., 2001). Currently, besides food and feed, there are numerous applications of microalgae such as the production of biodiesel feedstock, for biological carbon dioxide sequestration and production of bioactive molecules.

1.1 Biodiesel Feedstock

The accumulation of atmospheric greenhouse gases (GHG) as a result of human activities and industrialization is regarded to be the principal cause of global warming with carbon dioxide accounting for 68% of total GHG emissions (Ho et al., 2011). The burning of fossil fuels is the major cause of elevated atmospheric carbon dioxide levels with power plant flue gas accounting for more than one-third of global carbon dioxide emissions (Stewart and Hessami, 2005). Hence, sustainable energy production is a critical national economic issue. Although electricity can be generated from renewable sources in many ways, biofuels are currently the only option for the replacement of conventional liquid fuels in combustion engine vehicles (Mata et al., 2010).

Microalgae represent an important alternative to agricultural crops for second-generation biofuel production. There are several advantages of using microalgae for the production of biofuel. Compared to plants, microalgae have higher growth rates (e.g., one to three doublings per day) and do not compete for resources with food production as they can be grown in closed photobioreactors or open ponds on non-arable land using seawater or nutrient-rich agricultural wastewater (Chisti, 2007; Dismukes et al., 2008; Downing et al., 2002; Ho et al., 2011; Kumar et al., 2010).

Biodiesel is a biofuel alternative to petroleum-based diesel fuel. It is produced by the transesterification of triacylglycerols (TAG) with methanol to yield the corresponding monoalkyl fatty acid esters. As many microalgal species are able to accumulate significant amounts of lipid with the content of some species exceeding 80% of their dry cell weight (Banerjee et al., 2002; Chisti, 2008), microalgae are potential candidates for the production of biodiesel feedstock. In comparison, most agricultural crops used in biodiesel production have lipid yields of less than 5% of biomass (Chisti, 2008).

The lipid composition of microalgae is dependent on the physiological state of the cells. Under optimal growth conditions, algae synthesize fatty acids principally for esterification into glycerol-based membrane lipids (polar lipids). The formation of polar lipids is also induced by low-light intensities. In response to unfavorable environmental or stress conditions for growth, such as nutrient depletion (nitrogen limitation in particular), algae alter their lipid biosynthetic pathways toward the formation and accumulation of neutral lipids, mainly in the form of TAG in their stationary growth phase. Under the stress conditions, the flow of fixed carbon is also diverted from protein to lipid production (Rodolfi et al., 2009).

Similar to fatty acids in higher plants, the most common microalgal fatty acids have chain lengths that range from C16 to C18 and are either saturated or unsaturated. In general, saturated and monounsaturated algal fatty acids are predominant. For example, the major fatty acids of C16:0 and C16:1 are observed in the Bacillariophyceae; C16:0 and C18:1 in the Chlorophyceae, Euglenophyceae, Eustigmatophyceae, and Prasinophyceae; C16:0, C16:1 and C18:1 in the Chrysophyceae, Prymnesiophyceae, and cyanobacteria; C14:0, C16:0, and C16:1 in the Xanthophyceae (Cobelas and Lechado, 1988; Hu et al., 2008).

In contrast to higher plants, some algae and cyanobacteria possess the ability to produce medium-chain fatty acids (e.g., C10, C12, and C14) as the predominant species, whereas others synthesize very long-chain fatty acids (C20) (Hu et al., 2008). The filamentous cyanobacterium *Trichodesmium erythraeum* was found to produce C10 fatty acids comprising 27–50% of the total fatty acids, while C14 fatty acids accounted for almost 70% of the total fatty acids produced by the haptophyte *Prymnesium parvum* (Lee and Loeblich, 1971; Parker et al., 1967).

Microalgae also produce long-chain polyunsaturated fatty acids (PUFAs) such as those in the ω3 and ω6 families. PUFAs are essential in the diets of humans and animals as they are not able to synthesize these fatty acids (Gill and Valivety, 1997). In addition, PUFAs of the ω3 and ω6 families such as eicosapentaenoic (EPA), docosahexaenoic (DHA), and arachidonic acid are pharmacologically important in the treatment of chronic inflammation and the prevention of diseases like hypertension and coronary heart disease (Mata et al., 2010).

1.2 Biological Carbon Dioxide Sequestration

Microalgae can also be used directly in photosynthetic carbon dioxide sequestration. Photosynthesis is a natural process that converts sunlight, water, and carbon dioxide to produce oxygen and organic matter such as sugar. Microalgae have high carbon dioxide fixation efficiencies (10–50 times greater than plants) and can fix carbon dioxide from various sources, including gaseous carbon dioxide from the atmosphere and industrial flue gases, and in the form of soluble carbonates (Kumar et al., 2010; Sydney et al., 2010). Furthermore, some microalgae strains can tolerate high carbon dioxide, NO, and SO_2 concentrations as well as high temperatures, allowing efficient capture of carbon dioxide directly from power plant flue gas without the need for pretreatment (Li et al., 2008b; Ono and Cuello, 2007; Wang et al., 2008; Zeiler et al., 1995). Hence, biological carbon dioxide mitigation by microalgae with carbon dioxide being converted to organic matter through photosynthesis has been proposed as one of the strategies for atmospheric carbon dioxide capture (Ho et al., 2011; Sydney et al., 2010).

Several studies have quantified the carbon dioxide fixation abilities of various microalgae to evaluate the potential of microalgae in biological sequestration of carbon dioxide via photosynthesis (Table 1, adapted from Ho et al. (2011)).

Extracellular products synthesized by microalgae can also be used as carbon sinks. It was found that *Dunaliella tertiolecta* releases extracellular glycerol continuously under normal growth conditions, which translates into a five times larger capacity to perform carbon dioxide removal as compared to cellular material alone. Thus, extracellular glycerol may function as a sink for carbon fixation (Chow et al., 2013).

1.3 Pigments and Other Bioactive Molecules

Several valuable compounds such as pigments, antioxidants, and other biologically active molecules can be extracted from microalgae. Phycocyanin is a phycobiliprotein complex produced by cyanobacteria, which is used as a natural dye in food and cosmetics (Spolaore et al., 2006). Carotenoids such as β-carotene and astaxanthin are used as food colorants as well as antioxidant supplements in human nutraceuticals and animal feed (Spolaore et al., 2006). β-Carotene is produced by the green halophilic flagellate *D. salina* while astaxanthin is produced by the freshwater chlorophyte *Haematococcus pluvialis* (Lorenz and Cysewski, 2000; Metting, 1996). Other bioactive molecules produced by microalgae include sterols and vitamins (Mata et al., 2010; Spolaore et al., 2006).

1.4 Human Food and Animal Feed

The high protein contents of various microalgae make them a potential source of protein for human and animal nutrition. Microalgae are also a source of essential amino acids and vitamins in human and animal diets (Guil-Guerrero et al., 2004). In aquaculture, microalgae such as *Chlorella* and *Tetraselmis* are often fed directly to mollusks and shrimp larvae or used indirectly as food for the live prey fed to fish larvae (Brown et al., 1997; Muller-Feuga, 2000).

As compared to fish oils, which are the main dietary sources of long-chain PUFAs, the use of PUFAs from microalgae is more environmentally sustainable as fish stocks will be depleted. DHA produced by the dinoflagellate *Crypthecodinium cohnii* has been added to infant formula, while EPA extracted from *Nannochloropsis* has the potential to be used as a nutritional supplement (Spolaore et al., 2006).

2. DEVELOPMENT OF MICROALGAE CULTURE SYSTEMS

Culture systems are critical in maximizing the productivity of microalgae cultures. Microalgae can be cultivated in open ponds or enclosed photobioreactors. Three parameters—productivity per unit reactor volume $(g\ L^{-1}\ day^{-1})$, productivity per unit of ground area occupied by the reactor $(g\ m^{-2}\ day^{-1})$, and productivity per unit of reactor illuminated surface area $(g\ m^{-2}\ day^{-1})$—are used to evaluate performance achieved by open ponds and photobioreactors (Richmond, 2004).

An advantage of utilizing an open culture system for microalgae cultivation is its low operating cost. Additionally, open ponds have a larger production capacity as compared to closed systems and may be more durable

TABLE 1   Carbon Dioxide Fixation Ability and Biomass Productivity of Microalgae

Microalgal species	Biomass productivity (mg L^{-1} day^{-1})	CO$_2$ consumption rate (mg L^{-1} day^{-1})	Operation mode	References
Chlorella vulgaris	40	75	Batch	Scragg et al. (2002)
Scenedesmus obliquus	85	160	Batch	de Morais and Costa (2007c)
Scenedesmus obliquus	105	198	Batch	de Morais and Costa (2007b)
Scenedesmus obliquus	140	263	Serial	de Morais and Costa (2007a)
Spirulina sp.	200	376	Serial	de Morais and Costa (2007a)
Spirulina sp.	210	394	Batch	de Morais and Costa (2007b)
Scenedesmus sp.	217.5	408.9	Batch	Yoo et al. (2010)
Scenedesmus sp.	188	460.8	Batch	Jin et al. (2006)
Nannochloropsis sp.	270	508	Batch	Negoro et al. (1991)
Scenedesmus obliquus	292.5	549.9	Batch	Ho et al. (2010)
Chlorella vulgaris	273	612	Batch	Jin et al. (2006)
Chlorella sp.	335	700.2	Batch	Chiu et al. (2009)
Chlorella sp.	381.8	717.8	Batch	Chiu et al. (2009)
Chlorella sp.	171	857	Batch	Chiu et al. (2008)
Chlorococcum littorale	530	900	Batch	Kurano et al. (1995)
Botryococcus braunii	900	1000	Batch	Murakami and Ikenouchi (1997)
Chlorella sp.	610	1147	Semi-batch	Chiu et al. (2009)
Chlorella sp.	700	1316	Batch	Sakai et al. (1995)
Synechocystis aquaticus	590	1500	Batch	Murakami and Ikenouchi (1997)
Chlorella sp.	940	1767	Batch	Sung et al. (1999)
Chlorella vulgaris	150	3450	Batch	Fan et al. (2008)
Aphanothece microscopica Nägeli	1250	5435	Batch	Jacob-Lopes et al. (2009)

(Mata et al., 2010; Richmond, 1987). However, it has been reported that more energy is required to homogenize nutrients in ponds. Another drawback of an open system is its susceptibility to weather conditions, with little control of water temperatures, evaporation, and lighting. Because of its open nature, this system is also more prone to contamination from other algae and bacteria. Although open culture systems have the potential to produce large quantities of microalgae, extensive land area is required. Furthermore, mass transfer of carbon dioxide to the culture is low due to the low levels of carbon dioxide in the atmosphere (0.03−0.06%), and this limitation may result in the slow growth of microalgae (Mata et al., 2010).

On the other hand, photobioreactors are flexible systems that can be optimized according to the biological and physiological characteristics of the algae species. Unlike open ponds, photobioreactors offer better control over culture conditions and growth parameters such as pH, temperature, carbon dioxide, and oxygen gas mixing. The closed nature of a photobioreactor also minimizes the likelihood of culture contamination as direct exchange of contaminants with the environment is limited. In addition, light penetration into the culture can be optimized to achieve maximum productivity (Mata et al., 2010; Richmond, 1987).

Despite the advantages, photobioreactors are prone to various problems such as overheating, biofouling, oxygen accumulation, difficulty in scaling up, high cost of operation, cell damage by shear stress, and deterioration of material used for the photo stage (Mata et al., 2010). As a result of these limitations, photobioreactors are not expected to have a significant impact in the near future on any product or process that can be attained in large outdoor ponds (Mata et al., 2010; Richmond, 1987).

Therefore, a combination of a closed culture system with an open pond system may be the way forward in enhancing the cultivation of microalgae (Richmond, 1987). Productivity of an outdoor culture of the filamentous cyanobacterium *Arthrospira platensis* was increased when cultivated in an integrated device coupling a raceway pond with a flat alveolar panel, which maintained a near-optimum temperature regimen for growth of the microalgae. The total productivity (g reactor^{-1} day^{-1}) obtained in the integrated system was 15% higher than the sum of the productivities of the pond and panel systems when operated separately (Pushparaj et al., 1997).

3. PHOTOSYNTHETIC PRODUCTIVITY IN MICROALGAL MASS CULTURE

Although the theoretical solar-to-biomass conversion efficiency of microalgae has been estimated to be 8–10% with a maximum productivity of 77 g-biomass m^{-2} day^{-1}, it has been observed in outdoor cultures that actual algal biomass productivities do not exceed 4 g L^{-1} day^{-1} with short-term areal productivities ranging from 20 to 40 g-biomass m^{-2} day^{-1}. This translates to a 3% solar-to-biomass conversion efficiency, which is much lower than the theoretical solar-to-biomass conversion efficiency (Melis, 2009). The low productivity of outdoor cultures implies that the absorbed solar energy was not efficiently converted into biomass. Several factors such as light saturation, slow rate of carbon dioxide fixation, and slow response to varying solar irradiance could have limited the conversion of absorbed light energy into biomass (Green and Durnford, 1996).

4. OVERCOMING BIOLOGICAL LIMITATIONS IN MICROALGAL CULTURE

In photosynthetic microalgae, photons for photosynthesis are trapped by a light-harvesting antenna array consisting of a core light-harvesting antenna complex that includes a reaction center and other peripheral antenna pigment protein complexes (Green and Durnford, 1996). In response to growth irradiance, the light-harvesting complex changes the abundance of specific components in the photosynthetic apparatus (Laroche et al., 1991). This requires the organism to sense differences in light intensity for conversion of the signal to biochemical information. The information is subsequently transferred to regulatory elements responsible for gene expression, resulting in changes in the abundance of light-harvesting chlorophyll protein complexes (Teramoto et al., 2002).

For maximum absorption of light for survival, microalgae are genetically predisposed to assembling large antenna complexes due to light-limiting conditions in the natural environment, such as in the water column in a pond. Pulsed light-emitting diode experiments have demonstrated an efficient uptake of instantaneous but discontinuous irradiance flux of 5000 μmol m^{-2} s^{-1}, which is almost three times that of the solar irradiance measured at noon, without flux saturation (Tennessen et al., 1995). This reduces single-cell solar energy conversion efficiency as the chlorophyll antenna absorbs more photons than can be processed by the reaction center. Furthermore, the large chlorophyll antenna prevents light from penetrating and reaching cells deeper in the culture. Lastly, excess photons may be lost during nonphotochemical quenching, which can downregulate photosynthetic machinery or activities (Holt et al., 2004). These factors result in light saturation at relatively low-light intensities as well as damage of its associated reaction centers, thus reducing the number of available reaction centers for photosynthesis. Hence, in a dense culture, algal cells at the surface would overabsorb incoming photons and dissipate the excess excitation energy and the process may become photoinhibited. On the other hand, cells further from the top layer would be shaded and receive insufficient light for photosynthesis (Naus and Melis, 1991).

To improve light harvesting and photosynthetic efficiency, it has been suggested that a reduction in size of the functional chlorophyll antenna would allow greater transmittance of light into a dense culture (to reduce the cell harvesting cost), reduce the number of photons collected by surface cells, and reduce photoinhibition (Beckmann et al., 2009; Mussgnug et al., 2007; Nakajima et al., 1998; Polle et al., 2003; Radmer and Kok, 1977). Microalgal cells can reduce their chlorophyll antenna size as a protective mechanism when exposed to extended durations of high irradiance (Kim et al., 1993). The reduction in chlorophyll antenna size can also be achieved through genetic engineering approaches such as UV mutagenesis, direct modification of chlorophyll biosynthesis genes, or the

alteration of chlorophyll antenna regulatory genes (Michel et al., 1983; Tanaka et al., 2001). Some genes such as *tla1* have been identified to be important in chlorophyll antenna regulation. When the *tla1* gene was disrupted by DNA insertional mutagenesis, the mutant demonstrated a smaller antenna size with a two-fold higher maximum photosynthetic rate (Polle et al., 2003).

The decrease in photosynthesis efficiency due to light saturation may also result in a slow rate of carbon dioxide fixation. Thus, another strategy to improve the photosynthetic efficiency is to increase the rate of photosynthetic carbon assimilation through genetic modification of the carbon fixation pathways. The Calvin cycle occurs in the stroma of chloroplasts and is responsible for photosynthetic carbon assimilation. A key enzyme in the cycle is sedoheptulose-1,7-bisphosphatase (SBPase). SBPase catalyzes the dephosphorylation of sedoheptulose-1,7-bisphosphate (SBP) in the regeneration of ribulose-1,5-bisphosphate (RuBP) from glyceraldehyde-3-phosphate (PGAL) (Raines, 2003). It has been suggested that the catalytic activity of SBPase may limit the rate of carbon fixation (Raines et al., 2000). In the halophilic green microalga *Dunaliella bardawil*, the overexpression of *Chlamydomonas reinhardtii* SBPase (CrSBPase) resulted in a 50−100% increase in photosynthetic performance in addition to increased total organic carbon content and glycerol production (Fang et al., 2012).

In addition to modifying microalgal cells by genetic engineering, culture systems can also be optimized to attenuate the effect of light saturation. The distribution of photosynthetically active radiation (PAR) can be varied by orientating the photobioreactor at various angles to the sun (Hu et al., 1998a; Lee and Low, 1991; Pirt et al., 1983; Tredici et al., 1991). At an inclination of 45° when the sun is directly overhead, the PAR measured at the surface of the bioreactor remained constant throughout the day at approximately half the maximum irradiance measured horizontally. This diluted the sun irradiance impinging on the surface of the photobioreactor as the irradiance was distributed over a large surface area when the sun approached midday, preventing the irradiance from reaching photosynthetically saturated levels (Lee and Low, 1991; Pirt et al., 1983). Besides adjusting for diurnal fluctuations in PAR, the tilt angle can also be varied based on seasonal variations of PAR. In summer and winter, a tilt angle of 10–30° and a larger angle of 60°, respectively, resulted in maximum productivity for that season (Hu et al., 1998a).

The depth of culture in a photobioreactor can also be optimized to increase photosynthetic efficiency. In a photosynthetic culture, the light energy should be available at all times in order to maintain the growth of cells. For a given light source and culture depth, an optimal cell density (OCD) exists where the highest photosynthetic efficiency and areal biomass yield (g m^{-2} day^{-1}) were achieved (Hu and Richmond, 1994). At a cell density below OCD, the culture was light saturated, whereas at a cell concentration above OCD, cell shading resulted in nonproductive dark fraction and increased maintenance energy requirements (Lee and Pirt, 1981). Therefore, a reduction in the depth of a culture will significantly increase both volumetric (g L^{-1} day^{-1}) and areal productivity (g m^{-2} day^{-1}) (Hu et al., 1996, 1998b,c).

5. IMPROVE RESPONSE TO VARYING SOLAR IRRADIANCE

Solar irradiance varies in intensity during the day reaching the maximum near noon; it also varies throughout the year, with the highest solar irradiance occurring in summer. However, increases in the specific growth rate of microalgal cultures lag behind rapid increases in light intensity (Lee and Low, 1991, 1992; Post et al., 1984; Prézelin and Matlick, 1980). It was observed that in an outdoor *Chlorella* culture, the specific growth rate of a light-limited culture

increased by only 20% despite a five-fold increase in the solar irradiance received by the culture from early morning to noon (Lee and Low, 1992). Thus, the photosynthetic conversion efficiency in the early morning could not be maintained. The inability to respond to rapid increases in solar irradiance has also been reported in a laboratory study in which the transitional responses of the marine diatom *Thalasiosira weisflogii* to increases in light intensities were investigated. It was noted that the specific growth rate of the diatom remained relatively constant for approximately 10 h after an eight-fold increase in light intensity, followed by a sharp increase in the specific growth rate within the next few hours (Post et al., 1984). Hence, it is likely that many microalgae are not capable of responding rapidly to increases in solar irradiance, resulting in a more gradual increase in biomass output rate as compared to the rate of increase in growth irradiance. Consequently, there is a rapid decrease in the bioenergetic growth yield in the morning (Lee and Low, 1991).

Photobioreactor design can play a key role in mitigating the slow response of microalgae to varying solar irradiance. It is recognized that there is a "flashing light" effect when microalgal cells are exposed to intermittent light–dark cycles in which light energy captured in the photic zone is able to sustain the dark carbon dioxide fixation for a brief period of time (Vejrazka et al., 2012). This implies that as long as an algal cell is returned to the photic zone before the ATP and NADPH reserves are exhausted in the dark zone, continuous photosynthesis is possible in the culture. It was observed that a light–dark cycle of frequency of 100 Hz resulted in a 35% higher biomass yield than that obtained in continuous light (Vejrazka et al., 2012). Turbulence facilitates cycling of cells between photic and dark zones. It was observed that photosynthetic activity of *Chlorella* cultures and the biomass productivity of *Spirulina* cultures were dependent on the stirring speed or aeration rate (Hu and Richmond, 1996). Hence, an optimal level of stirring or aeration should be considered in photobioreactor design in order to achieve the maximum photosynthetic efficiency and biomass production.

6. ADVANCES IN MICROALGAL FEEDSTOCK-BASED BIODIESEL PRODUCTION

Despite the enormous potential of microalgal feedstock-based biodiesel, there are unique challenges in this field that should be addressed before large-scale production is possible. Recent advances in molecular biology coupled with a greater understanding of the lipid production processes in microalgae have allowed genetic manipulation of growth and lipid metabolism. Critical engineering breakthroughs related to algal mass culture and downstream processing have also been demonstrated.

6.1 Selection and Genetic Engineering of Algal Strains

For industrial-scale production of algal feedstock-based biodiesel to be feasible, the ideal microalgal strain must be highly productive with a constitutional lipid accumulation and fatty acids that mimic conventional diesel. The occurrence and the extent to which TAG is produced are species/strain specific and are ultimately controlled by the genome of individual organisms. Hence, an important component of the design of algal biofuel production processes is the selection and genetic engineering of algal strains.

A limitation of lipid accumulation in the microalgal cell through nitrogen deficiency is that lipid productivity is offset by lower productivities attained under nutrient (nitrogen) shortage (Rodolfi et al., 2009). Hence, it is desirable to genetically engineer microalgal

cells to produce lipid without compromising growth.

Lipid productivity has also been improved by metabolic engineering of lipid catabolism. Recently, it was reported that the targeted knockdown of a multifunctional lipase/phospholipase/acyltransferase increased neutral lipid yields without affecting growth in the diatom *Thalassiosira pseudonana* (Trentacoste et al., 2013).

6.2 Optimizing Fatty Acid Compositions to Produce Biodiesel with Desired Properties

In addition to the selection and genetic engineering of algal strains with high lipid productivity, fatty acid compositions of microalgae should also be optimized. This is because the properties of biodiesel are largely determined by the structure of its component fatty acid esters. The most important characteristics include ignition quality (i.e., cetane number), cold-flow properties, and oxidative stability. Of the range of fatty acids found in nature, saturated medium-chain fatty acids (C8–C14) are ideal for biodiesel with superior oxidative stability (Durrett et al., 2008). On the other hand, biodiesel from PUFAs has good cold-flow properties but is particularly susceptible to oxidation (Hu et al., 2008). Hence, it is important to modulate the species of fatty acids produced by microalgae so that biodiesel with the desired properties can be produced.

Temperature has been found to have a major effect on the degree of fatty acid saturation in microalgae. A general trend toward increasing fatty acid unsaturation with decreasing temperature and increasing saturated fatty acid content with increasing temperature has been observed in many algae and cyanobacteria (Lynch and Thompson, 1982; Murata et al., 1975; Raison, 1986; Renaud et al., 2002; Sato and Murata, 1980). For example, both chloroplast and microsomal phospholipids fatty acid unsaturation increased during acclimation to low temperature in *Dunaliella salina* (Lynch and Thompson, 1982).

Microalgal fatty acid chain lengths can also be altered for the production of biodiesel. This has been demonstrated by expressing plant FatB thioesterases (TEs) from the California bay plant *U. californica* in the diatom *Phaeodactylum tricornutum* and cyanobacterium *Synechocystis* sp. PCC6803 with the goal of short circuiting fatty acyl chain elongation so that shorter-chain fatty acids (C12 and C14) can be produced (Liu et al., 2011; Radakovits et al., 2011). However, these efforts were met with limited success.

6.3 Secretion of Fatty Acids into External Medium

The high cost of harvest and biomass recovery associated with microalgal mass cultures is one of the obstacles to the implementation of large-scale microalgal biodiesel production. To facilitate harvesting and avoid expensive biomass recovery processes currently applied in algal biofuel systems, a fatty acid secretion strategy was employed in the cyanobacterium *Synechocystis* sp. PCC 6803. Fatty acid secretion yields were increased by weakening the S layer and peptidoglycan wall of the cell. In addition, acyl–acyl carrier protein TEs from the California bay plant *Umbellularia californica* were heterogeneously expressed. The mutant *Synechocystis* sp. PCC 6803 strains were able to overproduce fatty acids (C10–C18) and secrete them into the medium, which could then be easily skimmed off the surface of the culture (Liu et al., 2011). Hence, fatty acid–secreting cyanobacteria are a promising technology for renewable biodiesel production.

7. CONCLUSIONS

Microalgae are industrially valuable microorganisms, especially in the production of biodiesel feedstock. However, there are several biological and culture system limitations that result in low productivity of microalgae cultures. Hence, there is an urgent need to enhance the yield and productivity of microalgae mass cultures. With the advancement of molecular biology and engineering, these limitations can now be better understood and overcome through approaches such as genetic engineering and culture system design. It is hoped that these biotechnological advances will increase the productivity of microalgae cultures and transform microalgae cultivation technology into a sustainable industry with environmental and societal benefits.

References

Banerjee, A., Sharma, R., Chisti, Y., Banerjee, U.C., 2002. *Botryococcus braunii*: a renewable source of hydrocarbons and other chemicals. Crit. Rev. Biotechnol. 22, 245–279.

Beckmann, J., Lehr, F., Finazzi, G., Hankamer, B., Posten, C., Wobbe, L., Kruse, O., 2009. Improvement of light to biomass conversion by de-regulation of light-harvesting protein translation in *Chlamydomonas reinhardtii*. J. Biotechnol. 142, 70–77.

Brown, M.R., Jeffrey, S.W., Volkman, J.K., Dunstan, G.A., 1997. Nutritional properties of microalgae for mariculture. Aquaculture 151, 315–331.

Chisti, Y., 2007. Biodiesel from microalgae. Biotechnol. Adv. 25, 294–306.

Chisti, Y., 2008. Biodiesel from microalgae beats bioethanol. Trends Biotechnol. 26, 126–131.

Chiu, S.-Y., Kao, C.-Y., Chen, C.-H., Kuan, T.-C., Ong, S.-C., Lin, C.-S., 2008. Reduction of CO_2 by a high-density culture of *Chlorella* sp. in a semicontinuous photobioreactor. Bioresour. Technol. 99, 3389–3396.

Chiu, S.-Y., Tsai, M.-T., Kao, C.-Y., Ong, S.-C., Lin, C.-S., 2009. The air-lift photobioreactors with flow patterning for high-density cultures of microalgae and carbon dioxide removal. Eng. Life Sci. 9, 254–260.

Chow, Y.Y.S., Goh, S.J.M., Su, Z., Ng, D.H.P., Lim, C.Y., Lim, N.Y.N., Lin, H., Fang, L., Lee, Y.K., 2013. Continual production of glycerol from carbon dioxide by *Dunaliella tertiolecta*. Bioresour. Technol. 136, 550–555.

Cobelas, M.A., Lechado, J.Z., 1988. Lipids in microalgae. A review. I. Biochemistry. Grasas Aceites 40, 118–145.

Dismukes, G.C., Carrieri, D., Bennette, N., Ananyev, G.M., Posewitz, M.C., 2008. Aquatic phototrophs: efficient alternatives to land-based crops for biofuels. Curr. Opin. Biotechnol. 19, 235–240.

Downing, J., Bracco, E., Green, F., Lundquist, T., Zubieta, I., Oswald, W., 2002. Low cost reclamation using the advanced integrated wastewater pond systems technology and reverse osmosis. Water Sci. Technol. 45, 117—125.

Durrett, T.P., Benning, C., Ohlrogge, J., 2008. Plant triacylglycerols as feedstocks for the production of biofuels. Plant J. 54, 593—607.

Fan, L.-H., Zhang, Y.-T., Zhang, L., Chen, H.-L., 2008. Evaluation of a membrane-sparged helical tubular photobioreactor for carbon dioxide biofixation by *Chlorella vulgaris*. J. Membr. Sci. 325, 336—345.

Fang, L., Lin, H.X., Low, C.S., Wu, M.H., Chow, Y., Lee, Y.K., 2012. Expression of the *Chlamydomonas reinhardtii* sedoheptulose-1,7-bisphosphatase in *Dunaliella bardawil* leads to enhanced photosynthesis and increased glycerol production. Plant Biotechnol. J. 10, 1129—1135.

Gill, I., Valivety, R., 1997. Polyunsaturated fatty acids, part 1: occurrence, biological activities and applications. Trends Biotechnol. 15, 401—409.

Green, B.R., Durnford, D.G., 1996. The chlorophyll-carotenoid proteins of oxygenic photosynthesis. Annu. Rev. Plant Phys. 47, 685—714.

Guil-Guerrero, J.L., Navarro-Juárez, R., López-Martınez, J.C., Campra-Madrid, P., Rebolloso-Fuentes, M., 2004. Functional properties of the biomass of three microalgal species. J. Food Eng. 65, 511—517.

Ho, S.H., Chen, C.Y., Lee, D.J., Chang, J.S., 2011. Perspectives on microalgal CO_2-emission mitigation systems — a review. Biotechnol. Adv. 29, 189—198.

Ho, S.H., Chen, W.M., Chang, J.S., 2010. *Scenedesmus obliquus* CNW-N as a potential candidate for CO(2) mitigation and biodiesel production. Bioresour. Technol. 8725—8730.

Holt, N.E., Fleming, G.R., Niyogi, K.K., 2004. Toward an understanding of the mechanism of nonphotochemical quenching in green plants. Biochemistry 43, 8281—8290.

Hu, Q., Faiman, D., Richmond, A., 1998a. Optimal tilt angles of enclosed reactors for growing photoautotrophic microorganisms outdoors. J. Ferment. Bioeng. 85, 230—236.

Hu, Q., Guterman, H., Richmond, A., 1996. A flat inclined modular photobioreactor for outdoor mass cultivation of photoautotrophs. Biotechnol. Bioeng. 51, 51—60.

Hu, Q., Kurano, N., Kawachi, M., Iwasaki, I., Miyachi, S., 1998b. Ultrahigh-cell-density culture of a marine green alga *Chlorococcum littorale* in a flat-plate photobioreactor. Appl. Microbiol. Biotechnol. 49, 655—662.

Hu, Q., Richmond, A., 1994. Optimizing the population density in *Isochrysis galbana* grown outdoors in a glass column photobioreactor. J. Appl. Phycol. 6, 391—396.

Hu, Q., Richmond, A., 1996. Productivity and photosynthetic efficiency of *Spirulina platensis* as affected by light intensity, algal density and rate of mixing in a flat plate photobioreactor. J. Appl. Phycol. 8, 139—145.

Hu, Q., Sommerfeld, M., Jarvis, E., Ghirardi, M., Posewitz, M., Seibert, M., Darzins, A., 2008. Microalgal triacylglycerols as feedstocks for biofuel production: perspectives and advances. Plant J. 54, 621—639.

Hu, Q., Yair, Z., Richmond, A., 1998c. Combined effects of light intensity, light-path and culture density on output rate of *Spirulina platensis* (Cyanobacteria). Eur. J. Phycol. 33, 165—171.

Jacob-Lopes, E., Revah, S., Hernández, S., Shirai, K., Franco, T.T., 2009. Development of operational strategies to remove carbon dioxide in photobioreactors. Chem. Eng. J. 153, 120—126.

Jensen, G.S., Ginsberg, D.I., Drapeau, M.S., 2001. Blue-green algae as an immuno-enhancer and biomodulator. J. Am. Nutraceut. Assoc. 3, 24—30.

Jin, H.F., Lim, B.R., Lee, K., 2006. Influence of nitrate feeding on carbon dioxide fixation by microalgae. J. Environ. Sci. Health A Tox. Hazard Subst. Environ. Eng. 41, 2813—2814.

Kim, J.H., Nemson, J.A., Melis, A., 1993. Photosystem II reaction center damage and repair in *Dunaliella salina* (green alga) (analysis under physiological and irradiance-stress conditions). Plant Physiol. 103, 181—189.

Kumar, A., Ergas, S., Yuan, X., Sahu, A., Zhang, Q., Dewulf, J., Malcata, F.X., van Langenhove, H., 2010. Enhanced CO_2 fixation and biofuel production via microalgae: recent developments and future directions. Trends Biotechnol. 28, 371—380.

Kurano, N., Ikemoto, H., Miyashita, H., Hasegawa, T., Hata, H., Miyachi, S., 1995. Fixation and utilization of carbon dioxide by microalgal photosynthesis. Energy Convers. Manage. 36, 689—692.

Laroche, J., Mortain-Bertrand, A., Falkowski, P.G., 1991. Light intensity-induced changes in cab mRNA and light-harvesting complex II apoprotein levels in the unicellular chlorophyte *Dunaliella tertiolecta*. Plant Physiol. 97, 147—153.

Lee, R.F., Loeblich, A.R., 1971. Distribution of 21:6 hydrocarbon and its relationship to 22:6 fatty acid in algae. Phytochemistry 10, 593—602.

Lee, Y.K., Low, C.S., 1991. Effect of photobioreactor inclination on the biomass productivity of an outdoor algal culture. Biotechnol. Bioeng. 38, 995—1000.

Lee, Y.K., Low, C.S., 1992. Productivity of outdoor algal cultures in enclosed tubular photobioreactor. Biotechnol. Bioeng. 40, 1003—1006.

Lee, Y.K., Pirt, S.J., 1981. Energetics of photosynthetic algal growth: influence of intermittent illumination in short (40 s) cycles. J. Gen. Microbiol. 124, 43—52.

Li, Y., Horsman, M., Wang, B., Wu, N., Lan, C., 2008a. Effects of nitrogen sources on cell growth and lipid accumulation of green alga *Neochloris oleoabundans*. Appl. Microbiol. Biotechnol. 81, 629—636.

Li, Y., Horsman, M., Wu, N., Lan, C.Q., Dubois-Calero, N., 2008b. Biofuels from microalgae. Biotechnol. Prog. 24, 815–820.

Liu, X.-Y., Sheng, J., Curtiss, R., 2011. Fatty acid production in genetically modified cyanobacteria. Proc. Natl Acad. Sci. USA 108, 6899–6904.

Lorenz, R.T., Cysewski, G.R., 2000. Commercial potential for *Haematococcus* microalgae as a natural source of astaxanthin. Trends Biotechnol. 18, 160–167.

Lynch, D.V., Thompson, G.A., 1982. Low temperature-induced alterations in the chloroplast and microsomal membranes of *Dunaliella salina*. Plant Physiol. 69, 1369–1375.

Mata, T.M., Martins, A.A., Caetano, N.S., 2010. Microalgae for biodiesel production and other applications: a review. Renewable Sustainable Energy Rev. 14, 217–232.

Melis, A., 2009. Solar energy conversion efficiencies in photosynthesis: minimizing the chlorophyll antennae to maximize efficiency. Plant Sci. 177, 272–280.

Metting Jr., F.B., 1996. Biodiversity and application of microalgae. J. Ind. Microbiol. 17, 477–489.

Michel, H., Tellenbach, M., Boschetti, A., 1983. A chlorophyll *b*-less mutant of *Chlamydomonas reinhardtii* lacking in the light-harvesting chlorophyll complex but not in its apoproteins. BBA-Bioenerg. 725, 417–424.

de Morais, M.G., Costa, J.A.V., 2007a. Biofixation of carbon dioxide by *Spirulina* sp. and *Scenedesmus obliquus* cultivated in a three-stage serial tubular photobioreactor. J. Biotechnol. 129, 439–445.

de Morais, M.G., Costa, J.A.V., 2007b. Carbon dioxide fixation by *Chlorella kessleri*, *C. vulgaris*, *Scenedesmus obliquus* and *Spirulina* sp. cultivated in flasks and vertical tubular photobioreactors. Biotechnol. Lett. 29, 1349–1352.

de Morais, M.G., Costa, J.A.V., 2007c. Isolation and selection of microalgae from coal fired thermoelectric power plant for biofixation of carbon dioxide. Energy Convers. Manage. 48, 2169–2173.

Muller-Feuga, A., 2000. The role of microalgae in aquaculture: situation and trends. J. Appl. Phycol. 12, 527–534.

Murakami, M., Ikenouchi, M., 1997. The biological CO_2 fixation and utilization project by rite (2)—screening and breeding of microalgae with high capability in fixing CO_2. Energy Convers. Manage. 38 (Suppl.), S493–S497.

Murata, N., Troughton, J.H., Fork, D.C., 1975. Relationships between the transition of the physical phase of membrane lipids and photosynthetic parameters in *Anacystis nidulans* and lettuce and spinach chloroplasts. Plant Physiol. 56, 508–517.

Mussgnug, J.H., Thomas-Hall, S., Rupprecht, J., Foo, A., Klassen, V., McDowall, A., Schenk, P.M., Kruse, O., Hankamer, B., 2007. Engineering photosynthetic light capture: impacts on improved solar energy to biomass conversion. Plant Biotechnol. J. 5, 802–814.

Nakajima, Y., Tsuzuki, M., Ueda, R., 1998. Reduced photoinhibition of a phycocyanin-deficient mutant of *Synechocystis* PCC 6714. J. Appl. Phycol. 10, 447–452.

Naus, J., Melis, A., 1991. Changes of photosystem stoichiometry during cell growth in *Dunaliella salina* cultures. Plant Cell Physiol. 32, 569–575.

Negoro, M., Shioji, N., Miyamoto, K., Micira, Y., 1991. Growth of microalgae in high CO_2 gas and effects of SO_x and NO_x. Appl. Biochem. Biotechnol. 28–29, 877–886.

Ono, E., Cuello, J.L., 2007. Carbon dioxide mitigation using thermophilic cyanobacteria. Biosyst. Eng. 96, 129–134.

Parker, P.L., Van Baalen, C., Maurer, L., 1967. Fatty acids in eleven species of blue-green algae: geochemical significance. Science 155, 707–708.

Pirt, S.J., Lee, Y.K., Walach, M.R., Pirt, M.W., Balyuzi, H.H.M., Bazin, M.J., 1983. A tubular bioreactor for photosynthetic production of biomass from carbon dioxide: design and performance. J. Chem. Tech. Biotech. B 33, 35–58.

Polle, J.E.W., Kanakagiri, S.D., Melis, A., 2003. tla1, a DNA insertional transformant of the green alga *Chlamydomonas reinhardtii* with a truncated light-harvesting chlorophyll antenna size. Planta 217, 49–59.

Post, A.F., Dubinsky, Z., Wyman, K., Falkowski, P.G., 1984. Kinetics of light-intensity adaptation in a marine planktonic diatom. Marine Biol. 83, 231–238.

Prézelin, B.B., Matlick, H.A., 1980. Time-course of photoadaptation in the photosynthesis-irradiance relationship of a dinoflagellate exhibiting photosynthetic periodicity. Marine Biol. 58, 85–96.

Pushparaj, B., Pelosi, E., Tredici, M., Pinzani, E., Materassi, R., 1997. An integrated culture system for outdoor production of microalgae and cyanobacteria. J. Appl. Phycol. 9, 113–119.

Radakovits, R., Eduafo, P.M., Posewitz, M.C., 2011. Genetic engineering of fatty acid chain length in *Phaeodactylum tricornutum*. Metab. Eng. 13, 89–95.

Radmer, R., Kok, B., 1977. Photosynthesis: limited yields, unlimited dreams. Bioscience 27, 599–605.

Raines, C., 2003. The Calvin cycle revisited. Photosynth. Res. 75, 1–10.

Raines, C.A., Harrison, E.P., Ölçer, H., Lloyd, J.C., 2000. Investigating the role of the thiol-regulated enzyme sedoheptulose-1,7-bisphosphatase in the control of photosynthesis. Physiol. Plant. 110, 303–308.

Raison, J.K., 1986. Alterations in the physical properties and thermal responses of membrane lipids: correlations with acclimation to chilling and high temperature. In: St Joh, J.B., Berlin, E., Jackson, P.G. (Eds.), Frontiers of Membrane Research in Agriculture. Rowman and Allanheld, Totowa, NJ, pp. 383–401.

Renaud, S.M., Thinh, L.-V., Lambrinidis, G., Parry, D.L., 2002. Effect of temperature on growth, chemical composition and fatty acid composition of tropical Australian microalgae grown in batch cultures. Aquaculture 211, 195–214.

Richmond, A., 1987. The challenge confronting industrial microagriculture: high photosynthetic efficiency in large-scale reactors. Hydrobiologia 151–152, 117–121.

Richmond, A., 2004. Handbook of Microalgal Culture: Biotechnology and Applied Phycology. Blackwell Science Ltd.

Rodolfi, L., Zittelli, G.Z., Bassi, N., Padovani, G., Biondi, N., Bonini, G., Tredici, M.R., 2009. Microalgae for oil: strain selection, induction of lipid synthesis and outdoor mass cultivation in a low-cost photobioreactor. Biotechnol. Bioeng. 102, 100–112.

Sakai, N., Sakamoto, Y., Kishimoto, N., Chihara, M., Karube, I., 1995. *Chlorella* strains from hot springs tolerant to high temperature and high CO_2. Energy Convers. Manage. 36, 693–696.

Sato, N., Murata, N., 1980. Temperature shift-induced responses in lipids in the blue-green alga, *Anabaena variabilis*: the central role of diacylmonogalactosylglycerol in thermo-adaptation. Biochim. Biophys. Acta 619, 353–366.

Scragg, A.H., Illman, A.M., Carden, A., Shales, S.W., 2002. Growth of microalgae with increased calorific values in a tubular bioreactor. Biomass Bioenergy 23, 67–73.

Spolaore, P., Joannis-Cassan, C., Duran, E., Isambert, A., 2006. Commercial applications of microalgae. J. Biosci. Bioeng. 101, 87–96.

Stewart, C., Hessami, M.-A., 2005. A study of methods of carbon dioxide capture and sequestration – the sustainability of a photosynthetic bioreactor approach. Energy Convers. Manage. 46, 403–420.

Sung, K.D., Lee, J.S., Shin, C.S., Park, S.C., 1999. Isolation of a new highly CO_2 tolerant fresh water microalga *Chlorella* sp. KR-1. Renew. Energy 16, 1019–1022.

Sydney, E.B., Sturm, W., de Carvalho, J.C., Thomaz-Soccol, V., Larroche, C., Pandey, A., Soccol, C.R., 2010. Potential carbon dioxide fixation by industrially important microalgae. Bioresour. Technol. 101, 5892–5896.

Tanaka, R., Koshino, Y., Sawa, S., Ishiguro, S., Okada, K., Tanaka, A., 2001. Overexpression of chlorophyllide a oxygenase (CAO) enlarges the antenna size of photosystem II in *Arabidopsis thaliana*. Plant J. 26, 365–373.

Tennessen, D., Bula, R., Sharkey, T., 1995. Efficiency of photosynthesis in continuous and pulsed light emitting diode irradiation. Photosynth. Res. 44, 261–269.

Teramoto, H., Nakamori, A., Minagawa, J., Ono, T., 2002. Light-intensity-dependent expression of Lhc gene family encoding light-harvesting chlorophyll-a/b proteins of photosystem II in *Chlamydomonas reinhardtii*. Plant Physiol. 130, 325–333.

Tredici, M.R., Carlozzi, P., Chini Zittelli, G., Materassi, R., 1991. A vertical alveolar panel (VAP) for outdoor mass cultivation of microalgae and cyanobacteria. Bioresour. Technol. 38, 153–159.

Trentacoste, E.M., Shresthra, R.P., Smith, S.R., Gle, C., Hartmann, A.C., Hildebrand, M., Gerwick, W.H., 2013. Metabolic engineering of lipid catabolism increases microalgal lipid accumulation without compromising growth. Proc. Natl Acad. Sci. USA 110, 19748–19753.

Vejrazka, C., Janssen, M., Streefland, M., Wijffels, R.H., 2012. Photosynthetic efficiency of *Chlamydomonas reinhardtii* in attenuated, flashing light. Biotechnol. Bioeng. 109, 2567–2574.

Wang, B., Li, Y., Wu, N., Lan, C., 2008. CO_2 bio-mitigation using microalgae. Appl. Microbiol. Biotechnol. 79, 707–718.

Yoo, C., Jun, S.-Y., Lee, J.-Y., Ahn, C.-Y., Oh, H.-M., 2010. Selection of microalgae for lipid production under high levels carbon dioxide. Bioresour. Technol. 101, S71–S74.

Zeiler, K.G., Heacox, D.A., Toon, S.T., Kadam, K.L., Brown, L.M., 1995. The use of microalgae for assimilation and utilization of carbon dioxide from fossil fuel-fired power-plant flue-gas. Energy Convers. Manage. 36, 707–712.

Stability of Valuable Components during Wet and Dry Storage

Lieselot Balduyck[1,2], Koen Goiris[3], Charlotte Bruneel[1,2], Koenraad Muylaert[4], Imogen Foubert[1,2]

[1]KU Leuven Kulak, Research Unit Food and Lipids, Kortrijk, Belgium; [2]KU Leuven, Leuven Food Science and Nutrition Research Centre (LFoRCe), Leuven, Belgium; [3]KU Leuven Technology Campus Ghent, Laboratory of Enzyme, Fermentation and Brewing Technology, Ghent, Belgium; [4]KU Leuven Kulak, Research Unit Aquatic Biology, Kortrijk, Belgium

1. INTRODUCTION

Microalgae are an evolutionary old group comprising very different lineages, which makes them extremely diverse from a biochemical point of view. Microalgae thus contain many different unique and valuable components that are not often found in terrestrial plants. Each of these components has its own specific importance and applications. Depending on the characteristics of these components, stability during storage can be a problem (Gordon, 2001; Ryckebosch et al., 2011). Storage of microalgae differs fundamentally from storage of lipid-rich seeds. After harvesting, microalgae appear as a paste with a dry weight of only 5–25%, which can pose several problems during storage. These stability problems are also reported in fruit oils, for example, olive oil and palm oil, where the harvested fruit also contains high amounts of water (Dijkstra and Segers, 2007).

Microalgal cells consist of 10–30% lipids, depending on the species and the nutritional and environmental factors during culturing (Spolaore et al., 2006; Ryckebosch et al., 2011). Because of these substantial amounts of lipids, microalgae can possibly have a role in the supply of biodiesel. In addition, long-chain omega-3 polyunsaturated fatty acids (ω-3 LC-PUFA) are found in microalgae, representing the only plant source of these nutritionally interesting lipids. Several studies have demonstrated the health-promoting functions of ω-3 LC-PUFA, particularly eicosapentaenoic acid (C20:5n-3) and docosahexaenoic acid (DHA; C22:6n-3) (Kagan et al., 2013). These fatty acids protect against cardiovascular diseases (Breslow, 2006), can have beneficial effects against inflammations (Calder, 2011, 2013), and can have positive effects against dementia and depression (Nemets et al., 2001; Freund-Levi et al., 2008; Sublette et al., 2011). Commercially used ω-3 LC-PUFA (in supplements or functional foods) normally originates from fish, but this will be a problem in the future because of the global decline in wild-harvest fish stocks. Microalgae and other alternative sources of ω-3 LC-PUFA are thus necessary (Ryckebosch et al., 2012). For both biodiesel and food applications, the stability of the lipids is of major importance, although thus far there has been little research on this issue.

A second important group of valuable components in microalgae are the antioxidants, which protect microalgae against photo-oxidative damage during photosynthesis (Stahl and Sies, 2003; Goiris et al., 2012). Especially autotrophic microalgae, which are exposed to light, contain important amounts of antioxidants (Duval et al., 2000; Kovácik et al., 2010). The extensive amount of antioxidants in autotrophic microalgae can improve the stability of the lipids in the microalgae and in the extracted oil (Ryckebosch et al., 2013). However, some of these antioxidants are quite thermolabile, and care must thus be taken to minimize losses of antioxidants during processing.

A few other components of microalgae can also be considered valuable, for example, some essential amino acids and proteins. Essential amino acids from microalgae can act as precursors to hormones and can have functions in metabolic pathways such as the citric acid and urea cycle (Welladsen et al., 2014). Phycobiliproteins, a group of proteins involved in photosynthesis, are important components mainly produced by *Arthrospira* and *Porphyridium*. These proteins have several applications, for example, in cosmetics, food pigments, and fluorescent labels in analytical techniques. Phycobiliproteins also have medical applications, as they have some hepatoprotective, anti-inflammatory, and antioxidant properties (Plaza et al., 2009; Spolaore et al., 2006).

The next section will focus on the stability of lipids and antioxidants in microalgal biomass and oil during processing and storage.

2. STABILITY OF LIPIDS DURING STORAGE OF MICROALGAL BIOMASS AND OIL

A first stability problem than can arise with lipids during storage of microalgae is lipolysis, a hydrolytic process mostly caused by endogenous lipases, during which free fatty acids (FFA) are released from lipids. FFA are not desirable in food applications because of their rancid flavor. In biodiesel applications, the presence of FFA is also unfavorable because saponification reactions take place between FFA and alkaline catalysts during transesterification. Consequently, the corresponding soaps are formed, leading to a reduced amount of catalyst present and several downstream separation problems (Greenwell et al., 2009; Halim et al., 2012).

A second stability problem is oxidation of lipids, during which hydroperoxides are formed (primary oxidation products), which are later decomposed to a variety of secondary oxidation products such as aldehydes, ketones, and alcohols. Oxidation of fatty acids is promoted by high temperatures and the presence of light and can be retarded by the presence of antioxidants. Unsaturated fatty acids are preferably oxidized, especially those containing double bonds, separated from each other by one methylene group, as is the case for ω-3 LC-PUFA (Frankel, 1980; Gordon, 2001; Gunstone et al., 2007; Knothe, 2007). Oxidation of lipids can be initiated by free radicals (autoxidation) or singlet oxygen (photo-oxidation) and can also be catalyzed by the enzyme lipoxygenase (LOX) (Knothe, 2007). Nuñez et al. (2002) isolated LOX from *Chlorella* by precipitation with $(NH_4)_2SO_4$ and further purification. The purified LOX produced 9- and 13- hydroperoxide derivatives from linoleic acid and had an optimal pH of 7.5. Cutignano et al. (2011) showed the presence of oxylipins in marine diatoms. These oxygenated derivatives of PUFA are most often formed by sequential action of several enzymes, among others lipases and LOX, but some can also be formed chemically (Mosblech et al., 2009).

Lipid oxidation can have negative effects on product quality, for both food and biodiesel applications. Secondary oxidation products, many of which are volatile components, can cause different off-flavors in food, depending on the fatty acid composition of the lipids and the oxidation conditions. Aldehydes are the most prevalent volatile components in microalgae and can give desirable as well as rancid flavors. Shorter chain linear aldehydes are often formed by chemical lipid oxidation, whereas branched and aromatic aldehydes are typically the result of enzymatic lipid oxidation (Van Durme et al., 2013). Besides the development of rancid flavors, oxidation can also cause bleaching of food by reactions with pigments, reduction of nutritional quality by loss of ω-3 LC-PUFA and vitamins, and formation of potential toxic products, for example, 4-hydroxynonenal. For these reasons, oxidation must be minimal in food applications (Kubow, 1992; Gordon, 2001). In biodiesel applications, oxidation can cause reduced fuel quality by formation of deposits in tanks, fuel systems, and filters during decomposition of lipids (Singer and Rühe, 2014). Polymerization reactions between hydroperoxides can also increase viscosity (Dunn, 2005; Knothe, 2007).

In many studies, oxidation of lipids is measured by comparing the fatty acid profile and the amount of ω-3 LC-PUFA in it. However, more accurate and sensitive methods to quantify lipid oxidation measure primary and secondary oxidation products. Primary oxidation products (hydroperoxides) are determined by peroxide value or level of conjugated dienes or trienes, while thiobarbituric acid value and *p*-anisidine value are used as measures for secondary oxidation products. The latter methods, however, lack sensitivity and specificity; therefore, it is better to quantify secondary volatile oxidation products by gas chromatography (Frankel, 1993; Shahidi and Zhong, 2005).

The stability of lipids during the storage of microalgae strongly depends on the moisture content and consequently on the stage in the production process where storage is performed. There are two stages of microalgae storage: (1) storage of microalgal paste with a dry weight of 5−25%, called "wet storage," and (2) storage of dried microalgae with a dry weight of 90−95%, called "dry storage."

2.1 Stability of Lipids during Wet Storage of Microalgal Biomass

2.1.1 Lipolysis

Ryckebosch et al. (2011) showed that after 2 days of wet storage at 4 °C of centrifuged microalgal paste of *Phaeodactylum*, pronounced lipolysis was observed, which was not the case when the paste was dried immediately after harvest (Figure 1(b)). As a consequence, the total lipid content also decreased significantly (Figure 1(a)). It was thus concluded that even short-term storage of fresh wet biomass can have detrimental effects on biomass quality and should be avoided where possible.

Budge and Parrish (1999) showed that boiling water treatment of harvested diatoms could be used to avoid elevated amounts of FFA, as this treatment was supposed to inactivate lipases. The formed FFA seemed to originate more from phospholipids than from triacylglycerols. This is in agreement with the studies of Bergé et al. (1995) and Pernet and Tremblay (2003), which also indicate that phospholipids in microalgae are more sensitive to lipolysis than triacylglycerols. A possible explanation is the presence of lipases that specifically act on phospholipids, also called phospholipases.

Despite the importance of the lipases for hydrolytic stability, they have been sparsely studied. Terasaki and Itabashi (2002) investigated the features of galactolipases in *Chattonella marina*. The enzymes have an optimal temperature

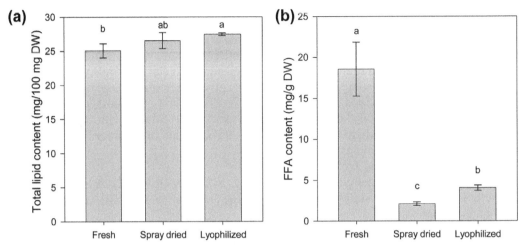

FIGURE 1　Influence of short-term storage, spray drying, and lyophilization of *Phaeodactylum tricornutum* on the total lipid content (a) and FFA content (b). *Reprinted with permission from Ryckebosch, E., Muylaert, K., Eeckhout, M., Ruyssen, T., Foubert, I., 2011. Influence of drying and storage on lipid and carotenoid stability of the microalga Phaeodactylum tricornutum. J. Agric. Food Chem. 59, 11063–11069. Copyright (2011) American Chemical Society.*

of 20–25 °C and preferentially act on monogalactosyldiacylglycerol (MGDG), but also on phosphatidylcholine. This confirmed the results of Cho and Thomson (1986), who studied lipases from the chloroplast of *Dunaliella salina*. The enzymes also preferentially attacked MGDG, which is distinct from other lipases studied in green plants. However, further research on lipases in microalgae is required for an insight into optimal conditions to minimize lipase activity during wet storage.

Cell wall disruption is performed during the production process to improve the extraction of lipids or other valuable components. Pernet and Tremblay (2003), however, showed that cell wall disruption can have negative effects on stability. They showed increased production of FFA during wet storage after two disruption processes (ultrasonication and grinding) than after one of these disruption processes. They explained this by the enhanced contact between enzymes and substrates if the degree of cell disruption increases.

2.1.2 Oxidation

Ryckebosch et al. (2011) evaluated the degree of oxidation in fresh wet microalgal paste compared to dried biomass of *Phaeodactylum tricornutum* by determining conjugated dienes and trienes of PUFA with a spectrophotometric method. They showed that short-term storage (48 h) at 4 °C showed no significant differences in the degree of oxidation when compared to biomass that was immediately freeze-dried or spray-dried. It was concluded that short-term storage of microalgal paste contributes no significant oxidation problems.

Welladsen et al. (2014) evaluated the oxidative stability on a longer term of two microalgal concentrates, *Nannochloropsis* and *Dunaliella*, and two other diatom concentrates, *Melosira dubia* and *Entomoneis punctulata*. Lipid oxidation was examined by measuring the loss of n-3 PUFA content in the fatty acid profile. They found a big difference in the oxidation of n-3 PUFA between the two chlorophytes, *Nannochloropsis* and *Dunaliella*, on the one hand and the two diatoms, *M. dubia* and *E. punctulata*, on the other hand. The two chlorophyte concentrates retained more than 95% of n-3 PUFA during 2 months of storage in a refrigerator, while it was only 10% in the other two diatoms. It was hypothesized that this difference is due to a difference in antioxidant content or in cell wall structure. Natural antioxidants, present in *Nannochloropsis* and *Dunaliella*, could have protected the lipids against oxidation. On the other hand, the silica frustules of the diatoms are possibly more permeable to oxygen than the cell wall of chlorophytes, resulting in increased lipid oxidation in diatoms.

2.2 Stability of Lipids during Dry Storage of Microalgal Biomass

The drying step has a crucial role in the stability of valuable components in microalgae and also in food in general, as it reduces the moisture content and water activity. However, water activity is a better predictor of stability in food products, since it is a measure of the availability of water for reactions occurring in food products. Drying, and consequently reducing water activity, diminishes the growth of microorganisms and retards enzymatic reactions because of the reduced contact between enzyme and substrate (Figure 2). Some enzymes are also inactivated during the drying process, depending on the time and temperature used. Enzymes that are particularly important in lipid stability are lipases and LOX. Autoxidation is also dependent on water activity, since the oxidation rate decreases when drying until a water activity of about 0.2, but increases again when water activity is lower than 0.2 (Figure 2) (Belitz et al., 2009; Kerr, 2013). Disadvantages of the drying process are the promotion of lipid oxidation because of the high temperatures and the inactivation of antioxidants (Knothe et al., 2007; Lim and Murtijaya, 2007; Nindo et al., 2003). The drying technique and process parameters may have

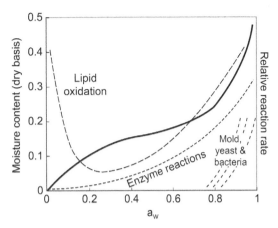

FIGURE 2 Relative rates of deteriorative reactions in foods as a function of water activity (Kerr, 2013). *Reprinted from Kerr, M.,2013. Food drying and evaporation processing operations. In: Kutz, M.(Ed.), Handbook of Microalgae: Biotechnology Advances, Second ed. pp. 317−354. Copyright (2013), with permission from Elsevier.*

an important influence on the properties and stability of dried microalgae during storage, although relatively little research has been done on this topic.

2.2.1 Lipolysis

Esquivel et al. (1993) showed that hot air drying as well as freeze-drying of *P. tricornutum* and *Chaetoceros* sp. caused a loss of 70% of the lipids. This was, however, not observed by Babarro et al. (2001) after freeze-drying of *Isochrysis galbana*, by Ryckebosch et al. (2011) after freeze-drying and spray drying of *P. tricornutum* and by Balasubramanian et al. (2013) after hot air drying, sun drying, and freeze-drying of *Nannochloropsis*. A possible explanation is the storage of the wet biomass before drying detailed by Esquivel et al. (1993), since it has been shown that even short-term storage of wet biomass can cause substantial lipid losses (Ryckebosch et al., 2011).

The drying process can possibly affect the extent of lipolysis occurring during dry storage. Balasubramanian et al. (2013) observed a different amount of FFA in biomass depending on the drying technique. FFA content was three times higher for sun-dried biomass than for freeze-dried and oven-dried biomass. On the other hand, freeze-dried biomass was shown to be more prone to lipolysis during storage than spray-dried biomass (Ryckebosch et al., 2011). The latter authors hypothesized that the higher temperatures during spray-drying caused denaturation of the lipases, while this was not the case during freeze-drying.

2.2.2 Oxidation

Next to lipolysis, oxidation of lipids can be dependent on the drying process. Oliveira et al. (2010) observed an influence of process parameters (temperature and sample thickness) of hot air drying on the degree of lipid oxidation of *Spirulina*—higher temperatures and lower thickness yielded less oxidation. Morist et al. (2001) detected no significant differences in the fatty acid profile of *Spirulina platensis* after spray drying. Ryckebosch et al. (2011) observed more primary and secondary oxidation products during the storage of spray-dried *P. tricornutum* compared to the storage of freeze-dried biomass. This may be linked to the temperature dependence of the oxidation mechanism, as spray drying causes higher heat exposure than freeze-drying. Moreover, antioxidants, which are often thermolabile components, are also more degraded during spray drying, compromising the protection of lipids against oxidation.

2.3 Stability of Lipids in Extracted Microalgal Oil

Ryckebosch et al. (2013) observed a different impact of lipolysis and oxidation during storage of oil extracted from *Isochrysis*, *Nannochloropsis gaditana*, and *Phaeodactylum*. While lipolysis was an important problem in the storage of wet and dry biomass, no FFAs were produced in the extracted oil during an accelerated storage test at 37 °C. It was suggested that lipases

were not active in the oil medium nor were co-extracted with the lipids. Oxidation, on the other hand, posed a problem in the extracted microalgal oils. Because of the substantial amounts of unsaturated fatty acids, the lipids are more prone to oxidation, especially if they are exposed to air and/or light.

Important factors in the production process that influence the oxidative stability of the lipids are the extraction solvent and eventually purification steps after extraction. Ryckebosch et al. (2013) observed that oils extracted with hexane showed a lower oxidative stability than oils extracted with hexane/isopropanol (Figure 3(a)). Two explanations are hypothesized for this observation. Firstly, other antioxidants were extracted with different solvents. Polyphenols, for example, are better extracted with a more polar extraction solvent (Dai and Mumper, 2010). As they are present in microalgae in amounts in the same order of magnitude as carotenoids (Li et al., 2007; Goiris et al., 2012), but have a higher antioxidant capacity (Podsedek, 2007), they contribute strongly to the antioxidant capacity of microalgae (Goiris et al., 2012). A second possible explanation is the higher extractability of glycolipids and phospholipids in hexane/isopropanol than in hexane. It has already been shown that these polar lipids have a higher oxidative stability than triacylglycerols (Song et al., 1997; Lyberg et al., 2005; Yamaguchi et al., 2012) and that addition of phospholipids improves the oxidative stability of triacylglycerols (Nwosu et al., 1997; Khan and Shahidi, 2000).

Ryckebosch et al. (2013) also found that the oxidative stability of microalgal oils was much better than that of several commercial oils rich in ω-3 LC-PUFA, such as fish and tuna oil and oil from heterotrophic microalgae (DHA-S oil) (Figure 3(b)). A possible explanation is the presence of endogenous antioxidants in the autotrophic microalgal oils. Although antioxidants are added during the production process of the commercial oils, they still seem to be quite unstable.

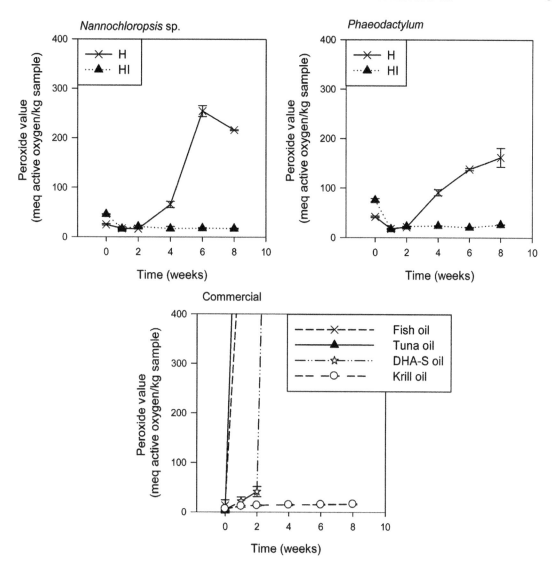

FIGURE 3 Peroxide value (in mequiv active oxygen/kg oil; mean ± SD; $n = 2$) as a function of storage time at 37 °C of *Nannochloropsis* sp. oil and *Phaeodactylum* oil, obtained with hexane/isopropanol (HI) and hexane (H), and the commercial omega-3 LC-PUFA oils (Ryckebosch et al., 2013). *Reprinted with permission from Ryckebosch, E., Bruneel, C., Termote-Verhalle, R., Lemahieu, C., Muylaert, K., Van Durme, J., Goiris, K., Foubert, I., 2013. Stability of omega-3 LC-PUFA-rich photoautotrophic microalgal oils compared to commercially available omega-3 LC-PUFA oils. J. Agric. Food Chem. 61, 10145–10155. Copyright (2013) American Chemical Society.*

3. ANTIOXIDANT PROTECTION AND STABILITY DURING STORAGE

Microalgae reduce (lipid) oxidation in living cells using two strategies: antioxidant enzymes and antioxidant molecules. The most important antioxidant enzymes in microalgae are superoxide dismutase, catalase, and peroxidases such as ascorbate peroxidase and glutathione peroxidase. However, when microalgae are harvested and dried, enzymatic activity is greatly reduced due to inactivation during the drying process and low water activity in the dry biomass, and thus enzymatic protection in dry biomass is limited (Lesser, 2006).

On the other hand, both in living cells and dried biomass, several antioxidant molecules protect the cell content against oxidation. An important class of low-molecular-weight antioxidants, present in microalgae, are antioxidant vitamins. High levels of vitamin C or ascorbic acid have been reported in *Chlorella* sp. (Running et al., 2002), *Dunaliella* sp. (Barbosa et al., 2005), *Chaetoceros calcitrans*, and *Skeletonema costatum* (Brown et al., 1998, 1999). Due to the thermolabile nature of ascorbic acid, care must be taken during the processing of microalgae in order to minimize losses of ascorbic acid. As ascorbic acid is hydrophilic, it is not transferred into the oil when extracted and is therefore of low importance for the stability of processed oils. On the other hand, the fat-soluble vitamin E (tocopherol) is also produced in high amounts by microalgae such as *Dunaliella tertiolecta* and *Tetraselmis suecica* (Carballo-Cardenas et al., 2003). Tocopherols are more thermostable and lipophilic and are therefore important for the stability of microalgal oils. Next to antioxidant vitamins, carotenoids (both primary and secondary) are another important group of lipophilic antioxidants that protect LC-PUFA. Important microalgal carotenoids include lutein, astaxanthin, fucoxanthin, and β-carotene. Carotenoids are efficient quenchers of singlet oxygen and are therefore efficient in preventing the initiation of lipid oxidation. During processing, care must be taken to limit exposure to light as these carotenoids are prone to photochemical bleaching. Also, thermal stress during spray drying has been shown to reduce carotenoid content in *P. tricornutum* biomass (Ryckebosch et al., 2011).

Although phenolics are well-studied antioxidant components in higher plants, the acknowledgment of their presence in microalgae is fairly recent. Li et al. (2007) and Hajimahmoodi et al. (2010) screened microalgae for polyphenol content and antioxidant activity and found large variations between species. Microalgae can further produce some remarkable polyphenolic antioxidant molecules such as marennine (structure not yet fully elucidated) in the diatom *Haslea ostrearia* (Pouvreau et al., 2008), purpurogallin in the extremophile snow algae *Mesotaenium berggrenii* (Remias et al., 2012), or even butylated hydroxytoluene, the well-known food additive (E321) that is usually obtained synthetically (Babu and Wu, 2008). Phenolics are efficient antioxidant molecules and exhibit a wide range of chemical structures and polarities. Many phenolics show good thermal stability.

Next to the common antioxidant components that are found in other plants, some microalgae produce specific types of antioxidants such as the phycobiliproteins, for example, phycocyanin and phycoerythrin in rhodophytes and cyanobacteria (Huang et al., 2007), dimethylsulfide/dimethylsulfoxide (Sunda et al., 2002), and sulfated polysaccharides (Tannin-Spitz et al., 2005). Thermostability of phycobiliproteins can vary significantly, but in most cases they should survive thermal stress during spray drying (Pumas et al., 2011).

4. CONCLUSIONS

Stability of valuable components in microalgae is an often understudied topic, but is of a major importance for several applications.

Nonstable lipids lead to off-flavors and possibly toxic components when microalgae are applied in food, whereas they also lead to problems in the production process of biodiesel. The most important reactions leading to problems with stability of lipids are oxidation and lipolysis.

It has already been shown that even short-term storage of wet biomass has to be avoided because pronounced lipolysis takes place, although this may be eliminated by adding boiling water. The stability of the lipids in dried biomass is higher than that in wet biomass, but dependent on the drying technique and its process parameters. In microalgal oil, lipolysis does not seem to be a problem, and the oxidative stability is much better than in several commercial oils rich in ω-3 LC-PUFA, although this depends on, for example, extraction solvent.

Many different types of antioxidants, both antioxidant enzymes and antioxidant molecules, are present in microalgal cells to protect them against oxidative stress. Although enzymatic activity of the antioxidant enzymes is reduced by drying, antioxidant molecules can protect both living cells and dried biomass against (lipid) oxidation. Examples of these antioxidant molecules are carotenoids, polyphenols, and antioxidant vitamins.

Given the importance of the stability of valuable components in microalgae, a lot more research must be done before high-quality microalgal products can be produced and applied in several industries.

Acknowledgments

This work was financially supported by the agency for Innovation by Science and Technology, Brussels, Belgium (IWT strategic research grant L. Balduyck).

References

Babarro, J.M.F., Fernandez Reiriz, M.J., Labarta, U., 2001. Influence of preservation techniques and freezing storage time on biochemical composition and spectrum of fatty acids of *Isochrysis galbana* clone T-ISO. Aquacult. Res. 32, 565–572.

Babu, B., Wu, J.-T., 2008. Production of natural butylated hydroxytoluene as an antioxidant by freshwater phytoplankton. J. Phycol. 44 (6), 1447–1454.

Balasubramanian, R.K., Doan, T.T.Y., Obbard, J.P., 2013. Factors affecting cellular lipid extraction from marine microalgae. Chem. Eng. J. 215–216, 929–936.

Barbosa, M.J., Zijffers, J.W., Nisworo, A., Vaes, W., van Schoonhoven, J., Wijffels, R.H., 2005. Optimization of biomass, vitamins, and carotenoid yield on light energy in a flat-panel reactor using the A-stat technique. Biotechnol. Bioeng. 89 (2), 233–242.

Belitz, H.-D., Grosch, W., Schieberle, P., 2009. Food Chemistry, fourth ed. Springer-Verlag, Berlin.

Bergé, J.-P., Gouygou, J.-P., Dubacq, J.-P., Durand, P., 1995. Reassessment of lipid composition of the diatom *Skeletonema costatum*. Phytochemistry 39 (5), 1017–1021.

Breslow, J.L., 2006. N-3 fatty acids and cardiovascular disease. Am. J. Clin. Nutr. 83, 1477S–1482S.

Brown, M.R., Skabo, S., Wilkinson, B., 1998. The enrichment and retention of ascorbic acid in rotifers fed microalgal diets. Aquacult. Nutr. 4, 151–156.

Brown, M.R., Mular, M., Miller, I., Farmer, C., Trenerry, C., 1999. The vitamin content of microalgae used in aquaculture. J. Appl. Phycol. 11, 247–255.

Budge, S.M., Parrish, C.C., 1999. Lipid class and fatty acid composition of *Pseudo-nitzschia multiseries* and *Pseudo-nitzschia pungens* and effects of lipolytic enzyme deactivation. Phytochemistry 52, 561–566.

Calder, P.C., 2011. Fatty acids and inflammation: the cutting edge between food and pharma. Eur. J. Pharmacol. 668, S50–S58.

Calder, P.C., 2013. Omega-3 polyunsaturated fatty acids and inflammatory processes: nutrition or pharmacology? Br. J. Clin. Pharmacol. 75, 645–662.

Carballo-Cardenas, E.C., Minh Tuan, P., Janssen, M., Wijffels, R.H., 2003. Vitamin E (a-tocopherol) production by the marine microalgae *Dunaliella tertiolecta* and *Tetraselmis suecica* in batch cultivation. Biomol. Eng. 20 (4–6), 139–147.

Cho, S.H., Thomson, G.A., 1986. Properties of a fatty acyl hydrolase preferentially attacking monogalactosyldiacylglycerols in *Dunaliella salina* chloroplasts. Biochim. Biophys. Acta 878, 353–359.

Cutignano, A., Lamari, N., D'ippolito, G., Manzo, E., Cimino, G., Fontana, A., 2011. Lipoxygenase products in marine diatoms: a concise analytical method to explore the functional potential of oxylipins. J. Phycol. 47, 233–243.

Dai, J., Mumper, R.J., 2010. Plant phenolics: extraction, analysis and their antioxidant and anticancer properties. Molecules 15, 7313–7352.

Dijkstra, A.J., Segers, J.C., 2007. Production and refining of oils and fats. In: Gunstone, F.D., Harwood, J.L., Dijkstra, A.J. (Eds.), The Lipid Handbook, third ed. CRC Press, Boca Raton, pp. 143–251.

Dunn, R.O., 2005. Effects of antioxidants on the oxidative stability of methyl soyate (biodiesel). Fuel Process. Technol. 86, 1071–1085.

Duval, B., Shetty, K., Thomas, W.H., 2000. Phenolic compounds and antioxidant properties in the snow alga Chlamydomonas nivalis after exposure to UV light. J. Appl. Phycol. 11, 559–566.

Esquivel, B.C., Lobina, D.V., Sandoval, F.C., 1993. The biochemical composition of two diatoms after different preservation techniques. Comp. Biochem. Physiol. 105B (2), 369–373.

Li, H.-B., Cheng, K.-W., Wong, C.-C., Fan, K.-W., Chen, F., Jiang, Y., 2007. Evaluation of antioxidant capacity and total phenolic content of different fractions of selected microalgae. Food Chem. 102, 771–776.

Frankel, E.N., 1980. Lipid oxidation. Prog. Lipid Res. 19, 1–22.

Frankel, E.N., 1993. In search of better methods to evaluate natural antioxidants and oxidative stability in food lipids. Trends Food Sci. Technol. 4, 220–225.

Freund-Levi, Y., Basun, H., Cederholm, T., Faxén-Irving, G., Garlind, A., Grut, M., Vedin, I., Palmblad, J., Wahlund, L., Eriksdotter-Jönhagen, M., 2008. Omega-3 supplementation in mild to moderate Alzheimer's disease: effects on neuropsychiatric. Int. J. Geriatr. Psychiatry 23 (2), 161–169.

Goiris, K., Muylaert, K., Fraeye, I., Foubert, I., De Brabanter, J., De Cooman, L., 2012. Antioxidant potential of microalgae in relation to their phenolic and carotenoid content. J. Appl. Phycol. 24 (6), 1477–1486.

Gordon, M.H., 2001. The development of oxidative rancidity in food. In: Pokorny, J., Yanishlieva, N., Gordon, M. (Eds.), Antioxidants in Food. Practical Applications. CRC Press, Boca Raton, pp. 7–20.

Greenwell, H.C., Laurens, L.M., Shields, R.J., Lovitt, R.W., Flynn, K.J., 2009. Placing microalgae on the biofuels priority list: a review of the technological challenges. J. R. Soc. Interface 7 (46), 703–726.

Gunstone, G., Kenar, J.A., Gunstone, F.D., 2007. Chemical properties. In: Gunstone, F.D., Harwood, J.L., Dijkstra, A.J. (Eds.), The Lipid Handbook, third ed. CRC Press, Boca Raton, pp. 535–587.

Hajimahmoodi, M., Faramarzi, M.A., Mohammadi, N., Soltani, N., Oveisi, M.R., Nafissi-Varcheh, N., 2010. Evaluation of antioxidant properties and total phenolic contents of some strains of microalgae. J. Appl. Phycol. 22, 43–50.

Halim, R., Harun, R., Danquah, M.K., Webley, P.A., 2012. Microalgal cell disruption for biofuel development. Appl. Energy 91, 116–121.

Huang, Z., Guo, B.J., Wong, R.N.S., Jiang, Y., 2007. Characterization and antioxidant activity of selenium-containing phycocyanin isolated from Spirulina platensis. Food Chem. 100 (3), 1137–1143.

Kagan, M.L., West, A.L., Zante, C., Calder, P.C., 2013. Acute appearance of fatty acids in human plasma—a comparative study between polar-lipid rich oil from the microalgae Nannochloropsis. Lipids Health Dis. 12 (102), 1–10.

Kerr, W.L., 2013. Food drying and evaporation processing operations. In: Kutz, M. (Ed.), Handbook of Farm, Dairy and Food Machinery Engineering, second ed. Elsevier, London, pp. 317–354.

Khan, M.A., Shahidi, F., 2000. Tocopherols and phospholipids enhance the oxidative stability of borage and evening primrose triacylglycerols. J. Food Lipids 7 (3), 143–150.

Knothe, G., 2007. Some aspects of biodiesel oxidative stability. Fuel Process. Technol. 88, 669–677.

Knothe, G., Kenar, J.A., Gunstone, F.D., 2007. Chemical properties. In: Gunstone, F.D., Harwood, J.L., Dijkstra, A.J. (Eds.), The Lipid Handbook, third ed. CRC Press, Boca Raton, pp. 535–587.

Kovácik, J., Klejdus, B., Backor, M., 2010. Physiological responses of Scenedesmus quadricauda (Chlorophyceae) to UV-A and UV-C light. J. Photochem. Photobiol. 86, 612–616.

Kubow, S., 1992. Routes of formation and toxic consequences of lipid oxidation products in foods. Free Radical Biol. Med. 12, 63–81.

Lesser, M.P., 2006. Oxidative stress in marine environments: biochemistry and physiological ecology. Annu. Rev. Physiol. 68, 253–278.

Lim, Y.Y., Murtijaya, J., 2007. Antioxidant properties of Phyllanthus amarus extracts as affected by different drying methods. LWT − Food Sci. Technol. 40, 1664–1669.

Lyberg, A.-M., Fasoli, E., Adlercreutz, P., 2005. Monitoring the oxidation of docosahexaenoic acid in lipids. Lipids 40 (9), 969–979.

Morist, A., Montesinos, J.L., Cusido, J.A., Godia, F., 2001. Recovery and treatment of Spirulina platensis cells cultured in a continuous photobioreactor to be used as food. Process Biochem. 37, 535–547.

Mosblech, A., Feussner, I., Heilmann, I., 2009. Oxylipins: structurally diverse metabolites from fatty acid oxidation. Plant Physiol. Biochem. 47, 511–517.

Nemets, B., Stahl, Z., Belmaker, R.H., 2001. Addition of omega-3 fatty acid to maintenance medication treatment for recurrent unipolar depressive disorder. Am. J. Psychiatry 159, 477–479.

Nindo, C.I., Sun, T., Wang, S.W., Tang, J., Power, J.R., 2003. Evaluation of drying technologies for retention of physical quality and antioxidants in asparagus (Asparagus officinalis, L.). LWT − Food Sci. Technol. 36, 507–516.

Nuñez, A., Savary, B.J., Foglia, T.A., Piazza, G.J., 2002. Purification of lipoxygenase from *Chlorella*: production of 9- and 13-hydroperoxide derivatives of linoleic acid. Lipids 37 (11), 1027–1032.

Nwosu, C.V., Boyd, L.C., Sheldon, B., 1997. Effect of fatty acid composition of phospholipids on their antioxidant properties and activity index. J. Am. Oil Chem. Soc. 74 (3), 293–297.

Oliveira, E.G., Duarte, J.H., Moraes, K., Crexi, V.T., Pinto, L.A.A., 2010. Optimisation of *Spirulina platensis* convective drying: evaluation of phycocyanin loss and lipid oxidation. Int. J. Food Sci. Technol. 45, 1572–1578.

Pernet, F., Tremblay, R., 2003. Effect of ultrasonication and grinding on the determination of lipid class content of microalgae harvested on filters. Lipids 38 (11), 1191–1195.

Plaza, M., Herrero, M., Cifuentes, A., Ibanez, E., 2009. Innovative natural functional ingredients from microalgae. J. Agric. Food Chem. 57, 7159–7170.

Podsedek, A., 2007. Natural antioxidants and antioxidant capacity of *Brassica* vegetables: a review. LWT – Food Sci. Technol. 40, 1–11.

Pouvreau, J.-B., Morançais, M., Taran, F., Rosa, P., Dufossé, L., Guérard, F., Pondaven, P., 2008. Antioxidant and free radical scavenging properties of marennine, a blue-green polyphenolic pigment from the diatom *Haslea ostrearia* (Gaillon/Bory) Simonsen responsible for the natural greening of cultured oysters. J. Agric. Food Chem. 56 (15), 6278–6286.

Pumas, C., Vacharapiyasophon, P., Peerapornpisal, Y., Leelapornpisid, P., Boonchum, W., Ishii, M., Khanongnuch, C., 2011. Thermostablility of phycobiliproteins and antioxidant activity from four thermotolerant cyanobacteria. Phycol. Res. 59 (3), 166–174.

Remias, D., Schwaiger, S., Aigner, S., Leya, T., Stuppner, H., Lütz, C., 2012. Characterization of an UV- and VIS-absorbing, purpurogallin-derived secondary pigment new to algae and highly abundant in *Mesotaenium berggrenii* (Zygnematophyceae, Chlorophyta), an extremophyte living on glaciers. FEMS Microbiol. Ecol. 79 (3), 638–648.

Running, J.A., Severson, D.K., Schneider, K.J., 2002. Extracellular production of L-ascorbic acid by *Chlorella protothecoides*, Prototheca species, and mutants of *P. moriformis* during aerobic culturing at low pH. J. Ind. Microbiol. Biotechnol. 29 (2), 93–98.

Ryckebosch, E., Muylaert, K., Eeckhout, M., Ruyssen, T., Foubert, I., 2011. Influence of drying and storage on lipid and carotenoid stability of the microalga *Phaeodactylum tricornutum*. J. Agric. Food Chem. 59, 11063–11069.

Ryckebosch, E., Bruneel, C., Muylaert, K., Foubert, I., 2012. Microalgae as an alternative source of omega-3 long chain polyunsaturated fatty acids. Lipid Technol. 24 (6), 128–130.

Ryckebosch, E., Bruneel, C., Termote-Verhalle, R., Lemahieu, C., Muylaert, K., Van Durme, J., Goiris, K., Foubert, I., 2013. Stability of omega-3 LC-PUFA-rich photoautotrophic microalgal oils compared to commercially available omega-3 LC-PUFA oils. J. Agric. Food Chem. 61, 10145–10155.

Sublette, M.E., Ellis, S.P., Geant, A.L., Mann, J.J., 2011. Meta-analysis of the effects of eicosapentaenoic acid (EPA) in clinical trials in depression. J. Clinical Psychiatry 72, 1577–1584.

Shahidi, F., Zhong, Y., 2005. Lipid oxidation: measurement methods. In: Shahidi, F. (Ed.), Bailey's Industrial Oil and Fat Products, sixth ed. John Wiley & Sons, Hoboken, New Jersey, pp. 357–385.

Singer, P., Rühe, J., 2014. On the mechanism of deposit formation during thermal oxidation of mineral diesel and diesel/biodiesel blends under accelerated conditions. Fuel 133, 245–252.

Song, J.-H., Inoue, Y., Miyazawat, T., 1997. Oxidative stability of docosahexaenoic acid-containing oils in the form of phospholipids, triacylglycerols, and ethyl esters. Biosci. Biotechnol. Biochem. 61 (12), 2085–2088.

Stahl, W., Sies, H., 2003. Antioxidant activity of carotenoids. Mol. Aspects Med. 24, 345–351.

Spolaore, P., Joannis-Cassan, C., Duran, E., Isambert, A., 2006. Commercial applications of microalgae. J. Biosci. Bioeng. 101 (2), 87–96.

Sunda, W., Kieber, D.J., Kiene, R.P., Huntsman, S., 2002. An antioxidant function for DMSP and DMS in marine algae. Nature 418, 317–320.

Tannin-Spitz, T., Bergman, M., Van-Moppes, D., Grossman, S., Arad, S.(M.), 2005. Antioxidant activity of the polysaccharide of the red microalga *Porphyridium* sp. J. Appl. Phycol. 17 (3), 215–222.

Terasaki, M., Itabashi, Y., 2002. Free fatty acid level and galactolipase activity in a red tide flagellate *Chattonella marina* (Raphidophyceae). J. Oleo Sci. 51 (3), 213–218.

Van Durme, J., Goiris, K., De Winne, A., De Cooman, L., Muylaert, K., 2013. Evaluation of the volatile composition and sensory properties of five species of microalgae. J. Agric. Food Chem. 61, 10881–10890.

Welladsen, H., Kent, M., Mangott, A., Yan, L., 2014. Shelf-life assessment of microalgae concentrates: effect of cold preservation on microalgal nutrition profiles. Aquaculture 430, 241–247.

Yamaguchi, T., Sugimura, R., Shimajiri, J., Suda, M., Abe, M., Hosokawa, M., Miyashita, K., 2012. Oxidative stability of glyceroglycolipids containing polyunsaturated fatty acids. J. Oleo Sci. 61 (9), 505–513.

Application of Microalgae Protein to Aquafeed

A. Catarina Guedes[1], Isabel Sousa-Pinto[1,2], F. Xavier Malcata[3,4]

[1]Interdisciplinary Centre of Marine and Environmental Research (CIIMAR/CIMAR), University of Porto, Porto, Portugal; [2]University of Porto, Department of Biology, Faculty of Sciences, Porto, Portugal; [3]University of Porto, Department of Chemical Engineering, Porto, Portugal; [4]LEPABE—Laboratory of Process Engineering, Environment, Biotechnology and Energy, Porto, Portugal

1. INTRODUCTION

Fisheries are the most important sources of feedstock for fish meal. Only a small percentage of global fish production is indeed channeled to human consumption, with the remainder being used for fish and animal feed. However, the proportion of fish processed into fish meal is not likely to grow due to an increasing demand for fish products in such emerging economies as China.

In 2011, for the first time ever fish farming exceeded beef production worldwide, and it is currently growing five times faster than beef production. This increased demand, along with limits on wild fish catch, has driven fish meal prices up by almost three-fold in the past decade. Fish meal is a protein-rich food obtained after processing fish caught in the wild, and sets the basis for any balanced formulation used in commercial aquaculture. For the reasons mentioned above, a problem in current fish farming is the fact that fish food can account for up to 30–60% of its operating costs and is anticipated to keep on rising as worldwide fish meal production has plateaued for several years now. There is a definite need for a new source of nutritious fish food. Therefore, the aquaculture industry has recently succeeded in reducing the inclusion rates of fish meal and fish oil in feeds of farmed aquatic animals. However, due to the increase in production of all farmed species, there is still a growing demand for such ingredients (Naylor et al., 2009). Fish meal is the principal source of protein in commercial aquatic feeds. As a result of the steep increase in the price of fish meal and the decline in fishery resources going into fish meal production, finding and testing alternate protein and lipid sources are important for the aquatic feed industry (Kiron et al., 2012).

Increased production in aquaculture is dependent on overall economic management, improved water management, better feeding strategies, more environmentally friendly feeds, genetically fit stocks, improved health management, and integration with agriculture. Although aquaculture dates back to early human history in Asia, Europe, and the Pacific islands (New and

93

Wagner, 2000), it is only in the last few decades that aquaculture has begun to catch up with the science of feed manufacture and nutrition—almost 40% of all aquaculture production is now firmly dependent on commercial feed.

This is especially true of high-value carnivorous species like shrimp, salmon, and trout, whose feed contains large portions of marine inputs in the form of fish meal (Alvarez et al., 2007). The percentage of farms using commercial feeds varies from 100% for salmon and trout to 83% in marine shrimp and 38% in carp farms (Hemaiswarya et al., 2011).

Therefore, the potential use of unconventional feed ingredients, such as microalgae, as feed inputs to replace high-cost feed stuffs has been increasing. Any satisfactory alternative feed ingredients must be able to supply comparable nutritional value at competitive cost. Conventional land-based crops, especially grains and oilseeds, have been feasible alternatives due to their low costs, and have proved successful for some applications when used as substitutes for a portion of the fish meal. But even when these plant-based substitutes can support good growth, they can cause significant changes in the nutritional quality of the final fish produced. As microalgal protein is of good quality, with amino acid profiles comparable to those of other reference food proteins (Becker, 2007), it could be a plausible alternative to fish meal protein. Therefore, microalgae, the source of most photosynthetically fixed carbon in the food web of aquatic animals (Kwak and Zedler, 1997), may be an ideal replacement for fish meal in aquatic feeds. Microalgae are also a more reliable and less volatile source of protein, and their availability is not dependent on fish captures. This provides industry with better control of their costs and supports a potential for future investment due to the reduction of risk in aquaculture farming operations.

Microalgae can be cultivated in areas unsuitable for plants (with less or no seasonality), and, compared to plants, some species have several folds higher production. Since they utilize sunlight energy more efficiently, their potential for production of valuable compounds or biomass is widely recognized, and they can be used to enhance the nutritional value of feed. Microalgae combine properties typical of higher plants with biotechnological attributes proper to microbial cells. In other words, they can reproduce fast in the liquid medium with simple nutritional requirements (they do not require an organic carbon source, like fungi or bacteria do) and accumulate interesting metabolites. However, of particular importance is the possibility to increase the production of biomass or desired compounds by manipulating cultivation conditions. For example, the lack of nitrogen compounds in the medium of microalga *Chlorella* can lead to lipid accumulation of 85% in biomass (Blazencic, 2007), or mixotrophic cultivation of microalga *Spirulina* can increase productivity five-fold (Chen and Zhang, 1997). Considering their intrinsic enormous biodiversity and the ongoing developments in genetic engineering, microalgae represent one of the most promising biological resources for new products and applications (Pulz and Gross, 2004).

Currently, microalgae have been used in aquaculture as food additives, fish meal and oil replacement, coloring of salmonids, inducers of biological activities, and enhancers of nutritional value of zooplankton fed to fish larvae and fry. Therefore, the use of microalgae as an additive in aquaculture has received a lot of attention due to the positive effect on weight gain, increased triglyceride and protein deposition in muscle, improved resistance to disease, improved taste and consistency of flesh, decreased nitrogen output into the environment, increased omega-3 fatty acid content, physiological activity, starvation tolerance, carcass quality, and increase in the rate of growth of aquatic species due to better digestibility (Becker, 2004; Fleurence, 2012). Their importance in aquaculture is thus not surprising, since they are natural food for these organisms.

Therefore, microalgae provide a way of controlling the nature and amount of nutrients given to aquatic species, as emphasized in Table 1 displaying the nutritional characteristics of selected microalgal strains.

By choosing microalgae with the highest protein content, manufacturers can reduce the overall pricing of their feed by mixing microalgae biomass with other feed sources such as soy or corn, while maintaining digestibility so as to assure maximum growth of aquatic crops. The digestibility determines the metabolic processes that induce growth in aquatic species for a given set of physical and chemical conditions. The better the digestibility, the faster the growth; this is the reason for reports of increased production on aquaculture operations after switching from mostly corn/soybean feed to microalgal feed.

Microalgae are utilized in aquaculture as live feeds for all growth stages of bivalve mollusks (e.g., oysters, scallops, clams, and mussels), for the larval/early juvenile stages of abalone, crustaceans, and some fish species, and for zooplankton used in aquaculture food chains. Over the last four decades, several hundred microalgae species have been tested as food, but probably less than 20 have gained

TABLE 1   Chemical Composition of Selected Microalgae, Expressed on a Dry Matter Basis (%)

Class	Microalga	Proteins	Carbohydrates	Lipids	Nucleic acids
Chlorophyceae	Chlamydomonas rheinhardii	48	17	21	−
	Chlorella pyrenoidosa	57	26	2	−
	Chlorella vulgaris	51−58	12−17	14−22	4−5
	Dunaliella bioculata	49	4	8	−
	Dunaliella salina	57	32	6	−
	Scenedesmus dimorphus	8−18	21−52	16−40	−
	Scenedesmus obliquus	50−56	10−17	12−14	3−6
	Scenedesmus quadricauda	47	−	1.9	−
Cyanophyceae	Anabaena cylindrica	43−56	25−30	4−7	−
	Aphanizomenon flos-aquae	62	23	3	−
	Spirulina maxima	60−71	13−16	6−7	3−4.5
	Spirulina platensis	46−63	8−14	4−9	2−5
	Synechococcus sp.	63	15	11	5
Euglenophyceae	Euglena gracilis	39−61	14−18	14−20	−
Labyrinthulomycetes	Schizochytrium sp.	−	−	50−77	−
Porphyridiophyceae	Porphyridium cruentum	28−39	40−57	9−14	−
Prasinophyceae	Tetraselmis maculata	52	15	3	−
Prymnesiophyceae	Prymnesium parvum	28−45	25−33	22−38	1−2
Zygnematophyceae	Spirogyra sp.	6−20	33−64	11−21	−

Adapted from Becker (2007), Chisti (2007), Kovač et al. (2013), FAO (1997), Lakshmanasenthil et al. (2013).

widespread use in aquaculture (Brown, 2002). The most frequently used species are *Chlorella, Tetraselmis, Isochrysis, Pavlova, Phaeodactylum, Chaetoceros, Nannochloropsis, Skeletonema,* and *Thalassiosira* genera. Combination of different algal species provides better balanced nutrition and improves animal growth better than a diet composed of only one algal species (Spolaore et al., 2006). To be suitable for aquaculture, a microalgal strain has to meet various criteria, such as ease of culturing, lack of toxicity, high nutritional value with correct cell size and shape, and digestible cell wall to make nutrients available (Raja et al., 2004; Patil et al., 2007). Protein and vitamin content is a major factor determining the nutritional value of microalgae. In addition, polyunsaturated fatty acid (e.g., eicosapentaenoic, EPA; arachidonic acid, AA; and docosahexaenoic acid, DHA) content is of major importance.

2. NUTRITIONAL FEATURES OF MICROALGAE

Several factors can contribute to the nutritional value of a microalga, including its cell size and shape, digestibility (related to cell wall structure and composition), production of toxic compounds, biochemical composition (e.g., nutrients, enzymes, and toxins if present) and requirements of the animal feeding on microalga. Together with differences among species and method of production, this explains the variability in the amount of protein, lipids, and carbohydrates, i.e., 12–35%, 7.2–23%, and 4.6–23%, respectively. This level of fluctuation can be influenced by culture conditions (Brown et al., 1997); for example, rapid growth and high lipid production can be achieved by somehow stressing the culture.

Microalgae vary appreciably in their biochemical composition, even when grown under standard conditions. Gross composition differs between species, but for many species this is not the major factor determining food value. The protein quality of all microalgae is high. Sugar composition is variable, and in some instances may affect the nutritional value. The essential PUFAs (EPA and DHA) are key nutrients in animal nutrition, and most microalgae are rich in one or both such acids. Chlorophytes, however, lack these acids, and this contributes to their low food value. Microalgae are rich sources of two key vitamins, ascorbic acid and riboflavin, but some species lack specific vitamins (Brown et al., 1997). Because microalgae may be limited in one or more key nutrients, mixed-microalgal diets provide a better balance, and are thus normally used in mariculture.

The biochemical composition of microalgae can be manipulated readily by changing the growth conditions, but the effects vary from one species to another. Knowledge of how species respond to different environments is useful to mariculturists, who may then grow microalgae to optimize the level of specific nutrient(s) needed by the feeding animal.

Microalgae also serve as relevant zooplankton food, since important microalgal nutrients (e.g., fatty acids and vitamins) are transferred to higher trophic levels via zooplankton intermediates (Brown et al., 1997).

Since early reports that demonstrated biochemical differences between microalgae in gross composition (Parsons et al., 1961) and fatty acids (Webb and Chu, 1983), many studies have attempted to correlate the nutritional value of microalgae with their biochemical profile (Richmond, 2004; Durmaz, 2007). However, results from feeding experiments encompassing microalgae differing in a specific nutrient are often difficult to interpret because of the confounding effects of other microalgal nutrients. Nevertheless, analysis of literature data, including experiments where algal diets have been supplemented with compounded diets or emulsions, permits a few conclusions to be reached (Knauer and Southgate, 1999).

The high protein content of various microalgal species is one of the main reasons to consider them as unconventional sources of proteins, as microalgae are capable of synthesizing all amino acids, including essential ones. Moreover, the average quality of most examined algae is similar, or superior, to conventional plant proteins (Becker, 2007). For example, dried *Spirulina* biomass contains all essential amino acids and ca. 68% of biomass is proteins, which is three-fold the content in beef. Another microalga, *Chlorella*, contains ca. 50—60% of proteins, whose quality is comparable to those of yeast, soy flour, and milk powder (Blazencic, 2007). Milovanovic et al. (2012) investigated the protein content in dry biomass of several cyanobacterial strains originating from Serbia, and their results showed that all of them possess very high values, ranging from 42.8% to 76.5%.

The amino acid composition of proteins of microalgae is very similar between species (Brown, 1991) and relatively unaffected by growth phase and light conditions (Brown et al., 1993a,b). Furthermore, the composition of essential amino acids in microalgae is very similar to that of protein from oyster larvae. Hence, it is unlikely the protein quality is a factor contributing to the differences in nutritional value of microalgal species.

Certain algal polysaccharides are of pharmacological importance, as they act on the stimulation of the human immune system (Pulz and Gross, 2004) or possess potential antiviral activity (Hemmingson et al., 2006). Oil content in microalgae can exceed 80% by weight of dry biomass, while levels of 20—50% are quite common (Chisti, 2007).

PUFAs derived from microalgae, i.e., DHA, EPA, and AA, are known to be essential for various larvae (Langdon and Waldock, 1981; Sergeant et al., 1997). The fatty acid content showed systematic differences according to taxonomic group, although there were examples of significant differences between microalgae from the same class. Because of this PUFA deficiency,

chlorophytes generally have low nutritional value and are not suitable as single species diet (Brown et al., 1997).

While the importance of PUFAs is widely recognized, the quantitative requirements of larval or juvenile animals feeding directly on microalgae are not well established (Knauer and Southgate, 1999).

The content of vitamins can vary between microalgae; however, to put the vitamin content of microalgae into context, data should be compared with the nutritional requirements of the consuming animal. Unfortunately, nutritional requirements of larval or juvenile animals that feed directly on microalgae are still poorly understood. The requirements of the adult are far better known (e.g., for marine fish and prawns) (Conklin, 1997) and might serve as a guide for the larval animal. These data suggest that a carefully selected, mixed-algal diet should provide adequate concentrations of vitamins for aquacultured food chains.

Sterols (Knauer et al., 1999), minerals (Fabregas and Herrero, 1986), and pigments also may contribute to nutritional differences of microalgae.

Microalgal species can vary significantly in their nutritional value, and this may also change under different culture conditions (Brown et al., 1997).

Nevertheless, a carefully selected mixture of microalgae can offer an excellent nutritional package for larval animals, either directly or indirectly (through enrichment of zooplankton). Microalgae found to have good nutritional properties—either as monospecies or within a mixed diet—include *Chaetoceros calcitrans*, *Chaetoceros muelleri*, *Pavlova lutheri*, *Isochrysis* sp., *Tetraselmis suecica*, *Skeletonema costatum*, and *Thalassiosira pseudonana* (Enright et al., 1986; Thompson et al., 1993; Brown et al., 1997).

Microalgae grown to late logarithmic growth phase typically contain 30—40% protein, 10—20% lipid, and 5—15% carbohydrate (Fujii et al., 2010). In the stationary phase, the proximate composition of microalgae can change significantly; for example, when nitrate is limiting, carbohydrate levels can double at the expense of protein (Liang et al., 2009). A strong correlation between the proximate composition of microalgae and their nutritional value is not apparent, though algal diets with high levels of carbohydrate have been reported to produce the best growth of juvenile oysters, *Ostrea edulis* (Ponis et al., 2006). Larval scallops, *Patinopecten yessoensis*, provided PUFAs in adequate proportions. In contrast, high dietary protein provided best growth for juvenile mussels, *Mytilus trossulus*, and Pacific oysters, *Crassostrea gigas* (Knuckey et al., 2002).

Microalgae can improve production of larvae, although the exact mechanism of action is unclear. Theories put forward include: (1) light attenuation (shading effects), with a beneficial effect; (2) maintenance of nutritional quality of zooplankton; (3) growth-promoting substances, such as vitamins provided by the algae; (4) probiotic effects. The mechanism is most likely a combination of several of these possibilities. While microalgae provide food for zooplankton, they also help stabilize and improve quality of the culture medium. For numerous freshwater and seawater animal species, the introduction of phytoplankton to rearing ponds leads indeed to much better results in terms of survival, growth, and transformation index (Muller-Feuga, 2000). The reasons for this are not entirely known, but may include water quality improvement and stabilization by algal oxygen production and pH, action of some excreted biochemical compounds along with induction of behavioral processes (like initial prey catching), and regulation of bacterial population, probiotic effects, and stimulation of immunity (Raja and Hemaiswarya, 2010).

Microalgal biomass may be considered as a multicomponent antioxidant system, which is generally more effective due to interactions between various antioxidant components (Gouveia et al., 2008). The most powerful water-soluble antioxidants found in algae are polyphenols, phycobiliproteins, and vitamins (Plaza et al., 2008). Being photosynthetic organisms, microalgae are exposed to light and high oxygen concentrations—and in cultures with high cell density in closed photobioreactors, oxygen concentrations can be very high. Such conditions promote accumulation of highly effective antioxidative scavenger complexes to protect cells, e.g., the antioxidative potential of *Spirulina platensis* can increase 2.3-fold during oxygen stress (Pulz and Gross, 2004). For functional feed, the radical scavenging capacity of microalgal products is of growing interest, especially in the beverage market segment (Pulz and Gross, 2004).

Synthetic colors used in the food industry are mainly coal tar derivatives, although they have been banned in many countries. Since the world trend for colorants is to replace artificial with natural ones, and extraction from plants requires greater amount of biomass, microalgae appear to be a good alternative. Besides chlorophylls, other types of pigments can be found in microalgae, such as carotenoids and phycobiliproteins. Some microalgae are a rich source of carotenoids, used as natural feed colorants, additive for feed, vitamin supplements, and health food products. The most important carotenoid is β-carotene and is most active as provitamin A; it is used as colorant, provitamin, additive to multivitamin preparations, and health food product under the antioxidant claim. The natural form of this pigment has a stronger effect than the synthetic one, from which it is much more easily absorbed by the body. Although the richest known food source of this carotenoid is *Spirulina*, the most important microalga for natural production on a large scale is *Dunaliella salina*, with an accumulation up to 16%/dry weight (del Campo et al., 2007). There is increasing evidence suggesting that astaxanthin surpasses the antioxidant benefits of β-carotene, vitamin C, vitamin E, and many xanthophylls. This carotenoid is used in aquaculture as a pigmentation source, as well as in feed industries. Although the natural form cannot compete commercially with the synthetic one, it is preferred for a few particular applications, and *Haematococcus pluvialis* represents a rich source suitable for cultivation on a large scale (Spolaore et al., 2006). Phycobiliproteins are a group of pigments with commercial value found only in algae, but not in all divisions of algae. Although they are used as natural dyes, there is also evidence of many health-promoting properties. Phycocyanin is one of the most promising commercial substances in *Spirulina*; it has been shown to possess antioxidant, anti-inflammatory, neuroprotective, and hepatoprotective activity, but also appears to be a potential chemotherapeutic, as well as hypocholesterolemic, agent (Gantar and Svircev, 2008).

Microalgal pigments, as well as the whole biomass, can be used to color feed and also to improve textural parameters. It has been shown that phycocyanin significantly improves emulsions' rheological properties, with a linearly increasing trend on phycocyanin concentration (Batista et al., 2006). On the other hand, due to the antioxidant properties of natural pigments, it is possible to improve resistance to oil oxidation that is particularly advantageous in high fat products like emulsions (Gouveia et al., 2008). The incorporation of microalgal biomass in emulsions by Gouveia et al. (2006) resulted in a wide range of appealing colors, from green to orange and pink, and also enhanced resistance to oxidation.

Another relevant issue is that marine environments are filled with bacteria and viruses that can attack fish and shellfish, and thus potentially devastate aquaculture farms. Bacteria and viruses can also attack single-celled microalgae, so these microorganisms have developed biochemical mechanisms for self-defense; such mechanisms involve secretion of compounds that inhibit bacterial growth or viral attachment. For instance, compounds synthesized by *Skeletonema costatum*, and partially purified from its organic extract, exhibited activity against aquacultured bacteria because of their fatty acids longer than 10 carbon atoms

in chain length, which apparently induce lysis of bacterial protoplasts (Guedes and Malcata, 2012).

The ability of fatty acids at large to interfere with bacterial growth and survival has been known for some time, and recent structure—function relationship studies have suggested that said ability lies on both their chain length and degree of unsaturation. Cholesterol and other compounds can antagonize antimicrobial features (Mendiola et al., 2007), so both composition and concentration of free lipids should be taken into account (Benkendorff et al., 2005). The activity of extracts of *Phaeodactylum tricornutum* against *Vibrio* spp. was attributed to EPA, a compound synthesized de novo by diatoms (Smith et al., 2010); this PUFA is found chiefly as a polar lipid species in structural cell components (e.g., membranes) and is toxic to grazers (Jüttner, 2001), as well as a precursor of aldehydes, with deleterious effects upon such consumers as copepods (d'Ippolito et al., 2004). Similarly, unsaturated and saturated long-chain fatty acids isolated from *Skeletonema costatum* (Naviner et al., 1999) and organic extracts from *Euglena viridis* (Das et al., 2005) display activity against that bacterial genus.

3. USE OF MICROALGAE IN FORMULATED FEED FOR AQUACULTURE

Many different microalgae already play a vital role in aquaculture. It is widely known that addition of microalgae to larval fish culture tanks confers a number of benefits, such as preventing bumping against the walls of the tanks (Battaglene and Cobcroft, 2007), enhancing predation on zooplankton (Rocha et al., 2008), and enhancing nutritional value of zooplankton (Van Der Meeren et al., 2007), as well as improving larval digestive (Cahu et al., 1998) and immune (Spolaorea et al., 2006) functions—such as lipid metabolism (Güroy et al.,

2011), antiviral and antibacterial action, improved gut function (Michiels et al., 2012) and stress resistance (Nath et al., 2012; Sheikhzadeh et al., 2012)—besides providing a source of protein, amino acids, fatty acids, vitamins, minerals and other biologically active phytochemicals (Becker, 2004; Pulz and Gross, 2004; Gouveia et al., 2008). Furthermore, it has also been shown that larvae of some fish benefit greatly from direct ingestion of microalgae.

It is not surprising that the biochemical compositions of certain marine microalgae are well matched to the nutritional requirements of some marine fish. Larval feeds probably deserve the most attention in efforts to discover how algae can best be used in fish feeds—because microalgae are a natural component of the diet of many larval fish, either consumed directly or acquired from the gut contents of prey species (such as rotifers and copepods).

The role of microalgae in aquaculture hatcheries may thus be summarized as follows:

- All developmental stages of bivalve mollusks directly rely on microalgae as feed source. Bivalve hatcheries therefore cultivate a range of microalgal strains for brood stock conditioning, and larval rearing and feeding of newly settled spat.
- Farmed gastropod mollusks (e.g., abalone) and sea urchins require a diet of benthic diatoms when they first settle out from the plankton, prior to transferring to their juvenile diet of macroalgae.
- The planktonic larval stages of commercially important crustaceans (e.g., penaeid shrimps) are initially fed on microalgae, followed by zooplanktonic live prey.
- The small larvae of most marine finfish species and some freshwater fish species also initially receive live prey, often in the presence of background microalgae. Depending on whether these microalgae are allowed to bloom within the fish larval rearing tanks or are added from external cultures, this is

referred to as the "green water" or "pseudo-green water" rearing technique.
- The zooplanktonic live prey referred to above are microscopic filter feeders that are themselves commonly fed on microalgae, although inert formulated feeds have been developed as a more convenient diet form for hatcheries (Shields and Lupatsch, 2012).

3.1 Microalgal Strains Used in Aquaculture Hatcheries

Often the algae chosen for fish feeding studies appear to have been selected largely for convenience, because they are low cost and commercially available. For example, *Spirulina*, *Chlorella*, and *Dunaliella* spp. can be produced by low-cost open-pond technologies and are marketed as dry powders, with their nutritional profiles being well documented.

In more traditional, extensive forms of aquaculture, adventitious populations of microalgae are bloomed in ponds or large tanks that act as mesocosms, in which the aquaculture species occupy the highest trophic level. By contrast, intensive aquaculture hatcheries cultivate individual strains of microalgae in separate reactors, and administer these regularly to the farmed species.

In recent years, there has been great interest in the potential of algae as a biofuel feedstock, and it has often been proposed that the protein portion remaining after lipid extraction might be a useful input for animal feed (Chen et al., 2010). However, the algae chosen for biofuel production may not be optimal for use as a feed input, and the economic pressure for the lowest-cost methods of fuel production is likely to result in protein residues with contamination that makes them unfit for use as feed (Hussein et al., 2013).

Conversely, the high-value microalgae that are used in shellfish and finfish hatcheries are generally produced in closed culture systems to exclude contaminating organisms, and they cannot be dried before use without adversely affecting their nutritional and physical properties—thus greatly reducing their value as feed. Inevitably, their production costs are higher, but their exceptional nutritional value justifies the extra cost.

Just as it would be senseless to arbitrarily substitute one conventional crop plant for another (e.g., potatoes for soybeans) in specific feed formulation, the particular attributes of each alga must be carefully considered. In addition to the protein/amino acid profile, lipid/PUFA/sterol profile and pigment content, there are additional important issues.

The type and quantity of extracellular polysaccharides, which are very abundant in certain algae, may interfere with nutrient absorption, or conversely be useful binding agents in forming feed pellets. The thick cell walls of such microalgae as *Chlorella* can prevent absorption of nutrients. On the other hand, depending on growth and processing conditions, algae can contain high concentrations of trace elements that may be detrimental.

Further careful study of the properties of algae will be necessary in order to optimally exploit the great potential offered by this diverse group of organisms. But it is already anticipated that microalgae will play an important role in the effort to move formulation of fish feed "down the food chain" to a more sustainable future.

Only a small number of microalgal strains are routinely cultured in aquaculture hatcheries, based on practical considerations of strain availability, ease of culture, cell physical characteristics, nutritional composition, digestibility, and absence of toxins or irritants (Muller-Fuega et al., 2003a,b; Muller-Feuga, 2004; Tredici et al., 2009; Anon, 2010; Guedes and Malcata, 2012).

A nonexhaustive list of the most commonly used strains and their typical areas of application in aquaculture is depicted in Table 2, for example, *Chlorella* or *Scenedesmus* fed to *Tilapia* (Tartiel et al., 2008), *Chlorella* fed to Korean rockfish (Bai et al., 2001), *Spirulina* fed to sea bream (Mustafa and Nakagawa, 1995) and *Nannochloropsis-Isochrysis* combination fed to Atlantic cod (Walker et al., 2009, 2010).

While scientific studies have demonstrated the ability to manipulate the nutritional composition of individual microalgal strains, hatchery operators do in practice focus on maintaining uninterrupted supplies of microalgae by avoiding system crashes or culture contamination. Delivery of a balanced diet to the aquaculture species is generally achieved by supplying a mixture of different microalgal strains, guided by typical nutritional profiles available for these strains (Brown et al., 1997; Guedes and Malcata, 2012).

3.2 Supplement for Nutritional Enhancement

3.2.1 Pigments

From the point of view of consumers, the concepts of sustainable, "chemical free," and organic farming have become quite appealing, including use of natural forms of pigments instead of synthetically produced ones.

The carotenoids are a class of yellow, orange, or red naturally occurring pigments that are distributed everywhere in the living world. Only fungi, algae, and higher plants are able to synthesize carotenoids de novo, so animals rely on the pigment or closely related precursor to be supplied in their diets—that in nature would have passed on through the food chain.

A few microalgae are used as sources of pigments in fish feeds. Farmed salmonid fish therefore require supplementation of dietary astaxanthin to achieve the pink color of the fillet. Synthetic carotenoids are mainly used for this purpose in commercial aquaculture, although microalga-derived carotenoids can also effectively impart pigmentation (Choubert and Heinrich, 1993; Soler-Vila et al., 2009). *Haematococcus* is used to produce astaxanthin that is responsible for the pink color of the flesh of

TABLE 2   Groups, Genera, and Species of Major Microalgal Strains Used in Aquaculture and Their Areas of Application

Group	Genus	Species	Area of application[a]
Cyanobacteria	*Spirulina*	*platensis*	FFI
	Spirulina	sp.	FFI
Chlorophyta	*Tetraselmis*	*suecica, chui*	B, CL
	Chlorella	sp., *vulgaris, minutissima, virginica, grossii*	R, FFI
	Dunaliella	sp., *tertiolecta, salina*	FFI
	Haematococcus	*pluvialis*	FFI
	Scenedesmus	sp.	FFI
Eustigmatophyceae	*Nannochloropsis*	sp., *oculata*	R, GW, FFI
Labyrinthulea	*Schizochytrium*	sp.	RAD
	Ulkenia	sp.	RAD
Bacillariophyta	*Chaetoceros*	*calcitrans, gracilis*	B, CL
	Skeletonema	*costatum*	B, CL
	Thalassiosira	*pseudonana*	B, CL
	Nitzschia	sp.	GU
	Navicula	sp.	GU
	Amphora	sp.	GU
Haptophyta	*Pavlova*	*lutheri*	B
	Isochrysis	*galbana, add. galbana "Tahiti"* (T-iso)	B, GW, FFI
Dinophyta	*Crypthecodinium*	*cohnii*	RAD

[a] *FFI formulated feed ingredient; B bivalve mollusks (larvae/postlarvae/brood stock), C crustacean larvae (shrimps, lobsters); R rotifer live prey; RAD rotifer and Artemia live prey (dry product form); GU gastropod mollusks and sea urchins; GW "green water" for finfish larvae.*
Mustafa and Nakagawa (1995); Bai et al. (2001); Tartiel et al. (2008); Walker et al. (2009, 2010); Shields and Lupatsch (2012).

salmon, and is typically used for organically certified salmon production.

Hematococcus pluvialis has been shown to successfully enhance the reddish skin coloration of red porgy, *Pagrus pagrus* (Chatzifotis et al., 2011), and also of penaeid shrimp, *Litopenaeus vannamei* (Parisenti et al., 2011). Both natural and synthetic sources of carotenoids have been successfully used to augment the yellow skin coloration in gilthead sea bream (Gomes et al., 2002; Gouveia et al., 2002). *Chlorella* sp. and

Spirulina sp. are used as a source of other carotenoids that such fish as ornamental koi can convert to astaxanthin and other brightly colored pigments (Gouveia and Rema, 2005; Sergejevová and Masojídek, 2012; Zatkova et al., 2011). *Dunaliella* produces large amounts of β-carotene.

3.2.2 Vitamins and Minerals

Microalgae also contain vitamins that can be beneficial to the health of the consumer (either human or other animal). However, the content of vitamins varies widely among microalgae. Ascorbic acid shows the greatest variation, that is, 16-fold ($1-16$ mg g^{-1} dry weight) (Brown and Miller, 1992). Concentrations of other vitamins typically show a two- to four-fold difference between species, that is, β-carotene from 0.5 to 1.1 mg g^{-1}, niacin from 0.11 to 0.47 mg g^{-1}, α-tocopherol from 0.07 to 0.29 mg g^{-1}, thiamin from 29 to 109 μg g^{-1}, riboflavin from 25 to 50 μg g^{-1}, pantothenic acid from 14 to 38 μg g^{-1}, folates from 17 to 24 μg g^{-1}, pyridoxine from 3.6 to 17 μg g^{-1}, cobalamin from 1.8 to 7.4 μg g^{-1}, biotin from 1.1 to 1.9 μg g^{-1}, retinol less than 2.2 μg g^{-1}, and vitamin D less than 0.45 μg g^{-1} (Seguineau et al., 1996; Brown et al., 1999). Some species of *Chlorella* genus contain more vitamins than most cultivated plants (Kovač et al., 2013). *Spirulina* genus also contains over 10-fold more β-carotene than any other food, including carrots (Mohammed et al., 2011), and more vitamin B12 compared to any fresh plant or animal food source. Compared to green algae, spinach, and liver, this genus represents the richest source of vitamin E, thiamine, cobalamine, biotin, and inositol (Gantar and Svircev, 2008). Several microalgal species produce α-tocopherol (the most biologically active form of vitamin E) to very high concentrations. Rodriguez-Zavala et al. (2010) found that the production of α-tocopherol in heterotrophically grown microalga *Euglena gracilis* after 120 h reaches 3.7 ± 0.2 mg g^{-1}, which, compared to sunflower, soybean, olive, and corn (some of the most common natural sources of vitamin E), is ca. 13-, 18-, 95-, and 56-fold higher, respectively. If reported high biomass yields of microalga *T. suecica* were reached, it would compete with *E. gracilis* for commercial α-tocopherol production (Carballo-Cardenas et al., 2003).

As mentioned above, the vitamin content of algal biomass can vary significantly among species. This variation is greatest for ascorbic acid (vitamin C), from 1 to 16 mg g^{-1} dry weight (Brown and Miller, 1992), although this may be due to differences in processing, drying, and storage of algae, as ascorbic acid is very sensitive to heat (Shields and Lupatsch, 2012); other vitamins typically show a two- to four-fold variation between species (Brown et al., 1999).

Moreover, every species of microalgae exhibits low concentrations of at least one vitamin, so a careful selection of a mixed microalgal diet would be necessary to provide all vitamins to cultured animals feeding directly on microalgae.

The aforementioned information highlights the drawbacks associated with supplying essential micronutrients via natural sources, i.e., there is too much variability arising from the combined effects of different algal species, growing seasons, culture conditions, and processing methods to reliably supply the required micronutrients in a predetermined fashion. Accordingly, microalgal biomass offers a supplement, rather than complete replacement, for manufactured minerals or vitamins in animal feeds.

On the other hand, microalgae exhibit a potential as mineral additives to replace inorganic mineral salts commonly used by the animal feed industry. It has been suggested that natural forms are more bioavailable to the animal than synthetic forms, and can even be altered or manipulated through bioabsorption (Doucha et al., 2009).

Mineral-rich microalgae have been incorporated in commercial salmon feeds to 15% in lieu of manufactured vitamin and mineral premixes (Kraan and Mair, 2010). Final tests suggested that salmon fed with microalgal feeds were healthier and more active, and their flavor and texture were improved.

3.2.3 Taurine

Often overlooked as a nutrient, taurine is a nonprotein sulfonic acid that is sometimes lumped with amino acids when discussing nutrition issues. It is an essential nutrient for carnivorous animals, including some fish, but is not found in any land plants. Although taurine has been scarcely investigated when compared with other amino acids, it has been reported to appear in significant quantities in certain microalgae, for example, the green flagellate *Tetraselmis* (Al-Amoudia and Flynn, 1989), the red unicellular alga *Porphyridium* (Flynn and Flynn, 1992), the dinoflagellate *Oxyrrhis* (Flynn and Fielder, 1989), and the diatom *Nitzschia* (Jackson et al., 1992).

3.2.4 Lipids

Farmed fish and shellfish offer rich sources of long-chain PUFAs, owing to inclusion of fish meal and fish oil in formulated aquafeeds. PUFAs are crucial to human health and play an important role in prevention and treatment of coronary heart disease, hypertension, diabetes, arthritis, and other inflammatory and autoimmune disorders. Due to global shortages of fish oil and fish meal since they currently cannot be synthesized in the laboratory, researchers have increasingly looked for alternative sources of lipids for use in fish feeds (Miller et al., 2008), for example, microalgae. Furthermore, there are many problems associated with using fatty fish as a source of PUFAS: (1) mercury and polychlorinated biphenyl levels are often unacceptable for most consumers (especially children and pregnant women, as the

developing nervous system is quite susceptible to even low levels of these contaminants); (2) unpleasant odor; (3) unsuitable for vegetarians; and (4) sustainability of fish as basic source (many species have been caught almost to extinction) (Spolaore et al., 2006; Cannon, 2009).

Unlike terrestrial crops, microalgae can directly produce PUFAs such as AA (20:4n-6) (*Porphyridium*), EPA (20:5n-3) (*Nannochloropsis, Phaeodactylum, Nitzschia, Isochrysis, Diacronema*), and docosahexaenoic acid (DHA, 22:6n-3) (*Crypthecodinium, Schizochytrium*). While most of these algae are not suitable for direct human consumption, they may indirectly boost their nutritional value for humans if added to animal feeds. However, few studies have been carried out to date to evaluate microalgal lipids in feeds for farmed fish (Atalah et al., 2007; Ganuza et al., 2008).

Lipids in microalgal cells play roles as both energy storage molecules and in formation of biological membranes, and can account for as high as 70% of dry weight in some marine species (Table 3). Under rapid growth conditions, such lipid levels can drop to 14—30% dry weight, a level more appropriate for aquaculture. These

TABLE 3 Oil Contents of Some Microalga Strains

Class	Microalgae	Oil content (% of dry matter basis)
Bacillariophyceae		
	Phaeodactylum tricornutum	20—30
	Cylindrotheca sp.	16—37
	Nitzschia sp.	45—47
Chlorophyceae		
	Dunaliella bioculata	8
	Dunaliella primolecta	23
	Dunaliella salina	6
	Scenedesmus dimorphus	16—40
	Scenedesmus obliquus	12—14
	Scenedesmus quadricauda	1.9

TABLE 3   Oil Contents of Some Microalga Strains—cont'd

Class	Microalgae	Oil content (% of dry matter basis)
	Chlamydomonas reinhardtii	21
	Nannochloris sp.	20–35
	Nannochloropsis sp.	31–68
	Chlorella pyrenoidosa	2
	Chlorella sp.	28–32
	Chlorella vulgaris	14–22
	Neochloris oleoabundans	35–54
Cyanophyceae		
	Aphanizomenon flosaqua	3
	Spirulina maxima	6–7
	Spirulina platensis	4–9
	Synechococcus sp.	11
Dinophyceae		
	Crypthecodinium cohnii	20
Euglenophyceae		
	Euglena gracilis	14–20
Labyrinthulomycetes		
	Schizochytrium sp.	50–77
Porphyridiophyceae		
	Porphyridium cruentum	9–14
Chlorodendrophyceae		
	Tetraselmis maculata	3
	Tetraselmis sueica	15–23
Coccolithophyceae		
	Isochrysis sp.	25–33
	Prymnesium parvum	22–38
Trebouxiophyceae		
	Botryococcus braunii	25–75
Conjugatophyceae		
	Spirogyra sp.	11–21

Becker (2007), Chisti (2007), Kovač et al. (2013).

lipids are composed of PUFA as DHA, EPA, and AA (Brown, 2002), and to high levels; most species possess EPA within 7—34% (Brown, 2002). There is a substantial literature devoted to analysis of PUFA content of microalgae, particularly those used in aquaculture, because they have long been recognized as the best source of such essential nutrients for production of zooplankton necessary for feeding of larval fish, as well as filter-feeding shellfish.

As emphasized above, PUFAs derived from microalgae (e.g., DHA, EPA, α-linoleic acid (ALA), and AA) are essential for various larvae (Sargent et al., 1997); and the proportions of these important PUFAs in 46 strains of microalgae have been clearly shown (Volkman et al., 1989; Dunstan et al., 1993). The fatty acid content showed systematic differences according to taxonomic group, although there were examples of significant differences between microalgae from the same class. Most microalgal species have moderate to high percentages (7—34%) of EPA. Prymnesiophytes, *Pavlova* sp., *Isochrysis* sp., and cryptomonads are relatively rich in DHA (0.2—11%), while eustigmatophytes, *Nannochloropsis*, and diatoms have the highest percentages of AA (0—4%). Chlorophytes, *Dunaliella* and *Chlorella*, are deficient in both C20 and C22 PUFAs, although some species exhibit small amounts (3.2%) of EPA. Because of such a PUFA deficiency, chlorophytes generally have low nutritional value, and thus are not suitable as single species diet (Brown, 2002). PUFA-rich microalgae, such as *Pavlova* sp. and *Isochrysis* sp., can be fed to zooplankton to enrich them in DHA.

EPA has been found in a wide variety of marine microalgae, including Bacillariophyceae (diatoms), Chlorophyceae, Chrysophyceae, Cryptophyceae, Eustigamatophyceae, and Prasinophyceae. *Nannochloropsis* species are widely used as feed in aquaculture and have accordingly been proposed for commercial production of EPA (Apt and Behrens, 1999). A high proportion of EPA in *Porphyridium propureum* has also been reported (Wen and Chen, 2003; Martínez-Fernández and Paul, 2007). The diatoms *P. tricornutum* and *Nitzschia laevis* have been intensively investigated for their EPA potential (Wen and Chen, 2003). Unlike a large number of EPA-containing microalgae, only a few microalgal species have demonstrated industrial production potential (Raja et al., 2007). This is mainly due to a poor specific growth rate and low cell density of microalgae when grown under conventional photoautotrophic conditions. The requirement for DHA fatty acid in marine fish and shrimp nutrition has been established via feeding diets, both rich and deficient in these lipids (Hemaiswarya et al., 2011).

Many shellfish producers are aware that the sterol profile of feed lipids is of critical importance, but much less attention has been paid to the importance of the sterol profile in fish feeds. Aside from alterations in the normal sterol profile of the final fish, the possible endocrine effects of plant phytosterols in fish feeds (e.g., soy phytohormones) have yet to be thoroughly investigated (Pickova and Mørkøre, 2007).

3.3 Source of Protein

Proteins are present in animal cells and tissues, and are continuously used to replace dying body cells and supply building body tissues (e.g., ligaments, hair, hooves, skin, organs, and muscle are indeed partially formed by protein). Apart from their important role as a basic structural unit, they are also needed for metabolism, hormone, antibody, and DNA production. When proteins are fed in excess, they are converted into energy and fat.

Proteins are complex organic compounds with high molecular weight. As with carbohydrates and fats, they contain carbon, hydrogen, and oxygen, but in addition they all contain nitrogen and generally sulfur. Proteins are composed of amino acids, arranged in a linear chain and folded into a globular form, which are released upon hydrolysis by enzymes, acids, or alkalis. Although over 200 amino acids have been isolated from biological materials, only 20 of these are commonly found as components of proteins. Within all known amino acids, 10 are classified as essential (EAA) because animals cannot synthesize them, and they must therefore be supplied in the diet. Although nonessential amino acids (NEAA) are not classified as dietary essential nutrients, they are necessary because they perform many essential functions at cellular or metabolic levels. In fact, they are termed dietary nonessential nutrients only because body tissues can synthesize them on demand. From a feed formulation viewpoint, it is important to know that the NEAA's cystine and tyrosine can be synthesized within the body from the EAA's methionine and phenylalanine respectively, and, consequently, the dietary requirement for these EAA is dependent on the concentration of the corresponding NEAA in the diet.

The requirements of amino acids in animals are well defined in various sets of recommendations, such as those of National Research Council in the United States. Requirements vary depending on the species and age of animals. For instance, the dietary EAA for fish and shrimp are threonine, valine, leucine, isoleucine, methionine, tryptophan, lysine, histidine, arginine, and phenylalanine.

In both aquaculture and agriculture, producers commonly rely on formulated feeds to ensure optimal growth, health, and quality of the farmed animal(s). Given the economic importance of feeds and feeding, nutritionists need to develop nutritionally balanced diets using commonly available raw ingredients. Once there are reliable data on the nutrient and energy requirements of the target species for a given production performance, specific feeds can be formulated and the feeding regimen established.

Typical compositions of feed and feed/gain ratio are summarized in Table 4 for several aquatic animal species; this table just provides an overview, as different feed formulations are used depending on the production stage of the

TABLE 4 Typical Composition of Formulated Feeds for Several Species of Commercial Fish (on % of Feed Basis) and Feed/Gain Ratio

	% Crude protein	% Crude lipid	% Crude carbo-hydrate	Metabolizable energy (MJ/kg)
Salmon	37.0	32.0	15	21.0
Sea bream	45.0	20.0	20	19.1
Tilapia	35.0	6.0	40	13.5
Shrimp	35.0	6.0	40	13.5

Adapted from Shields et al. (2012).

target species. Since protein is generally one of the most expensive feed ingredients, targeted rations are used and the amounts of protein in the diet are reduced as animals grow. As can be easily concluded, feeds for aquatic animals are more energy and nutrient dense than those for terrestrial animals; therefore, fish need to be fed less to support each unit of growth.

Traditionally, fish meal and fish oil have been a substantial component of feeds in aquaculture, yet this source is finite. With fish meal and fish oil prices increasing, there has been a growing interest in partial or complete replacement of fish meal by alternative protein sources of either animal or plant origin. Raw materials other than fish meal are selected for their nutritive value, balance of amino acids, digestibility of proteins, quality of fatty acids and other lipids, absence of antinutritional factors, availability and cost—and lipid-rich algal biomass is being considered as an alternative ingredient for the future (Lupatsch, 2009).

Fish meal is widely used in feeds, largely thanks to its substantial content of high-quality proteins containing all essential amino acids. A critical shortcoming of crop plant proteins commonly used in the formulation of fish feeds is their deficiency in certain amino acids, such as lysine, methionine, threonine, and tryptophan, whereas analyses of amino acid content of numerous algae have found that they generally contain all the essential amino acids despite significant variability. For example, analyses of microalgae have found high contents of essential amino acids, as exemplified by a comprehensive study of 40 species of microalgae from seven algal classes (Brown et al., 1997).

To help in assessing algae as a potential source of protein and energy in the form of carbohydrates and lipids, data in Table 5 allow comparison of the typical nutritional profiles of commercially available aquafeed ingredients with some selected microalgae.

Most figures published on the concentration of algal proteins are based on estimations of crude protein, and such other constituents of microalgae as nucleic acids, amines, glucosamides, and cell wall materials that contain nitrogen; this can result in overestimation of the true protein content (Becker, 2007).

The nonprotein nitrogen can reach 12% in *Scenedesmus obliquus*, 11.5% in *Spirulina*, and 6% in *Dunaliella*. Even with this overestimation, the nutritional value of microalgae is high, with average quality being equal, sometimes even

above that of conventional plant proteins (Becker, 2007).

Although microalgal varieties vary in their proximate composition, their amino acid and fatty acid profiles (Table 6) suggest that they can be made into valuable ingredients for aquatic animal feeds. Furthermore, the amino acid composition of the protein is relatively unaffected by growth phase and light conditions (Brown et al., 1993a, 1993b). Aspartate and glutamate occur to the highest levels (7.1–12.9%), whereas cysteine, methionine, tryptophan, and histidine occur to the lowest concentrations (0.4–3.2%), with other amino acids ranging 3.2–13.5% (Brown et al., 1997).

Based on the reported amino acid requirements of species studied (Wilson, 2002), the microalgal products are able to provide most of the essential amino acids.

In addition to quantifying the gross composition of feed ingredients, knowledge of their digestibility is needed to assess nutritional value. Digestibility trials are usually carried out in vivo by adding an indigestible marker to the

TABLE 5　Typical Composition of Commercially Available Feed Ingredients and Microalgae Species (Per Dry Matter)

	% Crude protein	% Crude lipid	% Crude carbohydrate[a]	% Ash	Gross energy (MJ/kg)
Fish meal	63.0	11.0	—	15.8	20.1
Poultry meal	58.0	11.3	—	18.9	19.1
Corn-gluten	62.0	5.0	18.5	4.8	21.3
Soybean	44.0	2.2	39.0	6.1	18.2
Wheat meal	12.2	2.9	69.0	1.6	16.8
Spirulina	58.0	11.6	10.8	13.4	20.1
Chlorella	52.0	7.5	24.3	8.2	19.3
Tetraselmis	27.2	14.0	45.4	11.5	18.0
Schizochytrium[b]	12.5	40.2	38.9	8.4	25.6

[a] Carbohydrates calculated as difference % DM − (% protein + % lipid + % ash).
[b] Commercial product, martek biosciences.
Adapted from Shields et al. (2012).

TABLE 6 Amino Acid Profile of Various Microalgae as Compared to Conventional Protein Source (g Per 100 Protein)

Source	Ile	Leu	Val	Lys	Phe	Tyr	Met	Cys	Try	Thr	Ala	Arg	Asp	Glu	Gly	His	Pro	Ser
Egg	6.6	8.8	7.2	5.3	5.8	4.2	3.2	2.3	1.7	5.0	—	6.2	11.0	12.6	4.2	2.4	4.2	6.9
Soybean	5.3	7.7	5.3	6.4	5.0	3.7	1.3	1.9	1.4	4.0	5.0	7.4	1.3	19.0	4.5	2.6	5.3	5.8
Chlorella vulgaris	3.8	8.8	55.5	8.4	5.0	3.4	2.2	1.4	2.1	4.8	7.9	6.4	9.0	11.6	5.8	2.0	4.8	4.1
Dunaliella Bardawil	4.2	11.0	5.8	7.0	5.8	3.7	2.3	1.2	0.7	5.4	7.3	7.3	10.4	12.7	5.5	1.8	3.3	4.6
Scenedesmus obliquus	3.6	7.3	6.0	5.6	4.8	3.2	1.5	0.6	0.3	5.1	9.0	7.1	8.4	10.7	7.1	2.1	3.9	3.8
Spirulina maxima	6.0	8.0	6.5	4.6	4.9	3.9	1.4	0.4	1.4	4.6	6.8	6.5	8.6	12.6	4.8	1.8	3.9	4.2
Spirulina platensis	6.7	9.8	7.1	4.8	5.3	5.3	2.5	0.9	0.3	6.2	9.5	7.3	11.8	10.3	5.7	2.2	4.2	5.1
Aphanizomenon sp.	2.9	5.2	3.2	3.5	2.5	—	0.7	0.2	0.7	3.3	4.7	3.8	4.7	7.8	2.9	0.9	2.9	2.9

feed at a known amount, collecting fecal matter by a suitable method, and analyzing the ratio between nutrient and marker in the fecal matter. Very few of the required digestibility trials have been completed with microalgal biomass to date, partly due to the limited availability of material. In this context, a digestibility trial with carnivorous mink, a model used for salmon and other farmed monogastric species, was recently reported by Skrede et al. (2011). Three microalgae—Nannochloropsis oceanica, P. tricornutum, and Isochrysis galbana—were included at levels up to 24% (dry weight) in the feed. The protein digestibilities determined by linear regression for N. oceanica, P. tricornutum, and I. galbana were found to be 35.5%, 79.9%, and 18.8%, respectively. The algae used had been freeze-dried prior to the trial, and those authors hypothesized that the cell wall of the diatom P. tricornutum may have been more easily broken down by digestive processes than the others (Shields and Lupatsch, 2012).

Elsewhere, digestibility coefficient of solar-dried Spirulina biomass has been tested for Arctic char and Atlantic salmon at 30% dietary inclusion level (Burr et al., 2011). Protein digestibility ranged between 82% and 84.7% for the two fish species, respectively. These relatively high digestibility coefficients compare favorably with terrestrial plant ingredients, thus confirming the high potential of Spirulina as a protein source for farmed fish.

In addition to digestibility measurements, in vivo growth trials need to be carried out in which the novel feed ingredient is supplied in suf ficient amounts. Even with seemingly nutritionally adequate diets, poor weight gain may be found in practice, because of low palatability of the test ingredient, and therefore reduced feed intake. Coutinho et al. (2006) found that supplementing feeds for goldfish fry with freeze-dried biomass of I. galbana, as a substitute for fish meal protein, had a negative effect on growth and survival (Coutinho et al., 2006). Aside from the question of palatability, one of the reasons may have been that the feeds were not isonitrogenous—dietary protein levels decreased with increasing algae inclusion level, and it is known that protein is a limiting factor, especially in the small, fast-growing larval stages.

In contrast, Nandeesha et al. (2001) reported improved growth rates for Indian carp fry with increasing levels of S. platensis in feeds. Palmegiano et al. (2005) reported that sturgeon fed Spirulina-based feeds even outperformed those receiving fish meal—based diets.

Contradictory results were reported by Olvera-Novoa et al. (1998), where *Spirulina*-supplemented feeds depressed growth performance of tilapia fry. A more recent study by Walker and Berlinsky (2011) tested the nutritional value of a *Nannochloropsis* sp. and *Isochrysis* sp. mix for juvenile Atlantic cod. These authors described decreased feed intake, and subsequently reduced growth with increasing algal inclusion. They concluded that reduced palatability of algal meal caused deterioration in cod growth.

As noted above, the costs of fish meal and fish oil are steadily increasing. Hence, if a source of protein- or lipid-rich algal meal came to the market at an affordable price, the animal feed industry would certainly consider using it based on existing evidence of the nutritional value of algal biomass. However, all categories of microalgal products are currently much higher in cost than the commodity feedstuffs used in aquafeed (Shields and Lupatsch, 2012). In conclusion, currently, fish meal and plant protein concentrates constitute the vast majority of feed for farmed fish. However, evaluating alternative protein sources, such as microalgae, in fish diets is critical to the economic success of the aquaculture industry. Demand for fish meal is high—and thus one of the main reasons for existence of the aquaculture industry; and plant protein concentrates are expensive, while fish are unable to absorb much of the protein from soybean meal in their diets. In addition, the use of fish for aquaculture feed shapes a public perception of aquaculture as depleting ocean resources.

The high protein content of microalgae and their amino acid profiles are the main reason for them being considered valuable components of fish diets. Microalgae are increasingly incorporated in aquafeeds to replace plant protein concentrates. Research has demonstrated that in *Tilapia* diets, substituting soybean meal for 6% of fish meal or 10% of meat-and-blood meal, can be accomplished without compromising fish growth. Microalgae are also carriers of high-value molecules that produce numerous antioxidant and antimicrobial compounds, although they still need to be screened for their activity and utility in fish diet formulations. The benefits of introducing algae as protein source, feed additive, and source of antibacterial compounds might further increase aquaculture's usefulness in human food production.

As pointed out by Glencross et al. (2007), ingredient digestibility and palatability are essential pieces of information to evaluate alternate ingredients in aquafeeds. Detailed investigations, covering these and other aspects, will be reviewed next.

The following short-term studies will help to explain the potential of microalgal meals as replacements for fish meal in the feeds of zooplankton, bivalve mollusks, gastropod mollusks, and echinoderms, Atlantic salmon, common carp, and shrimp.

3.3.1 Zooplankton

Although larvae of aquaculture species are predatory rather than filter-feeding (e.g., finfish larvae and decapod crustacean larvae), the most common husbandry strategy is to feed with zooplanktonic live prey rather than formulated inert diets.

Microalgae have an important role in aquaculture as a means of enriching zooplankton for feeding fish and larvae (Chakraborty et al., 2007). As such, the microalgal and zooplankton strains with the correct size and nutritional content have been identified for each major aquaculture species. The zooplankton most commonly used are rotifers *Brachionus plicatilis* and *Artemia salina* (Chakraborty et al., 2007). To a much lesser extent, cladocerans (*Moina macrocorpa*, *Daphnia* sp.) and copepods (*Euterpina acutifrons*, *Tigriopus japonicus*) are used. For zooplankton to grow and reproduce in the hatchery microalgal food, it is necessary to provide it to the newly hatched rotifers or *artemia* (*Nauplii*) until such zooplankton reach the desired size. Just prior to harvesting, zooplankton may be given a boost of microalgae or a formulated emulsion to pack the gut of the organism. This increases its nutritional value to the target culture species feeding on it.

Hatchery production of rotifers was initially based on feeding with live microalgae and/or baker's yeast. Commonly used microalgal strains for this purpose are *Nannochloropsis* sp., *Tetraselmis* sp., *P. lutheri*, and *I. galbana* (Conceição et al., 2010). Commercial off-the-shelf formulations have been developed and are now widely used as alternatives to live microalgae and yeasts. Depending on their specific formulation, these products are intended to optimize growth and reproduction of rotifers and/or to enhance their final nutritional composition before feeding larvae. This latter process is widely referred to as "enrichment." Even where hatcheries have adopted such artificial feeds for mass rotifer cultivation, it is common to retain rotifer master cultures on live microalgae, as this simplifies hygiene assurance and lessens the probability of culture crashing. Hatcheries prefer to use *Artemia* whenever possible because the cysts are purchased as a dry dormant phase. Upon immersion in seawater, the cysts hatch and can be ingested immediately. In addition to providing protein and energy, they provide other key nutrients such as vitamins, essential PUFAs, pigments, and sterols that are transferred through the food chain. For instance, rotifers fed with microalgae become rapidly enriched with ascorbic acid. After 24 h, rotifers fed on *Isochrysis* sp. and *Nannochloropsis oculata* contained 2.5 and 1.7 mg g^{-1} dry wt, respectively, whereas rotifers fed on baker's yeast alone were deficient in ascorbate and contained only 0.6 mg g^{-1} dry wt (Brown, 2002). Finfish hatcheries producing algae to feed rotifers or *Artemia* typically use the following: *Chlorella* sp., *Chlamydomonas* sp., *N. oculata*, *Tetraselmis tetrathele*, and *Tetraselmis chuii*.

The use of microalgae in fish hatcheries is required for both producing live prey and maintaining quality of the larvae-rearing medium (Spolaore et al., 2006). The use of small live plankton feeder preys, such as the rotifer *Brachionus plicatilis*, is still a prerequisite for success in hatcheries of small larvae finfish like sea breams. These preys can be raised on yeast-based artificial feeds, but this is much less efficient than with phytoplankton. Microalgae are interesting on three levels: (1) quick (7−13 days) recovery of rotifer populations, whereas yeasts take 20−35 days; (2) improved nutritional quality of live prey; and (3) lower bacterial contamination, especially from *Vibrio*. For numerous fresh- and sea-water animals, introduction of phytoplankton in rearing ponds leads to much better results in terms of survival, growth, and transformation index than in clear water. In the case of sea bream, this condition has become an economic need, but the reasons behind the positive role of microalgae in the larvae-rearing ponds of fish, as well as shrimp, have not been completely elucidated (Richmond, 2004). There is no doubt that the water quality is improved and stabilized by oxygen production and pH stabilization, but this fails to provide a complete explanation. The action of some excreted biochemical compounds is generally mentioned, as well as the induction of behavioral processes like initial prey catching; yet other positive functions (such as regulating bacterial population, conveying probiotic effects, and stimulating immunity) have also been suggested, although they are not sufficiently understood (Hemaiswarya et al., 2011).

3.3.2 Bivalve Mollusks

Intensive rearing of bivalves has so far relied on production of live microalgae that comprises an average of 30% of the operating costs in a bivalve hatchery. The relative microalgal requirements of the various stages of bivalve culture depend on whether the operation aims at mass production of larvae for remote setting or growing millions of seed until planting size (Lavens and Sorgeloos, 1996). In either case, juveniles consume the largest volumes of microalgal culture to respond to a demand of large biomass with high weight species. Bivalve hatcheries rely on a broad range of microalgal species, such as *Chaetoceros* sp., *Chlorella minutissima*, *Gomphonema* sp., *I. galbana*, *Nitzschia* sp., *Pavlova* sp., *P. tricornutum*, *Skeletonema* sp., *T. pseudonana*, and *Tetraselmis subcordiformis*. The microalgal species that were reported in an international survey among hatchery operators in 1995 are *Isochrysis* sp., *C. gracilis*, *C. calcitrans*, and *T. suecica* (Hemaiswarya et al., 2011).

3.3.3 Gastropod Mollusks and Echinoderms

Unlike bivalve mollusks, the larvae of abalone (gastropoda) and some species of sea urchin (echinoidea) do not require microalgae during their planktonic phase, relying instead on internal yolk reserves for energy. This simplifies hatchery rearing procedures (no microalgae required), yet abalone and urchins do initially graze on benthic microalgae (those living on surfaces) when they settle out from the plankton (Azad et al., 2010).

Natural assemblages of benthic diatoms are typically encouraged to grow as a feed source, by preexposing artificial substrates or macroalgal germlings to unfiltered seawater, upon which the microalgae grow (Heasman and Savva, 2007). This natural colonization process becomes limiting at higher abalone stocking densities, where the rate of algal growth can be outpaced by grazing (Dyck et al., 2011). The addition of cultured diatoms, such as *Navicula* sp., *Nitzschia* sp., and *Amphora* sp. (Viçose et al., 2012), offers greater control for intensive abalone nurseries, although challenges exist in optimizing their methods of cultivation and deployment.

3.3.4 Atlantic Salmon

A study conducted by Kiron et al. (2012) showed that growth performance, feed performance, and body composition of salmon fed with microalgae-based (5% and 10% protein replacement levels) and those fed with control feed were not different ($P > 0.05$).

The experimental feeds prepared by replacing a portion of the fish meal protein with microalgal protein contained cellulose as inert filler. Cellulose at levels up to 150 g kg^{-1} in the feeds of salmonids did not have any influence on digestibility of the main nutrients (Aslaksen et al., 2007; Hansen and Storebakken, 2007). The inclusion of the algae did not reveal any statistically significant difference in the growth data.

Protein efficiency ratio (PER) of fish fed with low levels of algae was close to that of the control group, thus indicating that replacement with algal protein may not have affected the rate of protein utilization. However, both microalgae at the higher inclusion levels seemed to lower the PER values, although the effect was not statistically significant.

No significant differences in growth or feed performance were observed for microalgae-based feeds relative to controls, at either 5% or 10% replacement level. Atlantic salmon, being carnivorous, may not be able to tolerate high amounts of plant materials in its feed (Torstensen et al., 2008). However, in a recent study on another carnivorous fish, Atlantic cod, replacement of 15% and 30% of fish meal protein with a microalgal mix of *Nannochloropsis* sp. and *Isochrysis* sp. caused a significant growth reduction at the higher replacement level (Walker and Berlinsky, 2011).

3.3.5 Common Carp

In the study conducted by Kiron et al. (2012), the performance parameters of the groups of common carp that received microalgae-based feeds (25% and 40% protein replacement) did not differ significantly from those of fish that were offered the control fish meal—based feed. Microalgal protein was used to replace fish meal in the experimental feeds, and cassava was incorporated to balance the nutrient composition. The inclusion of cassava up to 45% in the feeds of carp fingerlings was found to enhance both carbohydrate and protein digestion (Ufodike and Matty, 1983).

Common carp are omnivorous and can digest substantial amounts of carbohydrate from plants, and may utilize the energy from this component more effectively than carnivorous fishes. This ability is reflected in the growth rates attained by carp receiving algae at higher levels, even though not statistically supported.

Other varieties of microalgae have been employed in feeding trials on carp, but direct comparisons are not attempted here because there are wide differences in biochemical profile between microalgae. Atack et al. (1979) reported poor feed conversion for fingerling mirror carp (*C. carpio*) fed cyanobacterial protein (*Spirulina maxima*) as compared to casein- or petroyeast-protein feeds of similar protein and energy values (2.50 vs 1.39 for casein and 1.55 for petroyeast).

The digestibility of the algae was also lower (87.1%) compared to casein (93%) and petroyeast (96.6%). The PER of the microalgal feed was 1.15 against 2.08 for petroyeast and 2.48 for casein.

Most studies pertaining to application of algae in feeds of carps have focused on freshwater algae; and Kiron et al. (2012) authored the first report on the potential of commercially produced marine microalgal protein as replacement of fish meal protein in carp feeds.

3.3.6 Shrimp

In shrimp farming production, microalgae are necessary from the second stage of larval development (zoea) and in combination with zooplankton from the third stage (myses). Naturally occurring microalgal blooms are encouraged in large ponds with low water exchange, where larvae are introduced. Sometimes fertilizers and bacteria are added to induce more favorable conditions. This production system, with poor control of microalgae, provides a good part of shrimp production (López Elías et al., 2003). On the other hand, large-sized hatcheries require highly paid technicians, multimillion dollar investments, and highly controlled medium conditions. The observed trend is toward specialized production, particularly with supply of postlarvae in the hands of big centralized hatcheries. They open a pathway to new techniques, especially genetic selection of strains with stronger immunity.

In a study conducted by Kiron et al. (2012), whiteleg shrimp readily accepted all feeds tested, thus demonstrating the palatability of the new ingredients. Shrimp that were fed algae-based feeds (25% and 40% protein replacement) did not differ from the control fish meal—fed group, in terms of growth and feed performance. However, some differences were noted in their body proximate composition. Inclusion of 37% tapioca in the diets contributed to better growth and feed conversion ratio in Indian white prawn *Penaeus (Fenneropenaeus) indicus* (Ali, 1988).

There is hardly any information on the use of microalgae as a dry feed component for shrimps, though there are ongoing efforts to replace fish meal protein with microalgal proteins. Furthermore, beneficial impact of algal inclusion on shrimp health has been reported recently—feed diets supplemented with marine algal meals rich in DHA and AA produced significant improvement in immune responses (Nonwachai et al., 2010). The evidence from these studies indicates that microalgal meal that is capable of replacing fish meal protein also has the potential to improve health of shrimp. However, the latter aspect needs to be validated through additional studies.

4. ALTERNATIVES TO FRESH MICROALGAE

As mentioned in previous sections, marine microalgae have been the traditional food component in finfish and shellfish aquaculture, for example, for larval and juvenile animals; they are indeed essential in the hatchery and nursery of bivalves, shrimp, and some finfish cultures. Microalgae are also used to produce zooplankton—typically rotifers, which are in turn fed to freshly hatched carnivorous fish. As the aquaculture industry expands, and since microalgal biomass cultivation on site may represent up to 30% of the operating costs, there is a demand for marine microalgae that cannot be met by the conventional methods used in hatcheries, thus enforcing the use of substitutes with mediocre results and several associated problems.

Despite the obvious advantages of live microalgae in aquaculture, the current trend is to avoid using them because of their high cost and difficulty of production, concentration, and storage (Borowitzka, 1997; Becker, 2004). Alternatives that are potentially more cost-effective have been investigated, including nonliving food, viz microalga pastes (centrifu-gation, flocculation, foam fractionation, and filtration), dried microalgae, microencapsulates, cryopreservation, lipid emulsions (Coutteau et al., 1996), bacteria, or yeasts (Robert and Trintignac, 1997; Knauer and Southgate, 1999). However, nonliving diets generally give lower growth and higher mortalities compared to those fed with live microalgae (Ponis et al., 2003). Products other than live microalgae must be exempt from contamination and nontoxic. Hence, several criteria should be addressed for a substitute of live microalgae as a diet in aquaculture.

Bacteria can provide only a part of the metabolic requirements in aquaculture, by supplying a few organic molecules and vitamins. Under conditions close to those found in rearing facilities, the bacterial input represents less than 15% of the microalgal contribution for mollusk larvae and juveniles of many species (Brown et al., 1996; Langdon and Bolton, 1984). Yeasts were also investigated as an alternative food source, but poor results were observed (Coutteau et al., 1993; Robert and Trintignac, 1997). Therefore, these two alternatives do not appear suitable to replace live microalgae.

An alternative diet with an apparently better potential is microalgal pastes or concentrates (Nell and O'Connor, 1991; McCausland et al., 1999; Heasman et al., 2000); these are prepared by centrifugation (ca. 1:500 concentration) or flocculation (ca. 1:100 concentration). Commercially available concentrates offer a convenient source of microalgae for aquaculture hatcheries. This area of microalgal product development was recently reviewed by Tredici et al. (2009), including the technologies involved in concentrating and stabilizing microalgae, and descriptions of a range of commercially available products (with prices).

The practice of concentrating live microalgae originated in local use within individual hatcheries, typically using disk-stack centrifuges or membrane filters (Molina-Grima et al., 2003). This practice is still used in some large hatcheries, although commercial concentrates have become widely adopted.

From an aquaculture hatchery perspective, the key desired attributes for microalgal concentrates are:

- high cell concentration without damage to cells;
- suitable nutritional composition;
- acceptable shelf life (to maintain nutritional quality and avoid spoilage) using standard cold storage methods, thus avoiding use of preservatives that would be harmful to live prey or larvae;
- hygienic and free from pathogens;
- clumping-free and easy to suspend uniformly in water; and
- commonly available and affordable.

Two main categories of product have emerged: (1) concentrates of those microalgal strains that are particularly favored for aquaculture, and (2) industrial biotechnology strains, such as heterotrophically produced *Chlorella* sp., available at higher volume/lower price but with a more limited scope of application, such as in the production of live prey (Shields and Lupatsch, 2012).

The advantage of such concentrates is that they can be used "off the shelf," thus contributing favorably to the cost-efficiency in hatcheries. On the other hand, the lower nutritional value of most dried microalgae compared to live feed and the poor availability of commercial dried products, appear as main disadvantages. Spray-dried microalgae and microalgae paste may be useful to replace up to 50% of live microalgae (Guedes and Malcata, 2012).

Alternative processes have meanwhile been developed that are potentially less damaging to cells, including foam fractionation (Csordas and Wang, 2004), flocculation (Millamena et al., 1990; Poelman et al., 1997), and filtration (Rossingol et al., 1999). However, a common disadvantage encountered is that the harvested cells are difficult to disaggregate back to single cells, a requirement to feed them to filter-feeding species such as bivalves (Knuckey et al., 2006).

The artificial or nonliving diets are rarely applied in the routine feeding process of bivalves, and are mostly considered as a backup food source. The feasible alternative to live microalgae is freeze-dried forms, since they maintain the original cell shape and texture. Air-dried or spray-dried microalgae shrink and shrivel due to high processing temperature, and this decreases product quality. Freeze-dried products are easy to use, maintain, and store, and many research articles have shown that there is no difference in using live algae or freeze-dried algae. For many applications, freeze-dried microalgae even give higher yields (Lubzen et al., 1995; Yamasaki et al., 1989).

Cryopreservation has been thoroughly adopted by culture collections to preserve strains, but may also find application in aquaculture (Tzovenis et al., 2004). Viable cryostorage of biological specimens has been achieved following various protocols of cooling/thawing rates and cryoprotectant addition, which have been developed and tuned more or less empirically (Karlson and Toner, 1996). Temperatures used for cryostorage are well below freezing—down to −196 °C in liquid helium when biological specimens are to be stored without limit (Mazur, 1984). While cryostorage is generally thought to be innocuous to the cell, the events occurring upon freezing or thawing can lead to severe damage and, ultimately, cell death. Moreover, cryoprotectants that enhance cell viability at cryogenic temperatures are usually toxic at physiological temperatures (Fahy, 1986)—an obstacle overcome by reducing exposure time or temperature of incubation prior to cryopreservation (Fahy et al., 1990). Knowledge of cryoprotectant tolerance levels for microalgae is still limited (Taylor and Fletcher, 1999), as well as for early larval stages and zooplankton that are cultivated and rely on availability of microalgae for growth. In general, cryopreservation is a technique possessing a high potential for culture collections, and may offer a solution for reliable supply of microalgae in aquaculture. For instance, marine microalgae used in aquaculture were successfully cryopreserved at 4, −20, and −80 °C using common cryoprotectants (i.e., methanol, dimethylsulfoxide, propylene glycol, and polyvinylpyrrolidone), with promising results at least for *C. minutissima, Chlorella stigmatophora, I. galbana,* and *Dunaliella tertiolecta* (Tzovenis et al., 2004).

Several products based on thraustochytrids (i.e., microorganisms whose taxonomy may be related to certain algal classes), from the genus *Schizochytrium*, have been marketed. These products possess high concentrations of DHA (Barclay and Zeller, 1996), and so have been applied as alternatives to commercial oil enrichment for zooplankton fed to larvae. As direct feeds, most such products have a lower nutritional value than mixtures of microalgae, yet some performed well as components of a mixed diet with live microalgae (Robert and Trintignac, 1997; Langdon and Önal, 1999).

In general, substitutes of live microalgae should exhibit an appropriate physical behavior, and this constitutes a significant challenge; in particular, they should not aggregate or easily break apart. Drying microalgae can cause, due to oxidation, a loss of PUFAs (Dunstan et al., 1992), which are essential components for larval growth (Brown et al., 1996); and the poor performance reported for dried microalgae was associated chiefly with the difficulty of keeping cells in suspension without disintegrating them, so as to avoid said oxidation (Coutteau and Sorgeloos, 1992). Moreover, when cell walls are broken, a high fraction of water-soluble components cannot be ingested by the organism and may consequently interfere with the water quality of aquaculture (Dhont and van Stappen, 2003). Therefore, pathogenic bacterial proliferation may occur and cause costly production losses. There are similar difficulties when using microalgal paste, because the preparation procedures (i.e., centrifugation, flocculation, or filtration) and/or preservation techniques (i.e., additives or freezing) must ensure that cell wall integrity is essentially preserved.

Products other than live microalgae must obviously be free of bacterial contamination and toxicity. The use of bacteria as a food source in hatcheries consequently seems inappropriate, since physical and chemical treatments are often used to limit bacterial contamination that would be responsible for drastic larval mortality (Elston, 1990).

In conclusion, mitigated or unsuccessful results when using nonliving microalgae has driven one to consider live microalgae as the first choice in aquaculture feeding. Only partial replacement thereof has proven successful in studies using preserved nonliving algae (Donaldson, 1991), microencapsulated diets (Langdon and Bolton, 1984), or spray-dried algae (Zhou et al., 1991); and no whole replacement can be recommended, despite intensive research efforts (Langdon and Önal, 1999). Novel solutions to totally replace microalgae in aquaculture diets cannot at present be widely adopted (Guedes and Malcata, 2012).

5. BENEFITS AND SHORTCOMINGS

5.1 Advantages

The development of the aquaculture industry worldwide is constrained by the supply of fish meal produced from fish caught in the wild. However, the use of farmed microalgae reduces the cost of fish meal, which can be as high as 35% of a farm's total operating costs.

In a general way, the advantages of microalgae feeding to aquaculture species includes increased growth, increased health benefits, and improved taste, as the seafood would present a taste that is equivalent to fish and shrimp caught in the wild—thus increasing the value of aquaculture products in food markets. Moreover, microalgae have a great potential for use in sustainable aquaculture, as they are not only a source of protein and lipids, but they are phototrophic and can thus produce these directly from sunlight.

CO_2 affects the metabolic processes that induce growth in aquatic species; it is expelled by the gills of the fish when they take oxygen from water and use it in their metabolic processes. Just like with any other animal, excessive concentrations of CO_2 in the medium (in this case water) induce CO_2 poisoning that stresses fish and hampers growth. CO_2 poisoning is easy to observe when fish swim to the surface of the water and gasp for air. When microalgae are added to the production pond or tank, the CO_2 exhaled by the fish is taken by the algae and emits oxygen back to the water. This reduces the stress of the fish and increases their metabolic growth rates. The use of green water systems provides the additional advantage of providing food for some herbivorous filtering species (such as rotifers-copepods-artemia systems, tilapia, or oysters). Hence, microalgae management allows the safe use of microalgae to remove CO_2 from ponds, thus reducing stress and mortality and increasing growth rate of aquatic species. Producing 100 tons of algal biomass also fixes roughly 183 tons of carbon dioxide, which has obvious implications in the light of the current climate change.

Unlike terrestrial plants that require fertile land or irrigation, microalgae can grow in a wide range of habitats (Tamaru et al., 2004) and are not nutritionally imbalanced with regard to amino acid content (as is soybean). Furthermore, microalgae are rich in omega 3, so fish fed with algal-based food would bring about much better health benefits than fish fed with soy or corn-based fish meal.

Regarding microalgal production, their cultivation can be done independently of external conditions, as they more efficiently convert solar energy compared to higher plants, do not require fertile soil, produce a wide range of substances, can be used for different applications, and some species reproduce very fast—so these organisms constitute a remarkable source of biomass and certain compounds. An advantage of particular importance is the possibility to regulate and define metabolism and production of a target compound by manipulating the cultivation conditions. While a mixture of different species opens up many possibilities, their use in feed can also address the issue of using certain plants that are, in the first place, used as a food. They can also lower the price of animal feeding (Kovač et al., 2013).

The use of microalgae will allow use of more available and reliable feed, and this will free the aquaculture industry to keep growing and meet current and future demands.

5.2 Constraints

Although microalgae are able to enhance the nutritional content of conventional feed preparations, and thus to positively affect the health of animals prior to commercialization, microalgal material must be analyzed for the presence of toxic compounds to prove their harmlessness. Some constituents of microalgal biomass may represent constraints on their incorporation in feeds, like nucleic acids, toxins, and heavy-metal components.

Other issues that might constrain microalgal incorporation as dietary supplements are their digestibility, and the high amount of salt in marine microalgal species. Regarding microalgal digestibility, the cellulosic cell wall of most microalgae strains poses problems in biomass digestion; it is not digested by nonruminant animals (e.g., chickens, turkeys, pigs, pets, horses). From the tests performed thus far, the overall digestibility of alga carbohydrates is good, and there seems to be no limitation in using dried microalgae as a whole—but this has to be evaluated for each alga identified as a potential feed material (Becker, 2004). If a certain algal strain has low digestibility levels, the biomass may anyway be incorporated as feed material, but needs effective treatment to disrupt the cell wall and thus make the algal constituents accessible to digestive enzymes. The processes employed may be physical (e.g., boiling), high temperature drying, and, to a certain extent, even sun drying, as well as chemical methods (e.g., autolysis) (Becker, 2004).

There are still some obstacles, such as the powder-like consistency of the dried biomass and applications to feed manufacture, production costs, and pests and pathogens that may affect large-scale algal cultivation sustainability.

Many feeding trials are still required, as most research has focused on improving the nutritional value of rotifers and not microalgae as a potential replacement of fish meal and fish oil. Specifically, quite a number of animal nutrition studies have claimed a "super food" status for algae, and these have been publicized and reviewed, but the challenge remains to support such claims on scientifically based evidence. Results from experimental studies can be difficult to interpret, as several compounds in algae can have confounding effects (Shields and Lupatsch, 2012). Unfortunately, it has rarely been possible to determine the particular nutritional factors responsible for these beneficial effects, either because no attempt was made to do so, or because of poor design of the study. There is also interest in storing microalgal pastes and thus extend shelf life (2–8 weeks). More and better-designed studies are necessary before it is possible to gain a good understanding of how algae can best be used in fish feeds.

On the other hand, the successful use of the "green water" CO_2 control in aquaculture systems for carnivorous species has been difficult to achieve, given that if algal density increases over a threshold (because is not being consumed), it dies and takes the oxygen out of the water during its decay. In order to prevent algae density from reaching this threshold, aquaculture producers are required to do frequent water changes, and thus take algae away or harvest them via costly methods.

The production costs of microalgae are still too high to compete with traditional protein sources for aquaculture (Becker, 2007). However, there has been a shift away from typical systems, such as outdoor ponds and raceways, to large-scale photobioreactors that have a much higher surface area-to-volume ratio, which could potentially reduce production costs (Brown, 2002). However, this is only achievable through economies of scale.

6. FUTURE PERSPECTIVES

The high production costs of microalgae remain a constraint to many hatcheries. Despite efforts over several decades to develop cost-effective artificial diets to replace microalgae as hatchery feeds, on-site microalgal production remains a critical factor for most marine hatcheries. Improvements in alternative diets may continue, but production costs of microalgae may also decrease due to uptake of new technology by hatcheries. Therefore, it is unlikely that microalgae will be totally replaced, at least on the medium term. A wide selection of microalgal species is already available to support the aquaculture industry. However, specific applications and industrial subsectors demand novel species with improved nutritional quality or growth characteristics that are compatible with attempts to improve hatchery efficiency and yield.

Assuming sufficient quantities of algal biomass have become available at a suitable price, alga producers and animal feed manufacturers will still need to take into account the potentially large variations in proximate composition (proteins, lipids, fatty acids, minerals) and digestibility encountered among different microalgal strains and growing conditions. Efforts are needed to ensure a more consistent composition of algal biomass so that manufacturers can readily incorporate this new feedstuff alongside existing ingredients in formulated feeds. To improve their digestibility, some types of algal biomass may require additional processing steps (over and above those applied to conventional feedstuffs), which will add further to their cost (Shields and Lupatsch, 2012).

Although genetic engineering of microalgae has been studied in its application for biofuel production and bioremediation of heavy metals, there is scarce research on its application in aquaculture. The insertion of genes determining the nutritional parameters into microalgae can increase the quality of fish in aquaculture. A combined effort to standardize a genetically modified microalga coupled to a controlled bioprocess system will lead to an upliftment in the status of aquaculture (Hemaiswarya et al., 2011).

Finally, a better understanding of the mechanism of green water systems, both in intensive and extensive culture, will aid in optimizing usage of microalgae in larval cultures. A broader range of microalgae species, especially mixtures and including species rich in protein, should thus be assessed in green water systems (Brown, 2002).

Finally, it is important to reflect on the importance of the biorefinery concept. For more than 40 years, aquaculturists have devised robust methods for culturing a diverse range of photo-trophic microalgal strains with high nutritional value, but these are more susceptible to crashes and contamination than those extremophiles that are mass cultured for other purposes in open ponds or raceways (e.g., *Spirulina* sp., *Dunaliella* sp., *Haematococcus* sp.) (Shields and Lupatsch, 2012).

Aquaculture-associated technologies (e.g., affordable closed PBRs) for culturing "sensitive" microalgal strains may add value to the current microalgal biotechnology agenda of biofuels, and high-value biomass extracts through integrated biorefineries. It is expected that significant benefits will return to the aquaculture sector via current biotechnology investments, in the form of more efficient microalgal production systems and greater availability of high-quality microalgal biomass (and extracts for use as hatchery feeds).

Acknowledgments

A postdoctoral fellowship (ref. SFRH/BPD/72777/2010), supervised by authors F.X.M. and I.S.P., was granted to author A.C.G., by Fundação para a Ciência e Tecnologia (FCT, Portugal) under the auspices of POPH/FSE.

References

Al-Amoudia, O.A., Flynn, K.J., 1989. Effect of nitrate-N incorporation on the composition of the intracellular amino acid pool of N-deprived Tetraselmis marina. Br. Phycol. J. 24, 53—61.

Ali, A.S., 1988. Water stability of prawn feed pellets prepared using different binding materials with special reference to tapioca. Indian J. Fish. 35, 46—51.

Alvarez, J.S., Llamas, A.H., Galindo, J., Fraga, I., Garca, T., Villarreal, H., 2007. Substitution of fishmeal with soybean meal in practical diets for juvenile white shrimp Litopenaeus schmitti. Aquacul. Res. 38, 689—695.

Anon, 2010. Report on biology and biotechnology of algae with indication of criteria for strain selection. In: Report of the AquaFUELS FP7 Project, Deliverable 1.4 (download 2.7.12). http://www.aquafuels.eu/attachments/079_D%201.4%20Biology%20Biotechnology.pdf.

Apt, K.E., Behrens, P.W., 1999. Commercial developments in microalgal biotechnology. J. Phycol. 35, 215—226.

Aslaksen, M.A., Kraugerud, O.F., Penn, M., Svihus, B., Denstadli, V., Jørgensen, H.Y., Hillestad, M., Krogdahl, A., Storebakken, T., 2007. Screening of nutrient digestibilities and intestinal pathologies in Atlantic salmon, Salmo salar, fed diets with legumes, oilseeds, or cereals. Aquacult. 272, 541—555.

Atack, T.H., Jauncey, K., Matty, A.J., 1979. The utilization of some single cell proteins by fingerling mirror carp (Cyprinus carpio). Aquaculture 18, 337—348.

Atalah, E., Hernández Cruz, C.M., Izquierdo, M.S., Rosenlund, G., Caballero, M.J., Valencia, A., Robaina, L., 2007. Two microalgae crypthecodinium cohnii and Phaeodactylum tricornutum as alternative source of essential fatty acids in starter feeds for sea bream (Sparus aurata). Aquaculture 270, 178—185.

Azad, A.K., McKinley, S., Pearce, C.M., 2010. Factors influencing the growth and survival of larval and juvenile echinoids. Rev. Aquacult. 2, 121—137.

Bai, S.C., Koo, J.-W., Kim, K.W., Kim, S.K., 2001. Effects of Chlorella powder as a feed additive on growth performance in juvenile Korean rockfish, Sebastes schlegeli (Hilgendorf). Aquacult. Res. 32, 92—98.

Barclay, W., Zeller, S., 1996. Nutritional enhancement of n-3 and n-6 fatty acids in rotifers and Artemia nauplii by feeding spray-dried Schizochytrium sp. J. World Aquacult. Soc. 27, 314—322.

Batista, A.P., Raymundo, A., Sousa, I., Empis, J., 2006. Rheological characterization of coloured oil-in-water food emulsions with lutein and fycocyanin added to the oil and aqueous phases. Food Hydrocolloids 20, 44—52.

Battaglene, S.C., Cobcroft, J.M., 2007. Advances in the culture of striped trumpeter larvae: a review. Aquaculture 268, 195—208.

Becker, E.W., 2004. Microalgae for aquaculture. The nutritional value of microalgae for aquaculture. In: Richmond, A. (Ed.), Handbook of Microalgal Culture: Biotechnology and Applied Phycology. Blackwell, Oxford, pp. 380—391.

Becker, E.W., 2007. Microalgae as a source of protein. Biotec. Adv. 25, 207—210.

Benkendorff, K., Davis, A.R., Rogers, C.N., Bremner, J.B., 2005. Free fatty acids and sterols in the benthic spawn of aquatic molluscs, and their associated antimicrobial properties. J. Exp. Mar. Biol. Ecol. 316, 29—44.

Blazencic, J., 2007. Sistematika Algi. NNK Internacional, Beograd.

Borowitzka, M.A., 1997. Microalgae for aquaculture: opportunities and constraints. J. Appl. Phycol. 9, 393—401.

Brown, M.R., 2002. Nutritional value of microalgae for aquaculture. In: Cruz-Suárez, L.E., Ricque-Marie, D., Tapia-Salazar, M., Gaxiola-Cortés, M.G., Simoes, N. (Eds.), Avances en Nutrición Acuícola VI. Memorias del VI Simposium Internacional de Nutrición. Acuícola. 3 al 6 de Septiembre del 2002. Cancún. Quintana Roo, México.

Brown, M.R., Barret, S.M., Volkman, J.K., Nearhos, S.P., Nell, J., Allan, G.L., 1996. Biochemical composition of new yeasts and bacteria evaluated as food for bivalve aquaculture. Aquaculture 143, 341–360.

Brown, M.R., 1991. The amino acid and sugar composition of 16 species of microalgae used in mariculture. J. Exp. Mar. Biol. Ecol. 145, 79–99.

Brown, M.R., Dunstan, G.A., Jeffrey, S.W., Volkman, J.K., Barrett, S.M., LeRoi, J.M., 1993a. The influence of irradiance on the biochemical composition of the prymnesiophyte *Isochrysis* sp. (clone T-ISO). J. Phycol. 29, 601–612.

Brown, M.R., Garland, C.D., Jeffrey, S.W., Jameson, I.D., Leroi, J.M., 1993b. The gross and amino acid compositions of batch and semi-continuous cultures of *Isochrysis* sp. (clone T.ISO), Pavlova lutheri and Nannochloropsis oculata. J. Appl. Phycol. 5, 285–296.

Brown, M.R., Jeffrey, S.W., Volkman, J.K., Dunstan, G.A., 1997. Nutritional properties of microalgae for mariculture. Aquaculture 151, 315–331.

Brown, M.R., Miller, K.A., 1992. The ascorbic acid content of eleven species of microalgae used in mariculture. J. Appl. Phycol. 4, 205–215.

Brown, M.R., Mular, M., Miller, I., Trenerry, C., Farmer, C., 1999. The vitamin content of microalgae used in aquaculture. J. Appl. Phycol. 11, 247–255.

Burr, G.S., Barrows, F.T., Gaylord, G., Wolters, W.R., 2011. Apparent digestibility of macronutrients and phosphorus in plant derived ingredients for Atlantic Salmon, *Salmo salar* and arctic charr, *Salvelinus alpinus*. Aquacult. Nutr. 17, 570–577.

Cahu, C., Zambonino-Infante, J.L., Escaffre, A.M., Bergot, P., Kaushik, S., 1998. Preliminary results on sea bass *Dicentrarchus labrax* larvae rearing with compound diet from first feeding, comparison with carp *Cyprinus carpio* larvae. Aquaculture 169, 1–7.

Cannon, D., 2009. From fish oil to microalgae oil a win shift for humans and our habitat. Explore 5, 299–303.

Carballo-Cardenas, E.C., Tuan, P.M., Jans-sen, M.H.M., Wijffels, R.H., 2003. Vitamin E (α-tocopherol) production by the marine microalgae *Dunaliella tertiolecta* and *Tetraselmis suecica* in batch cultivation. Biomol. Eng. 20, 139–147.

Chakraborty, R.D., Chakraborty, K., Radhakrishnan, E.V., 2007. Variation in fatty acids composition of *Artemia salina* nauplii enriched with microalgae and baker's yeast for use in larviculture. J. Agric. Food Chem. 55, 4043–4051.

Chatzifotis, S., vaz Juan, I., Kyriazi, P., Divanach, P., Pavlidis, M., 2011. Dietary carotenoids and skin melanin content influence the coloration of farmed red porgy *Pagrus pagrus*. Aquacult. Nutr. 17, e90–e100.

Chen, S., Chi, Z., O'Fallon, J.V., Zheng, Y., Chakraborty, M., Laskar, D.D., 2010. System integration for producing microalgae as biofuel feedstock. Biofuels. 1, 889–910.

Chen, F., Zhang, Y., 1997. High cell density mixotrophic culture of *Spirulina platensis* on glucose for phycocyanin production using a fed-batch system. Enzym. Microb. Tech. 20, 221–224.

Chisti, Y., 2007. Biodiesel from microalgae. Biotech. Adv. 25, 294–306.

Choubert, G., Heinrich, O., 1993. Carotenoid pigments of the green alga *Haematococcus pluvialis*: assay on rainbow Trout, *Oncorhynchus mykiss*, pigmentation in comparison with synthetic astaxanthin and canthaxanthin. Aquacult. 112, 217–226.

Conceição, L.E.C., Yúfera, M., Makridis, P., Morais, S., Dinis, M.T., 2010. Live feeds for early stages of fish rearing. Aquacult. Res. 41, 613–640.

Conklin, D.E., 1997. Vitamins. In: D'Abramo, L.R., Conkin, D.E., Akiyama, D.M. (Eds.), Crustacean Nutrition, Advances in World Aquaculture, Vol. 6. World Aquaculture Society, pp. 123–149.

Coutinho, P., Rema, P., Otero, A., Pereira, O., Fábregas, J., 2006. Use of biomass of the Marine microalga *isochrysis galbana* in the nutrition of goldfish (*Carassius auratus*) larvae as source of protein and vitamins. Aquacult. Res. 37, 793–798.

Coutteau, P., Dravers, M., Leger, P., Sorgeloos, P., 1993. Manipulated yeast diets and dried algae as a partial substitute for live algae in the juvenile rearing of the Manila clam *Tapes philippinarum* and the Pacific oyster *Crassostrea gigas*. Spec. Publ. Eur. Aquatic Soc. Ghent, Belgium 18, pp. 523–531.

Coutteau, P., Sorgeloos, P., 1992. The use of algal substitutes and the requirement for live algae in the hatchery and nursery rearing of bivalve molluscs: an international survey. J. Shellfish Res. 11, 467–476.

Coutteau, P., Castell, J.D., Ackman, R.G., Sorgeloos, P., 1996. The use of lipid emulsions as carriers for essential fatty acids in bivalves: a test case with juvenile *Placopecten magellanicus*. J. Shellfish Res. 15, 259–264.

Csordas, A., Wang, J.-K., 2004. An integrated photobioreactor and foam fractionation unit for the growth and harvest of *Chaetoceros* spp. in open systems. Aquacult. Eng. 30, 15–30.

d'Ippolito, G., Tucci, S., Cutignano, A., Romano, G., Cimino, G., Miralto, A., Fontana, A., 2004. The role of complex lipids in the synthesis of bioactive aldehydes of the marine diatom *Skeletonema costatum*. Biochim. Biophys. Acta 1686, 100–107.

Das, K., Pradhan, J., Pattnaik, P., Samantaray, B.R., Samal, S.K., 2005. Production of antibacterials from the freshwater alga *Euglena viridis* (Ehren). World J. Microb. Biot. 21, 45–50.

Del Campo, J.A., González, M.G., Guerrero, M.G., 2007. Outdoor cultivation of microalgae for carotenoid production: current state and perspectives. Appl. Microbiol. Biotechnol. 74, 1163–1174.

Dhont, J., van Stappen, G., 2003. Live Feeds in Marine Aquaculture. Blackwell Science, pp. 65−121.

Donaldson, J., 1991. In: Proceedings of US-Asia Workshop, Honolulu, HA, January 28−31, the Oceanic Institute, HA, pp. 229−236.

Doucha, J., Lívanský, K., Kotrbáček, V., Zachleder, V., 2009. Production of chlorella biomass enriched by selenium and its use in animal nutrition: a review. Appl. Microbiol. Biotechnol. 83, 1001−1008.

Dunstan, A., Volkman, J.K., Jeffrey, S.W., Barret, S.M., 1992. Biochemical composition of microalgae from the green algal classes Chlorophyceae and Prasinophyceae. J. Exp. Mar. Biol. Ecol. 161, 115−134.

Dunstan, G.H., Volkman, J.K., Barret, S.M., Garland, C.D., 1993. Changes in the lipid composition and maximization of the polyunsaturated fatty acid content of three microalgae grown in mass culture. J. Appl. Phycol. 5, 71−83.

Durmaz, Y., 2007. Vitamin E (α-tocopherol) production by the marine microalgae Nannochloropsis oculata (Eustigmatophyceae) in nitrogen limitation. Aquaculture 272, 717−722.

Dyck, M., Roberts, R., Jeff, A., 2011. Assessing alternative grazing-tolerant algae for nursery culture of abalone, Haliotis iris. Aquaculture 320, 62−68.

Elston, A., 1990. Mollusc Diseases, Guide for the Shellfish Farmer. Washington Sea Grant Program, Seattle, WA, p. 73.

Enright, C.T., Newkirk, G.F., Craigie, J.S., Castell, J.D., 1986. Evaluation of phytoplankton as diets for juvenile Ostrea edulis L. J. Exp. Mar. Biol. Ecol. 96, 1−13.

Fabregas, J., Herrero, C., 1986. Marine microalgae as a potential source of minerals in fish diets. Aquaculture 51, 237−243.

Fahy, M., 1986. The relevance of cryoprotectant 'toxicity' to cryobiology. Cryobiology 23, 1−13.

Fahy, M., Lilley, T.H., Lindsell, H., Douglas, M.J., Meryman, H.T., 1990. Cryoprotectant toxicity and cryoprotectant toxicity reduction: in search of molecular mechanisms. Cryobiology 27, 247−268.

FAO, 1997. Renewable Biological Systems for Alternative Sustainable Energy Production. FAO Agricultural Services Bulletin (Chapter 6). Version 128.

Fleurence, J., Morançais, M., Dumay, J., Decottignies, P., Turpin, V., Munier, M., Garcia-Bueno, N., Jaouen, P., 2012. What are the prospects for using seaweed in human nutrition and for marine animals raised through aquaculture? Trends Food Sci. Tech. 27, 57−61.

Flynn, K.J., Fielder, J., 1989. Changes in intracellular and extracellular amino acids during the predation of the chlorophyte Dunaliella primolecta by the heterotrophic dinoflagellate Oxyrrhis marina and the use of the glutamine/glutamate ratio as an indicator of nutrient status in mixed populations. Mar. Ecol. Prog. Ser. 53, 117−127.

Flynn, K.J., Flynn, K., 1992. Non-protein free amines in microalgae: consequences for the measurement of intracellular amino acids and of the glutamine/glutamate ratio, Mar. Ecol. Prog. Ser. 89, 73−79.

Fujii, K., Nakashima, H., Hashidzume, Y., Uchiyama, T., Mishiro, K., Kadota, Y., 2010. Potential use of the astaxanthin-producing microalga, Monoraphidium sp. GK12, as a functional aquafeed for prawns. J. Appl. Phycol. 22, 363−369.

Gantar, M., Svircev, Z., 2008. Microalgae and cyanobacteria: food for thought. J. Phycol. 44, 260−268.

Ganuza, E., Benítez-Santana, T., Atalah, E., Vega-Orellana, O., Ganga, R., Izquierdo, M.S., 2008. Cryptheco-dinium cohnii and Schizochytrium sp. as potential substitutes to fisheries-derived oils from sea bream (Sparus aurata) Microdiets. Aquaculture 277, 109−116.

Glencross, B.D., Booth, M., Allan, G.L., 2007. A feed is only as good as its ingredients—a review of ingredient evaluation strategies for aquaculture feeds. Aquacult. Nutr. 13, 17−34.

Gomes, E., Dias, J., Silva, P., et al., 2002. Utilization of natural and synthetic sources of carotenoids in the skin pigmentation of gilthead sea bream (Sparus aurata). Eur. Food Res. Technol. 214, 287−293.

Gouveia, L., Batista, A.P., Sousa, I., Ray-mundo, A., Bandarra, N.M., 2008. Microalgae in novel food products. In: Papadoupoulos, K.N. (Ed.), Food Chemistry Research Developments. Nova Science Publishers, New York, pp. 75−112.

Gouveia, L., Raymundo, A., Batista, A.P., Sousa, I., Empis, J., 2006. Chlorella vulgaris and Haematococcus pluvialis biomass as colouring and antioxidant in food emulsions. Eur. Food Res. Technol. 222, 362−367.

Gouveia, L., Choubert, G., Gomes, E., et al., 2002. Pigmentation of gilthead sea bream, Sparus aurata, using Chlorella vulgaris (Chlorophyta, Volvocales) microalga. Aquacult. Res. 33, 987−993.

Gouveia, L., Rema, P., 2005. Effect of microalgal biomass concentration and temperature on ornamental goldfish (Carassius auratus) skin pigmentation. Aquacult. Nutr. 11, 19−23.

Guedes, A.C., Malcata, F.X., 2012. In: Muchlisin, Z.A. (Ed.), Nutritional Value and Uses of Microalgae in Aquaculture, Aquaculture, ISBN 978-953-307-974-5, pp. 59−78. InTech, Available from: http://www.intechopen.com/books/aquaculture/nutritional-value-and-uses-of-microalgae-in-aquaculture.

Güroy, D., Güroy, B., Merrifield, D.L., Ergün, S., Tekinay, A.A., Yiğit, M., 2011. Effect of dietary Ulva and spirulina on weight loss and body composition of rainbow trout, Oncorhynchus mykiss (Walbaum), during a starvation period. J. Anim. Physiol. Anim. Nutr. 95, 320−327.

Hansen, J.Ø., Storebakken, T., 2007. Effects of dietary cellulose level on pellet quality and nutrient digestibilities in rainbow trout (*Oncorhynchus mykiss*). Aquaculture 272, 458–465.

Heasman, M., Diemar, J., O'Connor, W., Sushames, T., Foulkes, L., 2000. Development of extended shelf-life micro-algae concentrate diets harvested by centrifugation for bivalve molluscs—a summary. Aquacult. Res. 31, 637–659.

Heasman, M., Savva, N., 2007. Manual for Intensive Hatchery Production of Abalone. NSW Department of Primary Industries.

Hemaiswarya, S., Raja, R., Kumar, R.R., Ganesan, V., Anbazhagan, C., 2011. Microalgae: a sustainable feed source for aquaculture. World J. Microbiol. Biotechnol. 27, 1737–1746.

Hemmingson, J.A., Falshaw, R., Furneaux, R.H., Thompson, K., 2006. Structure and anti-viral activity of the galactofucan sulfates extracted from Undaria pinnatifida (Phaeophyta). J. Appl. Phycol. 18, 185–193.

Hussein, E.E.S., Dabrowski, K., El-Saidy, D.M.S.D., Lee, B.J., 2013. Enhancing the growth of Nile tilapia larvae/juveniles by replacing plant (gluten) protein with algae protein. Aquacult. Res. 44, 937–949.

Jackson, A.E., Ayer, S.W., Laycock, M.V., 1992. The effect of salinity on growth and amino acid composition in the marine diatom *Nitzschia pungens*. Can. J. Bot. 70, 2198–2201.

Jüttner, F., 2001. Liberation of 5,8,11,14,17-eicosapentaenoic acid and other polyunsaturated fatty acids from lipids as a grazer defense reaction in epiphithic diatom biofilms. J. Phycol. 37, 744–755.

Karlson, J.O.M., Toner, M., 1996. Long-term storage of tissues by cryopreservation: critical issues. Biomaterials 17, 243–256.

Kiron, V., Phromkunthong, W., Huntley, M., Archibald, I., De Scheemaker, G., 2012. Marine microalgae from biorefinery as a potential feed protein source for Atlantic salmon, common carp and whiteleg shrimp. Aquacult. Nutr. 18, 521–531.

Knauer, J., Southgate, P.C., 1999. A review of the nutritional requirements of bivalves and the development of alternative and artificial diets for bivalve aquaculture. Rev. Fish. Sci. 7, 241–280.

Knauer, J., Barrett, S.M., Volkman, J.K., Southgate, P.C., 1999. Assimilation of dietary phytosterols by Pacific oyster *Crassostrea gigas* spat. Aquacult. Nutr. 5, 257–266.

Knuckey, R.M., Brown, M.R., Barrett, S.M., Hallegraeff, G.M., 2002. Isolation of new nanoplanktonic diatom strains and their evaluation as diets for the juvenile Pacific oyster. Aquaculture 211, 253–274.

Knuckey, R.M., Brown, M.R., René, R., Frampton, M.F.D., 2006. Production of microalgal concentrates by flocculation and their assessment as aquaculture feeds. Aquacul. Eng. 35, 300–313.

Kovač, D.J., Simeunović, J.B., Babić, O.B., Mišan, A.Č., Milovanović, I.L., 2013. Algae in food and feed. Food Feed Res. 40, 21–31.

Kraan, S., Mair, C., 2010. Seaweeds as ingredients in aquatic feeds. Int. Aquafeed 13, 10–14.

Kwak, T.J., Zedler, J.B., 1997. Food web analysis of southern California coastal wetlands using multiple stable isotopes. Oecologia 110, 262–277.

Lakshmanasenthil, S., Vinothkumar, T., Geetharamani, D., Maruthupandi, T., 2013. Influence of microalgae in enrichment of *Artemia salina* for aquaculture feed enhancement. J. Algal Biomass Utln. 4, 67–73.

Langdon, C.J., Önal, E., 1999. Replacement of living microalgae with spray-dried diets for the marine mussel *Mytilus galloprovincialis*. Aquaculture 180, 283–294.

Langdon, C.J., Bolton, E.T., 1984. A microparticulate diet for suspension-feeding bivalve mollusc, *Crassostrea virginica* (Gmelin). J. Exp. Mar. Biol. Ecol. 89, 239–258.

Langdon, C.J., Waldock, M.J., 1981. The effect of algal and artificial diets on the growth and fatty acid composition of *Crassostrea gigas* spat. J. Mar. Biol. Assoc. UK 61, 431–448.

Lavens, P., Sorgeloos, P., 1996. In: Lavens, P., Sorgeloos, P. (Eds.), Manual on the Production and Use of Live Food for Aquaculture. FAO Fisheries Technical Paper. Rome. pp. 36–39.

Liang, H., Gong, W.J., Chen, Z.L., Tian, J.Y., Qi, L., Li, G.B., 2009. Effect of chemical preoxidation coupled with in-line coagulation as a pretreatment to ultrafiltration for algae fouling control. Desalin. Water Treat. 9, 241–245.

López-Elías, J.A., Voltolina, D., Chavira-Ortega, C.O., Rodríguez, B.B., Sáenz-Gaxiola, L.M., Esquivel, B.C., Nieves, M., 2003. Mass production of microalgae in six commercial shrimp hatcheries of the Mexican northwest. Aquacultural Eng. 29, 155–164.

Lubzens, E., Gibson, O., Zmora, O., Sukenik, A., 1995. Potential advantages of frozen algae (*Nannochloropsis* sp.) for rotifer (*Brachionus plicatilis*) culture. Aquaculture 133, 295–309.

Lupatsch, I., 2009. Quantifying nutritional requirements in aquaculture—the factorial approach. In: Burnell, G., Allan, G. (Eds.), New Technologies in Aquaculture: Improving Production Efficiency. Quality and Environmental Management, Cambridge, pp. 417–439.

Martínez-Fernández, E., Paul, C., 2007. Southgate use of tropical microalgae as food for larvae of the black-lip pearl oyster *Pinctada margaritifera*. Aquaculture 263, 220–226.

Mazur, P., 1984. Freezing of living cells: mechanisms and implications. Am. J. Physiol. 247, 125–142.

McCausland, A., Brown, M.R., Barrett, S.M., Diemar, J.A., Heasman, M.P., 1999. Evaluation of live and pasted microalgae as supplementary food for juvenile Pacific oysters (*Crassostrea gigas*). Aquacult. Res. 174, 323–342.

Mendiola, J.A., Torres, C.F., Martín-Alvarez, P.J., Santoyo, S., Toré, A., Arredondo, B.O., Señoráns, F.J., Cifuentes, A., Ibáñez, E., 2007. Use of supercritical CO_2 to obtain extracts with antimicrobial activity from Chaetoceros muelleri microalga. A correlation with their lipidic content. Eur. Food Res. Technol. 224, 505−510.

Michiels, J., Skrivanova, E., Missotten, J., Ovyn, A., Mrazek, J., De Smet, S., Dierick, N., 2012. Intact brown seaweed (*Ascophyllum nodosum*) in diets of weaned piglets: effects on performance, gut bacteria and morphology and plasma oxidative status. J. Anim. Physiol. Anim. Nutr. 96, 1101−1111.

Millamena, M., Aujero, E.J., Borlongan, I.G., 1990. Techniques on algae harvesting and preservation for use in culture as larval food. Aquacult. Eng. 9, 295−304.

Miller, M.R., Nichols, P.D., Carter, C.G., 2008. n-3 oil sources for use in aquaculture—alternatives to the unsustainable harvest of wild fish. Nutr. Res. Rev. 21, 85−96.

Milovanovic, I., Misan, A., Saric, B., Kos, J., Mandic, A., Simeunovic, J., Kovac, D., 2012. Evaluation of protein and lipid content and determination of fatty acid profile in selected species of cyanobacteria. In: Proceedings of the 6th Central European Congress on Food, CEFood2012, Novi Sad, Serbia.

Mohammed, M.K., Mohd, M.K., 2011. Production of carotenoids (antioxidants/colourant) in *Spirulina platensis* in response to indole acetic acid (IAA). IJEST 3, 4973−4979.

Molina-Grima, E., Belarbi, E.H., Acién-Fernández, F.G., Robles Medina, A., Chisti, Y., 2003. Recovery of microalgal biomass and metabolites: process options and economics. Biotech. Adv. 20, 491−515.

Muller-Feuga, A., 2000. The role of microalgae in aquaculture: situation and trends. J. Appl. Phycol. 12, 527−534.

Muller-Feuga, A., 2004. Microalgae for aquaculture: the current global situation future trends. In: Richmond, A. (Ed.), Handbook of Microalgal Culture: Biotechnology and Applied Phycology. Oxford, pp. 352−364.

Muller-Feuga, A., Moal, J., Kaas, R., 2003a. The microalgae of aquaculture. In: Støttrup, J.G., McEvoy, L.A. (Eds.), Live Feeds in Marine Aquaculture. Oxford, pp. 206−252.

Muller-Feuga, A., Robert, R., Cahu, C., Robin, J., Divanach, P., 2003b. Uses of microalgae in aquaculture. In: Støttrup, J.G., McEvoy, L.A. (Eds.), Live Feeds in Marine Aquaculture. Oxford, pp. 253−299.

Mustafa, M.G., Nakagawa, H., 1995. A review: dietary benefits of algae as an additive in fish feed. Israeli J. Aquacult. Bamidgeh. 47, 155−162.

Nandeesha, M.C., Gangadhara, B., Manissery, J.K., Venkataraman, L.V., 2001. Growth performance of two indian major carps, catla (*Catla catla*) and Rohu (*Labeo rohita*) fed diets containing different levels of spirulina platensis. Bioresour. Technol. 80, 117−120.

Nath, P.R., Khozin-Goldberg, I., Cohen, Z., et al., 2012. Dietary supplementation with the microalgae *Parietochloris incisa* increases survival and stress resistance in guppy (*Poecilia reticulate*) fry. Aquacult. Nutr. 18, 167−180.

Naviner, M., Bergé, J.P., Durand, P., le Bris, H., 1999. Antibacterial activity of the marine diatom *Skeletonema costatum* against aquacultural pathogens. Aquaculture 174, 15−24.

Naylor, R.L., Hardy, R.W., Bureau, D.P., Chiu, A., Elliott, M., Farrell, A.P., Forster, I., Gatlin, D.M., Goldburg, R.J., Hua, K., Nichols, P.D., 2009. Feeding aquaculture in an era of finite resources. Proc. Natl. Acad. Sci. 106, 15103−15110.

Nell, J.A., O'Connor, W.A., 1991. The evaluation of fresh algae and stored algal concentrates as a food source for Sydney rock oyster, *Saccostrea commercialis* (Iredale and Roughley) larvae. Aquaculture 99, 277−284.

New, M.B., Wagner, C.V., 2000. Freshwater Prawn Culture. Blackwell Science, pp. 1−11, Oxford.

Nonwachai, T., Purivirojkul, W., Limsuwan, C., Chuchird, N., Velasco, M., Dhar, A.K., 2010. Growth, nonspecific immune characteristics, and survival upon challenge with *Vibrio harveyi* in Pacific white shrimp (*Litopenaeus vannamei*) raised on diets containing algal meal. Fish Shellfish Immunol. 29, 298−304.

Olvera-Novoa, M.A., Dominguez-Cen, L.J., Olivera-Castillo, L., 1998. Effect of the use of the Microalgae Spirulina maxima as fish meal replacement in diets for Tilapia, *Oreochromis mossambicus* fry. Aquacult. Res. 29, 709−715.

Palmegiano, G.B., Agradi, E., Forneris, G., Gai, F., Gasco, L., Rigamonti, E., Benedetto, S., Zoccarato, I., 2005. *Spirulina* as a nutrient source in diets for growing Sturgeon (*Acipenser baeri*). Aquacult. Res. 36, 188−195.

Parisenti, J., Beirão, L.H., Maraschin, M., Mouriño, J.L., Do Nascimento-Vieira, F., Bedin, L.H., Rodrigues, E., 2011. Pigmentation and carotenoid content of shrimp fed with haematococcus pluvialis and soy lecithin. Aquacult. Nutr. 17, e530−e535.

Parsons, T.R., Stephens, K., Strickland, J.D.H., 1961. On the chemical composition of eleven species of marine phytoplankters. J. Fish. Res. Board Can. 18, 1001−1016.

Patil, V., Källqvist, T., Olsen, E., Vogt, G., Gislerød, H.R., 2007. Fatty acid composition of 12 microalgae for possible use in aquaculture feed. Aquacul. Int. 15, 1−9.

Pickova, J., Mørkøre, T., 2007. Alternate oils in fish feeds. Eur. J. Lipid Sci. Technol. 109, 256−263.

Plaza, M., Cifuentes, A., Ibanez, E., 2008. In the search of new functional food ingredients from algae. Trends Food Sci. Tech. 19, 31−39.

Poelman, E., de Pauw, N., Jeurissen, B., 1997. Potential of electrolytic flocculation for recovery of microalgae. Resour. Conserv. Recy. 19, 1–10.

Ponis, E., Probert, I., Véron, B., Mathieu, M., Robert, R., 2006. New microalgae for the Pacific oyster *Crassostrea gigas* larvae. Aquaculture 253, 618–627.

Ponis, E., Robert, R., Parisi, G., 2003. Nutritional value of fresh and concentrated algal diets for larval and juvenile Pacific oysters (*Crassostrea gigas*). Aquaculture 221, 491–505.

Pulz, O., Gross, W., 2004. Valuable products from biotechnology of microalgae. Appl. Microbiol. Biotechnol. 65, 635–648.

Raja, R., Anbazhagan, C., Lakshmi, D., Rengasamy, R., 2004. Nutritional studies on *Dunaliella salina* (Volvocales, Chlorophyta) under laboratory conditions. Seaweed Res. Utili. 26, 127–146.

Raja, R., Hemaiswarya, S., 2010. In: Watson, R.R., Zibadi, S., Preedy, V.R. (Eds.), Microalgae and Immune Potential a Chapter in Dietary Components and Immune Function—prevention and Treatment of Disease and cancer. Humana Press/Springer, USA, ISBN 978-1-60761-060-1, pp. 517–529.

Raja, R., Hemaiswarya, S., Rengasamy, R., 2007. Exploitation of Dunaliella for b-carotene production. Appl. Microbiol. Biotechnol. 74, 517–523.

Richmond, A., 2004. Handbook of Microalgal Culture: Biotechnology and Applied Phycology. Blackwell Science Ltd, pp. 1–544.

Robert, R., Trintignac, P., 1997. Substitutes for live microalgae in mariculture: a review. Aquat. Living Resour. 10, 315–327.

Rocha, R.J., Ribeiro, L., Costa, R., Dinis, M.T., 2008. Does the presence of microalgae influence fish larvae prey capture? Aquacult. Res. 39, 362–369.

Rodriguez-Zavala, J.S., Ortiz-Cruz, M.A., Mendoza-Hernandez, G., Moreno-Sanchez, R., 2010. Increased synthesis of α-tocopherol, paramylon and tyrosine by *Euglena gracilis* under conditions of high biomass production. J. Appl. Microbiol. 109, 2160–2172.

Rossingol, N., Vandanjon, L., Jaouen, P., Quéméneur, F., 1999. Membrane technology for the continuous separation microalgae/culture medium: compared performances of cross-flow microfiltration and ultrafiltration. Aquacult. Eng. 20, 191–208.

Sargent, J.R., McEvoy, L.A., Bell, J.G., 1997. Requirements, presentation and sources of polyunsaturated fatty acids in marine fish larval feeds. Aquaculture 155, 117–127.

Seguineau, C., Laschi-Loquerie, A., Moal, J., Samain, J.F., 1996. Vitamin requirements in great scallop larvae. Aquacult. Int. 4, 315–324.

Sergejevová, M., Masojídek, J., 2012. *Chlorella* biomass as feed supplement for freshwater fish: sterlet, *Acipenser ruthenus*. Aquacult. Res. 44, 157–159.

Sheikhzadeh, N., Tayefi-Nasrabadi, H., Oushani, A.K., Enferadi, M.H., 2012. Effects of haematococcus pluvialis supplementation on antioxidant system and metabolism in rainbow Trout (*Oncorhynchus mykiss*). Fish Physiol. Biochem. 38, 413–419.

Shields, R.J., Lupatsch, I., 2012. Algae for aquaculture and animal feeds. Technikfolgenabschätzung—Theorie und Praxis. 21. Jg., Heft 1, Juli 2012.

Skrede, A., Mydland, L.T., Ahlstrøm, Ø., Reitan, K.I., Gislerød, H.R., Øverland, M., 2011. Evaluation of microalgae as sources of digestible nutrients for monogastric animals. J. Anim. Feed Sci. 20, 131–142.

Smith, V.J., Desbois, A.P., Dyrynda, E.A., 2010. Conventional and unconventional antimicrobials from fish, marine invertebrates and microalgae. Mar. Drugs 8, 1213–1262.

Soler-Vila, A., Coughlan, S., Guiry, M.D., Kraan, S., 2009. The red alga Porphyra dioica as a fish-feed ingredient for rainbow Trout (*Oncorhynchus mykiss*): effects on growth, feed efficiency and carcass composition. J. Appl. Phycol. 21, 617–624.

Spolaore, P., Joannis-Cassan, C., Duran, E., Isambert, A., 2006. Commercial applications of microalgae. J. Biosci. Bioeng. 101, 87–96.

Tamaru, C.S., Pang, L., Ako, H., 2004. Ornamental fish feeds - part 1. Effect of different maturation diets on the spawning of armored catfish, *Corydoras aeneus*. Int. Aquafeed. 7, 33–35.

Tartiel, M.B., Ibrahim, E.M., Zeinhom, M.M., 2008. In: Partial Replacement of Fish Meal with Dried Microalga (*Chlorella* Spp and Scenedesmus Spp) in Nile Tilapia (*Oreochromis niloticus*) Diets. 8th International Symposium on Tilapia in Aquaculture, pp. 801–811.

Taylor, R., Fletcher, R.L., 1999. Cryopreservation of eukaryotic algae—a review of methodologies. J. Appl. Phycol. 10, 481–501.

Thompson, P.A., Guo, M.X., Harrison, P.J., 1993. The influence of irradiance on the biochemical composition of three phytoplankton species and their nutritional value for larvae of the Pacific oyster (*Crassostrea gigas*). Mar. Biol. 117, 259–268.

Torstensen, B.E., Espe, M., Sanden, M., Stubhaug, I., Waagbø, R., Hemre, G.I., Fontanillas, R., Nordgarden, U., Hevrøy, E.M., Olsvik, P., Berntssen, M.H.G., 2008. Novel production of Atlantic salmon (*Salmo salar*) protein based on combined replacement of fish meal and fish oil with plant meal and vegetable oil blends. Aquaculture 285, 193–200.

Tredici, M.R., Biondi, N., Ponis, E., Rodolfi, L., Zittelli, G.C., 2009. Advances in microalgal culture for aquaculture feed and other uses. In: Burnell, G., Allan, G. (Eds.), New Technologies in Aquaculture: Improving Production Efficiency. Quality and Environmental Management, Cambridge, pp. 611–676.

Tzovenis, I., Triantaphyllidis, G., Naihong, X., Chatzinikolaou, E., Papadopoulou, K., Xouri, G., Tafas, T., 2004. Cryopreservation of marine microalgae and potential toxicity of cryoprotectants to the primary steps of the aquacultural food chain. Aquaculture 230, 457–473.

Ufodike, E.B.C., Matty, A.J., 1983. Growth responses and nutrient digestibility in mirror carp (*Cyprinus carpio*) fed different levels of cassava and rice. Aquaculture 31, 41–50.

Van Der Meeren, T., Mangor-Jensen, A., Pickova, J., 2007. The effect of green water and light intensity on survival, growth and lipid composition in Atlantic cod (*Gadus morhua*) during intensive larval rearing. Aquaculture 265, 206–2173.

Viçose, G.C., Viera, M.P.P., Huchette, S., Izquierdo, M.S., 2012. Improving Nursery performances of Haliotis tuberculata coccinea: nutritional value of four species of benthic diatoms and Green macroalgae germlings. Aquaculture 334–337, 124–131.

Volkman, J.K., Jeffrey, S.W., Nichols, P.D., Rodgers, G.I., Garland, C.D., 1989. Fatty acid and lipid composition of 10 species of microalgae used in mariculture. J. Exp. Mar. Biol. Ecol. 128, 219–240.

Walker, A.B., Fournier, H.R., Neefus, C.D., Nardi, G.C., Berlinsky, D.L., 2009. Partial replacement of fish meal with laver *Porphyra* spp. in diets for Atlantic Cod. N. Am. J. Aquacult. 71, 39–45.

Walker, A.B., Sidor, I.F., O'Keefe, T., Cremer, M., Berlinsky, D.L., 2010. Effects of replacement of fish meal protein by microalgae on growth, feed intake, and body composition of Atlantic Cod. N. Am. J. Aquacult. 72, 343–353.

Walker, A.B., Berlinsky, D.L., 2011. Effects of partial replacement of fish meal protein by microalgae on growth, feed intake, and body composition of Atlantic cod. N. Am. J. Aquacult. 73, 76–83.

Webb, K.L., Chu, F.E., 1983. Phytoplankton as a food source for bivalve larvae. In: Pruder, G.D., Langdon, C.J., Conklin, D.E. (Eds.), Proceedings of the Second International Conference on Aquaculture Nutrition: Biochemical and Physiological Approaches to Shellfish Nutrition. Louisiana State University, Baton Rouge, LA, pp. 272–291.

Wen, Z.Y., Chen, F., 2003. Heterotrophic production of eicosapentaenoic acid by microalgae. Biotechnol. Adv. 21, 273–294.

Wilson, R.P., 2002. Amino acids and proteins. In: Halver, J.E., Hardy, R.W. (Eds.), Fish Nutrition. Academic Press, CA, pp. 143–179.

Yamasaki, S., Tanabe, K., Hirata, H., 1989. Efficiency of chilled and frozen *Nannochloropsis* sp. (marine *Chlorella*) for culture of rotifer. Mem. Fac. Fish Kagoshima Univ. 38, 77–82.

Zatkova, I., Sergejevová, M., Urban, J., Vachta, R., Štys, D., Masojídek, J., 2011. Carotenoid-enriched microalgal biomass as feed supplement for freshwater ornamentals: albinic form of wels catfish (*Silurus glanis*). Aquacult. Nutr. 17, 278–286.

Zhou, B., Liu, W., Qu, W., Tseng, C.K., 1991. Application of *Spirulina* mixed feed in the breeding of bay scallop. Bioresour. Technol. 38, 229–232.

From the Ancient Tribes to Modern Societies, Microalgae Evolution from a Simple Food to an Alternative Fuel Source

Suphi S. Oncel[1], Ayse Kose[1], Fazilet Vardar[1], Giuseppe Torzillo[2]

[1]Ege University, Engineering Faculty, Department of Bioengineering, Izmir, Turkey;
[2]CNR-Istituto per lo Studio degli Ecosistemi Sezione di Firenze, Firenze, Italy

1. INTRODUCTION

From the flying bird to the swimming fish the human eye has captured each scene and stored it in its brain for evaluation. All these data sooner or later play some role in human life. Such observations from nature have been key to human progress in many scientific areas. Microalgal research is not an exception, in this regard. Microalgae blooms, far before they were taxonomically defined, captured the attention of the first human communities because they were consumed by animals for food (Becker, 1994).

The history of algae utilization goes back to ancient tribes mainly as traditional and folk medicines for their antifebrile, antiedema, diuretic, and expectorant functions as well as to prevent goiter, nephritic diseases, catarrh, and as anthelmintics. The oldest knowledge about algae (mainly macroalgae) utilization is known to be 2700 BC according to the list of species in

the Chinese materia medica by the emperor Shen-Nung (Niang and Hung, 1984; Pallela and Kim, 2011; Pohl et al., 1979; Wang et al., 2014). Pedanius Dioscorides, a Greek-origin Roman physician and botanist, reported the use of algae as medicine in his five-volume book *De Materia Medica*, some of the first medical records in Europe (Hoppe, 1979). Ancient Egyptians and Romans used algae for cosmetic purposes, in Asian countries they were mainly used as food and medicine, and in Europe they were used as fertilizer (Hoppe, 1979; Fox, 1996; Costa et al., 2003). A Chinese proverb *"Big fish eat little fish, little fish eat shrimp, shrimp eat mud"* is dedicated to algae and their importance in the food pyramid and a source of medicine is highlighted by Schwimmer and Schwimmer (Schwimmer and Schwimmer, 1955).

Early records of the consumption of microalgae to back to the ninth century where Kanem Empire used *Arthrospira* as a food source in

Africa (Barzanti and Gualtieri, 2006). Aztecs also harvested *Arthrospira* as a food source in the thirteenth century on a different continent, America, later reported to the world by the first colonialists (Belay, 2013). This is good proof of the importance of observation in life, in different continents at different times, but doing the same thing with the help of nature (Abdulgader et al., 2000; Sánchez et al., 2003).

In 1940, P. Dangeard made the first scientific report mentioning the use of algae as a food source in Africa around Lake Chad, which was the first spark for the microalgal industry (Ciferri, 1983; Sánchez et al., 2003; Belay, 2013). This is because even if studies on microalgae like *Chlorella* increased the scientific popularity with its first unialgal culture achieved by Beijerinck in 1890, and its cultivation started in 1919 in Otto Warburg's lab, the target was to use them as a model for physiology and photosynthesis research not as a potential food or fuel source (Beijerinck, 1890; Warburg, 1919; Soeder, 1980; Zallen, 1993; Sánchez et al., 2003; Grobbelaar, 2012).

Even though algae utilization has a strong historical background, commercial production was not started until the 1950s; one of the driving forces was World War II and the following food crisis (Borowitzka, 2013). The first attempts to grow microalgae in mass cultures were quite sporadic because of World War II, but then accelerated when studies were united with the efforts of John S. Burlew who edited the famous book *Algal Culture from Laboratory to Pilot Plant* (Burlew, 1953), which triggered research work around the globe, particularly in the United States, Germany, Israel, Czech Republic, Japan, Thailand, France, and Italy focusing on the cultivation of microalgae and investigating their chemical composition and industrial applications (Chaumont et al., 1988; Chaumont, 1993; Sánchez et al., 2003; Spoalore et al., 2006; Becker, 2013; Belay, 2013).

Before World War II, commercial production started with *Spirulina* during the 1870s in Mexico (Sosa Texcoco S.A.) and with *Chlorella* during the

1890s in Japan (Spoalore et al., 2006). In 1977, *Spirulina* production was also started in Thailand. By the 1980s, 46 companies were actively producing more than 1000 kg/month of microalgae, mostly *Chlorella*. In 1996, 2000 tons of *Chlorella* was marketed in Japan. New companies were also started to produce microalgae in Israel, the United States, and later China and India (Borowitzka, 1997; Fernandez et al., 1997). In a relatively short period of time, rapid progress was established by the microalgae industry (Tamiya, 1957).

Microalgae as a source evolved from three unique paths: (1) history of the ancient tribes and traditions, (2) scientific progress related to technology, and (3) engineering work in the environment, chemistry, and biology fields (Soeder, 1980). The aim of this book chapter is to focus on the progress and engineering of the microalgae industry, with a special emphasis on the current big debate of using microalgae as an alternative fuel source.

2. DISCOVERIES IN PHOTOSYNTHESIS AS A TOOL TO UNDERSTAND MICROALGAL METABOLISM FOR BIOFUEL DEVELOPMENT: A LITERATURE SURVEY FOR A HISTORICAL SNAPSHOT

Photosynthesis, formerly known as *assimilation* (Gest, 2002), has a strong historical background. The very first hypothesis about photosynthesis comes from ancient Greece, but the tree and soil experiment of Jan van Helmont in 1648 is the most memorable event. After van Helmont's studies, Joseph Priestly, John Woodward, Jan Ingen-Housz, and Antoine-Laurent Lavoisier made critical contributions to research on photosynthesis from a physiological point of view (Huzisige and Ke, 1993; Govindjee and Krogmann, 2004; Blankenship, 2010; Yu et al., 2012).

In 1818, the chlorophyll pigment was isolated from plant leaves and referred to as *light harvesting machinery* in photosynthesis. Chloroplasts were discovered in 1881 as photosynthetic apparatus of individual cells. The C^{14} isotope labeling technique to prove that O_2 evolution comes from H_2O not from CO_2 (Ruben et al., 1939; Ruben and Kamen, 1941; Gest, 1997, 2005) was another breakthrough. Calvin Melvin and his colleagues published numerous studies highlighting the origin of photosynthesis and stated that CO_2 is fixed into organic structures in cells (Calvin and Benson, 1948; Benson and Calvin, 1948; Stepka and Benson, 1948; Benson et al., 1949; Calvin, 1949; Badin and Calvin, 1950; Calvin et al., 1950, 1951). Later named as the Calvin–Benson cycle, these studies highlighted the metabolism and pathways of inorganic and organic substances in both plant and microalgae metabolism.

The evolution of photosynthesis in microalgae is also considered as key in terms of finding certain metabolic pathways and common ancestors with other photosynthetic organisms (Burris, 1977; Xiong, 2006). Regarding combined studies of carbon utilization (de Saussure, 1804; Kortschak et al., 1965; Hatch and Slack, 1966; Benson, 2002; Bassham, 2005; Yu et al., 2012), the Calvin–Benson cycle was later used for various carbon assimilation modes in microalgal metabolism (Sager and Granick, 1953; Kratz and Myers, 1955; Goldman et al., 1971; Chen, 1996; Perez-Garcia et al., 2011; Pires et al., 2012; Wang et al., 2010). With this metabolism, storage chemicals were further evaluated for various applications, for example, biodiesel production from microalgal lipids (Sheenah et al., 1998a, 1998b; Chisti, 2007; Oncel, 2013) and the role of starch for biohydrogen production (Kruse et al., 2005), and cell composition for bioethanol and biogas production (Samson and LeDuy, 1982; Mussatto et al., 2010). Highlights in the Calvin–Benson cycle further encouraged studies on the relation of carbon nutrition and valuable products (Perez-Garcia et al., 2011). Besides work on various growth and nutrition

modes, for example, heterotrophic and photomixotrophic, research applied to microalgal metabolism resulted in important findings; for example, algae were able to assimilate certain carbon sources under dark using pure organic carbon and nitrogen sources, wastewater, and industrial by-products (Theiarult, 1965; Kawaguchi, 1980; Ogawa and Aiba, 1981; Kaplan et al., 1986; Chen, 1996; Yang et al., 2002; Perez-Garcia et al., 2011).

Going back to 1919, the relation of photosynthesis with microalgae was researched by Otto Heinrich Warburg. He used microalgae, *Chlorella*, because of its fast-growing, nonmotile, and simple life cycle properties (Zallen, 1993). Warburg's study was important because he was able to understand the number of quantas required to possess photosynthesis and he also proposed the concepts of light utilization efficiency and light to chemical energy (Warburg, 1919, 1920; Warburg and Negelein, 1923; Nickelsen, 2007). Warburg also used the terms *photolysis* and *acceptors* in discussing photosynthesis reactions. His studies of photolysis are now widely used in microalgal biohydrogen production (biophotolysis of water into electrons, protons, and oxygen) (Myers, 1974; Benemann, 1997).

3. MICROALGAE, A MILESTONE IN LIGHT TO FUEL CONCEPT

Andrei Segeyevic Famintsyn, a Russian botanist, was the first scientist to use artificial light in plant research (1868) and to culture algae in a simple culture medium including inorganic salts (1871) (Andersen, 2005); however, the modern culture of microalgae started after Martinus Beijerinck established unialgal *Chlorella vulgaris* cultures in the laboratory (Beijerinck, 1890; Andersen, 2005; Borowitzka and Moheimani, 2013). E. J. Allen published detailed research (Allen and Nelson, 1910) on culture methods including both previous studies and his own experiments. After this, knowledge of the physiology, biochemistry, and metabolism of microalgae started to accumulate thanks to the efforts of many scientists (Bewicke and Potter, 1993; Toerien et al., 1987). Ernst Pringsheim, in 1921, was the first to introduce acetate in culture medium as an organic carbon source for heterotrophic growth of algae (Andersen, 2005). Myers and Clark (1944) developed a system for continuous culture of *Chlorella*, which enables the maintenance of culture density at a constant value by adding fresh medium. Myers also conducted experiments on the effects of high-light intensity on the photosynthesis of *Chlorella* (Myers and Burr, 1940). At the same time, studies on the effects of nitrogen and phosphate on the growth medium of algae were done by Chu (1942) using several species such as *Botryococcus braunii* (Kützing, 1849), a promising algal hydrocarbon feedstock (Watanabe and Tanabe, 2013).

The first steps into large-scale production of microalgae were taken during the late 1940s and early 1950s with the efforts of von Witsch in Germany (1948), PM Cook in the United States (1950—1951), and Gummert in Germany (1953) (Andersen, 2005) with *Chlorella* strains. First examples of outdoor large-scale production included both closed tubular reactors (the first one in 1951 in the United States) (Borowitzka and Moheimani, 2013) and open ponds (Richmond, 1999; Zittelli et al., 2013). The major issues with these processes, first described at this time and which are still being discussed, include temperature control, prevention of contamination, choice of flow rate, and the effect of climate on cultures (Pulz, 2001; Carvolha et al., 2006).

Pruess et al. (1954) published a detailed report using several different strains of algae including *Chlorella*, *Scenedesmus*, *Chlamydomonas*, and *Chlorococcum* cultured in different volumes starting from 100 mL in 500-mL flasks to 12 L in 20-L glass bottles and up to 300—2500 L in deep tank fermenters. They presented the dry weight yields obtained from cultures (maximum value being 2.4 g L^{-1} per day with *Chlorella vulgaris*)

and chemical composition analysis of algal cells grown in 20-L glass bottles and two batches of *Chlorella pyrenoidosa* grown in 2000-L tank fermenters.

As the knowledge about microalgae physiology and metabolism expanded, large-scale cultivation of microalgae continued in the early 1960s through 1970s. Information about large-scale cultivation of algae for biofuel production using high-rate algal ponds was first presented in the literature by Oswald and Golueke (1960). Mayer et al. (1964) developed a 2000-L, 1-m deep, continuously stirred culture unit with its one side (facing south) and top transparent, in which *Chlorella vulgaris*, *C. pyrenoidosa*, *C. ellipsoidea*, *Chlamydomonas*, and *Phaeodactylum* were cultivated. A yield of 13 g dry weight m^{-2} illuminated area per day was reported in addition to solubility of gases in tank, photosynthetic rates, and the effects of stirring speed, temperature, and light intensity. Several patents on algae cultivation process and production units (Buisson et al., 1969; Masahito, 1969), separation of algae (Ort, 1970), and industrial cultivation of algae (Minoru et al., 1967a, 1967b) were obtained in the late 1960s.

During the same period, the concept of thin-layer culture (cascade) to optimize the light utilization efficiency (Setlik et al., 1970) was studied in Trebon (Czech Republic). Although, cascade belongs to the open systems, it presents some advantages of the closed photobioreactor (PBR), namely the possibility to grow cultures at very high densities (10—15 g L^{-1}), which dramatically reduces the cost of harvesting. This system, however, has been tested only with *Chlorella* cultures.

Goldman and Rhyter (1977) argued the feasibility of large-scale algae systems solely for energy production and proposed smaller scale systems capable of wastewater treatment, waste recycling, and protein, fertilizer, and drug production (Becker, 2007). They also noted the importance of culture mode (batch, continuous, or semicontinuous), source of nutrients, surface area and depth of culture, mixing method, and residence time. Mass production experiments using two continuously mixed 2000-L ponds with 2.3-m width and 0.5-m depth reported algal yields of 12.7 and 19 g dry weight m^{-2} day (Goldman and Ryther, 1975), and Goldman published a review on outdoor mass cultures of algae and compiled biomass yields obtained in previous experiments by various researchers (Goldman, 1979). In order to be economically feasible, Goldman emphasized the exploitation of other applications such as production of chemicals, wastewater treatment, and aquaculture, in addition to bioconversion of solar energy by microalgae.

First reports on closed PBR systems for mass cultivation of algae came in the mid-1980s (Lee, 1986) and were followed by several articles on PBR design in the late 1980s. Weissman et al. (1988) compared an open raceway pond to a horizontal glass tubular reactor. Miyamoto et al. (1988) presented a vertical tubular reactor with an inner diameter of 5 cm and a length of 2.3 m, which holds about 4 L of culture medium.

By the late 1970s and early 1980s, after almost 40 years since Gaffron and Rubin first demonstrated the production of hydrogen by *Chlamydomonas* (Gaffron and Rubin, 1942), photosynthetic hydrogen production received some attention, and several articles, both research (Neil et al., 1976; Miyamoto et al., 1979) and review (Mitsui, 1976; Weaver et al., 1979), on the production of hydrogen using algae, were published. Stuart and Gaffron (1972) also published an article on hydrogen production by several strains of microalgae (*Chlamydomonas, Scenedesmus, Chlorella,* and *Ankistrodesmus*).

The next section discusses in detail microalgal biofuels—biodiesel, bioethanol, biogas, and the latest trend in sustainable biofuels, biohydrogen.

3.1 Biodiesel

The idea to use transportation fuels from renewable sources came from the early thoughts of Rudolph Diesel (Patil et al., 2008). According to Diesel, he stated that during his university years, at a thermodynamics class, he had the very first seeds of his idea after learning about the Carnot theorem (Diesel, 1913; Knothe, 2010). At the 1900 World Fair in France, the first diesel engine was exhibited by a French Company named Otto; its engine was designed by Rudolf Diesel. Diesel used peanut oil to run the engine and it worked! (Knothe, 2005).

Apart from the first and second generation biodiesel sources, microalgal lipids were also evaluated as a promising and sustainable biofuel source after the 1970s oil crisis and *food vs fuel* debate. Developments in microalgal biodiesel production have taken two different routes: (1) screening of suitable strains with high oil-producing properties considering nutrition as a key, and (2) screening the metabolic basis of microalgae lipid synthesis metabolism for strain improvement (Wang et al., 2014).

Heterotrophic production from *Chlorella pyrenidosa* was achieved in 1977. Before this study, the ability of *Chlamydomonas* to live in dark conditions was also known (Sager, 1955). The results pointing out lipid accumulation motivated further studies. With heterotrophic production techniques algae could be used in conventional fermenters (Endo et al., 1977; Xu et al., 2006; Yang et al., 2000).

From the 1940s to today, the history of biodiesel as one of the most promising commercial microalgal biofuels (Ackman et al., 1968; Benemann et al., 1978; Aaronson, 1973; Banerjee et al., 2002; Borowitzka, 1988; Brown et al., 1996; Certik and Shimizu, 1999; Grossman, 2005; Hu et al., 2008) has shown the following: (1) nutrient stress is a critical strategy to enhance the lipid storage as TGA (Borowitzka, 2013); (2) screening of potential species for microalgal biodiesel production can be promising; however, the lipid content of microalgae is more or less the same and further improvements should be required (Sheehan et al., 1998a, 1998b); (3) not only is the lipid content an important parameter but also biomass productivity is critical to develop feasible production (Hu et al., 2008); (4) the cell structure of microalgae is the barrier for extraction, thus an efficient extraction technology and downstream process requires further research (Chisti, 2007); and (5) for a reliable process, genetic engineering tools could be necessary (Radakovits et al., 2010).

3.2 Bioethanol

Bioethanol from renewable sources began to be discussed as a reliable transportation fuel after concerns were raised about global warming and an oil crisis (Bai et al., 2008). The raw materials for bioethanol production are based on crops mostly of sugarcane and corn (Oncel, 2013). Cellulosic materials have also been introduced to the technology (Sun and Cheng, 2002). Apart from microalgal bioethanol, plant material–derived bioethanol was introduced in to the transportation sector in the late nineteenth century (Solomon et al., 2007). Henry Ford launched a new car that worked with pure bioethanol or blends with gasoline (Mussatto et al., 2010). The country of Brazil developed the ProAlcool project in 1975 (Soccol et al., 2005) and currently leads the world in ethanol export potential, even compared to the United States, Japan, and Europe. Brasil also launched another project, OVEG, in 1983 aimed at biodiesel production from vegetable oils. The success of the bioethanol program was a solution to the 1970s fuel crisis; however, it introduced another crisis related to competition with food sources (Soccol et al., 2005; Mussatto et al., 2010).

In theory, microalgae are a reliable feedstock that does not compete with food sources, besides accumulating a higher content of carbohydrates in the form of starch (Chisti, 2008) However, it is not realistic to cultivate algae to use as substrate for ethanol production. Approaches to increase the carbohydrate content in microalgae for ethanol fermentation are another point to discuss (Hirano et al., 1997); however, rather than combining bulk cultivation with ethanol fermentation, bio-refinery concept is more

applicable. Oil extracted microalgae waste biomass is one of the promising sources for this purpose (Oncel, 2013).

3.3 Biogas

The first laboratory study of anaerobic digestion using algae biomass as a substrate was done by Golueka using *Scenesmus* sp. and *Chlorella vulgaris* for waste treatment processes (Golueka and Oswald, 1959). In later studies, open ponds were used to treat wastewater with microalgae, and microalgal biomass has been used for anaerobic digestion (Oswald and Golueka, 1960; Golueka and Oswald, 1965). This work can be considered as an important milestone for the development of algal studies in terms of integrating bioprocesses and also for using the mass cultivation concept. Anaerobic digestion of microalgae was more like a literature knowledge until the 1970 fuel criris. However similar to other alternative fuels, biogas began to be considered again as a possibility, with a special emphasis on the biorefinery concept (Ward et al., 2014).

In the case of biogas, microalgal biomass is itself the main substrate (Samson and LeDuy, 1982). Thus the cellular composition of biomass directly affects the yield of anaerobic digestion (Droop, 1983; Sialve et al., 2009). The complex structure of algae made it hard to estimate substrate ratios for digestion, so basic equations were adapted from conventional strategies (Symon and Buswells, 1933; Harris and Adams, 1979; Sialve et al., 2009).

Currently, rather than being a separate process, biogas production from microalgae is considered as a step in the biorefinery concept (Mussgnug et al., 2010; Oncel, 2013). The disrupted and oil-extracted waste biomass still contains high concentrations of proteins, carbohydrates, and other compounds. Utilization of waste for biogas production could be a more feasible pathway for the waste treatment policy, too. Until now, diverse research on microalgal biogas production has been conducted. Commercially important strains such as *Spirulina* sp., *Chlorella* sp., *Tetraselmis*, *Scenesmus*, *Dunaliella salina*, *Arthrospira* and their biogas production capacities have been investigated. A brief description of these results appears in the latest review related to anaerobic digestion of microalgal biomass (Ward et al., 2014).

3.4 Biohydrogen

Biohydrogen, one of the latest trends in algal biofuel studies, appears first in the literature in 1942 (Gaffron and Rubin, 1942). Today the known metabolism of algal hydrogen production is activation of certain enzymes called as hydrogenases under anaerobic conditions (Meyer, 2007). Algal biohydrogen production has been evaluated regarding the following aspects: (1) the sensitivity of hydrogenase enzymes to oxygen; (2) the relation of inhibitors with hydrogen production mechanism; (3) microalgae species-specific hydrogen production and screening of certain strains as hydrogen-producing machineries; (4) genetic basis of hydrogen production both in terms of enzymes and mechanistic models; (5) regulation of cell metabolism to shift their storage metabolism; (6) the effect of nutrition to PSII system; (7) evaluation of biophotolysis; (8) genetic engineering tools for an efficient production; (9) commercialization and scale-up; (10) immobilization for sustainable biohydrogen production; and (11) evolutionary development of hydrogen-producing machinery, which were highlighted in a series of comprehensive reports (Ghirardi et al., 1997; Wykoff et al., 1998; Melis et al., 2000; Kosourov et al., 2002; Forestier et al., 2003; Posewitz et al., 2004; Ghirardi et al., 2005; Kruse et al., 2005; Meyer, 2007; Laurinavichene et al., 2008; Kosourov and Seibert, 2009; Matthew et al., 2009; Faraloni and Torzillo, 2010; Gaffron, 1939; Kruse and Hankamer, 2010; Oh et al., 2011; Srirangan et al., 2011; Meuser et al., 2012; Oncel and Sabankay, 2012; Scoma et al., 2012; Oncel, 2013; Torzillo and Seibert, 2013; Oncel and Kose, 2014).

The restrictive nature of microalgal biohydrogen production requires anaerobic environments. The history of anaerobiosis aiming to produce microalgal biohydrogen is achieved via using inert gasses to purge oxygen, using PSII inhibitors to lock down the photo-damage repair system; the addition of O_2 scavengers is also another approach (Torzillo and Seibert, 2013). However, even to sustain a laboratory study, these procedures are impractical, expensive, and have no chance for scaling up. In 1998, Wykoff presented the sulfur-deprived PSII blocking, and later Melis et al. introduced a two-stage protocol separating aerobic phases and anaerobic phases with sulfur-deprived conditions (Melis et al., 2000).

Photobiological hydrogen production is the direct product of metabolic activation, which has encouraged further developments. The rise in the cost of energy, along with the carbon balance-related crisis, has been a driving force in the need for sustainable green energy, and thus the introduction of a clean fuel aiming for a clean future. Besides renewable energy features, combustion characteristics are also another strong point to encourage further studies (e.g., by an on-site fuel cell system linked to the electricity grid) (Oncel and Vardar, 2011; Oncel, 2013). Biotechnology of algal biohydrogen is a young branch of research that requires further interest. After a two-stage protocol, environmental conditions, nutrient regimes, possible algae strains with high biohydrogen production capacity were screened. However, still none of them achieved the introduction of an oxygen tolerant process. After 2000 the interest on biohydrogen production shifted to genetic engineering (Posewitz et al., 2004; Kruse et al., 2005; Beer et al., 2009; Srirangan et al., 2011; Kose and Oncel, 2014; Mathews and Wang, 2009) and since 2000 transcriptome, metabolome, system biology, and omics became the new areas of focus to study microalgal biohydrogen production (Rupprecht, 2009).

4. GENETIC ENGINEERING TOOLS TO DESIGN CONCEPTUAL STUDIES: TIME TRAVEL FOR GENETIC EVOLUTION OF MICROALGAL BIOFUEL RESEARCH

There are over 40,000 microalgae species known in current databases (Safi et al., 2014); however, around 100 species are attractive for cyanobacteria, green algae, and diatom studies (Pulz and Gross, 2004). Genetic engineering and system biology can be a new-generation research tool for further studies to make algal biofuels feasible (Rupprecht, 2009) but also as simulations for outdoor or large-scale productions, which are another possibility for fuel efficiency.

Currently most of the studies are conducted with the aid of molecular genetics tools. "Strain improvement" as a known concept has also been introduced into algal research, in addition to screening new promising strains. Microalgae with three genomes (chloroplast, nucleus, and mitochondria) can be seen as a pool in terms of genetic material (Walker et al., 2005; Neupert et al., 2008). However, today there is not adequate knowledge about algal genomics to go beyond this point (Radakovits et al., 2010). *Chlamydomonas* is used as a model for most of the algal genetic studies, providing good feedback; it has been suggested that future studies on algal genomes should be accelerated (Merchant et al., 2007).

The life cycle of *Chlamydomonas* was described back in 1876. Since then, *Chlamydomonas* was used as model microalga because of its high growth rates and ability to grow in dark conditons (heterotrophic metabolism), and lately it is the most studied genome representative for algal studies (Sager, 1955). Mutation and transgene studies unfortunately took a long time to be able to present a permanent transformant, but in 1988 the first stable chloroplast transformation was introduced (Radakovits et al., 2010).

In the case of biodiesel enhanced lipid accumulation via overexpression of lipid synthesis in microalgae is the desired target to reach. Because the downstream processing is expensive, approaches must be developed to reduce the process costs and also enhance the yield. With this in mind, in 1994 and 1996 the successful plasmid transformation with diatoms *Cyclotella cryptica* and *Navicula saprophila* was achieved in terms of enhancing lipid synthesis. With these studies, the role of ACCase enzyme was also clarified and lipid accumulation was enhanced in mutant strains (Dunahay et al., 1996). Blocking metabolic pathways, overexpression of related gene products or enzymes, altering lipid profiles, regulation of lipid catabolism, and inhibition of lipid oxidation have been studied. Other promising approaches for feasible biofuel production are also considered to be in situ biofuel production and secretion of lipid bodies as extracellualar compounds (Rosenberg et al., 2008).

Another crucial biofuel, biohydrogen, is also the focus of genetic studies (Posewitz et al., 2005). The biohydrogen metabolism within the cell is composed of complex reactions, and according to the previous studies, each approach in molecular metabolism gives positive results to upgrade biohydrogen studies. The following studies have been done: starch metabolism (Kruse et al., 2005; Esquível et al., 2006; Chochois et al., 2009); introduction of foreigner genes to enhance carbon utilization (Doebbe et al., 2007); role of electron transport chain and electron pools, the role of D1 protein (Faraloni and Torzillo, 2010); the effect of light-harvesting antenna size and the photosynthetic light utilization efficiency (Tetali et al., 2007); the role of Rubisco in the accumulation of electron pool (Hemschemeier et al., 2008); the molecular evolution of hydrogenase enzymes (Meyer, 2007; Kim and Kim, 2011; Oh et al., 2011); and the oxygen sensitivity of hydrogenase enzymes (Ghirardi et al., 2006; Nagy et al., 2007; Boyd et al., 2009; Meyer, 2007).

Using genetic engineering tools, researchers can now search and finalize the genome sequences of other microalgae that have potential for biofuels in addition to the completed genomes of *Nannochloropsis gaditana, Chlamydomonas reinhardtii, Chlorella variabilis, Micromonas pusila, Ostreococcus tauri, Ostreococcus lucimarinus, Thalassiosira psudonana, Emiliana huxleyii, Fragilariopsis cylindrus, Aerococcus aophagefferens, Cyanidioschyzon merolae,* and *Phaeodactylum tricornutum* (Oncel, 2013). Molecular-based focusing on the microalgae will accelerate the progress for future studies of biofuel production with enhanced productivities targeting commercial-scale applications.

5. GLOBAL PROJECTS

The era of microalgal biofuels started with the recognition of the risky balance of fossil fuel supply and demand forcing an economic imbalance. Even though the potential of microalgae was mentioned by various outstanding researchers decades ago, awareness of the risk has accelerated for further investments on reliable and green futures for society. Thus, various projects have been conducted to find potential sources, production technologies, and commercialization of microalgal biofuels.

5.1 Aquatic Species Program

Aquatic Species Program (ASP) is one of the glorious algal biofuel projects that was launched in 1978 by the US Department of Energy's Office of Fuels Development to develop renewable transportation fuels from algae (Borowitzka, 2013). The program first focused on biohydrogen production (until 1982), then shifted to screening potential species for biodiesel production from microalgae; the program ended in 1996 because of budget cuts (overall funding, $25 million) (Pienkos and Darzins, 2009). The aim of the project was not only to find microalgae species but also to define production strategies, including cultivation systems such as open ponds and PBRs. During ASP, 3000 strains were isolated

from saline, brackish, and freshwater sources. The strains were screened according to their environmental needs and lipid accumulation capacity. Selection of 300 strains, mostly green algae and diatoms, which were promising in terms of biodiesel production for commercial scales, increased scientific curiosity worldwide (Sheehan et al., 1998a, 1998b).

The 18 years of study also accomplished the evaluation of microalgal biofuels, mainly biodiesel. A final report suggested the following (Sheehan et al., 1998a, 1998b): (1) oil accumulation in the algal cells attained with nitrogen starvation does not increase the total oil productivity; (2) there is little prospect for any alternative for a more productive culture system to achieve low cost, for example, closed PBRs as an alternative to open ponds; (3) difficulties in maintaining axenic cultures for more than 2 weeks or months; (4) even with aggressive assumptions about biological productivity, the prospective costs for biodiesel are still higher than fossil fuels. However, the report also suggested that genetic engineering tools for algal biofuels are a trusted asset for further development, strain improvement, strain resistance, and overall yield. Besides being able to compete with fossil fuels or at least obtain an efficient commercialization, biorefinery concept with zero waste strategy was also another aspect during the project. However, the budget cut and finalization of the project is a demotivational result for the progress of algal biofuels, the only competitive fuel source from biomass.

5.2 The RITE

In 1990, the Japanese Ministry of International Trade and Industry launched an innovative R&D program including projects at the Research Institute of Innovative Technology for Earth (RITE) to develop technologies for biological fixation of CO_2 via photosynthesis (Michiki, 1995; Borowitzka, 2013). Although the project was not concerned with biofuel production, it contributed a great deal to the development of large-scale microalgae production. The major research themes addressed by the RITE projects were: (1) highly efficient photosynthetic bacteria and microalgae with high CO_2 fixation capability, which identified 13 strains including *Chlorococcum* sp., *Galdieria* sp., *Chlorella* sp., *Synechococcus* sp., and *Scedesmus* sp.; (2) development of devices and technologies to utilize sunlight with maximum efficiency, which is the major bottleneck to developing sustainable production; (3) development of large-scale PBRs for CO_2 fixation using dense microalgal culture; and (4) development of a technology to produce biomass, oil for biodiesel, proteins, and other valuable compounds (Usui and Ikenouchi, 1997).

5.3 Carbon Trust

Carbon Trust was also one of the leading projects to develop a new carbon economy in the scope of algal biofuels for sustainable future. The project was launched in 2001 by the government of the United Kingdom, with £30 million of investment. The project mostly concentrated on screening potential strains for biodiesel production, enhancing solar energy conversion, designing techno-economic production systems for an algal fuel process facility, and designing open ponds for mass cultivation. The project also acted as a funding pool for algal investments. But in 2011 the government cut 40% of the budget, resulting in job losses and also loss in trust to algal biofuels is damaged in public eye (http://www.theguardian.com/environment/2011/feb/14/carbon-trust-funding-cut).

5.4 BioMARA (The Sustainable Fuels from Marine Biomass)

The aim of the BioMARA project was to find alternative marine resources for biofuel production for commercialization. The project was launched in April 2009 by the Scottish Association for Marine Science, supported by partner organizations from Scotland, Northern Ireland, and the Republic of Ireland (http://www.biomara.org). Project objectives were to define high oil-producing microalgae strains, define genes responsible for oil synthesis, marine seaweeds and marine biomass for anaerobic digestion, and design techno-economical and feasible PBRs for microalgae cultivation to produce bioethanol and biogas. Total budget of the project was 6 million € for 7 years. So far, the research team has screened 2700 microalgae strains from culture collections in order to define promising strains for lipid production. The team has published a few papers, mostly dealing with climate change and global warming risks, alternative species for lipid production and strain screening, strain improvement, reviews about biogas production, techno-economic development , and research on biogas production from seaweeds (Allan, 2011; Callaway et al., 2012; Day et al., 2012; Hughes et al., 2012; Roleda et al., 2013; Dave et al., 2013; Slocombe et al., 2013a, 2013b; Vanegas and Bartlett, 2013).

5.5 OMEGA (Offshore Membrane Enclosures for Growing Algae)

OMEGA was established by NASA between 2009 and 2012. The system consisted of light-penetrating plastic bag PBRs with specially designed osmosis membranes (Hughes et al., 2014). Algae cultures within the bags floating on the water surface used sunlight and nutrients in water to release clean water and oxygen back into the water environment. In other studies, lipids extracted from harvested biomass were used as feedstock for biodiesel and waste biomass was also used as fertilizers. The prototype studies started with 1−2 L of PBRs and scaled up to 110, 1600 L; respectively. The very first findings reported 14.1 ± 1.3 g dry biomass per square meter of PBR surface area per day. The project was not a leading one in terms of innovative algae cultivation, but it established the usefulness of various equipment and apparatus to measure temperature, photosynthetic active radiation, and oxygen–carbon dioxide levels. In addition, the ecological aspects of the floating PBR on the sea surface were also discussed and evaluated (Trent, 2012; Hughes et al., 2014).

6. INTRODUCING A FEASIBLE APPROACH TO ALGAL BIOFUEL ECONOMY: CURRENT STATUS FOCUSING ON PHOTOBIOREACTORS

The strategy to decrease the cost of production of algal biomass is complex. One of the main issues is the strain selection, which should be based on the following aspects: (1) strains with high productivity under high light (i.e., not affected by light saturation and photo inhibition); (2) large cell with thin membrane to facilitate cell rupture; (3) insensitivity to high oxygen concentration; (4) resistance to shear stress; (5) cell must flocculate easily; (6) robust cells resistant to grazing; and (7) the oil should be excreted outside cells (hydrogen is an example) (Wijffels and Barbosa, 2010). Another key issue is the optimization of the PBR design. As a guideline, an optimal PBR design should have the following characteristics:

- From open ponds to sophisticated PBR designs with various modifications, mostly focused on light distribution to PBR surface and light utilization from microalgal cultures. Unfortunately, lack of a well-defined and optimized strategy also blocks the pathway for fuel production. Considering the sun as a free and sustainable light source, natural sunlight

utilization for illumination should be accomplished with increasing the illumination area of PBR, considering light dilution as a challenge.
- Construction materials for PBRs should be designed to allow light penetration from PBR surface into the culture broth, where microalgae harvest light. However, light/dark zones are the main problems to achieving desired light utilization. Culture mobilized with the mixing within the PBR is supposed to transfer within the PBR visiting light and dark zones. It is advised to keep dark volume-to-total volume ratio below 0.05 (Torzillo and Seibert, 2013).
- Surface-to-volume ratio (S/V) is a critical value for an efficient scale-up strategy synchronized with mixing time (Oncel and Kose, 2014). A high S/V ratio reduces the light path and also is an important parameter in terms of volumetric productivity, considering light utilization. Mixing time also has various effects on microalgal metabolism, light distribution into PBR, light/dark cycle frequency, and heat and mass transfer properties affecting bulk biomass production. Also, it is stated that mixing time has a significant effect on H_2 production, too (Oncel and Sabankay, 2012). Reduction in the mixing time is the key to a better transfer. To reach effective mass transfer coefficient turbulent mixing is advised. To achieve a turbulent regime, Reynolds number is the criteria (Torzillo and Seibert, 2013).
- The utilization of PBRs in terms of orientation is another point. For outdoor cultivation the surface meets with illumination is crucial. One-sided illumination and two-sided illumination differ in terms of light penetration into the culture. In the case of one-sided illumination, inadequate mixing will result in a loss of culture and a decrease in the overall biomass accumulation. The geographical region is a parameter for the capture of solar irradiation. East/west

orientation panel and tubular PBRs intercept more light than south—north orientation (Sierra et al., 2008; Torzillo and Seibert, 2013).

7. CONCLUSIONS

Efforts to produce biofuels from microalgae are still worthy, yet investors should realize that microalgal technology is not ready and may take another 10—15 years to become economically viable. Developing technologies that are able to reduce the cost of production are important not only for biofuel applications but also for other purposes, such as food and feed, which should also be considered by investors (Zittelli et al., 2013).

Results of research over 60 years show that to develop commercial algal biofuel production, the first thing to do must be to redirect studies over challenges and also screen new strains that may be promising for algal biofuels. Today the number of strains with identified genomes has increased (Radakovits et al., 2010). The genome, transcriptome, and metabolome studies combining system biology tools are another concept under development (Nguyen et al., 2008; Matthew et al., 2009; Rupprecht, 2009).

Compared to when microalgal biotechnology was first introduced, encouragement of the researchers kept alive the aim of finding a suitable supporter to fossil-derived fuel sources. With the know-how and a deep evaluation, one can claim that the microalgal biotechnology will go further and find a feasible platform to penetrate into global energy market. It can be also stated that the previous studies are a guide to go further.

Keeping in mind the immense gap in the investment for the research and development compared to the multibillion dollars for the fossil fuel industry, microalgal researchers were trying to take stronger steps with limited financial sources.

We can't wait for "Scotty to beam us up" like in the popular series of "Star Trek", to a greener future in a few seconds but we have to work and change our destiny by using lessons from the past. Knowing that with today's technology and knowledge, microalgal biofuels may have a long way to go but it is not impossible to reach the desired target of a sustainable and greener future.

References

Aaronson, S., 1973. Effect of incubation temperature on the macromolecular and lipid content of the phytoflagellate *Ochromonas danica*. J. Phycol. 9, 111—113.

Abdulqader, G., Barsanti, L., Tredici, M.R., 2000. Harvest of *Artrosphira platensis* from Lake Kossorom (Chad) and its household usage among the Nanembu. J. Appl. Phycol. 12, 493—498.

Ackman, R.G., Tocher, C.S., McLachlan, J., 1968. Marine phytoplankter fatty acids. J. Fish. Res. Board Can. 25, 1603—1620.

Allan, G.J., 2011. The Regional Economic Impacts of Biofuels: A Review of Multisectoral Modelling Techniques and Evaluation of Applications. No 2011-74, SIRE Discussion Papers. Scottish Institute for Research in Economics (SIRE).

Allen, E.J., Nelson, E.W., 1910. On the artificial culture of marine plankton organisms. J. Mar. Biol. Assoc. U.K. 8 (05), 421—474.

Andersen, R., 2005. Algal Culturing Techniques. Academic Press.

Badin, E.J., Calvin, M., 1950. The Path of Carbon in Photosynthesis IX. Photosynthesis, Photoreduction, and the Hydrogen-oxygen-carbon Dioxide Dark Reaction. UCRL-591.

Bai, F.W., Anderson, W.A., Moo-Young, M., 2008. Ethanol fermentation technologies from sugar and starch feedstocks. Biotechnol. Adv. 26, 89—105.

Banerjee, A., Sharma, R., Chisti, Y., Banerjee, U.C., 2002. *Botryococcus braunii*: a renewable source of hydrocarbons and other chemicals. Crit. Rev. Biotechnol. 22, 245—279.

Barsanti, L., Gualtieri, P., 2006. Algae: Anatomy, Biochemistry, and Biotechnology. CRC Press Taylor & Francis Group, pp. 251—256.

Bassham, J.A., 2005. Mapping the carbon reduction cycle: a personal retrospective. Discov. Photosynth. 815—832.

Becker, E.W., 1994. Microalgae: Biotechnology and Microbiology, Cambridge Studies in Biotechnology, vol. 10. Cambridge University Press Cambridge, p. 293.

Becker, E.W., 2007. Micro-algae as a source of protein. Biotechnol. Adv. 25, 207.

Becker, E.W., 2013. Microalgae for Human and Animal Nutrition. Handbook of Microalgal Culture: Applied Phycology and Biotechnology, second ed. Wiley, pp. 461—503.

Beer, L.L., Boyd, E.S., Peters, J.W., Posewitz, M.C., 2009. Engineering alga for biohydrogen and biofuel production. Curr. Opin. Biotechnol. 20, 264–271.

Beijerinck, M.W., 1890. Kulturversuche mit Zoochloren, Lichenengonidien und anderen niederen Algen. Bot. Z. 48, 725–785.

Belay, A., 2013. Biology and Industrial Production of *Arthrospira* (*Spirulina*). Handbook of Microalgal Culture: Applied Phycology and Biotechnology, second ed. Wiley, pp. 339–358.

Benemann, J.R., 1997. Feasibility analysis of photobiological hydrogen production. Int. J. Hydrogen Energy 22 (10/11), 979–987.

Benemann, J.R., Pursoff, P., Oswald, W.J., 1978. Engineering Design and Cost Analysis of a Large-Scale Microalgae Biomass System. Final Report to the US Department of Energy. NTIS# HCP/T1605-01 UC-61, pp. 1–91.

Benson, A.A., 2002. Following the path of carbon in photosynthesis: a personal story. Photosynth. Res. 73 (1–3), 31–49.

Benson, A.A., Bassham, J.A., Calvin, M., Goodale, T.C., Haas, V.A., Stepka, W., 1949. The Path of Carbon in Photosynthesis V. Paper Chromatography and Radioautography of the Products. Report Number AECU-356, UCRL-363.

Benson, A.A., Calvin, M., 1948. The Path of Carbon in Photosynthesis III. UCRL-133.

Bewicke, D., Potter, B.A., 1993. Chlorella: The Emerald Food. Ronin Publishing.

Blankenship, R.E., 2010. Early evolution of photosynthesis. Plant Physiol. 154, 434–438.

Borowitzka, A.M., 2013. Energy from microalgae: a short history. Algae Biofuels Energy 5, 1–15.

Borowitzka, M., 1988. Fats, oils and hydrocarbons. In: Borowitzka, M.A., Borowitzka, L.J. (Eds.), Microalgal Biotechnology. Cambridge University Press, Cambridge, UK, pp. 257–287.

Borowitzka, M., 1997. Microalgae for aquaculture: opportunities and constraints. J. Appl. Phycol. 9, 393–401.

Borowitzka, M.A., Moheimani, N.R., 2013. Algae for Biofuels and Energy. Springer, Netherlands.

Boyd, E.S., Spear, R.J., Peters, J.W., 2009. [FeFe] hydrogenase genetic diversity provides insight into molecular adaptation in a saline microbial mat community. Appl. Environ. Microbiol. 75 (13), 4620–4623.

Brown, M.R., Dunstan, G.A., Norwood, S.J., Miller, K.A., 1996. Effects of harvest stage and light on the biochemical composition of the diatom *Thalassiosira pseudonana*. J. Phycol. 32, 64–73.

Buisson, A., Van Landeghem, H., Rebeller, M., Trambouze, P., 1969. Process for the Culture of Algae and Apparatus Therefor: Google Patents.

Burlew, J.S., 1953. Algal Culture: From Laboratory to Pilot Plant. Carnegie Institution of Washington, Washington, DC, pp. 1–357.

Burris, J.E., 1977. Photosynthesis, photorespiration, and dark respiration in eight species of algae. Mar. Biol. 39, 371–379.

Callaway, R., Shinn, A.P., Grenfell, S.E., Bron, J.E., Burnell, G., Cook, E.J., et al., 2012. Review of climate change impacts on marine aquaculture in the UK and Ireland. Aquat. Conserv. 22 (3), 389–421.

Calvin, M., 1949. The Path of Carbon in Photosynthesis VI. Report Number AECU-433, UCRL-387.

Calvin, M., Bassham, J.A., Benson, A.A., Kawaguchi, S., Lynch, V.H., Stepka, W., Tolbert, N.E., 1951. The Path of Carbon in Photosynthesis XIV. UCRL-1386.

Calvin, M., Bassham, J.A., Benson, A.A., Lynch, V., Ouellet, C., Schou, L., Stepka, W., Tolbert, N.E., 1950. The Path of Carbon in Photosynthesis X. Carbon Dioxide Assimilation in Plants. UCRL-658.

Calvin, M., Benson, A.A., 1948. The Path of Carbon in Photosynthesis. AECD-1912; UCRL-65.

Carvalho, A.P., Meireles, L.A., Malcata, F.X., 2006. Microalgal reactors: a review of enclosed system designs and performances. Biotechnol. Prog. 22, 1490–1506.

Certik, M., Shimizu, S., 1999. Biosynthesis and regulation of microbial polyunsaturated fatty acid production. J. Biosci. Bioeng. 87, 1–14.

Chaumont, D., 1993. Biotechnology of algal biomass production: a review of systems for outdoor mass culture. J. Appl. Phycol. 5, 593–604.

Chaumont, D., Thepenier, C., Gudin, C., Junjas, C., 1988. Scaling up a tubular photoreactor for continuous culture of *Porphyridium cruentum* from laboratory to pilot plant (1981–1987). In: Stadler, T., Mollion, J., Verdus, M.C., Karamanos, Y., Morvan, H., Christiaen, D. (Eds.), Algal Biotechnology. Elsevier Applied Science, London, pp. 199–208.

Chen, F., 1996. High cell density culture of microalgae in heterotrophic growth. Trends Biotechnol. 14, 412–426.

Chisti, Y., 2007. Biodiesel from microalgae. Biotechnol. Adv. 25, 294–306.

Chisti, Y., 2008. Biodiesel from microalgae beats bioethanol. Trends Biotechnol. 26 (3), 126–131.

Chochois, V., Dauvillée, D., Beyly, A., Tolleter, D., Cuiné, S., Timpano, H., Ball, S., Cournac, L., Peltier, G., 2009. Hydrogen production in *Chlamydomonas*: photosystem II-dependent and -independent pathways differ in their requirement for starch metabolism. Plant Physiol. 151 (2), 631–640.

Chu, S., 1942. The influence of the mineral composition of the medium on the growth of planktonic algae: part I. Methods and culture media. J. Ecol. 30, 284–325.

Cifferi, O., 1983. *Spirulina*, the edible organism. Microbiol. Rev. 47, 551.

Costa, J.A.V., Colla, L.M., Duarte, P., 2003. *Spirulina platensis* growth in open raceway ponds using fresh water supplemented with carbon, nitrogen and metal ions. Z. Naturforschung C-A J. Biosci 58, 76–80.

Dave, A., Huang, Y., Rezvani, S., McIlveen-Wright, D., Novaes, M., Hewitt, N., 2013. Techno-economic assessment of biofuel development by anaerobic digestion of European marine cold-water seaweeds. Bioresour. Technol. 135, 120–127.

Day, J.G., Slocombe, S.P., Stanley, M.S., 2012. Overcoming biological constraints to enable the exploitation of microalgae for biofuels. Bioresour. Technol. 109, 245–251.

Diesel, E.R., 1913. Die Entstehung des Dieselmotors. Springer Berlin Heidelberg, pp. 267–367.

Doebbe, A., Rupprecht, J., Beckmann, J., Mussgnug, J.H., Hallmann, A., Hankamer, B., 2007. Functional integration of the HUP1 hexose symporter gene into the genome of *C.reinhardtii* impacts on biological H$_2$ production. J. Biotechnol. 131, 27–33.

Droop, M.R., 1983. 25 years of algal growth kinetics. A personal view. Bot. Mar. 26, 99–112.

Dunahay, T.G., Jarvis, E.E., Dais, S.S., Roessler, P.G., 1996. Manipulation of microalgal lipid production using genetic engineering. In: Seventeenth Symposium on Biotechnology for Fuels and Chemicals. Humana Press, pp. 223–231.

Endo, H., Hosoya, H., Koibuchi, T., 1977. Growth yield of *Chlorella regularis* in dark-heterotrophic continuous cultures using acetate. J. Fermentat. Technol. 55, 369–379.

Esquível, M.G., Pinto, T.S., Marín-Navarro, J., Moreno, J., 2006. Substitution of tyrosine residues at the aromatic cluster around the ßA-ßB loop of rubisco small subunit affects the structural stability of the enzyme and the in vivo degradation under stress conditions. Biochemistry 45, 5745–5753.

Faraloni, C., Torzillo, G., 2010. Phenotypic characterization and hydrogen production in *Chlamydomonas reinhardtii* QB-binding D1-protein mutants under sulfur starvation: changes in Chl fluorescence and pigment composition. J. Phycol. 46, 788–799.

Fernandez, F.G.A., Camacho, F.G., Perez, J.A.S., Sevilla, J.M.F., Grima, E.M., 1997. Modelling of biomass productivity in tubular photobioreactors for microalgal cultures: effects of dilution rate, tube diameter, and solar irradiance. Chem. Eng. Sci. 58 (6), 605–616.

Forestier, M., King, P., Zhang, L., Posewitz, M., Schwarzer, S., Happe, T., 2003. Expression of two [Fe]-hydrogenases in *Chlamydomonas reinhardtii* under anaerobic conditions. Eur. J. Biochem. 270, 2750–2758.

Fox, R.D., 1996. Spirulina: Production and Potential. Edisud, Aix-en-Provence, France, p. 232.

Gaffron, H., 1939. Reduction of CO$_2$ with H$_2$ in green plants. Nature 143, 204–205.

Gaffron, H., Rubin, J., 1942. Fermentative and photochemical production of hydrogen in algae. J. Gen. Physiol. 26, 219–240.

Gest, H., 1997. A "misplaced chapter" in the history of photosynthesis research; the second publication (1796) on plant processes by Dr Jan Ingen-Housz, MD, discoverer of photosynthesis. Photosynth. Res. 53, 65–72.

Gest, H., 2002. History of the word photosynthesis and evolution of its definition. Photosynth. Res. 73, 7–10.

Gest, H., 2005. Samuel Ruben's contribution to research on photosynthesis and bacterial metabolism with radioactive carbon. Photosynth. Res. 80, 77–83.

Ghirardi, M.L., Togasaki, R.K., Seibert, M., 1997. Oxygen sensitivity of algal H$_2$ production. In: Appl. Biochem. Biotechnol. Proceedings of the Eighteenth Symposium on Biotechnology for Fuels and Chemicals, 5–9 May 1996, Gatlinburg, TN, vol. 63–65, pp. 141–151. NREL Report No. TP-450-21371.

Ghirardi, M.L., King, P.W., Posewtiz, M.C., Maness, P.C., Fedoro, A., Kim, K., Cohen, J., Schulten, K., Seibert, M., 2005. Approaches to developing biological H2-photoproducing organisms and processes. Biochem. Soc. Transact. 33, 70–72.

Ghirardi, M.L., Cohen, J., King, P., Schulten, K., Kim, K., Seibert, M., 2006. [FeFe]- hydrogenases and photobiological hydrogen production. SPIE 6340, U257–U262.

Goldman, J.C., 1979. Outdoor algal mass cultures—I. Appl. Water Res. 13 (1), 1–19.

Goldman, J.C., Porcella, D.B., Middlebrooks, J.E., Toerien, D.F., 1971. The effect of carbon on algal growth—its relationship to eutrophication. Water Res. 6, 637–679.

Goldman, J.C., Ryther, J.H., 1975. Mass production of marine algae in outdoor cultures. Nature 254, 594–595.

Goldman, J.C., Ryther, J.H., 1977. Mass production of algae—bioengineering aspects. Biol. Sol. Energy Convers. 367–378.

Golueke, C.G., Oswald, W.J., 1959. Biological conversion of light energy to the chemical energy of methane. Appl. Microbiol. 7 (4), 219–227.

Golueke, C.G., Oswald, W.J., 1965. Harvesting and processing of sewage grown planktonic algae. J. Water Pollut. Control Fed. 37, 471–498.

Govindjee, Krogmann, D., 2004. Discoveries in oxygenic photosynthesis (1727–2003): a perspective. Photosynth. Res. 80, 15–57.

Grobbelaar, J.U., 2012. Microalgae mass culture: the constraints of scaling-up. J. Appl. Phycol. 24 (3), 315–318.

Grossman, A., 2005. Paths toward algal genomics. Plant Physiol. 137, 410–427.

Harris, R.F., Adams, S.S., 1979. Determination of the carbon-bound electron composition of microbial cells and metabolites by dichromate oxidation. Appl. Environ. Microbiol. 37, 237–243.

Hatch, M.D., Slack, C.R., 1966. Photosynthesis by sugar-cane leaves: a new carboxylation reaction and the pathway of sugar formation. Biochem. J. 101, 103–111.

Hemschemeier, A., Fouchard, S., Cournac, L., Peltier, G., Happe, T., 2008. Hydrogen production by *Chlamydomonas reinhardtii*: an elaborate interplay of electron sources and sinks. Planta 227 (2), 397–407.

Hirano, A., Ueda, R., Hirayama, S., Ogushi, Y., 1997. CO_2 fixation and ethanol production with microalgal photosynthesis and intracellular anaerobic fermentation. Energy 22, 137–142.

Hoppe, H.A., 1979. Marine algae and their products and constituents in pharmacy. In: Hoppe, H.A., Levring, T., Tanaka, Y. (Eds.), Marine Algae in Pharmaceutical Sciences. Walter de Gruyter, Berlin New York, pp. 25–119.

Hu, Q., Sommerfeld, M., Jarvis, E., Ghirardi, M., Posewitz, M., Seibert, M., Darzins, A., 2008. Microalgal triacylglycerols as feedstocks for biofuel production: perspectives and advances. Plant J. 54 (4), 621–639.

Hughes, A.D., Kelly, M.S., Black, K.D., Stanley, M.S., 2012. Biogas from macroalgae: is it time to revisit the idea. Biotechnol. Biofuels 5, 86.

Hughes, S.N., Tozzi, S., Harris, L., Hamsen, S., Young, C., Rask, J., Toy-Choutka, S., Clark, C., Cruickshank, M., Fennie, H., Kuo, J., Trent, J.D., 2014. Interactions of marine mammals and birds with offshore membrane enclosures for growing algae. Aquat. Biosyst. 10 (3), 1–13.

Husizige, H., Ke, B., 1993. Dynamics of the history of photosynthesis. Photosynth. Res. 38, 185–209.

Kaplan, D., Richmond, A.E., Dubinsky, Z., Aaronson, S., 1986. Algal Nutrition. Handbook for Microalgal Mass Culture. CRC Press, Boca Raton, FL, pp. 147–198.

Kawaguchi, K., 1980. Microalgae production system in Asia. In: Shelef, G., Soeder, C.J. (Eds.), Algae Biomass: Production and Use. Elsevier/North-Holland Biomedical Press, pp. 25–33.

Kim, D.H., Kim, M.S., 2011. Hydrogenases for biohydrogen production. Bioresour. Technol. 102, 8423–8431.

Knothe, G., 2005. The history of vegetable oil-based diesel fuels. In: Knothe, G., Van Gerpen, J., Krahl, J. (Eds.), The Biodiesel Handbook. AOCS Press, pp. 12–24.

Knothe, G., 2010. Biodiesel and renewable diesel: a comparison. Prog. Energy Combust. 36, 364–373.

Kortschak, H.P., Hartt, C.E., Burr, G.O., 1965. Carbon dioxide fixation in sugarcane leaves. Plant Physiol. 40, 209–213.

Kose, A., Oncel, S., 2014. Biohydrogen production from engineered microalgae *Chlamydomonas reinhardtii*. Adv. Energy Res. 2 (1), 1–9.

Kosourov, S., Tsygankov, A., Seibert, M., Ghirardi, M.L., 2002. Sustained hydrogen photoproduction by *Chlamydomonas reinhardtii*: effects of culture parameters. Biotechnol. Bioeng. 78 (7), 731–740.

Kosourov, S.N., Seibert, M., 2009. Hydrogen photoproduction by nutrient-deprived *Chlamydomonas reinhardtii* cells immobilized within thin alginate films under aerobic and anaerobic conditions. Biotechnol. Bioeng. 102, 50–58.

Kratz, A.W., Myers, J., 1955. Nutrition and growth of several blue-green algae. Am. J. Bot. 42 (3), 282–287.

Kruse, O., Hankamer, B., 2010. Microalgal hydrogen production. Curr. Opin. Biotechnol. 21, 238–243.

Kruse, O., Rupprecht, J., Bader, K.P., Thomas-Hall, S., Schenk, P.M., Finazzi, G., Hankamer, B., 2005. Improved photobiological H_2 production in engineered green algal cells. J. Biol. Chem. 280, 34170–34177.

Kützing, F.T., 1849. Species Algarum. FA Brockhaus. Leipzig, Germany.

Laurinavichene, T.V., Kosourov, S.N., Ghirardi, M.L., Seibert, M., Tsygankov, A.A., 2008. Prolongation of H_2 photoproduction by immobilized, sulfur-limited *Chlamydomonas reinhardtii* cultures. J. Biotechnol. 134, 275–277.

Lee, Y.K., 1986. Enclosed bioreactors for the mass cultivation of photosynthetic microorganisms: the future trend. Trends Biotechnol. 4 (7), 186–189.

Masahito, T., 1969. Process of Cultivating Algae: Google Patents.

Mathews, J., Wang, G., 2009. Metabolic pathway engineering for enhanced biohydrogen production. Int. J. Hydrogen Energy 34, 7404–7416.

Matthew, T., Zhou, W., Rupprecht, J., Lim, L., Thomas-Hall, S.R., Doebbe, A., Kruse, O., Hankamer, B., Marx, U.C., Smith, S.M., Schenk, P.M., 2009. The metabolome of *Chlamydomonas reinhardtii* following induction of anaerobic H_2 production by sulfur depletion. Curr. Opin. Biotechnol. 284, 23415–23425.

Mayer, A., Zuri, U., Shain, Y., Ginzburg, H., 1964. Problems of design and ecological considerations in mass culture of algae. Biotechnol. Bioeng. 6 (2), 173–190.

Melis, A., Zhang, L., Forestier, M., Ghirardi, M.L., Seibert, M., 2000. Sustained photobiological hydrogen gas production upon reversible inactivation of oxygen evolution in the green alga *Chlamydomonas reinhardtii*. J. Plant Physiol. 122, 127–135.

Merchant, S.S., Prochnik, S.E., Vallon, O., Harris, E.H., Karpowicz, S.J., Witman, G.B., Terry, A., Salamov, A., Fritz-Laylin, L.K., Marechal-Droudard, L., 2007. The *Chlamydomonas* genome reveals the evolution of key animal and plant functions. Science 318 (5848), 245–250.

Meuser, J.E., D'amo, S., Jinkerson, R.E., Mus, F., Yang, W., Ghirardi, M.L., Seibert, M., Grossman, A.R., Posewitz, M.C., 2012. Genetic disruption of both *Chlamydomonas reinhardtii* [FeFe] hydrogenases: insight into the role of HYDA2 in H_2 production. Biochem. Biophys. Res. Commun. 417, 704–709.

Meyer, J., 2007. [FeFe] hydrogenases and their evolution: a genomic perspective. J. Cell. Mol. Life Sci. 64, 1063–1084.

Michiki, H., 1995. Biological CO_2 fixation and utilization project. Energy Convers. Manage. 36, 701–705.

Minoru, S., Hiroshi, E., Kei, N., 1967a. Method of Industrial Cultivation of Unicellular Green Algae such as Chlorella: Google Patents.

Minoru, S., Hiroshi, E., Kei, N., 1967b. Method of New Industrial Cultivation of Unicellular Green Algae such as Chlorella: Google Patents.

Mitsui, A., 1976. Bioconversion of solar energy in salt water photosynthetic hydrogen production systems. In: Paper Presented at the 1st World Hydrogen Energy Conference, vol. 2.

Miyamoto, K., Hallenbeck, P.C., Benemann, J.R., 1979. Hydrogen production by the thermophilic alga *Mastigocladus laminosus*: effects of nitrogen, temperature, and inhibition of photosynthesis. Appl. Environ. Microbiol. 38 (3), 440–446.

Miyamoto, K., Wable, O., Benemann, J.R., 1988. Vertical tubular reactor for microalgae cultivation. Biotechnol. Lett. 10 (10), 703–708.

Mussatto, S.I., Dragone, G., Guimarães, P.M.R., Silva, J.P.A., Carneiro, L.M., Roberto, I.C., Vicente, A., Domingues, L., Teixeira, J.A., 2010. Technological trends, global market, and challenges of bio-ethanol production. Biotechnol. Adv. 28, 817–830.

Mussgnug, J.H., Klassen, V., Schlüter, A., Kruse, O., 2010. Microalgae as substrates for fermentative biogas production in a combined biorefinery concept. J. Biotechnol. 150, 51–56.

Myers, J., 1974. Conceptual developments in photosynthesis, 1924–1974. Plant Physiol. 54, 420–426.

Myers, J., Burr, G., 1940. Studies on photosynthesis: some effects of light of high intensity on *Chlorella*. J. Gen. Physiol. 24 (1), 45.

Myers, J., Clark, L., 1944. Culture conditions and the development of the photosynthetic mechanism: II. An apparatus for the continuous culture of *Chlorella*. J. Gen. Physiol. 28 (2), 103.

Nagy, L.E., Meuser, J.E., Plummer, S., Seibert, M., Ghirardi, M.L., King, P.W., Ahmann, D., Posewitz, M.C., 2007. Application of gene shuffling for the rapid generation of novel [FeFe]-hydrogenase libraries. Biotechnol. Lett. 29, 421–430.

Neil, G., Nicholas, D.J.D., Bockris, J.O.'M., McCann, J.F., 1976. The photosynthetic production of hydrogen. Int. J. Hydrogen Energy 1 (1), 45–48.

Neupert, J., Karcher, D., Bock, R., 2008. Generation of *Chlamydomonas* strains that efficiently express nuclear transgenes. Plant J. 57, 1140–1150.

Nguyen, A.V., Thomas-Hall, S.R., Malnoe, A., Timmins, M., Mussgnug, J.H., Rupprecht, J., Kruse, O., Hankamer, O., Schenk, P.M., 2008. Transcriptome for photobiological hydrogen production induced by sulfur deprivation in the green alga *Chlamydomonas reinhardtii*. Eukaryot. Cell 7 (11), 1965–1979.

Niang, L.L., Hung, X., 1984. Studies on biological active compounds of the algae from yellow sea. In: Eleventh International Seaweed Symposium Developments in Hydrobiology, vol. 22, pp. 168–170.

Nickelsen, K., 2007. Otto Warburg's first approach to photosynthesis. Photosynth. Res. 92, 109–120.

Ogawa, T., Aiba, S., 1981. Bioenergetic analysis of mixotrophic growth in *Chlorella vulgaris* and *Scenedesmus acutus*. Biotechnol. Bioeng. 23, 1121–1132.

Oh, Y.K., Raj, S.M., Jung, G.Y., Park, S., 2011. Current status of the metabolic engineering of microorganisms for biohydrogen production. Bioresour. Technol. 102, 8357–8367.

Oncel, S., 2013. Microalgae for a macroenergy world. Renew. Sust. Energ. Rev. 26, 241–264.

Oncel, S., Kose, A., 2014. Comparison of tubular and panel type photobioreactors for biohydrogen production utilizing *Chlamydomonas reinhardtii* considering mixing time and light intensity. Bioresour. Technol. 151, 265–270.

Oncel, S., Sabankay, M., 2012. Microalgal biohydrogen production considering light energy and mixing time as the key features for scale up. Bioresour. Technol. 121, 228–234.

Oncel, S., Vardar-Sukan, F., 2011. Application of proton exchange membrane fuel cells for the monitoring and direct usage of biohydrogen produced by *Chlamydomonas reinhardtii*. J. Power Sources 196, 46–53.

Ort, J.E., 1970. Unbalanced Culture Method of Algae Production. US patent No: 3521400.

Oswald, W.J., Golueke, C.G., 1960. Biological Transformation of Solar Energy. In: Advances in Applied Microbiology, vol. 2. Academic Press, pp. 223–262.

Paella, R., Kim, S.K., 2011. Impact of marine micro- and macroalgal consumption on photoproduction. Adv. Food Nutr. Res. 64, 287–295.

Patil, V., Tran, K.-Q., Giselrød, H.R., 2008. Towards sustainable production of biofuels from microalgae. Int. J. Mol. Sci. 9, 1188–1195.

Perez-Garcia, O., Escalante, F.M., de-Bashan, L.E., Bashan, Y., 2011. Heterotrophic cultures of microalgae: metabolism and potential products. Water Res. 45 (1), 11–36.

Pienkos, P.T., Darzins, A.L., 2009. The promise and challenges of microalgal-derived biofuels. Biofuels Bioprod. Biorefin. 3 (4), 431–440.

Pires, J.C.M., Alvim-Ferraz, M.C.M., Martins, F.G., Simões, M., 2012. Carbon dioxide capture from flue gases using microalgae: engineering aspects and biorefinery concept. Renew. Sust. Energy Rev. 16, 3043–3053.

Pohl, P., Zurheide, F., Hoppe, H.A., Levring, T., Tanaka, Y., 1979. Marine Algae in Pharmaceutical Science. In: Marine Algae in Pharmaceutical Science, vol. 1. Walter de Gruyter, pp. 25–26.

Posewitz, M.C., King, P.W., Smolinski, S.L., Smith, R.D., Ginley, A.R., Ghirardi, M.L., 2005. Identification of genes required for hydrogenase activity in *Chlamydomonas reinhardtii*. Biochem. Soc. Trans. 33, 102–104.

Posewitz, M.C., Smolinski, S.L., Kanakagiri, S., Melis, A., Seibert, M., Ghirardi, M.L., 2004. Hydrogen photoproduction is attenuated by disruption of an isoamylase gene in *Chlamydomonas reinhardtii*. Plant Cell 16, 2151–2163.

Pruess, L., Arnow, P., Wolcott, L., Bohonos, N., Oleson, J., Williams, J., 1954. Studies on the mass culture of various algae in carboys and deep-tank fermentations. Transfer 2, 2.

Pulz, O., 2001. Photobioreactors: production systems for phototrophic microorganisms. Appl. Microbiol. Biotechnol. 57, 287–293.

Pulz, O., Gross, W., 2004. Valuable products from biotechnology of microalgae. Appl. Microbiol. Biotechnol. 65, 635–648.

Radakovits, R., Jinkerson, R.E., Darzins, A., Posewitz, M.C., 2010. Genetic engineering of algae for enhanced biofuel production. Eukaryot. Cell 8, 486–501.

Richmond, A., 1999. Physiological principles and modes of cultivation in mass production of phototrophic microalgae. In: Cohan, Z. (Ed.), Chemicals from Microalgae. Taylor & Francis, London, pp. 353–386.

Roleda, M.Y., Slocombe, S.P., Leakey, R.J., Day, J.G., Bell, E.M., Stanley, M.S., 2013. Effects of temperature and nutrient regimes on biomass and lipid production by six oleaginous microalgae in batch culture employing a two-phase cultivation strategy. Bioresour. Technol. 129, 439–449.

Rosenberg, J.N., Oyler, G.A., Wilkinson, L., Betenbaugh, M.J., 2008. A green light for engineered algae: redirecting metabolism to fuel a biotechnology revolution. Curr. Opin. Biotechnol. 19, 430–436.

Ruben, S., Kamen, M.D., 1941. Long lived radioactive carbon: C14. Phys. Rev. 59, 349–354.

Ruben, S., Kamen, M.D., Hassid, W.Z., de Vault, D.C., 1939. Photosynthesis with radio-carbon. Science 90, 510–511.

Rupprecht, J., 2009. From system biology to fuel—*Chlamydomonas reinhardtii* as a model for a systems biology approach to improve biohydrogen production. J. Biotechnol. 142, 10–20.

Safi, C., Zebib, B., Merah, O., Pontalier, P.Y., Vaca-Garcia, C., 2014. Morphology, composition, production, processing and applications of *Chlorella vulgaris*: a review. Renew. Sust. Energy Rev. 35, 265–278.

Sager, R., 1955. Inheritance in the green alga *Chlamydomonas reinhardi*. Genetics 40, 476–489.

Sager, R., Granick, S., 1953. Nutritional studies with *Chlamydomonas reinhardi*. Ann. N.Y. Acad. Sci. 56, 831–838.

Samson, R., LeDuy, A., 1982. Biogas production from anaerobic digestion of *Spirulina maxima* algal biomass. Biotechnol. Bioeng. 24, 1919–1924.

de Saussure, N.T., 1804. Recherches chimique sur la végétation.

Sánchez, M., Castillo, B.J., Rozo, C., Rodríguez, I., 2003. *Spirulina (Arthrospira)*: an edible microorganism. a review. Universitas Scientiarum 8 (1), 1–16.

Schwimmer, M., Schwimmer, D., 1955. The Role of Algae and Plankton in Medicine. Grune & Stratton, NY.

Scoma, A., Gianelli, L., Faraloni, C., Torzillo, G., 2012. Outdoor H_2 production in a 50-L tubular photobioreactor by means of a sulfur-deprived culture of the microalga *Chlamydomonas reinhardtii*. J. Biotechnol. 157, 620–627.

Setlik, I., Veladimir, S., Malek, I., 1970. Dual purpose open circulation units for large scale culture of algae in temperate zones. I. Basic design considerations and scheme of pilot plant. Algol. Stud. 1, 111–164.

Sheehan, J., Dunahay, T., Benemann, J., Roessler, P., 1998a. A Look Back at the U.S. Department of Energy's Aquatic Species Program—Biodiesel from Algae. National Renewable Energy Laboratory, Golden, Colorado. NREL/TP-580–24190, pp. 1–328.

Sheehan, J., Dunahay, T., Benemann, J., Roessler, P., 1998b. A Look Back at the US Department of Energy's Aquatic Species Program: Biodiesel from Algae, vol. 328. National Renewable Energy Laboratory, Golden.

Sialve, B., Bernet, N., Bernard, O., 2009. Anaerobic digestion of microalgae as a necessary step to make microalgae biodiesel sustainable. Biotechnol. Adv. 27, 409–416.

Sierra, E., Acién, F.G., Fernández, J.M., García, J.L., González, C., Molina, E., 2008. Characterization of a flat plate photobioreactor for the production of microalgae. Chem. Eng. J. 138 (1–3), 136–147.

Slocombe, S.P., Ross, M., Thomas, N., McNeill, S., Stanley, M.S., 2013a. A rapid and general method for measurement of protein in micro-algal biomass. Bioresour. Technol. 129, 51–57.

Slocombe, S.P., Zhang, Q., Black, K.D., Day, J.G., Stanley, M.S., 2013b. Comparison of screening methods for high-throughput determination of oil yields in microalgal biofuel strains. J. Appl. Phycol. 25 (4), 961–972.

Soccol, C.R., Vanderberghe, L.P.S., Costa, B., Woiciechowski, A.L., Carvalho, J.C., Medeiros, A.B.P., Francisco, A.M., Bonomi, L.J., 2005. Brazilian biofuel program: an overview. J. Sci. Ind. Res. 64, 897–904.

Soeder, C.J., 1980. Massive cultivation of microalgae: results and prospects. Hydrobiologia 72, 197–209.

Solomon, B.D., Barnes, J.R., Halvorsen, K.E., 2007. Grain and cellulosic ethanol: history, economics, and energy policy. Biomass Bioenergy 31, 416–425.

Spolaore, P., Cassan, C.J., Duran, E., Isambert, A., 2006. Commercial applications of microalgae. J. Biosci. Bioeng. 101, 87–96.

Srirangan, K., Pyne, M.E., Chou, C.P., 2011. Biochemical and genetic engineering strategies to enhance hydrogen production in photosynthetic algae and cyanobacteria. Bioresour. Technol. 102, 8559–8604.

Stepka, W., Benson, A.A., 1948. The path of carbon in photosynthesis II. Amino acids. UCRL-119.

Stuart, T.S., Gaffron, H., 1972. The mechanism of hydrogen photoproduction by several algae. Planta 106 (2), 101–112.

Sun, Y., Cheng, J., 2002. Hydrolysis of lignocellulosic materials for ethanol production: a review. Bioresour. Technol. 83, 1–11.

Symons, G.E., Buswell, A.M., 1933. The methane fermentation of carbohydrates. J. Am. Chem. Soc. 55, 2028–2036.

Tamiya, H., 1957. Mass culture of algae. Annu. Rev. Plant Physiol. 8 (1), 309–334.

Tetali, S.D., Mitra, M., Melis, A., 2007. Development of the light-harvesting chlorophyll antenna in the green alga *Chlamydomonas reinhardtii* is regulated by the novel Tla1 gene. Planta 225 (4), 813–829.

Theriault, R.J., 1965. Heterotrophic growth and production of xanthophylls by *Chlorella pyrenoidosa*. Appl. Microbiol. 13, 402–416.

Toerien, D.F., Grobbelaar, J.U., Walmsley, R.D., 1987. Management of Autotrophic Mass Cultures of Micro-algae. Elsevier B.V., Amsterdam.

Torzillo, G., Seibert, M., 2013. Hydrogen Production by Microalgae. Handbook of Microalgal Culture: Applied Phycology and Biotechnology, second ed. pp. 417–444.

Trent, J., 2012. Offshore Membrane Enclosures for Growing Algae (OMEGA): A Feasibility Study for Wastewater to Biofuels (Energy Research and Development Division Final Project Report).

Usui, N., Ikenouchi, M., 1997. The biological CO_2 fixation and utilization project by RITE(1)—highly effective photobioreactor system. Energy Convers. Manage. 38, 487–492.

Vanegas, C.H., Bartlett, J., 2013. Green energy from marine algae: biogas production and composition from the anaerobic digestion of Irish seaweed species. Environ. Technol. 34 (15), 2277–2283.

Walker, T.L., Collet, C., Purton, S., 2005. Algal transgenics in the genomic era. J. Phycol. 41, 1077–1093.

Wang, H., Liu, G., Ruan, R., Liu, Y., 2014. Biofuel from microalgae: current status, opportunity and challenge. In: International Conference on Material and Environmental Engineering (ICMAEE 2014). Atlantis Press, pp. 25–28.

Wang, L.A., Min, M., Li, Y.C., Chen, P., Chen, Y.F., Liu, Y.H., Wang, Y.K., Yuan, R., 2010. Cultivation of green algae *Chlorella* sp. in different wastewaters from municipal wastewater treatment plant. Appl. Biochem. Biotechnol. 162, 1174–1186.

Warburg, O., 1919. Über die geschwindigkeit der kohlensäurezusammensetzung in lebenden zellen. Biochem. Z. 100, 230–270.

Warburg, O., 1920. Über die Geschwindigkeit der photochemischen kohlensaeurezersetzung in lebenden zellen. Biochem. Z. 103, 188–217.

Warburg, O., Negelein, E., 1923. Über den Einfluss der Wellenlänge auf den Energieumsatz bei der Kohlensäureassimilation. Zeit. Physik. Chem. 106, 191–218.

Ward, A.J., Lewis, D.M., Green, F.B., 2014. Anaerobic digestion of algae biomass: a review. Algal. Res. 5, 204–214.

Watanabe, M.M., Tanabe, Y., 2013. Biology and Industrial Potential of *Botryococcus Braunii*. Handbook of Microalgal Culture, pp. 369–387.

Weaver, P., Lien, S., Seibert, M., 1979. Photobiological Production of Hydrogen: A Solar Energy Conversion Option. Solar Energy Research Inst. Golden, CO.

Weissman, J.C., Goebel, R.P., Benemann, J.R., 1988. Photobioreactor design: mixing, carbon utilization, and oxygen accumulation. Biotechnol. Bioeng. 31 (4), 336–344.

Wijffels, R.H., Barbosa, M.J., 2010. An outlook on microalgal biofuels. Science 329, 796–799.

Wykoff, D.D., Davies, J.P., Melis, A., Grossman, A.R., 1998. The regulation of photosynthetic electron transport during nutrient deprivation in *Chlamydomonas reinhardtii*. Plant Physiol. 117, 129–139.

Xiong, J., 2006. Photosynthesis: what color was its origin. Genome Biol. 7, 2455.

Xu, H., Miao, X., Wu, Q., 2006. High quality biodiesel production from a microalga *Chlorella prototothecoides* by heterotrophic growth in fermenters. J. Biotechnol. 126 (4), 499–507.

Yang, C., Hua, Q., Shimizu, K., 2000. Energetics and carbon metabolism during growth of microalgal cells under photoautotrophic, mixotrophic and cyclic light-autotrophic/dark-heterotrophic conditions. Biochem. Eng. J. 6, 87–102.

Yang, C., Hua, Q., Shimizu, K., 2002. Integration of the information from gene expression and metabolic fluxes for the analysis of the regulatory mechanisms in *Synechocystis*. Appl. Microbiol. Biotechnol. 58, 813–822.

Yu, J.J., Wang, M.H., Xu, M., Ho, Y.S., 2012. A bibliometric analysis of research papers published on photosynthesis: 1992–2009. Photosynthetica 50 (1), 5–14.

Zallen, D.T., 1993. The "light" organism for the job: green algae and photosynthesis research. J. Hist. Biol. 26 (2), 269–279.

Zittelli, G.C., Biondi, N., Rodolfi, L., Tredici, M.R., 2013. Photobioreactors for Mass Production of Microalgae. Handbook of Microalgal Culture: Applied Phycology and Biotechnology, second ed. Blackwell, pp. 225–266.

Biofuel Production from Microalgae

Kurniadhi Prabandono[1], Sarmidi Amin[2]

[1]Ministry for Marine Affairs and Fisheries—Republic of Indonesia; [2]Agency for Assessment and
Application Technology (BPPT), Indonesia

1. INTRODUCTION

Microalgae as biomass has chemical composition, which varies depending on algae used, and it can be rich in proteins, lipids or have a balance composition of proteins, carbohydrates, and lipids (Raja et al., 2013). The first unialgal cultures were achieved by Beijerinck in 1890 with *Chlorella vulgaris*, and the use of such cultures for studying plant physiology was developed by Warburg in the early 1900s (Anonymous, beam). After World Wars I and II, scientists continued their research for sustainable food sources (Edward, 2010).

Mass culture of microalgae really began to be a focus of research after 1948 at Stanford (USA), Essen (Germany), and Tokyo (Anonymous, beam). The first attempt in the United States to translate the biological requirements for algal growth into engineering specifications for a large-scale plant was made at the Stanford Research Institute during 1948—1950 (Demirbas, 2010). In 1952, an Algae Mass Culture Symposium was held at Stanford University, California. One important outcome of the symposium was the 1953 publication of *Algae Culture from Laboratory to Pilot Plant* edited by J. S. Burlew (Boriwitzka, 2013). The concept of using algae as a fuel was first proposed by Meier in 1955

for production of methane gas from carbohydrate fractional cells. This idea was further developed by Oswald and Golueke in 1960 (Demirbas and Demirbas, 2010).

Production of methane gas from algae received a big push during the energy crisis of the 1970s (Anonymous, history). The production of microalgal lipids that can be synthesized into biodiesel has been of interest to the US Department of Energy (DOE) since the 1970s (Weldy and Huesemann, 2013). But because of financial constraints and cheap oil in 1995, this production was terminated (Anonymous, history). Then it was continued again by the National Renewable Energy Laboratory and DOE; the product was commercial grade fuel from triglyceride-rich microalgae (Danielo, 2005).

Recently, researchers and companies are not only focusing on algae as a source for powering cars and trucks but also as a source for renewable aviation fuel. Biojet fuel has already been successfully tested in both commercial and military aircraft. The fuel has been praised for having a low flash point and sufficiently low freezing point, issues that have been problematic for other biofuels (Gouveia, 2011). Lufthansa (German airline) with Algae Tec (Australian company) in 2012 signed a collaboration to build an industrial-scale plant to produce aviation fuels from algae (Bussiness, Green, 2012).

1.1 Biofuel Production

Bio-oil from microalgae can be used directly as fuel or chemically transesterified into biodiesel. Other microalgae biofuels such as ethanol and methane are produced as organic substrates and can be fermented by microbes under anaerobic conditions. Hydrogen can be produced by photosynthetic algae and cyanobacteria under certain nutrient- or oxygen-depleted conditions, and by bacteria and archae utilizing organic substrates under anaerobic conditions (Drapcho et al., 2008). The conversion technology of microalgae biomass can be classified into two categories: biochemical and thermochemical. Figure 1 shows the energy conversion process from microalgae to biofuel (Tsukahara and Sawayama, 2005). Spolaore et al. (2006) suggested that microalgae can also produce valuable coproducts such as proteins and residual biomass after oil extraction, which may be used as feed or fertilizer.

Gasification is the conversion process of biomass into CH_4, H_2, CO_2, and ammonia. The product gas from this process can be burned directly or used as a fuel gas engine, or can be used as a feedstock (synthetic gas) (McKendry, 2003). But in the case of microalgae biomass that has high water content (80–90%) (Patil et al., 2008), the gasification process is not suitable. It is better through a biological conversion process (McKendry, 2003) or liquefaction process (FAO, oil production). Direct hydrothermal liquefaction in subcritical water condition is a technology that can be employed to convert wet biomass material to liquid fuel (Patil et al., 2008). The separation scheme is presented in Figure 2 (Minowa et al., 1995; FAO, oil production; Minowa and Sawayama, 1999; Yang et al., 2004; Murakami et al., 1990; Itoh et al., 1994; Amin, 2009). The liquefaction is performed in an aqueous solution of alkali or alkaline earth salt at about 300 °C and 10 MPa without a reducing gas such as hydrogen and/or carbon monoxide (Minowa et al., 1995).

Pyrolysis is conversion of biomass to biofuel, charcoal, and gaseous fraction by heating the biomass in the absence of air to around 500 °C

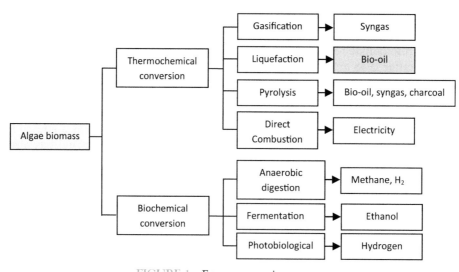

FIGURE 1   Energy conversion process.

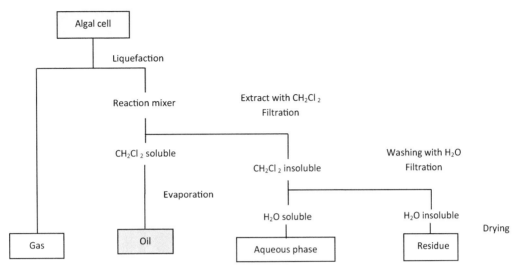

FIGURE 2   Separation scheme for liquefies microalgal cells.

(McKendry, 2003; Miao et al., 2004) or by heating in the presence of a catalyst (Agarwal, 2007), at high heating rate (103—104 K/s) and with short gas residence time to crack into short-chain molecules and then be rapidly cooled to liquid (Qi et al., 2007). Low heating rate and long residence time may increase the energy input (Agarwal, 2007), and fast pyrolysis processes for biomass have attracted a great deal of attention for maximizing liquid yields (Miao et al., 2004).

The advantage of fast pyrolysis is that it can directly produce a liquid fuel and also produce biogas. A conceptual fluidized bed fast pyrolysis system is shown in Figure 3 (Bridgwater and Peacocke, 2000). However, since microalgae usually have a high moisture content, a drying process requires much heating energy (Yang et al., 2004). Organic compounds such as carbohydrates, proteins, and lipids, in natural microorganisms, break the complex carbon into smaller substances through the digestion process. There are two types of digestion processes: aerobic and anaerobic (Suyog, 2011). The anaerobic digestion process can occur at mesophilic temperature (35—45 °C) or thermophilic temperature (50 °C).

Hydrolysis is the first step of organic matter enzymolyzed externally by extracellular enzymes, cellulose, amylase, protease, and lipase. Bacteria decompose long chains of complex organic matter such as carbohydrates, proteins, and lipids into small chains. During hydrolysis, complex organic matter (polymers) are converted into glucose, glycerol, purines, and pyridines (Al Seadi et al., 2008). Proteins are split into peptides and amino acids (Sorathia et al., 2012). In the first step, the complicated molecules like polymers, fats, proteins, and carbohydrates are converted to monomers, fatty acids (FAs), amino acids, and saccharides (Alexopoulos, 2012).

The second step is acidification, converting the intermediates of fermenting bacteria into acetic acid, hydrogen, and carbon dioxide. The acidification step has two parts: acidogenesis and acetogenesis. The third step is decomposition of compounds having a low molecular weight. They utilize hydrogen, carbon dioxide, and acetic acid to form methane and carbon dioxide, with other small traces of H_2S, H_2, and N_2, etc. (Sorathia et al., 2012). According to Al Seadi et al. (2008), 70% of the formed

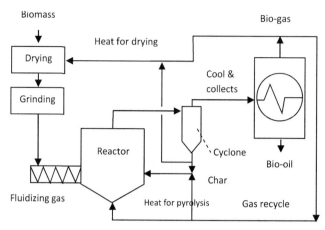

FIGURE 3   Fast pyrolysis principles.

methane originates from acetate and 30% is produced from conversion of hydrogen and carbon dioxide.

Continuous and batch are two types of processes for anaerobic fermentation. Gasification converts the biomass to syngas by means of partial oxidation with air, oxygen, and/or steam at high temperature, typically in the range of 800—900 °C. A flow diagram of a microalgae system for fuel production by low temperature catalytic gasification of biomass is shown in Figure 4 (Tsukahara and Sawayama, 2005). A novel energy production system using microalgae with nitrogen cycling combined with low-temperature catalytic gasification of the microalgae has been proposed (Minowa and Sawayama, 1999). Elliot and Sealock (1999) have also developed a low-temperature catalytic gasification of biomass with high moisture content. Biomass with high moisture is gasified directly to methane-rich fuel gas without drying. In addition, nitrogen in biomass is converted to ammonia during the reaction (Minowa and Sawayama, 1999).

2. MICROALGAE

There are two main populations of algae: filamentous and phytoplankton. Microalgae are unicellular, photosynthetic microorganisms, living in saline or freshwater environments, that convert sunlight, water, and carbon dioxide to algal biomass (Ozkurt, 2009). The three most important classes of microalgae in terms of abundance are the diatoms (*Bacillariophyceae*), the green algae (*Chlorophyceae*), and the golden algae (*Chrysophyceae*). The cyanobacteria or blue—green algae (*Cyanophyceae*) are also referred to as microalgae, that is, *Spirulina* (*Arthrospira platensis* and *Arthrospira maxima*).

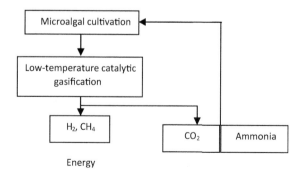

FIGURE 4   Flow diagram of microalgal system for fuel production by gasification.

2.1 Source of Energy

Algae have been suggested as good candidates for fuel production because of their higher photosynthetic efficiency, higher biomass production, and faster growth compared to other energy crops (Miao and Wu, 2004). The productivity rates of microalgae are higher than most other plants—up to $12.5\,kg\,m^{-2}\,year^{-1}$ (Shay, 1993). According to Shay (1993) and Minowa et al. (1995), among the biomasses, microalgae usually have a higher photosynthetic efficiency than other biomasses. This biomass is a highly promising resource according to Hall in Sawayama et al. (1999) and Minowa et al. (1995). Microalgae also have potential in the production of gaseous biofuel. According to Benemann (in Liewellyn and Skill, 2007; www.nbu.ac.uk), hydrogen can be produced via a number of photobiological processes either using hydrogenase through direct or indirect photolysis or using nitrogenase. Microalgae also can be converted to synthesis gas by means of partial oxidation with air, oxygen, and/or steam at high temperature, typically in the range of 800–900 °C (Amin, 2009). *Chlorella*, *Dunaliella*, *Chlamydomonas*, *Scenedesmus*, and *Spirulina* are known to contain a large amount (>50% of the dry weight, dw) of starch and glycogen, useful as raw materials for ethanol production (John et al., 2011). Biojet fuel can be also produced from microalgae.

2.2 Cultivation of Microalgae

Microalgae cultivation using sunlight energy can be carried out in open ponds or photobioreactors (PBRs). There are three distinct algae production mechanisms: photoautotrophic, heterotrophic, and mixotrophic.

2.2.1 Open Ponds

Algae can be cultured in open ponds (such as raceway-type ponds and lakes) and PBRs. Open ponds are less expensive, but they are highly vulnerable to contamination by other microorganisms. The water temperature is important condition for microalgae culture, because most type of algae grow well at temperatures from 17 °C to 22 °C. The lower temperatures will not usually kill the algae but will reduce their growth rate, and above 27 °C, most types of algae will die (Laing, 1991).

Open pond cultivation can exploit unusual conditions that suit only specific algae. For instance, *Spirulina* sp. thrives in water with a high concentration of sodium bicarbonate and *Dunaliella salina* grows in extremely salty water; these unusual media exclude other types of organisms, allowing the growth of pure cultures in open ponds (Wikipedia, algaculture). The open ponds raceway type usually is paddle-wheel driven for circulation of the water with nutrients, no more than 30 cm deep (Janssen, 2002).

2.2.2 Closed Bioreactors

A PBR is a bioreactor that incorporates a light source. Because PBR systems are closed, the cultivator must provide all nutrients, including CO_2. A PBR can operate in batch or continuous mode. The grower provides sterilized water, nutrients, air, and carbon dioxide at the correct rates. This allows the reactor to operate for long periods (Wikipedia, algaculture). There are two types of closed systems: flat panel PBR and tubular PBR with air-lift column.

2.3 Harvesting and Extraction

The most common microalgae harvesting methods include gravity sedimentation, flocculation, centrifugation, filtration and microscreening, flotation (sometimes with an additional step with a combination of flocculation and flotation), and electrophoresis techniques. The selection of harvesting technique is dependent on the properties of the microalgae. Microalgae harvesting can generally be divided into a two-step process, including (Brenann and Owende, 2010):

- Bulk harvesting is used to separate microalgal biomass from bulk suspension.
- Thickening is concentrated in the slurry with filtration and centrifugation.

Gravity sedimentation is commonly applied for separating microalgae in water. Flocculation is frequently used to increase the efficiency of gravity sedimentation, but addition of flocculations is currently not a method of choice for cheap and sustainable production (Schenk et al., 2008). Autoflocculation also occurs as a result of interrupting the CO_2 supply to an algal system (Dermibas, 2010). There are two main classifications of flocculants: inorganic and organic.

Filtration is often applied at a laboratory scale, but in application on a large scale it has problems, such as membrane clogging, the formation of compressible filter cake, and high maintenance costs. Centrifugation is a preferred method, especially for producing extended shelf-life concentrates for aquaculture; however, this method is time-consuming and costly. Centrifugation is a very useful secondary harvesting method to concentrate an initial slurry (10−20 g/L) to an algal paste (100−200 g/L) and could possibly be used in combination with oil extraction (Schenk et al., 2008). Flotation is a gravity separation process in which air or gas bubbles are attached to solid particles, which then carry them to the liquid surface. Zhang et al. (2010) developed an efficient technology for harvesting of algal biomass using membrane filtration.

After separation from the culture medium algal, biomass must be quickly processed because it can spoil in only a few hours in a hot climate (Mata et al., 2010). According to Liu et al. (2011), after harvesting, chemicals in biomass may be subjected to degradation induced by the process itself and also by internal enzymes in algal cells. For example, lipase contained in the cells can rapidly hydrolyze cellular lipids into free fatty acids, which are not suitable for biodiesel production.

Oil extraction from dried biomass can be performed in two steps—mechanical crushing followed by solvent extraction. Oil extraction from algal cells can also be facilitated by osmotic shock or ultrasonic treatment to break the cells. In other equipment for cell disruption, besides the mechanical crushing, there are cell homogenizer bead mills, ultrasound, autoclave and spray drying, and nonmechanical crushing that uses freezing, organic solvents, osmotic shock, and acid, base and enzyme reactions (Mata et al., 2010). Lee et al. (2010) investigated several methods for effective lipid extraction from microalgae, including autoclaving, bead-beating, microwaves, sonication, and 10% NaCl solution, to identify the most effective cell disruption method.

Recent research of bio-oil production by pyrolysis of biomass has received much interest (Huang et al., 2010; Miao and Wu, 2004). Pyrolysis was first used for production of bio-oil or biogas from lignocellulose. However, such technology may be more suitable for microalgae because of the lower temperature required for pyrolysis and the higher quality oils obtained (Bridgwater et al., 1999). Fast pyrolysis is a new technology (see Figure 3) that produces bio-oil in the absence of air at atmospheric pressure with a relatively low temperature (450−550 °C) and a high heating rate (10^3−10^4 °C s^{-1}) as well as short gas residence time to crack into short-chain molecules and be cooled to liquid rapidly (Bridgwater et al., 1999; Bridgwater and Peacocke, 2000). Fast pyrolysis has proved to be a promising way to produce bio-oils compared to slow pyrolysis (Miao and Wu, 2004) for the following reasons: (1) less bio-oils were produced from slow pyrolysis; (2) the viscous bio-oils from slow pyrolysis are not suitable for liquid fuel; and (3) the fast pyrolysis process is time saving and requires less energy compared to the slow pyrolysis process. It was reported that the experiment was completed at 500 °C with a heating rate of 600 °C s^{-1}, a sweep gas (N_2) flow rate of 0.4 m^3/h, and a vapor residence time of 2–3 s.

Liquefaction process can be also used to produce biofuel directly without the need of drying microalgae (Minowa et al., 1995). It was reported that *Dunaliella tertiolecta* cells with 78.4% water content convert to oils directly. The yield of oil reached 37% of the total organic matter. Sawayama et al. (1999) investigated the energy balance and CO_2 mitigating effect of a liquid fuel production process from *Botryococcus braunii* using thermochemical liquefaction.

3. LIPID PRODUCTION

Microalgae contain proteins, carbohydrates, and lipids as shown in Table 1, which displays the lipid content and FA profile of microalgae. Lipids can be converted into biodiesel, carbohydrates to ethanol and H_2, and proteins as raw material of biofertilizer. Lipid is a general name for plants and animal products that are structurally esters of higher FAs. The FAs are a variety of monobasic acids such as palmatic (C16:0), stearic (C18:0), oleic (C18:1) (Klass, 1998), linoleic (C18:2), and linolineic (C18:3). The FAs can be classified as medium chain (C10–C14), long chain (C16–C18) and very long chain (>C20) species and FA derivates. However, under unfavorable environmental conditions, many algae alter their lipid biosynthetic pathways to the formation and accumulation of neutral lipids (20–50% dw), mainly in the form of triglycerides (TAGs). For biodiesel production, neutral lipids have to be extracted (Alcaine, 2010).

Lipids can be divided into two main groups: (1) the storage lipid (neutral or nonpolar lipid) and (2) structural (membrane or polar lipid).

TABLE 1   Composition of Selected Microalgae, % Dry Matter Basis

Strain	Protein	Carbohydrates	Lipid
Scenedesmus obliquus	50–55	10–15	12–14
Scenedesmus quadricauda	40	12	1.9
Scenedesmus dimorphus	8–18	21–52	16–40
Chlamydomonas rheinhardii	48	17	21
Chlorella vulgaris	51–58	12–17	14–22
Chlorella pyrenoidosa	57	26	2
Spirogyra sp.	6–20	33–64	11–21
Dunaliella bioculata	49	4	8
Dunaliella salina	57	32	6
Euglena gracilis	39–61	14–18	14–20
Prymnesium parvum	28–45	25–33	22–38
Tetraselmis maculata	52	15	3
Porphyridium cruentum	28–39	40–57	9–14
Spirulina plantesis	46–63	8–14	4–9
Spirulina maxima	60–71	13–16	6–7
Synechoccus sp.	63	15	11
Anabaena cylindrical	43–56	25–30	4–7

Adapted from Raja et al. (2013).

The group of neutral lipids is formed by triacylglycerols or TAGs, steryl esters, and wax esters (Lang, 2007). TAGs generally serve as energy storage in microalgae that once extracted can be converted into biodiesel through transesterification reactions. This neutral lipid bears a common structure of a triple ester where usually three long-chain FAs are coupled to a glycerol molecule. Transesterification displaces glycerol with small alcohols or methanol. Structural lipids typically have a high content of polyunsaturated fatty acids (PUFAs), which are also essential nutrients for aquatic animals and humans. Polar lipids and sterols are important structural components of cell membranes that act as a selective permeable barrier for cells and organelles (Sharma et al., 2012).

The oil content of some microalgae exceeds 80% of the dw of algae biomass (Patil et al., 2008; Christi, 2007). According to Shay (1993), algae can accumulate up to 65% of total biomass as lipids, but according to Oilgae, some algae have only about 15–40% (dw), whereas palm kernel has about 50%, copra has about 60%. Oil content itself can be estimated to be 64.4% of total lipid component (Hill and Feinberg, 1984). Microalgal oil can be produced through either biological conversion to lipid, hydrocarbon, or thermochemical liquefaction of algal cells. Direct extraction of microalgal lipid appears to be a more efficient methodology for obtaining energy from these organisms than fermentation of algal biomass to produce either methane or ethanol (FAO). D. tertiolecta with a moisture content of 78.4 wt% were converted directly into oil by chemical liquefaction at around 300 °C and 10 MPa. The oil yield was about 37%. The oil obtained at a reaction temperature of 340 °C and holding time 60 min had a viscosity of 150–330 mPas and a heating value of 36 MJ/kg (Minowa et al., 1995).

3.1 Lipid Profile

Properties of biodiesel can be predicted from the FA profile of lipid feedstock. Table 2 displays total lipid and FA profile (Barman et al., 2012), but just selecting the major FAs such as palmitic (C16:0), oleic (C18:1), and linoleic (C18:2), and minor FAs such as myristic (C14:0) and palmitoleic (C16:1).

3.2 Lipid Production Pathway

The production pathway of lipids consists of cultivation, harvesting, drying, cell disruption, and extraction of algal oil or lipid separation (Beal et al., 2013). The selective microalgae were cultivated in a PBR or in outdoor culture. After harvesting, the algae were concentrated by using a centrifugal separator or other methods. Beal et al. (2013) reported that the concentrated algae were exposed to electromechanical pulsing in a process designed to lyse the cell. After lysing, the algae were again left in a dark container at room temperature (~ 24 °C) overnight (for about 20 h). Another researcher (Alcaine, 2010) reported to dry the algae after concentration process. Several methods have been employed to dry the algae, including spray drying, freeze-drying, or sun drying, and it follows the cell disruption of microalgae. Several methods can be used such as bead mill, homogenizer, freezing, and sonication, and then algal oil can be extracted using chemical or mechanical methods.

3.3 Optimization of Algal Production

It is possible to increase the concentration of the lipid content by optimizing the growth-determining factors. The flue gases from industrial power plants are rich in CO_2 and can be used as fertilizer for algae cultivation. Wastewater from domestic or industrial sources contains essential salts and minerals for growth of algae. Chinnasamy et al. (2010) reported that their study to evaluate the feasibility of algal biomass and biodiesel production was conducted using wastewater containing 85–90% carpet industry effluent with 10–15% municipal sewage. About 63.9% of algal oil obtained from this cultivation could be converted into biodiesel.

Light from sun or from artificial light is needed for the photosynthesis process in open ponds or closed PBRs. The ponds are kept shallow because of the need to keep the algae exposed to sunlight and the limited depth to which sunlight can penetrate the pond water. Race way ponds are about 15–35 cm deep to ensure adequate exposure to sunlight. A tubular PBR has tubes that are small and limited in diameter (0.2 m or less) to allow light penetration to the center of tube (Dermibas, 2010).

TABLE 2   Total Lipid Content and Fatty Acid Profile of Microalgae

		Fresh water microalgae					
Genus		*Synechocystis pevalekii*	*Nostoc ellipsosporum*	*Spirulina plantensis*	*Spirogyra orientalis*	*Chlorococcum infusionum*	*Rhizoclonium fontinale*
Total lipid (%)		9 ± 2	7 ± 1.5	8.5 ± 2	21 ± 2.5	11.34 ± 1	7.5 ± 1.4
FATTY ACID (FA)							
Saturated FA	C14:0	1.0	12.0	13.27	0.8	8.5	6.49
	C16:0	34.2	22.05	21.1	34.08	25.62	29.36
Monounsaturated FA	C16:1	3.8	6.6	9.32		5.46	2.26
	C18:1	29.8	17.85	11.27	6.4	15.66	22.13
Polyunsaturated FA	C18:2	14.9	2.5	0.24	6.7	7.5	—

		Brackish water microalgae						
Genus		*Lynghya birgei*	*Rhizoclonium riprium*	*Pithophora cleveana*	*Cladophora crystallina*	*Cheatomorpha gracilis*	*Enteromorpha intestinalis*	*Polysiphonia mollis*
Total lipid (%)		12 ± 2.8	8.65 ± 1	19 ± 2	23 ± 1.8	16 ± 0.5	12 ± 1	7.8 ± 1.6
FATTY ACID (FA)								
Saturated FA	C14:0	3.2	7.8	9.4	8.8	13.0	9.96	7.5
	C16:0	59.6	37.4	37.0	39.1	52.4	22.57	44.7
Monounsaturated FA	C16:1	2.3	3.1	3.1	2.8	6.0	9.7	3.7
	C18:1	13.1	17.3	34.3	31.8	15.9	18.49	10.3
Polyunsaturated FA	C18:2	9.1	17.2		11.6	3.6	14.83	—

		Marine water microalgae							
Genus		*Phormidium valderianum*	*Phormidium tenne*	*Rhizoclonium africanum*	*Ulva lactuca*	*Navicula minima*	*Catenella repens*	*Geledium pasillum*	*Ceramium manorensis*
Total lipid (%)		7.8 ± 2.8	8.01 ± 2.2	7.2 ± 2.7	11 ± 1	16.23 ± 0.59	8 ± 1.5	9.7 ± 2.8	8 ± 1.9
FATTY ACID (FA)									
Saturated FA	C14:0	1.9	4.8	6.5	0.9	2.7	5.3	2.0	2.4
	C16:0	35.8	32.9	30.2	45.2	26.4	56.2	36.2	60.6
Monounsaturated FA	C16:1	16.1	24.3	9.3	2.0	18.7	7.6	9.2	2.0
	C18:1	23.8	15.2	20.0	10.8	25.3	10.4	25.6	17.4
Polyunsaturated FA	C18:2	2.5	2.9	5.3	4.89	2.2	7.0	7.1	9.6

Adapted from Barman et al. (2012).

Microalgae are typically grown under autotrophic conditions. In autotrophic growth, algae utilize sunlight as an energy source and carbon dioxide as a carbon source. Certain species of algae can also grow under heterotrophic conditions, where algae utilize a reduced organic compound as their carbon and energy source (Drapcho et al., 2008). Miao and Wu (2004) and Xu et al. (2006) studied green algae *Chlorella protothecoides* and found that lipid content in heterotrophic conditions could be as high as 55%, which was four times higher than in autotrophic conditions under similar conditions. *Chlorella vulgaris* showed an increase in biomass production when the organism was grown under mesotrophic conditions.

There has been a wide range of studies carried out to identify and develop efficient lipid induction techniques in microalgae such as nutrient stress, osmotic stress, radiation, pH, temperature, heavy metals, and other chemicals. In addition, several genetic strategies for increased triacylglyceride production and inducibility are currently being developed (Sharma et al., 2012).

Stress growth conditions can often be used to increase the formation of natural lipid. The stress was caused by either the use of nutrient-deficient media or the addition of excess salt to nutrient-enriched media. The combination of both nutrient deficiency and salt enrichment appears to enhance lipid formation with *Isochrysis* sp. but to reduce it with *D. salina*. Interestingly, the free glycerol content can apparently be quite a bit higher for *Dunaliella* sp. *Botryococcus braunii* exhibited relatively high lipid content under each set of growth conditions, but the highest was 54.2% (dw) under nutrient-deficient growth conditions (Klass, 1998). Kishimoto et al. (1994), using CO_2 as a carbon source, found that the microalgae growth reached 1.0 g/L under non-sterilized conditions after 1 week.

Nannochloropis sp., when grown in media under nitrogen-limited conditions, will increase the lipid content from 28% to over 50%. Lipid productivity was demonstrated to have a maximum of 150 mg/L-day at 5—6% cell nitrogen and was apparently independent of the light supply and cell density when the initial nitrogen concentration exceeded 25 mg/L (Klass, 1998). Shiffrin and Chrisholm (in Weldy and Huesemann, 2013) reported that *Monallantus salina* produced as much as 72% lipid in nitrogen-deficient conditions. Xin et al. (2010) compared the freshwater microalgae *Scenedesmus* sp. with 11 species of high lipid content and reported that *Scenedesmus* sp. showed the best ability to adapt to growth in secondary effluent, and that it had the highest lipid content at 31—33%. Benemann et al. (2013) reported their study on eight strains of microalgae, and all of microalgae were subjected to nitrogen limitation in batch cultures.

4. PROPERTIES OF MICROALGAE-DERIVED BIODIESEL

The most common FA profile of algae consists mainly of palmatic (C16:0), stearic (C18:0), oleic (C18:1) (Klass, 1998), linoleic (C18:2), and linolineic acids (C18:3) (Knothe, 2009). These FAs have a direct impact on the chemical and physical properties of biofuel. Along with these major FAs, there are varieties of FAs in minor amounts that have some influence on fuel property. The minor FAs are C8:0, C10:0, C12:0, C13:0, C14:0, C15:0, C15:1, C16:1, C18:3, C18:4, C20:0, C20:2, C20:5, C20:6, C22:6, C24:0, and C24:1. However, there is no single FA that is responsible for any particular fuel property (Islam et al., 2013a). Microalgal lipids are predominantly polyunsaturated (C16:2, C18:2, C20:2, C16:3, C18:3, C20:3) and therefore they are more prone to oxidation. This is a serious issue with biodiesel while in storage.

Gunstone and Hilditch in Schenk et al. (2008) measured the relative rate of oxidation for the methyl ester of oleic (C18:1), linoleic (C18:2), and linolenic (C18:3) acids to be 1:12:25. It is therefore preferable that the level of PUFAs in biodiesel is kept to a minimum. In contrast, higher levels of polyunsaturated fats lower the cold filter plugging point; the temperature at which the fuel solidifies and blocks the fuel filter of an engine.

Table 3 presents the melting point of the major FAs. It can be seen that the more unsaturated the oil is, the lower the melting point (Knothe et al., 2005). Therefore, the colder climates require a higher unsaturated lipid content to enable the fuel perform at low temperature. A good-quality biodiesel should have a 5: 4:1 mass FA ratio of C16:1, C18:1, and C14:0, as recommended by Schenk et al. (2008). Of the nine microalgae species investigated by Islam et al. (2013b), the fatty acid methyl ester (FAME) composition of *Nannochlopsis oculata* was closest to the recommended ratio of 5.1:3.5: 1, but eicosapentaenoic acid (EPA) is also present in appreciable quantities (fourth most dominant FA). From eight marine water microalgae species investigated by Barman et al. (2012), the FAME composition of *Phoramidium tennue* was closest to the recommended ratio of 5: 3.16: 1 (Table 2). According to Knothe (2008), the ideal biodiesel feedstock would be composed entirely of C16:1 and C18:1 (monounsaturated FAs). In practice, a biodiesel feedstock should have high concentrations of C16:1 and C18:1 with less variation in the FA profile.

The Cetane number (CN) is another measure describing the combustion quality of diesel fuel during compression ignition. Higher CNs have shorter ignition delay periods than lower CN fuel. Ramos et al. (2009) proposed the equation to calculate some critical parameters like CN, iodine value, oxidation stability, and cold filter plugging point. Ramirez-Verduzco et al. (2012) developed a calculation method of physical properties of methyl esters and an average absolute deviation. They concluded that CN, viscosity, and heating value increase because of the increase of molecular weight, and decrease as the number of double bonds increases.

TABLE 3 Profiles of Fatty Acids

Fatty acid	Cetane number	Melting point (°C)	Ester MP
Caprylic (8:0)	33.6	16.7	Etyl, −43 °C
Capric (10:00)	47.7	31.6	Etyl, −20 °C
Lauric (12:00)	61.4	44.2	Etyl, −1.8 °C
Myristic (14:0)	66.2	54.4	Etyl, 12.3 °C
Palmitic (16:00)	74.5	62.9	Etyl, 24 °C
Palmitoleic (16:ω7)	45	−0.1	NA
Stearic (18:00)	86.9	69.6	Methyl, 39 °C
Oleic (18:1ω9)	55	14	Methyl, −20 °C
Linoleic (18:2ω6)	36	−5	Methyl, −35 °C
Linolenic (18:3ω3)	28	−11	Methyl, −57 °C
Gadoleic (20:1ω9)	82	23	NA
Arachidonic (20:4ω6)	NA	−50	NA

Sumber: Knothe et al. (2005).

5. SUMMARY

Biofuel is a biomass product that can be used as a substitute for petroleum fuels. Biofuel from microalgae can be processed by using thermochemical and biochemical conversion. The thermochemical process can be divided into gasification, liquefaction, pyrolysis, and direct combustion; meanwhile the biochemical process can be divided into anaerobic digestion, fermentation, and photobiological process. By using the gasification process, the biomass produces CH_2, H_2, CO_2, and ammonia. Liquid fuel can be produced by using the liquefaction process. Pyrolysis is conversion of biomass to biofuel, charcoal, and gaseous fraction. The anaerobic digestion process can produce methane and H_2, meanwhile the fermentation process can produce ethanol, and the photobiological process can produce hydrogen.

References

Agarwal, A.K., 2007. Biofuel (alcohol and biodiesel) application as fuel for internal combustion engine. Prog. Ener. Comb. Sci. 33, 233–271.

Alexopoulos, S., 2012. Biogas system: basic, biogas multifuction, principle of fermentation and hybrid application with a solar tower for the treatment of waste animal manure. JESTR 5, 48–55.

Alcaine, A.A., 2010. Biodiesel from Microalgae. Final degree project. Royal School of Technology Kungliga Tekniska Högskolan, Sweden.

Al Seadi, T., Rutz, D., Prassl, H., Köttner, M., Finsterwalder, T., Volk, S., Janssen, R., 2008. More about anaerobic digestion. In: Al Seadi (Ed.), Biogas Handbook. University of Southern Denmark, Esbjerg, pp. 21–29.

Amin, S., 2009. Review on biofuel oil and gas production process from microalgae. Ener. Conv. Manage. 50, 1834–1840.

Anonymous. History of Algae as Fuel. http://allaboutalgae.com/history (accessed 01.11.13).

Anonymous. Renewable Aviation Fuel. http://allaboutalgae.com/aviation-fuel (accessed 01.11.13).

Anonymous. Biotechnological and Environmental Applications of Microalgae. www.bsv.murdoch.edu.au/groups/beam/BEAM-App10.html (accesssed 01.11.13).

Barman, N., Satpati, G.G., Sen Roy, S., Khatoon, N., Sen, R., Kanjilat, S., Prasad, R.B.N., Pal, R., 2012. Mapping algae of Sundarban origin as lipid feedstock for potential biodiesel application. Algal Biom. Utln. 3 (2), 42–49.

Beal, C.M., Hebner, R.E., Romanovicz, D., Mayer, C.C., Connelly, R., 2013. Progression of lipid profile and cell structure in a research-scale production pathway for algal biocrude. Renew. Ener. 50, 86–93.

Benemann, J.R., Tellett, D.M., Suen, Yu., Hubbard, J., Tornabene, T.G. Chemical Profile of Microalgae with Emphasis on Lipids. Final Report to the Solar Energy Research Institute. School of Applied Biology. George Institute of Technology. www.nrel.gov/docs/legosti/old115418.pdf (accessed 05.11.13).

Borowitzka, M.A., 2013. Energy from microalgae: a short history. In: Borowitzka, M.A., Moheimani, N.R. (Eds.), Algae for Biofuel and Energy, Development in Applied Phycology, 5. Springer Science.

Brennan, L., Owende, P., 2010. Biofuel from microalgae—a review of technologies for production processing, and extraction of biofuels and co-products. Renew. Sust. Ener. Rev. 14, 557–577.

Bridgwater, A.V., Peacocke, G.V.C., 2000. Fast pyrolysis processes for biomass. Renew. Sust. Ener. Rev. 4, 1–73.

Bridgwater, A.V., Meier, D., Radlein, D., 1999. An overview of fast pyrolysis of biomass. Org. Geochem. 30, 1479–1493.

Business Green, 2012. Lufthansa Airline Signs Deal with Australia Company to Build Industrial-scale Plant in an Unnamed Country. www.businessgreen.com/bg/news/lufthansa-to-land-algae-jet-fuel-plant-in-europe (accessed 01.11.13).

Chinnasamy, S., Bhatnagar, A., Hunt, R.W., Das, K.C., 2010. Microalgae cultivation in a wastewater dominated by carpet mill effluents for biofuel applications. Biores. Tech. 101, 3097–3105.

Christi, Y., 2007. Biodiesel from microalgae. Biotech. Adv. 25, 294–306.

Danielo, O., 2005. An algae-base fuel. Biofutur 255.

Demirbas, A., 2010. Use of algae as biofuel sources. Ener. Conv. Manage. 51, 2738–2749.

Demirbas, A., Demirbas, M.F., 2010. Algae Energy. Springer, pp. 97–133.

Drapcho, C.M., Nhuan, N.P., Walker, T.H., 2008. Biofuel Engineering Process Technology. McGraw Hill, 197–259.

Edward, M., 2010. Algae History and Politics. www.algaeindustrymagazine.com/part-3-algae-history-and-politics (accessed 01.11.13).

Elliot, D.C., Sealock, L.J., 1999. Chemical processing in high-pressure aqueous environments: low temperature catalytic gasification. Trans. IchemE 74, 563–566.

FAO, Oil Production, FAO Corp Doc Repository. www.fao.org/docrep/w7241e/w7241e Oh.htm (accessed 02.05.07).

Gouviea, L., 2011. Microalgae as a Feedstock for Biofuels. Springer.

Hill, A.M., Feinberg, D.A., April, 1984. Fuel from microalgae lipid products. In: Presented at the energy from biomass: building on a generic technology base. Second Tech. Rev. Meet., pp. 23–25. Portland-Oregon.

Huang, G.H., Chen, F., Wei, D., Zhang, X.W., Chen, G., 2010. Biodiesel production by microalgal biotechnology. Appl. Ener. 87, 38–46.

Islam, M.A., Magnusson, M., Brown, R.J., Ayoko, G.A., Nabi, M.N., Heimann, K., 2013a. Microalgal species selection for biodiesel based on fuel properties derived from fatty acid profiles. Energies 6, 5676–5702.

Islam, M.A., Ayoko, G.A., Brown, R., Stuart, D., Heimann, K., 2013b. Influence of fatty acid structure on fuel properties of algae derived biodiesel. Procedia Eng. 56, 591–596.

Itoh, S., Suzuki, A., Nakamura, T., Yokoyama, S., 1994. Production of heavy oil from sewage sludge by direct thermochemical liqufaction. Desalination 98, 127–133.

Janssen, M.G.J., 2002. Cultivation of Microalgae: Effect of Light/Dark Cycles on Biomass Yield. Thesis of Wageningen Univesity, Ponsen & Looijen BV. Netherlands.

John, R.P., Anisha, G.S., Nampoothiri, K.M., Pandey, A., 2011. Micro and macroalgal biomass: a renewable source for bioethanol. Biores. Tech. 102, 186–193.

Kishimoto, M., Okakura, T., Nagoshima, H., Minowa, T., Yokoyama, S., Yamaben, K., 1994. CO_2 fixation oil production using microalgae. Ferment. Bioeng. 76 (6), 479–482.

Klass, D.L., 1998. Biomass for Renewable Energy, Fuels, and Chemicals. Elsevier Academic Press, 333–344.

Knothe, G., 2009. Improving biodiesel fuel properties by modifying fatty ester composition. Ener. Envir. Sci. 2, 759–766.

Knothe, G., 2008. Designer biodiesel: optimizing fatty ester composition to improve fuel properties. Ener. & Fuel 22, 1358–1364.

Knothe, G., Van Gerpen, J., Krahl, J., 2005. The Biodiesel Handbook. AOCS Press, pp. 277–280.

Lang, I., 2007. New Fatty Acids, Oxylipins and Volatiles in Microalgae. Dissertation Georg-August-University Göttingen.

Laing, I., 1991. Cultivation of Marine Unicellular Algae. Laboratory Leaflet Number 67, Lowestoff.

Lee, J.L., Yoo, C., Jun, S.Y., Ahn, C.Y., Oh, H.M., 2010. Comparison of several methods for effective lipid extraction from microalgae. Biores. Tech. 101, 575–577.

Liewellyn, C., Skill, S., 2007. Biofuel Production from Marine Algae. www.nbu.ac.uk/biota/Archieve_ marine BD/9251.htm (accessed 21.07.08).

Liu, J., Huang, J., Chen, F., 2011. Microalgae as feed stock for biodiesel production. In: Stoytcheva, Margarita (Ed.), Biodiesel—Feedstock & Processing Technologies. In Tech, pp. 133–160.

Mata, T.M., Martins, A.A., Caetano, N.S., 2010. Microalgae for biodiesel production and other application. A review. RSER 14, 217–232.

Miao, X., Wu, Q., 2004. High yield bio-oil production from fast pyrolysis by metabolic controlling of *Chlorella prototothecoides*. Biotechnology 110 (1), 85–93.

Miao, X., Wu, Q., Yang, C., 2004. Fast pyrolysis of microalgae to produce renewable fuels. Anal. Appl. Pyrolysis 71, 855–863.

Minowa, T., Yokoyama, S., Kishimoto, M., Okakurat, T., 1995. Oil production from algal cells of *Dunalliela tertiolecta* by direct thermochemical liquefaction. Fuel 74 (12), 1735–1738.

McKendry, P., 2003. Energy production from biomass (part 2): conversion technologies. Biores. Tech. 83, 47–54.

Minowa, T., Sawayama, S., 1999. A novel microalgal system for energy production with nitrogen cycling. Fuel 78, 1213–1215.

Minowa, T., Yokoyama, S., Kishimoto, M., Okakurat, T., 1995. Oil production from algal cells of *Dunaliella tertiolecta* by direct thermochemical liquifaction. Fuel 74 (12), 1735–1738.

Murakami, M., Yokoyama, S., Ogi, T., Koguchi, K., 1990. Direct liquefaction of activated sludge from aerobic treatment of effluents from the corn starch industry. Biomass 23, 215–228.

Ozkurt, I., 2009. Qualifying of safflower and algae for energy. Energy education science and technology part A. Energy Sci. Res. 23, 145–151.

Patil, V., Tran, K.Q., Giselrod, H.R., 2008. Toward sustainable production of biofuel from microalgae. Int. J. Mol. Sci. 9, 1188–1195.

Qi, Z., Lie, C., Tiejun, W., Ying, X., 2007. Review of biomass pyrolysis oil properties and upgrading research. Energy Conv. Manage. 48, 87–92.

Raja, A., Vipin, C., Aiyappan, A., 2013. Biological importance of marine algae—an overview. Int. J. Curr. Microbiol. App. Sci. 2 (5), 222–227.

Ramos, M.J., Fernandez, C.M., Casas, A., Rodrigulz, L., Perez, A., 2009. Influence of fatty acid composition of raw materials on biodiesel properties. Biores. Tech. 100 (1), 261–268.

Ramirez-Verduzco, L.F., Rodriguez, J.E.R., Jacob, A.R.J., 2012. Predicting cetane number, kinematic viscosity, density & higher heating value of biodiesel from its fatty acid methyl ester composition. Fuel 91, 102–111.

Sawayama, S., Minowa, T., Yokoyama, S.Y., 1999. Possibility of renewable energy production and CO_2 mitigation by thermochemical liquefaction of microalgae. Biom. Bioenergy 17, 33–39.

Sharma, K.K., Schuhmann, H., Schenk, P.M., 2012. High lipid induction in microalgae for biodiesel production. Energies 5, 1532–1553.

Shay, E.G., 1993. Diesel fuel from vegetable oils. Status &opportunities. Biom. Bioenergy 4 (4), 227–242.

Schenk, P., Thomas-Hall, S., Stephens, E., Marx, U., Mussgnug, J., Posten, C., Kruse, O., Hankamer, B., 2008. Second generation biofuel: high efficiency microalgae for biodiesel production. Bioenergy Res. 1, 20–43.

Sorathia, H.S., Rathod, P.P., Sorathia, A.S., July–September, 2012. Biogas generation and factors affecting the biogas generation—a review study. Int. J. Adv. Eng. Technol. III (III), 72–78.

Spolaore, P., Casson, C.J., Duran, E., Isambert, A., 2006. Commercial application of microalgae. Biores. Bioeng. 101 (2), 87–96.

Suyog, V.I.J., 2011. Biogas Production from Kitchen Waste Seminar Report, National Institute of Technology, Roukela.

Tsukahara, K., Sawayama, S., 2005. Liquid fuel production using microalgae. Jpn. Petro. Inst. 48 (5), 251–259.

Weldy, C.S., Huesemann, M., 2013. Lipid production by *Dunaliella salina* in batch culture: effect of nitrogen limitation and light intensity. US. DOE J. Undergraduate Res. 115–122. www.scied.science.doe.gov (accessed 01.11.13).

Wikipedia. Pond and Bioreactor Cultivation Methods. http://en.wikipedia.org/wiki/Alga culture, 115-122. (accessed 29.09.13).

Xin, L., Hong-yin, H., Jia, Y., 2010. Lipid accumulation and nutrient removal properties of a newly isolated freshwater microalgae, *Scenedesmus* sp. LX1, growing in secondary effluent. New Biotech. 27 (1), 59–63.

Xu, H., Miao, X., Wu, Q., 2006. High quality biodiesel production from a microalgae *Chlorella protothecoides* by heterotrophic growth in fermenters. Biotechnology 126, 499–507.

Yang, Y.F., Feng, C.P., Inamori, Y., Maekawa, T., 2004. Analysis of energy conversion characteristics in liquefaction of algae. Res. Cons. Recyc. 43, 21–33.

Zhang, X., Hu, Q., Sommerfeld, M., Puruhito, E., Chen, Y., 2010. Harvesting algal biomass for biofuel using ultra filtration membranes. Biores. Tech. 100, 5297–5304.

11

Biohydrogen from Microalgae, Uniting Energy, Life, and Green Future

Suphi S. Oncel

Ege University, Engineering Faculty, Department of Bioengineering, Izmir, Turkey

1. INTRODUCTION

In 1990, an astonishing photograph of our planet Earth, taken from a distance of 6 billion kilometer —nearly at the border of our solar system, by the space probe "Voyager I" which had started its journey in 1977—became one of the landmarks of man's great achievements. This photograph, which was the idea of the well-known scientist Carl Edward Sagan, showed once again the vulnerability and small size of Earth in the vast ocean of space. Today, Voyager I is somewhere in the universe, 1.925×10^{10} km from Earth according to the real-time odometer at the NASA Jet Propulsion Lab Website (voyager.jpl.nasa.nasa.gov/where/index.html), and still running into deep space as the farthest manmade spacecraft from Earth. On the other hand, the tiny "Pale blue dot," the phrase used for the photograph, is still in its place as the only home for humans. But the question is, "Are we aware enough of the vulnerability of this planet as being the one and only home for humans?"

Rising awareness after the environmental and economic crises through the nineteenth and twenty-first centuries forced societies to shift their attention to renewable energy sources (Amponsah et al., 2014) to fulfill the needs for a sustainable and environment friendly future (Wijffels and Barbosa, 2010; Oncel, 2013).

Fossil fuels are the flagship of the energy system for today's world with a share of more than 70% in the energy market (Parmar et al., 2011). Attempts to increase renewable fuels' share in the energy spectrum, which can be tracked with reports published by leading oil companies (Bentham, 2012; Dudley, 2014) and energy agencies (Muth, 2013; Conti, 2014; Hoeven, 2014) considering post fossil fuel scenarios, accelerated the research activities and also encouraged the flow of the new players to the sector.

Among the other alternatives of renewables, biobased sources with the ability to produce various biofuels like bioethanol, biomethanol, biodiesel, biomethane, and biohydrogen are attractive for the energy market (Schenk et al., 2008; Parmar et al., 2011; Singh et al., 2011; Shamsul et al., 2014; Nasir Uddin and Daud, 2014). Starting with the first generation followed by the second, third, and even the fourth, new concepts for biofuel technologies with the ultimate aim of fulfilling the sustainable, safe, diversified, environmentally friendly, and energy efficient demand of the energy future are the driving force of the progress (Naik et al., 2010; Oncel, 2013; Chaubey et al., 2013).

With regard to the importance of biofuels, this review aims to cover progress in the concept of microalgal biohydrogen production, highlighting theory, technology, economy, and ethics.

2. A NEW PARTNERSHIP THAT CAN CHANGE THE GAME: MICROALGAE AND BIOFUELS

Microalgae are a group of diversified microorganisms that can convert solar energy to cellular components by fixing CO_2 through photosynthesis. Even if their classification differs depending on the specifications of the microorganisms, the term *microalgae* usually covers all microscopic prokaryotic cyanobacteria and eukaryotic algae without the focus on their taxonomy (Pulz, 2001; Tomaselli, 2004). This simple terminology is quite convenient and accepted by researchers, keeping in mind their differences, especially in the area for microalgal bioproducts and biofuels (Pulz, 2001; Schenk et al., 2008; Oncel, 2013).

The history of the human use of microalgae goes back to human observations of wild animals feeding on microalgae. Mimicking this observation opened the doors of microalgal biotechnology and how microalgal biomass came to be used in many areas (Spoehr, 1976; Abdulqader et al., 2000; Olaizola, 2003; Pulz and Gros, 2004; Spolaore et al., 2006; Oncel, 2013) and today is reflected in the stage of energy.

Because of increasing demand on fuel utilization and rising CO_2 levels, microalgae have become more important due to their advantages over higher plants especially with regard to their potential for high photosynthetic efficiency, ability to utilize salty and brackish waters, no direct competition with food crops, potential for continuous supply of biofuels by harvesting year long, ability to sequester CO_2, and ability to be used in waste water systems both for treatment and biofuel production (Oncel, 2013). The US Aquatic Species Program accomplished an extensive study to screen potential microalgal species for transportation fuels (Sheehan et al., 1998).

Comprehensive reviews (Chisti, 2007; Schenk et al., 2008; Singh et al., 2011; Kruse and Hankamer, 2010; Parmar et al., 2011; Oncel, 2013; Uggetti et al., 2014) and reports (Weaver et al., 1979; Sheehan et al., 1998; Puri, 2009; Ryan et al., 2009; Lundquist et al., 2010) have covered the potential of microalgal biofuels in detail, showing that competition with fossil fuels is the main challenge for future. The key is to be realistic about their advantages and disadvantages while to be optimistic in their potential as a source of biofuel.

Microalgal biodiesel and bioethanol have attracted most of the attention as an alternative transportation fuel. But because their combustion still forms CO_2 to reach the desired reduction levels, researchers have also considered biohydrogen from microalgae as a zero emission fuel.

After the discovery of hydrogen by Henry Cavendish and naming by Antoine Lavoisier, hydrogen came to the stage of science with its unique properties like non-toxicity, high energy density, carbon-free combustion and flexibility to be applied in economy with various production processes (Midilli et al., 2005a,b; Parmar et al., 2011). Technically, even if hydrogen is the most abundant element in the environment because it is not in the molecular form, its production is energy consuming (Oncel, 2013; Sakurai et al., 2013; Torzillo et al., 2014). The target of building a hydrogen economy and rapid developments in the technology like fuel cell cars have become the accelerators for production of microalgal biohydrogen.

3. MICROALGAL BIOHYDROGEN PRODUCTION

Similar to some other scientific breakthroughs, biohydrogen from microalgae was initially investigated without the aim of utilization for a specific reason like energy. The first report of biohydrogen was published in 1896 by Jackson and Ellms about filamentous cyanobacteria *Anabaene*, which produced biohydrogen (Benemann, 1998; Lopes Pinto et al., 2002; Prince and Kheshgi, 2005; Torzillo and Seibert, 2013). Later, Gaffron and Rubin reported the ability of green microalgae *Scenedesmus obliquus* to produce biohydrogen in the dark (Gaffron and Rubin, 1942). The productivity was not remarkable until the introduction of a two-stage process (Melis et al., 2000), which was based on the separation of biomass production from biohydrogen production under sulfur-depleted anaerobic conditions accomplished by a sustainable production of 140 mL L^{-1} in 80 h by using *Chlamydomonas reinhardtii* culture in flat photobioreactors (Melis et al., 2000). This achievement caused a great deal of excitement among scientists.

Considering the production of biohydrogen from microalgae two strategies can be followed: (1) direct biohydrogen production by microalgae, which is the main focus of this review (Table 1) and (2) the utilization of microalgal biomass as a substrate for other microorganisms, which is the strength of microalgae in a biorefinery approach (Table 2).

3.1 How do Microalgae Produce Biohydrogen?

Microalgae are photosynthetic microorganisms that can harvest solar energy by special molecules (chlorophyll, carotenoid, phycobilin) to fix CO_2 and produce carbohydrates. Photosynthesis can be classified basically into two stages: (1) the light-dependent stage and (2) the light-independent (dark) stage. Light reactions takes place in the thylakoid membranes that contain light-harvesting antenna, photosystem I (PSI), photosystem II (PSII), cytochrome b_6f, and ATP synthase complexes. Photosystems also have reaction centers P700 (PSI) and P680 (PSII) where specific wavelengths of light can be harvested. The light reactions are responsible for the supply of NADPH$_2$ and ATP to fix inorganic carbon during the Calvin–Benson Cycle of the dark stage (Walker, 2009; Antal et al., 2011; Masojidek et al., 2013). On the other hand, photorespiration is a light-independent reaction that converts organic carbon back to CO_2. During respiration, the CO_2-fixing Rubisco, which actually has the ability to bind to both O_2 and CO_2, acts as an oxygenase. Rubisco fixes according to the partial pressure changes of CO_2 and O_2; if the pressure is in favor of CO_2 it will fix CO_2, and if not, it fixes O_2. Photorespiration is an unfavorable process that decreases the efficiency of photosynthesis in microalgae, because it consumes energy but does not produce any sugars. But because it can decrease oxygen levels in the culture, it can be important for hydrogen production (Miyake, 1998; Zhu et al., 2008; Masojidek et al., 2013; Oncel, 2013; Torzillo and Seibert, 2013). This prelude is the first step to understand metabolic activities for biohydrogen generation.

Hydrogen plays an important role in photosynthesis as being the vehicle in the photosynthetic transport grid where plants and microalgae can utilize water as the source of hydrogen (Antal et al., 2011; Polander and Barry, 2012). Different routes that are active in the hydrogen evolution process of microalgae are related to the dissipation of excess electrons of the electron transport chain when the main electron acceptors CO_2 and O_2 are absent or the other processes like Calvin cycle, photorespiration, and Mehler reactions used as sinks are blocked (Ghysels and Franck, 2010; Antal et al., 2011). The key of these routes is the transfer of electrons to hydrogen evolving enzymes, whether supplied directly from water splitting or indirectly from organic materials such as starch (Kruse and Hankamer, 2010).

TABLE 1   Microalgal Biohydrogen Production, a Brief Survey of Literature

Microalgae	Cultivation	Process	PBR type	Culture medium	Hydrogen production procedure	Time	H₂ production	References
Aphanothece halophytica	Batch-photomixotrophic	Lab	Vial (10 mL)	BG-11 + salt solution (containing various amounts of $NaCO_3$, $MgSO_4$, NaCl, Fe^{3+}, Ni^{2+}, sugars)	*Under argon flushing *Under different light intensities	7–21 days	Up to 13.8 μmol mg $Chla^{-1}$ h^{-1} with N-free medium containing 0.4 μM Fe^{3+}	Taikhao et al. (2013)
Anabaena variabilis (CCAP 1403/4B)	Continuous-photoautotrophic	Lab	Stirred (300 mL)	Nitrogen-free Allen–Arnon	*Nitrogen-free Allen–Arnon medium *Under vacuum for 15 min *Increased light intensity for 5 h at the 6th day of cultivation	35 days	12 – 14 mL g_{DW}^{-1} h^{-1}	Markov et al. (1997)
A. variabilis (ATCC 29413)	Continuous-photoautotrophic	Lab	Vial (14 mL)	Allen–Arnon (containing Na_2MoO_4 or Na_3VO_4 or neither)	Under argon flushing	80–100 h	Up to 5 nmol μg $Chla^{-1}$ h^{-1}at pH levels of 7–9 with Na_3VO_4 added cultures	Tsygankov et al. (1997)
Anabaena sp. PCC 7120 (3 hydrogenase mutants from PCC 7120)	Batch-photoautotrophic	Lab	Sealed polystyrene cuvettes (1 cm light path, 4.7 mL capacity)	BG 11	*Nitrogen-free BG 11 medium *Under anaerobic conditions *Various light intensities	32–40 h	Up to; 0.8–1 mmol per cuvette (4.7 mL)	Masukawa et al. (2002)
Anabaena sp. PCC 7120 and its mutant AMC 414	Continuous-photoautotrophic	Outdoor	Tubular-coiled (4.35 L)	BG 11	*Nitrogen-free BG 11 medium *Under argon atmosphere	7 days	14.9 mL h^{-1} L^{-1}_{PBR} (373 mL total) with the mutant	Lindblad et al. (2002)
A. variabilis (ATCC 29413)	Batch-photoautotrophic	Lab	Panel (500 mL capacity)	BG 11	*Nitrogen free BG 11 medium *Under anaerobic conditions *Increased light intensity	50 h	Up to; 40–50 mL	Yoon et al. (2006)
Chlamydomonas reinhardtii (C137 mt+)	Batch-photomixotrophic	Lab	Stirred glass bottles (1.2 L)	TAP	TAP-S medium	80 h	140 mL	Melis et al. (2000)
C. reinhardtii (CC124)	Batch-photomixotrophic, synchronous, and asynchronous growth,	Lab	Stirred glass bottles (1.2 L)	TAP	TAP-S medium (transferred by centrifugation)	140 h	*102 mL (with 4 h synchronized cultures) *86 mL (with unsynchronous cultures)	Tsygankov et al. (2002)

Organism	Culture mode	Scale	Reactor	Medium	Conditions	Duration	H_2 production	Reference
C. reinhardtii (Dang 137C mt+)	Batch-photomixotrophic	Lab	Stirred glass bottles (500 mL)	TAP	*TAP-S medium transfer by centrifugation *TAP-S medium inoculation by TAP culture (10% inoculation)	Up to 140 h	175 mL L^{-1} (with centrifuged cultures, under 20-40 µE m^{-2} s^{-1} average light intensity)	Laurinavichene et al. (2004)
C. reinhardtii (CC124)	Continuous-photomixotrophic	Lab	Stirred glass bottles (1050 mL)	TAP-S (90 µmol sulfate added)	*TAP-S medium, two stage chemostat with aerobic stage and anaerobic stage	4000 h	Up to: 0.58 mL h^{-1} L$_{PBR}$$^{-1}$	Federov et al. (2005)
C. reinhardtii (Dang 137C mt+ and a nonmotile mutant CC 1036 pf18 mt+)	Batch-photomixotrophic-immobilized	Lab	Panel (160 mL)	TAP (0.46 mM sulfate replete)	*TAP-S medium *TAP medium with limiting sulfate (10–20 mM) *Anaerobic conditions	23 days	45 mL day^{-1} (380 mL total)	Laurinavichene et al. (2006)
C. reinhardtii (Dang 137C mt+)	Batch-photoautotrophic, mixotrophic, and heterotrophic	Lab	Flat glass bottles (1.5 L volume)	HS (CO$_2$ bubbling); TAP with or without CO$_2$ bubbling	*Sulfur-free medium *Anaerobic	60–80 h	*1.1 ± 0.4 mmol L^{-1} (~2.8 mL (hL)$^{-1}$) in Photoautotrophic cultures *4.5 ± 1.6 mmol L^{-1} (~4.0 mL (hL)$^{-1}$) in Photoheterotrophic cultures *0.9 ± 0.8 mmol L^{-1} (~6.9 ml (hL)$^{-1}$) in photomixotrophic cultures	Kosourov et al. (2007)
C. reinhardtii (Dang 137C mt+)	Batch-photoautotrophic	Lab	Flat glass bottles (1.5 L)	High salt	*HS sulfur-free medium *Argon purging during the first 24 h of deprivation	100 h	*71 ± 3 mL L^{-1} (175 µE m^{-2} s^{-1} light intensity, pH 7.7) *52 ± 2 mL L^{-1} (420 µE m^{-2} s^{-1} light intensity, pH 7.4)	Tolstygania et al. (2009)
C. reinhardtii (CC124)	Semi-continuous-photomixotrophic	Lab	Stirred tank (2.5 L)	TAP	*TAP-S medium *Various dilutions *Anaerobic conditions	127 days	1108 mL	Oncel and Sukan (2009)
C. reinhardtii (CC124)	Batch-photomixotrophic-immobilized	Lab	Vials (75 mL)	TAP	*TA-S-P (no sulfate, no phosphate) medium *Alginate entrapped harvested cells flushed with or without argon *Anaerobic conditions	160–180 h	0.30 ± 0.02 mol m^{-2} Chl (12.5 µmol mg^{-1} Chl h^{-1})	Kosourov and Seibert (2009)

Continued

TABLE 1   Microalgal Biohydrogen Production, a Brief Survey of Literature—cont'd

Microalgae	Cultivation	Process	PBR type	Culture medium	Hydrogen production procedure	Time	H_2 production	References
C. reinhardtii (CC124)	Batch-photomixotrophic	Lab	Glass bottles (325 mL)	TAP	*TAP-S or TAP-N *DCMU mixed ethanol addition *Dark anaerobic fermentation	192 h	*3.95 µmol × 10^6 cells^{-1} h^{-1} with TAP-S *3.4 µmol × 10^6 cells^{-1} h^{-1} with TAP-N	Philipps et al. (2012;
C. reinhardtii (CC124)	Batch-photomixotrophic	Lab	Stirred tank (1, 2.5, 5 L)	TAP	*TAP-S medium transfer by centrifugation *Various mixing time *Various light energy *Scale-up	192 h	Max with 1.22 kJ s^{-1} m^{-3} light energy and 2.5 min mixing time: *1 L PBR: 1.53 ± 0.05 mL L^{-1} h^{-1} *2.5 L PBR: 1.32 ± 0.05 mL L^{-1} h^{-1} *5 L PBR: 1.02 ± 0.05 mL L^{-1} h^{-1}	Oncel and Sabankay (2012)
C. reinhardtii (CC124)	Batch-photomixotrophic	Lab	Tubular (110 L)	TAP	*TAP-S medium transfer by centrifugation *Modified with silica nanoparticle to enhance scattering	48 h	3121.5 ± 178.9 mL (0.6 mL L^{-1} h^{-1})	Giannelli and Torzillo (2012)
C. reinhardtii (CC400)	Batch-photomixotrophic	Lab	Glass bottles (100 mL)	TAP	*TAP-S medium transfer by centrifugation *Wild type and fnr-RNAi mutants tested	120 h	Up to 500 µL mg Chla^{-1} h^{-1} with mutants	Sun et al. (2013)
C. reinhardtii (CC849 and Iba mutants)	Batch-photoheterotrophic-co-culture	Lab	Glass bottles (60 mL)	TAP	*TAP grown Chlamydomonas and YEM grown Bradyrhizobium japonicum was washed and transferred to TAP-S for co-culture *Various inoculation ratios tested	21 days	Max 300 µmol with 849 + B. japonicum co-culture (100:1)	Shuangxiu et al. (2012)

Species	Culture type	Scale	Reactor	Medium	Conditions	Duration	Results	Reference
C. reinhardtii (CC849)	Batch-photoheterotrophic-co-culture	Lab	Glass bottles (60 mL)	TAP	*TAP grown Chlamydomonas and YEM grown bacteria L2, L3, L4 (Genus: Stenotrophomonas, Microbacterium, Pseudomonas) were washed and transferred to TAP-S for co-culture *Various inoculation ratios tested	13 days	Max 61.5 μmol with 849 + L4 co-culture (80:1)	Li et al. (2013)
C. reinhardtii (C238)	Batch-photomixotrophic-mixed-culture	Lab	Flat glass bottles (1 L)	Modified Bristol medium	*Microalgae was mixed cultivated with Rhodospirillum rubrum NCIB 8255 *Transferred to tubes (10 mL) for dark tests *Transferred to membrane reactor (15 mL) for L/D cycles (12:12) test	12 h (Dark) 156 h (L/D)	Max 2.39 μmol mg^{-1} DW 12 h^{-1} (total mass base) during dark Max ~3 μmol mg^{-1} DW 12 h^{-1} with L/D cycle Both at 8:2 population ratio	Miyamoto et al. (1987)
C. reinhardtii (Stm6Glc4 mutant)	Batch-photomixotrophic	Lab	Stirred tank (2 L)	TAP	*TAP medium with different sulfur concentrations selected (5–70 mg L^{-1}) *11 mg L^{-1} initial sulfur concentration applied for PBR experiments *At 120 h culture sealed and hydrogen phase started *Various light intensities were tested (80–721 μE m^{-2} s^{-1})	660 h	650 mL g^{-1} dry biomass with 280 μE m^{-2} s^{-1} intensity	Lehr et al. (2012)

Continued

TABLE 1   Microalgal Biohydrogen Production, a Brief Survey of Literature—cont'd

Microalgae	Cultivation	Process	PBR type	Culture medium	Hydrogen production procedure	Time	H_2 production	References
C. reinhardtii (CC124 mt⁻) and 4 mutants (CC4348,CC2890, CC1354,CC3723)	Batch-photoheterotrophic	Lab	Flat glass bottles (125 mL)	TAP	*Cells suspended in TAP medium under dark anaerobic conditions for 3 h with nitrogen flushing *Different lights were tested (White, Red690 nm, no light) *Different light–dark cycles tested (15:15 h:h) *TAP-S cultures tested for comparison	18–48	Up to 8.81 mL³ g⁻¹ DW with CC3723 mutant	Hoshino et al. (2013)
C. reinhardtii (CC124)	Batch-photoheterotrophic	Lab	Glass bottles (130 mL)	TAP	*Cells flushed with nitrogen *CCCP (15 µmol L⁻¹) added	12 h	Up to 20 mmol mg⁻¹ Chl with CCCP-treated cultures	Yang et al. (2014)
C. reinhardtii (Dang 137C mt+)	Batch-photomixotrophic	Lab	Stirred tank (0.5 L) and flat glass bottles (0.55 L)	TAP	*TAP-S, TA-P or TA-S-P *TA-P also tested in flat PBRs (sealed after 80 h, initial Chl concentration diluted to ~1.5 mg L⁻¹)	140 h (With tank) 300 h (With flat PBR)	Up to 80 mL L⁻¹	Batyrova et al. (2012)
C. reinhardtii (rbcS-T60-3 mt⁻ mutants; Y67A,Y68A,Y72A)	Batch-heterotrophic	Lab	Shake flasks (80 mL)	TAP	*Cells in dark harvested and transferred to S deplete and S replete medium	120 h	*Maximum 709.24 µmol L⁻¹ with Y67A mutant under S-deplete medium *Maximum 433.23 µmol L⁻¹ with Y67A mutant under S-replete medium	Pinto et al. (2013)

Organism	Mode	Condition	Reactor	Medium	Notes	Duration	Production	Reference
C. reinhardtii (CC124)	Batch-photomixotrophic	Lab	Tubular and flat panel (5 L)	TAP	*TAP-S medium transfer by centrifugation *Mixing time and light intensity as the comparison factors	120 h	*1.3 ± 0.05 mL L^{-1} h^{-1} with panel type PBR *1.05 ± 0.05 mL L^{-1} h^{-1} with tubular type PBR	Oncel and Kose (2014)
C. reinhardtii (CC124)	Batch-photomixotrophic	Outdoor	Tubular (50 L)	TAP	*TAP-S medium transfer by centrifugation *Effect of outdoor solar light on the outdoor acclimated cultures	75 h	*No hydrogen produced with cultures directly transferred from laboratory *930 mL H$_2$ under solar light with acclimated cultures *930 ± 100 mL under artificial light	Scoma et al. (2012)
C. reinhardtii (WT and D1- mutants)	Batch-photomixotrophic	Lab	Flat glass bottles (1.1 L)	TAP	*TAP-S medium transfer by centrifugation	Up to 240 h	*0.47 ± 0.11 mL L^{-1} h^{-1} with WT *1.54 ± 0.31 mL L^{-1} h^{-1} with D240 *2.60 ± 0.18 mL L^{-1} h^{-1} with D2390-40	Faraloni and Torzillo (2010)
C. reinhardtii (CC124)	Batch-photomixotrophic	Lab	Flat glass bottles (1.1 L)	TAP and olive oil mill wastewater (OMWW)	*TAP-S medium transfer by centrifugation *Effect of OMWW concentration on hydrogen	120 h	*1.03 ± 0.18 mL L^{-1} h^{-1} with TAP grown culture *1.29 ± 0.14 mL L^{-1} with TAP-OMWW grown culture	Faraloni et al. (2011)
C. reinhardtii (CC124) (WT and D1-mutants)	Batch-photomixotrophic	Lab	Flat glass bottles (1.1 L)	TAP	*TAP-S medium transfer by centrifugation *Effect of D1 mutations on hydrogen production	Up to 280 h	Mean production rates between: *0.3 ± 0.11 mL L^{-1} h^{-1} with WT *5.15 ± 0.11 mL L^{-1} h^{-1} with A250L	Torzillo et al. (2014)
Chlorella pyrenoidosa (C-101)	Batch-photoautotrophic	Lab	Bubble column (650 mL)	Modified Bristol	Nitrogen flushing under dark for ">20" hours	60−65 h	6.9 × 10^{-2} m^3 kg^{-1} cell	Kojima and Lin (2004)
Chlorella vulgaris (MSU 01)	Batch-photomixotrophic	Lab	Stirred (500 mL)	Modified BG11 and MJ	*Anaerobic conditions *Corn stalk as carbon source in the medium	6−7 days	26 mL	Amutha and Murugesan (2011)

Continued

TABLE 1   Microalgal Biohydrogen Production, a Brief Survey of Literature—cont'd

Microalgae	Cultivation	Process	PBR type	Culture medium	Hydrogen production procedure	Time	H₂ production	References
Chlorella protothecoides	Batch-photoheterotrophic	Lab	Stirred vessels (650 mL)	TAP (with various NH₄Cl concentrations; 0.35–7 mM)	*Harvested, washed and transferred to TAP-S or TAP-N medium under continuous illumination	140 h	Max 233.7 mL L⁻¹ with TAP-S cultures (0.35 mM NH₄Cl)	He et al. (2012)
Chlorella vulgaris (NIER-10003)	Batch-photomixotrophic-immobilized	Lab	Glass bottle (1 L)	MA	*Harvested, washed, and transferred to MA-S *Nitrogen flushed *Various carbon sources, their concentrations and medium pH were tested *Optical fibers tested	40–120 h	Maximum volume of 1315 mL L⁻¹ with pH = 8, sucrose = 5 mg L⁻¹ Maximum rate of 24 mL L⁻¹ h⁻¹ was with fructose	Rashid et al. (2013)
Chlorella protothecoides	Batch-photoheterotrophic	Lab	Glass tubes (20 mL)	TAP	*Harvested, and transferred to TAP-S or TAP-N medium with/ without DCMU tested	120 h	Up to ~140 mL L⁻¹ (with simultaneous nitrogen and sulfur limitation)	Zhang et al. (2014)
Microcystis flos-aquae	Batch-photomixotrophic	Lab	Glass bottle (60 mL)	BG 11	*Sealed samples (30 mL) were placed under dark or light	26 days	6.11 μmol mg Chla⁻¹ h⁻¹	Wei et al. (2013)
Platymonas subcordiformis	Batch-photoautotrophic	Lab	Serum bottle (295 mL)	Sea water medium	*Sulfur deprive medium *Incubation under dark *Anaerobic conditions *Continuous illumination afterward	50 h	11720 nL h⁻¹ (with Seawater-S medium)	Guan et al. (2004a)

Species	Mode		Vessel	Medium	Conditions	Duration	Yield	Reference
Platymonas subcordiformis	Batch-photoautotrophic	Lab	Serum bottle (295 mL)	Sea water medium	*Dark incubation *Anaerobic nitrogen atmosphere *Addition of specific effectors (DCMU, DCCD, DBMIB, and CCCP) *Continuous illumination afterward	8 h	0.339 mL h^{-1} L^{-1} (1.44 mL); based on 1 × 10^6 cells mL^{-1} at 15 µM CCCP added culture after 8 h of illumination	Guan et al. (2004b)
Platymonas subcordiformis	Batch-photoautotrophic	Lab	Torus (1.5 L)	Defined mineral	*Dark incubation *Anaerobic nitrogen atmosphere *15 µM CCCP addition *Continuous illumination afterward	4–6 days	7.20 mL h^{-1} (236.6 ± 7.0 mL)	Ji et al. (2010)
Platymonas helgolandica var. tsingtaoensis	Batch-photoautotrophic	Lab	Jars (130 mL)	F2	*Anaerobic *Illuminated-S medium *Dark CCCP/DCMU or both added medium	25 h	*0.002 mmol L^{-1} with −S medium *0.160 mmol L^{-1} with CCCP medium *0.014 mmol L^{-1} with DCMU medium *0.290 mmol L^{-1} with CCCP + DCMU medium	Zhang et al. (2012)
Spirulina platensis (NIES-46)	Batch-photoautotrophic	Lab	Erlenmeyer flasks (60 mL)	SOT	*Nitrogen-free medium *Under dark *Anaerobic conditions	20 h	2 µmol mg$^{-1}_{DW}$	Aoyama et al. (1997)
Synechocystis sp. (PCC 6803 and its mutant NDH-1 complex deficient M55)	Batch-photoautotrophic-encapsulated	Lab	Vials (2 mL)	BG 11	*Nitrogen atmosphere *Under cycled light and dark exposure	5 days	0.005–0.045 mM	Dickson et al. (2009)
Tetraspora sp. (CU2551)	Batch-photomixotrophic	Lab	Vials	TAP (with, or without, 0.5 mM β-mercaptoethanol for 24 h)	*Sulfur and Nitrogen deprived medium *Argon flushed anaerobic atmosphere	>24 h	17.3–161.7 µmol mg$^{-1}_{Chla}$ h^{-1}	Maneeruttanarungro et al. (2010)

TABLE 2   Integrated Biohydrogen Processes With a Special Emphasis on Microalgal Biomass as a Substrate for Fermentation

Microalgae	System	Max hydrogen production	References
Anabaena sp. PCC7120	Fermentation of hydrogen produced residual algal biomass by Enterobacter aerogenes	$0.0114 \, kg \, kg^{-1}$ biomass	Ferreria et al. (2012)
Arthrospira (Spirulina) platensis, Nannochloropsis sp. and Dunaliella tertiolecta	Anaerobic fermentation of thermal pretreated algal biomass by immobilized Clostridium acetobutylicum	$8.5 \, mmol \, L^{-1} \, day^{-1}$	Efremenko et al. (2012)
Arthrospira (Spirulina) platensis	Fermentation of pretreated (boiling, bead milling, ultrasonication, enzymatic hydrolysis) wet biomass by anaerobic activated sludge	$92 \, mL \, g^{-1} \, DW$	Cheng et al. (2012)
Arthrospira maxima	Enzymatic hydrolysis and fermentation of algal biomass by activated sludge	$78.7 \, mL \, g^{-1}$	Cheng et al. (2011)
Chlorella vulgaris ESP6	Dark fermentation of acid or alkaline/enzyme pretreated algal biomass hydrolysate by Clostridium butyricum CGS5	$246 \, mL \, L^{-1} \, h^{-1}$	Liu et al. (2012)
Chlorella sorokiniana	Dark fermentation of HCl-heat pretreated algal biomass by Enterobacter cloacae IIT-BT08	$148 \, mL \, L^{-1} \, h^{-1}$	Kumar et al. (2013)
Chlorella sp.	Simultaneous hydrolysis and fermentation of algal biomass (Chlorella sp. ESP-6 and commercial Chlorella powder) by sewage sludge consortia	$1.01 \pm 0.09 \, mmol \, g^{-1}$ powder	Ho et al. (2013)
Chlorella vulgaris and Dunaliella tertiolecta	Fermentation of algal biomass by anaerobic activated sludge	$12.6 \, mL \, g^{-1}$ vs. from D. tertiolecta	Lakaniemi et al. (2011)
Thalassiosira weissflogii	Dark fermentation of algal biomass with the thermophilic bacterium Thermotoga neapolitana	$18.1 \pm 5 \, mL \, L^{-1} \, h^{-1}$	Dipasquale et al. (2012)
Chlamydomonas reinhardtii (IAM C-238), Chlorella pyrenoidosa (IAM C-212), and Dunaliella tertiolecta (ATCC 30909)	Fermentation of algal biomass by Lactobacillus amylovorous (ATCC 33620) and Rhodobacter sphaeroides RV, Rhodobacter capsulata (ATCC 11166), Rhodospirillum rubrum (ATCC 11170), Rhodovullum sulfidophilus (ATCC 35886), Rhodobium marinum (ATCC 35675)	$8 \, mol \, mol^{-1}$ starch–glucose equivalent from C. reinhardtii biomass with R. marinum	Ike et al. (1997)

Microalgae	Process	Value	Reference
C. reinhardtii (IAM C-238) and Dunaliella tertiolecta (ATCC 30909)	Fermentation of algal biomass by Lactobacillus amylovorous and Rhodobium marinum (A-501) mixed culture	2.47 ± 0.21 mmol h^{-1} L^{-1} with D. tertiolecta biomass	Kawaguchi et al. (2001)
C. reinhardtii (UTEX 90)	Fermentation of algal biomass by C. butyricum to hydrogen and organic acids followed by the photodissimilation of organic acids to hydrogen by R. sphaeroides	8.30 mol mol^{-1} starch–glucose equivalent algal biomass	Kim et al. (2006)
Chlamydomonas (MGA 161)	Fermentation of algal biomass by Rhodovulum sulfidophilum (W-1S)	0.1875 mmol L^{-1} h^{-1}	Miura et al. (1997)
C. reinhardtii (UTEX 90)	Fermentation of heat-HCl and enzymatic pre-treated microalgal biomass by hyperthermotolerant eubacterium T. neopolitana	227 ± 9.4 mL L^{-1} h^{-1}	Nguyen et al. (2010)
Nannochloropsis sp.	Dark fermentation of algal biomass extraction residues by E. aerogenes	60.6 mL g$^{-1}_{DW}$	Nobre et al. (2013)
Scenedesmus sp.	Fermentation of algal biomass by anaerobic digested sludge	3.84 mL h^{-1} (19.53 mL g$^{-1}_{Volatile\ solids}$)	Zhiman et al. (2011)
Scenedesmus obliquus	Fermentation of algal biomass by C. butyricum	2.9 ± 0.3 mol mol$^{-1}_{algal\ sugars}$	Ferreria et al. (2013)
S. obliquus (YSW15)	Fermentation of algal biomass by anaerobic digester consortia	120.3 mL L^{-1} h^{-1}	Choi et al. (2011)
S. obliquus	Fermentation of algal biomass by bacteria (E. aerogenes ATCC 13048 and C. butyricum DSM 10702)	113.1 mL g$^{-1}_{VS}$ C. butyricum	Batista et al. (2014)
Spirulina platensis, Anabaena variabilis, and Chlorella sp.	Processing of algal biomass by an integrated system of methanogenic culture, R. capsulatus and Thermohydrogenium kirishi	1080 mL L^{-1} day^{-1} with Chlorella sp. Biomass by R. capsulatus	Teplyakov et al. (2002)

The first route is the direct biophotolysis of water into hydrogen and oxygen where the water-splitting PSII and ferredoxin-reducing PSI act cooperatively. The second route is the indirect biophotolysis where the electrons from glycolysis are transferred to the linear electron transport chain and utilized only by a PSI active hydrogen evolution pathway. A third route other than these light-dependent routes is the dark fermentation of the decarboxylated pyruvate from glycolysis by pyruvate ferredoxin oxidoreductase (Beer et al., 2009; Maness et al., 2009; Ghysels and Franck, 2010; Torzillo et al., 2014).

Applications that can show important enhancement with regard to biohydrogen productivities in microalgae (Lopes Pinto et al., 2002; Mathews and Wang, 2009; Oncel and Kose, 2014; Torzillo et al., 2014) include the proper changes in several culture parameters such as adjusting salinity, percentages of sparged gasses (CO_2, Ar), nutrient levels, illumination patterns as well as partial pressure of head space.

3.2 Enzymes in Microalgal Biohydrogen Production

Hydrogen is catalyzed by specific enzymes that differ according to the taxonomic cluster of the hydrogen-producing microorganisms, microalgae, or cyanobacteria (Melis, 2002; Lopes Pinto et al., 2002; Lindblad, 2004; Simmons et al., 2014). Cyanobacteria have NiFe-type uptake hydrogenase (used for energy conservation and hydrogen oxidation, so not wanted in a sustainable hydrogen production process), NiFe-type bidirectional hydrogenase (used for energy conservation and maintaining intracellular redox balance), and nitrogenase (used for hydrogen production and nitrogen fixation). Eukaryotic microalgae have FeFe hydrogenases utilized in hydrogen production, which are used for both energy conservation and fermentation (Prince and Kheshgi, 2005; Mathews and Wang, 2009; Srirangan et al., 2011;

Tamburic et al., 2012; Catalanotti et al., 2013; Sakurai et al., 2013).

Considering the input of 2 ATP per electron in the nitrogenase-based hydrogen production efficiency decreases nearly by half compared to the ATP-free hydrogenase-based production (also the advantage of high turnover potential of hydrogenases (up to $10^6 \, s^{-1}$) should be mentioned) both for direct and indirect biophotolysis. Apart from this disadvantage, nitrogenases are less sensitive to oxygen, and because of the hydrolysis of the ATP they can generate hydrogen up to 50 atm pressure, while hydrogenases cannot, where the free energy is relatively small (Kruse et al., 2005; Prince and Kheshgi, 2005; Oncel, 2013). Even if the efficiencies (hydrogenase: 10–13%; nitrogenase: 6%) and the mechanisms (hydrogenase: 4 photons/produced H_2; nitrogenase: 15 photons/produced H_2) are different, the important commonality for both of these enzyme classes is the sensitivity to oxygen, which shaped the strategies for biohydrogen production by microalgae in a way that anaerobiosis should be accomplished (Kruse et al., 2005; Prince and Kheshgi, 2005; Mathews and Wang, 2009).

Cyanobacteria (unicellular, filamentous, or colonial) have different strategies for hydrogen production. A heterocyst-forming group that is able to fix nitrogen can separate the oxygen from hydrogen evolution spatially with the help of the differentiated heterocyst. Heterocysts are constructed by an oxygen impermeable glycolipid layer with the lack of oxygen-evolving PSII and have high respiration rates, which are able to consume the produced oxygen easily and act as a safety chamber against the negative effects of oxygen (Maness et al., 2009; Srirangan et al., 2011; Sakurai et al., 2013). The nonheterocyst group does not have the ability to shield the effects of oxygen from vegetative cells but can still fix nitrogen and produce hydrogen by the on–off control of the photosynthesis with a specialized mechanism, special storage granules (cyanophycin) and nitrogenase-comprising group of cells

(Sakurai et al., 2013). On the other hand, some unicellular cyanobacteria can produce hydrogen with the indirect separation of the oxygenic processes under light and dark cycles. Oxygen from photosynthesis continued by the nitrogen fixation in the day followed by the oxygen consuming respiration ended up in the production of biohydrogen at night (Mathews and Wang, 2009; Srirangan et al., 2011; Sakurai et al., 2013).

Similarly, eukaryotic microalgae can produce hydrogen directly by utilizing their FeFe hydrogenase enzymes during photosynthesis, but the production is very transient due to the sensitivity to concomitantly produced oxygen from the splitting of water or by the separation of the oxygen evolution by the day and night shift with the help of the oxygen-consuming respiration (Oncel, 2013; Torzillo and Seibert, 2013). Table 1 summarizes the microalgae and cyanobacteria species in terms of biohydrogen production, giving an idea about the state of the art and research focuses.

4. MICROALGAL CULTURE SYSTEMS: OPEN AND CLOSED

The first step of a sustainable biohydrogen production is microalgae cultivation, which needs special production systems with optimum conditions. Optimum conditions for light, temperature, nutrients, and pH should be met to prevent inhibition or cell death (Long et al., 1994; Ort, 2001; Pulz, 2001; Posten, 2009; Zitelli et al., 2013). According to the atmospheric interaction, culture systems can be classified as either open systems (Figure 1) or closed systems (Figures 2–4).

4.1 Open Systems

Open systems target the outdoor production with minimal investment and having a special emphasis on the resistant species like *Spirulina*, *Dunaliella*, or *Chlorella* even if they can be constructed in greenhouses or with artificial illumination systems (mostly lab or pilot scales) (Pulz and Gross, 2004; Chisti, 2007). These systems can be subdivided according to their presence of mixing strategy, as unmixed and mixed.

Unmixed ponds are primitive systems where the mixing and aeration is established by the help of the wind acting on the water surface or sometimes by manpower in smaller scales (Becker, 1995; Grobbelaar, 2009; Borowitzka and Moheimani, 2013). Today the biggest commercial *Dunaliella* production in the world, owned by BASF, has a productivity of about $1 \, g \, m^{-2} \, day^{-1}$ and produces β-carotene; it uses unmixed natural lagoons in Australia (Borowitzka, 2013).

Mixed open systems (Figure 1) having an average surface-to-volume ratio around $10 \, m^{-1}$ have found greater application in commercial productions (Pulz, 2001; Chisti, 2007; Masojidek et al., 2013). Raceway ponds are one of the highly recognized mixed systems in the industry both in wastewater treatment and microalgae cultivation (Stephenson et al., 2010). They are shallow (around 0.3 m deep), rectangular-shaped ponds divided by a separator wall to define the flow channels (Chisti, 2007; Borowitzka and Moheimani, 2013). The world's largest facility of raceways is owned by the Earthrise Farms and covers an area of about $440,000 \, m^2$ (Chisti, 2007). Another important design, circular ponds, can be scaled up to 50-m diameter and $50,000 \, m^2$ in area and are mainly used in Japan, Taiwan, and Indonesia (Lee, 2001; Borowitzka and Moheimani, 2013; Costa and Morais, 2014).

Another design that is modified with the aim of contributeing high light utilization and turbulent mixing regime is the cascade (also called thin layer cascade or inclined) open systems. Cascade systems that were developed in Trebon, Czech Republic, in the 1960s have the advantage of reaching higher culture densities (up to $35 \, g \, L^{-1}$) and higher surface area-to-volume ratios ($100 \, m^{-1}$) (Becker, 1995; Masojidek et al.,

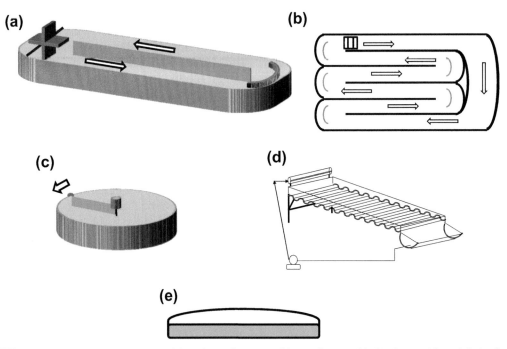

FIGURE 1   Open systems: Raceway type (a), multi-grid raceway (b), circular type (c), thin layer with undulating base (d), covered pond (e).

2011; Borowitzka and Moheimani, 2013; Zitelli et al., 2013).

4.2 Closed Systems (Photobioreactors)

Closed systems, technically termed *photobioreactors* (PBRs), are systems where the culture is partially or completely isolated from the surrounding environment. PBRs can be constructed outdoors to interact with the environment for heat exchange and light harvesting to be cost-effective, while isolating the culture to prevent contamination.

Compared to open ponds, PBRs have diversified designs that have the ability to produce more species other than the resistant ones. Especially in the area of energy, thousands of different species can be investigated with the help of the PBRs, enabling the opportunity to find alternative high producers. PBRs can be used for heterotrophic, mixotrophic, or photoautotrophic cultivation with different modes such as continuous, batch, or semicontinuous to investigate various process parameters (Becker, 1995; Ogbonna and Tanaka, 1997; Ogbonna et al., 1995; Pulz, 2001; Oncel and Sukan, 2008, 2009; Oncel and Sabankay, 2012; Fernandez et al., 2013; Oncel and Kose, 2014).

PBR designs focus on two main units: illuminated area (solar receiver or solar stage) and the culture reservoir. Solar stage is the naturally or artificially illuminated zone that targets light harvesting with the awareness of the optimum levels and inhibition risks. On the other hand, the culture reservoir unit, which actually houses a smaller volume relative to the illuminated area, comprises the degassing, control, mixing, and circulation systems. Depending on the design, these units can be separated or unite according to the needs of the process. Considering their major

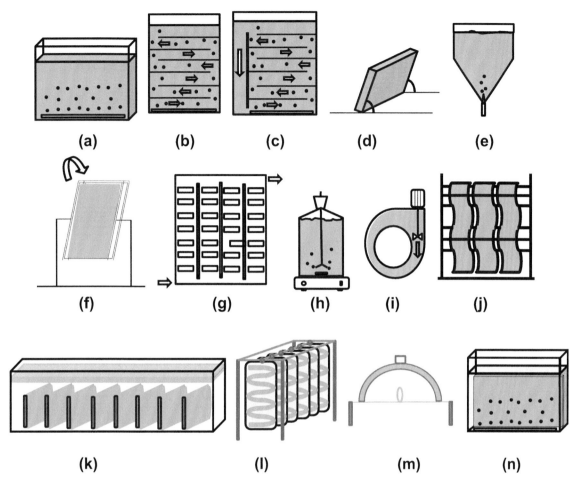

FIGURE 2 Panel type PBRs: Vertical flat panel (a), with baffles (b), airlift type (c), inclined (d), V-shaped (e), rocking (f), integrated compartment (g), Roux-type (h), Torus type (i), accordion type (j), submerged bag type (k), with pressed tubular grid (l), dome type (m), immobilized (n).

designs, PBRs can be classified as flat panels (Figure 2), tubular (Figure 3), and fermenter types (tanks and columns) (Figure 4).

Flat panel (or flat plate) PBRs are compact vertical designs that are widely used in microalgae cultivation both in the laboratory and outdoors (Tredici and Zitelli, 1998; Pulz, 2001; Oncel and Kose, 2014). They are basically shear friendly hydraulically, pneumatically, or mechanically mixed (with/without baffles). PBRs having high surface-to-volume ratio for better light harvesting. With the modular design, they have the ability to be oriented according to the sun, which makes them applicable outdoors (Ugwu et al., 2008; Xu et al., 2009; Wijffels and Barbosa, 2010). From cheap plastic materials to costly glass, various materials increase the chance for the economic competitiveness of these PBRs for specific targets like biofuels to high-value chemicals (Zitelli et al., 2013). Technically, the height

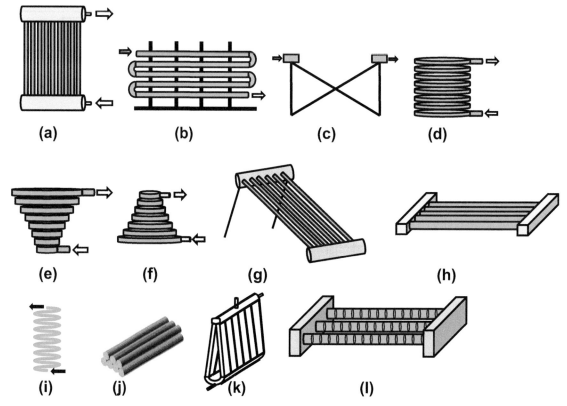

FIGURE 3 Tubular PBRs: Vertical with manifolds (a), fence type with u-bends (b), α-type (c), helical (d), conical (e), pyramid (f), inclined (g), horizontal (h), strongly curved (i), multi stack (j), Vertical loop like (k), with static mixers (l).

and width of the panels are around 1.5×2.5 m with an optical path of 0.1 m to avoid high construction loads and allow for better light utilisation (Posten, 2009; Fernandez et al., 2013).

Tubular PBRs are the other well-known category with various subtypes that are developed and applied for both laboratory and commercial scales. They are basically a horizontal- or vertical-oriented stack of tubes attached by U-bends or manifolds, in which culture circulation and mixing are accompanied by pumps or airlifts (Grima et al., 2003; Wongluang et al., 2013; Zitelli et al., 2013). Similar to the flat panels, tubular PBRs have surface-to-volume ratio up to 80 m^{-1} (some very high ratios up to 2000 m^{-1} have also been reported) and are

generally constructed by transparent tubes of ≤ 0.1 m in diameter for light harvesting; the length of the tubes should be well designed to prevent O_2 accumulation or the need for high pumping power (Tredici and Zitelli, 1998; Pulz, 2001; Chisti, 2007).

Fermenter-type PBRs can be classified as the stirred tanks and aerated columns. The concept of design actually depends on standard bioreactors used in bioprocesses usually modified with mixing and illumination systems for microalgae cultivation. Airlift and bubble columns are the vertical PBRs with no moving mechanical parts (unless modified with internal mixers) having mass transfer coefficients about 0.006 s^{-1} and low power inputs for mixing by supplying

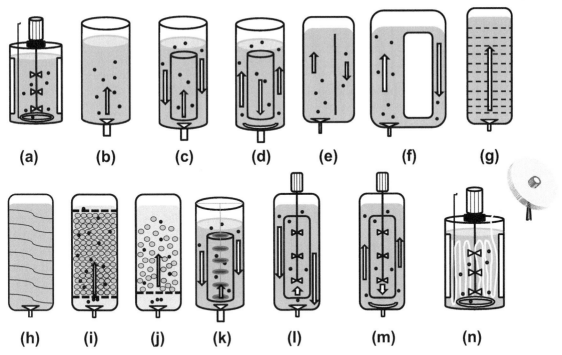

FIGURE 4 Fermenter type PBRs: Stirred tank (a), bubble column (b), internal loop draft tube sparged airlift (c), anulus sparged (d), divided column (split cylinder airlift) (e), external loop (f), perforated plate column (g), static mixer (h), packed bed (i), fluidized bed (j), draft tube baffled airlift (k), mechanical mixer adapted draft tube sparged (l), annulus sparged (m), Collector adapted stirred tank (internal illumination by optic fibers) (n).

aeration rates of 0.25 vvm (Posten, 2009; Xu et al., 2009; Dasgupta et al., 2010). Column PBRs can be constructed as simple polyethylene bags or in more sophisticated designs up to 500 L working capacity (Oncal and Sukan, 2008; Xu et al., 2009; Zitelli et al., 2013; Oncel, 2014; Pirouzi et al., 2014) Stirred tank bioreactors, which can also be used as PBRs, are mechanically mixed, sparger or nozzle-aerated glass or steel tanks usually used in laboratory scales with a typical height-to-diameter ratio up to 3 for an adequate mixing (Shuler and Kargı, 1992).

5. CHALLENGES

The term *challenge* can be defined as passing each production bottleneck to reach the ultimate goal of sustainable, environment friendly, clean, and feasible supply of energy with an equal share of sources and wealth for a better future. The major bottlenecks in photobiological hydrogen production are the sensitivity to oxygen and limited utilization of light. Processes should be modified biologically or mechanically for sustainable biohydrogen production.

5.1 Sensitivity to Oxygen

To overcome oxygen sensitivity some strategies like applying PSII inhibitors (DCMU, SAL, DBMIB, C1-CCP), purging inert gas, or using light—dark cycles have been tested. But because they are not practical or are even damaging to the cells, especially in the case of inhibitors, these techniques are not preferred for sustainable production (Melis, 2002; Melis and Happe, 2001; Oncel, 2013; Torzillo and Seibert, 2013).

5.1.1 Two-Stage Approach

To avoid hydrogenase inactivation by oxygen, a two-stage process that combines the direct and indirect biophotolysis pathways with the model microalgae *Chlamydomonas reinhardtii* targeting the separation of oxygenic reactions from hydrogen evolution becomes the primary focus (Melis et al., 2002; Antal et al., 2011). The key is to block the PSII repair cycle under light by sulfur deprivation based on the fact that sulfur is essential for the amino acids (cysteine and methionine) of the D1 protein in the PSII reaction centers (Melis et al., 2002; Melis and Happe, 2001; Melis, 2009; Sakurai et al., 2013). After a lag time in sealed cultures, residual PSII activity drops the respiration limit of mitochondria triggering anaerobiosis. Suffocating cells stop proliferation and start to breakdown proteins, especially Rubisco, resulting in an increase in storage compounds like starch and lipids and a shift to hydrogen metabolism for survival (Melis et al., 2002; Ghysels and Franck, 2010; Sakurai et al., 2013). Electrons needed during the hydrogen production are supplied from water splitting by the residual PSII activity and from the breaking down of the stored starch linked with bidirectional FeFe hydrogenase (which also has hydrogen uptake activity) related to PSI activity (Ghysels and Franck, 2010; Antal et al., 2011; Peters et al., 2013). After reaching maximum levels within 3–4 days (Melis et al., 2002; Kosourov et al., 2007; Oncel and Sabankay, 2012), hydrogen production starts to decrease and ceases, even if some amount can be detected up to 30 days later (Oncel and Sukan, 2009), and levels are not meaningful to continue. The reason for the stop mechanism is probably due to the negative effect of sulfur deprivation as well as the building up of the potential inhibiting products like ethanol and formate (Ghysels and Franck, 2010). Various applications like semicontinuous, continuous, or immobilized cultures with different light regimes and light cycles are being investigated to increase the biohydrogen levels by using the two-stage process (Aparico et al., 1985; Miura et al., 1997; Tsygankov et al., 2002; Laurinavichene et al., 2004, 2006, 2008; Kim et al., 2006; Kosourov et al., 2007; Oncel and Sukan, 2011).

A similar approach by using nitrogen deprivation was also investigated in nitrogen fixing cyanobacteria, which resulted in higher biohydrogen productivities (Masukawa et al., 2002). The target is to stimulate the formation of the heterocysts, which are only 10–20% of the vegetative cells, in order to prevent the oxygen stress on the nitrogenases by the spatial separation of the oxygen evolving reactions from enzymatic reaction evolving biohydrogen (Mathews and Wang, 2009; Sakurai et al., 2013).

5.1.2 Molecular Approach

Genetic modification is another approach used to circumvent the oxygen stress. Considering the key role of the D1 in biohydrogen production, various D1 mutant strains of *Chlamydomonas reinhardtii* with different amino acid deletions produced in longer periods (up to 13 days) have reached maximum daily productivities of $7.1 \, mL \, L^{-1} h^{-1}$ (Faraloni and Torzillo, 2010; Torzillo et al., 2014). Also to target the higher accumulation of starch to support the anaerobic metabolism, a new *Chlamydomonas* mutant Stm6 (locked in state 1) with random gene insertion reached higher productivities up to 13 times relative to the wild type due to its higher respiration rates (Mathews and Wang, 2009; Antal et al., 2011; Torzillo et al., 2014). The modifications are done on an enzymatic basis both for maturation proteins or mature enzyme itself in order to achieve more tolerant enzyme active site to oxygen (Mathews and Wang, 2009; Kim and Kim, 2011; Oh et al., 2011; Srirangan et al., 2011). Other mutants with low sulfate permease, which is responsible for the transfer of the sulfate to chloroplasts, are potentially high-producer candidates (Chen et al., 2005; Antal et al., 2011).

For cyanobacteria, heterocyst formation is enhanced by the overexpression or inactivation of related genes, and the mutants targeting these genes like hetR, hetN, patS, patA, show effective results in high heterocyst frequency (Buikema and Haselkorn, 2001; Mathews and Wang, 2009; Sakurai et al., 2013). Also a type I NADH deficient mutant of *Synechocystis* showed a higher production potential (Sakurai et al., 2013), again good proof of the potential for mutation.

5.2 Light Utilization

With regard to light utilization by microalgae, a major limitation was related to the efficiency of photosynthesis to convert solar energy into biomass, which is lower compared to its theoretical potential ($\approx 36\%$) keeping in mind that 48 photons (40 kcal/photon) are needed to fix 6 carbon dioxide to form glucose ($\Delta G^{o}_{glucose} = 686$ kcal/mol) (Atkinson and Mavituna, 1983; Pirt, 1986; Walker, 2009). On the other hand, in conversion of solar energy to biohydrogen, considering direct biophotolysis, the losses due to the reflection and scattering, radiation outside photosynthetic active radiation region and non-photochemical reactions decrease the efficiency to a theoretical limit of $13 \pm 1\%$. But taking into account the two-step process like sulfur deprivation, addition of cell maintenance and down regulation of PSII steps, will further decrease to only $1 \pm 0.5\%$ (Long et al., 2006; Walker, 2009; Torzillo and Seibert, 2013). These values can even fluctuate with the dynamic nature of the outside environment as well as the cell itself (Table 3).

5.2.1 Antenna Size Truncation

A strategy to increase the efficiency is related with the antenna size of the chloroplasts in microalgae. Microalgae cells target to harvest as many photons as possible for survival, but because their capacity to utilize light is limited, this results in a wasted energy up to 90% (Melis, 2009; Kruse and Hankamer, 2010). This energy

loss can be explained by the lower electron transport rate compared to the photon capture rate by antennas (Mathews and Wang, 2009). Even if large antennas can contain 600 chlorophyll molecules in PSII and PSI reaction centers, a minimum number of 132 chlorophylls to be enough for survival makes the idea realistic (Melis et al., 1999; Melis, 2009). Antenna size truncation will serve as a dilution effect and increase the photosynthetic efficiency up to three-fold (Melis, 2009); a list of truncated chlorophyll antenna transformants tla1, tlaX, and tla-CW^{+} of *Chlamydomonas* are reported with tla1 reaching higher biohydrogen productivities when immobilized (Mathews and Wang, 2009; Melis, 2009; Torzillo and Seibert, 2013).

5.2.2 PBR Design

Conventional PBRs should be modified when the topic comes to biohydrogen production with a special emphasis on leakage risks and ability to scale-up (Table 4).

Mixing should be sufficient enough for the uniform illumination of the culture considering shear effects and wall growth variations depending on the strains, without disturbing anaerobic environment (Dasgupta et al., 2010; Zitelli et al., 2013). Construction materials of all the ports, fittings, and seals should be gas tight and also nonpermeable to prevent hydrogen escape as well as be nontoxic against inhibition risk on the cells (Skjnes et al., 2008; Dasgupta et al., 2010; Torzillo and Seibert, 2013). Materials should be selected according to the needs of the designs—some should be rigid and some flexible with a good ability to be attached and detached (Chaumont, 1993). Also, produced hydrogen should be collected easily through gas-tight ports for storage or direct utilization without increasing the head space biohydrogen partial pressure, which decreases the productivity (Oncel and Sukan, 2011; Kosourov et al., 2012; Oncel and Kose, 2014).

Apart from these general specifications, PBRs will need to meet the expectancy for high

TABLE 3   Energy Flow from the Sun to Biohydrogen

	Loss	Remaining	Efficiency	Energy ($\times 10^{12}$ W)
RADIATION ENERGY EMITTED BY THE SUN: 3.8×10^{26} W				
Total radiation energy reaching to earth		100		173,000
Reflected by the atmosphere	6	94	0.94	162,620
Absorbed by the atmosphere	16	78	0.83	134,975
Reflected by the clouds	20	58	0.74	99,882
Absorbed by the clouds	3	55	0.95	94,888
Reflected by land and oceans	4	51	0.93	88,246
Absorbed by land and oceans	51	0		88,246
ENERGY ABSORBED BY LAND AND OCEANS: 88246×10^{12} W **ROUTE 1: HYDROGEN PRODUCTION**				
Energy outside PAR	25.5	25.5	0.50	44,123
Reflected and transmitted light	2.5	23	0.9	39,711
Absorbed by nonphotosynthetic pigments	0.9	22.1	0.96	38,123
Photochemical inefficiencies	4.2	17.9	0.81	30,880
Hydrogen production	10.6	7.3	0.41	12,661
ENERGY FOR HYDROGEN: 12661×10^{12} W ($\approx 14.35\%$ CONVERSION) **ROUTE 2: TWO STAGE HYDROGEN PRODUCTION (BIOMASS THAN HYDROGEN)**				
Energy outside PAR	25.5	25.5	0.50	44,123
Reflected and transmitted light	2.5	23	0.9	39,711
Absorbed by nonphotosynthetic pigments	0.9	22.1	0.96	38,123
Photochemical inefficiencies	4.2	17.9	0.81	30,880
Carbohydrate synthesis	11.4	6.5	0.36	11,117
Photorespiration	1.75	4.75	0.73	8115
Dark respiration	1.7	3.05	0.64	5194
PSII downregulation during sulfur deprivation	2.62	0.43	0.14	727
ENERGY FOR HYDROGEN: 727×10^{12} W ($\approx 0.82\%$ CONVERSION)				

Modified from: Atkinson and Mavituna (1983), Cengel and Turner (2005), Long et al. (2006), Zhu et al. (2008), Walker (2009), Torzillo et al. (2014), http://asd-www.larc.nasa.gov/erbe/components2.gif.

productivity in outdoor cultures where the control of temperature and light plays a vital role. Specific temperature control units can be constructed inside or outside the PBRs to prevent temperature fluctuations both on the culture and biohydrogen stages. Modular designs oriented properly according to the sun, with a flexible adaptation for different strains, having high surface-to-volume ratio and illuminated area ground surface area ratio will serve as an advantage for outdoor productions (Posten, 2009; Dasgupta et al., 2010; Olivieri et al., 2014; Torzillo and Seibert, 2013).

5.3 Scale-up

Scaling-up is another challenge that actually covers all of the process from laboratory to

TABLE 4 Microalgae Production Systems, Comments on the Key Features for Biohydrogen

Type	Mixing	Illumination	Temperature	Large-scale applications	Commercial	Biohydrogen
OPEN SYSTEMS: OUTDOOR TARGETED NATURAL WATER RESERVOIRS OR ARTIFICIAL PONDS TYPICALLY CONSTRUCTED BY CONCRETE OR JUST EXCAVATED AND COMPRESSED EARTH COVERED BY WATER RESISTANT COATINGS USUALLY FOCUSING ON PHOTOAUTOTROPHIC CULTIVATION OF TOLERANT MICROALGAE						
Unmixed ponds	*No mechanical mixing *Mixing done by wind and convection flows	Utilization of direct sun light	No control, depend on the climate (thermal convection and radiation mechanisms)	Large areas of natural lagoons or reservoirs	*Production of biomass for feed, food, chemicals, and liquid biofuels *Competitive especially if constructed in convenient climates	Has potential as a biomass supplier for hydrogen phase
Mixed ponds (raceway ponds and circular ponds)	Various types of mechanical mixing (paddle wheels, horizontal arms, water jets, airlifts, propellers, pumps, Archimedes screws, and drag boards) modified with baffles	*Utilization of direct sun light or artificial illumination as a support (usually lab scale) *Modified with white colored base to distribute light inside the culture *Special awnings or reflecting covers are used for light dilution	*Usually no control, depend on the climate *Some limited applications constructed in green houses are temperature controlled *Special awnings or reflecting covers are used for sun protection	*Large ponds with multimixers *Modular designs integrated for larger culture volumes	*Production of biomass for feed, food, chemicals, and liquid biofuels *Competitive especially if constructed in convenient climates	*Has potential as a biomass supplier for hydrogen phase *Has potential if modified with hydrogen impermeable covers and construction materials (but not feasible and technically not easy)
PBRS: SPECIAL CONSTRUCTIONS BY TRANSPARENT MATERIALS IN ORDER TO CULTIVATE VARIOUS MICROALGAE PHOTOAUTOTROPHICALLY, PHOTOMIXOTROPHICALLY, OR HETEROTROPHICALLY, BY SERVING A CONTROLLED LIVING ENVIRONMENT FREE OF CONTAMINANTS						
Tubular (horizontal, fence, conical, helical, pyramid, α-type)	Various types of mixing systems (hydraulic, pneumatic or mechanical) modified with baffles and static mixers	*Utilization of direct sun light or artificial illumination *Modified with light scattering particles, optical fibers *Modified with white colored base to distribute light inside the culture *Special awnings or reflecting covers *Multi-stack formation for light dilution	*Depend on climate at outdoors *Some constructed in green houses for temperature control *Special awnings or reflecting covers are used for sun protection *Conventional heat exchange systems are used with internal (concentric tubes) or external (jackets) attachments *Water sprays or pools	*Long tubes attached by manifolds or U-fittings *Modular designs integrated for larger culture volumes	*Production of biomass for feed, food, chemicals and all biofuels *Competitive in harsh climates where open systems are not applicable *Competitive in the case of hydrogen production	*Has a potential as a biomass supplier for hydrogen phase *Has potential if modified with hydrogen impermeable construction materials with a special emphasis on hydrogen sealed mixing and degas systems

Continued

TABLE 4   Microalgae Production Systems, Comments on the Key Features for Biohydrogen—cont'd

Type	Mixing	Illumination	Temperature	Large-scale applications	Commercial	Biohydrogen
Flat panels (vertical, baffled, airlift, rotating, swing, V-shaped, Roux-type)	Various types of mixing systems (hydraulic, pneumatic or mechanical) modified with baffles and static mixers	*Utilization of direct sun light or artificial illumination *Modified with light scattering particles, optical fibers *Tilt and turning assemblies for better illumination *Special awnings or reflecting covers *Close-up formation for light dilution	*Depend on climate at outdoors *Some applications constructed in green houses for temperature control *Special awnings or reflecting covers are used for sun protection *Conventional heat exchange systems are used with cooling fingers or jackets *Water sprays or pools	*Modular designs integrated for larger culture volumes	*Production of biomass for feed, food, chemicals, and all biofuels *Competitive in harsh climates where open systems are not applicable *Competitive in the case of hydrogen production	*Has potential as a biomass supplier for hydrogen phase *Has potential if modified with hydrogen impermeable construction materials
Tank (stirred tanks, bubble columns, internal airlifts, external airlifts, divided columns, mechanical mixer integrated columns)	Various types of mixing systems (hydraulic, pneumatic, or mechanical) modified with baffles	*Utilization of direct sun light or artificial (internal or external) illumination *Modified with light-scattering particles, optical fibers with special collectors like Fresnel and Himawari	*Depend on climate at outdoors *Conventional heat exchange systems are used with cooling fingers, coils, or jackets	*Modular designs integrated for larger culture volumes	*Production of biomass to be used for feed, food, chemicals, and all biofuels *Competitive in harsh climates where open systems are not applicable *Competitive for hydrogen-targeted productions	*Has potential as a biomass supplier for hydrogen phase *Has potential if modified with hydrogen impermeable construction materials *Has potential if conventional fermentation systems can be modified for hydrogen production

INTEGRATED SYSTEMS (OPEN PONDS AND PBRS): TARGETS TO INTEGRATE THE ADVANTAGES OF BOTH SYSTEMS AND UNITE IN HYDROGEN PRODUCTION. HAS AN ADVANTAGE ESPECIALLY IN THE INTEGRATION OF TWO-STEP PROCESS WHERE BIOMASS CAN BE PRODUCED IN COST-EFFICIENT OPEN PONDS AND THEN TRANSFERRED TO HYDROGEN PHASE INSIDE THE SPECIAL PBRS

Type	Mixing	Illumination	Temperature	Large-scale applications	Commercial	Biohydrogen
Open ponds and PBRs	Various types of mixing systems from each concept	Utilization of direct sunlight or different artificial illumination systems	*Depend on climate at outdoors *Can be controlled with conventional strategies used for both concepts	Flexible application according to the design for example larger volumes in open ponds and smaller volumes in PBRs	Increase competitiveness in broader climates with the integration of both concepts	Has potential if modified well with a good integration of the systems
PBRs and PBRs	Various types of mixing systems from each design	Utilization of direct sunlight or different artificial illumination systems	*Depend on climate at outdoors *Can be controlled with conventional strategies used for both concepts	Flexible application according to each design, for example, aeration at panel type illumination at tubular type. Or heterotrophic cultivation at fermenter type and photoautotrophic at panel type	Increased competitiveness in broader climates with the integration of both concepts that can be applied for numerous microalgae species	Has potential if modified well with a good integration of the systems

outdoors. Scale-up criteria of microalgae production can be defined as the projection from unit volume to the total volume. Power input, k_La, dissolved oxygen, surface-to-volume ratio, light supply coefficient, light–dark cycle frequency, mixing time, Reynolds number, dissolved CO_2, blade tip velocity, or optical path length are some of the key factors for an efficient scale-up strategy (Oncel, 2013). Even if scale-up subsumes a volume increase, the modular construction approach can be a circumventing way to pass the construction and maintenance challenges such as durability, gas build-up, cleaning and sterilization, temperature control, uniform illumination, orientation, leakage, etc. by portioning to smaller volumes. This gives elasticity in operation against contamination and productivity losses with the ability to respond locally to a single module without disturbing all of the system (Posten, 2009; Xu et al., 2009). Each PBR design has its own pros and cons for scale-up, but for biohydrogen where the mixing is the limiting step, flat panels or tank-type PBRs with easier adaptation to a mechanical mixing system will be advantageous relative to long, tubular PBRs.

6. COMMERCIALIZATION AND ECONOMY

Commercialization of microalgal biofuels depends on the interactive relations between the market and the production technologies. Experience from the conventional food, feed, cosmeceutical and pharmaceutical industries provide information that can be valuable if properly directed to biofuel sector. From the economic aspect of the microalgae in these sectors, the cost of production of the raw algal biomass changes from 1 to 7 $/kg can increase up to 1000 $/kg depending on the production system and downstream processes for the desired product. But it is noteworthy to mention that the higher the cost of production in biotechnology usually pays a higher revenue, like the price

tag of high purity pigments such as astaxanthin, which can rise up to 10,000 $/kg (Oncel, 2013).

With regard to biohydrogen, the two-stage process has captured the attention of scientists and entrepreneurs. The concern about the two-stage process is the need for a culture shift, which on a large scale adds an extra cost for harvest and transfer of the culture addition to the high cost PBRs. For commercial scale, a process with efficient light conversion for high productivities other than the low cost PBRs will be important, keeping in mind the alternative uses of microalgae. There are two routes for the produced biohydrogen: to store or to use immediately. Storage of hydrogen is a challenging step for applications because it needs energy and special systems (Mao et al., 2012; Durbin and Jugroot, 2013; Dutta, 2014). There are several methods like storage: in high-pressure gas cylinders, as liquefied hydrogen by cryogenic applications, by adsorption on carbon structures, by absorption in host metals and metallic hydrides, in complex compounds such as borohydrides, and by metals and complexes together with water (Züttel, 2004; Kim et al., 2005; Zhou, 2005; Chen and Zhou, 2008). Direct usage of biohydrogen by proton exchange membrane fuel cells can have potential for commercial applications as the direct conversion to electricity, which can be used as a support in the processes or can be transferred to a main electric network without any storage step (Oncel and Sukan, 2011; Dutta, 2014).

The dynamic energy market dominated by fossil fuels is a hard game to enter. Changes in fossil fuel prices and increasing cost of production pushes the sector for alternatives. To have an idea about the transportation sector in which biohydrogen can have a chance, the costs including lifting and finding of crude oil depending on the oil field are in the range of 16–52 $/barrel, which can rise up to 75 $/barrel in the Arctic fields (Wullf, 2014). Considering the production, which is dominated by the onshore or offshore fields in Middle East and America, the average cost can be roughly estimated as 34 $/barrel (Bawks, 2011). With the average value of 103 $/barrel (West Texas and Brent; average price of crude oil for July 2014) (http://www.oil-price.net/) the cost of production actually equals 33% of selling price. But this percentage will decrease depending on included costs of refining, distribution, marketing, and taxes (http://www.eia.gov/petroleum/gasdiesel/). Even if all countries have the same global market prices, because of taxes or sometimes subsidies, the retail prices can have a sharp fluctuation all over the globe between 1.59 $/barrel (in Venezuela) up to 408.6 $/barrel (in Norway) (http://www.globalpetrolprices.com/gasoline_prices/). Compared to the estimated values of the microalgal green diesel with a cost of production ranging between 413 $/barrel and 863 $/barrel depending on the production strategy (open ponds or PBRs) even the highest retail price of fuel is still lower than the cost of microalgal fuel; this is a good sign of the competitiveness in the fuel market, which will be very harsh (Oncel, 2013). These values are important to know in order to have a clear view of the challenges in this field.

An estimation of the biohydrogen production by microalgae without the storage costs can be 1.38 $/kg, which is lower than the estimate of USDE and NREL having set a target to 2.99 $/kg biohydrogen by using oxygen-tolerant microalgae (James et al., 2009; Sakurai et al., 2013). The capital cost in biohydrogen production can be summarized as the total costs of subcomponents like PBRs (42%), hydrogen collection assembly (34%), microalgae feed assembly (4%), recycle and pumping systems (13%), and other costs like control systems, consumables (7%), which shows the vital role of the main infrastructure of specific PBRs on the cost determination for the retail price (James et al., 2009). Apart from microalgal oils, biohydrogen has a potential advantage that will help attain a feasible price tag because it is a shorter process path with three major steps for production: biomass production,

harvesting, and transferring to anaerobic environment. Assuming each step has a high yield of 90%, the biohydrogen production gives a role throughput of %73 (RT (three steps) = $(90/100)^3$), while an increase in process steps, like in biodiesel, shows the negative effect more clearly, other than the addition of extra costs for downstream and extraction processes.

The target of direct biophotolysis that has a potential solar conversion efficiency of 13.4% while biomass production is 9.4% will also support realization of feasible prices. On the other hand, an ultimate target of direct biohydrogen production with oxygen tolerant species will make possible the usage of cheaper ponds that will reduce the high costs, serving as an advantage like microalgal oils for biodiesel.

Scale-up should also be considered because it is the main step for building up the market. Today different companies having various prices according to their projections are the main question mark in front of realization. Especially in an emerging market like microalgal biofuels, to blend the scientific knowledge with industry will need the proper scale-up approaches and their real-life applications supported with all optimization possibilities to reach a realistic price.

7. FUTURE

For the future of microalgal biohydrogen two basic areas of progress may act as the game changers: (1) the progress in microalgae biotechnology and (2) the process of production. Considering microalgal biotechnology, there is still long way to go, with thousands of microalgae species to screen having higher oxygen tolerance and productivities for biohydrogen production. Under optimal conditions, estimations of 20 g hydrogen per m^2 of culture area can be produced with the future progress in genetic modifications incorporating omics and systems biology with special emphasis on a single cell to a culture having a reduced amount of chlorophyll per antenna, high photosynthetic efficiency, high respiration rates, high carbohydrate accumulation ability to survive under stress conditions like anaerobic cultures, and high xanthophyll cycle (Melis, 2000; Kruse and Hankamer, 2010; Torzillo et al., 2014). There will be screening for truncated antenna, D1 protein, starch producing, thermotolerant or oxygen tolerant microalgae or their mutants with the ability to survive in fluctuating light outdoors, form flocs, be susceptible for genetic modifications, grow fast in waste or saline waters, with stable and robust nature. Using specified protocols like elimination of the electron sinks or the regulation of photosynthesis respiration equilibrium and analyses with the finalization of the genome sequences, each species can further be investigated, and this will catalyze the progress for a competitive biohydrogen production (Chisti, 2007; Melis, 2009; Xu et al., 2009; Wijffels and Barbosa, 2010; Ghysels and Franck, 2010; Ortiz-Marquez et al., 2013; Oncel, 2013; Kose and Oncel, 2014; Scoma et al., 2014).

In addition, new genes from other organisms will play a role in the progress of the construction of the ultimate producer. Good examples of progress (Sakurai et al., 2013) include the case of HUP1 hexose uptake protein from *Chlorella kessleri* introduced to *Chlamydomonas*, which gave the ability to assimilate glucose in dark heterotrophic environment (Mathews and Wang, 2009); using marine phage PsaJF subunit in *Synechocytis*, which enabled its transformant to accept electrons from alternative sources (Mazor et al., 2012); and introducing the hydrogenase enzymes from *Clostridium pasteurianum* into *Synechococcus* sp., which resulted in the active FeFe hydrogenase production.

The future of microalgal biohydrogen will be tied to new materials and innovative designs that will make biohydrogen effective and cost efficient. Durable, transparent, and hydrogen-tight materials with practical construction techniques need to be developed.

Considering the mixing strategies, modular PBRs with modified mixer systems that can be applied to different scales and designs will be helpful for real-life applications. Compact PBRs with decreased land usage should be selected both for economy and light dilution modified with special sterilization and cleaning systems to prevent contamination and fouling. Special illumination systems based on LEDs, optical fibers, and solar collectors (Fresnel lenses, Himawari lenses), or light scattering particles inside the culture may also act as an enhancer for light utilization (Posten, 2009; Scoma et al., 2012; Zitelli et al., 2013). Using renewable sources like solar or wind to support the energy requirements of the production systems will also be very important for the future applications to be more competitive (Mohammed et al., 2014; Nasir Uddin and Daud, 2014). Also, integration of open systems and PBR systems can decrease the cost of the production.

Sustainable, environment friendly, and feasible biohydrogen economy targeting on the PBR design and construction with special emphasis on materials and all the system assemblies should target a reduction in the capital cost for both biomass ($100 \$/m^2$) and for biofuel ($15 \$/m^2$) production to be competitive (Chen et al., 2011). The SWOT analysis of microalgal biohydrogen production gives an idea about the future considerations and studies (Table 5).

An idealized process should cover the "light to fuel concept" integrated with biorefinery, which totals the process with all its subunits like potential alternatives for nutrients, by-products, and especially other potential fuels (methane, ethanol, biodiesel) or high value chemicals (pigments, lipids, vitamins, etc.) and even wastes (Figure 5). Some LCA analysis showed that the wastewater treatment technologies integrated with microalgae for eco-friendly treatment and biofuel production at the same time will result in feasible production costs (Oncel, 2013). Also carbon credits will play a

TABLE 5   SWOT Analysis Focusing on Biohydrogen With Specific Comments for Progress

		Strengths	Weaknesses	Opportunities	Threats	Key points for progress
Production	Biohydrogen	Clean and sustainable source for fuel and energy	Limited applications with regards to the need of new technologies for storage and usage that are still in progress	Increasing awareness for daily life applications like using fuel cell vehicles with the integration to conventional fuel grid	Immature technologies with high costs blocking the interest of the consumers and the industry	Applications in daily life to catch the attention of the public with a competitive price
	Microalgae	Various potential species, some with defined genetic codes	Few species are studied Lack of knowledge in the cellular metabolism of microalgal bioprocesses	Screening of new species out of the vast collection of nature with a special emphasis on high productivity and outdoor large-scale productions	Productivity fluctuations between lab and large scales	Selection of high production species that can live in salt or wastewaters with durable and sustainable nature especially with high oxygen tolerance
	PBR	Various designs	Need of more experience on outdoor large-scale applications	Already established conventional types with a potential to modify for hydrogen production	Hydrogen leakage risks	New materials and designs for outdoor large scale hydrogen production
	Illumination	Usage of various sources like sun or artificial light	Risk of productivity loss by the lack of proper illumination especially in large scales	Focusing on low energy consuming illumination systems	Need of high energy for illumination especially in case of artificial illumination	Integration with renewable sources as a support for the energy consumption
	Mixing	Application of various systems for lab and pilot scales	Risk of productivity loss by the improper mixing especially in large scales	Flexible mixing systems with the ability to integrate different PBR designs having low energy consumption	High energy need for large scales	Integration of new mixing systems with renewable sources as a support for the energy consumption
	Temperature	Easy adaptation of conventional systems like heat exchangers and control units	Need of a proper control especially outdoor large scales where the fluctuations are very dynamic depending on the environment	New systems to minimize the energy consumption with an ease on the integration to the PBRs	High energy cost to control (heat or cool) and sterilize all the systems especially in large scales	Integration with renewable systems as a support of energy and high efficiency temperature control systems well applied to large scales

Nutrients	Usage of various nutrient sources	Utilization of fossil fuel-based fertilizers or depleting sources like phosphate	Utilization of recycle streams from various industries or crops	Risk of environmental problems eutrophication if not handle properly	Integration to wastewater or recycle streams for large-scale applications
Scale-up	Experience from the conventional microalgae productions	Limited experience to transfer up to outdoors where the large-scale PBRs will be more complex compared to open pond systems used in industry	Proper procedures and methods to apply for a scale-up strategy with the ability to modify and adapt according to microalgae	Productivity losses due to dynamic outdoor conditions and dynamic responses in large culture volumes	Practical and reliable strategies for scale-up with an easy application potential in industry
Energy requirement	Quite a direct production eliminating high energy consuming downstream and refining processes	Usage of artificial illumination and mixing systems	Focusing on low energy-consuming systems with an ease on the integration with renewable sources	Energy gap between the produced hydrogen and consumption leading a dependence on fossil fuels	Integration with renewable sources to support the energy consumption
Environment	Clean combustion gasses	Necessity to use the limited sources for water and land	Integration with recycle streams and wastewaters	Waste streams from the process and risk of strain escape especially if genetically modified	Wastewater usage
Economy	Potential of hydrogen as a sustainable and environment friendly alternative with a high flexibility to integrate to different sectors	Limited daily life applications with a need for a well-defined market	New legislation and politics to catch the attention of investors and entrepreneurs to build up the industry with the catalysis of carbon credits and government subsidies	Lack of interest by the governments and end users based on the experience	Building up a sustainable market with a special emphasis on new technologies, biorefinery concept, and opportunities for new jobs
Ethical Issues	A new perspective putting the human and environment in the center	Dependence on limited sources like water, land, and also fossil fuels	Equally shared knowledge and experience to build up a new industry for all humanity	Lack of knowledge and experience with a risk of monopolization	Focusing on knowledge transfer and a well-defined market, giving equal opportunities to all the societies to spread the wealth

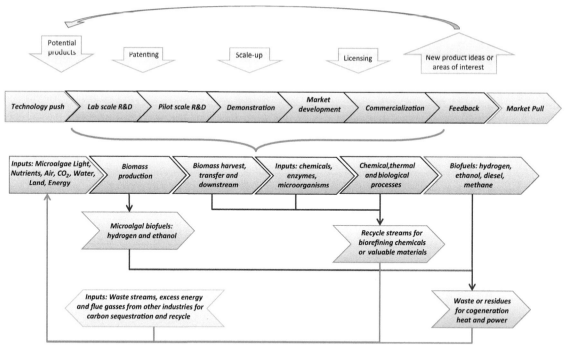

FIGURE 5   Techno-economic flow of microalgal biofuels with a biorefinery approach. *Modified from: King (2010), Isaka (2013), Oncel (2013).*

role in the cost reduction, especially with the new legislation forcing countries to reduce emissions. Organizations with well-defined networks of raw materials and products will be the key players that are successful with the concept (Parmar et al., 2011; Koller et al., 2012; Kondo et al., 2013; Lakaniemi et al., 2013; Oncel, 2013).

Future strategies to share knowledge between industries will support the development of the microalgal biohydrogen technologies. A radical but useful attempt that was announced recently (http://www.teslamotors.com/blog/all-our -patent-are-belong-you) by the electric car maker Tesla about open access to patents for the new players that are interested in the field has already positively affected acceleration of the technology (with a resulting positive reaction on financial stock exchanges).

8. ETHICAL ISSUES

Today even if the wild competition for energy forces industries to forget or ignore ethical issues, increasing awareness is also forcing policy makers to consider new ethical principles. Construction of a better future should use justice, solidarity, sustainability, and stewardship as the building blocks and stick them all with the cement of equity and human rights (Oncel, 2013).

The ability of microalgae to be cultivated without interfering with potential food crops and limited clean water by integrating with different production routes and processes serves as a flexibility to build a bio-economy that is eco-friendly, sustainable, safe, and fits with ethical values. Jobs in renewable energy are estimated as 6.5 millon in 2013 and 1.4 milion of them are on liquid biofuels (Ferroukhi et al., 2013, 2014) microalgal biofuels can also support the wealth and share between the societies with a special emphasis on inter-generational equity and resilience by elevating the chance for labor and fair trade (McGraw, 2009; Buyx and Tait, 2011; Oncel, 2013).

9. CONCLUSIONS

Microalgal biohydrogen with its unique advantages can be a valuable alternative regarding energy needs of the future. Efforts to reach higher productivities, searching for new species, and investigating alternative production methods still need further research in order for biohydrogen to have a sustainable share in the energy sector. On the other hand aiming direct biohydrogen production with reaching higher solar efficiencies by using cost efficient production systems will continue to be the target.

Societies should focus on alternatives like biohydrogen without exaggeration but with optimism. Todays' reality is related to labor and progress catalyzed by the massive economy driven by the fossil fuels. Fossil fuels accelerated the industrial age, but massive consumption makes it impossible for sustainability unless new solutions and supports are added to the energy stream. It is not realistic to convert to a bio-economy immediately, but alternatives like microalgal biohydrogen can replace the use of some of the fossil fuels by being incorporated step by step in a planned and foreseeable future; the ultimate target is for such energy alternatives to become direct substitutes for our dwindling fossil fuels.

References

Abdulqader, G., Barsanti, L., Mario, R., 2000. Harvest of *Arthrospira platensis* from Lake Kossorom (Chad) and its house hold usage among the Kanembu. J. Appl. Phycol. 12, 493–498.

Amponsah, N.Y., Troldborg, M., Kington, B., Aalders, I., Hough, R.L., 2014. Greenhouse gas emissions from renewable energy sources: a review of lifecycle considerations. Renewable Sustainable Energy Rev. 39, 461–475.

Amutha, K.B., Murugesan, A.G., 2011. Biological hydrogen production by the algal biomass *Chlorella vulgaris* MSU 01 strain isolated from pond sediment. J. Biotechnol. 102, 194–199.

Antal, T.K., Krendeleva, T.E., Rubin, A.B., 2011. Acclimation of green algae to sulfur deficiency: underlying mechanisms and application for hydrogen production. Appl. Microbiol. Biotechnol. 89, 3–15.

Aoyama, K., Uemura, I., Miyake, J., Asada, Y., 1997. Fermentative metabolism to produce hydrogen gas and organic compounds in a cyanobacterium, *Spirulina platensis*. J. Ferment. Bioeng. 83, 17–20.

Aparico, P.J., Azuara, M.P., Antonio, B., Fernandez, V.M., 1985. Effect of light intensity and oxidized nitrogen sources on hydrogen production by *Chlamydomonas reinhardtii*. Plant Physiol. 78, 803–806.

Atkinson, B., Mavituna, F., 1983. Biochemical Engineering and Biotechnology Handbook, p. 1118. The Nature Press, New York, NY.

Batista, A.P., Moura, P., Marques, P.A.S.S., Ortigueira, J., Alves, L., Gouveia, L., 2014. *Scenedesmus obliquus* as feedstock for biohydrogen production by *Enterobacter aerogenes* and *Clostridium butyricum*. Fuel 117, 537–543.

Batyrova, K.A., Anatoly, A., Tsygankov, S., Kosourov, N., 2012. Sustained hydrogen photoproduction by phosphorus- deprived *Chlamydomonas reinhardtii* cultures. Int. J. Hydrogen Energy 37, 8834–8839.

Bawks, B., 2011. Performance Profiles of Major Energy Producers 2009. US Energy Information Administration. DOE/EIA-0206(09), p. 70.

Becker, E.W., 1995. Large scale application, chapter 10. pp. 65–171. In: Microalgae Biotechnology and Microbiology. Cambridge University Press, Cambridge, UK, p. 293.

Beer, L.L., Boyd, E.S., Peters, J.W., Posewitz, M.C., 2009. Engineering algae for biohydrogen and biofuel production. Curr. Opin. Biotechnol. 20, 264–271.

Benemann, J.R., 1998. The technology of biohydrogen. In: Zaborsky, O.R. (Ed.), Biohydrogen. Plenum Press, New York, pp. 19–30.

Bentham, J., 2012. 40 Years of Shell Scenarios, p. 39.

Borowitzka, A.M., 2013. Energy from microalgae: a short history. In: Borowitzka, M.A., Moheimani, N.R. (Eds.), Algae for Biofuels and Energy. Springer Dordrecht Heidelberg, New York, London, pp. 1–15.

Borowitzka, M.A., Moheimani, N.R., 2013. In: Borowitzka, M.A., Moheimani, N.R. (Eds.), Open Pond Culture Systems in Algae for Biofuels and Energy. Springer Dordrecht Heidelberg, New York, London, pp. 133–152.

Buikema, W.J., Haselkorn, R., 2001. Expression of the *Anabaena hetR* gene from a copper-regulated promoter leads to heterocyst differentiation under repressing conditions. PNAS 98 (5), 2729—2734.

Buyx, A., Tait, J., 2011. Biofuels: ethical issues, a guide to the report. Nuffield council on bioethics. Science 332, 540—541.

Catalanotti, C., Yang, W., Posewitz, M.C., Grosmann, A.R., 2013. Fermentation metabolism and its evolution in algae. Front. Plant Sci. 150, 1—17.

Cengel, Y.A., Turner, R.H., 2005. Fundamentals of Thermal Fluid Sciences, second ed. McGraw Hill International Edition, Singapore. pp. 974—975.

Chaubey, R., Sahu, S., James, O.O., Maity, S., 2013. A review on development of industrial processes and emerging techniques for production of hydrogen from renewable and sustainable sources. Renewable Sustainable Energy Rev. 23, 443—462.

Chaumont, D., 1993. Biotechnology of algal biomass production: a review of systems for outdoor mass culture. J. Appl. Phycol. 5, 593—604.

Chen, C.Y., Liu, C.H., Lo, Y.C., Chang, J.S., 2011. Perspectives on cultivation strategies and photobioreactor designs for photo-fermentative hydrogen production. Bioresour. Technol. 102, 8484—8492.

Chen, H.C., Newton, A.J., Melis, A., 2005. Role of SulP, a nuclear-encoded chloroplast sulfate permease, in sulfate transport and H_2 evolution *in Chlamydomonas reinhardtii*. Photosynth. Res. 84, 289—296.

Chen, P., Zhou, M., 2008. Recent progress on hydrogen storage. Mater. Today 11 (12), 36—43.

Cheng, J., Xia, A., Song, W., Su, H., Zhou, J., Cen, K., 2012. Comparison between heterofermentation and autofermentation in hydrogen production from *Arthrospira (Spirulina) platensis* wet biomass. Int. J. Hydrogen Energy 37, 6536—6544.

Cheng, J., Zhang, M., Song, W., Xia, A., Zhou, J., Cen, K., 2011. Cogeneration of hydrogen and methane from *Arthrospira* maxima biomass with bacteria domestication and enzymatic hydrolysis. Int. J. Hydrogen Energy 36, 1474—1481.

Chisti, Y., 2007. Biodiesel from microalgae. Biotechnol. Adv. 25, 294—306.

Choi, J.A., Hwang, J.H., Dempsey, B.A., Shanab, R.A.I., Min, B., Song, H., Lee, D.S., Kim, J.R., Cho, Y., Hongi, S., Jeon, B.H., 2011. Enhancement of fermentative bioenergy (ethanol/hydrogen) production using ultrasonication of *Scenedesmus obliquus* YSW15 cultivated in swine wastewater effluent. Energy Environ. Sci. 4, 3513—3520.

Conti, J.J., 2014. Annual Energy Outlook 2014 with Projections to 2040. US Energy Information Administration. DOE/EIA-0383. p. 269.

Costa, J.A.V., Morais, M.G., 2014. An open pond system for microalgal cultivation. In: Pandey, A., Lee, D.-J., Chisti, Y., Soccol, C.R. (Eds.), Biofuels from Algae. Elsevier Press, Burlington, USA, pp. 1—22.

Dasgupta, C.N., Gilbert, J.J., Lindblad, P., Heirdorn, T., Borgvang, S.A., Skjanes, K., Das, D., 2010. Recent trends on the development of photobiological processes and photobioreactors for the improvement of hydrogen production. Int. J. Hydrogen Energy 35, 10218—10238.

Dickson, D.J., Page, C.J., Ely, R.L., 2009. Photobiological hydrogen production from *Synechocystis sp.* PCC 6803 encapsulated in silica sol gel. Int. J. Hydrogen Energy 34, 204—215.

Dipasquale, L., D'Ippolito, G., Gallo, C., Vella, F.M., Gambacorta, A., Picariello, G., 2012. Hydrogen production by the thermophilic eubacterium *Thermotoga neapolitana* from storage polysaccharides of the CO2-fixing diatom *Thalassiosira weissflogii*. Int. J. Hydrogen Energy 37, 12250—12257.

Dudley, B., 2014. BP Energy Outlook 2035 bp.com/energyoutlook #BPstats. p. 96.

Durbin, D.J., Malardier-Jugroot, C., 2013. Review of hydrogen storage techniques for on board vehicle applications. Int. J. Hydrogen Energy 38, 14595—14617.

Dutta, S., 2014. A review on production, storage of hydrogen and its utilization as an energy resource. J. Ind. Eng. Chem. 20, 1148—1156.

Efremenko, E.N., Nikolskaya, A.B., Lyagin, I.V., Senko, O.V., Makhlis, T.A., Stepanov, N.A., 2012. Production of biofuels from pretreated microalgae biomass by anaerobic fermentation with immobilized *Clostridium acetobutylicum* cells. Bioresour. Technol. 114, 342—348.

Faraloni, C., Torzillo, G., 2010. Phenotypic characterization and hydrogen production in *Chlamydomonas reinhardtii* QB-binding D1-protein mutants under sulfur starvation: changes in Chl fluorescence and pigment composition. J. Phycol. 46, 788—799.

Faraloni, C., Ena, A., Pintucci, C., Torzillo, G., 2011. Enhanced hydrogen production by means of sulfur-deprived *Chlamydomonas reinhardtii* cultures grown in pretreated olive mill wastewater. Int. J. Hydrogen Energy 36, 5920—5931.

Fedorov, A.S., Kosourov, S., Ghirardi, M.I., Seibert, M., 2005. Continuous hydrogen photoproduction by *Chlamydomonas reinhardtii* using a novel two-stage, sulfate-limited chemostat system. Appl. Biochem. Biotechnol. 121, 403—412.

Fernandez, F.G.A., Sevilla, J.M.F., Molina, E., 2013. Photobioreactors for the production of microalgae. Sci. Biotechnol. 12, 131—151.

Ferreira, A.F., Marques, A.C., Batista, A.P., Marques, P.A.S.S., Gouveia, L., Silva, C.M., 2012. Biological hydrogen production by *Anabaena* sp.—yield, energy and CO_2

analysis including fermentative biomass recovery. Int. J. Hydrogen Energy 37, 179–190.

Ferreira, A.F., Ortigueira, J., Alves, L., Gouveia, L., Moura, P., Silva, C.M., 2013. Energy requirement and CO_2 emissions of bioH_2 production from microalgal biomass. Biomass Bioenergy 49, 249–259.

Ferroukhi, R., Khalid, A., Pena, A.L., Renner, M., 2014. Renewable Energy and Jobs, Annual Review 2014. International Renewable Energy Agency, p. 12.

Ferroukhi, R., Lucas, H., Renner, M., Lehr, U., Breitschopf, B., Lallement, D., Petrick, K., 2013. Renewable Energy and Jobs. International Renewable Energy Agency, p. 144.

Gaffron, H., Rubin, J., 1942. Fermentative and photochemical production of hydrogen in algae. J. Gen. Physiol. 26, 219–240.

Ghysels, B., Franck, F., 2010. Hydrogen photo-evolution upon S deprivation stepwise: an illustration of microalgal photosynthetic and metabolic flexibility and a step stone for future biotechnological methods of renewable H_2 production. Photosynth. Res. 106, 145–154.

Giannelli, L., Torzillo, G., 2012. Hydrogen production with the microalga *Chlamydomonas reinhardtii* grown in a compact tubular photobioreactor immersed in a scattering light nanoparticle suspension. Int. J. Hydrogen Energy 37, 16951–16961.

Grima, E.M., Belarbi, E.H., Fernandez, F.G.A., Medina, A.R., Chisti, Y., 2003. Recovery of microalgal biomass and metabolites: process options and economics. Biotechnol. Adv. 20, 491–515.

Grobbelaar, J.U., 2009. Factors governing algal growth in photobioreactors: the "open" versus "closed" debate. J. Appl. Phycol. 21, 489–492.

Guan, Y., Deng, M., Yu, X., Zhang, W., 2004a. Two-stage photo-biological production of hydrogen by marine green alga *Platymonas subcordiformis*. Biochem. Eng. J. 19, 69–73.

Guan, Y., Zhang, W., Deng, M., Jin, Y.X., 2004b. Significant enhancement of photo-biological H_2 evolution by carbonylcyanide *m*-chlorophenylhydrazone in the marine green alga *Platymonas subcordiformis*. Biotechnol. Lett. 26, 1031–1035.

He, M., Li, L., Zhang, L., Liu, J., 2012. The enhancement of hydrogen photoproduction in *Chlorella protothecoides* exposed to nitrogen limitation and sulfur deprivation. Int. J. Hydrogen Energy 37, 16903–16915.

Ho, K.L., Lee, D.J., Suc, A., Chang, J.S., 2013. Biohydrogen from lignocellulosic feedstock via one-step process. Int. J. Hydrogen Energy 37, 15569–15574.

Hoeven, M., 2014. Special Report, World Energy Investment Outlook. International Energy Agency, p. 190.

Hoshino, T., Johnson, D.J., Scholz, M., Cuello, J.L., 2013. Effects of implementing PSI-light on hydrogen production via biophotolysis in *Chlamydomonas reinhardtii* mutant strains. Biomass Bioenergy 59, 243–252.

Ike, A., Toda, N., Hirata, K., Miyamoto, K., 1997. Hydrogen photoproduction from CO_2-fixing microalgal biomass: application of lactic acid fermentation by *Lactobacillus amylovorus*. J. Ferment. Bioeng. 84, 428–433.

Isaka, M., 2013. Intellectual Property Rights, the Role of Patents in Renewable Energy Technology Innovation. International Renewable Energy Agency, IRENA, p. 34.

James, B.D., Baum, G.N., Perez, J., Baum, K.N., 2009. Technoeconomic Boundary Analysis of Biological Pathways to Hydrogen Production. Subcontract Report NREL/SR-560-46674.

Ji, C.F., Legrand, J., Pruvost, J., Chen, Z.A., Zhang, W., 2010. Characterization of hydrogen production by *Platymonas subcordiformis* in torus photobioreactor. Int. J. Hydrogen Energy 35, 7200–7206.

Kawaguchi, H., Hashimoto, K., Hirata, K., Miyamoto, K., 2001. H_2 production from algal biomass by a mixed culture of *Rhodobium marinum* A-501 and *Lactobacillus amylovorus*. J. Biosci. Bioeng. 91, 277–282.

Kim, D.-H., Kim, M.-S., 2011. Hydrogenases for biohydrogen production. Bioresour. Technol. 102, 8423–8431.

Kim, J.-H., Shim, J.-H., Cho, Y.W., 2005. On the reversibility of hydrogen storage in Ti- and Nb- catalyzed $Ca(BH_4)_2$. J. Power Sources 181, 140–143.

Kim, M.S., Baek, J.S., Yun, Y.S., Sim, S.J., Park, S., Kim, S.C., 2006. Hydrogen production from *Chlamydomonas reinhardtii* biomass using a two-step conversion process: anaerobic conversion and photosynthetic fermentation. Int. J. Hydrogen Energy 31, 812–816.

King, D., 2010. The future of industrial biorefineries. World Economic Forum, p. 40.

Kojima, E., Lin, B., 2004. Effect of partial shading on photoproduction of hydrogen by *Chlorella*. J. Biosci. Bioeng. 97, 317–321.

Koller, M., Salerno, A., Tuffner, P., Koinigg, M., Bochzelt, H., Schober, S., Pieber, S., Schnitzer, H., Mittelbach, M., Braunegg, G., 2012. Characteristics and potential of micro algal cultivation strategies: a review. J. Cleaner Prod. 37, 377–388.

Kondo, A., Ishii, J., Hara, K.Y., Hasunuma, K., Matsuda, F., 2013. Development of microbial cell factories for biorefinery through synthetic bioengineering. J. Biotechnol. 163, 204–216.

Kose, A., Oncel, S., 2014. Biohydrogen production from engineered microalgae *Chlamydomonas reinhardtii*. Adv. Energy Res. 2 (1), 1–9.

Kosourov, S., Patrusheva, E., Ghirardi, M.L., Seibert, M., Tsygankov, A., 2007. A comparison of hydrogen photoproduction by sulfur-deprived *Chlamydomonas reinhardtii* under different growth conditions. J. Biotechnol. 128, 776–787.

Kosourov, S.N., Batyrova, K.A., Petushkova, E.P., Tsygankov, A.A., Maria, L.G., Seibert, M., 2012.

Maximizing the hydrogen photoproduction yields in *Chlamydomonas reinhardtii* cultures: the effect of the H_2 partial pressure. Int. J. Hydrogen Energy 37, 8850−8858.

Kosourov, S.N., Seibert, M., 2009. Hydrogen photoproduction by nutrient-deprived *Chlamydomonas reinhardtii* cells immobilized within thin alginate films under aerobic and anaerobic conditions. Biotechnol. Bioeng. 102, 50−58.

Kruse, O., Hankamer, B., 2010. Microalgal hydrogen production. Curr. Opin. Biotech. 21, 238−243.

Kruse, O., Rupprecht, J., Bader, K.P., Thomas-Hall, S., Schenk, P.M., Finazzi, G., Hankamer, B., 2005. Improved photobiological H_2 production in engineered green algal cells received for publication. J. Biol. Chem. 280 (40), 34170−34177.

Kumar, K., Roy, S., Das, D., 2013. Continuous mode of carbon dioxide sequestration by *C. sorokiniana* and subsequent use of its biomass for hydrogen production by *E. cloacae* IIT-BT 08. Bioresour. Technol. 145, 116−122.

Lakaniemi, A.M., Hulatt, C.J., Thomas, D.N., Tuovinen, O.H., Puhakka, J.A., 2011. Biogenic hydrogen and methane production from *Chlorella vulgaris* and *Dunaliella tertiolecta* biomass. Biotechnol. Biofuels 4 (34), 1−12.

Lakaniemi, A.M., Tuovinen, O.H., Puhakka, J.A., 2013. Anaerobic conversion of microalgal biomass to sustainable energy carriers—A reveiw. Bioresour. Technol. 135, 222−231.

Laurinavichene, T., Tolstygina, I., Tsygankov, A., 2004. The effect of light intensity on hydrogen production by sulfur-deprived *Chlamydomonas reinhardtii*. J. Biotechnol. 114, 143−151.

Laurinavichene, T.V., Fedorov, A.S., Ghirardi, M.L., Seibert, M., Tsygankov, A., 2006. Demonstration of sustained hydrogen photoproduction by immobilized, sulfur-deprived *Chlamydomonas reinhardtii* cells. Int. J. Hydrogen Energy 31, 659−667.

Laurinavichene, T.V., Kosourov, S.N., Ghirardi, M.L., Seibert, M., Tsygankov, A.A., 2008. Prolongation of H_2 photoproduction by immobilized, sulfur-limited *Chlamydomonas reinhardtii* cultures. J. Biotechnol. 134, 275−277.

Lee, Y.K., 2001. Microalgal mass culture systems and methods: their limitations and potential. J. App. Phycol. 13, 307−315.

Lehr, F., Morweiser, M., Sastre, R.R., Kruse, O., Posten, C., 2012. Process development for hydrogen production with *Chlamydomonas reinhardtii* based on growth and product formation kinetics. J. Biotechnol. 162, 89−96.

Li, X., Huang, S., Yu, J., Wang, Q., Wu, S., 2013. Improvement of hydrogen production of *Chlamydomonas reinhardtii* by co-cultivation with isolated bacteria. Int. J. Hydrogen Energy 38, 10779−10787.

Lindblad, P., 2004. The potential of using cyanobacteria as producers of molecular hydrogen. In: Miyake, Y., Igarashi, Y., Rögner, M. (Eds.), Biohydrogen III-Renewable Energy Systems by Biological Solar Energy Conversion. Elsevier Ltd, Amsterdam, pp. 75−82.

Lindblad, P., Christensson, K., Lindberg, P., Fedorov, A., Pinto, F., Tsygankov, A., 2002. Photoproduction of H2 by wild type *Anabaena* PCC 7120 and a hydrogen uptake deficient mutant: from laboratory experiments to outdoor culture. Int. J. Hydrogen Energy 27, 1271−1281.

Liu, C.H., Chang, C.Y., Cheng, C.L., Lee, D.Y., Chang, J.S., 2012. Fermentative hydrogen production by *Clostridium butyricum* CGS5 using carbohydrate-rich micro-algal biomass as feedstock. Int. J. Hydrogen Energy 37 (20), 15458−15464.

Long, S.P., Humpries, S., Falkowski, P.G., 1994. Photoinhibition of photosynthesis in nature. Annu. Rev. Plant Physiol. Plant Mol. Biol. 45, 633−662.

Long, S.P., Zhu, X.G., Naidu, S.L., Ort, D.R., 2006. Can improvement in photosynthesis increase crop yields? Plant Cell Environ. 29, 315−330.

Lopes Pinto, F.A., Troshinaa, O., Lindblad, P., 2002. A brief look at three decades of research on cyanobacterial hydrogen evolution. Int. J. Hydrogen Energy 22, 1209−1215.

Lundquist, T.J., Woertz, I.C., Quinn, N.W.T., Benemann, J.R., 2010. A Realistic Technology and Engineering Assessment of Algae Biofuel Production. Energy Biosciences Institute, University of California, Berkeley, California.

Maneeruttanarungroj, C., Lindblad, P., Incharoensakdi, A., 2010. A newly isolated green alga, *Tetraspora* sp. CU2551, from Thailand with efficient hydrogen production. Int. J. Hydrogen Energy 35, 3193−3199.

Maness, P.C., Yu, J., Eckert, C., Maria, L., Features, G., 2009. Photobiological hydrogen production − prospects and challenges. Microbe. 4, 6.

Mao, S.S., Shen, S., Guo, L., 2012. Nanomaterials for renewable hydrogen production, storage and utilization. Prog. Nat. Sci. Mat. Int. 22 (6), 522−534.

Markov, S.A., Thomas, A.D., Bazin, M.J., Hall, D.O., 1997. Photoproduction of hydrogen by cyanobacteria under partial vacuum in batch culture or in a photobioreactor. Int. J. Hydrogen Energy 22, 521−524.

Masojidek, J., Kopecky, J., Giannelli, L., seibert, G., 2011. Productivity correlated to photobiochemical performance of *Chlorella* mass cultures grown outdoors in thin-layer cascades. J. Ind. Microbiol. Biotechnol. 38, 307−317.

Masojidek, J., Torzillo, G., Koblizek, M., 2013. Photosynthesis in microalgae. In: Richmond, A., Hu, Q. (Eds.), Culture: Applied Phycology and Biotechnology, Second ed. John Wiley & Sons, Ltd, Oxford. http://dx.doi.org/10.1002/9781118567166.ch2.

Masukawa, H., Mochimaru, M., Sakurai, H., 2002. Hydrogenases and photobiological hydrogen production utilizing nitrogenase system in cyanobacteria. Int. J. Hydrogen Energy 27, 1471−1474.

Mathews, J., Wang, G., 2009. Metabolic pathway engineering for enhanced biohydrogen production. Int. J. Hydrogen Energy 34, 7404–7416.

Mazor, Y., Toporik, H., Nelson, N., 2012. Temperature-sensitive PSII and promiscuous PSI as a possible solution for sustainable photosynthetic hydrogen production. Biochim. Biophys. Acta 1817, 1122–1126.

McGraw, L., 2009. The Ethics of Adoption and Development of Algae-based Biofuels. Prepared under the Outline Framework of WG9 in the Ethics of Climate Change in Asia and the Pacific (ECCAP) Project. RUSHSAP, UNESCO, 83.

Melis, A., Neidhardt, J., Benemann, J.R., 1999. *Dunaliella salina* (chlorophyta) with small chlorophyll antenna sizes exhibit higher photosynthetic productivities and photon use efficiencies than normally pigmented cells. J. App. Phycol. 10, 515–525.

Melis, A., Happe, T., 2001. Hydrogen production. Green algae as a source of energy. Plant Physiol. 127, 740–748.

Melis, A., 2002. Green alga hydrogen production: progress, challenges and prospects. Int. J. Hydrogen Energy 27, 1217–1228.

Melis, A., 2009. Solar energy conversion efficiencies in photosynthesis: minimizing the chlorophyll antenna to maximize efficiency. Plant Sci. 177, 272–280.

Melis, A., Zhang, L., Forestier, M., Ghirardi, M.L., Seibert, M., 2000. Sustained photobiological hydrogen gas production upon reversible inactivation of oxygen evolution in the green alga *Chlamydomonas reinhardtii*. J. Plant Physiol. 122, 127–135.

Midilli, A., Ay, M., Dincer, I., Rosen, M.A., 2005a. On hydrogen and hydrogen energy strategies. II: future projections affecting global stability and unrest. Renewable Sustainable Energy Rev. 9, 273–287.

Midilli, A., Ay, M., Dincer, I., Rosen, M.A., 2005b. On hydrogen and hydrogen energy strategies. I: current status and needs. Renewable. Sustainable. Energy Rev. 9, 255–271.

Miura, Y., Akano, T., Fukatsu, K., Miyasaka, H., Mizoguchi, T., Yagi, K., 1997. Stably sustained hydrogen production by biophotolysis in natural day/night cycle. Energy Convers. Manage. 38, 533–537.

Miyake, J., 1998. The science of biohydrogen, an energetic view. In: Zaborsky, O.R. (Ed.), Biohydrogen. Plenum Press, New York, pp. 7–18.

Miyamoto, K., Ohta, S., Nawa, Y., Mori, Y., Miura, Y., 1987. Hydrogen production by a mixed culture of a green alga, *Chlamydomonas reinhardtii* and a photosynthetic bacterium, *Rhodospirilum rubrum*. Agric. Biol. Chem. 51, 1319–1324.

Mohammed, Y.S., Mustafa, M.W., Bashir, N., 2014. Hybrid renewable energy systems for off-grid electric power: review of substantial issues. Renewable Sustainable Energy Rev. 35, 527–539.

Muth, J., 2013. Hat-Trick 2030, an Integrated Climate and Energy Framework. European Renewable Energy Council, p. 24.

Naik, S.N., Goud, V.V., Rout, P.K., Dalai, A.K., 2010. Production of first and second generation biofuels: a comprehensive review. Renewable. Sustainable. Energy Rev. 14, 578–597.

Nasir Uddin, Md., Wan Daud, W.M.A., 2014. Technological diversity and economics: coupling effects on hydrogen production from biomass. Energy Fuels 28, 4300–4320.

Nguyen, T.A.D., Kim, K.R., Nguyen, M.T., Kim, M.S., Kim, D., Sim, S.J., 2010. Enhancement of fermentative hydrogen production from green algal biomass of *Thermotoga neapolitana* by various pretreatment methods. Int. J. Hydrogen Energy 35, 13035–13040.

Nobre, B.P., Barragán, V.F.B.E., Oliveira, A.C., Batista, A.P., Marques, P.A.S.S., Mendes, R.L., Sovová, H., Palavra, A.F., Gouveia, L., 2013. A biorefinery from *Nannochloropsis* sp. microalga – extraction of oils and pigments. Production of biohydrogen from the leftover biomass. Bioresour. Technol. 135, 128–136.

Ogbonna, J.C., Tanaka, H., 1997. Industrial size photobioreactors. Chemtech. 27, 43–49.

Ogbonna, J.C., Yada, H., Tanaka, H., 1995. Light supply coefficient: a new engineering parameter for photobioreactor design. J. Ferment. Bioeng. 80, 369–376.

Oh, Y.-K., Raj, S.M., Jung, G.Y., Park, S., 2011. Current status of the metabolic engineering of microorganisms for biohydrogen production. Bioresour. Technol. 102, 8357–8367.

Olaizola, M., 2003. Commercial development of microalgal biotechnology: from the test tube to the marketplace. Biomol. Eng. 20, 459–466.

Olivieri, G., Salatino, P., Marzocchella, A., 2014. Advances in photobioreactors for intensive microalgal production: configurations, operating strategies and applications. J. Chem. Technol. Biotechnol. 89, 178–195.

Oncel, S., 2014. Focusing on the optimization for scale up in airlift bioreactors and the production of *Chlamydomonas reinhardtii* as a model microorganism. Ekoloji 23, 20–32.

Oncel, S., 2013. Microalgae for a macro energy world. Renewable Sustainable Energy Rev. 26, 241–264.

Oncel, S., Kose, A., 2014. Comparison of tubular and panel type photobioreactors for biohydrogen production utilizing *Chlamydomonas reinhardtii* considering mixing time and light intensity. Bioresour. Technol. 151, 265–270.

Oncel, S., Sabankay, M., 2012. Microalgal biohydrogen production considering light energy and mixing time as the two key features for scale-up. Bioresour. Technol. 121, 228–234.

Oncel, S., Sukan, F.V., 2008. Comparison of two different pneumatically mixed column photobioreactors for the cultivation of *Arthrospira platensis* (*Spirulina platensis*). Bioresour. Technol. 99, 4755–4760.

Oncel, S., Sukan, F.V., 2009. Photo-bioproduction of hydrogen by *Chlamydomonas reinhardtii* using a semi-continuous process regime. Int. J. Hydrogen Energy 34, 7592−7602.

Oncel, S., Sukan, F.V., 2011. Effect of light intensity and the light: dark cycles on the long term hydrogen production of *Chlamydomonas reinhardtii* by batch cultures. Biomass Bioenerg. 35, 1066−1074.

Ort, D.R., 2001. When there is too much light. Plant Physiol. 125, 29−32.

Ortiz-Marquez, J.C.F., Nascimento, M.D., Zehr, J.P., Curatti, L., 2013. Genetic engineering of multispecies microbial cell factories as an alternative for bioenergy production. Trends Biotechnol. 31 (9), 521−529.

Parmar, A., Singh, N.K., Pandey, A., Ganansounou, E., Madamwar, D., 2011. Cyanobacteria and microalgae: a positive prospect for biofuels. Bioresour. Technol. 10163−10172.

Peters, J.W., Boyd, E.S., D'Adamo, S., Mulder, D.W., Therien, J., Posewitz, M.C., 2013. Hydrogenases, nitrogenases, Anoxia, and H_2 production in water-oxidizing phototrophs. In: Pandey, A., Lee, D.-J., Chisti, Y., Soccol, C.R. (Eds.), Biofuels from Algae. Elsevier Press, Burlington, USA, pp. 37−75.

Philipps, G., Happe, T., Hemschemeier, A., 2012. Nitrogen deprivation results in photosynthetic hydrogen production in *Chlamydomonas reinhardtii*. Planta 235, 729−745.

Pinto, T.S., Malcata, F.X., Arrabaca, J.D., Silva, J.M., Spreitzer, R.J., Esquível, M.G., 2013. Rubisco mutants of *Chlamydomonas reinhardtii* enhance photosynthetic hydrogen production. Appl. Microbiol. Biotechnol. 97, 5635−5643.

Pirouzi, A., Nosrati, M., Shojaosadati, S.A., Shakhesi, S., 2014. Improvement of mixing time, mass transfer, and power consumption in an external loop airlift photobioreactor for microalgae cultures. Biochem. Eng. J. 87, 25−32.

Pirt, J., 1986. The thermodynamic efficiency (quantum demand) and dynamics of photosynthetic growth. New Phytol. 102, 3−37.

Polander, B.C., Barry, B.A., 2012. A hydrogen bonding network plays a catalytic role in photosynthetic oxygen evolution. PNAS 109, 6112−6117.

Posten, C., 2009. Design principles of photobioreactors for cultivation of microalgae. Eng. Life Sci. 9 (3), 165−177.

Prince, R.C., Kheshgi, H.D., 2005. The photobiological production of hydrogen: potential efficiency and effectiveness as a renewable fuel. Crit. Rev. Microbiol. 31, 19−31.

Pulz, O., Gross, W., 2004. Valuable products from biotechnology of microalgae. Appl. Microbiol. Biotechnol. 65, 635−648.

Pulz, O., 2001. Photobioreactors: production systems for phototrophic microorganisms. Appl. Microbiol. Biotechnol. 57, 287−293.

Puri, L., 2009. The Biofuels Market: Current Situation and Alternative Scenarios. United Nations Conference on Trade and Development, United Nations.

Rashid, N., Lee, K., Han, J., Gross, M., 2013. Hydrogen production by immobilized *Chlorella vulgaris*: optimizing pH, carbon source and light. Bioprocess Biosyst. Eng 36, 867−872.

Ryan, C., Bright, T., Green, L.L.C., Hartley, A., 2009. The Promise of Algae Biofuels. NRDC.

Sakurai, H., Masukawa, H., Kitashima, M., Inoue, K., 2013. Phtobiological hydrogen production: bioenergetics and challenges for its practical application. J. Photochem. Photobiol. C 17, 1−25.

Schenk, P.M., Thomas-Hall, S.R., Stephens, E., Marx, U.C., Mussgnug, J.H., Posten, C., Kruse, O., Hankamer, B., 2008. Second generation biofuels: high-efficiency microalgae for biodiesel production. Bioenergy Res. http://dx.doi.org/10.1007/s12155-008-9008-8.

Scoma, A., Durante, L., Bertin, L., Fava, F., 2014. Acclimation of hypoxia in *Chlamydomonas reinhardtii*: can biophotolysis be the major trigger for long term H2 production? New Phytol. 1−11. http://dx.doi.org/10.1111/nph.12964.

Scoma, A., Giannelli, L., Faraloni, C., Torzillo, G., 2012. Outdoor H2 production in a 50-L tubular photobioreactor by means of a sulfur-deprived culture of the microalga *Chlamydomonas reinhardtii*. J. Biotechnol. 157, 620−627.

Shamsul, N.S., Kamarudin, S.K., Rahman, N.A., Kofli, N.T., 2014. An overview on the production of bio-methanol as potential renewable energy. Renewable Sustainable Energy Rev. 33, 578−588.

Sheehan, J., Dunahay, T., Benemann, J., Roessler, P., 1998. A Look Back at the US Department of Energy's Aquatic Species Program—Biodiesel from Algae.

Shuangxiu, W., Li, X., Yu, J., Wang, Q., 2012. Increased hydrogen production in co-culture of *Chlamydomonas reinhardtii* and *Bradyrhizobium japonicum*. Bioresour. Technol. 123, 184−188.

Shuler, M.L., Kargı, F., 1992. Bioprocess Engineering, Basic Concepts. Parntice Hall, New Jersey, USA, 272−273.

Simmons, T.R., Berggren, G., Bacchi, M., Fontecave, M., Artero, V., 2014. Mimicking hydrogenases: from biomimetics to artificial enzymes. Coord. Chem. Rev. 270−271, 127−150.

Singh, A., Nigam, P.S., Murphy, J.D., 2011. Mechanism and challenges in commercialization of algal biofuels. Bioresour. Technol. 102, 26−34.

Skjanes, K., Knutsen, G., Kallqvist, T., Lindblad, P., 2008. H2 production from marine and freshwater species of green algae during sulfur deprivation and considerations for bioreactor design. Int. J. Hydrogen Energy 33, 511−521.

Spoehr, H.A., 1976. The need for a new source food. In: Burlew, J.S. (Ed.), Algal Culture from Laboratory to Pilot Plant, fifth ed. Carnegie Institution of Washington Publication, Washington, USA, pp. 24−28.

Spolaore, P., Cassan, C.J., Duran, E., Isambert, A., 2006. Commercial applications of microalgae. J. Biosci. Bioeng. 101, 87–96.

Srirangan, K., Michael, E.P., Chou, C.P., 2011. Biochemical and genetic engineering strategies to enhance hydrogen production in photosynthetic algae and cyanobacteria. Bioresour. Technol. 102, 8589–8604.

Stephenson, A.L., Kazamia, E., Dennis, J.S., Howe, C.J., Scott, S.A., Smith, A.G., 2010. Life-cycle assessment of potential algal biodiesel production in the United Kingdom: a comparison of raceways and air-lift tubular bioreactors. Energy Fuels 24, 4062–4077.

Sun, Y., Chen, M., Yang, H., Zhang, J., Kuang, T., Huang, F., 2013. Enhanced H2 photoproduction by down-regulation of ferredoxin-NADPD+ reductase (FNR) in the green alga Chlamydomonas reinhardtii. Int. J. Hydrogen Energy 38 (36), 16029–16037.

Taikhao, S., Junyapoon, S., Incharoensakdi, A., Phunpruch, S., 2013. Factors affecting biohydrogen production by unicellular halotolerant cyanobacterium Aphanothece halophytica. J. Appl. Phycol. 25, 575–585.

Tamburic, B., Fessehaye, W., Maitland, G.C., Hellgardt, K., 2012. A novel nutrient control method to deprive green algae of sulphur and initiate spontaneous hydrogen production. Int. J. Hydrogen Energy 37, 8988–9001.

Teplyakov, V.V., Gassanova, L.G., Sostina, E.G., Slepova, E.V., Modigell, M., Netrusov, A.I., 2002. Lab-scale bioreactor integrated with active membrane system for hydrogen production: experience and prospects. Int. J. Hydrogen Energy 27, 1149–1155.

Tolstygina, I.V., Antal, T.K., Kosourov, S.N., Krendeleva, T.E., Rubin, A.B., Tsygankov, A.A., 2009. Hydrogen production by photoautotrophic sulfur-deprived Chlamydomonas reinhardtii pre-grown and incubated under high light. Biotechnol. Bioeng. 102, 1055–1061.

Tomaselli, L., 2004. The microalgal cell. In: Richmond, A. (Ed.), Handbook of Microalgal Culture Applied Phycology and Biotechnology. Blacwell-Science, IA, pp. 3–19.

Torzillo, G., Scoma, A., Faraloni, C., Gianelli, L., 2014. Advances in the biotechnology of hydrogen production with the microalga Chlamydomonas reinhardtii. Crit. Rev. Biotechnol. 1–12. http://dx.doi.org/10.3109/07388551.2014.900734.

Torzillo, G., Seibert, M., 2013. Hydrogen production by microalgae. In: Richmond, A., Hu, Q. (Eds.), Handbook of Microalgal Culture: Applied Phycology and Biotechnology Second Edition. Elsevier, pp. 21–36.

Tredici, M.R., Zittelli, G.C., 1998. Efficiency of sunlight utilization: tubular versus flat photobioreactors. Biotechnol. Bioeng. 57 (2), 187–197.

Tsygankov, A., Kosourov, S., Seibert, M., Ghirardi, M.L., 2002. Hydrogen photoproduction under continuous illumination by sulfur deprived, synchronous Chlamydomonas reinhardtii cultures. Int. J. Hydrogen Energy 27, 1239–1244.

Tsygankov, A.S., Serebryakova, L.T., Svesnikov, D.A., Rao, K.K., Gogotov, I.N., Hall, D.O., 1997. Hydrogen photoproduction by three different nitrogenases in whole cells of Anabaena variabilis and the dependence on pH. Int. J. Hydrogen Energy 22, 859–867.

Uggetti, E., Sialve, B., Trably, E., Steyer, J.-P., 2014. Integrating microalgae production with anaerobic digestion: a biorefinery approach. Biofuels Bioprod. Biorefin. 8, 516–529.

Ugwu, C.U., Ayogi, H., Uchiyama, H., 2008. Photobioreactors for mass cultivation of algae. Bioresour. Technol. 99, 4021–4028.

Walker, D.A., 2009. Biofuels, facts, fantasy and feasibility. J. Appl. Phycol. 21, 509–517.

Weaver, P., Lien, S., Seibert, M., 1979. Photobiological Production of Hydrogen—a Solar Energy Conversion Option. Solar Energy Research Institute, US Department of Energy.

Wei, L., Li, X., Yi, J., Yang, Z., Wang, Q., Ma, W., 2013. A simple approach for the efficient production of hydrogen from Taihu Lake Microcystis spp. blooms. Bioresour. Technol. 139, 136–140.

Wijffels, R.H., Barbosa, M.J., 2010. An outlook on microalgal biofuels. Science 329, 796–799.

Wongluang, P., Chisti, Y., Srinophakun, T., 2013. Optimal hydrodynamic design of tubular photobioreactors. J. Chem. Technol. Biotechnol. 88, 55–61.

Wullf, P., 2014. Credit Suisse 2014 Energy Summit.

Xu, L., Weathers, P.J., Xiong, X.-R., Liu, C.-Z., 2009. Microalgal bioreactors: challenges and opportunities. Eng. Life Sci. 9 (3), 178–189.

Yang, D., Zhang, Y., Barupal, D.K., Fan, X., Gustafson, R., Guo, R., Fiehn, O., 2014. Metabolomics of photobiological hydrogen production induced by CCCP in Chlamydomonas reinhardtii. Int. J. Hydrogen Energy 39, 150–158.

Yoon, J.H., Shin, J.H., Kim, M.S., Sim, S.J., Park, T.H., 2006. Evaluation of conversion efficiency of light to hydrogen energy by Anabaena variabilis. Int. J. Hydrogen Energy 31, 721–727.

Zhang, L., He, M., Liu, J., 2014. The enhancement mechanism of hydrogen photoproduction in Chlorella protothecoides under nitrogen limitation and sulfur deprivation. Int. J. Hydrogen Energy 39, 8969–8976.

Zhang, Y., Fan, X., Yang, Z., Wang, H., Yang, D., Guo, R., 2012. Characterization of H2 photoproduction by a new marine green alga, Platymonas helgolandica var. Tsingtaoensis. Appl. Energy 92, 38–43.

Zhiman, Y., Guo, R., Xu, X., Fan, X., Luo, S., 2011. Fermentative hydrogen production from lipid extracted microalgal biomass residues. Appl. Energy 88, 3468–3472.

Zhou, L., 2005. Progress and problems in hydrogen storage. Renewable Sustainable Energy Rev. 9, 395–408.

Zhu, X.-G., Long, S.P., Ort, D.R., 2008. What is the maximum efficiency with which photosynthesis can covert solar energy into biomass. Curr. Opin. Biotechnol. 19, 153–159.

Zittelli, G.C., Biondi, N., Rodolfi, L., Tredici, M.R., 2013. Photobioreactors for mass production of microalgae.

In: Richmond, A., Hu, Q. (Eds.), Handbook of Microalgal Culture Applied Phycology and Biotechnology, second ed. Blacwell-Science, West Sussex, UK, pp. 225–266.

Züttel, A., 2004. Hydrogen storage methods. Naturwissenschaften 91, 157–172.

Bioethanol Production from Microalgae

Man Kee Lam[1], Keat Teong Lee[2]

[1]Chemical Engineering Department, Universiti Teknologi PETRONAS, Bandar Seri Iskandar, Perak, Malaysia; [2]Low Carbon Economy (LCE) Research Group, School of Chemical Engineering, Universiti Sains Malaysia, Nibong Tebal, Pulau Pinang, Malaysia

1. INTRODUCTION

Recently, increasing demand and usage of fossil fuels have escalated global concerns about several environmental issues, such as air pollution, melting of Arctic ice, and the frequent occurrence of droughts (Lam et al., 2012). As the energy crisis is beginning to affect almost every part of the world due to rapid industrialization and population growth, the search for renewable energy sources is the key challenge in this century to stimulate more sustainable energy development for the future (John et al., 2011; Lam and Lee, 2012).

Bioethanol is an alternative biofuel to gasoline since they share similar physical and chemical properties (Dale, 2007; Demirbaş, 2000; Naik et al., 2010). Bioethanol is usually produced through yeast fermentation by using carbohydrate (e.g., glucose) as the main substrate. First-generation bioethanol is derived from food crops, specifically from sugary and starchy feedstock, such as sugar cane, sugar beet, corn, and wheat (Balat, 2009; Naik et al., 2010). Nevertheless, the sustainability of bioethanol production from food crops is heavily debated due to their priority for food purposes and requirement of large amount of agricultural land to pro-

duce the feedstock (Harun et al., 2010b). Therefore, lignocellulosic biomass is proposed as the second-generation feedstock for bioethanol production; it is the most abundant biomass in the world and does not compete with food commodities (Alvira et al., 2010). However, the main challenge of using lignocellulosic biomass for bioethanol production is the requirement of an additional pretreatment step to break down the complex structure of lignin, so that the cellulose and hemicellulose can be released for subsequent hydrolysis and fermentation (Cardona and Sánchez, 2007).

The production of third-generation bioethanol mainly refers to use of aquatic microorganisms as the main feedstock. Specifically, microalgae biomass has recently gained wide attention as a plausible renewable source for biofuels production. Microalgae are one of the oldest living microorganisms on Earth (Song et al., 2008) and represent a vast variety of photosynthetic species dwelling in diverse environments (John et al., 2011; Mata et al., 2010; Nigam and Singh, 2011). They can grow at an exceptionally fast rate, which is a hundred times faster than terrestrial plants, and they can double their biomass in less than one day (Tredici, 2010). Besides having high lipid content for biodiesel production, some microalgae species also can accumulate high content of carbohydrates within their cells (Figure 1) that can be used as a carbon source or substrate for the fermentation process to produce bioethanol (Harun et al., 2010b; Radakovits et al., 2010). In this chapter, studies reported in the literature on bioethanol production from microalgae biomass are comprehensively discussed, including species selection, carbohydrate hydrolysis and saccharification methods, and the fermentation process.

2. MICROALGAE SPECIES WITH HIGH CARBOHYDRATE CONTENT

Apart from lipids, certain microalgae species are able to accumulate large quantities of carbohydrate within their cells. The carbohydrate is usually stored at the outer layer of cell wall (e.g., pectin, agar, alginate), inner layer of cell wall (e.g., cellulose, hemicellulose), and inside the cell (e.g., starch) (Chen et al., 2013). Through hydrolysis reaction, the carbohydrate can be

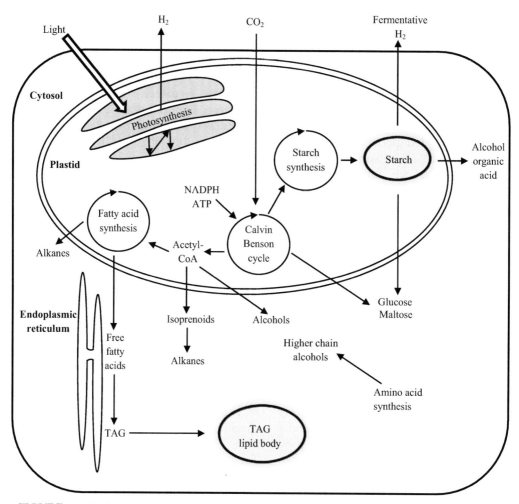

FIGURE 1   Metabolic pathways of microalgae for biofuel production. *Modified from Radakovits et al. (2010).*

hydrolyzed to fermentable sugar (e.g., glucose) for subsequent bioethanol production via fermentation process (Harun et al., 2010b). The carbohydrate contents for specific microalgae species and composition of fermentable sugars are shown in Tables 1 and 2, respectively.

As indicated in Table 1, *Porphyridium cruentum*, *Prymnesium parvum*, *Scenedesmus abundans*, *Scenedesmus dimorphus*, *Scenesdesmus obliquus*, *Spirogyra* sp., and *Tetraselmis suecica* are among the microalgae species that could accumulate high carbohydrate content and suitable to be cultivated for bioethanol production. The carbohydrate content ranges from 21% to 64% of their cell dry weight, which is considered high compared to lignocellulosic biomass. In addition, microalgae do not contain lignocellulosic compounds, in which mild pretreatment (hydrolysis) is already sufficient to release the carbohydrate for subsequent fermentation process and bioethanol production. In addition, the most abundant carbohydrate after hydrolysis process is glucose (Table 2), which is a preferred carbon source for conventional fermentation by *Saccromyces seriviea*.

3. CARBOHYDRATE METABOLISM

The accumulation of carbohydrates within the microalgae cells is mainly due to carbon fixation during photosynthesis process (Chen et al., 2013). Specifically, there are two stages in the photosynthetic reactions, known as light and dark reactions (Figure 2; Richmond and Hu, 2013). During light reactions, light energy is absorbed to split water to oxygen and producing chemical energy (NADPH and ATP). Subsequently, during dark reactions, NADPH and ATP are utilized to reduce CO_2 to carbohydrates via the Calvin−Benson cycle (Markou et al., 2012; Richmond and Hu, 2013). In the Calvin−Benson cycle, CO_2 fixation is catalyzed by ribulose-1,5-biphosphate carboxylase oxygenase (*Rubisco*) enzyme to form two three-carbon compounds

TABLE 1 Carbohydrate Content of Different Microalgae Species

Microalgae species	Carbohydrate content (% dry weight)	References
Chlamydomonas rheinhardii	17	Becker (1994)
Chlorella pyrenoidosa	26	Becker (1994)
Chlorella sp.	19	Phukan et al. (2011)
Chlorella vulgaris	12−17	Becker (1994)
Chloroccum sp.	32.5	Harun and Danquah (2011a)
Dunaliella bioculata	4	Becker (1994)
Dunaliella salina	32	Becker (1994)
Euglena gracilis	14−18	Becker (1994)
Isochrysis galbana	7.7−13.6	Fidalgo et al. (1998)
Isochrysis sp.	5.2−16.4	Renaud et al. (1991)
Mychonastes afer	28.4	Guo et al. (2013)
Nannochloropsis oculata	8	Biller and Ross (2011)
Porphyridium cruentum	40	Biller and Ross (2011)
Prymnesium parvum	25−33	Becker (1994)
Scenedesmus abundans	41	Guo et al. (2013)
Scenedesmus dimorphus	21−52	Becker (1994)
Scenesdesmus obliquus	15−51.8	Ho et al. (2012)
Spirogyra sp.	33−64	Becker (1994)
Synechoccus sp.	15	Becker (1994)
Tetraselmis maculate	15	Becker (1994)
Tetraselmis sp.	24	Schwenzfeier et al. (2011)
Tetraselmis suecica	15−50	Bondioli et al. (2012)

TABLE 2 Sugar Composition of Different Microalgae Species

Microalgae	Monomers (simple reducing sugar)								References
	Arabinose	Galactose	Glucose	Mannose	Xylose	Fucose	Rhamnose	Ribose	
Chaetoceros calcitrans	0.2	20.5	54.7	2.0	1.7	14.3	3.3	3.3	Brown (1991)
Chlamydomonas reinhardtii	3.2	4.5	74.9	2.3	–	0.7	1.5	–	Choi et al. (2010)
Chlorella vulgaris	–	–	98.0	–	–	–	–	–	Ho et al. (2013a)
Chlorococcum infusionum	–	8.9	46.8	15	29.3	–	–	–	Harun et al. (2011a)
Chroomonas Salina	0.08	2.7	87.5	2.3	1.6	3	0	2.5	Brown (1991)
Dunaliella tertiolecta	0.65	1.1	85.3	4.5	1	0	5.5	2	Brown (1991)
Isochrysis galbana	5.7	19	76.5	3.6	2.3	0.6	0.3	2	Brown (1991)
Nannochloropsis oculata	0	3.8	68.3	6.1	4.4	4.4	8.3	4.6	Brown (1991)
Pavlova salina	1.6	4.4	81.0	5.0	8.0	0.51	0	2.5	Brown (1991)
Scenesdesmus obliquus	5.0	13.0	56.0	19.0	7.0				Miranda et al. (2012a)
Spirulina platensis		2.6	54.4	9.3	7.0		22.3		Shekharam et al. (1987)
Tetraselmis chui	0.4	11.3	84.7	1.8	0	0	0.04	1.8	Brown (1991)

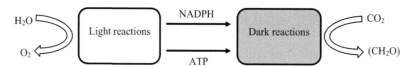

FIGURE 2 Light and dark reactions in photosynthetic reaction. *Modified from Richmond and Hu (2013).*

(3-phosphoglecerate) from a five-carbon compound (ribulose 1,5-biphosphate) (Markou et al., 2012; Richmond and Hu, 2013). One of the formed three-carbon compounds is then used as the substrate for carbohydrate synthesis, while the other is used to continue the next Calvin–Benson cycle (Markou et al., 2012). The carbohydrates are usually formed in the plastids as storage components (e.g., starch) or become the structural components in cell walls (e.g., cellulose, pectin, and sulfated polysaccharides) (Chen et al., 2013). The metabolic pathway of carbohydrates (starch and other sugars) production in microalgae cells is clearly illustrated in Figure 3.

4. PRETREATMENT AND SACCHARIFICATION OF MICROALGAE BIOMASS

A major portion of microalgae carbohydrate is usually entrapped within the microalgae cells. Therefore, a pretreatment step is usually required to disrupt the microalgae cell walls to release the carbohydrate. Carbohydrate is a

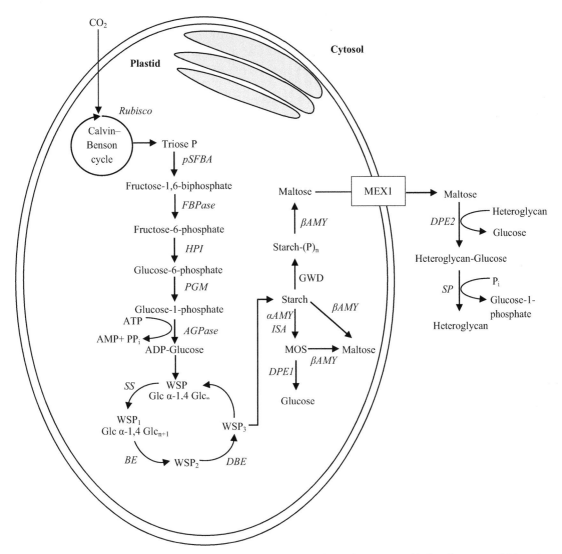

FIGURE 3 Metabolic pathway for carbohydrate synthesis in microalgae. Glucans are added to the water-soluble polysaccharide (WSP) by α-1, 4 glycosidic linkages (WSP₁) until a branching enzyme highly branched the ends (WSP₂). Some of these branches are trimmed (WSP₃), and this process is repeated until a starch granule is formed. Abbreviations: *αAMY*: α-amylase; *βAMY*: β-amylases; *AGPase*: ADP-glucose pyrophosphorylase; *BE*: branching enzymes; *DBE*: debranching enzymes; *DPE*: disproportioning enzyme (1 and 2) α-1,4 glucanotransferase; *FBPase*: fructose biphosphatase; Glc: glucose; *GWD*: glucan-water dikinases; *HPI*: hexose phosphate isomerase; *ISA*: isoamylases; MEX1: maltose transporter; MOS: maltose-oligosaccharides; *PGM*: plastidial phosphoglucomutase; *pSFBA*: plastidial sedulose/fructose-biphosphatase aldolase; P: phosphate; Pᵢ: inorganic phosphate; PPᵢ: pyrophosphate. *SP*: starch phosphorylases; *SS*: starch synthases. *Modified from Radakovits et al. (2010) and Subramanian et al. (2013).*

complex compound and a hydrolysis step is required to break it down to a monosaccharide, such as glucose, for the subsequent fermentation process. The following subsections will describe possible methods in pretreating and hydrolyzing carbohydrate from microalgae biomass.

4.1 Chemical Method

Chemical pretreatment using acid (e.g., H_2SO_4, HCl, and HNO_3) or alkali (NaOH, KOH, and Na_2CO_3) is a widely employed method to pretreat lignocellulosic biomass for bioethanol production (Bensah and Mensah, 2013; Sun and Cheng, 2002). This method is usually fast and relatively inexpensive compared to enzymatic pretreatment. However, due to the extreme reaction conditions required in this process (high temperature and pressure), degradation of carbohydrate is usually observed. The degraded carbohydrate, such as furfural (2-furaldehyde), 5-hydroxymethylfurfural (5-HMF), acetic acid, gypsum, vanillin, and aldehydes (4-hydroxybenzaldehyde) could directly inhibit yeast fermentation and subsequently reduce bioethanol yield (Bensah and Mensah, 2013; Chen et al., 2013). Thus, optimizing suitable reaction conditions (temperature, pressure, moisture content, and reaction reagent concentration) is the key factor in chemical pretreatment to reduce inhibitory effect and to improve operational efficiency (Chen et al., 2013).

Using chemical pretreatment method to produce sugar from microalgae biomass has been recently reported by many researchers (Harun and Danquah, 2011b; Ho et al., 2013a; Miranda et al., 2012b). Due to the simple cellular structure of microalgae and absence of lignin, only mild reaction conditions are required for simultaneous pretreatment (to release carbohydrate from inner cell wall) and hydrolysis reaction (to hydrolyze the complex carbohydrate molecules to simple fermentable sugars) (Ho et al., 2013a). As reported by Harun and Danquah (2011b), diluted H_2SO_4 concentration of 1–3%

v/v, reaction temperature of 140–160 °C, and reaction time of 15–30 min were sufficient to cause a rupture in the *Chlorococcum humicola* cell wall and to release the entrapped carbohydrate for bioethanol production. The study also reported that increasing the reaction temperature and reaction time beyond the optimum point could reduce the bioethanol yield, predominantly due to the degradation of carbohydrate. Similar findings were also reported by Ho et al. (2013a); 1% of diluted H_2SO_4 concentration, reaction temperature at 121 °C, and reaction time of 20 min could recover 96% of glucose yield from *Chlorella vulgaris* biomass. Table 3 compares the bioethanol yield from microalgae biomass using different hydrolysis methods.

Besides acids, the use of alkali has also been successfully demonstrated for the hydrolysis of microalgae carbohydrate (Harun et al., 2011a). Harun et al. (2011) used 0.75 w/v% NaOH at 120 °C for 30 min to obtain optimum glucose yield of 0.35 g/g biomass from *Chlorococcum infusionum*. The results indicated that NaOH is able to rupture the microalgae cell wall, causing the cell wall to lose its integrity and release the carbohydrate. Then, the carbohydrate was further broken down to simple fermentable sugar (glucose) by the presence of NaOH. However, a contradictory observation was reported by Miranda et al. (2012b) in which the use of NaOH as hydrolysis catalyst could not give high glucose yield from *Scenedesmus obliquus*. The reason given was that the high alkaline concentration used in the study caused severe sugar degradation. Thus, more research is still required to explore the potential of alkaline catalyst in pretreating and hydrolyzing microalgae carbohydrate.

4.2 Enzymatic Method

For enzymatic hydrolysis of carbohydrate (particularly referring to starch), α-d-$(1 \rightarrow 4)$-glucosidic linkages in starch will be hydrolyzed by α-amylase (liquefaction) in a random manner

TABLE 3   Bioethanol Yield from Microalgae Biomass via Different Hydrolysis Methods

Hydrolysis method	Microalgae	Total carbohydrate (%)	Sugar yield (g/g algae)	Bioethanol yield (g/g algae)	Reference
CHEMICAL					
H_2SO_4	*Chlorella vulgaris*	51	0.5	0.233	Ho et al. (2013a)
H_2SO_4	*Scenedesmus obliquus*	51.8	0.50–0.51	0.213	Ho et al. (2013c)
H_2SO_4	*Chlorococcum humicola*	32.5	–	0.52	Harun and Danquah (2011b)
H_2SO_4	*Scenedesmus obliquus*	50	–	0.202	Ho et al. (2013b)
NaOH	*Chlorococcum infusionum*	32.5	0.35	0.26	Harun et al. (2011a)
ENZYME					
α-amylase and amyloglucosidase	*Chlamydomonas reinhardtii*	59.7	0.561	0.235	Choi et al. (2010)
Pectinase	*Chlorella vulgaris*	22.4	0.177	0.07	Kim et al. (2014)
Cellulase	*Trichoderma reesei*	32.5	0.222	–	Harun and Danquah (2011a)
Endoglucanase, β-glucosidase and amylase	*Chlorella vulgaris*	51	0.461	0.178–0.214	Ho et al. (2013a)

to produce oligosaccharides (usually denoted as maltodextrin) with three or more α-(1 → 4)-linked D-glucose units as an intermediate product (Choi et al., 2010; Montesinos and Navarro, 2000; Richardson et al., 2000). Subsequently, starch saccharifying enzyme (amyloglucosidase) is introduced to the liquefied starch, in which the enzyme can act on both α-d-(1 → 4) and α-d-(1 → 6)-glucosidic linkages, and therefore maltodextrin is converted to simple reducing sugar (β-D-glucose) (Choi et al., 2010; Van Der Maarel et al., 2002). Enzymatic hydrolysis has several advantages over chemical hydrolysis, such as higher conversion yield, minimal by-product formation, mild operating condition, and low energy input (Taherzadeh and Karimi, 2007; Wald et al., 1984).

Several successful studies on the use of enzymes for hydrolysis of microalgae carbohydrate have been recently reported (Choi et al., 2010; Harun

and Danquah, 2011a; Ho et al., 2013a; Kim et al., 2014). Choi et al. (2010) studied sugar production from *Chlamydomonas reinhardtii* biomass by using α-amylase from *B. licheniformis* and amyloglucosidase from *Aspergillus niger* for liquefaction and saccharification, respectively. The optimum conditions for liquefaction process were identified as 0.005% α-amylase, 90 °C, and 30 min; whereas for saccharification, the optimum conditions were 0.2% amyloglucosidase, 55 °C, pH 4.5, and 30 min. In addition, the study also found that bioethanol yield produced from microalgae biomass (0.235 g/g biomass) is comparable to other cellulose and starch-based feedstock, such as corn stover (0.260 g/g biomass), cane bagasse (0.111 g/g biomass), wheat (0.308 g/g biomass), and corn (0.324 g/g biomass). However, as shown in Table 3, most of the studies concluded that diluted acid hydrolysis could attain higher sugar recovery and bioethanol yield from microalgae biomass than enzymatic method (Choi et al., 2010; Ho et al., 2013a). Besides the advantage of mild reaction conditions, enzymatic hydrolysis method is actually more expensive than chemical hydrolysis and faces difficulty in the recovery of enzymes, which makes the process apparently not economically feasible.

4.3 Mechanical and Other Alternative Pretreatments

Mechanical pretreatment usually refers to using physical force to reduce the particulate size of biomass, such as milling, grinding, and extrusion (Tedesco et al., 2014). Up to now, this pretreatment method is still rarely being explored for microalgae biomass except macroalgae (seaweed) (Schultz-Jensen et al., 2013; Tedesco et al., 2014). This is probably because microalgae biomass is a relatively new feedstock for bioethanol production, and, hence, commercial production of the biomass is still very limited and not sufficient for mechanical pretreatment process that requires a large amount of biomass. On the other hand, the production of macroalgae biomass is already very high, reaching up to 110,000 tons in the year 2005 (Goh and Lee, 2010). Some successful studies of applying mechanical pretreatment to release carbohydrate from macroalgae biomass were reported by Tedesco et al. (2014) and Schultz-Jensen et al. (2013).

Apart from mechanical pretreatment, using ultrasound technology is an alternative way to rupture microalgae for the release of carbohydrate (Zhao et al., 2013). Ultrasound is defined as sound with frequency beyond the capability of the human ear to respond. The normal sound frequency that can be detected by humans lies between 16 and 18 kHz; the frequency for ultrasound generally lies between 20 kHz and 100 MHz (Vyas et al., 2010). This high-frequency sound wave will compress and stretch the molecular spacing of a medium through which it passes. Thus, molecules will be continuously vibrated and cavities will be created. As a result, microfine bubbles are formed through sudden

expansion and collapse violently, generating energy for chemical and mechanical effects (Colucci et al., 2005; Luo et al., 2014). In a recent study, it was found that ultrasonic-assisted extraction (UAE) method could extract more glucose from *Chlorella* sp. biomass than conventional solvent extraction and fluidized bed extraction (Zhao et al., 2013). The highest glucose yield attained using UAE method was 36.9 g/100 g dry cell weight using the following extraction conditions: ultrasonic power of 800 W, extraction time of 80 min, flow rate of 1.52 L/min, and cell concentration of 0.3 g/L. It was observed that ultrasonication caused complete lysis of the nucleus membrane and extensive internal cell wall damage, thus, causing carbohydrate to be excreted to the exocellular medium (Choi et al., 2011; Jeon et al., 2013).

Supercritical fluid technology is a relatively new and alternative pretreatment technique. The basic principal of this technology is achieving a certain phase (supercritical) that is beyond the critical point of a fluid, in which meniscus separating the liquid and vapor phases disappears, leaving only single homogeneous phase (Sawangkeaw et al., 2010). Supercritical fluids that are currently being explored include ethylene, CO_2, ethane, methanol, ethanol, benzene, toluene, and water (Mendes et al., 2003; Sawangkeaw et al., 2010). Among these, supercritical CO_2 has received the most interest, typically in extraction of pharmaceutical and health-related products from microalgae (Jaime et al., 2007; Kitada et al., 2009). Research on the use of supercritical CO_2 to extract microalgae carbohydrate for bioethanol production has recently been reported (Harun et al., 2010a). Harun et al. (2010a) observed that after the pretreatment of supercritical CO_2, the cell wall of *Chlorococum* sp. was ruptured due to the high process temperature and pressure. This led to the release of the carbohydrate embedded in the cell wall. From the report, microalgae biomass with supercritical CO_2 gave 60% higher ethanol concentration for all samples than the biomass without pretreatment.

5. FERMENTATION PROCESS

Fermentation using sugar derived from microalgae as substrate for bioethanol production has been recently explored. There are two common methods for the fermentation process, namely separate hydrolysis and fermentation (SHF), and simultaneous saccharification and fermentation (SSF). The following subsections will discuss the SHF and SSF processes with emphasis on microalgae biomass.

5.1 SHF Process

For SHF, hydrolysis and fermentation process are carried out separately using two different reactors (Xiros et al., 2013). The advantage of this approach is that hydrolysis and fermentation can be carried out independently at their respective optimum conditions, such as pH, temperature, and time (Harun et al., 2011b; Xiros et al., 2013). Thus, during the hydrolysis process using acid, alkali, or enzyme, more substrate could be produced for subsequent fermentation process. However, this approach will face some limitations when acid or alkali is used in the hydrolysis step. Additional neutralizing and purification steps are required to neutralize the acid or alkali and to remove by-products formed during the hydrolysis step. On the other hand, if an enzyme is used during the hydrolysis step, the accumulation of glucose and cellobiose could lead to inhibition of cellulases that will progressively reduce the hydrolysis rate (Tengborg et al., 2001).

The potential of the SHF process in producing bioethanol from microalgae biomass has been revealed in recent studies (Guo et al., 2013; Harun and Danquah, 2011b; Harun et al., 2011a,b; Ho et al., 2013a,c; Kim et al., 2014). In a study carried out by Ho et al. (2013c), diluted acid hydrolysis on *Scenedesmus obliquuswas* biomass was carried out to obtain simple reducing sugar (mainly glucose) before proceeding to fermentation process by bacterium *Zymomonas mobilis*. Under the fermentation condition at 30 °C, initial pH of 6, and glucose concentration of 16.0–16.5 g/L, the maximum bioethanol yield produced was 0.213 g ethanol/g biomass (99.8% of the theoretical yield) within 4 h of fermentation time. The study also pointed out that microalgae biomass after acid hydrolysis process needed to be separated out from the hydrolysate. This is because the wet biomass will increase the viscosity in the hydrolysate and leads to mass transfer limitation during fermentation process, thereby reducing bioethanol production rate.

In another study reported by Harun et al. (2011a), SHF process was applied to produce bioethanol from *C. infusionum*. In that study, after hydrolyzing microalgae biomass with alkali, fermentation was carried out using conventional yeast *Saccharomyces cerevisiae*, at 30 °C and 200 rpm for 72 h. The highest bioethanol yield obtained was 26.1 wt% (g ethanol/g microalgae) resulting from biomass pretreatment with 0.75% (w/v) NaOH at 120 °C for 30 min. The study also found that fluctuation of glucose concentration was observed throughout the fermentation process. This could be because the bioethanol that was produced during the fermentation process was simultaneously being used as substrate by yeast (changing in metabolism) (Piškur et al., 2006) instead of solely depending on glucose.

5.2 SSF Process

For SSF, hydrolysis and fermentation occur simultaneously in the same reactor (Hahn-Hägerdal et al., 2006). Therefore, inhibition by glucose and cellobiose could be minimized because both compounds are directly converted to bioethanol by yeast (Xiros et al., 2013). SSF would also require lower enzyme loading, and higher bioethanol yield could be attained as compared to SHF (Lin and Tanaka, 2006). Other advantages of SSF are eliminating the needs of separate reactors for saccharification and fermentation, shorter fermentation time, and reduced risk of contamination by external microflora (Lin and Tanaka, 2006). However, the main drawback of SSF is the need to compromise operating conditions (e.g., temperature and pH) suitable for both saccharification and fermentation (Gupta and Demirbas, 2010). In addition, recycling of enzymes and yeast is difficult in SSF, which makes the process more challenging when considering scale-up for commercialization purposes (Olofsson et al., 2008).

Apparently, reports on using SSF process for producing bioethanol from microalgae biomass are still scatted in the literature (Harun et al., 2011b; Ho et al., 2013a). Ho et al. (2013a) studied the potential of SSF process by using *C. vulgaris* microalgae biomass as the substrate. The process was carried by using endoglucanase, β-glucosidase, and amylase as the saccharifying enzymes, while *Z. mobilis* was used as the fermenting bacterium. The maximum bioethanol concentration and bioethanol yield attained were 4.27 g/L and 92.3%, respectively, at biomass concentration of 20 g/L, temperature 30 °C, pH 6, and fermentation time of 60 h. In addition, the bioethanol attained through SSF was 15.5% higher than SSF process using the same saccharifying enzyme. However, when the SSF process was switched to using diluted H_2SO_4 as the hydrolysis catalyst, the bioethanol yield attained was 5% higher than SHF and required five-fold shorter operation time. Similar results were also reported by Harun et al. (2011b)—that SHF could attain higher bioethanol yield than SSF when using diluted acid as the hydrolysis catalyst. The reasons for SSF process not able to achieve the desired bioethanol yield are due to contamination during anaerobic treatment and uncompromised operating conditions for both saccharification and fermentation (Harun et al., 2011b).

6. CONCLUSIONS

Cultivating microalgae for biofuel production has revitalized the development of the renewable fuel industry. Besides the opportunity to utilize the microalgae lipids for biodiesel production, the carbohydrates embedded within the microalgae cells have high potential for bioethanol production. The carbohydrates, mainly consisting of starch, can be hydrolyzed to simple reducing sugars through chemical (e.g., diluted H_2SO_4 or NaOH) and enzymatic (e.g., amylase and amyloglucosidase) methods before proceeding to a yeast fermentation process. More research and studies on the life cycle energy and techno-economic analysis are still required to further justify the feasibility of bioethanol production from microalgae biomass.

Acknowledgments

The authors would like to acknowledge the funding given by Ministry of Higher Education (MOHE) Malaysia (long-term Research Grant (LRGS) No. 203/PKT/6723001) and Universiti Sains Malaysia (Research University Grant No. 814146, Postgraduate Research Grant Scheme No. 8044031, and USM Vice-Chancellor's Award) for this project.

References

Alvira, P., Tomás-Pejó, E., Ballesteros, M., Negro, M.J., 2010. Pretreatment technologies for an efficient bioethanol production process based on enzymatic hydrolysis: a review. Bioresour. Technol. 101, 4851–4861.

Balat, M., 2009. Bioethanol as a vehicular fuel: a critical review. Energy Sources Part A 31, 1242–1255.

Becker, E.W., 1994. Microalgae: Biotechnology and Microbiology. Cambridge University Press.

Bensah, E.C., Mensah, M., 2013. Chemical pretreatment methods for the production of cellulosic ethanol: technologies and innovations. Int. J. Chem. Eng. 2013, 1–21.

Biller, P., Ross, A.B., 2011. Potential yields and properties of oil from the hydrothermal liquefaction of microalgae with different biochemical content. Bioresour. Technol. 102, 215–225.

Bondioli, P., Della Bella, L., Rivolta, G., Chini Zittelli, G., Bassi, N., Rodolfi, L., Casini, D., Prussi, M., Chiaramonti, D., Tredici, M.R., 2012. Oil production by the marine microalgae *Nannochloropsis* sp. F&M-M24 and *Tetraselmis suecica* F&M-M33. Bioresour. Technol. 114, 567–572.

Brown, M.R., 1991. The amino-acid and sugar composition of 16 species of microalgae used in mariculture. J. Exp. Mar. Biol. Ecol. 145, 79–99.

Cardona, C.A., Sánchez, O.J., 2007. Fuel ethanol production: process design trends and integration opportunities. Bioresour. Technol. 98, 2415–2457.

Chen, C.-Y., Zhao, X.-Q., Yen, H.-W., Ho, S.-H., Cheng, C.-L., Lee, D.-J., Bai, F.-W., Chang, J.-S., 2013. Microalgae-based carbohydrates for biofuel production. Biochem. Eng. J. 78, 1–10.

Choi, J.A., Hwang, J.H., Dempsey, B.A., Abou-Shanab, R.A.I., Min, B., Song, H., Lee, D.S., Kim, J.R., Cho, Y., Hong, S., Jeon, B.H., 2011. Enhancement of fermentative bioenergy (ethanol/hydrogen) production using (ethanol/hydrogen) production using ultrasonication of Scenedesmus obliquus YSW15 cultivated in swine wastewater effluent. Energy Environ. Sci. 4, 3513−3520.

Choi, S.P., Nguyen, M.T., Sim, S.J., 2010. Enzymatic pretreatment of Chlamydomonas reinhardtii biomass for ethanol production. Bioresour. Technol. 101, 5330−5336.

Colucci, J.A., Borrero, E.E., Alape, F., 2005. Biodiesel from an alkaline transesterification reaction of soybean oil using ultrasonic mixing. J. Am. Oil Chem. Soc. 82, 525−530.

Dale, B.E., 2007. Thinking clearly about biofuels: ending the irrelevant 'net energy' debate and developing better performance metrics for alternative fuels. Biofuels, Bioprod. Biorefin. 1, 14−17.

Demirbaş, A., 2000. Conversion of biomass using glycerin to liquid fuel for blending gasoline as alternative engine fuel. Energy Convers. Manage. 41, 1741−1748.

Fidalgo, J.P., Cid, A., Torres, E., Sukenik, A., Herrero, C., 1998. Effects of nitrogen source and growth phase on proximate biochemical composition, lipid classes and fatty acid profile of the marine microalga Isochrysis galbana. Aquaculture 166, 105−116.

Goh, C.S., Lee, K.T., 2010. A visionary and conceptual macroalgae-based third-generation bioethanol (TGB) biorefinery in Sabah, Malaysia as an underlay for renewable and sustainable development. Renewable Sustainable Energy Rev. 14, 842−848.

Guo, H., Daroch, M., Liu, L., Qiu, G., Geng, S., Wang, G., 2013. Biochemical features and bioethanol production of microalgae from coastal waters of Pearl River Delta. Bioresour. Technol. 127, 422−428.

Gupta, R.B., Demirbas, A., 2010. Gasoline, Diesel and Ethanol Biofuels from Grasses and Plants. Cambridge University Press.

Hahn-Hägerdal, B., Galbe, M., Gorwa-Grauslund, M.F., Lidén, G., Zacchi, G., 2006. Bio-ethanol − the fuel of tomorrow from the residues of today. Trends Biotechnol. 24, 549−556.

Harun, R., Danquah, M.K., 2011a. Enzymatic hydrolysis of microalgal biomass for bioethanol production. Chem. Eng. J. 168, 1079−1084.

Harun, R., Danquah, M.K., 2011b. Influence of acid pre-treatment on microalgal biomass for bioethanol production. Process. Biochem. 46, 304−309.

Harun, R., Danquah, M.K., Forde, G.M., 2010a. Microalgal biomass as a fermentation feedstock for bioethanol production. J. Chem. Technol. Biotechnol. 85, 199−203.

Harun, R., Jason, W.S.Y., Cherrington, T., Danquah, M.K., 2011a. Exploring alkaline pre-treatment of microalgal biomass for bioethanol production. Appl. Energy 88, 3464−3467.

Harun, R., Liu, B., Danquah, M.K., 2011b. Analysis of Process Configurations for Bioethanol Production from Microalgal Biomass. Progress in Biomass and Bioenergy Production: In Tech.

Harun, R., Singh, M., Forde, G.M., Danquah, M.K., 2010b. Bioprocess engineering of microalgae to produce a variety of consumer products. Renew. Sust. Energ. Rev. 14, 1037−1047.

Ho, S.H., Chen, C.Y., Chang, J.S., 2012. Effect of light intensity and nitrogen starvation on CO_2 fixation and lipid/carbohydrate production of an indigenous microalga Scenedesmus obliquus CNW-N. Bioresour. Technol. 113, 244−252.

Ho, S.H., Huang, S.W., Chen, C.Y., Hasunuma, T., Kondo, A., Chang, J.S., 2013a. Bioethanol production using carbohydrate-rich microalgae biomass as feedstock. Bioresour. Technol. 135, 191−198.

Ho, S.H., Kondo, A., Hasunuma, T., Chang, J.S., 2013b. Engineering strategies for improving the CO_2 fixation and carbohydrate productivity of Scenedesmus obliquus CNW-N used for bioethanol fermentation. Bioresour. Technol. 143, 163−171.

Ho, S.H., Li, P.J., Liu, C.C., Chang, J.S., 2013c. Bioprocess development on microalgae-based CO_2 fixation and bioethanol production using Scenedesmus obliquus CNW-N. Bioresour. Technol. 145, 142−149.

Jaime, L., Mendiola, J.A., Ibáñez, E., Martin-Álvarez, P.J., Cifuentes, A., Reglero, G., Señoráns, F.J., 2007. β-Carotene isomer composition of sub- and supercritical carbon dioxide extracts. Antioxidant activity measurement. J. Agric. Food Chem. 55, 10585−10590.

Jeon, B.H., Choi, J.A., Kim, H.C., Hwang, J.H., Abou-Shanab, R.A.I., Dempsey, B.A., Regan, J.M., Kim, J.R., 2013. Ultrasonic disintegration of microalgal biomass and consequent improvement of bioaccessibility/bioavailability in microbial fermentation. Biotechnol. Biofuels 6, 37.

John, R.P., Anisha, G.S., Nampoothiri, K.M., Pandey, A., 2011. Micro and macroalgal biomass: a renewable source for bioethanol. Bioresour. Technol. 102, 186−193.

Kim, K.H., Choi, I.S., Kim, H.M., Wi, S.G., Bae, H.J., 2014. Bioethanol production from the nutrient stress-induced microalga Chlorella vulgaris by enzymatic hydrolysis and immobilized yeast fermentation. Bioresour. Technol. 153, 47−54.

Kitada, K., Machmudah, S., Sasaki, M., Goto, M., Nakashima, Y., Kumamoto, S., Hasegawa, T., 2009. Supercritical CO_2 extraction of pigment components with pharmaceutical importance from Chlorella vulgaris. J. Chem. Technol. Biotechnol. 84, 657−661.

Lam, M.K., Lee, K.T., 2012. Microalgae biofuels: a critical review of issues, problems and the way forward. Biotechnol. Adv. 30, 673−690.

Lam, M.K., Lee, K.T., Mohamed, A.R., 2012. Current status and challenges on microalgae-based carbon capture. Int. J. Greenhouse Gas Contr. 10, 456−469.

Lin, Y., Tanaka, S., 2006. Ethanol fermentation from biomass resources: current state and prospects. Appl. Microbiol. Biotechnol. 69, 627−642.

Luo, J., Fang, Z., Smith Jr., R.L., 2014. Ultrasound-enhanced conversion of biomass to biofuels. Prog. Energy Combust. Sci. 41, 56−93.

Markou, G., Angelidaki, I., Georgakakis, D., 2012. Microalgal carbohydrates: an overview of the factors influencing carbohydrates production, and of main bioconversion technologies for production of biofuels. Appl. Microbiol. Biotechnol. 96, 631−645.

Mata, T.M., Martins, A.A., Caetano, N.S., 2010. Microalgae for biodiesel production and other applications: a review. Renewable Sustainable Energy Rev. 14, 217−232.

Mendes, R.L., Nobre, B.P., Cardoso, M.T., Pereira, A.P., Palavra, A.F., 2003. Supercritical carbon dioxide extraction of compounds with pharmaceutical importance from microalgae. Inorganica Chimica Acta 356, 328−334.

Miranda, J.R., Passarinho, P.C., Gouveia, L., 2012a. Bioethanol production from Scenedesmus obliquus sugars: the influence of photobioreactors and culture conditions on biomass production. Appl. Microbiol. Biotechnol. 96, 555−564.

Miranda, J.R., Passarinho, P.C., Gouveia, L., 2012b. Pre-treatment optimization of Scenedesmus obliquus microalga for bioethanol production. Bioresour. Technol. 104, 342−348.

Montesinos, T., Navarro, J.M., 2000. Production of alcohol from raw wheat flour by amyloglucosidase and Saccharomyces cerevisiae. Enzyme. Microb. Technol. 27, 362−370.

Naik, S.N., Goud, V.V., Rout, P.K., Dalai, A.K., 2010. Production of first and second generation biofuels: a comprehensive review. Renewable Sustainable Energy Rev. 14, 578−597.

Nigam, P.S., Singh, A., 2011. Production of liquid biofuels from renewable resources. Prog. Energy Combust. Sci. 37, 52−68.

Olofsson, K., Bertilsson, M., Lidén, G., 2008. A short review on SSF − an interesting process option for ethanol production from lignocellulosic feedstocks. Biotechnol. Biofuels 1, 1−14.

Phukan, M.M., Chutia, R.S., Konwar, B.K., Kataki, R., 2011. Microalgae Chlorella as a potential bio-energy feedstock. Appl. Energy 88, 3307−3312.

Piškur, J., Rozpedowska, E., Polakova, S., Merico, A., Compagno, C., 2006. How did Saccharomyces evolve to become a good brewer? Trends Genet. 22, 183−186.

Radakovits, R., Jinkerson, R.E., Darzins, A., Posewitz, M.C., 2010. Genetic engineering of algae for enhanced biofuel production. Eukaryot. Cell 9, 486−501.

Renaud, S.M., Parry, D.L., Thinh, L.V., Kuo, C., Padovan, A., Sammy, N., 1991. Effect of light intensity on the proximate biochemical and fatty acid composition of Isochrysis sp. and Nannochloropsis oculata for use in tropical aquaculture. J. Appl. Phycol. 3, 43−53.

Richardson, S., Nilsson, G.S., Bergquist, K.E., Gorton, L., Mischnick, P., 2000. Characterisation of the substituent distribution in hydroxypropylated potato amylopectin starch. Carbohydr. Res. 328, 365−373.

Richmond, A., Hu, Q., 2013. Handbook of Microalgal Culture: Applied Phycology and Biotechnology. Wiley.

Sawangkeaw, R., Bunyakiat, K., Ngamprasertsith, S., 2010. A review of laboratory-scale research on lipid conversion to biodiesel with supercritical methanol (2001−2009). J. Supercrit. Fluids 55, 1−13.

Schultz-Jensen, N., Thygesen, A., Leipold, F., Thomsen, S.T., Roslander, C., Lilholt, H., Bjerre, A.B., 2013. Pretreatment of the macroalgae Chaetomorpha linum for the production of bioethanol − comparison of five pretreatment technologies. Bioresour. Technol. 140, 36−42.

Schwenzfeier, A., Wierenga, P.A., Gruppen, H., 2011. Isolation and characterization of soluble protein from the green microalgae Tetraselmis sp. Bioresour. Technol. 102, 9121−9127.

Shekharam, K.M., Venkataraman, L.V., Salimath, P.V., 1987. Carbohydrate composition and characterization of two unusual sugars from the blue green alga Spirulina platensis. Phytochemistry 26, 2267−2269.

Song, D., Fu, J., Shi, D., 2008. Exploitation of oil-bearing microalgae for biodiesel. Chin. J. Biotechnol. 24, 341−348.

Subramanian, S., Barry, A.N., Pieris, S., Sayre, R.T., 2013. Comparative energetics and kinetics of autotrophic lipid and starch metabolism in chlorophytic microalgae: implications for biomass and biofuel production. Biotechnol. Biofuels 6, 150.

Sun, Y., Cheng, J., 2002. Hydrolysis of lignocellulosic materials for ethanol production: a review. Bioresour. Technol. 83, 1−11.

Taherzadeh, M.J., Karimi, K., 2007. Enzymatic-based hydrolysis processes for ethanol. BioResources 2, 707−738.

Tedesco, S., Marrero Barroso, T., Olabi, A.G., 2014. Optimization of mechanical pre-treatment of Laminariaceae spp. biomass-derived biogas. Renew. Energy 62, 527−534.

Tengborg, C., Galbe, M., Zacchi, G., 2001. Reduced inhibition of enzymatic hydrolysis of steam-pretreated softwood. Enzyme Microb. Technol. 28, 835−844.

Tredici, M.R., 2010. Photobiology of microalgae mass cultures: understanding the tools for the next green revolution. Biofuels 1, 143−162.

Van Der Maarel, M.J.E.C., Van Der Veen, B., Uitdehaag, J.C.M., Leemhuis, H., Dijkhuizen, L., 2002. Properties and applications of starch-converting enzymes of the α-amylase family. J. Biotechnol. 94, 137−155.

Vyas, A.P., Verma, J.L., Subrahmanyam, N., 2010. A review on FAME production processes. Fuel 89, 1−9.

Wald, S., Wilke, C.R., Blanch, H.W., 1984. Kinetics of the enzymatic hydrolysis of cellulose. Biotechnol. Bioeng. 26, 221−230.

Xiros, C., Topakas, E., Christakopoulos, P., 2013. Hydrolysis and fermentation for cellulosic ethanol production. WIREs Energy Environ. 2, 633−654.

Zhao, G., Chen, X., Wang, L., Zhou, S., Feng, H., Chen, W.N., Lau, R., 2013. Ultrasound assisted extraction of carbohydrates from microalgae as feedstock for yeast fermentation. Bioresour. Technol. 128, 337−344.

Medicinal Effects of Microalgae-Derived Fatty Acids

Luísa Barreira[1], Hugo Pereira[1], Katkam N. Gangadhar[1,2], Luísa Custódio[1], João Varela[1]

[1]University of Algarve, CCMAR, Centre of Marine Sciences, Faro, Portugal;
[2]New University of Lisbon, Institute of Chemical and Biological Technology, Lisbon, Portugal

1. INTRODUCTION

Humans have long used microalgae as food, particularly as a source of protein (e.g., *Nostoc* in China and *Arthrospira* (syn. *Spirulina*) in Chad and Mexico; Milledge, 2011). However, the chemical composition of microalgae allows them to be valuable sources of other compounds with human interest. Microalgae also produce carbohydrates, mainly in the form of starch or glucose, and lipids, mainly in the form of triacylglycerols (TAG) (Spolaore et al., 2006). Additionally, they produce secondary metabolites, for example, pigments (e.g., carotenoids, phycobilins), phytosterols, and other compounds with applications in the pharmaceutical industry (e.g., dolastatin 10, an antitumoral drug produced by the cyanobacterium *Symploca* sp.; Borowitzka, 2013). Some of these secondary metabolites are already exploited commercially, such as β-carotene (*Dunaliella salina* in Australia, Israel, and the United States) or astaxanthin

(*Haematococcus pluvialis* in the United States and India; Spolaore et al., 2006).

Although their protein content was initially the driving force for mass production of microalgae, their content in polyunsaturated fatty acids (PUFAs) is currently recognized as key to their overall nutritional value. This is particularly true in the aquaculture field, where the biomass of *Chlorella*, *Tetraselmis*, *Isochrysis*, *Phaeodactylum*, *Nannochloropsis*, *Skeletonema*, and *Spirulina* is used as feed for larvae of invertebrate and vertebrate species with high commercial value (Roy and Pal, 2014). However, the importance of microalgae as major feedstocks for PUFA production is bound to increase even further in the near future. For example, diets rich in marine oily fish with health-promoting PUFA profiles have been highly encouraged by the World Health Organization (WHO) (FAO, 2010). However, the global depletion of fish stocks has made this recommendation highly unsustainable. It is thus clear that alternative sources

Handbook of Marine Microalgae
http://dx.doi.org/10.1016/B978-0-12-800776-1.00013-3

of health-promoting n-3 PUFA, such as eicosa-pentaenoic (EPA) and docosahexaenoic (DHA) acids, must be found (Adarme-Vega et al., 2014).

This chapter addresses the potential of microalgae as sources of PUFA, in particular of n-3 PUFA, as well as the cultivation of microalgae and how their content in PUFA can be modulated. How microalgae can be applied as a functional food, as well as sources of nutraceuticals and/or pharmaceutical drugs, will also be discussed.

2. MICROALGAE AS A SOURCE OF PUFA

2.1 Fatty Acid Profile of Microalgae

The fatty acid (FA) profile of different microalgae species has been extensively analyzed and reviewed by several authors searching for interesting species with applications not only in the aquaculture feed market but also in the search for and development of new microalgal strains for biodiesel production (Renaud et al., 1999; Pereira et al., 2011, 2013; Benemann, 2013). The FA profile is generally considered as a chemotaxonomic marker to define groups of various taxonomic ranks. In fact, different algal groups generally show different lipid compositions, though FA profiles can be species specific as well (Zhukova and Aizdaicher, 1995). Recently, Lang et al. (2011) published an extensive analysis of the FA profiles of more than 2000 strains of microalgae from the SAG culture (Culture Collection of Algae at Goettingen University). In this study, common FA distribution patterns were found in species belonging to the same phyla or classes; however, at lower taxonomic levels, FA contents could be rather variable (e.g., between closely related species or among multiple isolates of the same species; Lang et al., 2011). Typically, microalgae display FA profiles comprising molecules with a chain length of 12–24 carbons and several degrees of unsaturation or number of double bonds,

namely saturated (SFA; 0 double bonds), monounsaturated (MUFA; 1 double bond), and PUFA (≥ 2 double bonds). Though the FA profiles may vary considerably among strains, the most abundant FAs are generally those with an even number of carbons with chain lengths of 14, 16, and 18 carbons. Table 1 shows the profile of several microalgal strains belonging to different classes of the Bacillariophyta, Chlorophyta, Cryptophyta, Cyanobacteria, Dinophyta, Ochrophyta, Haptophyta, and Rhodophyta megagroups. The most common C14 is usually C14:0 (myristic acid), and its occurrence and abundance are highly variable among classes and even among species belonging to the same class. The highest abundances were found in Cyanobacteria, Bacillariophyta, Dinophyta, and Haptophyta (Table 1). Concerning C16 FA, besides C16:0, several unsaturated forms of C16 are often found, such as the MUFA C16:1 (palmitoleic acid) and the PUFA C16:2 n-6, C16:3 n-3, C16:3 n-6, and C16:4 n-3. In general, the C16:0 and the C16:1 are the most abundant FA with 16 carbons, being virtually ubiquitous in all microalgal megagroups. However, C16 PUFA are apparently more megagroup specific. For example, the C16:4 n-3 is almost restricted to the Bacillariophyta, whereas the C16:3 n-3 is more frequently detected among the Chlorophyta (Table 1). Looking at the C18 FA, the MUFA C18:1 n-9 and all the C18 PUFA appear to be more discriminative than C18:0 or C18:1 n-7. Specifically, C18:1 n-9 and the C18:2 n-6 are more frequently detected and abundant amid the Chlorophyta, while the C18:3 n-3 occurs at significant levels in Cyanobacteria, Cryptophyta, and Rhodophyta. Though the C18:4 n-3 PUFA can be found in several different megagroups, its abundance seems to be higher in Chlorophyta, in particular, in Nephrophyceae and Prasinophyceae strains (Table 1). The long-chain polyunsaturated fatty acids (LC-PUFA, carbon chain length ≥ 20) are notably abundant in some strains. C20:4 n-6 (arachidonic acid;AA) is commonly detected at high levels in diatoms

TABLE 1 Fatty Acid Composition of Microalgae Belonging to Different Phyla

Species	FA (% of total)																			
	12:0	14:0	14:1	16:0	16:1	16:2 n-6	16:3 n-3	16:3 n-4	16:4 n-3	18:0	18:1 n-7	18:1 n-9	18:2 n-6	18:3 n-3	18:3 n-6	18:4 n-3	20:4 n-6	20:5 n-3	22:6 n-3	Others
BACILLARIOPHYTA																				
Bacillariophyceae																				
Amphora sp. (Renaud et al., 1999)	—	9.7	0.3	19.6	28.0	—	—	1.1	—	1.7	0.2	0.5	3.4	0.6	2.7	2.4	4.9	13.9	0.3	10.7
Nitzchia spp. (Renaud et al., 1999)	—	7.4	0.1	24.0	34.2	—	—	6.8	—	0.6	1.5	0.5	0.6	0.1	0.5	0.4	8.0	8.4	0.3	6.6
Phaeodactylum tricornutum (Zhukova and Aizdaicher, 1995)	—	7.4	0.1	11.3	21.5	1.0	—	12.3	0.1	0.4	0.5	2.3	1.5	0.9	0.5	0.5	0.3	28.4	0.7	10.3
Coscinodiscophyceae																				
Chaetoceros muelleri (Zhukova and Aizdaicher, 1995)	—	15.0	—	17.3	30.0	1.6	—	7.8	0.1	0.8	0.5	1.4	0.7	0.3	1.1	0.8	1.7	12.8	0.8	7.3
Chaetoceros constrictus (Zhukova and Aizdaicher, 1995)	—	14.0	0.6	16.4	14.3	1.0	—	7.9	—	4.8	1.5	4.7	1.8	0.2	0.3	0.3	3.3	18.8	0.6	9.5
Skeletonema costatum (Zhukova and Aizdaicher, 1995)	—	12.7	0.4	9.4	19.0	0.9	—	10.2	—	2.2	—	2.6	1.6	0.2	—	2.9	0.2	15.4	2.3	20.0
Thalassiosira weissflogii (Lang et al., 2011 (and references therein))	—	8.8	—	36.6	28.7	—	—	—	—	—	14.0	—	—	—	—	—	—	—	—	11.9

Continued

TABLE 1 Fatty Acid Composition of Microalgae Belonging to Different Phyla—cont'd

Species	FA (% of total)																			
	12:0	14:0	14:1	16:0	16:1	16:2 n-6	16:3 n-3	16:3 n-4	16:4 n-3	18:0	18:1 n-7	18:1 n-9	18:2 n-6	18:3 n-3	18:3 n-6	18:4 n-3	20:4 n-6	20:5 n-3	22:6 n-3	Others
Fragilariophyceae																				
Fragilaria sp. (Renaud et al., 1999)	—	1.6	—	32.1	24.6	—	—	5.2	—	2.4	1.0	2.6	1.4	1.0	1.1	1.9	8.7	6.8	1.0	8.6
CHLOROPHYTA																				
Chlorophyceae																				
Chlorella sp. (Zhukova and Aizdaicher, 1995)	—	2.0	0.5	19.6	6.2	3.6	12.0	—	—	3.3	1.6	5.7	11.8	22.3	0.3	0.1	0.5	1.3	—	9.2
Dunaliella primolecta (Lang et al., 2011 (and references therein))	—	0.6	—	26.0	0.9	—	—	—	—	1.6	—	16.3	7.0	38.7	—	0.6	—	—	—	8.3
Dunaliella maritima (Zhukova and Aizdaicher, 1995)	—	0.4	0.1	11.8	2.7	1.1	2.7	—	22.6	0.4	0.4	2.1	4.1	42.6	3.2	1.3	—	—	—	4.5
Dunaliella salina (Zhukova and Aizdaicher, 1995)	—	0.5	0.1	17.8	0.8	1.5	2.1	0.3	18.2	1.5	0.6	2.8	6.1	36.9	2.5	0.7	—	0.1	—	7.5
Dunaliella tertiolecta (Zhukova and Aizdaicher, 1995)	—	0.3	0.1	10.3	4.0	2.0	3.2	—	23.9	0.3	0.5	1.7	5.2	38.7	3.2	1.3	0.3	0.4	—	4.6
Scenedesmus sp. (Custódio et al., 2014a)	—	1.3	—	20.5	19.0	1.9	2.4	—	—	0.8	—	0.9	10.3	39.3	—	—	—	—	—	3.6

Nephrophyceae

Taxon																				
Nephroselmis sp. (Renaud et al., 1999)	1.7	0.6	19.9	5.9	3.1	1.8	–	1.8	13.1	2.3	3.8	2.4	3.0	14.0	1.3	18.0	1.9	2.4	3.0	1.8
Prasinophyceae																				
Tetraselmis sp. (Custódio et al., 2014a)	0.7	–	24.9	7.9	1.7	1.9	–	–	–	0.4	–	27.1	21.3	–	–	–	2.9	9.4	–	1.8
Tetraselmis viridis (Zhukova and Aizdaicher, 1995)	0.8	0.2	15.9	3.3	0.3	1.2	–	–	19.9	0.8	3.1	4.6	3.1	15.0	0.2	13.3	0.2	6.7	–	11.4
Trebouxiophyceae																				
Botryococcus braunii (Custódio et al., 2014b)	–	–	18.1	9.7	2.2	3.1	–	–	–	0.8	–	35.8	23.2	3.4	–	–	–	–	–	3.7
Nannochloris sp. (Pereira et al., 2013)	–	–	32.5	12.1	–	3.8	–	–	–	2.3	–	33.8	15.4	–	–	–	–	–	–	0.1
Parietochloris incisa (Lang et al., 2011 (and references therein))	–	–	19.8	–	5.2	–	–	–	–	18.2	–	10.2	14.3	14.3	–	–	–	–	–	18.0
Picochlorum sp. (Pereira et al., 2013)	–	–	26.1	12.0	7.2	5.7	–	–	–	3.2	–	22.0	23.8	–	–	–	–	–	–	0.0
Ulvophyceae																				
Desmochloris sp. (Pereira et al., 2013)	–	–	33.6	–	–	1.3	–	–	–	1.9	–	28.2	28.6	2.2	–	–	–	–	–	4.2
CHRYPTOPHYTA																				
Cryptophyceae																				
Chroomonas salina (Zhukova and Aizdaicher, 1995)	5.0	0.3	13.5	2.0	–	–	–	3.0	–	3.0	2.9	2.3	1.2	10.8	–	30.3	2.8	12.9	7.1	5.9

Continued

TABLE 1 Fatty Acid Composition of Microalgae Belonging to Different Phyla—cont'd

Species	FA (% of total)																			
	12:0	14:0	14:1	16:0	16:1	16:2 n-6	16:3 n-3	16:3 n-4	16:4 n-3	18:0	18:1 n-7	18:1 n-9	18:2 n-6	18:3 n-3	18:3 n-6	18:4 n-3	20:4 n-6	20:5 n-3	22:6 n-3	Others
Cryptomonas sp. (Renaud et al., 1999)	–	1.0	–	11.9	3.8	–	–	–	–	1.5	1.1	1.1	0.6	25.1	–	30.7	0.2	12.0	6.6	4.4
Rhodomonas sp. (Renaud et al., 1999)	–	6.4	0.4	13.7	3.5	–	–	–	–	2.5	4.7	2.4	–	25.2	1.8	22.6	–	8.7	4.6	3.5
CYANOBACTERIA																				
Cyanophiceae																				
Nostoc commune (Lang et al., 2011 (and references therein))	–	–	–	25.3	24.1	–	–	–	–	–	–	–	12.5	38.1	–	–	–	–	–	0.0
Synechocystis sp. (Lang et al., 2011 (and references therein))	–	42.5	–	18.8	30.1	–	–	–	–	–	–	–	–	14.2	–	–	–	–	–	0.0
DINOPHYTA																				
Dinophyceae																				
Gymnodinium kowalevskii (Zhukova and Aizdaicher, 1995)	–	12.4	–	26.7	1.8	–	–	–	–	8.5	–	6.5	3.7	7.2	–	15.6	–	0.1	9.5	8.0
OCHROPHYTA																				
Eustigmatophyceae																				
Nannochloropsis oculata (Custódio et al., 2014b)	0.5	4.6	–	15.7	26.9	0.3	–	–	–	0.3	–	4.3	4.5	–	–	–	–	41.9	–	1.0
Raphidophyceae																				
Heterosigma akashiwo (Lang et al., 2011 (and references therein))	–	6.6	–	40.0	12.7	4.0	–	–	–	–	–	–	4.5	6.7	–	5.2	3.5	14.8	–	2.0

Pavlovophyceae

Species (reference)																				
Pavlova lutheri (Lang et al., 2011 (and references therein))	–	10.1	–	11.1	26.3	–	–	–	–	–	–	5.2	0.6	0.5	–	9.1	0.3	18.0	9.7	9.1
Pavlova salina (Zhukova and Aizdaicher, 1995)	–	13.1	–	15.1	30.4	0.6	0.1	0.4	–	1.0	0.7	3.1	1.5	–	2.2	4.2	3.7	19.1	1.5	3.3

Prymnesiophyceae

Species (reference)																				
Emiliania huxleyi (Lang et al., 2011 (and references therein))	–	18.8	–	10.3	–	–	–	–	–	10.8	–	42.2	–	–	–	8.7	–	–	9.2	0.0
Isochrysis galbana T-ISO (Custódio et al., 2014a)	–	18.3	–	14.4	20.5	0.7	–	–	–	0.3	–	15.2	12.1	–	–	–	0.7	2.8	12.7	2.3
Isochrysis sp. (Renaud et al., 1999)	–	17.3	0.2	12.0	3.1	–	–	–	–	1.1	1.1	6.9	4.0	5.7	1.0	19.0	–	0.9	9.9	17.8

RHODOPHYTA

Bangiophyceae

Species (reference)																				
Porphyridium cruentum (Zhukova and Aizdaicher, 1995)	–	0.5	–	28.6	1.1	–	0.2	–	–	0.8	0.7	1.3	8.2	0.4	0.3	–	27.6	21.1	–	9.2

Stylonemaphyceae

Species (reference)																				
Rhodosorus sp. (Renaud et al., 1999)	7.4	0.4	0.4	24.5	1.6	2.8	–	–	–	1.6	2.8	13.9	9.8	20.7	–	–	2.5	7.8	0.4	3.4

(Bacillariophyta). However, even higher contents can be found in the rhodophyte *Porphyridium cruentum* (27.6% of total FA). The distribution of the n-3 LC-PUFA, C20:5 n-3 (EPA), and C22:6 n-3 (DHA) is further discussed below.

2.2 n-3 PUFA in Microalgae

Microalgae are traditionally considered good sources of PUFA (Borowitzka, 2013). In fact, the data here collected and summarized in Table 2, in the form of sums and ratios of particular FA, agree with this assumption: about 62% of the microalgal strains here reviewed possess a ratio of Σ PUFA over Σ SFA higher than 1. Moreover, must of these PUFA are n-3 PUFA. In most strains under review, their n-3 PUFA content accounts for more than 20% of the total FA, and around 70% of the strains in Table 2 show a Σ n-3 PUFA/Σ n-6 PUFA ratio higher than 1. Strains with exceedingly high levels of n-3 PUFA (higher than 50% of total FA) can be found in Chlorophyta, in particular, those from the genus *Dunaliella* (*Dunaliella maritima*, *D. salina*, and *Dunaliella tertiolecta*), but also in the strains *Tetraselmis viridis* and *Nephroselmis* sp., and in Cryptophyta (*Chroomonas salina*, *Cryptomonas* sp., and *Rhodomonas* sp.). Strains with considerably low n-3 PUFA contents (<5% of total FA) are mainly found in Bacillariophyta (*Thalassiosira weissflogii*) and Chlorophyta (*Desmochloris* sp. and *Nannochloris* sp.). Considering just the levels of EPA and DHA, most of the Bacillariophyta, Cryptophyta, Ochrophyta, Haptophyta, and Rhodophyta reviewed in this study can be considered good sources of EPA, ranging from 6.8% (*Fragilaria* sp.) to 41.9% (*Nannochloropsis oculata*). The number of strains with high amounts of DHA is far less than those with EPA; nonetheless, Cryptophyta, Dinophyta, and Haptophyta microalgae appear to be good sources with DHA levels ranging from 4.6% (*Rhodomonas* sp.) to 12.7% (*Isochrysis galbana*; Table 1).

Although microalgal lipid profiles are undoubtedly an expedient way of deducing the potential of a microalga as a source of n-3 PUFA, one must keep in mind that the amount of total lipids in microalgae, and thus the real amount of these compounds in the microalgal biomass, may vary considerably among strains (Borowitzka, 2013; Roy and Pal, 2014). In this sense, higher amounts of EPA are probably delivered by diatoms and by *Nannochloropsis* and *Tetraselmis* microalgae (Martins et al., 2013; Roy and Pal, 2014). Concerning DHA production, *Isochrysis galbana* and *Pavlova lutheri*, which possess a lipid composition enriched in this n-3 PUFA (12.7% and 9.7%, respectively) and a total lipid content of 20−25% of total biomass dry weight, are estimated to produce about 2−3 g of DHA per 100 g of dry biomass (Fidalgo et al., 1998; Rodolfi et al., 2009).

3. MICROALGAL PUFA PRODUCTION

3.1 Microalgae Cultivation and PUFA Production

Large-scale production of microalgae was effectively established decades ago, being the most successful examples of the commercial production of *Chlorella*, *Spirulina*, and *D. salina* (Spolaore et al., 2006). Nowadays, mass production of microalgal biomass is performed autotrophically or heterotrophically, allowing a continuous supply of biomass and secondary metabolites (e.g., PUFA) from different commercial ventures (Martins et al., 2013).

Photoautotrophic microalgae use light, nutrients, and CO_2 to generate chemical energy through photosynthesis. The major advantage of photoautotrophic growth over heterotrophic is the environmental benefit of CO_2 fixation (grossly for each Kg of microalgae produced autotrophically 1.8 Kg of CO_2 is captured; Gouveia and Oliveira, 2009). Moreover,

TABLE 2 Fatty Acid Composition, IA and IT of Microalgae from Different Phyla and Classes

Species	Σ SFA	Σ MUFA	Σ PUFA	Σ PUFA/ Σ SFA	Σ n-3	Σ n-6	Σ n-6/ Σ n-3	IA	IT
BACILLARIOPHYTA									
Bacillariophyceae									
Amphora sp. (Renaud et al., 1999)	31.0	29.0	29.3	0.95	17.2	11.0	0.64	1.02	0.42
Nitzchia spp. (Renaud et al., 1999)	32.0	36.3	25.1	0.78	9.20	9.10	0.99	0.98	0.62
Phaeodactylum tricornutum (Zhukova and Aizdaicher, 1995)	19.1	24.4	46.2	2.42	30.6	3.30	0.11	0.70	0.17
Coscinodiscophyceae									
Chaetoceros muelleri (Zhukova and Aizdaicher, 1995)	33.1	31.9	27.7	0.84	14.8	5.10	0.34	1.49	0.50
Chaetoceros constrictus (Zhukova and Aizdaicher, 1995)	35.2	21.1	34.2	0.97	19.9	6.40	0.32	1.53	0.46
Skeletonema costatum (Zhukova and Aizdaicher, 1995)	24.3	22.0	33.7	1.39	20.8	2.70	0.13	1.32	0.29
Thalassiosira weissflogii (Lang et al., 2011 (and references therein))	45.4	42.7	0.00	0.00	0.00	0.00	—	1.68	2.13
Fragilariophyceae									
Fragilaria sp. (Renaud et al., 1999)	36.1	28.2	27.1	0.75	10.7	11.2	1.05	0.77	0.68
CHLOROPHYTA									
Chlorophyceae									
Chlorella sp. (Zhukova and Aizdaicher, 1995)	24.9	14.0	51.9	2.08	35.7	16.2	0.45	0.42	0.20
Dunaliella primolecta (Lang et al., 2011 (and references therein))	28.2	17.2	46.3	1.64	39.3	7.00	0.18	0.45	0.21
Dunaliella maritima (Zhukova and Aizdaicher, 1995)	12.6	5.30	77.6	6.16	69.2	8.40	0.12	0.16	0.06
Dunaliella salina (Zhukova and Aizdaicher, 1995)	19.8	4.30	68.4	3.45	58.0	10.1	0.17	0.27	0.11

Continued

TABLE 2   Fatty Acid Composition, IA and IT of Microalgae from Different Phyla and Classes—cont'd

Species	Σ SFA	Σ MUFA	Σ PUFA	Σ PUFA/ Σ SFA	Σ n-3	Σ n-6	Σ n-6/ Σ n-3	IA	IT
Dunaliella tertiolecta (Zhukova and Aizdaicher, 1995)	10.9	6.30	78.2	7.17	67.5	10.7	0.16	0.14	0.05
Scenedesmus sp. (Custódio et al., 2014a)	22.6	19.9	53.9	2.38	41.7	12.2	0.29	0.35	0.16
Nephrophyceae									
Nephroselmis sp. (Renaud et al., 1999)	23.9	12.7	61.6	2.58	52.3	9.30	0.18	0.36	0.14
Prasinophyceae									
Tetraselmis sp. (Custódio et al., 2014a)	26.0	35.0	37.2	1.43	11.3	25.9	2.29	0.38	0.40
Tetraselmis viridis (Zhukova and Aizdaicher, 1995)	17.5	11.2	59.9	3.42	56.1	3.80	0.07	0.27	0.09
Trebouxiophyceae									
Botryococcus braunii (Custódio et al., 2014b)	18.9	45.5	31.9	1.69	6.50	25.4	3.91	0.23	0.34
Nannochloris sp. (Pereira et al., 2013)	34.8	45.9	19.2	0.55	3.80	15.4	4.05	0.50	0.82
Parietochloris incisa (Lang et al., 2011 (and references therein))	38.0	10.2	33.8	0.89	14.3	19.5	1.36	0.45	0.65
Picochlorum sp. (Pereira et al., 2013)	29.3	34.0	36.7	1.25	5.70	31.0	5.44	0.37	0.59
Ulvophyceae									
Desmochloris sp. (Pereira et al., 2013)	35.5	28.2	32.1	0.90	3.50	28.6	8.17	0.56	0.91
CRYPTOPHYTA									
Cryptophyceae									
Chroomonas salina (Zhukova and Aizdaicher, 1995)	21.5	7.50	65.1	3.03	61.1	4.00	0.07	0.46	0.11
Cryptomonas sp. (Renaud et al., 1999)	14.4	6.00	75.2	5.22	74.4	0.80	0.01	0.20	0.05
Rhodomonas sp. (Renaud et al., 1999)	22.6	11.0	62.9	2.78	61.1	1.80	0.03	0.53	0.10

TABLE 2   Fatty Acid Composition, IA and IT of Microalgae from Different Phyla and Classes—cont'd

Species	Σ SFA	Σ MUFA	Σ PUFA	Σ PUFA/ Σ SFA	Σ n-3	Σ n-6	Σ n-6/ Σ n-3	IA	IT
CYANOBACTERIA									
Cyanophiceae									
Nostoc commune (Lang et al., 2011 (and references therein))	25.3	24.1	50.6	2.00	38.1	12.5	0.33	0.34	0.19
Synechocystis sp. (Lang et al., 2011 (and references therein))	61.3	30.1	14.2	0.23	14.2	0.00	0.00	4.26	1.06
DINOPHYTA									
Dinophyceae									
Gymnodinium kowalevskii (Zhukova and Aizdaicher, 1995)	47.6	8.30	36.1	0.76	32.4	3.70	0.11	1.72	0.43
OCHROPHYTA									
Eustigmatophyceae									
Nannochloropsis oculata (Custódio et al., 2014b)	21.1	31.2	46.7	2.21	41.9	4.80	0.11	0.44	0.14
Raphidophyceae									
Heterosigma akashiwo (Lang et al., 2011 (and references therein))	46.6	12.7	38.7	0.83	26.7	12.0	0.45	1.29	0.49
HAPTOPHYTA									
Pavlovophyceae									
Pavlova lutheri (Lang et al., 2011 (and references therein))	21.2	31.5	38.2	1.80	37.3	0.90	0.02	0.74	0.13
Pavlova salina (Zhukova and Aizdaicher, 1995)	29.2	34.2	33.3	1.14	24.9	8.00	0.32	1.01	0.30
Prymnesiophyceae									
Emiliania huxleyi (Lang et al., 2011 (and references therein))	39.9	42.2	17.9	0.45	17.9	0.00	0.00	1.42	0.53
Isochrysis galbana T-ISO (Custódio et al., 2014a)	33.0	35.7	29.0	0.88	15.5	13.5	0.87	1.35	0.46
Isochrysis sp. (Renaud et al., 1999)	30.4	11.3	40.5	1.33	35.5	5.00	0.14	1.57	0.25

Continued

TABLE 2   Fatty Acid Composition, IA and IT of Microalgae from Different Phyla and Classes—cont'd

Species	Σ SFA	Σ MUFA	Σ PUFA	Σ PUFA/ Σ SFA	Σ n-3	Σ n-6	Σ n-6/ Σ n-3	IA	IT
RHODOPHYTA									
Bangiophyceae									
Porphyridium cruentum (Zhukova and Aizdaicher, 1995)	29.9	3.10	57.8	1.93	21.7	36.1	1.66	0.50	0.35
Stylonematophyceae									
Rhodosorus sp. (Renaud et al., 1999)	33.9	18.7	44.0	1.30	28.9	12.3	0.43	0.56	0.25

photoautotrophic microalgae can grow in wastewater, recycling the nutrients present in the water. However, apart from bioremediation, nutrient availability is considered a serious constraint for industrial production of microalgae (Chisti, 2013). Large-scale production of photoautotrophic microalgae is commonly achieved in two distinct culture systems: open (e.g., open ponds and raceways) and closed systems (photobioreactors; PBRs).

Commercial production of microalgae in open systems has been fully established by different microalgae ventures (e.g., Qualitas Health, Aurora Algae) that use open ponds and raceways for mass cultivation (Figure 1), being the cultured biomass later processed into commercial oil enriched in PUFA (Martins et al., 2013). The main advantages of open systems over closed systems are the initial lower capital costs coupled with low maintenance and reduced energy requirements (Lee, 2001). The main drawback of open systems is their higher susceptibility to contamination, since cultures are directly exposed to the environment (Pulz, 2001). For this reason, only a narrow number of strains, which withstand specific culture conditions (e.g., high salinity, extreme pH) or display higher growth rates than competing microorganisms, can be cultivated in monoculture (Lee, 2001). One key example is the large-scale culture

of *D. salina* grown under high salinity, which prevents the proliferation of other microalgae (Borowitzka, 1999). Another serious limitation is that controlling and regulating the culture conditions are very difficult or virtually impossible (Pulz, 2001), which can be a serious disadvantage for most sensitive strains. Moreover, light penetration and the surface-to-volume ratio in open systems are lower than in closed systems, leading to lower volumetric productivities (Pienkos and Darzins, 2009).

FIGURE 1   Small-scale raceway for microalgae production. *Depicted image was kindly provided by Necton S.A. (Algarve, Portugal).*

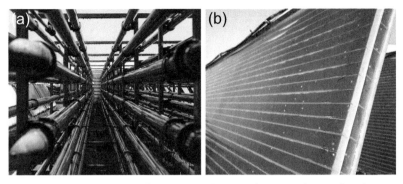

FIGURE 2   Examples of closed systems used for commercial mass cultivation of microalgae biomass at the facilities of Necton S.A. (Algarve, Portugal). (a) Tubular photobioreactor. (b) Flat panel flow through photobioreactor.

Microalgae cultivation in closed systems is performed in specialized systems commonly called PBR (Figure 2). To date, several PBRs have been developed and optimized for different microalgal strains. However, for large-scale commercial systems, they can be divided into three main categories: tubular (horizontal), air-lifts (vertical), and flat panel flow through PBR (Sierra et al., 2008; Kunjapur and Eldridge, 2010). PBRs display key advantages over open systems for large-scale production of microalgae; since the system is closed, cultures are less prone to contamination and culture conditions can easily be controlled and manipulated (Lee, 2001). PBRs are thus known to display significantly higher biomass productivities, also enabling the culture of sensitive strains (Ugwu et al., 2008). High investment and operational costs (e.g., capital, maintenance, and energy; Carvalho et al., 2006) are important economic constraints as far as their widespread use is concerned. However, further developments in PBR design, and implementation of large-scale facilities, mainly driven by the research in the field of biofuels, are expected to decrease these costs, enabling the culture of microalgal biomass for the exploitation of high-value products, such as PUFA.

Instead of atmospheric CO_2 and light, heterotrophic microalgae use dissolved organic compounds (e.g., glucose, acetate) as carbon and energy sources. Heterotrophic cultures can lead to significantly higher biomass and lipid productivities, being thus an alternative to photoautotrophic growth systems (O'Grady and Morgan, 2010). Large-scale production of heterotrophic microalgae is performed in conventional bioreactors/fermenters. Although this technology is expensive, it can be used for large-scale production of biomass containing high levels of valuable metabolites (Lee, 2001). Currently, most of the heterotrophic production of biomass is devoted to PUFA production (Bumbak et al., 2011). Nowadays, different commercial ventures (e.g., DSM-NP, Source-Omega) provide purified DHA-rich oil obtained from microalgae grown heterotrophically (Martins et al., 2013). Although heterotrophic growth presents promising advantages for PUFA production, there are still some drawbacks that need to be considered: (1) the addition of energy and of an organic carbon source to the media present significant costs; (2) only a restricted number of strains are able to grow heterotrophically; (3) there is a high risk of contamination; and (4) excess organic substrate might inhibit culture growth (Chen, 1996).

3.2 Modulation of PUFA Content

Microalgae biomass is known to present different biochemical profiles under different culture conditions. Lipid synthesis is affected by the imposition of unfavorable/stressful conditions in microalgal cells, causing considerable alterations in the lipid content as well as in the FA composition of the produced biomass (Hu et al., 2008). Therefore, the manipulation of cultivation parameters is a key strategy for the modulation of the biochemical profile in microalgae cells in order to obtain the target end product. The understanding of lipid biosynthesis mechanisms is essential for the modulation of FA production under mass cultivation of microalgae with the purpose of increasing the production of n-3 PUFA.

Lipids are present in different structures within microalgal cells, mainly in the form of structural (polar lipids) and storage lipids (nonpolar lipids). Structural lipids are present in cell membranes in the form of glycerol-based phospholipids and glycolipids and are known to contain high levels of PUFA. Conversely, storage lipids tend to accumulate SFA and MUFA in TAG, forming intracellular lipid droplets in the cytosol and/or in the chloroplast. There are, however, exceptions, as it has been found that a limited number of strains are able to accumulate significant amounts of PUFA in TAG as well (Khozin-Goldberg et al., 2002). TAGs are used by cells as an energy-rich carbon source when culture conditions are favorable for growth; under unfavorable/stressful conditions, when growth is reduced, cells tend to accumulate FAs in the form of TAG (up to 70% of total dried biomass) as an energy storage product (Hu et al., 2008).

Different culture parameters (nutrient levels, temperature, photon flux densities, pH, osmotic stress, among others) can be manipulated during the cultivation of microalgal strains in order to enhance the total lipid content of biomass (Sharma et al., 2012). Among these, limitation of nutrient levels is one of the most studied and most effective procedures for high lipid induction. Unfortunately, nutrient depletion as well as other stressful conditions (e.g., salinity,

light irradiance) generally result in decreased amounts of PUFA in the lipid fraction despite the overall increase in total lipids and TAG (Khozin-Goldberg et al., 2011; Guihéneuf and Stengel, 2013). Under these conditions, cells shift their metabolism in order to store energy in the form of TAG, which usually contain SFA and MUFA rather than PUFA. However, this lower unsaturation levels can be prevented if the right strain and experimental conditions are selected (Griffiths et al., 2012). Indeed, Guihéneuf and Stengel (2013) have recently reported that inorganic carbon availability is a major factor limiting the accumulation of n-3 PUFA under nutrient depletion in *P. lutheri*. In this study, bicarbonate addition improved biomass productivity under nutrient depletion, which was accompanied by improved accumulation of EPA and DHA in TAG. This result highlights how little is known about the way metabolic pathways leading to n-3 PUFA production in microalgae are regulated. Therefore, further research is urgently needed in order to understand how specific genomes respond to environmental cues, such as bicarbonate.

Temperature shifts are probably one of the most studied procedures for modulating the FA profile of living cells. Generally, FA profiles present higher degrees of unsaturation when the biomass is cultured under low temperatures, while higher temperatures generally increase the saturation of the lipid fraction (Sharma et al., 2012). Temperature shifts have successfully been used to increase the content of EPA and DHA in different microalgae (Jiang and Chen, 2000; Jiang and Gao, 2004). This change in saturation allows the cell to counteract alterations in membrane fluidity caused by the temperature shift, ensuring that cell integrity and physiological functions are maintained (Jiang and Chen, 2000). However, modulation of PUFA in large-scale production systems through the regulation of cultivation temperature can only be performed in closed systems and has never been reported.

3.3 Expedite Methods of PUFA Detection and Quantification

The detection and quantification of PUFA (and other FA) in microalgae and other sources of biomass are commonly performed by chromatographic techniques, mainly gas chromatography (GC) coupled with an appropriate detector, such as mass spectrometry (GC-MS) or flame ionization detector (GC-FID). However, alternative methods that are able to accurately determine total saturated and unsaturated FA composition in oils have successfully been established. Examples are mass spectroscopy, Raman spectroscopy, FTIR, and ^{1}H-NMR (Dong et al., 2013; Nuzzo et al., 2013; Vongsvivut et al., 2014). The latter methodologies were shown to possess several advantages over chromatographic techniques including speed, cost effectiveness, eco-friendliness, and suitability for routine research and industrial purposes. Nevertheless, for precise determination and purification of PUFA, chromatographic methods are still more accurate than spectroscopic analyses.

3.4 PUFA Extraction from Microalgae

In order to separate the PUFA from microalgae, crude lipids must first be extracted from the biomass. Various methods have been established for lipid extraction from microalgae: (1) mechanical (e.g., expeller, bead-beating, or high pressure homogenization) and (2) chemical (e.g., solvent, enzymatic, supercritical fluid, or ionic liquids)—the most common are solvent-based extractions, which are generally coupled with methods for cell disruption such as microwaves, ultrasonication and/or autoclave, and pressurized fluid extractions (Grimi et al., 2014 and references therein). After the extraction, lipids need to be converted into Fatty Acid Methyl Esters (FAME) by hydrolysis and/or transesterification using either chemical or enzymatic catalysts (Takisawa et al., 2013).

To concentrate and/or enrich in PUFA the extracts from natural sources such as fish or microalgal oils, several approaches have been established, such as thin-layer chromatography (TLC), molecular or fractional distillation, liquid—liquid extraction, membrane filtration, silver-based chromatography, urea complexation, supercritical fluid, supercritical fluid chromatography/or semipreparative supercritical fluid chromatography, metal ion complexation extraction, low-temperature crystallization, and lipase-catalyzed hydrolysis or transesterification (Nakano et al., 2008; Rubio-Rodríguez et al., 2010; Dillon et al., 2013 and references therein). Each method has innate advantages and disadvantages, and most of them are difficult to scale up or are energetically inefficient.

A well-known approach is based on the complexation of SFA/or esters with urea, which is a rather efficient and simple method (Hayes et al., 1998). Using this method, *Crypthecodinium cohnii* microalgae oil was enriched in DHA (from 47.4% to 97.1%), with a process yield of 32.5% of the weight of the parent oil (Senanayake, 2010). Still, urea can lead to the formation of carcinogenic ethyl or methyl carbamates, and it can only be applied for oils with low amounts of SFA (Baeza-Jimenez et al., 2014). Therefore, an integrated/or a combination of two or more of the described methods is advised. For example, Mendes et al. (2007) reported a four-step method to obtain DHA (99% pure) from *C. cohnii* oil, which involves saponification (i.e., liquid—liquid extraction for the separation of unsaponifiable matter), methylation, winterization (e.g., cooling) followed by urea complexation. However, it relies on the use of strong bases (e.g., NaOH) and large amounts of organic solvents to solubilize all components and to perform the necessary extraction steps (Baeza-Jimenez et al., 2014). Even though several methods exist, an efficient, simple, and inexpensive procedure for the production of a product enriched in n-3 PUFA (i.e., EPA and DHA) is still required to meet the future demand for highly pure/or EPA and DHA enriched products.

4. n-3 PUFA HEALTH BENEFITS

Health benefits of PUFA were first described in the 1930s. While experimenting with rats, Burr and Burr (1930) observed that rats fed with fat-free diet would develop skin, kidney, and ovary lesions leading to deficiencies in reproduction and often death. Such conditions could, however, be ameliorated by treatment with linoleic acid (PUFA) or complex unsaturated oils (e.g., corn oil or linseed oil, which are rich in PUFA) but not with oleic acid (MUFA) or saturated FAs. Burr concluded that linoleic acid, and possibly other PUFAs, were essential for healthy nutrition (Burr and Burr, 1930). From this study onwards an increasing amount of evidence suggests that PUFA and n-3 PUFA, in particular, produce several health benefits including: (1) prevention of cardiovascular diseases; (2) prevention and treatment of inflammatory conditions; and (3) aid brain development and function (Ruxton et al., 2004).

4.1 Cardiovascular Diseases

The role of n-3 PUFA in the prevention of cardiovascular diseases (CVD) is probably the most consensual. The meta-analysis concluded that n-3 PUFA could reduce overall mortality, mortality because of myocardial infarction (MI) and sudden death in patients with coronary heart disease (CHD). Several observational studies have related fish-eating habits (e.g., higher ingestion of n-3 PUFA) with lower rates of CVD, improvement in the high-density lipoprotein (HDL) to low-density lipoprotein (LDL) ratios, and low levels of blood triglycerides (Sidhu, 2003; Ruxton et al., 2004). Clinical trials have also demonstrated that dietary supplementation

with EPA and DHA are helpful in the primary prevention of CHD and post-MI, sudden cardiac death (SCD), heart failure (HF), atherosclerosis, and atrial fibrillation (AF; Lavie et al., 2009; Swanson et al., 2012).

Deficits in n-3 PUFA dietary intake, however, may not be the only cause of CVD. The Western diet has become increasingly rich in n-6 PUFA in detriment of the n-3 PUFA (Blasbalg et al., 2011). Whereas it is believed that the human race has evolved on a ~1 ratio between n-6 and n-3 PUFA, nowadays this ratio is close to 15:1 or 16:1 (Simopoulos, 2002). The PUFA composition of cell membranes is generally determined by the profile of the FA ingested, and, since mammalian cells cannot convert n-6 PUFA to n-3 PUFA due to a lack of the converting enzyme, the composition of cell membranes will reflect the ratio of n-3 PUFA/n-6 PUFA of the diet. The proportion of each FA in the membrane is bound to have an effect on its fluidity hence affecting its function (Candela et al., 2011). A diet enriched in EPA and DHA is able to replace the n-6 PUFA (e.g., arachidonic acid (AA)), in membranes of probably every cell (Figure 3; Simopoulos, 2002). Since the oxidation of AA may lead to the production of prostaglandins, thromboxanes, leukotrienes, hydroxy fatty acids, and lipoxins, which can, when in large quantities, contribute to the formation of thrombus and atheromas, a diet rich in n-6 PUFA may induce pro-thrombotic and pro-aggregatory processes and lead to CVD (Simopoulos, 1991; Candela et al., 2011). Recently, a behavioral study performed with men and women with MI, stroke, angina pectoris, coronary angioplasty, and coronary bypass associated the adherence to a Mediterranean-style diet (e.g., richer in n-3 than in n-6 PUFA) with lower all-cause mortality in individuals with CVD (Lopez-Garcia et al., 2013).

4.2 Inflammatory Disease

Inflammation is a process designed to deal with injury, invading pathogens, allergens, and

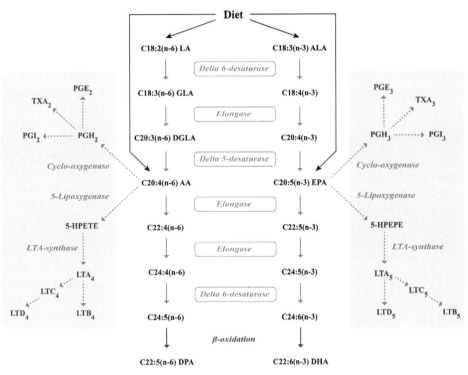

FIGURE 3   n-3 and n-6 PUFA biosynthetic pathway showing the oxidative metabolism of arachidonic acid (C20:4 n-6) and EPA (C22:5 n-3) by the cyclooxygenase and 5-lipoxygenase pathways.

toxins leading to damaged tissue repair (Rangel-Huerta et al., 2012). It involves the timely release of mediators and expression of receptors such as acute phase protein reactants (C-reactive protein; CRP), vasoactive amines, vasoactive peptides, lipid mediators (e.g., eicosanoids such as prostaglandin E_2 (PGE_2) and leukotriene B4 (LTB4), docosanoids and platelet-activating factors), cytokines (e.g., tumor necrosis factor (TNF-α), interleukins (IL-1, IL-6, IL-8)), chemokines (e.g., monocyte chemotactic protein 1 (MCP-1) and macrophage inflammatory protein 1 alpha (MIP-1a)) and proteolytic enzymes (Rangel-Huerta et al., 2012). Chronic inflammation diseases such as rheumatoid arthritis and inflammatory bowel disease and pathological inflammatory responses like endotoxic shock

and acute respiratory syndrome are characterized by high levels of TNF-α, IL-1, and IL-6 (Rangel-Huerta et al., 2012). n-3 and n-6 PUFA compete in the oxidative metabolism that leads to prostaglandin formation—higher dietary levels of n-6 PUFA lead to the formation of IL-1 and IL-6 and are associated with higher levels of thromboxane A_2 (TXA_2) and LTB4 (Figure 3). In fact, EPA and DHA have been reported to decrease circulating levels of CRP, TNF-α, IL-1, and IL-6 and hence may be related to anti-inflammatory processes (Swanson et al., 2012). Part of the anti-inflammatory effects of n-3 PUFA may be related to the inhibition of the 5-lipoxygenase pathway in neutrophils and monocytes that lead to lower levels of LTB4 (Simopoulos, 2002). In randomized controlled trials in humans, the consumption of n-3 LCPUFA decreased the plasma levels of pro-inflammatory n-6 eicosanoids and cytokines and was associated with reduced levels of some inflammatory biomarkers in plasma (James and Cleland, 1997; Rangel-Huerta et al., 2012).

4.3 Brain Development and Function

The brain is mostly made of fat and contains high amounts of AA and DHA. It is believed that these FA, DHA in particular, enable neuronal membranes fluidity and help to regulate neurotransmitters, both being central functions for optimal brain performance (Chang et al., 2009). Experiments with pregnant rats led to the conclusion that DHA deficiency during brain maturation reduces its plasticity and impairs brain function in adulthood (Bhatia et al., 2011). Therefore, it appears that n-3 PUFA are essential for fetal brain development, and since they must be supplied to the fetus through the mother, it is crucial that pregnant women include EPA and DHA in their diet (Ruxton et al., 2004; Food and Agriculture Organization of the United Nations, 2010). This has also been assessed through several randomized clinical trial (RCT), which link maternal DHA intake with several benefits in infants, namely: problem-solving skills (at 9 months old), better eye-and-hand coordination (at 2.5 years old), decreased incidence of asthma (at 16 years old) and allergies, and decreased release of maternal and fetal inflammatory markers, which may be related to preterm labor (see Swanson et al., 2012 and references therein).

4.4 Mental Health and Behavioral Disorders

The imbalance in FA composition of the brain has been linked to several other disorders related to mental health and behavior, including attention deficit and hyperactivity disorder (ADHD), depression, age-related cognitive decline, and neurodegenerative conditions such as Alzheimer's disease (AD) and Parkinson's disease (PD; Ruxton et al., 2004; Chang et al., 2009; Bradbury, 2011; Simopoulos, 2011). ADHD is characterized by hyperactive and impulsive behavior, and attention deficit, and researchers have found that these symptoms are often associated with lower levels of n-3 PUFA (Chang et al., 2009). However, RCTs have failed to demonstrate that such conditions may be reversed by treatment with EPA and DHA (Ruxton et al., 2004).

The possible link between depression and n-3 PUFA deficiency or imbalance between n-6 and n-3 PUFA may be provided by evidence in which the psychological stress in humans induces the release of pro-inflammatory cytokines such as TNFα and IL-6 (Simopoulos, 2011). However, Tiemeier et al. (2003), working with the elderly, found that depressed subjects had a significantly higher ratio of n-6 to n-3 PUFA, and that this relationship was not secondary to inflammation or atherosclerosis, indicating that there might be a direct link between fatty acid composition and mood. Clinical trials, however, provide conflicting results with evidence of both positive or absence of beneficial results from the intake of n-3 PUFA in the treatment of depression symptoms (Simopoulos et al., 2011).

Recently, DHA has been established as a precursor for the synthesis of a newly identified neuroprotectin, a docosanoid found in nervous tissues, which is thought to induce nerve regeneration. This discovery unveils new therapeutic opportunities in the treatment of neurodegenerative diseases like AD (Bradbury, 2011). In addition, epidemiological studies have found an inverse relationship between fish consumption and the risk of developing dementia or AD, indicating that relatively high intakes of DHA and EPA are linked to lower risks of dementia incidence or progression (Chang et al., 2009).

5. MICROALGAE AS FUNCTIONAL FOODS AND SOURCES OF PHARMACEUTICAL DRUGS

Considering some of the more well-established health benefits of n-3 PUFA ingestion mentioned in the last section, several organizations such as the WHO, Food and Agriculture Organization of the United Nations (FAO), and North American Treaty Organization (NATO) have proposed the following daily dietary intakes: (1) general adult population, 300—400 mg of EPA + DHA; (2) pregnant/lactating women, 200 mg of DHA; and (3) children (2—10 years), 100—250 mg of EPA + DHA (Simopoulos, 1989; FAO, 2010).

The amounts described are total daily amounts independent of their source: whole foods (e.g., oily fish, seeds), supplemented foods (e.g., designer eggs or meat enriched in n-3 PUFA), or dietary supplements (e.g., fish-oil capsules) (FAO, 2010; Plaza et al., 2009; Swanson et al., 2012). It is estimated that the ingestion of one to four portions of oily fish per week could represent an average daily intake of EPA plus DHA of 0.45—0.9 g, which is more than the amount proposed for the general adult population (Ruxton et al., 2004). Considering the close-to-depletion state of some populations of fish in the oceans, it is debatable if fish captures can sustain the global demand required to supply all the n-3 PUFA (Adarme-Vega et al., 2014). Moreover, safety concerns have been raised concerning the contamination of wild fish with metals (e.g., methyl mercury), persistent organic pollutants (e.g., polychlorinated biphenyls, dioxins, and organochloride pesticides), and other environmental contaminants (Verbeke et al., 2005). These issues have promoted efforts toward land-based production of alternative n-3 PUFA sources. These include farmed fish, oleaginous plants, and large-scale production of microalgae.

Marine microalgae seem to be one of the best alternatives since they are the primary producers of n-3 PUFA and therefore contain all the genes encoding the enzymes required for the synthesis of EPA and DHA (Vaezi et al., 2013). In Section 2.2, the best sources of microalgal n-3 PUFA were identified, based not only on their lipid profiles but also on their total lipid production, which resulted in higher net production of n-3 PUFA. The importance of microalgae, however, lies beyond a good source of n-3 PUFA; some microalgae strains also present balanced PUFA composition. Table 2 presents, in addition to total n-3 PUFA contents, the n-6 to n-3 PUFA ratios and the indexes of atherogenicity (IA) and of thrombogenicity (IT). The significance of a low n-6 to n-3 PUFA ratio was already amply discussed in the previous section and almost all of the microalgal strains here reviewed have very low amounts of n-6 compared to n-3 PUFA: over 75% of the strains have a ratio lower than 1% and 67% a ratio lower than 0.5. The IA and IT indexes may be used to estimate the lipid quality of several foods, in order to predict its cardiovascular-related health benefits (Ulbricht and Southgate, 1991). The IA is a ratio between the main classes of saturated and unsaturated FA. Since the first are considered pro-atherogenic and the latter anti-atherogenic, a low IA ratio is desirable (Garaffo et al., 2011). The IT is related to the tendency to form clots in the blood vessels and is defined as the relationship between the pro-thrombogenetic (saturated FA) and the anti-thrombogenetic (MUFA and n-3 and n-6 PUFA). Similarly to IA, a low value of IT is also desirable (Garaffo et al., 2011). The majority of the strains reviewed in this work (around 67%) have low IA values (e.g., lower than 1). The lowest values, however, are found in the Chlorophyta megagroup, in which all the strains reviewed have IA values close to or lower than 0.5. These IA values are similar to those of olive oil (0.14), sunflower oil (0.07), raw mackerel (0.28), or the Eskimos' diet (0.39), and considerably lower than those of lamb (1.00) or beef meat (0.70–0.74; Ulbricht and Southgate, 1991), or even of raw tuna (0.69–0.76; Garaffo et al., 2011). Concerning the IT index, the results are overwhelming: around 73% of the microalgal strains in Table 2 have an index of thrombogenicity lower than 0.5, suggesting their biomass is healthier than all types of meat and again comparable to other sources of n-3 PUFA such as olive oil (0.32), sunflower oil (0.28), raw mackerel (0.16; Ulbricht and Southgate, 1991), or raw tuna (0.27–0.28; Garaffo et al., 2011).

Despite the growing evidence of the future role of microalgae as sources of n-3 PUFA, few clinical studies are found in the literature in which microalgae or microalgal-derived oil supplements have been used. Nonetheless, a DHA-rich, but almost EPA-free, microalgal oil (*Ulkenia* sp.) supplementation was shown to improve some CHD risk factors such as plasma triacylglycerol (TG) and TG:HDL cholesterol ratio in a randomized trial (Geppert et al., 2006). Similar effects were observed with several independent human clinical trials in which both healthy and CVD and diabetes patients took supplements of *Spirulina* spp (Deng and Chow, 2010 and references therein).

Notwithstanding the lack of evidence supported by clinical trials involving microalgae or microalgal-oil supplementation, some companies currently produce a DHA-enriched product resulting from the blend of sunflower oil with oil extracted from the heterotrophic thraustochytrids *Crypthecodinium* (e.g., DSM-NP life's DHA™), *Schizochytrium* (DSM-NP life's DHA plus EPA™) and *Ulkenia* (Lonza DHAid™) and an EPA-enriched oil from the Chlorophyta *N. oculata* (e.g., Qualitas Health EicoOil™; Martins et al., 2013). Whole microalgal products have also reached the markets as "health foods"; these include *Chlorella* sp. (e.g., Taiwan Chlorella Manufacturing & Co. producing over 400 t/year

of dry algal biomass), or *Spirulina* (*Arthospira*) as a source of linolenic acid (e.g., DIC producing 300 t/year; Spolaore et al., 2006; Milledge, 2011). Most of these products are sold and advertised as preventing atherosclerosis and hypercholesterol, among others (Milledge, 2011).

A large number of investments are currently being made in the food industry. Consumers are increasingly aware of the importance of healthy eating habits, allowing for the expansion of markets such as that of "functional foods" (Bigliardi and Galati, 2013). "Functional foods" are defined by the Functional Food Center as "Natural or processed foods that contains [sic] known or unknown biologically-active compounds; which, in defined quantitative and qualitative amounts, provide a clinically proven and documented health benefit for the prevention, management, or treatment of chronic disease" (functionalfoodscenter.net); and their market is expected to display the highest growth index, especially in Europe and Asia (Borowitzka, 2013).

This chapter has provided a comprehensive picture of the microalgal content of n-3 PUFA and of the health benefits associated with the consumption of these compounds in addition to their placement in the markets as functional ingredients or functional foods. Several species are already commercially available and have proven to be nontoxic for human consumption, providing a safe and viable alternative to fish products as sources of health-promoting commodities as the n-3 PUFA.

References

Adarme-Vega, T.C., Thomas-Hall, S.R., Schenk, P.M., 2014. Towards sustainable sources for omega-3 fatty acids production. Curr. Opin. Biotechnol. 26, 14–18.

Baeza-Jimenez, R., No, D.S., Otero, C., Garcia, H.S., Lee, J.S., Kim, I.H., 2014. Lipase-catalyzed enrichment of γ-linolenic acid from evening primrose oil in a solvent-free system. J. Am. Oil Chem. Soc. 91, 1147–1153.

Benemann, J., 2013. Microalgae for biofuels and animal feeds. Energies 6, 5869–5886.

Bhatia, H.S., Agrawal, R., Sharma, S., Huo, Y.-X., Ying, Z., Gomez-Pinilla, F., 2011. Omega-3 fatty acid deficiency during brain maturation reduces neuronal and behavioral plasticity in adulthood. PLos One 6, e28451.

Bigliardi, B., Galati, F., 2013. Innovation trends in the food industry: the case of functional foods. Trends Food Sci. Technol. 31, 118–129.

Blasbalg, T.L., Hibbeln, J.R., Ramsden, C.E., Majchrzak, S.F., Rawlings, R.R., 2011. Changes in consumption of omega-3 and omega-6 fatty acids in the United States during the 20th century. Am. J. Clin. Nutr. 93, 950–962.

Borowitzka, M.A., 1999. Commercial production of microalgae: ponds, tanks, tubes and fermenters. J. Biotechnol. 70, 313–321.

Borowitzka, M.A., 2013. High-value products from microalgae—their development and commercialization. J. Appl. Phycol. 25, 743–756.

Bradbury, J., 2011. Docosahexaenoic acid (DHA): an ancient nutrient for the modern human brain. Nutrients 3, 529–554.

Bumbak, F., Cook, S., Zachleder, V., Hauser, S., Kovar, K., 2011. Best practices in heterotrophic high-cell-density microalgal processes: achievements, potential and possible limitations. Appl. Microbiol. Biotechnol. 91, 31–46.

Burr, G.O., Burr, M.M., 1930. On the nature and role of the fatty acids essential in nutrition. J. Biol. Chem. 86, 587–621.

Candela, C.G., López, L.M.B., Kohen, V.L., 2011. Importance of a balanced omega 6/omega 3 ratio for the maintenance of health. Nutritional recommendations. Nutr. Hosp. 26, 323–329.

Carvalho, A.P., Meireles, L.A., Malcata, F.X., 2006. Microalgal reactors: a review of enclosed system designs and performances. Biotechnol. Prog. 22, 1490–1506.

Chang, C.-Y., Ke, D.-S., Chen, J.-Y., 2009. Essential fatty acids and human brain. Acta Neurol. Taiwan. 18, 231–241.

Chen, F., 1996. High cell density culture of microalgae in heterotrophic growth. Trends Biotechnol. 14, 412–426.

Chisti, Y., 2013. Constraints to commercialization of algal fuels. J. Biotechnol. 167, 201–214.

Custódio, L., Soares, F., Pereira, H., Barreira, L., Vizetto-Duarte, C., Rodrigues, M.J., Rauter, A.P., Alberício, F., Varela, J., 2014a. Fatty acid composition and biological activities of Isochrysis galbana T-ISO, Tetraselmis sp. and Scenedesmus sp.: possible application in the pharmaceutical and functional food industries. J. Appl. Phycol. 26, 151–161.

Custódio, L., Soares, F., Pereira, H., Rodrigues, M.J., Barreira, L., Rauter, A.P., et al., 2014b. Botryococcus braunii and Nannochloropsis oculata extracts inhibit cholinesterases and protect human dopaminergic SH-SY5Y cells from H_2O_2-induced cytotoxicity. J. Appl. Phycol. http://dx.doi.org/10.1007/s10811-014-0369-4.

Deng, R., Chow, T.J., 2010. Hypolipidemic, antioxidant, and antiinflammatory activities of microalgae spirulina. Cardiovasc. Ther. 28, e33–e45.

Dillon, J.T., Aponte, J.C., Tarozo, R., Huang, Y., 2013. Purification of omega-3 polyunsaturated fatty acids from fish oil using silver-thiolate chromatographic material and high performance liquid chromatography. J. Chromatogr. A 1312, 18–25.

Dong, W., Zhang, Y., Zhang, B., Wang, X., 2013. Rapid prediction of fatty acid composition of vegetable oil by Raman spectroscopy coupled with least squares support vector machines. J. Raman Spectrosc. 44, 1739–1745.

Fidalgo, J.P., Cid, A., Torres, E., Sukenik, A., Herrero, C., 1998. Effects of nitrogen source and growth phase on proximate biochemical composition, lipid classes and fatty acid profile of the marine microalga Isochrysis galbana. Aquaculture 166, 105–116.

Food and Agriculture Organization of the United Nations, 2010. Fats and Fatty Acids in Human Nutrition: Report of an Expert Consultation. Food and Nutrition Paper 91. FAO, Rome.

Garaffo, M.A., Vassallo-Agius, R., Nengas, Y., Lembo, E., Rando, R., Maisano, R., Dugo, G., Giuffrida, D., 2011. Fatty acids profile, atherogenic (IA) and thrombogenic (IT) health lipid indices of raw roe of blue fin tuna (Thunnus thynnus L.) and their salted product "Bottarga". Food Nutr. Sci. 2, 736–743.

Geppert, J., Kraft, V., Demmelmair, H., Koletzko, B., 2006. Microalgal docosahexaenoic acid decreases plasma triacylglycerol in normolipidaemic vegetarians: a randomised trial. Br. J. Nutr. 95, 779–786.

Gouveia, L., Oliveira, C., 2009. Microalgae as a raw material for biofuels production. J. Ind. Microbiol. Biotechnol. 36, 269–274.

Griffiths, M.J., van Hille, R.P., Harrison, S.T.L., 2012. Lipid productivity, settling potential and fatty acid profile of 11 microalgal species grown under nitrogen replete and limited conditions. J. Appl. Phycol. 24, 989–1001.

Grimi, N., Dubois, A., Marchal, L., Jubeau, S., Lboka, N.I., Vorobiev, E., 2014. Selective extraction from microalgae Nannochloropsis sp. using different methods of cell disruption. Bioresour. Technol. 153, 254–259.

Guihéneuf, F., Stengel, D.B., 2013. LC-PUFA-enriched oil production by microalgae: accumulation of lipid and triacylglycerols containing n-3 LC-PUFA is triggered by nitrogen limitation and inorganic carbon availability in the marine haptophyte Pavlova lutheri. Mar. Drugs 11, 4246–4266.

Hayes, D.G., Bengttson, Y.C., Van Alstine, J.M., Setterwall, F., 1998. Urea complexation for the rapid, ecologically responsible fractionation of fatty acids from seed oil. J. Am. Oil Chem. Soc. 75, 1403–1409.

Hu, Q., Sommerfeld, M., Jarvis, E., Ghirardi, M., Posewitz, M., Seibert, M., Darzins, A., 2008. Microalgal triacylglycerols as feedstocks for biofuel production: perspectives and advances. Plant J. 54, 621—639.

James, M.J., Cleland, L.G., 1997. Dietary n-3 fatty acids and therapy for rheumatoid arthritis. Semin. Arthritis Rheum. 27, 85—97.

Jiang, H., Gao, K., 2004. Effects of lowering temperature during culture on the production of polyunsaturated fatty acids in the marine diatom *Phaeodactylum tricornutum* (bacillariophyceae). J. Phycol. 40, 651—654.

Jiang, Y., Chen, F., 2000. Effects of temperature and temperature shift on docosahexaenoic acid production by the marine microalga *Crypthecodinium cohnii*. J. Am. Oil Chem. Soc. 77, 613—617.

Khozin-Goldberg, I., Bigogno, C., Shrestha, P., Cohen, Z., 2002. Nitrogen starvation induces the accumulation of arachidonic acid in the freshwater green alga *Parietochloris incise* (Trebouxiophyceae). J. Phycol 38, 991—994.

Khozin-Goldberg, I., Iskandarov, U., Cohen, Z., 2011. LC-PUFA from photosynthetic microalgae: occurrence, biosynthesis, and prospects in biotechnology. Appl. Microbiol. Biotechnol. 91, 905—915.

Kunjapur, A.M., Eldridge, R.B., 2010. Photobioreactor design for commercial biofuel production from microalgae. Ind. Eng. Chem. Res. 49, 3516—3526.

Lang, I., Hodac, L., Friedl, T., Feussner, I., 2011. Fatty acid profiles and their distribution patterns in microalgae: a comprehensive analysis of more than 2000 strains from the SAG culture collection. BMC Plant Biol. 11, 124—139.

Lavie, C.J., Milani, R.V., Mehra, M.R., Ventura, H.O., 2009. Omega-3 polyunsaturated fatty acids and cardiovascular diseases. J. Am. Coll. Cardiol. 54, 585—594.

Lee, Y.-K., 2001. Microalgal mass culture systems and methods: their limitation and Potential. J. Appl. Phycol. 13, 307—315.

Lopez-Garcia, E., Rodriguez-Artalejo, F., Li, T.Y., Fung, T.T., Li, S., Willet, W.C., Rimm, E.B., Hu, F.B., 2013. The Mediterranean-style dietary pattern and mortality among men and women with cardiovascular disease. Am. J. Clin. Nutr. 99, 172—180.

Martins, D.A., Custódio, L., Barreira, L., Pereira, H., Ben-Hamadou, R., Varela, J., Abu-Salah, K.M., 2013. Alternative sources of *n*-3 long-chain polyunsaturated fatty acids in marine microalgae. Mar. Drugs 11, 2259—2281.

Mendes, A., da Silva, T.L., Reis, A., 2007. DHA concentration and purification from the marine heterotrophic microalga *crypthecodinium cohnii* CCMP 316 by winterization and urea complexation. Food Technol. Biotechnol. 45, 38—44.

Milledge, 2011. Commercial application of microalgae other than as biofuels: a brief review. Rev. Environ. Sci. Biotechnol. 10, 31—41.

Nakano, K., Maeta, R., Nagahama, K., Kato, S., 2008. Extraction of eicosapentaenoic acid ethyl ester using metallic copper and a copper (II) salt in water. Solvent Extr. Res. Dev. 15, 117—120.

Nuzzo, G., Gallo, C., d'Ippolito, G., Cutignano, A., Sardo, A., Fontana, A., 2013. Composition and quantitation of microalgal lipids by ERETIC ^{1}H NMR method. Mar. Drugs 11, 3742—3753.

O'Grady, J., Morgan, J., 2010. Heterotrophic growth and lipid production of *Chlorella protothecoides* on glycerol. Bioprocess Biosyst. Eng. 34, 121—125.

Pereira, H., Barreira, L., Custódio, L., Alrokayan, S., Mouffouk, F., Varela, J., Abu-Salah, K.M., Ben-Hamadou, R., 2013. Isolation and fatty acid profile of selected microalgae strains from the Red sea for biofuel production. Energies 6, 2773—2783.

Pereira, H., Barreira, L., Mozes, A., Florindo, C., Polo, C., Duarte, C.V., Custódio, L., Varela, J., 2011. Microplate-based high throughput screening procedure for the isolation of lipid-rich marine microalgae. Biotechnol. Biofuels 4, 61—72.

Pienkos, P.T., Darzins, A., 2009. The promise and challenges of microalgal-derived biofuels. Biofuels Bioprod. Biorefin. 3, 431—440.

Plaza, M., Herrero, M., Cifuentes, A., Ibáñez, E., 2009. Innovative natural functional ingredients from microalgae. J. Agric. Food Chem. 57, 7159—7170.

Pulz, O., 2001. Photobioreactors: production systems for phototrophic microorganisms. Appl. Microbiol. Biotechnol. 57, 287—293.

Rangel-Huerta, O.D., Aguilera, C.M., Mesa, M.D., Gil, A., 2012. Omega-3 long-chain polyunsaturated fatty acids supplementation on inflammatory biomakers: a systematic review of randomised clinical trials. Br. J. Nutr. 107, S159—S170.

Renaud, S.M., Thinh, L.V., Parry, D.L., 1999. The gross chemical composition and fatty acid composition of 18 species of tropical Australian microalgae for possible use in mariculture. Aquaculture 170, 147—159.

Rodolfi, L., Zittelli, G.C., Bassi, N., Padovani, G., Biondi, N., Bonini, G., Tredici, M.R., 2009. Microalgae for oil: strain selection, induction of lipid synthesis and outdoor mass cultivation in a low-cost photobioreactor. Biotechnol. Bioeng. 102, 100—112.

Roy, S.S., Pal, R., 2014. Microalgae in aquaculture: a review with special references to nutritional value and fish dietetics. Proc. Zool. Soc. http://dx.doi.org/10.1007/s12595-013-0089-9.

Rubio-Rodríguez, N., Beltrán, S., Jaime, I., de Diego, S.M., Sanz, M.T., Carballido, J.R., 2010. Production of omega-3 polyunsaturated fatty acid concentrates: a review. Innovative Food Sci. Emerging Technol. 11, 1—12.

Ruxton, C.H.S., Reed, S.C., Simpson, M.J.A., Millington, K.J., 2004. The health benefits of omega-3 polyunsaturated fatty acids: a review of the evidence. J. Hum. Nutr. Diet. 17, 449–459.

Senanayake, S.P.J.N., 2010. Methods of concentration and purification of omega-3 fatty acids. In: Separation, Extraction and Concentration Processes in the Food, Beverage and Nutraceutical Industries. Woodhead Publishing Limited, Cambridge, pp. 483–505.

Sharma, K.K., Schuhmann, H., Schenk, P.M., 2012. High lipid induction in microalgae for biodiesel production. Energies 5, 1532–1553.

Sidhu, K.S., 2003. Health benefits and potential risks related to consumption of fish or fish oil. Regul. Toxicol. Pharmacol. 38, 336–344.

Sierra, E., Acien, F.G., Fernandez, J.M., Garcia, J.L., Gonzalez, C., Molina, E., 2008. Characterization of a flat plate photobioreactor for the production of microalgae. Chem. Eng. J. 138, 136–147.

Simopoulos, A.P., 1989. Summary of the NATO advanced research workshop on dietary ω3 and ω6 fatty acids: biological effects and nutritional essentiality. J. Nutr. 119, 521–528.

Simopoulos, A.P., 1991. Omega-3 fatty acids in health and disease and in growth and development. Am. J. Clin. Nutr. 1991 (54), 438–463.

Simopoulos, A.P., 2002. The importance of the ratio of omega-6/omega-3 essential fatty acids. Biomed. Pharmacother. 56, 365–379.

Simopolous, A.P., 2011. Evolutionary Aspects of diet: the Omega-6/Omega-3 ratio and the brain. Mol. Neurobiol. 44, 203–215.

Spolaore, P., Joannis-Cassan, C., Duran, E., Isambert, A., 2006. Commercial applications of microalgae. J. Biosci. Bioeng. 101 (2), 87–96.

Swanson, D., Block, R., Mousa, S.A., 2012. Omega-3 fatty acids EPA and DHA: health benefits throughout life. Adv. Nutr. 3, 1–7.

Takisawa, K., Kanemoto, K., Miyazaki, T., Kitamura, Y., 2013. Hydrolysis for direct esterification of lipids from wet microalgae. Bioresour. Technol. 144, 38–43.

Tiemeier, H., van Tuijl, H.R., Hofman, A., Kiliaan, A.J., Breteler, M.M.B., 2003. Plasma fatty acid composition and depression are associated in the elderly: the Rotterdam study. Am. J. Clin. Nutr. 78, 40–46.

Ugwu, C.U., Aoyagi, H., Uchiyama, H., 2008. Photobioreactors for mass cultivation of algae. Bioresour. Technol. 99, 4021–4028.

Ulbricht, T.L.V., Southgate, D.A.T., 1991. Coronary heart disease: seven dietary factors. Lancet 338, 985–992.

Vaezi, R., Napier, J.A., Sayanova, O., 2013. Identification and functional characterization of genes encoding Omega-3 polyunsaturated fatty acid biosynthetic activities from unicellular microalgae. Mar. Drugs 11, 5116–5129.

Verbeke, W., Sioen, I., Pieniak, Z., Van Camp, J., De Henauw, S., 2005. Consumer perception versus scientific evidence about health benefits and safety risks from fish consumption. Public Health Nutr. 8, 422–429.

Vongsvivut, J., Mathew, R., Miller, McNaughton, D., Heraud, P., Barrow, C.J., 2014. Rapid discrimination and determination of polyunsaturated fatty acid composition in marine oils by FTIR spectroscopy and multivariate data analysis. Food Bioprocess Technol. 7, 2410–2422.

Zhukova, N.V., Aizdaicher, N.A., 1995. Fatty acid composition of 15 species of marine microalgae. Phytochemistry 39, 351–356.

Innovative Microalgae Pigments as Functional Ingredients in Nutrition

Efterpi Christaki, Eleftherios Bonos, Panagiota Florou-Paneri

Laboratory of Nutrition, School of Veterinary Medicine, Faculty of Health Sciences, Aristotle University of Thessaloniki, Thessaloniki, Greece

1. INTRODUCTION

Isolating and identifying bioactive compounds from marine microalgae have emerged as a recent trend, mainly to develop new functional foods. The interest in functional food ingredients derived from microalgae in relation to nutritional genomics (nutrigenomics and nutrigenetics) aims to minimize the risk of chronic diseases that have a genetic predisposition (Simopoulos, 2010; Cadoret et al., 2012; Freitas et al., 2012). From a biotechnological point of view, a great deal of interest has been expressed for the commercial large-scale production of microalgae; several different research fields are exploring the use of algal biomass.

Due to their abundant availability in the aquatic ecosystem, microalgae have the potential to become excellent sources of high-biological-value compounds with health benefits, such as lipids, polyunsaturated fatty acids, proteins, vitamins, minerals, and pigments, which can be used in different markets. Thus, microalgae represent an important and dynamic new area

in biotechnology (Gouveia et al., 2008; Christaki et al., 2011; Borowitzka, 2013).

2. MICROALGAE

For centuries, microalgae have been recognized as the basis of the food chain in aquatic ecosystems. They are primitive prokaryotic (e.g., cyanobacteria: blue-green algae) or eukaryotic (e.g., chlorophyta: green algae) photosynthetic microorganisms, which range in diameter from 0.2 to 2.0 μm (Richmond, 2004; Lordan et al., 2011). Because of their simple structure, they can grow rapidly and live in harsh environmental conditions such as heat, cold, anaerobiosis, salinity, and exposure to ultraviolet radiation (Richmond, 2004; Batista et al., 2013). Microalgae, as an alternative to higher plants, have the ability to be cultivated on nonarable land or seashores; consequently, they do not compete with food and feed cultivation. For these reasons, microalgal biomass is attracting worldwide attention to satisfy the so-called bioeconomy demand (Gouveia et al., 2008;

233

Christaki et al., 2013). Some edible microalgae are already commercialized and being used in biotechnology, including Chlorophyta (green algae), Rhodophyta (red algae), and Cyanobacteria (blue-green algae) (Spolaore et al., 2006; Christaki et al., 2010, 2011; Mulders et al., 2014).

The cultivation of microalgae can be performed in open-culture systems (e.g., lakes, lagoons, artificial ponds) and highly controlled closed-culture systems (photobioreactors) (Mata et al., 2010; Wijffels et al., 2013). Closed systems offer better options for the growth of most microalgae strains, providing the ability to protect the culture from contamination by undesirable microorganisms, control processing conditions, and allow higher productivity and better quality for the biomass; however, they are expensive. Consequently, much effort has been expended toward finding more cost-efficient modes of microalgal mass cultivation with regard to open systems. The main advantage of open systems is their use of sunlight to produce biomass and lower costs to build and operate. On the other hand, environmental factors are not fully controlled (Freitas et al., 2012; Barra et al., 2014). Recently, some new concepts for new bioreactors have been proposed (e.g., hybrid systems combining open ponds and photobioreactors, floating photobioreactors); most of them aim to reduce costs (Norsker et al., 2011; Zittelli et al., 2013).

Microalgae could be used as whole biomass without extraction because their bioactive compounds are more stable in their natural matrix than in the extracts (Gouveia and Empis, 2003; Christaki et al., 2013). Traditionally, microalgae pigments are extracted with solvents after breaking the cell walls through mechanical or thermal treatments (Freitas et al., 2012; Batista et al., 2013). In recent years, in order to obtain bioactive compounds from microalgae, some novel environmentally friendly and efficient selective extraction technologies have been developed, such as enzyme-assisted extraction, microwave-assisted extraction, ultrasound-assisted extraction, supercritical fluid extraction, and pressurized liquid extraction (Plaza et al., 2009; Kadam et al., 2013; Reyes et al., 2014). Therefore, to make microalgal biomass economically feasible, it is mandatory to develop the technology and to optimize the use of all produced substances (Vanthoor-Koopmans and Wijffels, 2013; Wijffels et al., 2013).

3. PIGMENTS FROM MICROALGAE

One of the most prominent characteristics of algae is their color, which is determined by their pigments. These pigments are colorful chemical substances that are part of the photosynthetic system of microalgae. They are distinguished into three classes: carotenoids (0.1−0.2% of dry weight on average, and up to 14% of dry weight for β-carotene in some species), chlorophylls (0.5−1.0% of dry weight), and phycobiliproteins (up to 8% of dry weight) (Gouveia et al., 2008; de Jesus Raposo et al., 2013; Markou and Nerantzis, 2013; Barra et al., 2014).

To be deemed suitable for producing pigments commercially, microalgae strains have to meet various criteria, such as ease of culture, lack of toxicity, high nutritional value, and presence of digestible cell walls to make the nutrients available. As shown in Table 1, the most frequently used species are *Dunaliella salina*, *Haematococcus pluvialis*, *Chlorella* spp., *Muriellopsis* spp., *Scenedesmus* spp., *Spirulina* (*Arthrospira*) spp., and *Porphyridium* spp. (Eonseon et al., 2003; Herrero et al., 2006; Spolaore et al., 2006; Hu et al., 2008; Mata et al., 2010; Borowitzka, 2013).

3.1 Carotenoids

Probably the most commercially valuable groups of algae pigments are the carotenoids. They are isoprenoid molecules, yellow to red in

TABLE 1 Commercially Interesting Pigments Across the Most Common Microalgae Groups

Microalgae	Pigments	References
Dunaliela salina	β-carotene, zeaxanthin, chlorophylls *a, b*	Plaza et al. (2009); Herrero et al. (2006); Hu et al. (2008)
Haematococcus pluvialis	astaxanthin, canthaxanthin, lutein, chlorophylls *a, b*	Plaza et al. (2009); Herrero et al. (2006); Jaime et al. (2010); Batista et al. (2013)
Chlorella spp.	astaxanthin, canthaxanthin, chlorophylls *a, b*	Wu et al. (2007); Mulders et al. (2014)
Scenedesmus spp.	lutein, β-carotene	Del Campo et al. (2007); Plaza et al. (2009)
Muriellopsis spp.	lutein	Del Campo et al. (2007); Fernandez-Sevilla et al. (2010)
Spirulina spp.	β-carotene, zeaxanthin, phycocyanin, allophycocyanin	Maoka (2011); Eriksen (2008); de Jesus Raposo et al. (2013)
Porphyridium spp.	phycoerythrin	Borowirka (2013); Rodriguez-Sanchez et al. (2012); Plaza et al. (2009)

color, and were some of the first pigments in nature. Because the carotenoids are lipid-soluble substances, they are absorbed with fats and enter the circulation in the blood bound to different lipoproteins (Lordan et al., 2011; Christaki et al., 2013). Humans and animals (with the exception of an aphid genus) are incapable of synthesizing carotenoids and hence require them in their diet (Moran and Jarvik, 2010; Christaki et al., 2013; Latowski et al., 2014). Chemically, carotenoids can be divided into two groups: the carotenes (hydrocarbons; e.g., α-carotene, β-carotene, lycopene) and the xanthophylls (oxygenated molecules; e.g., astaxanthin, lutein, canthaxanthin; Eonseon et al., 2003; Guedes et al., 2011; Christaki et al., 2013).

Another distinction of carotenoids can be made between primary and secondary carotenoids. Primary ones, such as β-carotene and lutein, are structural and functional components of the photosynthetic apparatus and are directly involved in photosynthesis. They have a light-harvesting role as well as a photoprotective one, which can be expressed either by filtering, quenching, or scavenging (Christaki et al., 2013; Skjanes et al., 2013; Latowski et al., 2014; Mulders et al., 2014).

Secondary carotenoids, including astaxanthin and canthaxanthin, are not bound to the photosynthetic apparatus. They are synthesized in the chloroplast and accumulate in the cytoplasm (Eonseon et al., 2003; Mulders et al., 2014). Some environmental and cultural factors, such as high light density, nutrient starvation, temperature changes, pH values, high salt concentration, and oxidative stress, exclusively induce the overproduction of secondary carotenoids as an adaptation to the above conditions. Thus, these carotenoids, produced in microalgae via carotenogenesis process, act only as photoprotective pigments with a filtering role so they can defend themselves (e.g., from harmful radiation; Lemoine and Schoefs, 2010; Takashi, 2011; Christaki et al., 2013; Skjanes et al., 2013; Mulders et al., 2014). Chlorophyta microalgae is the only group known to contain species that overproduce secondary carotenoids (Lamers et al., 2008; Mulders et al., 2014). Nevertheless, there is a lack of data about metabolic pathways in microalgae for the overproduction of both primary and secondary carotenoids, because the genes and enzymes involved are largely unpredictable (Takashi, 2011; Mulders et al., 2014).

β-Carotene was the first high-value product commercially produced from the green algae *Dunaliella salina*, which is its richest source (Herrero et al., 2006; Gouveia et al., 2008; Hu et al., 2008; Borowitzka, 2013; Dewapriya and Kim, 2014). Natural β-carotene is a mixture of all-*trans* and 9-*cis* isomers, which is rarely obtained in synthetic carotenoids. Although it is in competition with the less-expensive synthetic form, it is often preferred because it is considered to have anticancer activity (Gouveia et al., 2008; Lordan et al., 2011; Borowitzka, 2013). Furthermore, natural β-carotene produced by algae is absorbed better by the body compared to the synthetic form (Lordan et al., 2011; Skjanes et al., 2013).

Although β-carotene is a primary carotenoid, under stress conditions, it can act as a secondary carotenoid. For example, in the cultivation of *Dunaliela salina* under high salinity, light stress, and nitrogen starvation, β-carotene is accumulated to more than 14% of the microalgae dry weight, the highest content in the known sources (Del Campo et al., 2007; Lamers et al., 2012). Other researchers (Gomez and Gonzalez, 2005; Coesel et al., 2008) reported that the shift in low culture temperature of *Dunaliela* from 30 °C to 10 °C resulted in a twofold increase of β-carotene and 9-*cis*-β-carotene, while increasing the salt concentration from 4% to 9% had a 30-fold increase in the accumulation of β-carotene.

Astaxanthin derived from microalgae is gaining importance commercially. It can be produced under stress conditions from the green freshwater alga *Haematococcus pluvialis,* which contains up to 0.2–3.0% on a dry weight basis; this is currently the prime natural source of astaxanthin (Dufosse, 2009; Jaime et al., 2010; Chandi and Gill, 2011; Batista et al., 2013). This red-orange secondary xanthophyll is a powerful biological antioxidant that occurs in nature, protecting membranous phospholipids and other lipids against peroxidation (Hussein et al., 2006; Mata et al., 2010). Furthermore, astaxanthin has some additional advantages over other carotenoids: it is more stable, it can easily cross the blood–brain barrier, and it has high tinctorial properties (Dufosse, 2009). Astaxanthin is probably best known for eliciting the pinkish-red color in the flesh of salmonoids, shrimp, lobsters, and crayfish (Eonseon et al., 2003; Chu, 2012), representing 74–98% of their total pigments (Lordan et al., 2011).

Lutein is a yellow primary carotenoid found in microalgae species, such as *Scenedesmus* (4.5–5.5 mg/g) and *Chlorella* (4.6 mg/g) (Plaza et al., 2009). Another microalgae species, *Muriellopsis* spp., was found to possess high lutein content (up to 35 mg/l) that does not seem to be triggered under stress conditions, coupled with a high growth rate. Therefore, this microalga could be exploited for commercial production of this carotenoid (Fernandez-Sevilla et al., 2010; Guedes et al., 2011; Skjanes et al., 2013). Lutein, apart from being a strong antioxidant, is used as a natural colorant in the feed industry, in drugs, and in cosmetics (Wu et al., 2007; Plaza et al., 2009; de Jesus Raposo et al., 2013).

Canthaxanthin, a secondary carotenoid, is produced in large quantities by *Chlorella* spp. and *Scenedesmus* spp. under salt stress, nitrogen deprivation, and ultraviolet irradiation (Takaichi, 2011; Skjanes et al., 2013).

3.2 Chlorophylls

Chlorophyll, present in all higher plants, is a valuable bioactive constituent that can be extracted from the microalgal biomass. It is a green pigment that is ubiquitous in nature because it is responsible for the photosynthetic process (Hosikian et al., 2010; Stengel et al., 2011). Nowadays, there is a strong focus on the commercial production of dhlorophyll as a natural pigment in the food and feed industries, in pharmaceuticals, and in cosmetics (Hosikian et al., 2010; Stengel et al., 2011). Chlorophyll is a tetrapyrrole with a centrally bound magnesium ion (Pangestuti and Kim, 2011; Mulders et al., 2014).

There are several kinds of chlorophylls in microalgae—chlorophylls *a*, *b*, *c*, *d* and *f*—which have some small differences in their absorption spectra and consequently their tonality. Chlorophyll *a* has a blue-green color, chlorophyll *b* is a brilliant green, chlorophyll *c* is yellow-green, chlorophyll *d* is a brilliant/forest green, and chlorophyll *f* is emerald green (Chen et al.,

2010; Roy et al., 2011). Chlorophyll *a*, which is the major light-harvesting pigment, appears in all photosynthetic organisms. Chlorophyll *b* appears exclusively in chlorophyta and their descendants, whereas chlorophyll *c* appears exclusively in rhodophyta (Jeffrey and Wright, 2005).

Although chlorophyll is a natural pigment, there are some disadvantages associated with its use. For example, it is unstable in foods where it is incorporated under different pH conditions (Hosikian et al., 2010). Additionally, when the magnesium ion is lost, chlorophyll becomes pale and dusky colored (Humphrey, 2004).

It has been reported that the chlorophylls may not need to be overproduced to make a microalgal-based production process economically feasible (Mulders et al., 2014). Consequently, rapidly growing species like *Chlorella* can contain the two main types of chlorophylls (*a*, *b*), up to 4.5% of dry weight, and therefore could be the most attractive production material when grown under optimal conditions (Cuaresma et al., 2011; Miazek and Ledakowicz, 2013). On the other hand, it has been reported that chlorophyll content in microalgal biomass is reduced significantly under stress conditions (Markou and Nerantzis, 2013).

3.3 Phycobiliproteins

Phycobiliproteins are deep-colored, water-soluble proteins that are present mainly in cyanobacteria and rhodophyta. They capture light energy, which is then passed on to chlorophylls during photosynthesis. Phycobiliproteins are composed of proteins and covalently bound via cysteine amino acid chromophores called phycobilins, belonging to open-chain tetrapyrroles (Eriksen, 2008; de Jesus Raposo et al., 2013; Watanabe and Ikeuchi, 2013; Mulders et al., 2014). Phycobiliproteins that are not essential for the function of cells include phycocyanin (blue pigment), phycoerythrin (red pigment), and allophycocyanin (light-blue pigment),

which differ in their spectral properties (Wright and Jeffrey, 2006; Eriksen, 2008; Blot et al., 2009; Chu, 2012; Freitas et al., 2012; Mulders et al., 2014). In addition to being natural food colorants, these phycobiliproteins are strongly fluorescent markers and have antioxidant properties (Eriksen, 2008; Stengel et al., 2011; de Jesus Raposo et al., 2013). Also, the above-mentioned substances can neutralize the reactive oxygen species (ROS) due to their chemical structures and chelating properties, thus reducing oxidative stress (Roy et al., 2007; Rodriguez-Sanchez et al., 2012).

Phycocyanin from *Spirulina* spp. and phycoerythrin from *Porphyridium* spp. are two of the most well-known phycobiliproteins that have been produced commercially (Plaza et al., 2009; Rodriguez-Sanchez et al., 2012; Borowitzka, 2013).

The content of these protein-bound unique pigments found in microalgae is degraded under some stress conditions, such as phosphorus, nitrogen, and sulfur starvation (Eriksen, 2008; Hifney et al., 2013). Nevertheless, the increase in salt concentration up to 0.6 M resulted in an elevation of the total phycobiliprotein content in *Spirulina* spp., from 25% to 45% of dry matter, while a further increase to 0.9 M salt affected negatively the phycobiliprotein synthesis (Hifney et al., 2013).

4. BIOLOGICAL ACTIVITIES OF MICROALGAE PIGMENTS AND HEALTH BENEFIT EFFECTS

Several studies have focused on the use of natural pigments derived from microalgae, because they have health-promoting properties and a broad range of potential industrial applications. Consumers are becoming increasingly aware of the correlation between diet, health, and disease prevention; thus, microalgae represent an important and dynamic new area in biotechnology (Figure 1).

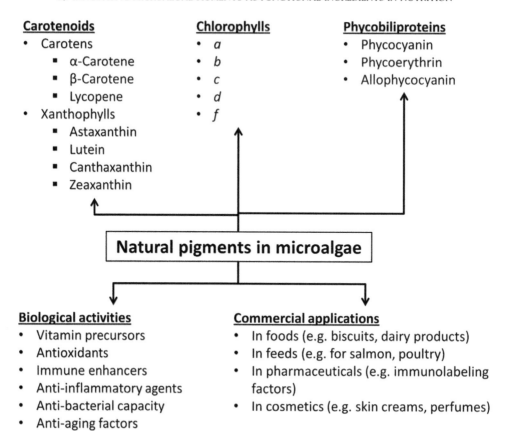

Carotenoids
- Carotens
  - α-Carotene
  - β-Carotene
  - Lycopene
- Xanthophylls
  - Astaxanthin
  - Lutein
  - Canthaxanthin
  - Zeaxanthin

Chlorophylls
- *a*
- *b*
- *c*
- *d*
- *f*

Phycobiliproteins
- Phycocyanin
- Phycoerythrin
- Allophycocyanin

Natural pigments in microalgae

Biological activities
- Vitamin precursors
- Antioxidants
- Immune enhancers
- Anti-inflammatory agents
- Anti-bacterial capacity
- Anti-aging factors
- Anti-obesity activity

Commercial applications
- In foods (e.g. biscuits, dairy products)
- In feeds (e.g. for salmon, poultry)
- In pharmaceuticals (e.g. immunolabeling factors)
- In cosmetics (e.g. skin creams, perfumes)

FIGURE 1   Distribution, biological activities, and commercial applications of natural pigments derived from microalgae.

Because humans and animals cannot synthesize pigments de novo, these are either provided by the diet or partly modified through metabolic reactions from precursor substances. For example, β-carotene can be metabolized to vitamin A (retinol). β-Carotene has a very high pro-vitamin A activity because every molecule of this compound produces two molecules of retinol (Liaanen-Jensen, 1998; Graham and Rosser, 2000; Christaki et al., 2013).

Regarding β-carotene and astaxanthin, and to a lesser degree other microalgae pigments, it must be noticed that they have strong antioxidant activity, which aims to mediate the harmful effects of free radicals by protecting the lipophilic part from lipid peroxidation or scavenging ROS in photo-oxidative processes (Stahl and Sies, 2003; Christaki et al., 2011; Lordan et al., 2011; Pangestuti and Kim, 2011). However, it has been reported that β-carotene might act as a pro-oxidant in the process of lipid peroxidation when there is a high oxygen pressure and high carotenoid amount (Polyakov et al., 2001; Stahl and Sies, 2003).

Apart from their high concentration in the microalgal biomass and the even larger quantities produced by controlling some environmental conditions, the antioxidants from microalgae can act synergistically to increase the positive effects on human health (Stahl and

Sies, 2003; Heydarizadeh et al., 2013). These pigments have greater antioxidant effects than vitamin E, but weaker than synthetic commercial antioxidants, such as butylated hydroxytoluene or butylated hydroxyanisole (Natrah et al., 2007; Pangestuti and Kim, 2011). The use of synthetic antioxidants is under strict regulation, especially in the European Union countries, due to their potential health hazards. Therefore, natural antioxidants can be used as safe alternatives in the industry (Spolaore et al., 2006; Gouveia et al., 2008).

Dietary intake of marine microalgae-derived antioxidants has shown the ability to protect organisms against various chronic disorders, such as cancer, diabetes, atherosclerosis, coronary disease, ischemic brain development, metabolic syndromes, gastrointestinal and liver diseases, as well as neurodegenerative diseases such as Alzheimer disease and Parkinson disease (Riccioni et al., 2011; Cadoret et al., 2012; Gouveia, 2014; Martins et al., 2014), or to ameliorate cognitive functions (Kidd, 2011). According to Guerin et al. (2003), astaxanthin could be effective against human prostatic hyperplasia and prostatic cancer through the enzyme 5-α-reductase, which is involved in the abnormal prostate growth. Other researchers have described the use of phycocyanin as a nephroprotector (Rodriguez-Sanchez et al., 2012) or a protector of human pancreatic cells (Chu, 2012).

In addition, the carotenoids lutein and zeaxanthin, due to their antioxidant capacity, can protect the eye macula from adverse photochemical reactions (Friedman et al., 2004), while β-carotene helps to prevent premature aging caused by ultraviolet radiation (Miyashita, 2009; Fernandez-Sevilla et al., 2010) or to treat skin melasma (Yaakob et al., 2014). Generally, β-carotene, apart from limiting photooxidative damage to the skin, provides protection against sunburns (erythema solare) (Eonseon et al., 2003; Stahl and Sies, 2003). Moreover, in marine animals, β-carotene is stored in the gonads, so it is essential for reproduction (Maoka, 2011).

Studies both in vitro and in vivo have shown the cancer-preventive effects of chlorophylls, with particular emphasis on their antimutagenic effects (Ferruzi and Blakeslee, 2007; Gouveia et al., 2008; Hosikian et al., 2010). The protective effects of carotenoids and phycobiliproteins on the development of cancerous tumors and leukemia have also been reported (Plaza et al., 2009; Vilchez et al., 2011; Heydarizadeh et al., 2013).

Microalgae pigments are able to boost the immune system (e.g., antibody production) and to act as anti-inflammatory agents against asthma, ulcers, arthritis, and muscle damage (by providing increased muscle endurance; Mata et al., 2010; Guedes et al., 2011; Stengel et al., 2011; de Jesus Raposo et al., 2013), as well as antiproliferative agents (Plaza et al., 2009; Lordan et al., 2011). In addition, some epidemiological data explain the above properties of carotenoids, which could directly act toward the DNA to regulate the production of RNA (Guerin et al., 2003; Hussein et al., 2006). Moreover, chlorophylls seem to be a good approach for the treatment of ulcers and postoperative wounds (Hosikian et al., 2010). Also, patients with pancreatitis can possibly be treated with chlorophyll *a* (Yaakob et al., 2014).

5. NEW TRENDS IN COMMERCIAL APPLICATIONS OF MICROALGAE PIGMENTS

Pigments from microalgae are now strongly demanded by the market as renewable natural color enhancers for foods and feeds, which simultaneously provide certain health benefits. Furthermore, these pigments, which are particularly strong dyes even at very low levels (parts per million), have important applications in the pharmaceutical industry (e.g., as fluorescence-based indicators, as biochemical tracers in immune assays) and in the cosmetic industry (e.g., as skin cream to stimulate collagen synthesis).

The development of foods with attractive appearances is an important goal in the food and feed industries. β-Carotene produced by microalgae serves as one of the most utilized food coloring agents in pasta, fruit juices, soft drinks, confectionary, margarine, dairy products, and salad dressings (Gouveia et al., 2008; Plaza et al., 2009; Christaki et al., 2011; Guedes et al., 2011; Christaki et al., 2013). Phycocyanin, with its blue color, can be used in various food products such as chewing gums, candies, dairy products, jellies, ice creams, and beverages; its color is stable in dry preparations, but it is sensitive to high temperatures and light (Gouveia et al., 2008; Dufosse, 2009). Phycoerythrin, with its red color, can be used for the pigmentation of confections, gelative desserts, and dairy products. This color is stable when phycoerythrine is incorporated in dry food preparations (stored under low humidity) and has a long shelf life at pH 6—7 (Dufosse et al., 2005). In addition, Chandi and Gill (2011) reported that novel food applications of pigments can include tomato ketchup, processed meats such as sausage and ham, and marine products such as fish paste and surimi. It is important to notice that microalgae pigments in foods besides the coloring and nutritional purposes can also cause significant changes in the rheological properties (Gouveia et al., 2008).

Carotenoids are also used as feed additives in the commercial rearing of many aquatic organisms to enhance the reddish color of the flesh of salmon, trout, and shrimp, as well as feed supplement for some types of zooplankton (e.g., rotifers and copepods (Plaza et al., 2009; Mata et al., 2010; Guedes et al., 2011)). Also, microalgae pigments could be used to enrich the yellowish color of egg yolk and chicken skin (Plaza et al., 2009; Mata et al., 2010; Guedes et al., 2011) or to improve the appearance of pet foods (Spolaore et al., 2006; Skjanes et al., 2013).

Laws and regulations concerning food additives and functional foods vary by country. For example, astaxanthin is recognized as a food colorant by the Food and Drug Administration in the United States, in Japan, and in some European countries, it but has not yet been approved in the European Union countries (Lordan et al., 2011; Borowitzka, 2013; Ambati et al., 2014). Pigments derived from microalgae are unique, so extensive assessment tests (e.g., acute toxicity, mutagenicity, teratogenicity, embryotoxicity, and reproductive toxicity) to confirm their safety for consumers need to be considered before their introduction into the market.

Apart from the conventional (nontransgenic) approach, the biotechnological production of pigments and other valuable substances from microalgae with the use of genetic engineering is a very tempting alternative. Such an approach may result in improvements of nutritional value and optimization of metabolite production with functional properties, which makes these substances better suited for commercial applications (Freitas et al., 2012; Htet et al., 2013; Rasala et al., 2014). Nevertheless, the use of transgenic microalgae faces some problems and challenges, at least in Europe, such as competitiveness, public acceptance, regulatory issues, and biosafety concerns (Freitas et al., 2012; Htet et al., 2013; Rasala et al., 2014). Generally, genetically modified food is viewed with a large degree of skepticism by consumers.

6. CONCLUSIONS

Microalgae pigments such as carotenoids, chlorophylls, and phycobiliproteins could be a leading natural resource for innovative potential functional ingredients in nutrition. Nowadays, these natural pigments are preferred over synthesized substances and are finding commercial applications, mainly in the food and feed industries. Nevertheless, some bottlenecks, such as high production costs and low yields, need to be solved before microalgae can be moved from niche markets to large-scale use.

References

Ambati, R.R., Moi, P.S., Ravi, S., Aswathanarayana, R.G., 2014. Astaxanthin: sources, extraction, stability, biological activities and its commercial applications—a review. Mar. Drugs 12, 128—152.

Barra, L., Chandrasekaran, R., Corato, F., Brunet, C., 2014. The challenge of ecophysiological biodiversity for biotechnological applications of marine microalgae. Mar. Drugs 12, 1641—1675.

Batista, A.P., Gouveia, L., Bandarra, N.M., Franco, J.M., Raymundo, A., 2013. Comparison of microalgal biomass profiles as novel functional ingredient for food products. Algal Res. 2, 164—173.

Blot, N., Wu, X.J., Thomas, J.C., Zhang, J., Garczarek, I., Bohm, S., Tu, J.M., Zhou, M., Ploscher, M., Eichacker, L., Partensky, F., Scheer, H., Zhao, K.H., 2009. Phycourobilin in trichromatic phycocyanin from oceanic cyanobacteria is formed post-translationally by a phycoerythrobilin lyase-isomerase. J. Biol. Chem. 284, 9290—9298.

Borowitzka, M.A., 2013. High-value products from microalgae—their development and commercialisation. J. Appl. Phycol. 25, 743—756.

Cadoret, J.P., Garnier, M., Saint-Jean, B., 2012. Microalgae, functional genomics and biotechnology. Adv. Bot. Res. 64, 285—341.

Chandi, G.K., Gill, B.S., 2011. Production and characterization of microbial carotenoids as an alternative to synthetic colors: a review. Int. J. Food Prop. 14, 503—513.

Chen, M., Schliep, M., Willows, R.D., Cai, Z.L., Neilan, B.A., Scheer, H., 2010. A red-shifted chlorophyll. Science 329, 1318—1319.

Christaki, E., Bonos, E., Giannenas, I., Florou-Paneri, P., 2013. Functional properties of carotenoids originating from algae. J. Sci. Food Agric. 93, 5—11.

Christaki, E., Florou-Paneri, P., Bonos, E., 2011. Microalgae: a novel ingredient in nutrition. Int. J. Food Sci. Nutr. 62, 794—799.

Christaki, E., Karatzia, M., Florou-Paneri, P., 2010. The use of algae in animal nutrition. J. Hellenic Vet. Med. Soc. 61, 267—276.

Chu, W.L., 2012. Biotechnological applications of microalgae. Int. E-J. Sci. Med. Educ. 6, S24—S37.

Coesel, S.N., Baumfartner, A.C., Teles, L.M., Ramos, A.A., Henriques, N.M., Cancela, L., Varela, J.C.S., 2008. Nutrient limitation is the main regulatory factor for carotenoid accumulation and for Psy and Pds steady state transcript levels in *Dunaliella salina* (Chlorophyta) exposed to high light and salt stress. Mar. Biotechnol. 10, 602—611.

Cuaresma, M., Janssen, M., Vilchez, C., Wijffels, R.H., 2011. Horizontal or vertical photobioreactors? How to improve microalgae photosynthetic efficiency. Biores. Technol. 102, 5129—5137.

de Jesus Raposo, M.F., de Morais, R.M.S.C., de Morais, A.M.M.B., 2013. Health applications of bioactive compounds from marine microalgae. Life Sci. 93, 479—486.

Del Campo, J.A., Garcia-Gonzalez, M., Guerrero, M.G., 2007. Outdoor cultivation of microalgae for carotenoid production: current state and perspectives. Appl. Microbiol. Biotechnol. 74, 1163—1174.

Dewapriya, D., Kim, S.K., 2014. Marine microorganisms: an emerging avenue in modern nutraceuticals and functional foods. Food Res. Int. 56, 115—125.

Dufosse, L., 2009. Microbial and microalgal carotenoids as colourants and supplements. In: Britton, G., Liaanen-Jensen, S., Pfander, H. (Eds.), Carotenoids, Nutrition and Health, Volume 5. Birkhauser Verlag, Basel, Switzerland.

Dufosse, L., Galaup, P., Yaron, A., Arad, S.M., Blanc, P., Murthy, N.C., Ravishankar, G.A., 2005. Microorganisms and microalgae as sources of pigments for food use: a scientific oddity or an industrial reality? Trends Food Sci. Technol. 16, 389—406.

Eonseon, J., Polle, J.E.W., Lee, H.K., Hyund, S.M., Chang, M., 2003. Xanthophylls in microalgae: from biosynthesis to biotechnological mass production and application. J. Microbiol. Biotechnol. 13, 165—174.

Eriksen, N.T., 2008. Production of phycocyanin—a pigment with applications in biology, biotechnology, foods and medicine. Appl. Microbiol. Biotechnol. 80, 1—14.

Fernandez-Sevilla, J.M., Fernandez, A.F.G., Grima, M.E., 2010. Biotechnological production of lutein and its applications. Appl. Microbiol. Biotechnol. 86, 27—40.

Ferruzi, M.G., Blakeslee, J., 2007. Digestion, absorption, and cancer preventive activity of dietary chlorophyll derivatives. Nutr. Res. 27, 1—12.

Freitas, A.C., Rodrigues, D., Rocha-Santos, T.A.P., Gomes, A.M.P., Duarte, A.C., 2012. Marine biotechnology advances towards applications in new functional foods. Biotechnol. Adv. 30, 1506—1515.

Friedman, D.S., O'Colmain, B.J., Muñoz, B., Tomany, S.C., McCarty, C., De Jong, P.T., Nemesure, B., Mitchell, P., Kempen, J., Congdon, N., 2004. Prevalence of age-related macular degeneration in the United States. Arch. Ophthalmol. 122, 564—572.

Gomez, P., Gonzalez, M., 2005. The effect of temperature and irradiance on the growth and carotenogenic capacity of seven strains of *Dunaliella salina* (Chlorophyta) cultivated under laboratory conditions. Biol. Res. 38, 151—162.

Gouveia, L., 2014. From tiny microalgae to huge biorefineries. Oceanography 2, 120.

Gouveia, L., Batista, A.P., Sousa, I., Raymundo, A., Bandarra, N.M., 2008. Microalgae in novel food products. In: Papadopoulos, K.N. (Ed.), Food Chemistry Research Developments. Nova Science Publishers Inc., Hauppauge, NY, pp. 75—112.

Gouveia, L., Empis, J., 2003. Relative stabilities of microalgal carotenoids in microalgal extracts, biomass and fish feed: effect of storage conditions. Innov. Food Sci. Emerging Technol. 4, 227—233.

Graham, R.D., Rosser, J.M., 2000. Carotenoids in staple foods: their potential to improve human nutrition. Food Nutr. Bull 21.

Guedes, A.C., Amaro, H.M., Malcata, F.X., 2011. Microalgae as sources of carotenoids. Mar. Drugs 9, 625—644.

Guerin, M., Huntley, M.E., Olaizola, M., 2003. *Haematococcus* astaxanthin: applications for human health and nutrition. Trends Biotechnol. 21, 210—216.

Herrero, M., Jaime, L., Martin-Alvarez, P.J., Cifuentes, A., Ibanez, E., 2006. Optimization of the extraction of antioxidant from *Dunaliella salina* microalga by pressurized liquids. J. Agric. Food Chem. 54, 5597—5603.

Heydarizadeh, P., Poirier, I., Loizeau, D., Ulmann, L., imouni, V., Schoefs, B., Bertrand, M., 2013. Plastids of marine phytoplankton produce bioactive pigments and lipids. Mar. Drugs 11, 3425—3471.

Hifney, A.F., Issa, A.A., Fawzy, M.A., 2013. Abiotic stress induced production of β-carotene, allophycocyanin and total lipids in *Spirulina sp.* J. Biol. Earth Sci. 3, B54—B64.

Hosikian, A., Lim, S., Halim, R., Danquah, M.K., 2010. Chlorophyll extraction from microalgae: a review on the process engineering aspects. Int. J. Chem. Eng. 2010, 11.

Htet, M.Z., Ling, L.Y., Yun, S.H., Rajee, O., 2013. Biofuel from microalgae—a review on the current status and future trends. Int. J. Adv. Biotechnol. Res. 4, 329—341.

Hu, C.C., Lin, J.T., Lu, F.J., Chou, F.P., Yang, D.J., 2008. Determination of carotenoids in *Dunaliella salina* cultivated in Taiwan and antioxidant capacity of the algal carotenoid extract. Food Chem. 109, 439—446.

Humphrey, A.M., 2004. Chlorophyll as a color and functional ingredient. J. Food Sci. Eng. 69, 422—425.

Hussein, G., Sankawa, U., Goto, H., Matsumoto, K., Watanabe, H., 2006. Astaxanthin, a carotenoid with potential in human health and nutrition. J. Nat. Prod. 69, 443—449.

Jaime, L., Rodríguez-Meizoso, I., Cifuentes, A., Santoyo, S., Suarez, S., Ibáñez, E., Señorans, F.J., 2010. Pressurized liquids as an alternative process to antioxidant carotenoids' extraction from *Haematococcus pluvialis* microalgae. LWT Food Sci. Technol. 43, 105—112.

Jeffrey, S.W., Wright, S.W., 2005. Photosynthetic pigments in marine microalgae: insights from cultures and the sea. In: Subba Rao, D.V. (Ed.), Algal Cultures Analogues of Blooms and Applications. Science Publishers, New Hampshire, pp. 33—90.

Kadam, S.U., Tiwari, B.K., O'Donnell, C.P., 2013. Application of novel extraction technologies for bioactives from marine algae. J. Agric. Food Chem. 61, 4667—4675.

Kidd, P., 2011. Astaxanthin, cell membrane nutrient with diverse clinical benefits and anti-aging potential. Altern. Med. Rev. 16, 355—364.

Lamers, P.P., Janssen, A.M., de Vos, R.G.H., BIno, R.J., Wijffels, R.H., 2008. Exploring and exploiting carotenoid accumulation in *Dunaliella salina* for cell-factory applications. Trends Biotechnol. 26.

Lamers, P.P., Janssen, A.M., de Vos, R.G.H., BIno, R.J., Wijffels, R.H., 2012. Carotenoid and fatty acid metabolism in nitrogen-starved *Dunaliella salina*, a unicellular green microalga. J. Biotechnol. 162, 21—27.

Latowski, D., Szymanska, R., Kazimierz, S., 2014. Carotenoids involved in antioxidant system of chloroplasts. In: Ahmad, P. (Ed.), Oxidative Damage to Plants: Antioxidant Networks and Signaling. Academic Press, Waltham, MA, pp. 289—319.

Lemoine, Y., Schoefs, B., 2010. Secondary ketocarotenoid astaxanthin biosynthesis in algae: a multifunctional response to stress. Photosynthesis Res. 106, 155—177.

Liaanen-Jensen, S., 1998. Carotenoids in food chain. In: Britton, G., Liaanen-Jensen, S., Pfander, H. (Eds.), Carotenoids: Biosynthesis and Metabolism, Vol. 3. Birkhauser, Basel, Switzerland, pp. 359—371.

Lordan, S., Paul Ross, R., Stanton, C., 2011. Marine bioactives as functional food ingredients: potential to reduce the incidence of chronic diseases. Mar. Drugs 9, 1056—1100.

Maoka, T., 2011. Carotenoids in marine animals. Mar. Drugs 9, 278—293.

Markou, G., Nerantzis, E., 2013. Microalgae for high-value compounds and biofuels production: a review with focus on cultivation under stress conditions. Biotechnol. Adv. 31, 1532—1542.

Martins, A., Veieira, H., Gaspar, H., Santos, S., 2014. Marketed marine natural products in the pharmaceutical and cosmeceutical industries: tips for success. Mar. Drugs 12, 1066—1101.

Mata, T.M., Martins, A.A., Caetano, N.S., 2010. Microalgae for biodiesel production and other applications: a review. Renew. Sust. Energ. Rev. 14, 217—232.

Miazek, K., Ledakowicz, S., 2013. Chlorophyll extraction from leaves, needles and microalgae: a kinetic approach. Int. J. Agric. Biol. Eng. 6, 107—115.

Miyashita, K., 2009. Function of marine carotenoids. Forum Nutr. 61, 136—146.

Moran, N.A., Jarvik, T., 2010. Lateral transfer of genes from fungi underlies carotenoids production in aphids. Science 328, 624—627.

Mulders, K.J.M., Lamers, P.P., Martens, D.E., Wijffels, R.H., 2014. Phototropic pigment production with microalgae: biological constraints and opportunities. J. Phycol. 50, 229–242.

Natrah, F., Yosoff, F.M., Shariff, M., Abas, F., Mariana, N.S., 2007. Screening of Malaysian indigenous microalgae for antioxidant properties and nutritional value. J. Appl. Phycol. 19, 711–718.

Norsker, N.H., Barbosa, M.J., Vermue, M.H., Wijffels, R.H., 2011. Microalgal production—a close look at the economics. Biotechnol. Adv. 29, 24–27.

Pangestuti, R., Kim, S.K., 2011. Biological activities and health benefit effects of natural pigments derived from marine algae. J. Funct. Foods 3, 255–266.

Plaza, M., Herrero, M., Cifuentes, A., Ibanez, E., 2009. Innovative natural functional ingredients from microalgae. J. Agric. Food Chem. 57, 7159–7170.

Polyakov, N.E., Leshina, T.V., Kovalova, T.A., Kispert, L.D., 2001. Carotenoids as scavengers of free radicals in a Fenton reaction: antioxidants or pro-oxidants? Free Rad. Biol. Med. 31, 398–404.

Rasala, B.A., Chao, S.S., Pier, M., Barrera, D.J., Mayfield, S.P., 2014. Enhanced genetic tools for engineering multigene traits into green algae. PLoS One 9, e94028.

Reyes, F.A., Mendiola, J.A., Ibanez, E., del Valle, J.M., 2014. Astaxanthin extraction from *Haematococcus pluvialis* using CO_2-expanded ethanol. J. Supercrit. Fluid 92, 75–83.

Riccioni, G., D'Orazio, N., Franceschelli, S., Speranza, L., 2011. Marine carotenoids and cardiovascular risk markers. Mar. Drugs 9, 1166–1175.

Richmond, A., 2004. Handbook of Microalgal Culture: Biotechnology and Applied Phycology. Blackwell Science Ltd, IA.

Rodriguez-Sanchez, R., Ortiz-Butron, R., Blas-Valdivia, V., Hernandez-Garcia, A., Cano-Europa, E., 2012. Phycobiliproteins or C-phycocyanin of *Arthrospira (Spirulina) maxima* protect against $HgCl_2$-caused oxidative stress and renal damage. Food Chem. 135, 2359–2365.

Roy, K.R., Arunasree, K.M., Dhoot, A., Aparna, R., Reddy, G.V., Vali, S., Reddanna, P., 2007. C-phycocyanin inhibits 2-acetylaminofluorene-induced expression of MDR1 in mouse macrophage cells: ROS mediated pathway determined via combination of experimental and in silico analysis. Arch. Biochem. Bioph. 459, 169–177.

Roy, S., Llewellyn, C.A., Egeland, E.S., Johnson, G., 2011. Phytoplankton Pigments: Characterization, Chemotaxonomy, and Applications in Oceanography. Cambridge University Press, New York, NY.

Simopoulos, A.P., 2010. Nutrigenetics/nutrigenomics. Annu. Rev. Public Health 31, 53–68.

Skjanes, K., Rebours, C., Lindblad, P., 2013. Potential for green microalgae to produce hydrogen, pharmaceuticals and other high value products in a combined process. Crit. Rev. Biotechnol. 33, 172–215.

Spolaore, P., Joannis-Cassan, C., Duran, E., Isambert, A., 2006. Commercial applications of microalgae. J. Biosci. Bioeng. 101, 87–96.

Stahl, W., Sies, H., 2003. Antioxidant activity of carotenoids. Mol. Aspect. Med. 24, 345–351.

Stengel, D.B., Connan, S., Popper, Z.A., 2011. Algal chemodiversity and bioactivity: sources of natural variability and implications for commercial application. Biotechnol. Adv. 29, 483–501.

Takaichi, S., 2011. Carotenoids in algae: distributions, biosynthesis and functions. Mar. Drugs 9, 1101–1118.

Takashi, M., 2011. Carotenoids in marine animals. Mar. Drugs 9, 278–293.

Vanthoor-Koopmans, M., Wijffels, R.H., 2013. Biorefinery of microalgae for food and fuel. Biores. Technol. 135, 142–149.

Vilchez, C., Forjan, E., Cuaresma, M., Bedmar, F., Garbayo, I., Vega, J.M., 2011. Marine carotenoids: biological functions and commercial applications. Mar. Drugs 9, 319–333.

Watanabe, M., Ikeuchi, M., 2013. Phycobilisome: achitecture of a light-harvesting supercomplex. Phytother. Res. 116, 265–276.

Wijffels, R.H., Kruse, O., Klaas, J.H., 2013. Potential of industrial biotechnology with cyanobacteria and eukaryotic microalgae. Curr. Opin. Biotechnol. 24, 405–413.

Wright, S.W., Jeffrey, S.W., 2006. Pigment markers for phytoplankton production. In: Volkman, J.K. (Ed.), Marine Organic Matter. Springer Verlag, Berlin, Heidelberg, pp. 71–104.

Wu, Z., Wu, S., Shi, X., 2007. Supercritical fluid extraction and determination of lutein in heterotrophically cultivated *Chlorella pyrenoidosa*. J. Food Proc. Eng. 30, 174–185.

Yaakob, Z., Ali, E., Zainal, A., Mohamad, M., Takriff, M.S., 2014. An overview: biomolecules from microalgae for animal feed and aquaculture. J. Biolog. Res. Thessaloniki 21, 6.

Zittelli, G.C., Biondi, N., Rodolfi, L., Tredici, M.R., 2013. Photobioreactors for mass production of microalgae. In: Richmond, A., Hu, Q. (Eds.), Handbook of Microalgae Culture: Applied Phycology and Biotechnology, Second ed. Wiley Blackwell, Chichester, pp. 225–266.

15

Application of Diatom Biosilica in Drug Delivery

Jayachandran Venkatesan[1,2], Baboucarr Lowe[1], Se-Kwon Kim[1,2]

[1]Pukyong National University, Department of Marine-Bio Convergence Science, Busan, South Korea;
[2]Marine Bioprocess Research Center, Pukyong National University, Busan, Korea

1. INTRODUCTION

Many unicellular organisms in nature make use of cell wall surfaces or vesicles formed by organelles as biomineralized stents to manufacture mineral tissues. Diatoms—one type of eukaryotic unicellular algae—have a mineralized shell that provides protection. There are 10,000 species of diatoms, and they have unique cell walls made of silica (hydrated silicon dioxide) called frustules (Hamm et al., 2003). The frustules, made of a composite of organic material and silica that is often intricately and ornately shaped, are not as stiff as calcite but are more flexible (Hildebrand, 2008). In diatoms, the silica is formed on the surface of the cell in a complex three-dimensional (3D) network that is only partially understood.

Each diatom species has a specific biosilica cell wall with regularly arranged slits or pores, ranging from 10 to 1000 nm in size. Biosilica morphogenesis takes place inside the diatom cell within a specialized membrane-bound compartment, termed the silica deposition vesicle. It has been postulated that the silica deposition vesicle contains a matrix of organic macromolecules that

not only regulate silica formation but also act as templates to mediate the growth of the frustules and the creation of holes and slits (i.e., nanopatterning; Chen et al., 2012). Poulsen et al. (2003) used biosilica-associated phosphoproteins, known as silaffins, to create a silica assembly with pores having nanodiameters.

Depending on the species of diatom and the growth conditions, frustules can display a wide range of different morphologies. It is possible to design and produce specific frustule morphologies that have potential applications in nanotechnology (Parkinson and Gordon, 1999). Diatoms secrete a hydrated silica cage ($SiO_2 \cdot nH_2O$) that is not as stiff as calcite; thus, it can undergo more deformation per unit load, making it more flexible. Young's modulus of the glass sponge spicule is $\sim 40\,GPa$ (Woesz et al., 2006) and the diatom frustule is 22.4 GPa (Hamm et al., 2003), whereas that for calcite is $\sim 76\,GPa$. Diatoms contain two valves with a regular set of perforations through which they filter the nourishment from the ocean. Several types of diatom structures are shown in Figure 1.

Biologists—and diatomists in particular—have long studied the properties of single-cell algae,

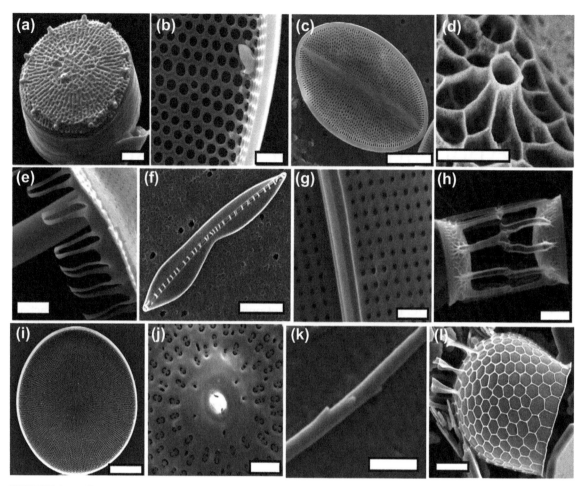

FIGURE 1   Different morphologies and diversity of diatoms. (a) Thalassiosira pseudonana, scale-1 μm, (b) close up of Cosci-nodiscus wailesii, scale-5 μm, (c) Cocconeis sp., scale-10 μm, (d) rimoportula from Thalassiosira weissflogii, scale-500 nm, (e) corona structure of Ditylum brightwellii, scale-2 μm, (f) Bacilaria paxillifer, scale-10 μm, (g) close up of pores in Gyrosigma balticum, scale-2 μm, (h) Skeletonema costatum, scale-2 μm, (i) valve of C. wailesii, scale-50 μm, (j) close up of pores in D. brightwellii, scale-2 μm, (k) seta of Chaetoceros gracilis, scale-1 μm, and (l) Stephanopyxis turris, scale-10 μm. *Reprinted with permission from Hildebrand, M., (2008). Diatoms, biomineralization processes, and genomics. Chemical Reviews. 108, 4855–4874. Copyright @ American Chemical Society.*

and engineers are just discovering how to exploit features unique to these organisms. Their uniform nanopore structure, microchannels, chemical inertness, and silica microcrystal structure suggest many nanoscale applications. Diatom-derived biosilica has been widely used for several applications, including drug delivery (Wee et al., 2005), biodetection (Li et al., 2014), molecular and particle separations (Losic et al., 2006a), silica immobilization (Poulsen et al., 2007), bone tissue engineering (Wang et al., 2012, 2013a), and nanotechnology (Gordon and Parkinson, 2005).

Progress toward the goal of creating frustules synthetically was made when the genome of the marine diatom *Thalassiosira pseudonana* was established (Armbrust et al., 2004), including novel genes for silicic acid transport and the formation of silica-based cell walls. Based on this, it was proposed that the first step is to identify cell wall synthesis genes involved in structure formation; the completed genome sequence of *T. pseudonana* may open the door for genomic and proteomic approaches to accomplish this (Hildebrand, 2005, 2008; Hildebrand et al., 2006). An approach that is also used in other organisms is to modify gene sequences or expression, introduce the modified genes into the diatoms, and monitor the effect on structure.

The unique morphologies and properties of the diatom silica shell—with its intricate, hierarchically organized 3D structures with nanoscale dimensions—have attracted considerable interest in materials science and nanotechnology in recent years. This chapter highlights emerging applications for diatom nanotechnology research in drug delivery (Losic et al., 2009; Kurkuri et al., 2011).

2. CULTURE AND HARVEST METHODS IN DIATOM PRODUCTION

Diatoms are the major contributors to phytoplankton blooms in lakes and in the sea; hence, they are prominent in aquatic ecosystems and the global carbon cycle (Hamm et al., 2003). Scientists have developed culturing techniques and carried out experimental designs to study the biological and ecological features of microalgae. Distinguished among them is the agar plate culturing method, which was first developed more than half a century ago.

Kei Kimura and Yuji Tomaru (2013) described a simple and easy agar plate culturing method for centric diatoms using the planktonic species *Chaetoceros tenuissimus Meunier* (Centrales), which had been difficult to culture on regular agar plates. The agar plates were prepared with 25 mL of 1% agar in medium enriched with 2 nM Na_2SeO_3 in a plastic petri dish. Five

milliliters of the same agar media, molten, was poured on to the mesh. After solidification, the mesh with extra fixed agar was removed from the agar plate. In this procedure, the nylon mesh created a dimpled surface on the smooth agar plate. Axenic clonal *C. tenuissimus* strains 2–10 were grown in modified SWM3 liquid medium under a 12-h light–dark cycle of ~110–150 mol of photon $m^2 s^{-1}$ using cool white fluorescent illumination at 15 °C. The cell growth on the dimpled plate was compared with that on an agar plate with a smooth surface (smooth plate). All plates were cultured under the conditions described above for 7 days. It was concluded that *C. Tenuissimus* grew well on agar plates, with a suggestion that the daubing method was not suitable for plate culturing of diatoms (Kimura and Tomaru, 2013).

Sanjay et al. (2013) used three different methods to culture diatoms: shake flash, polythene bag, and a photobioreactor culture. In the shake flask method, 100 mL of Chu medium was prepared and transferred to 500-mL conical flasks. Then, 10 mL of collected samples were transferred aseptically to the conical flasks. The flasks were kept in an incubator shaker with an illuminator at 20 °C and 120 rpm for 10 days. After 10 days, diatoms were identified using a stereomicroscope and subsequently isolated.

In the polythene bag culturing method, autoclavable polythene bags were used. A total of 200 mL of medium was transferred to each bag and sterilized at 121 °C with 15 lbs of pressure for 15 min. The bags were inoculated aseptically with 20 mL of a pure culture of *Navicula* and kept at 20 °C below the light source (6 W). Aerators with sterile filters were used for sufficient aeration for the growth of diatoms (Sanjay et al., 2013).

A vertical photobioreactor with a glass chamber was designed for culturing diatoms. The reactor was provided with a source of light (12 W), aerator, thermometer, inlet, and outlets. A total of 15 L of sterilized Chu medium was transferred to the reactor and 1.5 L of inoculum was added aseptically. The temperature was maintained at 20 °C, with sufficient aeration and light, and incubated for 15 days. The biomass was transferred to a preweighed clean petri plate and dried at 50 °C to determine the dry weight of the biomass, then stored under refrigerated conditions (Sanjay et al., 2013).

Lewin (1966) showed the boron requirement for 12 species of marine pennate diatoms, four species of centric diatoms, and eight other freshwater species. He concluded that boron was essential for the growth of most (probably all) diatoms.

3. ISOLATION AND CHARACTERIZATION OF DIATOM BIOSILICA

Silica is a material with many technological applications in biotechnology, including many industrial and synthetic production processes. Diatoms have delicate porous structures that are very beneficial in improving the absorbing ability of the biodetection field (Li et al., 2014). The chemical synthesis of silica in industrial applications requires extreme temperatures, pressures, and pH levels, as well as dangerous chemical compounds. In contrast, biological synthesis of silica in nature (e.g., in diatoms) occurs at ambient temperature/pressure and neutral pH.

Diatoms produce nanostructured silica as a component of the cell wall (Manurung et al., 2009). In living diatoms, the diatom's silica skeleton is covered with an organic envelope composed of polysaccharides and proteins. The high degree of complexity and hierarchical structure displayed by diatom silica walls is achieved under mild physiological conditions (Dolatabadi and de la Guardia, 2011).

Gnanamorthy et al. (2014) conducted a scanning electron microscopy (SEM) investigation of three diatom species: *Coscinodiscus* spp., *Cantharellus concinnus*, and *Odontella mobiliensis*. The authors revealed the elemental composition of

silica in *C. concinnus* (36%), *Coscinodiscus* spp. (30.71%), and *O. mobiliensis* (34.27%), and they suggested future investigations on the nanoporous diatom silica for applications in the antireflection coating of materials (Figure 2). The SEM stereoimaging technique was used to reconstruct the surface features of the diatomaceous frustules to quantitatively evaluate specimens. Geometrical parameters, such as volume and area, were given based on the reconstructed 3D image (Chen et al., 2010). Atomic force microscopy, histochemical analysis, infrared spectrometry, molecular spectroscopy, and confocal infrared microscopy are also used to study diatom bionanotribology (Gebeshuber et al., 2005).

In fossilized and recent diatoms, the complex patterns of silica shell are statistically stable and lightweight. Young's modulus E of the diatom silica of a pleura was 22.4 GPa, which is comparable to cortical bone (20 GPa) or medical dental composites (about 6–25 GPa). Like diatom silica, this is composed of inorganic particles associated with an organic matrix (Hamm et al., 2003).

The procedure for pore size modifications of two centric diatom species, *Coscinodiscus* sp. and *Thalassiosira eccentrica*, using atomic layer deposition of ultrathin films of titanium oxide (TiO_2) has been described. TiO_2 was deposited by sequential exposures to titanium chloride ($TiCl_4$) and water. These techniques confirmed the controlled reduction of pore sizes while preserving the shape of the diatom membrane pores. Pore diameters of diatom membranes can be further tailored for specific applications by varying the number of cycles and by changing their surface functionality (Losic et al., 2006b; De Stefano et al., 2008).

A diatom frustule is functionalized with an alkyl halide to allow the growth of a polymer from its surface via deactivation-enhanced atom transfer radical polymerization. The diatom core is partially dissolved to form a more translational platform. This method can be used to create an array of nanostructured composites derived from the species-specific

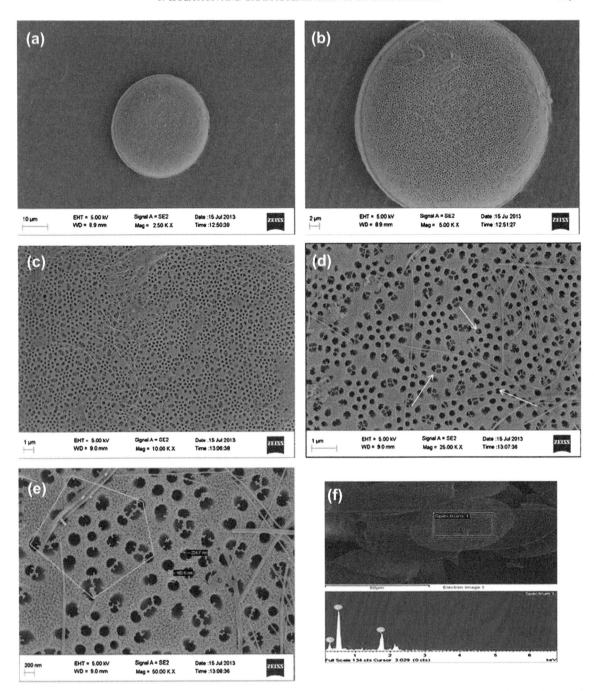

FIGURE 2 (a) Structure of *Cantharellus concinnus*. (b) Field emission scanning electron microscopy (FESEM) image (2 μm) of the surface showing the porous topography. (c) Well-arranged FESEM image of foramen surface. (d) Enlarged FESEM image of a foramen shows the details of pore organization (1 μm). (e) High-resolution FESEM image of one typical pentagonal pattern is marked on the pore array and star-shaped hyaline area of the surface of the diatom with nanoporous diameter and inter pore-distance details. (f) Corresponding EDS graph shows the silica element presence. *Reprinted with permission from Gnanamoorthy, P., Karthikeyan, V., Prabu, V. A., 2014. Field emission scanning electron microscopy (FESEM) characterisation of the porous silica nano-particulate structure of marine diatoms. Journal of Porous Materials, 1—9. Copyright @ Elsevier.*

diatom architecture (O'Connor et al., 2014). These modified structures have potential applications in antibody arrays and may have use in techniques such as immunoprecipitation. These silica structures are produced in diatoms using only light and minimal nutrients and, therefore, generate an exceptionally cheap and renewable material (Townley et al., 2008).

4. APPLICATION OF BIOSILICA IN DRUG-DELIVERY VEHICLES

Nature has developed an elegant, biologically based, self-assembling synthetic route to produce silica biomaterials with complex 3D porous structures, offering great potential to replace synthetic materials with suitable drug carriers for cost-effective delivery systems (Bariana et al., 2013b). The biocompatibility, biodegradability, and pore size of diatom biosilica are the most important features for use in drug delivery. The incorporation of nanocarriers into drug molecules can protect against degradation as well as offer possibilities and target release. Studies have proposed several methods of functionalizing cylindrical frustule for controlled two-step release (Wang et al., 2013b). Biosilicified nanostructured microshells from the marine diatom *Coscinodiscus wailesii* have been properly functionalized to bind a molecular probe that specifically recognizes a target analyte. Fluorescence measurements demonstrate that the antibodies used, even if linked to the amorphous silica surface of *C. wailesii* microshells, still efficiently recognize their antigens (De Stefano et al., 2008).

Diatomaceous earth (DE), which is naturally available silica that originated from fossilized diatoms, has been explored for use in drug-delivery applications as a potential substitute for synthetic silica materials. DE's particle zeta potential (Losic et al., 2010) shows that surface diatom silica structures are negatively charged in aqueous media in the pH range of 2−11;

accordingly, it favors the immobilization of the positively DOPA/Fe_3O_4 nanoparticles. In addition, experiments on nitrogen adsorption/desorption confirmed a moderately fair surface area of 18.5 ± 0.8 $m^2\,g^{-1}$ of hollow diatom frustules; this unique feature allows them to accommodate a large amount of drugs (Losic et al., 2010). To understand this better, Aw et al. (2011) conducted structural and spectroscopic analysis using scanning electron microscopy (SEM), Energy-dispersive X-ray spectroscopy (EDXS), nuclear magnetic resonance (NMR), and X-Ray Powder Diffraction (XRPD) characterizations of encapsulated drug molecules in diatom microcapsules. Drug molecules were physisorbed on the diatom silica surface, with a small proportion integrated into the diatom structure. There are many potential applications in diatom nanotechnology using its silica. However, characterization of the biosilica of diatoms will require much engineering to suit application models in biosensing, biophotonics, and drug delivery (Aw et al., 2011).

Various silica-based morphologies, from mesoporous nanoparticles to implantable microcarriers, have been synthesized and applied for drug-delivery purposes, with the aim of addressing common therapeutic problems, such as limited drug solubility leading to poor bioavailability and undesirable pharmacokinetics in drug release over weeks (as required for therapeutic implants; Aw et al., 2011).

The easy functionalization of diatom biosilica can offer protection, design, and control of drug release through pores or by covering an ultrathin polymer layer. Diatom structures in dopamine-modified iron oxide nanoparticles with magnetic properties showed an ability to work as magnetically guided drug-delivery microcarriers (Losic et al., 2010). This was accompanied by in vitro drug dissolution studies, oral delivery, implant delivery, and NMR characterization of encapsulated drug molecules in diatom microcapsules. Results demonstrated the effectiveness of diatom microstructures, with 12–14% drug loading capacity and sustained release for 2 weeks.

Zhang et al. (2013) also demonstrated diatom biosilica's remarkable features for application in drug delivery. They proved that DSMs have almost no toxicity in Caco-2, HT-29, HCT-116, and Caco-2/HT-29 cells at concentrations greater than 1000 $\mu g\ mL^{-1}$ (Zhang et al., 2013). According to Gordon et al., the physical and structural properties of porous silica capsules of diatoms are ideal microscale bodies for designing robotic devices for medical applications (Gordon et al., 2009).

Biosilica is much preferred over other materials because of its natural origin and nontoxicity with hydroxyl groups (Bayramoglu et al., 2013). Using silica hollow flower, Chen et al. (2014) confirmed the nontoxic sustained release of BMP-2, with much biocompatibility in osteoblasts of MC3T3-E1 cells. Thus, amorphous silica is applicable in system delivery for tissue regeneration (Chen et al., 2014).

Water-insoluble (indomethacin) and water-soluble (gentamicin) drugs were loaded in DE particles to study their drug release performances. In vitro drug release studies were performed over 1–4 weeks to examine the impact of the particle size and hydrophilic/hydrophobic functional groups. The release studies showed a biphasic pattern, comprising an initial burst release for 6 h followed by near-zero order sustained release (Bariana et al., 2013a).

Aw et al. (2012) proved the drug-delivery concept based on diatoms for implants and oral drug delivery; indomethacin, as a model of a poorly water-soluble drug, was investigated. The effectiveness of diatom silica for drug-delivery application was approximately 22 wt% drug loading capacity with sustained drug release over 2 weeks. A two-step drug release from diatom structures was observed: the first is a rapid release (over 6 h) attributed to the surface-deposited drug, whereas the second is a slow and sustained release over 2 weeks with zero-order kinetics, as a result of release from diatom pores and internal hollow structures (Aw et al., 2012; Figure 3).

Aw et al. (2013) also reported indomethacin delivery using diatoms. Surface modification of the diatoms was performed with two organosilanes, 3-aminopropyltriethoxy silane and N-(3-(trimethoxysilyl) propyl) ethylene diamine and phosphonic acids (2-carboxyethyl-phosphonic acid and 16-phosphono-hexadecanoic acid), providing organic surface hydrophilic and hydrophobic properties. Differences in the loading capacities of diatoms (15–24%) and release times (6–15 days) were observed, which were due to the presence of different functional groups on the surface. It was found that 2-carboxyethyl-phosphonic acid, 3-aminopropyltriethoxy silane, and N-(3-(trimethoxysilyl) propyl) ethylene diamine render diatom surfaces hydrophilic, due to a polar carboxyl functional group (COOH) and active amine species (NH and NH_2) that favor drug adsorption. Better encapsulation efficiency and prolonged release of drugs over the hydrophobic surface were created by 16-phosphono-hexadecanoic acid (Aw et al., 2013).

5. FUTURE DIRECTIONS

Diatoms have been known to play important roles on earth and in oceans as oxygen synthesizers and biomass sources. Before the late 1980s, diatom nanotechnology was a minor field of academic research, with little or no thought given to its applications. Today, however, the situation is quite different. Diatom nanotechnology has great potential for applications in many areas of research, such as in the design of complex drug vehicles, biosensors, biophotonics, and cell labeling.

The interdisciplinary approaches required in diatom nanotechnology research for creating useful technologies are quite promising. According to Richard Gordon, in basic research, diatoms are likely to contribute to the solution of one of the major unsolved biological problems: how the genome is involved in creation of form

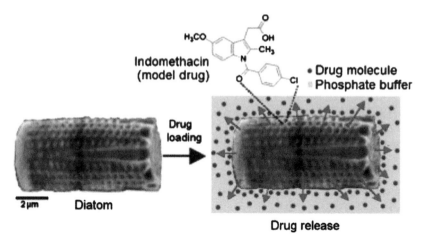

FIGURE 3   Scheme of drug release from diatom silica microshell. *Reprinted with permission from Aw, M. S., Simovic, S., Yu, Y., Addai-Mensah, J., Losic, D., 2012. Porous silica microshells from diatoms as biocarrier for drug-delivery applications. Powder Technology. 223, 52—58. Copyright @ Elsevier.*

or how form evolves (Gordon et al., 2009). We now have a few competing hypotheses for diatoms, and there is a great need for intracellular observation to resolve what is going on.

There is still work to be done to answer the following questions, which will widen the scope and application potentials for diatom nanotechnology in major biological and medical problems:

- What are the processes involved in the formation of diatom frustule?
- What role does biosilica play in the aggregate formation of diatom patterns, shapes, and sizes?
- What are the genetic characteristic and roles of specific functional molecules to ascertain diatom motility and speed?

In addition, the phylogenetic characterization of other diatom species is critical. For diatom biosilica features such as biocompatibility, pore size will be important for a wide range of applications in the future, such as drug-delivery vehicles. The diatom biosilica structure, which is less than 300 nm in thickness, shows better elasticity, such as the girdle band of most diatoms, the septum and thin valve of pinnate diatoms

smaller than 10 μm, the long central spine of some centric diatoms (e.g., *Ditylum* spp.), and some thin and long substructures of frustules (e.g., strutted processes of *Skeletonema* spp.). These may serve as testers, force actuators/sensors, optical parts, or filters in biomedical microelectromechanical systems and microfluidic systems (Wang et al., 2013b).

Almost all studies have shown that silica could be an excellent, biocompatible candidate for gene transfer into different cells and tissues. From published works, it is clear that surface-modified silica can improve the in vitro transfer of plasmid DNA and RNA into mammalian cells. Considering the chemical inertness and biocompatibility of silica, Dolatabadi and Guardia suggested that such technology might offer a new platform for nonviral gene delivery, as well as a contrasting agent in magnetic imaging (Dolatabadi and de la Guardia, 2011).

6. CONCLUSIONS

Diatom biosilica is a very important biomedical agent for the development and study of novel therapeutic agents, diagnostic techniques, and drug-delivery vehicles. There have been a great deal of scientific studies on this subject, and its prospects for industrial applications in the production of therapeutic agents are showing great signs of progress each day. Studies related to diatom biosilica will continue to attract more attention in nanotechnology and biomedical engineering, as well as materials science engineering.

Acknowledgments

This research was supported by a grant from the Marine Bioprocess Research Center of the Marine Biotechnology Program, funded by the Ministry of Oceans and Fisheries, Republic of Korea.

References

Armbrust, E.V., Berges, J.A., Bowler, C., Green, B.R., Martinez, D., Putnam, N.H., Zhou, S., Allen, A.E., Apt, K.E., Bechner, M., 2004. The genome of the diatom *Thalassiosira pseudonana*: ecology, evolution, and metabolism. Science 306 (5693), 79−86.

Aw, M.S., Simovic, S., Addai-Mensah, J., Losic, D., 2011. Silica microcapsules from diatoms as new carrier for delivery of therapeutics. Nanomedicine 6 (7), 1159−1173.

Aw, M.S., Simovic, S., Yu, Y., Addai-Mensah, J., Losic, D., 2012. Porous silica microshells from diatoms as biocarrier for drug delivery applications. Powder Technol. 223, 52−58.

Aw, M.S., Bariana, M., Yu, Y., Addai-Mensah, J., Losic, D., 2013. Surface-functionalized diatom microcapsules for drug delivery of water-insoluble drugs. J. Biomater. Appl. 28 (2), 163−174.

Bariana, M., Aw, M.S., Losic, D., 2013a. Tailoring morphological and interfacial properties of diatom silica microparticles for drug delivery applications. Adv. Powder Technol. 24 (4), 757−763.

Bariana, M., Aw, M.S., Kurkuri, M., Losic, D., 2013b. Tuning drug loading and release properties of diatom silica microparticles by surface modifications. Int. J. Pharm. 443 (1−2), 230−241.

Bayramoglu, G., Akbulut, A., Yakup Arica, M., 2013. Immobilization of tyrosinase on modified diatom biosilica: enzymatic removal of phenolic compounds from aqueous solution. J. Hazard Mater. 244, 528−536.

Chen, P.-Y., McKittrick, J., Meyers, M.A., 2012. Biological materials: functional adaptations and bioinspired designs. Prog. Mater. Sci. 57 (8), 1492−1704.

Chen, S., Shi, X., Osaka, A., Gao, H., Hanagata, N., 2014. Facile synthesis, microstructure and BMP-2 delivery of novel silica hollow flowers for enhanced osteoblast differentiation. Chem. Eng. J. 246, 1−9.

Chen, X., Ostadi, H., Jiang, K., 2010. Three-dimensional surface reconstruction of diatomaceous frustules. Anal Biochem. 403 (1−2), 63−66.

De Stefano, L., Lamberti, A., Rotiroti, L., De Stefano, M., 2008. Interfacing the nanostructured biosilica microshells of the marine diatom Coscinodiscus wailesii with biological matter. Acta Biomater. 4 (1), 126−130.

Dolatabadi, J.E.N., de la Guardia, M., 2011. Applications of diatoms and silica nanotechnology in biosensing, drug and gene delivery, and formation of complex metal nanostructures. TrAC Trends Anal. Chem. 30 (9), 1538−1548.

Gebeshuber, I.C., Stachelberger, H., Drack, M., 2005. Diatom bionanotribologybiological surfaces in relative motion: their design, friction, adhesion, lubrication and wear. J. Nanosci Nanotechnol. 5 (1), 79−87.

Gnanamoorthy, P., Karthikeyan, V., Prabu, V.A., 2014. Field emission scanning electron microscopy (FESEM) characterisation of the porous silica nanoparticulate structure of marine diatoms. J. Porous Mater. 21 (2), 225−233.

Gordon, R., Parkinson, J., 2005. Potential roles for diatomists in nanotechnology. J. Nanosci. Nanotechnol. 5 (1), 35−40.

Gordon, R., Losic, D., Tiffany, M.A., Nagy, S.S., Sterrenburg, F.A., 2009. The glass menagerie: diatoms for novel applications in nanotechnology. Trends Biotechnol. 27 (2), 116−127.

Hamm, C.E., Merkel, R., Springer, O., Jurkojc, P., Maier, C., Prechtel, K., Smetacek, V., 2003. Architecture and material properties of diatom shells provide effective mechanical protection. Nature 421 (6925), 841−843.

Hildebrand, M., 2005. Prospects of manipulating diatom silica nanostructure. J. Nanosci. Nanotechnol. 5 (1), 146−157.

Hildebrand, M., 2008. Diatoms, biomineralization processes, and genomics. Chem. Rev. 108 (11), 4855−4874.

Hildebrand, M., York, E., Kelz, J.I., Davis, A.K., Frigeri, L.G., Allison, D.P., Doktycz, M.J., 2006. Nanoscale control of silica morphology and three-dimensional structure during diatom cell wall formation. J. Mater. Res. 21 (10), 2689−2698.

Kimura, K., Tomaru, Y., 2013. A unique method for culturing diatoms on agar plates. Plankton Benthos Res. 8 (1), 46−48.

Kurkuri, M.D., Saunders, C., Collins, P.J., Pavic, H., Losic, D., 2011. Combining micro and nanoscale structures: emerging applications of diatoms. Micro Nanosyst. 3 (4), 277−283.

Lewin, J., 1966. Boron as a growth requirement for diatom. J. Phycol. 2 (4), 160−163.

Li, A., Cai, J., Pan, J., Wang, Y., Yue, Y., Zhang, D., 2014. Multi-layer hierarchical array fabricated with diatom frustules for highly sensitive bio-detection applications. J. Micromech. Microeng. 24 (2), 025014.

Losic, D., Mitchell, J.G., Voelcker, N.H., 2009. Diatomaceous lessons in nanotechnology and advanced materials. Adv. Mater. 21 (29), 2947–2958.

Losic, D., Rosengarten, G., Mitchell, J.G., Voelcker, N.H., 2006a. Pore architecture of diatom frustules: potential nanostructured membranes for molecular and particle separations. J. Nanosci. Nanotechnol. 6 (4), 982–989.

Losic, D., Triani, G., Evans, P.J., Atanacio, A., Mitchell, J.G., Voelcker, N.H., 2006b. Controlled pore structure modification of diatoms by atomic layer deposition of TiO_2. J. Mater. Chem. 16 (41), 4029–4034.

Losic, D., Yu, Y., Aw, M.S., Simovic, S., Thierry, B., Addai-Mensah, J., 2010. Surface functionalisation of diatoms with dopamine modified iron-oxide nanoparticles: toward magnetically guided drug microcarriers with biologically derived morphologies. Chem. Commun. 46 (34), 6323–6325.

Manurung, A.I., Pratiwi, A.R., Syah, D., Suhartono, M.T., 2009. Isolation and characterization of silaffin that catalyze biosilica formation from marine diatom Chaetoceros gracilis. HAYATI J. Biosci. 14 (3), 119.

O'Connor, J., Lang, Y., Chao, J., Cao, H., Collins, L., Rodriguez, B.J., Dockery, P., Finn, D.P., Wang, W., Pandit, A., 2014. Nano-structured polymer-silica composite derived from a marine diatom via deactivation enhanced atom transfer radical polymerization grafting. Small 10 (3), 469–473.

Parkinson, J., Gordon, R., 1999. Beyond micromachining: the potential of diatoms. Trends Biotechnol. 17 (5), 190–196.

Poulsen, N., Sumper, M., Kröger, N., 2003. Biosilica formation in diatoms: characterization of native silaffin-2 and its role in silica morphogenesis. Proc. Natl Acad. Sci. 100 (21), 12075–12080.

Poulsen, N., Berne, C., Spain, J., Kroeger, N., 2007. Silica immobilization of an enzyme through genetic engineering of the diatom Thalassiosira pseudonana. Angew. Chem. Int. Ed. 46 (11), 1843–1846.

Sanjay, K., Nagendra, P., Anupama, S., Yashaswi, B., Deepak, B., 2013. Isolation of diatom Navicula cryptocephala and characterization of oil extracted for biodiesel production. Afr. J. Environ. Sci. Technol. 7 (1), 41–48.

Townley, H.E., Parker, A.R., White-Cooper, H., 2008. Exploitation of diatom frustules for nanotechnology: tethering active biomolecules. Adv. Funct. Mater. 18 (2), 369–374.

Wang, X., Schröder, H.C., Wiens, M., Ushijima, H., Müller, W.E., 2012. Bio-silica and bio-polyphosphate: applications in biomedicine (bone formation). Curr. Opin. Biotechnol. 23 (4), 570–578.

Wang, X., Schröder, H.C., Feng, Q., Draenert, F., Müller, W.E., 2013a. The deep-sea natural products, biogenic polyphosphate (Bio-PolyP) and biogenic silica (Bio-Silica), as biomimetic scaffolds for bone tissue engineering: fabrication of a morphogenetically-active polymer. Mar. Drugs 11 (3), 718–746.

Wang, Y., Cai, J., Jiang, Y., Jiang, X., Zhang, D., 2013b. Preparation of biosilica structures from frustules of diatoms and their applications: current state and perspectives. Appl. Microbiol. Biotechnol. 97 (2), 453–460.

Wee, K.M., Rogers, T.N., Altan, B.S., Hackney, S.A., Hamm, C., 2005. Engineering and medical applications of diatoms. J. Nanosci. Nanotechnol. 5 (1), 88–91.

Woesz, A., Weaver, J.C., Kazanci, M., Dauphin, Y., Aizenberg, J., Morse, D.E., Fratzl, P., 2006. Micromechanical properties of biological silica in skeletons of deep-sea sponges. J. Mater. Res. 21 (8), 2068–2078.

Zhang, H., Shahbazi, M.-A., Mäkilä, E.M., da Silva, T.H., Reis, R.L., Salonen, J.J., Hirvonen, J.T., Santos, H.A., 2013. Diatom silica microparticles for sustained release and permeation enhancement following oral delivery of prednisone and mesalamine. Biomaterials 34 (36), 9210–9219.

Microalgal Nutraceuticals

J. Paniagua-Michel

Laboratory for Bioactive Compounds and Bioremediation, Department of Marine Biotechnology, Centro de Investigación Científica y de Educación Superior de Ensenada (CICESE), Ensenada, BC, México

1. INTRODUCTION

Marine and freshwater ingredients have been used for millennia. Consumers in modern societies have also embraced these dietary components, now termed "nutraceuticals," because of their recognized health benefits, as evidenced from scientific findings. The word *nutraceutical*, as originally coined, is a combination of the words *nutrition* and *pharmaceutical*. A nutraceutical is defined as a food or food product that provides health and medical benefits, including the prevention and treatment of disease (Kim, 2013). Essential nutrients can be considered nutraceuticals, but only when they confer additional attributes on the normal growth or maintenance of organisms (Wildman and Taylor, 2006). Contrary to pharmaceutical drugs, nutraceutical foods are not subject to testing and regulation protocols, which has resulted in their wide distribution.

The increasing world population requires adequate and proper nourishment of high nutritional value for specific and possibly chronic diseases (Bishop and Zubeck, 2012). Moreover, the increasing costs of health care, increasing life expectancy, and individuals' desire to improve their health quality (Plaza et al., 2009) have increased the demand for nutraceuticals in recent years. Algae have been recognized as having diverse nutritional components, along with simple and rapid growth characteristics. As autotrophic organisms, microalgae exhibit properties of nutraceuticals and have health benefits for the immune system to fight cancer and heart disease (Bishop and Zubeck, 2012). Nutraceuticals in aquaculture are used to enhance the growth of shellfish under intensive and controlled conditions because of the rudimentary immune system of shellfish. Microalgae also supply nutritional protein, pro-vitamin compounds, and fatty acids (FAs) to enhance the flesh color of fish and crustaceans, mainly salmonids and *Penaeidae*. The objective of this chapter is to review and analyze the use of nutraceuticals from microalgae in human and aquaculture applications. Their respective characteristics and effects for consumers are also discussed.

2. NUTRACEUTICAL: DEFINING THE TERM

There are several definitions of the term *nutraceutical*, which often is confused with or used as a synonym for *dietary supplement*, *functional foods*, and/or *medical food* (González-Sarrías et al., 2013). Such foods are regulated by the US Food and Drug Administration (FDA) under the authority of the Federal Food, Drug, and Cosmetic Act, even though they are not specifically defined by law. De Felice (1989) coined the term *nutraceutical*, which the American Nutraceutical Association has defined as follows: "A nutraceutical is any substance conceived as a food, or part of a food which provides medical or health benefits, as well as the prevention and treatment of a disease" (www.acronymfinder.com/American-Nutraceutical-Association-(ANA).html). Unlike pharmaceutical drugs, within the United States nutraceutical products are monitored as "dietary supplements."

Generally, nutraceuticals play a positive role in enhancing health and wellness in the consumers. Hence, health-promoting nutrients that are derived from food or food products to enhance the prevention or treatment of a disease and/or disorder (Bishop and Zubeck, 2012) are also considered to be nutraceuticals. There are more than 470 commercial nutraceuticals with health-promoting properties. The consumer trend in nutraceuticals is aimed at promoting health as well as treating potential neurodegenerative diseases, such as Alzheimer disease, heart disease, cancer, and Parkinson disease, among others.

3. MICROALGAE: THE NUTRACEUTICAL BENEFITS

Microalgae are a diverse group of autotrophic organisms that have the ability to grow rapidly, efficiently use light energy, fix atmospheric CO_2, and produce more biomass per surface than any natural source. The ability of microalgae to synthesize a variety of compounds confers nutraceutical properties, with an elevated potential for industrial exploitation (Bishop and Zubeck, 2012). Different types of microalgae are used for current commercial and potential nutraceuticals, such as *Chlamydomonas*, *Scenedesmus*, *Synechococcus*, *Dunaliella*, *Haematoccoccus*, *Chlorella*, and *Porphyridium*. Moreover, certain microalgae have the ability to accumulate essential elements, such as potassium, zinc, iodine, selenium, iron, manganese, copper, phosphorus, sodium, nitrogen, magnesium, cobalt, molybdenum, sulfur, and calcium, which can be used in nutraceutical formulae. Carotenogenic microalgae biosynthesize secondary carotenoids, such as canthaxanthin and astaxanthin, in lipid globules of the chloroplast plastids. Algae are also high producers of essential amino acids, as well as omega-6 (arachidonic acid) and omega-3 (docosahexaenoic acid (DHA), eicosapentaenoic acid (EPA)) FAs (Simoons, 1991). A lucrative market for algae and derived nutraceuticals has developed due to their abundant production of health-promoting molecules and respective nutritive contents. Microalgae can use solar energy to transform wastewater, CO_2, and some nutrients into useful biomass rich in lipids, sugars, proteins, carbohydrates, and other organic compounds. These microorganisms convert inorganic substances into colored biomass (Batista et al., 2013).

Microalgae are able to produce a rich diversity of critical nutrients to support human and animal health. Microalgae are found worldwide in numerous types of environments; therefore, they have enormous potential as nutraceuticals. Their properties and adaptations have resulted in benefits to different organisms further up the food web. For instance, the unique characteristics of many microalgae—including phycobilins, phycoerytrines, carotenoids, micronutrients, amino acids, and carbohydrates—have led to a number of compounds that have important benefits for human health (Figure 1). However,

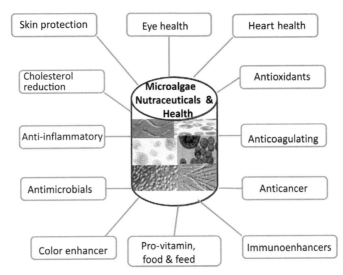

FIGURE 1   Different uses and applications of microalgal nutraceuticals.

despite the above-mentioned properties and benefits of microalgae compounds, their development is still in its infancy, although many products are currently available. In recent years, several products or extracts with microalgal origins have been used in nutritional supplements and cosmetics, such as *Dunaliella, Spirulina, Chlorella,* and *Haematococcus* (Gellenbeck, 2012).

The use and acceptance of new species and bioproducts must overcome many barriers and regulatory aspects, although they are not as strict as for pharmaceuticals. Among the many examples and applications of microalgae in the area of nutraceutical products (i.e., algal extracts and powders used for human nutritional supplementation), many are being produced for commercial use (Gellenbeck, 2012), as described in Table 1.

3.1 Beta-Carotene: A Nutraceutical from Extreme Algae

The richest natural source of the carotenoid beta-carotene is the microalgae *Dunaliella salina*, which thrives in hypersaline coastal water bodies; large-scale production also can be

performed in open raceway ponds. The massive accumulation of beta-carotene production by *D. salina* is inversely proportional to high salinity and nutrient-deprivation conditions. The cells

TABLE 1   Main Microalgal Products and Applications as Nutraceuticals

Species/group	Product	Application areas
Arthrospira platensis	Phycocyanin,	Health food
Chlorella vulgaris	Minerals, trace elements	Health food, food supplement, feeds
Dunaliella salina	Lutein, beta-carotene	Health food, food supplement, feeds
Haematococcus pluvialis	Astaxanthin	Health food, feeds
Porphyridium cruentum	Polysaccharides	Pharmaceuticals, cosmetics,
Isochrysis galbana	Fatty acids	Animal nutrition (aquaculture)
Phaedactylum tricornutum	Fatty acids, fucoxanthin	Nutrition, aquaculture
Lyngbya majuscule	Immune modulators	Health food, nutrition

from noninductive conditions are green; however, under stress conditions, cells turn red from a massive accumulation of carotenoids.

Of all the carotenoids, beta-carotene is the most prominent, accounting for 95% on a dry weight basis. The pigment accumulates in the interthylakoid spaces of the chloroplast. The current price output of this natural product on a pure basis is valued around US $1500 per kilogram, which can vary depending on the marketing and commercial demand for the product (Gellenbeck, 2012).

The nutraceutical use of this pigment is based on its function as a nontoxic vitamin A precursor, which is used in multivitamin and specialty formulations (Linan-Cabello et al., 2002). Because of its unique growth environment and halotolerance of saline conditions (up to 100 g NaCl/L) carotenogenic *Dunaliella* strains cannot use the same water as for agricultural and domestic uses.

The global production of *Dunaliella* is estimated to be 1200 tons dry weight per year (Gouveia et al., 1996). The major producers of *Dunaliella*, mainly for beta-carotene, are located in Israel, China, United States, and Australia: Betatene, Western Biotechnology, AquaCarotene Ltd, Cyanotech Corp., Nature Beta Technologies, and more (Benedetti et al., 2004).

Dunaliella produces numerous carotenoid pigments, with the most prominent being beta-carotene (up to 14% dry weight), along with smaller amounts of alpha-carotene, lutein, and lycopene (Desmorieux et al., 2005). The total carotenoid content of *Dunaliella* varies with growth conditions; it can yield around 400 mg beta-carotene/m^2 of cultivation area (Muller-Feuga, 2004).

Beta-carotene is an antioxidant that can trap reactive oxygen species involved in the aging process (Muller-Feuga, 2000; Borowitzka, 1991). Several studies have shown beta-carotene to prevent cancer of various organs, such as the lungs, stomach, cervix, pancreas, colon, rectum, breast, prostate, and ovaries, by means of antioxidant activity. Findings on the role of beta-carotene from *Dunaliella* as a nutraceutical have reported a positive influence on intracellular communication, immune response (Becker, 2004), and protection against many types of neoplasms (Aragão, 2004). Supplements of *Dunaliella* have also shown excellent hepatoprotective effects and reduced the occurrence of liver lesions (Chuntapa et al., 2003).

Despite the benefits and the advancement in the natural production of beta-carotene from *Dunaliella*, more than 90% of commercialized beta-carotene is produced synthetically. However, natural beta-carotene exhibits a higher bioavailability compared to the synthetically manufactured analog beta-carotene (Lio-Po, 2005). The amount of the antioxidant enzymes catalase, peroxidase, and superoxide dismutase has been reported as significantly greater in naturally produced beta-carotene from *Dunaliella* compared to synthetic. The health benefits from *Dunaliella* are numerous and diverse, and there are no risks associated with consumption of supplements containing this alga. When rats were fed 10% *Dunaliella* in their diets, they showed no significant negative effects, which is indicative of the safety of *Dunaliella* for human consumption (Rodolfi et al., 2003).

3.2 Astaxanthin: The Red Nutraceutical

This red pigment is produced by the freshwater alga *Haematococcus*. This unicellular, green alga is a common component of nutraceuticals, pharmaceuticals, cosmetics, aquaculture, and numerous food products (Ambati et al., 2014; Guerin et al., 2003). Astaxanthin has shown innovative anti-inflammatory and antioxidant applications in human nutrition. A theoretical analysis of the costs of production of astaxanthin was estimated at US $ 718 per kilogram (Li et al., 2011). The dry weight production of *Haematococcus* is around 300 tons annually in the United States, India, and Israel (Irianto and Austin, 2002). The annual world market of this pigment is estimated at $200 million, with 95% of this market consuming synthetically derived astaxanthin (Irianto and Austin, 2002; Guerin et al., 2003). The current price of astaxanthin is approximately $2500 per kilogram dry weight (but for human consumption astaxanthin from Haematococcus pluvialis can be sold over 7000 US$ per kg (Ambati et al., 2014)). Companies that commercially produce *Haematococcus*, predominantly for astaxanthin, include Cynotech Corporation, Parry Nutraceuticals, BioReal, Inc., Fuji Health Science, Aquasearch Inc., Valensa International, and Alga Technologies.

The astaxanthin level in *Haematococcus* (the largest natural source of astaxanthin) comprises 1.5–3% of its dry weight (Spolaore, 2006; Waldenstedt et al., 2003). Astaxanthin has shown antioxidant activity that is more than 10 times stronger than beta-carotene, and close to 1000 times more effective than vitamin E (Lorenz, 2000; Zittelli et al., 1999). Research demonstrated that this pigment is effective in decreasing arterial blood pressure, plasma levels of triglycerides, and nonesterified FAs (Robinson et al., 2005). Among the main metabolic functions of astaxanthin in humans, the following have been reported: protection against oxidation of essential polyunsaturated fatty acids (PUFAs), protection against ultraviolet radiation effects, and enhanced vision, immune response, pigmentation, and reproductive behavior (Yuan et al., 2011). Several studies have indicated antioxidative (Kim et al., 2009), anticancer (Gradelet et al., 1998), anti-inflammatory, and antibacterial activities and properties for this pigment.

To date, there have been no reports of negative consequences associated with the use of *Haematococcus* as a direct dietary supplement. In studies with rat models, no adverse effects were reported when consuming 5–18 g per kilogram per day (Jiang, 1999). The effects of astaxanthin extracts from *Haematococcus* in humans were investigated and did not reveal significant differences for a period of 8 weeks (Molina-Grima et al., 2003). After that time, exposure to doses of 20 mg/day for 4 weeks revealed no negative effects on blood biochemistry or hematology (Wen and Chen, 2003). *Haematococcus pluvialis* is generally regarded as safe by the FDA; since 1999, it has been approved for marketing as a new dietary ingredient in the United States (Kroes et al., 2003).

3.3 *Spirulina*: The Original Blue Nutraceutical

Spirulina is probably the best documented example of microalgae as a food, being used by ancient cultures such as the Aztecs in Mexico and the Kanembu tribes in Lake Chad, Africa (Paniagua-Michel et al., 1992). *Spirulina* was sold in the Tlatelolco market (Mexico City) and eaten with maize or a sauce made of chili peppers and tomatoes (Figure 2).

This cyanobacteria, through photosynthesis, converts sunlight into protein, FAs, carbohydrates, and nearly every other nutrient

FIGURE 2 Scheme representation of Aztec people activities harvesting tecuitlatl (*Spirulina*) at Texcoco Lake, Mexico.

essential to life. Its characteristic blue color is derived from the exotic pigment phycocyanin (PC). PC is a blue, light-harvesting pigment in cyanobacteria and in the two eukaryote algal genera, *Rhodophyta* and *Cryptophyta*. In fact, the bluish color of cyanobacteria is due to PC. The phycobiliproteins in this cyanophyte— allophycocyanin (APC) and c-phycocyanin (CPC)—can have several applications like nutraceuticals, mainly as colorants in food and fluorescent labels. These proteins harbor a tetrapyrrolic pigment, called phycobilin, which is covalently attached to their structure.

Sprirulina has been reported as a supplementary food and a rich source of nutrients, such as B vitamins, phycocyanin, chlorophyll, vitamin E, omega-6 FAs, and abundant minerals (Paniagua-Michel et al., 1992; Gershwin, 2008). According to the culture conditions, this cyanophyte can accumulate around 60—70% protein by weight, including many amino acids, and beta-carotene. Commercial presentations are packed with protein, essential amino acids, and gamma linolenic acid, also providing alpha linolenic acid, linoleic acid (LA), stearidonic acid, EPA, DHA, and arachidonic acid (AA). *Spirulina* is also a rich source of potassium, calcium, chromium, copper, iron, magnesium, manganese, phosphorus, selenium, sodium, and zinc. Commercial brands of *Spirulina* are used for the health promotion of weight loss, diabetes, high blood pressure, and hypertension (Iwata, 1990). There are some reports emphasizing its antiviral properties and anticancer effects (Mishima et al., 1998).

The commercial value of dried *Spirulina* biomass is valued at about US $15—50 per kilogram, depending on production and market-driven variables (Gellenbeck, 2012). Approximately 3000 tons dry weight is currently produced annually in the United States, Thailand, India, Taiwan, China, Pakistan, and Burma (Raja et al., 2007). *Spirulina* has been used as a weight loss supplement, as well as to control diabetes, high blood pressure, and hypertension. Reports have pointed out its positive uses in lowering cholesterol by increasing HDL levels, which can lead to healthy cardiovascular functions (de Caire et al., 1995). The antioxidant and anti-inflammatory properties of C-phycocyanine of *Spirulina* have been successfully demonstrated in mice models as well as in humans (Romay et al., 1998). The potential effects of *Spirulina* for the improvement of digestion, food absorption, and enhancement of the immune system to help fight infections have been investigated (Archer et al., 1985).

Despite all of the above-mentioned benefits of *Spirulina* as a nutraceutical, there are still questions concerning the assimilation, digestion, and bioconversion capacity and the antagonist effects of the other nutrients contained in *Spirulina* extracts or meals. Some findings pointed out the need for efficient processing and engineering techniques to formulate nutrients in a digestible way for humans. Potential side effects related to the consumption of *Spirulina* include allergic reactions, diarrhea, nausea, and vomiting.

3.4 *Chlorella*: The Green Nutraceutical

The green cellular microalga *Chlorella* is widely sold as a health food, food supplement, and nutraceutical (Morita, 1999). In the Far East, *Chlorella* has been used as an alternative medicine since ancient times. In China and the Orient, this Chlorophyte is considered as a traditional food similar to a nutraceutical. Nowadays, the microalgae *Chlorella* is produced and marketed as a health food supplement in many countries, like China, Japan, Europe, and the United States. Its estimated total production is around 2000 tons/year (Batista et al., 2013) of dried *Chlorella* in the United States, Japan, China, Taiwan, and Indonesia. Because of its content of nutrients and positive health effects, *Chlorella* is considered an important functional food and nutraceutical. Concerning its composition, *Chlorella*

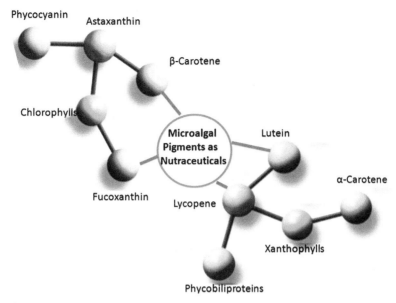

FIGURE 3   The main pigments contained in microalgae used as Nutraceuticals.

is composed of 55—60% protein, 1—4% chlorophyll, 9—18% dietary fiber, and numerous minerals and vitamins (Shim et al., 2008). Reports on the protein of *Chlorella* reveal all essential amino acids required for the nutrition of heterotrophic organisms. The detoxication of metals and pesticides performed by *Chlorella* is associated with porphyrin rings in chlorophyll or glutathione-induced pathway production by vitamin B12. *Chlorella* also accumulates large amounts of lutein, which has been associated with prevention and treatment of macular degeneration (Shibata et al., 2003). This Chlorophyte is also able to lower cholesterol levels and decrease blood pressure. The consumption of *Chlorella* significantly decreased the low-density lipoprotein and the cholesterol levels. As in the case of other microalgae, *Chlorella* consumption as a nutraceutical product could exhibit side effects (Gouveia et al., 2007). Among the registered effects, each commercial brand could exert different response among consumers, as cases of gastrointestinal diseases,

nausea, and vomiting. This Chlorophyte has been classified as a weak allergen. Figure 3 shows the different pigments contained in the diverse microalgal nutraceuticals.

3.5 Lipids and Omega Fatty Acids

During the last decade, efforts have been focused on the use of algal oils containing long-chain polyunsaturated fatty acids (LCPU-FAs) as nutraceuticals. Among, these, the omega-3 LCPUFAs: DHA and EPA are among the most prominent (Kirk and Behrens, 1999). Docosahexaenoic acid is an omega-3 LCPUFA with 22 carbon atoms and six methylene-interrupted *cis*-double bonds (22:6) (Figure 4). It is a dominant FA in neurological tissue, constituting 20—25% of the total FAs in the gray matter of the human brain and 50—60% in retina rod outer segments (Kirk and Behrens, 1999). In the heart muscle tissue and sperm cells it is also abundant. Limited storage of the n-3

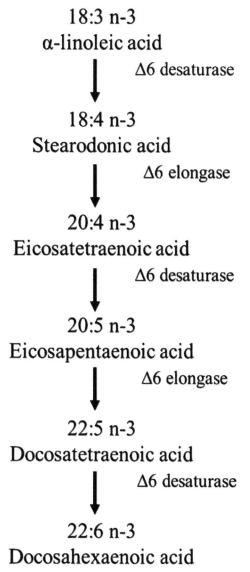

18:3 n-3
α-linoleic acid

| Δ6 desaturase

18:4 n-3
Stearodonic acid

| Δ6 elongase

20:4 n-3
Eicosatetraenoic acid

| Δ6 desaturase

20:5 n-3
Eicosapentaenoic acid

| Δ6 elongase

22:5 n-3
Docosatetraenoic acid

| Δ6 desaturase

22:6 n-3
Docosahexaenoic acid

FIGURE 4   Scheme for Δ6-pathway for biosynthesis of EPA and DHA polyunsaturated fatty acids. *From Adarme-Vega et al. (2012).*

fatty acids, such as DHA in adipose tissue of mammals suggests that an exogenous dietary source is needed (Arterburn, et al., 2006). Thus, adequate supplies of DHA must be obtained from dietary sources. Fish and

fish oils have long been recognized as the common sources of LCPUFAs, because of their enriched DHA and EPA content. Microalgae is considered the alternative of these FAs, because like humans, fish receive much of their LCPUFAs from dietary sources, which in this case are the primary producers in the oceanic environment (Robles-Medina et al., 1998). Actually, the main advantage of microalgae over fish oils is the lack of unpleasant odor. These FAs with methylene-interrupted double bonds in their structure are essential for normal cell function. The elucidation of the biological role of these FAs in obesity and cardiovascular diseases allowed their inclusion as nutraceutical (Cardozo et al., 2007; Sayanova and Napier, 2004). In humans and animals, the adequate dietary supply of DHA is an important issue because its precursor linolenic acid is very poor (Qiu, 2003).

Docosahexaenoic acid is also considered an essential nutrient during infancy, and is the natural nutritional source for a human infantits mother's breast milk. Several algae have been proposed for production of EPA, mainly diatoms like *Nitzschia* sp., *Navicula* sp., and *Phaeodactylum*. In addition, EPA is an LCPUFA, but with 20 carbons and five double bonds (20:5). One of the main biological roles of PUFAs is in cellular and tissue metabolism, as well as the regulation of membrane fluidity, electron and oxygen transport, as well as thermal adaptation (Funk, 2001). Particularly, the PUFA family (ω-3), named eicosapentaenoic acid (EPA, C20:5Δ5,8,11,14,17, 20:5 ω-3), is one of the most studied and considered effective FA. The five double bonds of the carboxy [Δ] terminus of the molecule, may exhibit also the condition with the last double bond at the third carbon from the methyl [ω] terminus (Petrie et al., 2010). EPA forms complex lipid molecules and is an important precursor of a group of eicosanoids, hormone-like substances such as prostaglandins, thromboxanes, and leucotrienes that are crucial in regulating developmental and regulatory physiology (Wen and Chen, 2003). EPA is common in a wide variety of marine microalgal classes, particularly Diatoms and some Chlorophyte are valuable nutraceuticals, specifically omega-3 FAs. These FAs are very important in products directed at maintaining heart health. FA purified can reach an approximate value of around US$260/L. As noted earlier, omega-3 oils, which are rich in long-chain ω3 PUFA, are another important group of bioactive compounds, mainly from marine sources. During the last couple of decades, the health benefits from seafoods rich in ω3 have been recognized as one of the most promising developments in human nutrition and disease prevention research (Shahidi 2009). Probably, long-chain ω3 PUFAs are the most used nutraceutical because of their effectiveness in prevention and treatment of coronary heart disease (Schmidt and Skou, 2000), hypertension (Howe, 2006), diabetes (Krishna Mohan and Das, 2001), arthritis and other inflammations (Babcock and Helton, 2000), autoimmune disorders, and cancers. Recent studies have pointed out the importance of FA in the maintenance and development of normal growth (Anderson et al., 1990). Moreover, the frequent consumption of ω3 PUFA can lower the rate of incidence and death from cardiovascular disease including ischemic heart disease, nonischemic myocardial heart disease, and hypertension (Arterburn et al., 2006).

4. NUTRACEUTICALS AND AQUACULTURE

In aquaculture activities and industries, microalgae have gained huge interest as a source of biomolecules and biomass for feed purposes. Microalgae as nutraceutical, presently, is having a great acceptation in the shellfish aquaculture industry, since marine invertebrates have rudimentary immune systems. Hence, in order to enhance the equivalent of defense mechanisms in these organisms, nutraceuticals must supply the required molecules having immunoenhancer properties. Moreover, the need for nutritional sources safer than commercial and traditional animal products has developed the interest particularly in microalgae. The nutritional composition supplied by nutraceuticals, generally is critical for the proper development or enhanced properties of organisms, because aquatic organisms when cultured in hatcheries or under controlled conditions generally are lacking in these nutrients. During the last decade, the supplementation of new immunostimulants in aquafeed has increased the possibility of safer production through aquaculture as well as for increasing yield of shellfish flesh (Robinson et al., 2005; Molina et al., 2003). Moreover, in the aquaculture feed, microalgae is a rich source of a number of biomolecules including chlorophyll, astaxanthin, lutein, beta-carotene, phycobiliprotein, and PUFAs, which can be considered nutraceutical compounds with relevant commercial importance. Most of the shellfish organisms are unable to synthesis carotenoids de novo, in such situation the exogenous diet is the exclusive source (Linan et al., 2002; Harrison, 1990). Hence, the actual commercial utilization of microalgae relies in the utilization of nutraceuticals and nutritional supplements, antioxidants, food coloring, immunoenhancers, and PUFAs (Spolaore et al., 2006). Microalgal species rich in long-chain PUFAs, mainly EPA (20:5ω3) and DHA (22:6ω3) could be considerd nutraceuticals. Salmonids, shrimp, lobster, seabream, goldfish, and koi carp under intensive rearing conditions are fed with carotenoid pigments in their diet, to attain their characteristic muscle color. Moreover, pigmenting effects, carotenoids, namely astaxanthin and canthaxanthin, exert benefits on animal health and welfare, promote larval development and provide growth and performance stimulatory effects in farmed fish and shrimp (Gouveia et al., 1996; Baker and

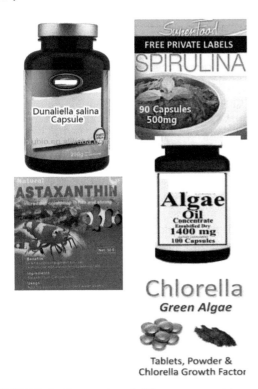

FIGURE 5   Presentation of different commercial nutraceuticals from microalgal origin.

shape according to the mouth of organisms; (2) high content of bioactive compounds, nutritional value, and a digestible cell wall; (3) high growth rates; (4) easy to cultivate; (5) non-toxic; (6) few and weak allergenic properties and side effects (Raja et al., 2004; Patil et al., 2007; Hemaiswarya et al., 2011). Table 2 shows different commercial applications of FAs from microalgae.

5. CONCLUSIONS AND RESEARCH NEEDS

In spite of the useful properties of microalgae as a natural material to be used as nutraceutical for human and animal applications, this area is just beginning. Still there are many considerations concerning the physiological and therapeutic doses to be specified to consumers, mainly on aspects of safety, allergenics, and side effects. Labels should play an important role in the specification of the amounts of all ingredients contained in the product. In USA FDA regulates nutraceuticals under the authority of the Federal Food, Drug, and Cosmetic Act, even though law does not specifically define them. In most of the cases, the use of a nutraceutical relies on the responsibility of the consumer, which needs to change in order to make safer its consumption. The role of

Gunther, 2004). Figure 5 shows the different commercial presentations of nutraceutical products from microalgae. In aquaculture, the following criteria are recommended for species selection as nutraceuticals: (1) cell size and

TABLE 2   Representative Fatty Acids from Microalgae, Uses and Potential Applications

PUFA	Structure	Potential application	Microorganism producer
g-Linolenic acid	18:3 w6, 9, 12	Infant formulas, NS	Arthrospira sp.
Arachidonic acid	20:4 w6, 12, 15	Infant formulas, NS	Phorphyridium sp.
Eicosapentaenoic acid	20:5 w3, 6, 9, 12, 15	Aquaculture, NS	Nannochloropsis, Phaedactylum
Docosahexaenoic acid	22:6 w3, 6, 9, 12, 15, 18	Infant formula NS, aquaculture	Crypthecodinium, Schizochytrium

NS, Nutritional supplements.

microalgae nutraceuticals in human and animal nutrition will continually increase aiming to define exact doses, allergenic, and side effects. Moreover, the need for major bioprospection programs for new microalgal species is evident, because of the small number of algal species used as nutraceutical for human and aquaculture. This task shall incorporate new microalgal species for specific and multipurpose tasks in the coming years. Strains with properties for the overproduction of active compounds in low cost media and versatile growth conditions need to be encouraged considering the recent advances in genomics, metabolic engineering, high throughput screening and derivative chemistry.

References

Adarme-Vega, C.T., Lim, D.K., Timmins, M., Vernen, F., Li, Y., Schenk, P.M., 2012. Microalgal biofactories: a promising approach towards sustainable omega-3 fatty acid production. Microbial Cell factories 11, 1–10.

Ambati, R.R., Phang, S.M., Ravi, S., Aswathanarayana, R.G., 2014. Astaxanthin: sources, extraction, stability, biological activities and its commercial applications. Mar Drugs 12, 128–152.

Anderson, G.J., Connor, W.E., Corliss, J.D., 1990. Docosahexaenoic acid is the preferred dietary n-3 fatty acid for the development of the brain and retina. Pediatr. Res. 27, 89–97.

Aragão, C., Conceição, L.E.C., Dinis, M.T., Fyhn, H.-J., 2004. Amino acid pools of rotifers and Artemia under different conditions: nutritional implications for fish larvae. Aquaculture 234, 429–445.

Archer, D.L., Glinsmann, W.H., 1985. Intestinal infection and malnutrition initiate acquired immune deficiency syndrome (AIDS). Nutr. Res. 5, 9–19.

Arterburn, L.M., Hall, E.B., Oken, H., 2006. Distribution, interconversion, and dose response of n-3 fatty acids in humans. Am. J. Clin. Nutr. 83, 1467–1476.

Babcock, T., Helton, W.S., 2000. Eicosapentaenoic acid (EPA): an antiinflammatory omega-3 fat with potential clinical applications. Nutrition 16, 1116–1118.

Baker, R., Gunther, C., 2004. The role of carotenoids in consumer choice and the likely benefits from their inclusion into products for human consumption. Trends Food Sci. Technol. 15, 484–488.

Batista, A.P., Gouveia, L., Bandarra, N.M., Franco, J.M., Raymundo, A., 2013. Comparison of microalgal biomass profiles as novel functional ingredient for food products. Algal Res. 2, 164–173.

Becker, W., 2004. Microalgae for aquaculture. The nutritional value of microalgae for aquaculture. In: Richmond, A. (Ed.), Handbook of Microalgal Culture. Blackwell, Oxford, pp. 380–391.

Benedetti, S., Benvenuti, F., Pagliarani, S., Francogli, S., Scoglio, S., Canestrari, F., 2004. Antioxidant properties of a novel phycocyanin extract from the blue-green alga Aphanizomenon flos-aquae. Life Sci. 75, 2353–2362.

Bishop, W.M., Zubeck, H.M., 2012. Evaluation of microalgae for use as nutraceuticals and nutritional supplements. J. Nutr. Food Sci. 2, 5–10.

Borowitzka, M.A., 1991. Microalgae for aquaculture: opportunities and constraints. J. Appl. Phycol. 9, 393–401.

de Caire, G.Z., de Cano, M.S., de Mule, C.Z., Steyerthal, N., Piantanida, M., 1995. Effect of Spirulina platensis on glucose, uric acid and cholesterol levels in the blood of rodents. Intern. J. Exp. Bot. 57, 93–96.

Cardozo, K.H.M., Guaratini, T., Barros, M., Falcão, V., Tonon, A.P., Lopes, N.P., Campos, S., Torres, M.A., Souza, A.O., Pinto, E., 2007. Metabolites from algae with economical impact. Comp. Biochem. Physiol. Part C 146, 60–78.

Zittelli, C.G., Lavista, F., Bastianini, A., Rodolfi, L., Vincenzini, M., Tredici, M.R., 1999. Production of eicosapentaenoic acid by nannochloropsis sp. cultures in outdoor tubular photobioreactors. J. Biotechnol. 70, 299–312.

Chuntapa, D., Powtongsook, S., Menasveta, P., 2003. Water quality control using Spirulina platensis in shrimp culture tanks. Aquaculture 220, 355–366.

Desmorieux, H., Decaen, N., 2005. Convective drying of Spirulina in thin layer. J. Food Eng. 66, 497–503.

De Felice, S., 1989. The Nutraceutical Revolution: Fueling a Powerful, New International Market. www.fimdefelice.org.

Funk, C.D., 2001. Prostaglandins and leukotrienes: advances in eicosanoids biology. Science 294, 1871–1875.

Gellenbeck, K.W., 2012. Utilization of algal materials for nutraceutical and cosmeceutical aplications—what do manufacturers need to know? J. Appl. Phycology 24, 309–313.

Gershwin, M.E., Belay, A., 2008. Spirulina in Human Nutrition and Health. CRC Press, Boca Raton, FL.

González-Sarrías, A., Larrosa, M., García-Conesa, M.T., Tomás-Barberán, T.M., Espín, J.C., 2013. Nutraceuticals for older people: facts, fictions and gaps in knowledge. Maturitas 75, 313–334.

Gouveia, L., Veloso, V., Reis, A., Fernandes, H., Novais, J., Empis, J., 1996. Evolution of pigment composition in Chlorella vulgaris. Bioresour. Technol. 57, 157–159.

Gouveia, L., Batista, A.P., Miranda, A., Empis, J., Raymundo, A., 2007. *Chlorella vulgaris* biomass used as colouring source in traditional butter cookies. Innovative Food Sci. Emerging Technologies 8, 433−436.

Gradelet, S., Le Bon, A.M., Berges, R., Suchelet, M., Astorg, P., 1998. Dietary carotenoids inhibit aflotoxin B1-induced liver preneoplastic foci and DNA damage on the rats: role of the modulation of aflatoxin B1 metabolism. Carcinogenesis 19, 403−411.

Guerin, M., Huntley, M.E., Olaizola, M., 2003. Haematococcus astaxanthin: applications for human health and nutrition. Trends Biotechnol. 21, 210−216.

Harrison, K.E., 1990. The role of nutrition in maturation, reproduction and embryonic development of decapod crustaceans: a review. J. Shellfish Res. 9, 1−28.

Hemaiswarya, S., Raja, R., Ravi Kumar, R., Ganesan, V., Anbazhagan, C., 2011. Microalgae: a sustainable feed source for aquaculture World. J. Microbiol. Biotechnol. 27, 1737−1746.

Howe, P.C., 2006. Dietary fats and hypertension focus on fish oil. Ann. New York Acad. Sci. 827, 339−352.

Irianto, A., Austin, B., 2002. Probiotics in aquaculture. J. Fish Dis. 25, 633−642.

Iwata, K., Inayama, T., Kato, T., 1990. Effects of *Spirulina platensis* on plasma lipoprotein lipase activity in fructose-induced hyperlipidemic rats. J. Nutr. Sci. Vitaminol (Tokyo) 36, 65−171.

Jiang, Y., Chen, F., Liang, S.-Z., 1999. Production potential of docosahexaenoic acid by the heterotrophic marine dinoflagellate *Crypthecodinium cohnii*. Process. Biochem. 34, 633−637.

Kim, S.K., 2013. Marine Nutraceuticals: Prospects and Perspectives, 2013. CRC Press, Boca Raton, FL.

Kim, S.H., Jean, D.I., Lim, Y.P., An, G., 2009. Weight gain limitations and liver protection by long term feeding of astaxanthin in murines. J. Korean Soc. Appl. Biol. Chem. 52, 180−185.

Kirk, E.A., Behrens, P.W., 1999. Commercial developments in microalgal biotechnology. J. Phycol 35, 215−226.

Kroes, R., Schaefer, E.J., Squire, R.A., Williams, G.M., 2003. A review of the safety of DHA45-oil. Food Chem. Toxicol. 41, 1433−1446.

Li, J., Zhu, D.J., Niu, J., Shen, S., Guangce, Wang., 2011. An economic assessment of astaxanthin production by large scale cultivation of *Haematococcus pluvialis*. Biotechnol. Adv. 29, 568−574.

Linan-Cabello, M.A., Paniagua-Michel, J., Hopkins, P.M., 2002. Bioactive roles of carotenoids and retinoids in crustaceans. Aquacult. Nutr. 8, 299−309.

Lio-Po, G.D., Leaño, E.M., Peñaranda, M.M.D., Villa-Franco, A.U., Sombito, C.D., Guanzon, N.G., 2005. Antiluminous Vibrio factors associated with the 'green water' growout culture of the tiger shrimp Penaeus monodon. Aquaculture 250, 1−7.

Lorenz, T.R., Cysewski, G.R., 2000. Commercial potential for Haematococcus microalgae as a natural source of astaxanthin. Trends Biotechnol. 18, 160−167.

Mishima, T., Murata, J., Toyoshima, M., Fujii, H., Nakajima, M., 1998. Inhibition of tumor invasion and metastasis by calcium spirulan (Ca-SP), a novel sulfated polysaccharide derived from a blue-green alga, *Spirulina platensis*. Clin. Exp. Metastasis 16, 541−550.

Mohan, K., Das, U.N., 2001. Prevention of chemically induced diabetes mellitus in experimental animals by polyunsaturated fatty acids. Nutrition 17 (2), 126−151.

Molina Grima, E., Belarbi, E.H., Acién Fernández, F.G., Robles Medina, A., Chisti, Y., 2003. Recovery of microalgal biomass and metabolites: process options and economics. Biotechnol. Adv. 20, 491−515.

Morita, K., Matsueda, T., Iida, T., Hasegawa, T., 1999. *Chlorella* accelerates dioxin excretion in rats. J. Nutr. 129, 1731−1736.

Muller-Feuga, A., 2000. The role of microalgae in aquaculture: situation and trends. J. Appl. Phycol. 12, 527−534.

Muller-Feuga, A., 2004. Microalgae for aquaculture. The current global situation and future trends. In: Richmond (Ed.), Handbook of Microalgal Culture. Blackwell, Oxford, pp. 352−364.

Paniagua-Michel, J., Dujardin, E., Sironval, C., 1992. Le Tecuitlatl, concentre de Spirulines des Aztecs. Bull Acad. Roy Sci. Belgium, pp. 10, 253−263.

Patil, V., Kallqvist, T., Olsen, E., Vogt, G., Gislerød, H.R., 2007. Fatty acid composition of 12 microalgae for possible use in aquaculture feed. Aquacul Int. 15, 1−9.

Petrie, J.R., Shrestha, P., Mansour, M.P., Nichols, P.D., Liu, Q., Singh, S.P., 2010. Metabolic engineering of omega-3 long-chain polyunsaturated fatty acids in plants using an acyl-CoA [delta] 6-desaturase with 33 3-preference from the marine microalga micromonas pusilla. Metab. Eng. 12 (3), 233−240.

Plaza, M., Herrero, M., Cifuentes, A., Ibáñez, E., 2009. Innovative natural functional ingredients from microalgae. Agric. Food Chem. 57 (16), 7159−7170.

Qiu, X., 2003. Biosynthesis of docohexanoic acid (DHA): two distinct pathways. Prostaglandisn, leukotrienes and essential fatty acids 68, 181−186.

Raja, R., Anbazhagan, C., Lakshmi, D., Rengasamy, R., 2004. Nutritional studies on *Dunaliella salina* (Volvocales, Chlorophyta) under laboratory conditions. Seaweed Res. Utili 26, 127−146.

Raja, R., Hemaiswarya, S., Rengasamy, R., 2007. Exploitation of *Dunaliella* for beta-carotene production. Appl. Microbiol. Biotechnol. 74, 517−523.

Robinson, C.B., Samocha, T.M., Fox, J.M., Gandy, R.L., McKee, D.A., 2005. The use of inert artificial commercial food sources as replacements of traditional live food items in the culture of larval shrimp, *Farfantepenaeus aztecus*. Aquaculture 245, 135−147.

Robles Medina, A., Molina Grima, E., Giménez Giménez, A., Ibáñez González, M.J., 1998. Downstream processing of algal polyunsaturated fatty acids. Biotechnol. Adv. 16, 517–580.

Rodolfi, L., Chini Zittelli, G., Barsanti, L., Rosati, G., Tredici, M.R., 2003. Growth medium recycling in *Nannochloropsis* sp. mass cultivation. Biomol. Eng. 20, 243–248.

Romay, C., Armesto, J., Remirez, D., González, R., Ledon, N., et al., 1998. Antioxidant and anti-inflammatory properties of C-phycocyanin from blue-green algae. Inflamm. Res. 47, 36–41.

Sayanova, O.V., Napier, J.A., 2004. Eicosapentaenoic acid: biosynthetic routes and the potential for synthesis in transgenic plants. Phytochemistry 65, 147–158.

Schmidt, E.B., Skou, H.A., 2000. Christensen JH, Dyerberg. N-3 fatty acids from fish and coronary artery disease: implications for public health. J. Public Health Nutr. 3 (1), 91–98.

Shahidi, F., 2009. Nutraceuticals and functional foods: whole versus processed foods. Trends Food Sci. Technol. 20, 376–387.

Shibata, S., Natori, Y., Nishihara, T., Tomisaka, K., Matsumoto, K., 2003. Antioxidant and anti-cataract effects of *Chlorella* on rats with streptozotocin induced diabetes. J. Nutr. Sci. Vitaminol (Tokyo) 49, 334–339.

Shim, J.Y., Shin, H.S., Han, J.G., Park, H.S., Lim, B.L., 2008. Protective effects of *Chlorella vulgaris* on liver toxicity in cadmium-administered rats. J. Med. Food 11, 479–485.

Simoons, F.J., 1991. Food in China: A Cultural and Historical Inquiry. CRC Press, Boca Raton, FL, USA.

Spolaore, P., Cassan, C.J., Duran, E., Isambert, A., 2006. Commercial applications of microalgae. J. Biosci. Bioeng. 101, 87–96.

Waldenstedt, L., Inborr, J., Hansson, I., Elwinger, K., 2003. Effects of astaxanthin-rich algal meal (*Haematococcus pluvalis*) on growth performance, caecal campylobacter and clostridial counts and tissue astaxanthin concentration of broiler chickens. Anim. Feed Sci. Technol. 108, 119–132.

Wen, Z.Y., Chen, F., 2003. Heterotrophic production of eicosapentaenoic acid by microalgae. Biotechnol. Adv. 21, 273–294.

Wildman, R.E.C., Taylor, C., 2006. Wallace Handbook of Nutraceuticals and Functional Foods, Second ed. CRC Press, Boca Raton, FL.

Yuan, J.P., Peng, J., Yin, K., Wang, J.H., 2011. Potential health promoting effects of astaxanthin: a high-value carotenoid mostly from microalgae. Mol. Nutr. Food Res. 55, 150–165.

Microalgae as a Novel Source of Antioxidants for Nutritional Applications

Koen Goiris[1], Koenraad Muylaert[2], Luc De Cooman[1]

[1]KU Leuven Technology Campus Ghent, Laboratory of Enzyme, Fermentation and Brewing Technology, Ghent, Belgium; [2]KU Leuven Kulak, Research Unit Aquatic Biology, Kortrijk, Belgium

1. OXIDATIVE STABILITY OF FOOD SYSTEMS

1.1 Lipid Oxidation

Along with microbial stability, oxidative stability is a determining factor in the overall shelf life of foodstuffs. In most cases, lipid oxidation determines the oxidative stability of foods, although other food components are also subject to oxidative transformations. Lipid oxidation is crucial for food stability as highly flavor-active aldehydes are formed, resulting in flavor deterioration. The most important mechanisms leading to lipid oxidation are spontaneous oxidation of lipids (auto-oxidation), light-induced oxidation (photo-oxidation), and lipoxygenase-mediated oxidation. The different stages in the process of auto-oxidation, which is a free-radical reaction, are given in Reactions (1)–(9):

Initiation:

$$LH + X^{\bullet} \rightarrow L^{\bullet} \qquad (1)$$

Propagation:

$$L^{\bullet} + O_2 \rightarrow LO_2^{\bullet} \qquad (2)$$

$$LO_2^{\bullet} + LH \rightarrow LOOH + L^{\bullet} \qquad (3)$$

Termination:

$$LO_2^{\bullet} + LO_2^{\bullet} \rightarrow LOOL + O_2 \qquad (4)$$

$$LO_2^{\bullet} + L^{\bullet} \rightarrow LOOL \qquad (5)$$

$$L^{\bullet} + L^{\bullet} \rightarrow LL \qquad (6)$$

Alternative Initiation:

$$LOOH \rightarrow LO^{\bullet} + HO^{\bullet} \qquad (7)$$

$$2\,LOOH \rightarrow LO^{\bullet} + LO_2^{\bullet} \qquad (8)$$

$$M^{n+} + LOOH \rightarrow LO^{\bullet} + HO^{-} + M^{(n+1)+} \qquad (9)$$

In the first step in Reaction (1), the lipid radical $L^{\bullet}$ is generated through the intervention of an initiator molecule that induces the abstraction of an α-methylenic hydrogen atom from the lipid LH. Once the free radical $L^{\bullet}$ is generated, Reactions (2) and (3) lead to a chain reaction; the lipid peroxide radical, which is formed in the fast reaction (2), attacks lipid molecules with formation of lipid hydroperoxides (3) and another lipid radical $L^{\bullet}$. The function of chain-breaking antioxidants (ArOH) is the reaction with lipid peroxide, which halts the chain reaction, as given in Reaction (10):

$$LO_2^{\bullet} + ArOH \rightarrow LOOH + ArO^{\bullet} \qquad (10)$$

Effective antioxidants must react slowly with the substrate LH but rapidly with the $LO_2^{•}$ and, secondly, the formed $ArO^{•}$ radical must be relatively stable. The weaker the OH bond in ArOH, the more effective the antioxidant will be in donating an H-atom. Well-known chain-breaking antioxidants that inhibit lipid peroxidation are tocopherols and phenolics (Wright et al., 2001).

There are two main pathways by which nonenzymatic antioxidants can deactivate radicals and prevent oxidative damage, namely hydrogen atom transfer (HAT) and single electron transfer (SET), which often occur in parallel. The dominant mechanism is predetermined by the properties of the antioxidant as well as the reaction environment—that is, solubility of the antioxidant and solvent used (Prior et al., 2005). The general reaction of the HAT mechanism, with AH being the hydrogen-donating antioxidant, is given by Reaction (11):

$$AH + X^{•} \rightarrow A^{•} + XH \qquad (11)$$

The second mechanism by which an antioxidant can deactivate radicals is single electron transfer, given by Reactions (12)–(14):

$$X^{•} + AH \rightarrow X^{-} + AH^{•+} \qquad (12)$$

$$AH^{•+} \overset{H_2O}{\rightleftharpoons} A^{•} + H_3O^{+} \qquad (13)$$

$$X^{-} + H_3O^{+} \rightarrow XH + H_2O \qquad (14)$$

Considering Reaction (13), it is obvious that the overall electron transfer mechanism is pH dependent and the reactivity of the antioxidant increases at higher pH values.

1.2 Commercial Antioxidants: Synthetic and Natural Alternatives

To reduce the oxidative adulteration of food or bulk oils, many synthetic and natural antioxidants are added during food processing. Synthetic antioxidants, such as butylated hydroxytoluene (BHT), butylated

hydroxyanisol (BHA), and t-butyl hydroquinone, are restricted in their applications and levels (<0.02% of lipid content) because their toxicological safety has been debated (Namiki, 1990). Therefore, they are replaced with natural antioxidants where possible. Examples of natural antioxidants are tocopherols, polyphenols, and carotenoids. Well-known sources of natural sources of food-grade antioxidants are rosemary (rosmarinic acid), tea (catechins), and grape (flavonoids).

2. ANTIOXIDANT ACTIVITY IN MICROALGAE

2.1 Reactive Oxygen Species: Formation and Physiological Role

The most important pro-oxidants in biological systems are reactive oxygen species (ROS), which are formed in varying physiological processes. In microalgae, ROS are continuously produced in chloroplasts, mitochondria, and peroxisomes (Figure 1). To avoid damage to cell components, production and scavenging of ROS must be strictly balanced; hence, antioxidant protective mechanisms must be in place.

2.1.1 Reactive Oxygen Species

In its ground state, molecular oxygen or triplet oxygen can be considered a biradical because it contains two unpaired electrons in parallel spin. Ground-state oxygen can be converted to much more reactive ROS forms by energy transfer or by electron transfer reactions, leading to radical ROS and nonradical ROS (Figure 2).

2.1.2 Formation Sites of Reactive Oxygen Species

In photosynthetic organisms, including microalgae, ROS are continuously produced as byproducts from various metabolic pathways (Apel and Hirt, 2004). Under light stress, excited

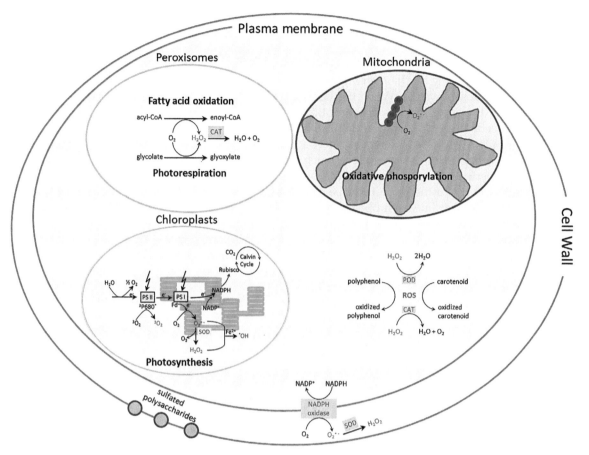

FIGURE 1 Cellular pathways of reactive oxygen species (ROS) in microalgae. *Based on Cirulis et al. (2013) and Laloi et al. (2006).*

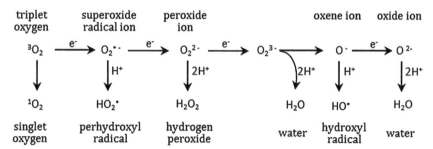

FIGURE 2 Generation of different ROS by energy transfer (production of singlet oxygen) or sequential univalent reduction of ground-state triplet oxygen. *Based on Apel and Hirt (2004).*

triplet chlorophyll from the photosystem II reaction center in the chloroplasts may transfer its excitation energy onto triplet ground-state oxygen, yielding the highly reactive singlet oxygen. A second ROS-generating mechanism originates in the electron transfer system. When the light-driven electron transport exceeds consumption of electrons needed for CO_2 fixation or NADP supply is limited, molecular oxygen can be reduced by photosystem I to superoxide, which is rapidly converted to hydrogen peroxide by superoxide dismutase. Furthermore, in the peroxisomes, recycling of glycolate from photorespiration and fatty acid oxidation are both processes that produce hydrogen peroxide, which in its turn is mitigated by catalase. A third endogenous ROS source is the generation of superoxide by ubiquinone in the electron transport chain during oxidative phosphorylation in the mitochondria. However, mitochondrial ROS production is much lower than the production in chloroplasts. A last source of superoxide is the activity of NADPH oxidase in the plasma membrane.

2.2 ROS Detoxification in Microalgae

To counteract the detrimental effects of ROS, all living organisms have several defensive systems at their disposal, both enzymatic and nonenzymatic. In this section, the most important ROS-associated enzymes and antioxidants found in microalgae are discussed.

2.2.1 Enzymatic Antioxidant Protection

Superoxide dismutase (EC 1.15.1.1) consists of a mixture of metalloproteins differentiated by their metal cofactor. The three isoforms common to plants (CuZn-SOD, Fe-SOD, and Mn-SOD) are also present in microalgae (Janknegt et al., 2009). Superoxide dismutase catalyzes the neutralization of superoxide radicals with the formation of hydrogen peroxide and oxygen.

Catalase (EC 1.11.1.6) catalyzes the conversion of hydrogen peroxide to water and oxygen. The catalase enzyme is sensitive to light, which

may affect the ability of photoautotrophic organisms to tolerate oxidative stress when exposed to high light irradiance.

Another group of enzymes that catalyze reduction of hydrogen peroxide to water are *peroxidases*. They differ from catalase in their requirement of an electron donor that subsequently becomes oxidized. *Ascorbate peroxidase* (EC 1.11.1.11), present in the stroma and thylakoids of chloroplasts, has a significantly lower K_m value for hydrogen peroxide than catalase and uses vitamin C as a specific electron donor (Asada and Akahashi, 1987). *Glutathione peroxidase* (EC 1.11.1.9) requires glutathione for the removal of hydrogen peroxide. Several isoforms have been detected in the microalgae *Chlamydomonas reinhardtii* (Dayer et al., 2008) and *Chlorella* sp. (Wang and Xu, 2012).

2.2.2 Nonenzymatic Antioxidant Protection

L-*Ascorbic acid* or vitamin C, abundant in photosynthetic organisms, can reduce many ROS. Vitamin C is present in both cytosol and chloroplast where it takes part in the ascorbate–glutathione cycle to remove hydrogen peroxide (Mallick and Mohn, 2000). Next to its vital role in the elimination of hydrogen peroxide, vitamin C also scavenges superoxide, hydroxyl radicals, and lipid hydroperoxides (Lesser, 2006). In the chloroplast, vitamin C plays a crucial role in the regeneration of membrane-bound carotenoids and tocopherols, thereby protecting the photosynthetic apparatus (Mallick and Mohn, 2000). High levels of vitamin C have been reported in *Chlorella* sp. (Running et al., 2002), *Dunaliella* sp. (Barbosa et al., 2005), *Chaetoceros calcitrans*, and *Skeletonema costatum* (Brown et al., 1998, 1999). Production of vitamin C is stimulated by high light exposure (Barbosa et al., 2005), allelochemicals such as ethyl 2-methyl acetoacetate (Yang et al., 2011), or ultraviolet stress (Abd El-baky et al., 2004).

Glutathione (GSH), a tripeptide (Glu-Cys-Gly) found in animals and photosynthetic organisms, acts as an antioxidant in many ways including reaction with superoxide, singlet oxygen, and hydroxyl radicals. Glutathione also acts as a chain-breaker of free-radical reactions and is crucial in the regeneration of ascorbate (Lesser, 2006). As a substrate for glutathione peroxidase, it donates the electrons necessary for decomposition of hydrogen peroxide (Kohen and Nyska, 2002).

Tocopherols are located in the lipid bilayers of cell membranes. The most widespread homologs in nature are four tocopherols and four tocotrienols: α, β, γ, and δ -tocopherol as well as α, β, γ, and δ -tocotrienol (Colombo, 2010). Tocopherols and tocotrienols have the same basic chemical structure characterized by a long chain attached at the 2-position of a chromane ring. However, tocotrienols contain three conjugated double bonds in the aliphatic side chain instead of the saturated C16 side chain found in tocopherols (Figure 3).

The most active antioxidant form is α-tocopherol, which is only synthesized in the chloroplasts of photosynthetic organisms. α-Tocopherol acts as an antioxidant through its ability to quench both singlet oxygen and (lipid) peroxides (Lesser, 2006; Mallick and Mohn, 2000). Although α-tocopherol is located in the membranes and the hydrophilic vitamin C is located in the liquid phase, vitamin C is able to reduce the tocopheroxyl radical, thereby recycling the active form of tocopherol in the chloroplast (Buettner, 1993; Niki, 1991).

α-Tocopherol is produced in high amounts by *Dunaliella tertiolecta* and *Tetraselmis suecica*. Carballo-Cardenas et al. (2003) demonstrated that production of α-tocopherol is highly variable throughout the growth cycle and that nutrient composition can be used to control its production in both species. In *Dunaliella salina*, α-tocopherol production is stimulated by UV-stress, nitrogen limitation, and salt stress (Abd El-baky et al., 2004). Another study showed that decreasing N-concentrations in the growth medium leads to increased α-tocopherol accumulation in *Nannochloropsis oculata*, but growth rate is reduced under these conditions (Durmaz, 2007).

The lipophilic *carotenoids* are produced de novo by photoautotrophs. In photosynthetic organisms, carotenoids are present in the pigment–protein complexes of the thylakoid membranes of chloroplasts, where they fulfill a dual function (Cogdell et al., 1994). Some carotenoids (especially keto-carotenoids such as fucoxanthin, Figure 4) act as accessory light-harvesting pigments by

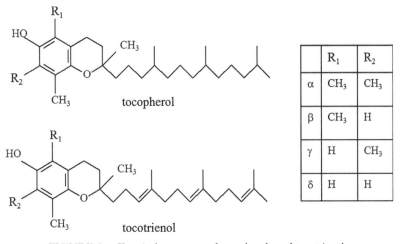

FIGURE 3   Chemical structures of tocopherols and tocotrienols.

transferring light energy of wavelengths that cannot be captured by chlorophylls (Takaichi, 2011). Next, carotenoids also have a protective function by dissipating excess energy and by quenching ROS that are produced during photosynthesis (Goss and Jakob, 2010). The xanthophyll cycle, that is, the cyclical interconversion of violaxanthin, antheraxanthin, and zeaxanthin in chlorophytes, provides zeaxanthin needed for dissipation of excess energy from excited chlorophylls in photosynthetic organisms. In diatoms and dinoflagellates, an alternative xanthophyll cycle exists where diadinoxanthin is converted into diatoxanthin (diatoms) or dinoxanthin (dinoflagellates).

Another mode of photoprotection by carotenoids is the quenching of excited triplet-state chlorophyll and singlet oxygen by β-carotene. The main mechanism in carotenoid photoprotection against singlet oxygen functions through electronic energy transfer as given by reaction (15) (Edge et al., 1997); however, chemical quenching with formation of carotenoid epoxides also occurs (Liebler, 1993).

$$^1O_2{}^* + CAR \rightarrow {}^3O_2 + {}^3CAR^* \qquad (15)$$

During physical quenching (reaction (15)), the carotenoid triplet state is produced through energy transfer. This excited state can return to the ground state by dissipating energy as heat or by translocation over the conjugated double bond system. Therefore, the ability to quench singlet oxygen increases with longer chain lengths of the conjugated system (Edge et al., 1997). Next to the ability of carotenoids to quench singlet oxygen, they can also react with free radicals. Carotenoids can react with peroxyl radicals and are involved in recycling of phenoxyl radicals and tocopheroxyl radicals, which are formed upon reaction of phenolic antioxidants and tocopherols with peroxyl and alkoxyl radicals (Burke et al., 2001; Burton, 1989; Edge et al., 1997). However, unlike quenching of singlet oxygen, which mainly leads to

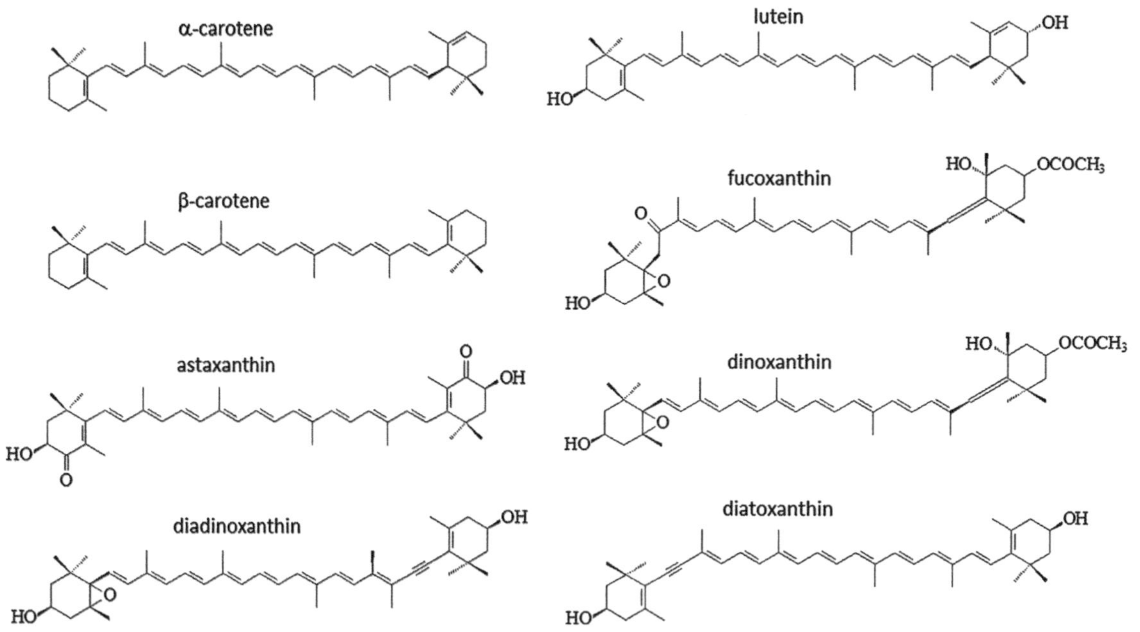

FIGURE 4   Chemical structures of some carotenoids occurring in microalgae.

energy dissipation as heat, the reactions of carotenoids (or any antioxidant) with free radicals will lead to electron transfer or addition reactions. Three reaction mechanisms describe the reaction of free radicals with carotenoids, that is, electron transfer, hydrogen atom transfer, and radical addition to the carotenoid (Martínez and Barbosa, 2008; Martínez et al., 2008). In order to scavenge free radicals, carotenoids can either donate or accept unpaired electrons. Usually, antioxidant molecules become oxidized by donating electrons to the free radical. However, carotenoids can also quench free radicals by accepting an unpaired electron, rendering it harmless by translocation over the conjugated side chain. In a comparative study, it was observed that the apolar lycopene as well as the xanthophylls were the most effective carotenoids in reducing ferric ions (FRAP assay) (Müller et al., 2011). In the same study, it was further demonstrated that the group of carotenes (lycopene, α- and β-carotene) were more efficient quenchers of the ABTS$^{\cdot+}$ radical than most of the xanthophylls and that keto-carotenoids were most efficient in scavenging peroxyl radicals, due to their more pronounced conjugated double bond systems.

In microalgae, a distinction is usually made between primary and secondary carotenoids. Whereas primary carotenoids are structural and functional components of the photosynthetic apparatus, and thus essential for survival, secondary carotenoids are produced at high levels when cells are exposed to specific environmental stimuli (Jin et al., 2003). At present, carotenoids (both primary and secondary) are the most commercialized products from microalgae. One such carotenoid that can be sourced from microalgae is lutein. Lutein is also found in the human retina where it acts both as an optical filter and as an antioxidant to protect long-chain polyunsaturated fatty acids and is therefore important for our eye's health (Rapp et al., 2000). In *Scenedesmus*, lutein production, as well as β-carotene production, can be stimulated

by increasing pH and temperature during cultivation (Guedes et al., 2011b). Further, Wei et al. (2008) showed that lutein content of heterotrophically grown *Chlorella protothecoides* increased in response to singlet oxygen, but was reduced when cells were exposed to hydroxyl radicals. Growth-limiting conditions, such as pH values of six or nine and a temperature of 33 °C, were found to stimulate carotenogenesis in the chlorophyte *Muriellopsis* sp. (Del Campo et al., 2000), which is currently the commercial source of lutein. Lutein content in this species is the highest in early stationary phase. Another microalgal pigment, sold as antioxidant, is the secondary carotenoid astaxanthin, produced by *Haematococcus pluvialis*. Accumulation of astaxanthin occurs when *H. pluvialis* cells are exposed to stress, induced by a combination of high light, high salt levels and nitrogen deprivation (Boussiba, 2000; Wang et al., 2003). The last microalgal pigment that is currently produced commercially for its antioxidant properties is β-carotene, extracted from the halophile *D. salina*. Carotenogenesis in this microalga is induced by Fe^{2+}, as well as by UV-stress. Also in *Chlorella vulgaris*, an increase in carotenoid content is observed upon metal exposure (Mallick, 2004). A good overview of the optimum conditions for carotenoid production is given by Guedes et al. (2011a).

A last group of antioxidant secondary metabolites are *polyphenols*, which are present, often at high levels, in virtually all plants (Pietta, 2000). Polyphenols comprise a structurally diverse group of components, including simple phenols, phenolic acids, flavonoids, tannins, and lignans. Polyphenols can inhibit lipid oxidation in different ways, that is, by directly scavenging HOCl, singlet oxygen, lipid peroxyl, superoxide and hydroxyl radicals, by metal chelation or by inhibiting lipoxygenase (Dugas et al., 2000; Rice-Evans et al., 1996; Salah et al., 1995). The general mechanism of radical scavenging by phenolics is given in Figure 5. In the first step, a hydrogen atom is donated to the radical, and an aroxyl radical is produced. This radical is

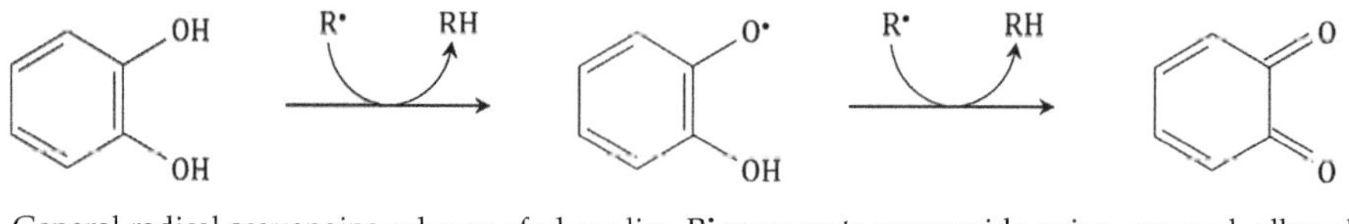

FIGURE 5   General radical scavenging scheme of phenolics. R˙ represents superoxide anion, peroxyl, alkoxyl, or hydroxyl radicals.

relatively stable through resonance and can interact with another radical yielding a quinone. This means that, in the general mechanism, one polyphenol molecule is able to quench two radical molecules. However, if high metal concentrations are present, the aroxyl radical can interact with oxygen, generating a quinone and a superoxide anion, rather than terminating the radical chain reactions. This mechanism is responsible for the undesired pro-oxidant effect of phenolics that can occur under specific conditions (Pietta, 2000).

Next to their direct radical scavenging properties, polyphenols are able to chelate metals, hereby reducing the oxidative stress as transition metals are involved in ROS generation (Pietta, 2000).

Although phenolics are well-studied antioxidant components in higher plants, the acknowledgment of their presence in microalgae is fairly recent. Li et al. (2007) and Hajimahmoodi et al. (2010) screened microalgae for polyphenol content and antioxidant activity and found large variations between samples. More recently, Goiris et al. (2012) indicated that the antioxidant potential of microalgae is not only determined by its carotenoid content but also other components, including phenolics, are important contributors to overall antioxidant activity. Next to the presence of simple phenols in microalgae (Klejdus et al., 2009; Onofrejová et al., 2010), the presence of flavonoids in microalgae has been acknowledged, albeit at low levels (Goiris et al., 2014; Klejdus et al., 2010). Microalgae can further produce some remarkable polyphenolic antioxidant molecules such as marennine in the diatom *Haslea ostrearia* (Pouvreau et al., 2008), purpurogallin in the extremophile snow algae

Mesotaenium berggrenii (Remias et al., 2012) or even BHT, the well-known food additive (E321), which was found in the chlorophyte *Botryococcus braunii* and the cyanobacteria *Cylindrospermopsis raciborskii*, *Microcystis aeruginosa*, and *Oscillatoria* sp. (Babu and Wu, 2008).

In microalgae, little is known about the response of phenolic components to environmentally induced oxidative stress. Duval et al. (2000) examined the effects of UV-exposure on antioxidant properties of *Chlamydomonas nivalis* and observed an increase in phenolics upon exposure to UV-C light. Another study on potential effects of UV on phenolic content was performed with *Scenedesmus quadricauda* (Kováčik et al., 2010). This study found no significant changes in total phenolic content when cells were exposed to elevated levels of UV-A but noticed a 50% decrease in the flavonols quercetin and kaempherol. When the cells were exposed to UV-C, these flavonols were not found in the biomass, suggesting breakdown of these components by UV-C. On the other hand, benzoic acids increased upon UV-A exposure and cinnamic acid decreased when cells were exposed to UV-A or UV-C. Others found that production of BHT (Babu and Wu, 2008) was stimulated under high light irradiation in *B. braunii*, *C. raciborskii*, *M. aeruginosa*, and *Oscillatoria* sp. Other studies described the influence of metal stress on the phenolic content in microalgae. Ulloa et al. (2012) found that phenolic content was stimulated by strontium addition to cultures of *T. suecica*. Also in *S. quadricauda* (Kováčik et al., 2010) and *Phaeodactylum tricornutum* (Rico et al., 2013), phenolic content was higher when cells were exposed to metals. A better understanding on how polyphenol concentrations change in response to oxidative stress will clarify the role of polyphenols as antioxidants in microalgae and learn how to maximize production for use in the food, feed, or chemical industry.

2.3 Other Antioxidants

Next to the antioxidant components commonly found in other plants, some microalgae produce specific types of antioxidants such as the phycobilin proteins phycocyanin in cyanobacteria (Benedetti et al., 2004; Huang et al., 2007; Thangam et al., 2013; Yoshikawa and Belay, 2008) and phycoerythrin in rhodophytes and cyanobacteria (Huang et al., 2007). Especially the cyanobacteria *Arthrospira platensis* is known to produce high amounts of phycocyanin (Oliveira et al., 2009). Also dimethylsulfide/dimethylsulfoxide (Sunda et al., 2002) and sulfated polysaccharides (Tannin-Spitz et al., 2005) contribute to the antioxidant pool of microalgae.

3. POTENTIAL OF MICROALGAL ANTIOXIDANTS TO REDUCE LIPID OXIDATION IN FOODSTUFF

3.1 Current Knowledge and Applications

Over the last decade, increasing evidence has been gathered on the potential of microalgal extracts for retarding lipid oxidation. For instance, Ranga Rao et al. (2006) studied the effect of crude acetone extracts of *B. braunii* on lipid peroxidation in model systems. The relatively high degree of inhibition of lipid peroxidation, in comparison with the synthetic antioxidant BHT, was ascribed to carotenoids and polyphenols in the extracts. Tannin-Spitz et al. (2005) studied the effect of sulfated polysaccharides from *Porphyridium cruentum* on lipid oxidation and found inhibition rates of up to 80% at a concentration of $10\,\mathrm{mg\,mL}^{-1}$. Benedetti et al. (2004) reported the use of a phycocyanin-rich extract from the cyanophyte

Aphanizomenon flos-aquae to reduce cupper-induced oxidation of plasma lipids and found that at a concentration of $1\,\mu M$ phycocyanin, oxidation of blood lipids was reduced by a factor three. Lee et al. (2010) assessed the antioxidant properties of the microalgae *Halochlorococcum porphyrae* and *Oltamannsiellopsis unicellularis* using both solvent extracts and enzymatic digests. Lipid peroxidation was strongly inhibited by all methanolic extracts, as well as the ethyl acetate fraction of *H. porphyrae* and the chloroform fraction of *O. unicellularis* which inhibited lipid peroxidation similar to α-tocopherol. In addition, some enzymatic digests exhibited effects similar to the synthetic antioxidant BHT. Several studies demonstrated the efficacy of ethanolic extracts of *Chlorella* sp. on lipid peroxidation and measured similar degrees of inhibition compared to BHT (Choochote et al., 2014; Rodriguez-Garcia and Guil-Guerrero, 2008). Another study by Natrah et al. (2007) screened 14 samples of Malaysian indigenous microalgae and identified six species, that is, *S. quadricauda*, *C. vulgaris*, *N. oculata*, *Tetraselmis tetrathele* and especially *Isochrysis galbana* and *C. calcitrans*, of which crude methanolic extracts inhibited the oxidation of linoleic acid to the same extent as the commercial antioxidants BHA and BHT. Recently, the use of whole biomass of *C. vulgaris* and *H. pluvialis* has been shown to retard lipid oxidation in bulk oils (Lee et al., 2013) as well as in food emulsions (Gouveia et al., 2005).

Although all studies mentioned above describe the activity against lipid oxidation in view of radical scavenging activity of the extracts, earlier work by Matsukawa et al. (1997) looked at the potential of microalgal extracts for inhibition of two important oxidative enzymes that are involved in oxidation of lipids and proteins, that is, lipoxygenase and tyrosinase, respectively. In this study, it was indicated that ethanol extracts contained efficient inhibitors of the lipoxygenase activity, whereas methanol extracts showed the highest tyrosinase inhibition, compared to ethanol and aqueous extracts.

3.2 Future Perspectives

Although the efficacy of microalgal antioxidants toward lipid oxidation has been proven, little data are available on how these novel antioxidant formulations perform compared to commercially available antioxidant products, both from a cost-perspective and from the antioxidant action in real food systems. Further, the active principles still need to be further characterized to allow standardization and commercialization. Thirdly, growth conditions should be optimized for maximal productivity of antioxidant components. Finally, a big hurdle that has to be taken before incorporating microalgal products in foodstuffs are the legislative constrictions concerning novel foods. Since only a few species are currently allowed for human consumption, many microalgae with high antioxidant potential still await approval before they can be used in foodstuff.

References

Abd El-baky, H.H., El-Baz, F.K., El-baroty, G.S., 2004. Production of antioxidant by the green alga *Dunaliella salina*. Int. J. Agri. Biol. 6 (1), 49–57.

Apel, K., Hirt, H., 2004. Reactive oxygen species: metabolism, oxidative stress, and signal transduction. Ann. Rev. Plant Biol. 55, 373–399.

Asada, K., Akahashi, M., 1987. Production and scavenging of active oxygen in photosynthesis. In: Kyle, D.J., Osmond, C.B., Arntzen, C.J. (Eds.), Photoinhibition. Elsevier, Amsterdam, The Netherlands, pp. 228–287.

Babu, B., Wu, J.-T., 2008. Production of natural butylated hydroxytoluene as an antioxidant by freshwater phytoplankton 1. J. Phycol. 44 (6), 1447–1454.

Barbosa, M.J., Zijffers, J.W., Nisworo, A., Vaes, W., van Schoonhoven, J., Wijffels, R.H., 2005. Optimization of biomass, vitamins, and carotenoid yield on light energy in a flat-panel reactor using the A-stat technique. Biotech. Bioeng. 89 (2), 233–242.

Benedetti, S., Benvenuti, F., Pagliarani, S., Francogli, S., Scoglio, S., Canestrari, F., 2004. Antioxidant properties of a novel phycocyanin extract from the blue-green alga *Aphanizomenon flos-aquae*. Life Sci. 75 (19), 2353–2362.

Boussiba, S., 2000. Carotenogenesis in the green alga *Haematococcus pluvialis*: cellular physiology and stress response. Phys. Plant. 108, 111–117.

Brown, M.R., Mular, M., Miller, I., Farmer, C., Trenerry, C., 1999. The vitamin content of microalgae used in aquaculture. J. App. Phycol. 11, 247–255.

Brown, M.R., Skabo, S., Wilkinson, B., 1998. The enrichment and retention of ascorbic acid in rotifers fed microalgal diets. Aquacult. Nutr. 4, 151–156.

Buettner, G.R., 1993. The pecking order of free radicals and antioxidants: lipid peroxidation, alpha-tocopherol, and ascorbate. Arch. Biochem. Biophys. 300, 535–543.

Burke, M., Edge, R., Land, E.J., Mcgarvey, D.J., Truscott, T.G., 2001. One-electron reduction potentials of dietary carotenoid radical cations in aqueous micellar environments. FEBS Lett. 500, 132–136.

Burton, G.W., 1989. Antioxidant action of carotenoids. J. Nutr. 119, 109–111.

Carballo-Cardenas, E.C., Minh Tuan, P., Janssen, M., Wijffels, R.H., 2003. Vitamin E (a-tocopherol) production by the marine microalgae *Dunaliella tertiolecta* and *Tetraselmis suecica* in batch cultivation. Biomol. Eng. 20 (4–6), 139–147.

Choochote, W., Suklampoo, L., Ochaikul, D., 2014. Evaluation of antioxidant capacities of green microalgae. J. Appl. Phycol. 26 (1), 43–48.

Cirulis, J.T., Scott, J.A., Ross, G.M., 2013. Management of oxidative stress by microalgae. Can. J. Physiol. Pharmacol. 91 (1), 15–21.

Cogdell, R.J., Gillbro, T., Andersson, P.O., Liu, R.S.H., Asato, A.E., 1994. Carotenoids as accessory light-harvesting pigments. Pure Appl. Chem. 66 (5), 1041–1046.

Colombo, M.L., 2010. An update on vitamin E, tocopherol and tocotrienol-perspectives. Molecules 15 (4), 2103–2113.

Dayer, R., Fischer, B.B., Eggen, R.I.L., Lemaire, S.D., 2008. The peroxiredoxin and glutathione peroxidase families in *Chlamydomonas reinhardtii*. Genetics 179 (1), 41–57.

Del Campo, J.a, Moreno, J., Rodríguez, H., Vargas, M.a, Rivas, J., Guerrero, M.G., 2000. Carotenoid content of chlorophycean microalgae: factors determining lutein accumulation in *Muriellopsis* sp. (Chlorophyta). J. Biotech. 76 (1), 51–59.

Dugas, A.J., Castañeda-Acosta, J., Bonin, G.C., Price, K.L., Fischer, N.H., Winston, G.W., 2000. Evaluation of the total peroxyl radical-scavenging capacity of flavonoids: structure-activity relationships. J. Nat. Prod. 63 (3), 327–331.

Durmaz, Y., 2007. Vitamin E (α-tocopherol) production by the marine microalgae *Nannochloropsis oculata* (Eustigmatophyceae) in nitrogen limitation. Aquaculture 272 (1–4), 717–722.

Duval, B., Shetty, K., Thomas, W.H., 2000. Phenolic compounds and antioxidant properties in the snow alga *Chlamydomonas nivalis* after exposure to UV light. J. Appl. Phycol. 11, 559–566.

Edge, R., McGarvey, D.J., Truscott, T.G., 1997. The carotenoids as anti-oxidants—a review. J. Photochem. Photobiol. B 41 (3), 189–200.

Goiris, K., Muylaert, K., Fraeye, I., Foubert, I., De Brabanter, J., De Cooman, L., 2012. Antioxidant potential of microalgae in relation to their phenolic and carotenoid content. J. App. Phycol. 24 (6), 1477−1486.

Goiris, K., Muylaert, K., Voorspoels, S., Noten, B., De Paepe, D., Baart, G.J.E., E Baart, G.J., 2014. Detection of flavonoids in microalgae from different evolutionary lineages. J. Phycol. 50 (3), 483−492.

Goss, R., Jakob, T., 2010. Regulation and function of xanthophyll cycle-dependent photoprotection in algae. Photosynth. Res. 106 (1−2), 103−122.

Gouveia, L., Raymundo, A., Batista, A.P., Sousa, I., Empis, J., 2005. *Chlorella vulgaris* and *Haematococcus pluvialis* biomass as colouring and antioxidant in food emulsions. Eur. Food Res. Technol. 222 (3−4), 362−367.

Guedes, A.C., Amaro, H.M., Malcata, F.X., 2011a. Microalgae as sources of carotenoids. Mar. Drugs 9 (4), 625−644.

Guedes, A.C., Amaro, H.M., Pereira, R.D., Malcata, F.X., 2011b. Effects of temperature and pH on growth and antioxidant content of the microalga *Scenedesmus obliquus*. Biotech. Progr. 27 (5), 1218−1224.

Hajimahmoodi, M., Faramarzi, M.A., Mohammadi, N., Soltani, N., Oveisi, M.R., Nafissi-Varcheh, N., 2010. Evaluation of antioxidant properties and total phenolic contents of some strains of microalgae. J. Appl. Phycol. 22 (1), 43−50.

Huang, Z., Guo, B.J., Wong, R.N.S., Jiang, Y., 2007. Characterization and antioxidant activity of selenium-containing phycocyanin isolated from *Spirulina platensis*. Food Chem. 100 (3), 1137−1143.

Janknegt, P., De Graaff, C.M., Van De Poll, W., Visser, R., Rijstenbil, J., Buma, A., 2009. Short-term antioxidative responses of 15 microalgae exposed to excessive irradiance including ultraviolet radiation. Eur. J. Phycol. 44 (4), 525−539.

Jin, E., Polle, J.E.W., Lee, H.K.U.M., Hyun, S.M.I.N., Chang, M.A.N., 2003. Xanthophylls in microalgae: from biosynthesis to biotechnological mass production and application. J. Microbiol. Biotech. 13, 165−174.

Klejdus, B., Kopecký, J., Benesová, L., Vacek, J., 2009. Solid-phase/supercritical-fluid extraction for liquid chromatography of phenolic compounds in freshwater microalgae and selected *cyanobacterial* species. J. Chrom. A 1216 (5), 763−771.

Klejdus, B., Lojková, L., Plaza, M., Snóblová, M., Stěrbová, D., 2010. Hyphenated technique for the extraction and determination of isoflavones in algae: ultrasound-assisted supercritical fluid extraction followed by fast chromatography with tandem mass spectrometry. J. Chrom. A 1217 (51), 7956−7965.

Kohen, R., Nyska, A., 2002. Oxidation of biological systems: oxidative stress phenomena, antioxidants, redox reactions, and methods for their quantification. Toxicol. Pathol. 30 (6), 620−650.

Kováčik, J., Klejdus, B., Backor, M., 2010. Physiological responses of *Scenedesmus quadricauda* (Chlorophyceae) to UV-A and UV-C light. Photochem. Photobiol. 86, 612−616.

Kováčik, J., Klejdus, B., Hedbavny, J., Bačkor, M., 2010. Effect of copper and salicylic acid on phenolic metabolites and free amino acids in *Scenedesmus quadricauda* (Chlorophyceae). Plant Sci. 178 (3), 307−311.

Laloi, C., Przybyla, D., Apel, K., 2006. A genetic approach towards elucidating the biological activity of different reactive oxygen species in *Arabidopsis thaliana*. J. Exp. Bot. 57 (8), 1719−1724.

Lee, S.-H., Lee, J.-B., Lee, K.-W., Jeon, Y.-J., 2010. Antioxidant properties of tidal pool microalgae, *Halochlorococcum porphyrae* and *Oltamannsiellopsis unicellularis* from Jeju Island, Korea. Algae 25 (1), 45−56.

Lee, Y.-L., Chuang, Y.-C., Su, H.-M., Wu, F.-S., 2013. Freeze-dried microalgae of *Nannochloropsis oculata* improve soybean oil's oxidative stability. Appl. Microbiol. Biotech. 97 (22), 9675−9683.

Lesser, M.P., 2006. Oxidative stress in marine environments: biochemistry and physiological ecology. Ann. Rev. Physiol. 68 (3), 253−278.

Li, H., Cheng, K., Wong, C., Fan, K., Chen, F., Jiang, Y., 2007. Evaluation of antioxidant capacity and total phenolic content of different fractions of selected microalgae. Food Chem. 102 (3), 771−776.

Liebler, D.C., 1993. Antioxidant reactions of carotenoids. Ann. N. Y. Acad. Sci. 691 (1), 20−31.

Mallick, N., 2004. Copper-induced oxidative stress in the chlorophycean microalga *Chlorella vulgaris*: response of the antioxidant system. J. Plant Physiol. 161 (5), 591−597.

Mallick, N., Mohn, F.H., 2000. Reactive oxygen species: response of algal cells. J. Plant Physiol. 157, 183−193.

Martínez, A., Barbosa, A., 2008. Antiradical power of carotenoids and vitamin E: testing the hydrogen atom transfer mechanism. J. Phys. Chem. B 112 (51), 16945−16951.

Martínez, A., Rodríguez-Gironés, M.a, Barbosa, A., Costas, M., 2008. Donator acceptor map for carotenoids, melatonin and vitamins. J. Phys. Chem. A 112 (38), 9037−9042.

Matsukawa, R., Dubinsky, Z., Masaki, K., Takeuchi, T., Karube, I., 1997. Enzymatic screening of microalgae as a potential source of natural antioxidants. Appl. Biochem. Biotech. 66 (3), 239−247.

Müller, L., Fröhlich, K., Böhm, V., 2011. Comparative antioxidant activities of carotenoids measured by ferric reducing antioxidant power (FRAP), ABTS bleaching assay (αTEAC), DPPH assay and peroxyl radical scavenging assay. Food Chem. 129 (1), 139−148.

Namiki, M., 1990. Antioxidants/antimutagens in food. Crit. Rev. Food Sci. Nutr. 29 (4), 273−300.

Natrah, F.M.I., Yusoff, F.M., Shariff, M., Abas, F., Mariana, N.S., 2007. Screening of Malaysian indigenous microalgae for antioxidant properties and nutritional value. J. App. Phycol. 19 (6), 711−718.

Niki, E., 1991. Action of ascorbic acid as a scavenger of active and stable oxygen radicals. Am. J. Clin. Nutr. 54, 1119S–1124S.

Oliveira, E.G., Rosa, G.S., Moraes, M.A., Pinto, L.A.A., 2009. Characterization of thin layer drying of *Spirulina platensis* utilizing perpendicular air flow. Bioresour. Technol. 100 (3), 1297–1303.

Onofrejová, L., Vasícková, J., Klejdus, B., Stratil, P., Misurcová, L., Krácmar, S., Vacek, J., 2010. Bioactive phenols in algae: the application of pressurized-liquid and solid-phase extraction techniques. J. Pharm. Biomed. Anal. 51 (2), 464–470.

Pietta, P.G., 2000. Flavonoids as antioxidants. J. Nat. Prod. 63 (7), 1035–1042.

Pouvreau, J.-B., Morançais, M., Taran, F., Rosa, P., Dufossé, L., Guérard, F., Pondaven, P., 2008. Antioxidant and free radical scavenging properties of marennine, a blue-green polyphenolic pigment from the diatom *Haslea ostrearia* (Gaillon/Bory) Simonsen responsible for the natural greening of cultured oysters. J. Agric. Food Chem. 56 (15), 6278–6286.

Prior, R.L., Wu, X., Schaich, K., 2005. Standardized methods for the determination of antioxidant capacity and phenolics in foods and dietary supplements. J. Agric. Food Chem. 53 (10), 4290–4302.

Ranga Rao, A., Sarada, R., Baskaran, V., Ravishankar, G.A., 2006. Antioxidant activity of *Botryococcus braunii* extract elucidated in vitro models. J. Agric. Food Chem. 54 (13), 4593–4599.

Rapp, L.M., Maple, S.S., Choi, J.H., 2000. Lutein and zeaxanthin concentrations in rod outer segment membranes from perifoveal and peripheral human retina. Invest. Ophthalmol. Vis. Sci. 41 (5), 1200–1209.

Remias, D., Schwaiger, S., Aigner, S., Leya, T., Stuppner, H., Lütz, C., 2012. Characterization of an UV- and VIS-absorbing, purpurogallin-derived secondary pigment new to algae and highly abundant in *Mesotaenium berggrenii* (Zygnematophyceae, Chlorophyta), an extremophyte living on glaciers. FEMS Microbiol. Ecol. 79 (3), 638–648.

Rice-Evans, C., Miller, N.J., Paganga, G., 1996. Structure-activity relationships of flavonoids and phenolic acids. Free Radic. Biol. Med. 20 (7), 933–956.

Rico, M., López, A., Santana-Casiano, J.M., González, A.G., González-Dávila, M., 2013. Variability of the phenolic profile in *Phaeodactylum tricornutum* diatom growing under copper and iron stress. Limnol. Oceanogr. 58 (1), 144–152.

Rodriguez-Garcia, I., Guil-Guerrero, J.L., 2008. Evaluation of the antioxidant activity of three microalgal species for use as dietary supplements and in the preservation of foods. Food Chem. 108 (3), 1023–1026.

Running, J.A., Severson, D.K., Schneider, K.J., 2002. Extracellular production of L-ascorbic acid by *Chlorella protothecoides*, *Prototheca* species, and mutants of *P. moriformis* during aerobic culturing at low pH. J. Ind. Microbiol. Biotechnol. 29 (2), 93–98.

Salah, N., Miller, N.J., Paganga, G., Tijburg, L., Bolwell, G.P., Rice-Evans, C., 1995. Polyphenolic flavanols as scavengers of aqueous phase radicals and as chain-breaking antioxidants. Arch. Biochem. Biophys. 322 (2), 339–346.

Sunda, W., Kieber, D.J., Kiene, R.P., Huntsman, S., 2002. An antioxidant function for DMSP and DMS in marine algae. Nature 418, 317–320.

Takaichi, S., 2011. Carotenoids in algae: distributions, biosyntheses and functions. Mar. Drugs 9 (6), 1101–1118.

Tannin-Spitz, T., Bergman, M., Van-Moppes, D., Grossman, S., Arad, S.(M.), 2005. Antioxidant activity of the polysaccharide of the red microalga *Porphyridium* sp. J. Appl. Phycol. 17 (3), 215–222.

Thangam, R., Suresh, V., Princy, W.a., Rajkumar, M., SenthilKumar, N., Gunasekaran, P., Kannan, S., 2013. C-Phycocyanin from *Oscillatoria tenuis* exhibited an antioxidant and in vitro antiproliferative activity through induction of apoptosis and G0/G1 cell cycle arrest. Food Chem. 140 (1–2), 262–272.

Ulloa, G., Otero, A., Sánchez, M., Sineiro, J., Núñez, M.J., Fábregas, J., 2012. Effect of Mg, Si, and Sr on growth and antioxidant activity of the marine microalga *Tetraselmis suecica*. J. App. Phycol. 24 (5), 1229–1236.

Wang, B., Zarka, A., Trebst, A., Boussiba, S., 2003. Astaxanthin accumulation in *Haematococcus pluvialis* (chlorophyceae) as an active photoprotective process under high irradiance. J. Phycol. 39, 1116–1124.

Wang, X., Xu, X., 2012. Molecular cloning and functional analyses of glutathione peroxidase homologous genes from *Chlorella* sp. NJ-18. Gene 501 (1), 17–23.

Wei, D., Chen, F., Chen, G., Zhang, X., Liu, L., Zhang, H., 2008. Enhanced production of lutein in heterotrophic *Chlorella protothecoides* by oxidative stress. Sci. China, Ser. C: Life Sci. 51 (12), 1088–1093.

Wright, J.S., Johnson, E.R., DiLabio, G.A., 2001. Predicting the activity of phenolic antioxidants: theoretical method, analysis of substituent effects, and application to major families of antioxidants. J. Am. Chem. Soc. 123 (6), 1173–1183.

Yang, C.-Y., Liu, S.-J., Zhou, S.-W., Wu, H.-F., Yu, J.-B., Xia, C.-H., 2011. Allelochemical ethyl 2-methyl acetoacetate (EMA) induces oxidative damage and antioxidant responses in *Phaeodactylum tricornutum*. Pestic. Biochem. Physiol. 100 (1), 93–103.

Yoshikawa, N., Belay, A., 2008. Single-laboratory validation of a method for the determination of c-phycocyanin and allophycocyanin in *Spirulina* (*Arthrospira*) supplements and raw materials by spectrophotometry. J. AOAC Int. 91 (3), 524–529.

18

Production of Biopharmaceuticals in Microalgae

Bernardo Bañuelos-Hernández, Josué I. Beltrán-López,
Sergio Rosales-Mendoza

Universidad Autónoma de San Luis Potosí, Laboratorio de Biofarmacéuticos Recombinantes,
Facultad de Ciencias Químicas, San Luis Potosí, Mexico

1. INTRODUCTION

Recombinant DNA and hybridoma technologies made it possible to produce, at large scale, proteins acting as drugs, which are referred to as biodrugs, biologics, or biopharmaceuticals (BFs). Many disorders are currently treated with such BFs, which comprise cytokines, monoclonal antibodies, clotting factors, hormones, and enzymes. BFs differ from traditional drugs in terms of manufacturing processes, structure, and action. BFs are much more complex due to their high molecular weight, specific tridimensional structure (or even quaternary structures), and heterogeneity given by the number of posttranslational modifications to which they can be subjected (Schellekens, 2004; WHO, 2013).

Because BFs are of key importance in the fight against a myriad of both infectious and noncommunicable diseases, they are typically produced, at a commercial scale, using mammalian cells, yeast, or bacteria as the expression host. The election of the production platform mainly depends on the requirements determined by the BF complexity (e.g., required specific glycosylation patterns or oligomeric structure formation). Therefore, BFs with high complexity are produced in appropriate eukaryotic platforms such as mammalian cells, whereas simpler molecules that do not require complex posttranslational modifications, such as insulin and some interleukins, are typically produced in bacterial systems. Production of BFs in mammalian cells is characterized by the most appropriate posttranslational processing, yielding high-quality proteins with optimum functionality and proper half-life (Almo and Love, 2014). However, the limitations of this system comprise the required scaling-up procedure and the potential contamination with pathogenic viruses or prions. The latter factor demands strict purification processes and quality control procedures, leading to costly BFs. On the other hand, although the production of recombinant proteins in bacteria cells is more economical than in mammalian cells, the system is limited by the lack of capability to perform complex posttranslational modifications and the frequent production of nonsoluble proteins that require refolding processes, which in fact do not always guarantee functionality (Overton, 2014; Hochkoeppler, 2013).

Given this outlook, the development of new BFs production platforms is a relevant aim in this area. Proper functionality, safety, and low cost are the main desired attributes, considering that low-income countries have an urgent need for these drugs to be more accessible. In recent years, progres in genetic modification procedures in combination with plant biotechnology has led to propose next-generation platforms for the production of BFs. Among them, plants have the potential to serve as a low-cost and robust platform. Microalgae have also been proposed as a potential platform because these organisms offer some of the strengths of the mammalian and bacterial systems. The characteristics identified in microalgae for the potential BFs production include: (1) high growth rates can be achieved in low-cost media culture, (2) efficient genetic modification technologies are available, (3) technologies for bioreactor-based production are available, (4) many species possess the machinery to perform complex post-translational modifications, and (5) numerous species are edible and thus may serve as oral delivery vehicles of certain BFs.

The exploration of this topic has been mainly performed using the freshwater microalga species *Chlamydomonas reinhardtii*, with promising results in terms of yield and functionality. A substantial number of BFs have been produced in *Chlamydomonas* thus far, comprising mainly vaccines and antibodies. These cases have generated a solid expectation of the potential impact of this technology in the pharmaceutical industry (Specht and Mayfield, 2014; Rosales-Mendoza, 2013; Specht et al., 2010). This chapter analyzes the current outlook on the use of microalgae as expression hosts of BFs, emphasizing the potential of marine algae as convenient platforms, in particular the species *Phaeodactylum tricornutum*, *Dunaliella salina*, and *Nannochloropsis* spp., for which genetic transformation procedures are currently available.

2. CURRENT OUTLOOK OF BFs PRODUCED IN MICROALGAE

Since genetic engineering tools are available for a number of algae species, the notion of using them as recombinant protein production platforms emerged in the early 1990s (Kindle, 1990; Kindle et al., 1991). Figure 1 presents a general outlook of the methodology that currently allows for the development of genetically engineered algae clones. Following this type of procedure, approximately 13 vaccine prototypes have been produced in microalgae during the last decade, targeting both infectious and noncommunicable pathologies of relevance in humans, such as hypertension (Soria-Guerra et al., 2014), hepatitis B virus (Geng et al., 2003), and human papillomavirus infection/cervical cancer (Demurtas et al., 2013), among others. In the veterinary field, these approaches have also been applied, such as in the case of the VP28 protein-based vaccine against the spread of white spot syndrome virus (WSSV), which is of relevance in shrimp farms (Feng et al., 2014a). Twelve of these candidate vaccines have been produced in *C. reinhardtii* and only one in *D. salina*. The characterization of most of these vaccines comprised the detection and quantification of the recombinant antigen in algae candidate clones and preclinical studies conducted in mice to evaluate their immunogenic potential in terms of antibody titers and polarization of the immune response (see Table 1).

On the other hand, antibodies have constituted another key target in this field. Although antibodies are effective therapeutics against a variety of human diseases, the high cost of such BFs remains a limiting issue for their massive use. *C. reinhardtii* has been used for the production of different types of antibodies, which range from large single-chain antibodies directed against the glycoprotein D of herpes simplex virus (HSV) (Mayfield et al., 2003) to immunotoxins that consist of chimeric antibodies comprising variable regions that recognize a

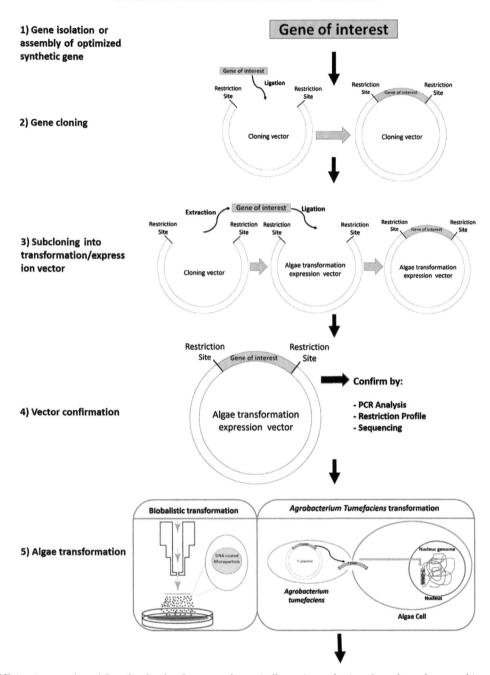

FIGURE 1 A general workflow for the development of genetically engineered microalgae clones for recombinant protein production.

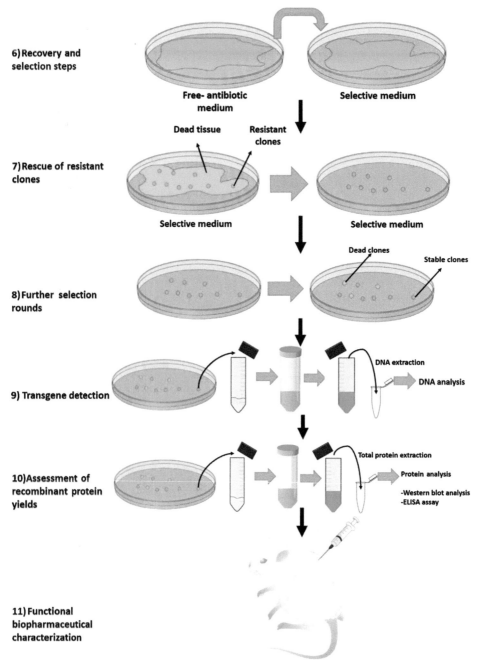

FIGURE 1   cont'd.

TABLE 1 An Overview of the Recombinant Vaccines Expressed in Microalgae Species

Target antigen	Expression host	Transformation method	Integration site	Yields	Route of administration	Adjuvant	Findings	References
							Features of vaccine	
White spot syndrome virus VP28 protein	*Dunaliella salina*	Glass beads	Nuclear genome	Approximately 780 µg/L of recombinant VP28 protein	Oral	—	Immunogenicity not assessed	Feng et al. (2014b)
Angiotensin II fused to hepatitis B virus capsid antigen (HBcAg)	*Chlamydomonas reinhardtii*	*Agrobacterium tumefaciens*	Nuclear genome	Up to 0.05% of TSP	—	—	Immunogenicity not assessed	Soria-Guerra et al. (2014)
Human papillomavirus type 16 E7 protein (attenuated mutant, E7GGG)	*C. reinhardtii*	Glass beads	Chloroplast genome	Up to 0.12% of TSP 0.02% of TSP for the His-tagged version	Subcutaneus	Quil A	Vaccination in mice induced high titers of specific IgGs	Demurtas et al. (2013)
Plasmodium falciparum surface protein Pfs25 fused to cholera toxin B subunit	*C. reinhardtii*	Biobalistic	Chloroplast genome	Up to 0.09% TSP	Intraperitoneal and oral	Aluminum salt	Intraperitonial immunization elicits IgG responses that recognized Pfs25. Oral vaccination elicits secretory IgA responses	Gregory et al. (2013)
C-terminal domain of the *P. falciparum* surface protein Pfs48/45	*C. reinhardtii*	Biobalistic	Chloroplast genome	Not quantified; detected by Western blot after affinity purification	—	—	Pfs48/45 protein was expressed and properly folded; immunogenicity not assessed	Jones et al. (2013)

Continued

TABLE 1   An Overview of the Recombinant Vaccines Expressed in Microalgae Species—cont'd

Target antigen	Expression host	Transformation method	Integration site	Yields	Route of administration	Adjuvant	Findings	References
							Features of vaccine	
P. falciparum surface proteins 25 (Pfs25) and 28 (Pfs28)	*C. reinhardtii*	Biobalistic	Chloroplast genome	Not quantified; detected by Western blot and Coomassie blue staining after affinity purification	Intraperitoneal	Freund's	Both candidates are immunogenic in mice inducing antibodies against the target pathogen form. Pfs25 elicits transmission blocking antibodies.	Gregory et al. (2012)
Chimeric protein comprising the granule bound starch synthase (GBSS) and either the C-terminal domains from the Apical Major Antigen (AMA1) or Major Surface Protein (MSP1)	*C. reinhardtii*	Glass beads	Nuclear genome	0.2–1 μg of protein per mg of purified starch	Oral Intraperitoneal	LTB Freund's	Immunogenic in mice inducing systemic protective immune responses.	Dauvillée et al. (2010)
Chimeric protein comprising the cholera toxin B subunit and the D2 fibronectin-binding domain of *Staphylococcus aureus*	*C. reinhardtii*	Biobalistic	Chloroplast genome	Up to 0.7% of TSP	Oral	—	Immunogenic in mice inducing mucosal and systemic immune responses. Induced immunoprotection against *S. aureus* challenge	Dreesen et al. (2010)

Protein	Organism	Transformation method	Genome	Expression level	Route	Adjuvant	Immunogenicity	Reference
VP28 protein of the White spot syndrome virus	C. *reinhardtii*	Biobalistic	Chloroplast genome	Variable, ranging from 0.2% to 20.9% of total cellular protein (0.1–10.5% TSP)	—	—	Immunogenicity not assessed	Surzycki et al. (2009)
Human glutamic acid decarboxylase 65	C. *reinhardtii*	Biobalistic	Chloroplast genome	0.25–0.3% of TSP	—	—	Showed a positive reactivity with sera from diabetic patients and is immunogenic in nonobese diabetic mice	Wang et al. (2008)
Structural protein E2 of the classical swine fever virus	C. *reinhardtii*	Biobalistic	Chloroplast genome	1.5–2% of TSP	Subcutaneous Oral	Freund's —	Immunogenic in mice inducing systemic immune responses only when subcutaneously administered.	He et al. (2007)
Hepatitis B surface antigen (HBsAg)	D. *salina*	Electroporation	Nuclear genome	1.64–3.11 ng/mg of TSP	—	—	Immunogenicity not assessed	Geng et al. (2003)
Chimeric protein comprising the cholera toxin B subunit foot and the mouth disease virus VP1 protein	C. *reinhardtii*	Biobalistic	Chloroplast genome	3–4% of TSP	—	—	Displays antigenic determinants from both components; immunogenicity not assessed	Sun et al. (2003)

relevant cell-specific target (e.g., CD22) fused to a toxin (e.g., gelonin toxin) that is intended as a therapeutic agent against neoplastic diseases (e.g., chronic B-cell lymphoid leukemia; Tran et al., 2013). Genetic engineering procedures at the chloroplast compartment have allowed for the successful assembly of these antibodies, which have shown proper reactivity (see Table 2).

Besides vaccines and antibodies, many recombinant proteins have been expressed successfully in microalgae, including reporter genes such as luciferase (Matsuo et al., 2006; Mayfield and Schultz, 2004; Minko et al., 1999), green fluorescent protein (GFP) (Franklin et al., 2002), and β-glucuronidase (Ishikura et al., 1999). In addition, selection markers such as aminoglycoside transferase (aphA) (Bateman and Purton, 2000) and the adenine aminoglycoside transferase (aadA) (Goldschmidt-Clermont, 1991) have been expressed properly, allowing for the selection of transformed clones.

Other BFs that have been expressed in microalgae include the Kunitz trypsin inhibitor of soybean (SKTI), which is used to downregulate inflammatory activity (Chai et al., 2013); the mammary-associated serum amyloid A protein (M-SAA), which stimulates mucin production in gut epithelial cells to protect the bowel in newborns (Manuell et al., 2007); and the human tumor necrosis factor-related apoptosis-inducing ligand (TRAIL) for cancer therapy (Yang et al., 2006). Although most of these proteins have been expressed in C. reinhardtii, a remarkable interest in expressing recombinant proteins in other promising microalgae species, such as P. tricornutum and D. salina, has been reflected in the recent literature (see Table 3).

3. OTHER MICROALGAE AS PLATFORMS TO PRODUCE BIOPHARMACEUTICALS

Although C. reinhardtii has been the main microalga applied for BFs production (Rasala and Mayfield, 2014; Mayfield et al., 2003), marine algae species are also being explored for this purpose. The attributes that justify the use of new species include efficient protein secretion mechanisms, higher yields, use of seawater to avoid interference with potable or agriculture water sources, and the feasibility of growing algae under full containment favoring biosafety. The following sections describe these cases and the potential for using them as new microalgae-based platforms in the biopharming field.

3.1 Phaeodactylum tricornutum

Phaeodactylum tricornutum is a diatom that has emerged in the biotechnology field for various applications, such as a biofuel precursor and recombinant protein expression host due to its biosynthetic capacity and high growth rates. Both biolistic and electroporation-mediated genetic engineering methods have been reported for P. tricornutum (Hempel et al., 2011; Hempel and Maier, 2012; Zaslavskaia et al., 2000; Apt et al., 1996; Niu et al., 2012; Miyahara et al., 2013). Initial attempts comprised the use of vectors containing selection markers targeting the nuclear genome via a biolistic method. Apt et al. (1996) showed that the sh-ble (phleomycin-resistance gene) and cat (chloramphenicol acetyltransferase) genes serve as appropriate selection markers. Another important genetic engineering tool applied in this species consists of the use of reporter genes. Zaslavskaia et al. (2000) developed P. tricornutum clones expressing both the GFP and GUS proteins, which are the main reporter genes used to monitoring either transient or stable transgene expression.

An important advance related to BFs production consisted of the production of functional antibodies in P. triconutum. Hempel et al. (2011) expressed a monoclonal human immunoglobulin (Ig) G antibody against the hepatitis B surface protein. Assembling and binding properties of the antibodies produced were positively

TABLE 2   An Overview of Antibodies Expressed in Microalgae Species

Description of antibody	Expression host	Transformation method	Integration site	Findings	Yields	References
Immunotoxin protein containing a ribosome inactivating protein named gelonin with variable regions of heavy and light chains from an antibody (scFv) that recognizes CD22	*Chlamydomonas reinhardtii*	Biobalistic	Chloroplast genome	Functional immunotoxin is expressed in the chloroplast of *C. reinhardtii*, having as perspective the use of therapies in neoplastic diseases such as B-cell chronic lymphocytic leukemia	αCD22 accumulates at approximately 0.7% of TSP. αCD22Gel accumulates at approximately 0.2−0.3% of TSP αCD22CH23Gel accumulates at approximately 0.1−0.2% of TSP	Tran et al. (2013)
Human monoclonal IgG antibody CL4mAb against the hepatitis B virus surface protein	*Phaeodactylum tricornutum*	Biobalistic	Nuclear genome	Functional IgG antibody was expressed and secreted into the culture medium.	IgG antibodies were secreted to the medium at levels of up to 2.5 μg/mL	Hempel and Maier (2012)
Human monoclonal IgG1 antibody CL4mAb against the hepatitis B virus surface antigen	*P. tricornutum*	Biobalistic	Nuclear genome	Functional antibody is expressed and assembled in the endoplasmatic reticulum of *P. tricornutum* using the endoplasmic reticulum retention signal (DDEL) at the C-terminus of both antibody chains and use a nitrate-inducible promoter system HBsAg was expressed and accumulated in *P. tricornutum* more efficient than in other plant systems, such as *N. tabacum*	Monoclonal IgG antibody is accumulated at up to 8.7% of TSP. HBsAg is accumulated at up to 0.7% TSP	Hempel et al. (2011)

Continued

TABLE 2 An Overview of Antibodies Expressed in Microalgae Species—cont'd

Description of antibody	Expression host	Transformation method	Integration site	Findings	Yields	References
Human IgG1 monoclonal antibody 83K7C against the PA83 anthrax antigen	C. reinhardtii	Biobalistic	Chloroplast genome	Human antibody was successfully expressed and the levels of expression are similar to the same antibody expressed in mammalian cells	Not quantified; detected by Western blot	Tran et al. (2009)
Single-chain variable regions antibody against the Herpes simplex virus glycoprotein D (HSV8-scFv)	C. reinhardtii	Biobalistic	Chloroplast genome	Functional immunotoxin protein is expressed in C. reinhardtii chloroplast	Antibody is accumulated at up to 0.25% of TSP	Mayfield and Franklin (2005)
Large single-chain (lsc) antibody directed against herpes simplex virus (HSV) glycoprotein D (HSV8-lsc)	C. reinhardtii	Biobalistic	Chloroplast genome	HSV8-lsc antibody is produced in a functional manner in C. reinhardtii chloroplast	Antibody is accumulated at >1% TSP	Mayfield et al. (2003)

TABLE 3 An Overview of Other Biopharmaceuticals and Reporter/Marker Genes Expressed in Microalgae Species

Description of recombinant protein	Expression host	Transformation method	Integration site	Findings	Yields	References
Soybean Kunitz trypsin inhibitor (SKTI)	*Dunaliela salina*	LiAc transformation method	Nuclear genome	*SKTI* gene was successfully expressed in *D. salina* cells as an approach to down regulate inflammatory activity.	0.68% of TSP	Chai et al. (2013)
Chloramphenicol acetyltransferase (CAT)	*P. tricornutum*	Electroporation	Nuclear genome	*CAT* gene was successfully expressed in *P. tricornutum* cells. An interesting strategy consisted of the use of an NaNO$_3$-inducible promoter.	Not quantified; detected by Western blot	Niu et al. (2012)
Mammary-associated serum amyloid A (M-SAA)	*Chlamydomonas reinhardtii*	Biobalistic	Chloroplast genome	Robust expression of bioactive M-SAA was achieved, with the potential to protect against intestinal bacterial and viral infection in newborns.	12.5% of TSP	Manuell et al. (2007)
Firefly luciferase	*C. reinhardtii*	Biobalistic	Chloroplast genome	Firefly luciferase was successfully used as a reporter gene that will be useful in chloroplast gene regulation studies	Not quantified	Matsuo et al. (2006)
Human tumor necrosis factor-related apoptosis-inducing ligand (TRAIL)	*C. reinhardtii*	Biobalistic	Chloroplast genome	*Chlamydomonas*-made TRAIL was functional, with possible use for therapies in viral disease and cancer.	0.43−0.67% of TSP	Yang et al. (2006)
Human metallothionine-2	*C. reinhardtii*	Biobalistic	Chloroplast genome	The hMT-2 human gene was integrated into the chloroplast genome of *C. reinhardtii* and successfully expressed in the transplastomic alga. hMT-2 transplastomic algae were resistant to UV-B radiation exposure.	Not quantified; visible by Western blot	Zhang et al. (2006)

Continued

TABLE 3 An Overview of Other Biopharmaceuticals and Reporter/Marker Genes Expressed in Microalgae Species—cont'd

Description of recombinant protein	Expression host	Transformation method	Integration site	Findings	Yields	References
Allophycocyanin	C. reinhardtii	Biobalistic	Chloroplast genome	Successful expression of polycistronic arrays of prokaryotic cyanobacteria in C. reinhardtii.	2%—3% (W/W) of TSP	Su et a. (2005)
Bacterial luciferase	C. reinhardtii	Biobalistic	Chloroplast genome	A codon-optimized bacterial luciferase gene was successfully used as a reporter gene that will be useful in chloroplast gene regulation studies.	Not quantified; detected by Western blot	Myfield and Schultz (2004)
Green fluorescent protein	C. reinhardtii	Biobalistic	Chloroplast genome	A codon-optimized green fluorescent protein gene was successfully used as a reporter gene that will be useful in chloroplast gene regulation studies.	0.5 of TSP	Franklin et al. (2002)
Aminoglycoside phosphotransferase	C. reinhardtii	Biobalistic	Chloroplast genome	The aphA-6 gene serves as a selectable marker that confers kanamycin and amikacin resistance in chloroplast-transformed Chlamydomonas clones.	Not quantified	Bateman and Purton (2000)
β-Glucuronidase	C. reinhardtii	Biobalistic	Chloroplast genome	Expression of GUS, a functional reporter gene using different promoters in C. reinhardtii.	Not quantified	Ishikura et al. (1999)
Renilla luciferase	C. reinhardtii	Biobalistic	Chloroplast genome	Renilla luciferase was successfully used as a reporter gene that will be useful in chloroplast gene regulation studies.	Not quantified; detected by Western blot	Minko et al. (1999)
Aminoglycoside adenine transferase (aadA)	C. reinhardtii	Biobalistic	Chloroplast genome	Functional aadA gene is expressed in C. reinhardtii and allowed for the selection of transformed clones.	Not quantified	Goldschmidt-Clermont (1991)

evidenced by enzyme-linked immunosorbent assay (ELISA). In terms of production, recombinant protein yields were of up to 8.7% of the total soluble protein, which represents approximately 1.6 mg of antibody per liter. The expression system used in this approach was driven by an inducible promoter from the nitrate reductase, which is activated by the presence of nitrate and downregulated by ammonia. Another characteristic of this system was given by the inclusion of an endoplasmic reticulum retention signal to avoid the complex glycosylation patterns that occur in the Golgi apparatus.

In 2012, Hempel and Maier made a modification in the expression system that removed the endoplasmic reticulum retention signal. This modification allowed for the antibody to be secreted by the diatom into the culture medium. Because *P. tricornutum* essentially does not secrete endogenous proteins, the secreted recombinant protein can be easily purified, which makes the production extremely cost efficient (Hempel and Maier, 2012). Importantly, the expression levels in this system were higher (2250 ng per mL) than those observed when the protein was subjected to endoplasmic reticulum retrieval.

3.2 Dunaliella salina

Dunaliella salina is a unicellular, biflagellate, naked green alga that is morphologically similar to *Chlamydomonas*, with the main difference being the lack of a rigid cell wall (Emeish, 2013). The first description of *D. salina* was made in 1832 by Dunal, who initially considered it as a variety of *Haematococcus salinus*. In 1905, Teodoresco designated it as a new genera of microalgae. The *Dunaliella* genus currently comprises many algae species (Oren, 2005; Borowitzka and Siva, 2007). An important feature of *D. salina* is the ability to be grown in high salt concentrations of up to 35% w/v (Brown and Borowitzka, 1979). This property diminishes the contamination problems that are frequently present in

other species, such as *Chlamydomonas*. In addition, *D. salina* is able to grow under high-light intensity, high temperatures, and a wide pH range (Cifuentes et al., 1996). Another important characteristic is the absence of a rigid cell wall, which creates a natural protoplast, facilitating the DNA transfer step during genetic transformation procedures (Ben-Amotz and Avron, 1990).

Several methods have been applied to transform *D. salina*: electroporation, particle bombardment, *Agrobacterium tumefaciens*, glass beads, and the lithium acetate/polyethylene glycol (LiAc/PEG)-mediated method, with the last two methods being the most efficient (Feng et al., 2014b; Feng et al., 2009). *D. salina* has been cultivated for several decades for pigment production at industrial levels (Olaizola, 2003; Gómez et al., 2003; Ben-Amotz, 1993), but in recent years it has also emerged as a new platform for BFs production. Hepatitis B surface antigen (HBsAg), VP28 from WSSV, and Soybean Kunitz trypsin inhibitor are important recombinant proteins produced in *D. salina*. These approaches are presented in the following sections.

3.2.1 HBsAg

The first nuclear stable transformation of *D. salina* was accomplished by Geng et al. 2003. Microalga was transformed by an electroporation method, obtaining approximately 50 colonies per plate in a period of 3 weeks. The *cat* gene that confers chloramphenicol resistance was used as a selection marker. The transformed colonies were subcultivated 60 times, and then the presence of the transgene was evaluated. All colonies were positive, which reflected a highly stable transgene insertion. The best productivity was of 3.11 ng HBsAg/mg of soluble protein (Geng et al., 2003). The expression of HBsAg is important because of the epidemiologic relevance of the hepatitis B virus. In addition, this antigen can also serve as an immunogenic carrier of genetically fused, nonrelated epitopes from other pathogens.

3.2.2 VP28 from WSSV

WSSV is a serious problem affecting shrimp yields (Lightner and Redman, 1998), and vaccination with plant-made vaccines has been applied to reduce lost production (Thagun et al., 2012). Feng et al. (2014b) obtained transgenic *D. salina* clones expressing the Vp28 protein of WSSV. A nuclear transformation was achieved through the glass bead method, and the vector pUΩ-GUS designed by Geng in 2003 was used to drive the expression. By means of ELISA, the Vp28 protein productivity was estimated at levels of up to 3.04 ng/mg of total protein in *D. salina* clones. In terms of production by volume of culture, approximately 780 µg of recombinant VP28 protein was produced per liter. Lysates from these transgenic microalgae clones were used to orally immunize shrimp, as was reported by Fu et al. (2010). This immunization approach was able to diminish shrimp mortality by 40% with respect to the nonvaccinated shrimp population (Feng et al., 2014b). It is envisioned that this algae-based vaccine will be a breakthrough in resolving aquaculture problems.

3.2.3 Soybean Kunitz Trypsin Inhibitor

Soybean Kunitz trypsin inhibitor (SKTI) is a serine proteinase inhibitor against the activities of both trypsin and chymotrypsin, which leads to anti-inflammatory and anticarcinogenic activities (Hsieh et al., 2010; Lippman and Matrisian, 2000; Ribeiro et al., 2010). Kobayashi et al. (2004) showed that SKTI can suppress ovarian cancer cell invasion by blocking urokinase upregulation. Trypsin inhibitors are mainly extracted from human urine, soybean, and pumpkin. To diminish production costs, the expression of SKTI in *D. salina* has been attempted. Using the expression vector pCAM2201, in which the transgene *skti* is under the control of the 35S promoter, Chai et al. (2013) transformed *D. salina* via the lithium acetate/polyethylene glycol method. Expression levels of SKTI in algae clones were calculated at about 0.68% of total soluble protein, and the stability of transgene insertion was positively evaluated after 35 subcultures (Chai et al., 2013).

3.3 Haematococcus pluvialis

Hematococus pluvialis is an autotrophic fresh water microalgae that has been applied for industrial production of pigments (Fábregas et al., 2000; Sarada et al., 2002; Olaizola, 2000). Although *H. pluvialis* is a freshwater microalgae, it is able to grow in extreme environments and survive extreme fluctuations in light, temperature, and salt concentration (reviewed by Saei et al. (2012)). An important feature in *H. pluvialis* is its codon usage, which is highly similar to codon usage in *Homo sapiens*, as reported by Saei et al. (2012). This is an important trait because codon optimization is not necessary in the case of human origin BFs to avoid the low productivity associated to codon usage issues.

The development of genetic engineering tools for *H. pluvialis* has been reported. Gutiérrez et al. (2012) have developed a chloroplast-transforming vector, designed using the endogenous 5′ rbcL as promoter and 3′ rbcL as terminator. These elements were used to express the *aadA* selection marker. The construct was introduced into the chloroplast through the biolistic method. The transgene insertion was stable through 40 subcultivation steps in three transgenic lines, although only one line became homoplastic after the selection rounds. Another important advantage of *H. pluvialis* is the availability of nuclear transformation protocols via agro-inoculation. Kathiresan et al. (2009) used agrobacterium and elements from the pCAMBIA vector to express the reporter genes *gfp* and *uidA*. The integration of the transgene was maintained after 2 years of subcultivation. These genetic engineering tools will be critical to explore the use of *H. pluvialis* as a new BF production platform.

3.4 *Nannochloropsis* spp.

Nannochloropsis spp. are microalgae living in freshwater and seawater that are related to diatoms and brown algae (Sukenik et al., 2009; Andersen et al., 1998). *Nannochloropsis* species have been used for several decades to produce nutraceuticals and feed supplements (Rodolfi et al., 2009). Genetic engineering tools for nuclear transformation have been recently developed for this species. Kilian et al. (2011) established a protocol for nuclear transformation of *Nannochloropsis* via homologous recombination with a high transformation efficiency. These tools increase the possibilities of implementing robust BF production platforms based in this species.

4. PERSPECTIVES

Green microalgae are potential expression hosts for the production of biomolecules of high biotechnological value because these organisms have unique advantages, such as low production costs, high growth rates, high yields, and the ability to accomplish posttranslational modifications. They are compatible with high biosafety procedures, with some of them considered Generally Recognized as Safe by the US Food and Drug Administration. Among green algae, *C. reinhardtii* is the eukaryotic green microalga mainly used as an expression host in this field. The adoption of these approaches by the industry highlights the potential of algae-based platforms. For example, Triton Algae Innovations is a company that produces proteins, enzymes, and other biologics in *Chlamydomonas*. One of the products currently in the market is the mammary-associated amyloid, which is contained in the colostrum of mammals and can be used to treat intestinal disease in livestock, companion animals, and humans. In addition, antibacterials, antioxidants, biosurfactants, DNA repair enzymes, antimicrobials, intestinal health proteins, growth factors, bone growth enhancers, and vaccines

are in the pipeline. All of these candidates will have implications on the development of pharmaceuticals, nutraceuticals, cosmetics, and human and animal health nutrition products; applications in agricultural and other retail markets are also envisioned.

Although current biologicals produced in *Chlamydomonas* are promising and interesting cases, relevant perspectives related to the expansion of this field can be also outlined. Improvements in distinct aspects can be identified as a need in the *Chlamydomonas*-based approaches. For example, *Chlamydomonas* cultures are prone to contamination, which causes frequency delays in clone propagation steps. On the other hand, many BFs require posttranslational modifications, such as glycosylation; in this case, following a nuclear-based expression, it is necessary to direct the protein to endoplastmic reticulum and Golgi apparatus or even the secretory pathway. However, nuclear-based expression in *Chlamydomonas* has shown generally low yields (Rasala et al., 2012). In addition, the use of seawater to cultivate algae is desirable in this process to avoid the use of freshwater, which is also used as potable water or for agricultural purposes. It is envisioned that these pitfalls would be overridden by the use of marine algae species with distinct characteristics.

The identified marine algae species in this chapter, for which culture and genetic modification tools are available, constitute relevant hosts to generate new BFs production platforms with improved features, including the following: low-contamination events due to cultivation under extreme culture conditions (e.g., pH, and high salinity); higher protein yields; the use of seawater instead fresh water; and the use of the secretion machinery to facilitate downstream processing in the case of BFs requiring parenteral administration. The coming years will be critical to assess the potential of these candidate species to determine their performance with a wide number of specific BFs.

References

Almo, S.C., Love, J.D., 2014. Better and faster: improvements and optimization for mammalian recombinant protein production. Curr. Opin. Struct. Biol. 7, 39–43.

Andersen, R.A., Brett, R.W., Potter, D., Sexton, J.P., 1998. Phylogeny of the Eustigmatophyceae based upon 18S rDNA, with emphasis on *Nannochloropsis*. Protist 149, 61–74.

Apt, K.E., KrothPancic, P.G., Grossman, A.R., 1996. Stable nuclear transformation of the diatom *Phaeodactylum tricornutum*. Mol. Gen. Genet. 252, 572–579.

Bateman, J.M., Purton, S., 2000. Tools for chloroplast transformation in *Chlamydmonas*: expression vectors and a new dominant selectable marker. Mol. Gen. Genet. 263, 404–410.

Ben-Amotz, A., 1993. Production of β-carotene and vitamin by the halotolerant algae *Dunaliella*. In: Ahaway, A., Zabrosky, O. (Eds.), Marine Biotechnology. Plenum Press, New York, pp. 411–417.

Ben-Amotz, A., Avron, M., 1990. The biotechnology of cultivating the halotolerant alga *Dunaliella*. Trends Biotechnol. 8, 121–126.

Borowitzka, M.A., Siva, C.J., 2007. The taxonomy of the genus *Dunaliella* (Chlorophyta, Dunaliellales) with emphasis on the marine and halophilic species. J. Appl. Phycol. 19, 567–590.

Brown, A.D., Borowitzka, L.J., 1979. Halotolerance of *Dunaliella*. In: Levandowsky, M., Hunter, S.H. (Eds.), Biochemistry and Physiology of Protozoa, second ed., vol. 1. Academic Press, NY, pp. 139–190.

Chai, X.F., Chen, H.X., Xu, W.Q., Xu, Y.W., 2013. Expression of soybean Kunitz trypsin inhibitor gene *SKTI* in *Dunaliella salina*. J. Appl. Phycol. 25, 139–144.

Cifuentes, A.S., Gonzalez, M., Parra, O., Zuñiga, M., 1996. Culture of strains of *Dunaliella salina* (Teodoresco 1905) in different media under laboratory conditions. Rev. Chil. Hist. Nat. 69, 105–112.

Dauvillée, D., Delhaye, S., Gruyer, S., Slomianny, C., Moretz, S.E., d'Hulst, C., Long, C.A., Ball, S.G., Tomavo, S., 2010. Engineering the chloroplast targeted malarial vaccine antigens in *Chlamydmonas* starch granules. PLoS One 5, e15424.

Demurtas, O.C., Massa, S., Ferrante, P., Venuti, A., Franconi, R., Giuliano, G.A., 2013. *Chlamydmonas*-derived humanpapillomavirus16 E7 vaccine induces specific tumor protection. PLoS One 8, e61473.

Dreesen, I.A., Charpin-El Hamri, G., Fussenegger, M., 2010. Heat-stable oral alga-based vaccine protects mice from *Staphylococcus aureus* infection. J. Biotechnol. 145, 273–280.

Emeish, S., 2013. Production of natural biopharmaceuticals from the microalgae living in the dead sea. J. Environ. Earth Sci. 3, 6–15.

Fábregas, J., Domínguez, A., Regueiro, M., Maseda, A., Otero, A., 2000. Optimization of culture medium for the continuous cultivation of the microalga *Haematococcus pluvialis*. Appl. Microbiol. Biotechnol. 53, 530–535.

Feng, S., Feng, W., Zhao, L., Gu, H., Li, Q., Shi, K., Guo, S., Zhang, N., 2014a. Preparation of transgenic *Dunaliella salina* for immunization against white spot syndrome virus in crayfish. Arch. Virol. 159, 519–525.

Feng, S., Li, X., Xu, Z., Qi, J., 2014b. *Dunaliella salina* as a novel host for the production of recombinant proteins. Appl. Microbiol. Biotechnol. 98, 4293–4300.

Feng, S., Xue, L., Liu, H., Lu, P., 2009. Improvement of efficiency of genetic transformation for *Dunaliella salina* by glass beads method. Mol. Biol. Rep. 36, 1433–1439.

Franklin, S., Ngo, B., Efuet, E., Mayfield, S.P., 2002. Development of a GFP reporter gene for *Chlamydmonas reinhardtii* chloroplast. Plant J. 30, 733–744.

Fu, L.L., Shuai, J.B., Xu, Z.R., Li, J.R., Li, W.F., 2010. Immune responses of *Fenneropenaeus chinensis* against white spot syndrome virus after oral delivery of VP28 using *Bacillus subtilis* as vehicles. Fish Shellfish Immunol. 28, 49–55.

Geng, D., Wang, Y., Wang, P., Li, W., Sun, Y., 2003. Stable expression of hepatitis B surface antigen gene in *Dunaliella salina* (Chlorophyta). J. Appl. Phycol. 15, 451–456.

Gregory, J.A., Li, F., Tomosada, L.M., Cox, C.J., Topol, A.B., Vinetz, J.M., Mayfield, S.P., 2012. Algae produced Pfs25 elicits antibodies that inhibit malaria transmission. PLoS One 7, e37179.

Gregory, J.A., Topol, A.B., Doerner, D.Z., Mayfield, S., 2013. Alga-produced cholera toxin-Pfs25 fusion proteins as oral vaccines. Appl. Environ. Microbiol. 79, 3917–3925.

Goldschmidt-Clermont, M., 1991. Transgenic expression of aminoglycoside adenine transferase in the chloroplast: a selectable marker of site-directed transformation of *Chlamydmonas*. Nucleic Acids Res. 19, 4083–4089.

Gómez, P.I., Barriga, A., Cifuentes, A.S., González, M.A., 2003. Effect of salinity on the quantity and quality of carotenoids accumulated by *Dunaliella salina* (strain CONC-007) and *Dunaliella bardawil* (strain ATCC 30861) Chlorophyta. Biol. Res. 36, 185–192.

Gutiérrez, C.L., Gimpel, J., Escobar, C., Marshall, S.H., Henríquez, V., 2012. Chloroplast genetic tool for the green microalgae *Haematococcus pluvialis* (chlorophyceae, volvocales). J. Phycol. 48, 976–983.

He, D.M., Qian, K.X., Shen, G.F., Zhang, Z.F., Li, Y.N., Su, Z.L., Shao, H.B., 2007. Recombination and expression of classical swine fever virus (CSFV) structural protein E2 gene in *Chlamydmonas reinhardtii* chroloplasts. Colloids Surf. B. Biointerfaces 55, 26–30.

Hempel, F., Lau, J., Klingl, A., Maier, U.G., 2011. Algae as protein factories: expression of a human antibody and the respective antigen in the diatom *Phaeodactylum tricornutum*. PLoS One 6, e28424.

Hempel, F., Maier, U.G., 2012. An engineered diatom acting like a plasma cell secreting human IgG antibodies with high efficiency. Microb. Cell. Fact. 11, 1—6.

Hochkoeppler, A., 2013. Expanding the landscape of recombinant protein production in *Escherichia coli*. Biotechnol. Lett. 35, 1971—1981.

Hsieh, C.C., Hernández-Ledesma, B., Jeong, H.J., Park, J.H., Ben, O., 2010. Complementary roles in cancer prevention: protease inhibitor makes the cancer preventive peptide lunasin bioavailable. PLoS One 5, e8890.

Ishikura, K., Takaoka, Y., Kato, K., Sekine, M., Yoshida, K., Shinmyo, A., 1999. Expression of a foreign gene in *Chlamydmonas reinhardtii* chloroplast. Biosci. Bioeng. 87, 307—314.

Jones, C.S., Luong, T., Hannon, M., Tran, M., Gregory, J.A., Shen, Z., Briggs, S.P., Mayfield, S.P., 2013. Heterologous expression of the C terminal antigenic domain of the malaria vaccine candidate Pfs48/45 in the green algae *Chlamydmonas reinhardtii*. Appl. Microbiol. Biotechnol. 97, 1987—1995.

Kathiresan, S., Chandrashekar, A., Ravishankar, A., Sarada, R., 2009. Agrobacterium-mediated transformation in the green alga *Haematococcus pluvialis* (Chlorophyceae, volvocales). J. Phycol. 45, 642—649.

Kilian, O., Benemann, C.S., Niyogi, K.K., Vick, B., 2011. High-efficiency homologous recombination in the oil-producing alga *Nannochloropsis* sp. Proc. Natl Acad. Sci. U.S.A. 108, 21265—21269.

Kindle, K.L., 1990. High-frequency nuclear transformation of *Chlamydomonas reinhardtii*. Proc. Natl Acad. Sci. USA 87, 1228—1232.

Kindle, K.L., Richards, K.L., Stern, D.B., 1991. Engineering the chloroplast genome: techniques and capabilities for chloroplast transformation in *Chlamydomonas reinhardtii*. Proc. Natl Acad. Sci. USA 88, 1721—1725.

Kobayashi, H., Suzuki, M., Kanayama, N., Terao, T., 2004. A soybean Kunitz trypsin inhibitor suppresses ovarian cancer cell invasion by blocking urokinase upregulation. Clin. Exp. Metastasis 21, 159—166.

Lightner, D.V., Redman, R.M., 1998. Shrimp diseases and current diagnostic methods. Aquaculture 164, 201—220.

Lippman, S.M., Matrisian, L.M., 2000. Protease inhibitors in oral carcinogenesis and chemoprevention. Clin. Cancer Res. 6, 4599—4603.

Manuell, A.L., Beligni, M.V., Elder, J.H., Siefker, D.T., Tran, M., Weber, A., McDonald, T.L., Mayfield, S.P., 2007. Robust expression of a bioactive mammalian protein in *Chlamydomonas* chloroplast. Plant Biotechnol. J. 5, 402—412.

Matsuo, T., Onai, K., Okamoto, K., Minagawa, J., Ishiura, M., 2006. Real-time monitoring of chloroplast gene expression by a luciferase reporter: evidence for nuclear regulation of chloroplast circadian period. Mol. Cell. Biol. 26, 863—870.

Mayfield, S.P., Franklin, S.E., Lerner, R.A., 2003. .Expression and assembly of a fully active antibody in algae. Proc. Natl. Acad. Sci. U.S.A. 100, 438—442.

Mayfield, S.P., Franklin, S.E., 2005. Expression of human antibodies in eukaryotic micro-algae. Vaccine 23, 1828—1832.

Mayfield, S.P., Schultz, J., 2004. Development of a luciferase reporter gene, luxCt, for *Chlamydmonas reinhardtii* chloroplast. Plant J. 37, 449—458.

Minko, I., Holloway, S.P., Nikaido, S., Carter, M., Odom, O.W., Johnson, C.H., Herrin, D.L., 1999. Renilla luciferase as a vital reporter for chloroplast gene expression in *Chlamydomonas*. Mol. Gen. Genet. 262, 421—425.

Miyahara, M., Aoi, M., Inoue-Kashino, N., Kashino, Y., Ifuku, K., 2013. Highly efficient transformation of the diatom *Phaeodactylum tricornutum* by multi-pulse electroporation. Biosci. Biotechnol. Biochem. 77, 874—876.

Niu, Y.F., Yang, Z.K., Zhang, M.H., Zhu, C.C., Yang, W.D., Liu, J.S., Li, H.Y., 2012. Transformation of diatom *Phaeodactylum tricornutum* by electroporation and establishment of inducible selection marker. BioTechniques 1—3.

Olaizola, M., 2003. .Commercial development of microalgal biotechnology: from the test tube to the marketplace. Biomol. Eng. 20, 459—466.

Olaizola, M., 2000. Commercial production of astaxanthin from *Haematococcus pluvialis* using 25,000-liter outdoor photobioreactors. J. Appl. Phycol. 12, 499—506.

Oren, A., 2005. A hundred years of *Dunaliella* research: 1905—2005. Saline Syst. 1, 1—14.

Overton, T.W., 2014. Recombinant protein production in bacterial hosts. Drug Discov. Today 19, 590—601.

Rasala, B.A., Mayfield, S.P., 2014. Photosynthetic biomanufacturing in green algae; production of recombinant proteins for industrial, nutritional, and medical uses. Photosynth. Res. http://dx.doi.org/10.1007/s11120-014-9994-7.

Rasala, B.A., Lee, P.A., Shen, Z., Briggs, S.P., Mendez, M., Mayfield, S.P., 2012. Robust expression and secretion of xylanase1 in *Chlamydomonas reinhardtii* by fusion to a selection gene and processing with the FMDV 2A peptide. PLoS One 7 (8), e43349.

Ribeiro, J.K., Cunha, D.D., Fook, J.M., Sales, M.P., 2010. New properties of the soybean trypsin inhibitor: inhibition of human neutrophil elastase and its effect on acute pulmonary injury. Eur. J. Pharmacol. 644, 238—244.

Rodolfi, L., Chini Zittelli, G., Bassi, N., Padovani, G., Biondi, N., Bonini, G., Tredici, M.R., 2009. Microalgae for oil: strain selection, induction of lipid synthesis and outdoor mass cultivation in a low-cost photobioreactor. Biotechnol. Bioeng. 102, 100—112.

Rosales-Mendoza, S., 2013. Future directions for the development of *Chlamydmonas*-based vaccines. Expert Rev. Vaccines 12, 1011—1019.

Saei, A.A., Ghanbari, P., Barzegari, A., 2012. Haematococcus as a promising cell factory to produce recombinant pharmaceutical proteins. Mol. Biol. Rep. 39, 9931–9939.

Sarada, R., Bhattacharya, S., Ravishankar, G.A., 2002. Optimization of culture conditions for growth of the green alga *Haematococcus pluvialis*. World J. Microbiol. Biotechnol. 18, 517–521.

Schellekens, H., 2004. When biotech protein go off-patent. Trends Biotechnol. 22, 406–410.

Soria-Guerra, R.E., Ramírez-Alonso, J.I., Ibáñez-Salazar, A., Govea-Alonso, D.O., Paz-Maldonado, L.M.T., Bañuelos-Hernández, B., Korban, S.S., Rosales-Mendoza, S., 2014. Expression of an HBcAg-based antigen carrying angiotensin II in *Chlamydmonas reinhardtii* as a candidate hypertension vaccine. Plant Cell, Tissue Organ Cult. 116, 133–139.

Specht, E.A., Mayfield, S.P., 2014. Algae-based oral recombinant vaccines. Front. Microbiol. 17, 5–60.

Specht, E.A., Miyake-Stoner, S., Mayfield, S.P., 2010. Microalgae come of age as a platform for recombinant protein production. Biotechnol. Lett. 32, 1373–1383.

Su, Z.L., Qian, K.X., Tan, C.P., Meng, C.X., Qin, S., 2005. Recombination and heterologous expression of allophycocyanin gene in the chloroplast of *Chlamydmonas reinhardtii*. Acta Biochim. Biophys. Sin 37, 709–712.

Sukenik, A., Beardall, J., Kromkamp, J.C., Kopecký, J., Masojídek, J., Bergeijk, S.V., Gabai, S., Shaham, E., Yamshon, A., 2009. Photosynthetic performance of outdoor *Nannochloropsis* mass cultures under a wide range of environmental conditions. Aquat. Microb. Ecol. 56, 297–308.

Sun, M., Qian, K., Su, N., Chang, H., Liu, J., Shen, G., 2003. Foot-and-mouth disease virus VP1 protein fused with cholera toxin B subunit expressed in *Chlamydmonas reinhardtii* chloroplast. Biotechnol. Lett. 25, 1087–1092.

Surzycki, R., Greenham, K., Kitayama, K., Dibal, F., Wagner, R., Rochaix, J.D., Ajam, T., Surzycki, S., 2009.

Factors effecting expression of vaccines in microalgae. Biologicals 37, 133–138.

Thagun, C., Srisala, J., Sritunyalucksana, K., Narangajavana, J., Sojikul, P., 2012. Arabidopsis-derived shrimp viral-binding protein, PmRab7 can protect white spot syndrome virus infection in shrimp. J. Biotechnol. 161, 88–97.

Tran, M., Henry, R.E., Siefker, D., Van, C., Newkirk, G., Kim, J., Bui, J., Mayfield, S.P., 2013. Production of anti-cancer immunotoxins in algae: ribosome inactivating proteins as fusion partners. Biotechnol. Bioeng. 110, 2826–2835.

Tran, M., Zhou, B., Pettersson, P.L., Gonzalez, M.J., Mayfield, S.P., 2009. Synthesis and assembly of a full-length human monoclonal antibody in algal chloroplasts. Biotechnol. Bioeng. 104, 663–673.

Wang, X.F., Brandsma, M., Tremblay, R., Maxwell, D., Jevnikar, M.A., Huner, N., Ma, S., 2008. A novel expression platform for the production of diabetes-associated autoantigen human glutamic acid decarboxylase (hGAD65). BMC Biotechnol. 8, e87.

World Health Organization, 2013. WHO Guidelines on the Quality, Safety, and Efficacy of Biotherapeutic Protein Products Prepared by Recombinant DNA Technology. World Health Organization. Technical Report Series, No. 814.

Yang, Z., Li, Y., Chen, F., Li, D., Zhang, Z., Liu, Y., Zheng, D., Wang, Y., Shen, G., 2006. Expression of human soluble TRAIL in *Chlamydmonas reinhardtii* chloroplast. Chin. Sci. Bull. 51, 1703–1709.

Zaslavskaia, L.A., Lippmeier, J.C., Kroth, P.G., Grossman, A.R., Apt, K.E., 2000. Transformation of the diatom *Phaeodactylum tricornutum* (Bacillariophyceae) with a variety of selectable marker and reporter genes. J. Phycol. 36, 379–386.

Zhang, Y.K., Shen, G.F., Ru, B.G., 2006. Survival of human metallothionein-2 transplastomic *Chlamydmonas reinhardtii* to ultraviolet B exposure. Acta Biochim. Biophys. Sin. 38, 187–193.

Nutritional and Pharmaceutical Properties of Microalgal *Spirulina*

Thanh-Sang Vo[1], Dai-Hung Ngo[1], Se-Kwon Kim[1,2]

[1]Marine Bioprocess Research Center, Pukyong National University, Busan, Republic of Korea;
[2]Specialized Graduate School Science and Technology Convergence, Pukyong National University, Department of Marine-Bio Convergence Science, Busan, Republic of Korea

1. GENERAL CHARACTERISTICS OF *SPIRULINA*

1.1 Morphology

Spirulina is symbiotic, multicellular, and filamentous blue-green microalgae with symbiotic bacteria that fix nitrogen from air. It is recognizable by the arrangement of the multicellular cylindrical trichomes in an open left-hand helix along the entire length. The blue-green nonheterocystous filaments, composed of vegetative cells that undergo binary fission in a single plane, show easily visible transverse cross-walls. The presence of gas-filled vacuoles in the cells, together with the helical shape of the filaments, results in floating mats. The trichomes have a length of 50–500 µm and a width of 3–4 µm. The trichomes, enveloped by a thin sheath, show more or less slightly pronounced constrictions at cross-walls. Although the helical shape of the trichome is considered to be a stable and constant property maintained in culture, there may be considerable variation in the degree of helicity between different strains of the same species and within the same strain. The body surface of *Spirulina* is smooth and without covering, so it easily digestible by simple enzymatic systems. Its main photosynthetic pigment is phycocyanin, which is blue in color. It also contains chlorophyll *a* and carotenoids. Some contain the pigment phycoythrin, giving a red or pink color. *Spirulina* are photosynthetic and therefore autotrophic (Capelli and Cysewski, 2010; Habib et al., 2008; Ali and Saleh, 2002).

1.2 Natural Habitat and Source

The largest *Spirulina* lakes are found in Central Africa around Lakes Chad and Niger, Lake Texcoco, and in East Africa along the Great Rift Valley. Lakes Bodou and Rombou in Chad have a stable monoculture of *Spirulina* dating back centuries. It is also a major species in Kenya's Lakes Nakuru and Elementeita and Ethiopia's Lakes Aranguadi and Kilotes. *Spirulina* thrives in alkaline lakes, where it is difficult or impossible for other microorganisms to survive (Kebede and Ahlgren, 1996). The algae population grows rapidly, reaches a maximum density, and then dies off when nutrients are exhausted. A new seasonal cycle begins when decomposed algae release their nutrients or when more nutrients flow into the lake. *Spirulina* is found in soil, marshes, freshwater, brackish water, seawater, and thermal springs. Alkaline, saline water (>30 g/L) with high pH (8.5–11.0) favors good production of *Spirulina*, especially where there is a high level of solar radiation at altitude in the tropics. *Spirulina* is an obligate photoautotroph; thus, it cannot grow in the dark on media containing organic carbon compounds. It reduces carbon dioxide in the light and assimilates mainly nitrates. The main assimilation product of *Spirulina* photosynthesis is glycogen. *Spirulina* shows an optimum growth between 35 and 39 °C (Richmond, 1986).

2. NUTRITIONAL VALUES OF *SPIRULINA*

2.1 *Spirulina* Biochemical Composition

Microalgae are considered to be the actual producers of some highly bioactive macromolecules in marine resources, including carotenoids, long-chain polyunsaturated fatty acids (PUFAs), proteins, chlorophylls, vitamins, and unique pigments (Kay, 1991; Pasquet et al., 2011). Notably, *Spirulina* is one of the more promising microalgae, being rich in proteins, essential amino acids, PUFAs, vitamins, minerals, and many phytonutrients. *Spirulina* has a high protein concentration, with 60–70% of its dry weight, depending upon the source (Phang et al., 2000). It is a complete protein, containing all of the essential amino acids, including leucine, isoleucine, and valine, although with reduced amounts of methionine, cystine, and lysine when compared with standard proteins such as those from meat, eggs, or milk.

Spirulina has a high amount of PUFAs, 1.5–2.0% of 5–6% total lipid. In particular,

Spirulina is rich in γ-linolenic acid (36% of total PUFAs), and also provides γ-linolenic acid (ALA), linoleic acid (LA, 36% of total), stearidonic acid, eicosapentaenoic acid, docosahexaenoic acid, and arachidonic acid (Li and Qi, 1997). Moreover, it also contains relatively high concentrations of vitamin B1 (thiamine), B2 (riboflavin), B3 (nicotinamide), B6 (pyridoxine), B9 (folic acid), B12 (cyanocobalamin), vitamin C, vitamin D, and vitamin E. Notably, all of the essential minerals (about 7%) are available in *Spirulina*, including potassium, calcium, chromium, copper, iron, magnesium, manganese, phosphorus, selenium, sodium, and zinc. Furthermore, *Spirulina* contains many pigments, including chlorophyll *a*, xanthophyll, beta-carotene, echinenone, myxoxanthophyll, zeaxanthin, canthaxanthin, diatoxanthin, 3-hydroxyechinenone, beta-cryptoxanthin, oscillaxanthin, c-phycocyanin, and allophycocyanin. In addition, *Spirulina* contains about 13.5% carbohydrates, which is mainly composed of glucose, along with rhamnose, mannose, xylose, galactose, and two unusual sugars, including 2-*O*-methyl-L-rhamnose and 3-*O*-methyl-L-rhamnose (Habib et al., 2008; Koru, 2009).

2.2 Use of *Spirulina* as a Human Food

Spirulina has been used as an additive in a variety of human foods and animal feeds. The current use of this resource has three precedents: traditional, scientific, and technological development, and the so-called green tendency (Koru, 2009). It has long been used as food in Mexico, dating back to the Aztec civilization approximately 400 years ago. It was reported that *Spirulina maxima* was harvested from the Lake Texcoco, dried and sold for human consumption in a Tenochtitlán market (Sánchez et al., 2003). It is still being used as food by the Kanembu tribe in the Lake Chad area of the Republic of Chad where it is sold as dried bread called "dihe" (Fox, 1996; Belay, 2002). In 1967,

Spirulina was established as a "wonderful future food source" in the International Association of Applied Microbiology (Sasson, 1997) and is now widely cultured throughout the world. It has been produced commercially for the last 30 years for food and specialty feeds. Commercial *Spirulina* are normally produced in large outdoor ponds under controlled conditions or produced directly from lakes. Current production of *Spirulina* worldwide is estimated to be about 3000 metric tons. Currently, more than 70 per cent of *Spirulina* market is for human consumption, mainly as health food because of its rich content of protein, essential amino acids, minerals, vitamins, and essential fatty acids. *Arthrospira platensis* and *Arthrospira maxima*, which are commonly used as food, dietary supplement, and feed supplement (Wikfors and Ohno, 2001). Nowadays, *Spirulina* has been marketed and consumed as a human food and has been approved as a food for human consumption by many governments, health agencies, and associations of almost countries (Becker and Venkataraman, 1984; Vonshak, 2002; Koru, 2009; Henrikson, 2010).

2.3 Safety Assurance for *Spirulina*

Beside nutritional values, several cyanobacteria contain a certain level of toxin that has become a major issue in public health due to the increased occurrence of toxic cyanobacterial blooms. These toxic blooms contain algae that produce hepatotoxins called microcystins (Carmichael, 1994). That toxin with low levels of exposure may have chronic effects in humans (Martinez, 2007). Therefore, the strict and reliable control of these toxins should be carried out in order to prevent serious public health problems. For this reason, *Spirulina* has been subjected to extensive safety studies with human, animals, and fish in many countries (Cevallos et al., 2008). However, there has been no report of cyanobacterial toxins in *Spirulina* species to date. Although *Spirulina* does not normally contain toxins but contamination of outdoor culture by other cyanobacteria is a possibility. Thus, it is very unlikely to be a problem in proper control and management of monoculture of *Spirulina*.

3. PHARMACEUTICAL PROPERTIES OF *SPIRULINA*

Spirulina is a great product which can be used to improve overall health due to many essential nutrients for human body. There are several companies producing *Spirulina*, and the product is sold as a food supplement in many health food stores around the world. Recently, more attention has been given to the study of the therapeutic effects of *Spirulina*. A number of published studies suggest significant therapeutic effects of *Spirulina*. Many preclinical studies and a few clinical studies have suggested several therapeutic effects ranging from reduction of cholesterol and cancer to enhancement of the immune system, an increase in intestinal lactobacilli, reduction of nephrotoxicity by heavy metals and drugs, and radiation protection (Belay et al., 1993; Blinkova et al., 2001; Khan et al., 2005). In addition, it has been experimentally proven, in vivo and in vitro, that *Spirulina* is effective for treating certain allergies, anemia, cancer, hepatotoxicity, viral and cardiovascular diseases, hyperglycemia, hyperlipidemia, immunodeficiency, and inflammatory processes (Chamorro et al., 2002).

3.1 Antioxidant Activities

The antioxidant properties of *Spirulina* and its extracts have attracted the attention of many researchers. According to Miranda and colleagues, the antioxidant activity of a methanolic extract of *Spirulina* was determined in vitro and in vivo (Miranda et al., 1998). The in vitro antioxidant capacity of the methanolic extract of *Spirulina* was related to reduction of oxidation on a brain homogenate with an IC_{50} value of 0.18 mg/mL. The in vivo antioxidant capacity was evaluated in plasma and liver of animals receiving a daily dose of 5 mg for 2 and 7 weeks. Upon treatment, the antioxidant capacity of plasma was 71% for the experimental group and 54% for the control group. The antioxidant effect was suggested to be associated with beta-carotene, tocopherol, and phenolic compounds working individually or in synergy in *Spirulina*. Moreover, the neuroprotective effect of *Spirulina* against the harmful effect of free radicals was evaluated by Tobon-Velasco et al. (2013). The pretreatment with 700 mg/kg/day of *Spirulina* resulted in the reduction of various indicators of toxicity such as nitric oxide levels, ROS formation, lipoperoxidation, and mitochondrial activity in rat injected with a single dose of 6-hydroxydopamine, 6-OHDA (16 μg/2 μL). *Spirulina* supplement possesses free radical scavenging function against gamma-irradiation-induced oxidative stress and tissue damage in rats (Makhlouf and Makhlouf, 2012). Aqueous extract of *Spirulina* has a protective effect against apoptotic cell death via scavenging free radicals (Chu et al., 2010). In addition, *Spirulina* has been shown to exert protective effects against oxidative stress caused by lead acetate in the rat liver and kidney (Ponce-Canchihuaman et al., 2010). Notably, the c-phycocyanin from *Spirulina* has been found to be a strong free radical scavenger (Huang et al., 2007). Romay et al. (1998a) showed that phycocyanin was able to scavenge hydroxyl ($IC_{50} = 0.91$ mg/mL) and alkoxyl ($IC_{50} = 76$ μg/mL) radicals with activity equal to 0.125 mg/mL of dimethylsulfoxide and 0.038 μg/mL of trolox. In addition, the hepatoprotective effect of phycocyanin was exhibited due to its radical scavenging activity, and thus reducing the hepatoxicity caused by carbon tetrachloride and R-(+)-pulegone (Vadiraja et al., 1998). Studies in vitro and in vivo have shown that the antioxidant components produced by *Spirulina* (Reddy et al., 2000a,b; Kulshreshtha et al., 2008) can prevent or delay oxidative

damage by reducing the accumulation of ROS (Zhang et al., 2011) through the activation of the antioxidant enzyme systems of catalase, SOD, and glutathione peroxidase (Thaakur and Jyothi, 2007). Recently, Bhat and Madayastha reported that c-phycocyanin from *Spirulina* effectively inhibited CCl4-induced lipid peroxidation in rat liver in vivo with an IC_{50} of 11.35 μM (Bhat and Madyastha, 2000).

3.2 Anti-inflammatory Activities

C-phycocyanin is a protein-bound pigment soluble in water and found in some blue-green microalgae such as *Spirulina*, which is used in many countries as a dietary supplement. According to Romay et al. (1998b), phycocyanin shows anti-inflammatory activity in four experimental models of inflammation. It reduced significantly and in a dose-dependent manner ear edema induced by arachidonic acid and tetradecanoyl phorbol acetate in mice as well as carrageenan-induced rat paw oedema in intact and adrenalectomized animals. Moreover, phycocyanin also exerted an inhibitory effect in the cotton pellet granuloma test. In the further study, Romay and colleagues have revealed that phycocyanin exerts inhibitory effects on TNF-α and NO production in serum of mice treated with lipopolysaccharide (Romay et al., 2001). Furthermore, phycoryanin treatment (50–200 mg/kg) inhibited in a dose-dependent manner edema as well as PGE_2 and LTB_4 levels in the mouse ear treated with arachidonic acid (Romay et al., 1999, 2000). In another study, anti-inflammatory activities of phycocyanin was found to be related to attenuation of myeloperoxidase activity, inhibition of inflammatory cell infiltration, and reduction to some extent in colonic damage in rats (González et al., 1999). Similarly, Remirez et al. (1999) have found that phycoryanin significantly reduced the levels of β-glucuronidase, inhibited cellular infiltration, reducing synovial hyperplasia and synovitis in a zymosan-induced arthritis model in mice. Notably, c-phycocyanin from *Spirulina* was determined as a selective inhibitor of cyclooxygenase−2 (COX-2) with a very low IC_{50} COX-2/IC_{50} COX-1 ratio (0.04). Interestingly, it was shown that the IC_{50} value obtained for COX-2 inhibition by phycocyanin was much lower (180 nM) as compared to those for celecoxib (255 nM) and rofecoxib (401 nM), the well-known selective COX-2 inhibitors (Reddy et al., 2000a,b). It was suggested that the anti-inflammatory activities of phycocyanin might be due, in part, to its selective COX-2 inhibitory, anti-oxidative and oxygen free radical scavenging properties.

3.3 Anti-allergic Activities

Consumption of various edible microalgae can modulate both adaptive and innate aspects of immunity (Price et al., 2002), which may provide protective effect against allergic responses. Indeed, *Spirulina* are able to decrease IgE antibody level, and increase IgG1 and IgA antibody production in the serum of the mice immunized with crude shrimp extract as an antigen. Moreover, an enhancement of IgA antibody production was observed in culture supernatant of lymphoid cells, especially in the spleen and mesenteric lymph node from mice treated with *Spirulina* extract for 4 weeks before antigen stimulation (Hayashi et al., 1998). In a clinical trial, *Spirulina* consumption resulted in significant amelioration in the symptoms and physical findings of allergic rhinitis patients compared with placebo, including nasal discharge, sneezing, nasal congestion, and itching (Cingi et al., 2008). The clinical effect of *Spirulina* on allergic rhinitis was also determined due to inhibiting the production of IL-4 and thus may suppress the differentiation of Th2 cells (Mao et al., 2005). Moreover, *Spirulina* has been recognized to be a great inhibitor of allergic reaction via suppressing anaphylactic shock, passive cutaneous anaphylaxis, and serum histamine levels in rats activated by compound 48/80 or anti-DNP IgE. Further, the in vitro experiment

has confirmed that *Spirulina* inhibited histamine release and TNF-α production from rat peritoneal mast cells (Yang et al., 1997; Kim et al., 1998). In addition, *Spirulina* exhibited the protective effect against the lead-induced increase in mast cells in the ovary during the oestrous cycle of rats (Karaca and Simşek, 2007).

Recently, phycocyanin has been found to be an inhibitor of different allergic responses such as histamine release from rat peritoneal mast cells, ear swelling in mice induced by OVA, and skin reactions in rats caused by histamine and compound 48/80 (Remirez et al., 2002). Moreover, phycocyanin was shown to enhance biological defense activity against infectious diseases through suppressing antigen-specific IgE antibody and thus reducing allergic inflammation in mice (Nemoto-Kawamura et al., 2004). In the most recent study, Chang et al. (2011) have evaluated the therapeutic potential of R-phycocyanin (R-PC), a novel phycobiliprotein, against allergic airway inflammation. Interestingly, R-PC treatment resulted in a decrease of endocytosis and augmentation of IL-12 production in mouse BMDCs. Additionally, R-PC-treated dendritic cells promote CD4+ T-cell stimulatory capacity and increase IFN-γ expression in CD4+ T cells. Meanwhile, intraperitoneal administration of R-PC suppressed OVA-induced airway hyperresponsiveness, serum levels of OVA-specific IgE, eosinophil infiltration, Th2 cytokine levels, and eotaxin in bronchoalveolar lavage fluid of mice. These findings implied that R-PC promoted activation and maturation of cultured dendritic cells, and enhanced the immunological function toward Th1 activity. Taken together, *Spirulina* and its phycobiliprotein are expected to be a useful foodstuff for regulating allergic inflammatory responses.

3.4 Anti-cancer Activities

Spirulina, either alone or in combination with certain other compounds, has been studied for anti-cancer activities, and its role and mechanisms of actions were well-described through various pathways. Experimental studies in animal models have demonstrated an inhibitory effect of Spirulina algae on oral carcinogenesis. The local injection of phycotene (extract of Spirulina and Dunaliella algae) 250 μg into DMBA (7, 12 dimethylbenz(a)anthracene)-induced squamous cell carcinomas of hamster causes tumor regression. Total tumor regression was found in 30% of animals, while partial tumor regression was found in the remaining 70% of animals (Schwartz and Shklar, 1987). Moreover, the algae extract delivered by mouth in continued dosages of 140 μm for 28 weeks presented a complete absence of gross tumors induce by DMBA (Schwartz et al., 1988). Twice weekly injection of Spirulina-Dunaliella algae into DMBA-induced squamous cell carcinomas of hamster significantly induced the production of tumor necrosis factor in macrophages, suggesting a possible mechanism of tumor destruction (Shklar and Schwartz, 1988).

Akao and collaborators have reported that the hot-water extract of Spirulina when taken orally in adult human enhances NK activation (Hirahashi et al., 2002). Moreover, ingestion of Spirulina confers both IFN-γ production and NK-mediated Rae-1-positive cell killing activity in mice, thus leading to retardation of implant tumor growth in mice (Akao et al., 2009). In addition, the chemoprevention of cancer by Spirulina has been demonstrated in dibutyl nitrosamine induced rat liver toxicity and carcinogenesis (Hamidah et al., 2009). Spirulina supplementation reduced the incidence of liver tumors from 80% to 20%. Reduction of p53 was significant along with inhibition of cell proliferation, increased p21 and decreased Rb expression levels at 48 h post-treatment. Spirulina also increased Bax and decreased Bcl-2 expression, indicating induction of apoptosis. Notably, selenium-containing phycocyanin (Se-PC) showed potent antiproliferative properties in human melanoma A375 cells and human breast adenocarcinoma MCF-7 cells via induction of apoptosis, accumulation of sub-G1 cell populations, DNA fragmentation, and nuclear condensation (Chen and Wong, 2008). Calcium spirulan (Ca-SP), a sulfated polysaccharide, was found to be effective in reducing the lung metastasis of B16-BL6 melanoma cells, by inhibiting the tumor invasion of basement membrane, thus contrbuting to the prevention of adhesion and migration of tumor cells (Mishima et al., 1998). On the other hand, c-phycocyanin showed anticancer effects on human chronic myeloid leukemia cell line (K562) due to a decrease (49%) in the proliferation of K562 cells and down-regulating antiapoptotic Bcl-2 with no alterations in propoptotic Bax (Subhashini et al., 2004).

3.5 Anti-viral and Anti-bacterial Activities

According to Hayashi and co-workers, the water extract of Spirulina platensis was shown to inhibit the replication in vitro of HSV-1 in HeLa cells within the concentration range of 0.08−50 mg/mL (Hayashi et al., 1993). Addition of the S. platensis extract (1 mg/mL) at 3 h before infection causes blockade of virus-specific protein synthesis at 50% effective inhibition dose (ED 50) value of 0.173 mg/mL without affecting host cell protein synthesis. Moreover, it was observed that food containing the S. platensis extract effectively prolonged the survival time of infected hamsters at doses of 100 and 500 mg/kg per day. Subsequently, Hayashi et al. isolated from S. platensis a novel sulfated polysaccharide, calcium spirulan (Ca-SP), which inhibits the replication in vitro of several enveloped viruses including HSV-1, human cytomegalovirus, measles virus, mumps virus, influenza A virus, and HIV-1 virus. The anti-HSV-1 activity of Ca-SP was determined to be fivefold higher than that of dextran sulfate (Hayashi et al., 1996a,b). In a similar trend, other bioactive components from S. platensis were explored due to anti-HSV activity. As expected, a polysaccharide with rhamnose as the main sugar component and a glycolipid sulphoquinovosyl diacylglycerol were found to be active against HSV-1 at IC 50 values of 21.32 and 6.8 lg/mL, respectively (Chirasuwan et al., 2007, 2009). While S. platensis was efficient for anti-HSV-1, S. maxima exposed inhibitory activity against HSV-2. A hot water extract of S. maxima showed appreciable suppression HSV-2 infection at the initial events of adsorption and penetration (ED 50, 0.069 mg/mL) (Hernández-Corona et al., 2002). Recently, Ayehunie et al. (1998) reported that an aqueous extract of S. platensis inhibited HIV-1 replication in human T-cell lines, peripheral blood mononuclear cells, and Langerhans cells. The extract inactivated HIV-1 infectivity directly when preincubated with virus before addition to human T-cell lines.

Microalgal cultures of Spirulina have displayed significant anti-bacterial activity against six Vibrio strains: Vibrio parahaemolyticus, Vibrio anguillarum, Vibrio splendidus, Vibrio scophthalmi, Vibrio alginolyticus, and Vibrio lentus (Kokou et al., 2012). Moreover, purified C-phycocyanin from S. platensis markedly inhibited the growth of some drug-resistant bacteria such as Escherichia coli, Klebsiella pneumoniae, Pseudomonas aeruginosa, and Staphylococcus aureus (Sarada et al., 2011).

3.6 Anti-diabetes and Anti-obesity Activities

Recently, Spirulina has also evidenced to possess anti-diabetes and anti-obesity activities. The water-soluble fraction of Spirulina was found to be effective in lowering the serum glucose level at fasting while the water-insoluble fraction suppressed the glucose level at glucose loading (Takai et al., 1991). Following 2 g/day for 21 days of Spirulina supplementation, it was observed that the fasting blood sugar level of patients was significantly decreased

(Mani et al., 1998). Meanwhile, Becker et al. (1986) have found that a supplementary diet of 2.8 g of Spirulina three times/day over 4 weeks resulted in a statistically significant reduction of body weight in obese outpatients. Spirulina has also been found to suppress high blood pressure in rats (Iwata et al., 1990).

3.7 Other Biological Activities

The protective role of dietary Spirulina in lead toxicity in Swiss albino mice was investigated by Shastri et al. (1999). Dietary Spirulina resulted in a significant increase of the survival time as compared with a control group without Spirulina. The crude ethanol precipitate of S. platensis was studied due to its radioprotective effect (Qishen et al., 1989). The extract caused a significant reduction of micronucleus frequencies induced by γ-radiation. It was concluded that the protective extract probably acts as a DNA-stabilizing factor, and they ruled out the possibility of a radical scavenging mechanism. Moreover, Spirulina exhibited a significant reduction of the low-density lipoprotein to high-density lipoprotein ratio after 4 months of supplementation (Mittal et al., 1999). Spirulina supplementation decreased the levels of plasma lipid concentration and reduced blood pressure by promoting vasodilation and restricting vasoconstriction (Schwartz et al., 1988).

4. CONCLUSION

Spirulina is well known as a non-toxic and nutritious food with various health beneficial effects. It is believed that Spirulina is a potential source of wonder food supplement and alternative medicine. However, the harvest of Spirulina may be contaminated with harmful elements and toxic substances that can potentially cause several serious human health effects. Moreover, Spirulina may contain toxins called microcystins, which accumulate in the liver and cause cancer or other liver diseases. Thus, the extensive studies of *Spirulina* will contribute to the development of safe health food.

References

Akao, Y., Ebihara, T., Masuda, H., Saeki, Y., Akazawa, T., Hazeki, K., Hazeki, O., Matsumoto, M., Seya, T., 2009. Enhancement of antitumor natural killer cell activation by orally administered *Spirulina* extract in mice. Cancer Sci. 100, 1494–1501.

Ali, S., Saleh, A.M., 2002. *Spirulina*—an overview. Int. J. Pharm. Pharm. Sci. 4, 9–15.

Ayehunie, S., Belay, A., Baba, T.W., Ruprecht, R.M., 1998. Inhibition of HIV-1 replication by an aqueous extract of *Spirulina platensis* (*Arthrospira platensis*). J. Acquir. Immune Defic. Syndr. 18, 7–12.

Becker, E.W., Venkataraman, L.V., 1984. Production and utilization of the blue-green alga *Spirulina* in India. Biomass 4, 105–125.

Becker, E.W., Jakover, B., Luft, D., Schmuelling, R.M., 1986. Clinical and biochemical evaluations of the alga *Spirulina* with regard to its application in the treatment of obesity: a double-blind cross-over study. Nutr. Rep. Int. 33, 565–574.

Belay, A., 2002. The potential application of *Spirulina* (*Arthrospira*) as a nutritional and therapeutic supplement in health management. JANA 5, 1–24.

Belay, A., Ota, Y., Miyakawa, K., Shimamatsu, H., 1993. Current knowledge on potential health benefits of *Spirulina*. J. Appl. Phycol. 5, 235–241.

Bhat, V.B., Madyastha, K.M., 2000. C-phycocyanin: a potent peroxyl radical scavenger *in vivo* and *in vitro*. Biochem. Biophys. Res. Commun. 1, 20–25.

Blinkova, L.P., Gorobets, O.B., Baturo, A.P., 2001. Biological activity of *Spirulina*. Zh. Mikrobiol. Epidemiol. Immunobiol. 2, 114–118.

Capelli, B., Cysewski, G.R., 2010. Potential health benefits of *Spirulina* microalgae. Nutra. Foods 9, 19–26.

Carmichael, W., 1994. The toxins of cyanobacteria. Scientific Am. 279, 78–86.

Cevallos, C.G., Barrón, B.L., Vázquez-Sánchez, J., 2008. Toxicologic studies and antitoxic properties of *Spirulina*. In: Gershwin, M.E., Belay, A. (Eds.), *Spirulina* in Human Nutrition and Health. CRC Press Taylor & Francis Group, Boca Raton, FL, 33487-2742, 42–65.

Chamorro, G., Salazar, M., Araujo, K.G., dos Santos, C.P., Ceballos, G., Castillo, L.F., 2002. Update on the pharmacology of *Spirulinao* (*Arthrospira*), an unconventional food. Arch. Latinoam. Nutr. 52, 232–240.

Chang, C.J., Yang, Y.H., Liang, Y.C., Chiu, C.J., Chu, K.H., Chou, H.N., Chiang, B.L., 2011. A novel phycobiliprotein alleviates allergic airway inflammation by modulating immune responses. Am. J. Respir. Crit. Care Med. 183, 15–25.

Chen, T., Wong, Y.S., 2008. *In vitro* antioxidant and antiproliferative activities of selenium-containing phycocyanin from selenium-enriched *Spirulina platensis*. J. Agric. Food Chem. 56, 4352–4358.

Chirasuwan, N., Chaiklahan, R., Kittakoop, P., Chanasattru, W., Ruengjitchatchawalya, M., Tanticharoen, M., Bunnag, B., 2009. Anti HSV-1 activity of sulphoquinovosyl diacylglycerol isolated from *Spirulina platensis*. Sci. Asia 35, 137–141.

Chirasuwan, N., Chaiklahan, R., Ruengjitchatchawalya, M., Bunnag, B., Tanticharoen, M., 2007. Anti HSV-1 activity of *Spirulina platensis* polysaccharide. Kasetsart J. (Nat. Sci.) 41, 311–318.

Chu, W.L., Lim, Y.W., Radhakrishnan, A.K., Lim, P.E., 2010. Protective effect of aqueous extract from *Spirulina platensis* against cell death induced by free radicals. BMC Complement. Altern. Med. 10, 53.

Cingi, C., Conk-Dalay, M., Cakli, H., Bal, C., 2008. The effects of *Spirulina* on allergic rhinitis. Eur. Arch. Otorhinolaryngol. 265, 1219–1223.

Fox, D.R., 1996. *Spirulina*: Production & Potential. Edisud, France, p. 232.

González, R., Rodriguez, S., Romay, C., 1999. Anti-inflammatory activity of phycocyanin extract in acetic acid-induced colitis in rats. Pharmacol. Res. 39, 55–59.

Habib, M.A.B., Parvin, M., Huntington, T.C., Hasan, M.R., 2008. A Review on Culture, Production and Use of Spirulina as Food for Humans and Feeds for Domestic Animals and Fish. FAO Fisheries and Aquaculture Circular. No. 1034. FAO, Rome, 33p.

Hamidah, A., Rustam, Z.A., Tamil, A.M., Zarina, L.A., Zulkifli, Z.S., Jamal, R., 2009. Prevalence and parental perceptions of complementary and alternative medicine use by children with cancer in a multi-ethnic Southeast Asian population. Pediatr. Blood Cancer 52, 70–74.

Hayashi, K., Hayashi, T., Kojima, I., 1996a. A natural sulfated polysaccharide, calcium spirulan, isolated from *Spirulina platensis*: *in vitro* and *ex vivo* evaluation of anti-herpes simplex virus and anti-human immunodeficiency virus activities. AIDS Res. Hum. Retroviruses 12, 1463–1471.

Hayashi, K., Hayashi, T., Morita, N., 1993. An extract from *Spirulina platensis* is a selective inhibitor of herpes simplex virus type 1 penetration into HeLa cells. Phytother. Res. 7, 76–80.

Hayashi, O., Hirahashi, T., Katoh, T., Miyajima, H., Hirano, T., Okuwaki, Y., 1998. Class specific influence of dietary *Spirulina platensis* on antibody production in mice. J. Nutr. Sci. Vitaminol. 44, 841–851.

Hayashi, T., Hayashi, K., Maedaa, M., Kojima, I., 1996b. Calcium spirulan, an inhibitor of enveloped virus replication, from a blue-green alga *Spirulina platensis*. J. Nat. Prod. 59, 83—87.

Henrikson, R., 2010. *Spirulina*: World Food, How This Micro Algae Can Transform Your Health and Our Planet. Ronore Enterprises, Inc., PO Box 909, Hana, Maui, Hawaii 96718 USA, ISBN 1453766987, p. 195.

Hernández-Corona, A., Nieves, I., Meckes, M., Chamorro, G., Barron, B.L., 2002. Antiviral activity of *Spirulina maxima* against herpes simplex virus type 2. Antivir. Res. 56, 279—285.

Hirahashi, T., Matsumoto, M., Hazeki, K., Saeki, Y., Ui, M., Seya, T., 2002. Activation of the human innate immune system by *Spirulina*: augmentation of interferon gamma production and NK cytotoxicity by oral administration of *Spirulina*. Int. Immunopharmac. 2, 423—434.

Huang, Z., Guo, B.J., Wong, R.N.S., Jiang, Y., 2007. Characterization and antioxidant activity of selenium-containing phycocyanin isolated from *Spirulina platensis*. Food Chem. 100, 1137—1143.

Iwata, K., Munakata, K., Inayama, T., Kato, T., 1990. Effect of *Spirulina platensis* on blood pressure in rats. Bull. Kagawa Nutr. Univ. 21, 63—70.

Karaca, T., Simşek, N., 2007. Effects of *Spirulina* on the number of ovary mast cells in lead-induced toxicity in rats. Phytother. Res. 21, 44—46.

Kay, R.A., 1991. Microalgae as food and supplement. Crit. Rev. Food Sci. Nutr. 30, 555—573.

Kebede, E., Ahlgren, G., 1996. Optimum growth conditions and light utilization efficiency of *Spirulina platensis* (= *Arthrospira fusiformis*) (Cyanophyta) from Lake Chitu, Ethiopia. Hydrobiology 332, 99—109.

Khan, Z., Bhadouria, P., Bisen, P.S., 2005. Nutritional and therapeutic potential of *Spirulina*. Curr. Pharm. Biotechnol. 6, 373—379.

Kim, H.M., Lee, E.H., Cho, H.H., Moon, Y.H., 1998. Inhibitory effect of mast cell-mediated immediate-type allergic reactions in rats by *Spirulina*. Biochem. Pharmacol. 55, 1071—1076.

Kokou, F., Makridis, P., Kentouri, M., Divanach, P., 2012. Anti-bacterial activity in microalgae cultures. Aquacult. Res. 43, 1520—1527.

Koru, E., 2009. *Spirulina* microalgae production and breeding in commercial. Turk. J. Agric. 11, 133—134.

Kulshreshtha, A., Zacharia, A.J., Jarouliya, U., Bhadauriya, P., Prasad, G.B., Bisen, P.S., 2008. *Spirulina* in health care management. Curr. Pharm. Biotechnol. 9, 400—405.

Li, D.M., Qi, Y.Z., 1997. *Spirulina* industry in China: present status and future prospects. J. Appl. Phycol. 9, 25—28.

Makhlouf, R., Makhlouf, I., 2012. Evaluation of the effect of *Spirulina* against gamma irradiation-induced oxidative stress and tissue injury in rats. Int. J. Appl. Sci. Eng. Res. 1, 152—164.

Mani, S., Iyer, U., Subramanian, S., 1998. Studies on the effect of *Spirulina* supplementation in control of diabetes mellitus. In: Subramanian, G., et al. (Eds.), Cyanobacterial Biotechnology. Science Publishers Inc., USA, pp. 301—304.

Mao, T.K., Van de Water, J., Gershwin, M.E., 2005. Effects of a *Spirulina*-based dietary supplement on cytokine production from allergic rhinitis patients. J. Med. Food 8, 27—30.

Martinez, G.A., 2007. Hepatotoxic cyanobacteria. In: Botana, M.L. (Ed.), Phycotoxins: Chemistry and Biochemistry. Blackwell Publishing, p. 259.

Miranda, M.S., Cintra, R.G., Barros, S.M., Mancini-Filho, J., 1998. Antioxidant activity of the microalga *Spirulina maxima*. Braz. J. Med. Biol. Res. 31, 1075—1079.

Mishima, T., Murata, J., Toyoshima, M., Fujii, H., Nakajima, M., Hayashi, T., Kato, T., Saiki, I., 1998. Inhibition of tumor invasion and metastasis by calcium spirulan (Ca-SP), a novel sulfated polysaccharide derived from a blue-green alga, *Spirulina platensis*. Clin. Exp. Metastasis 16, 541—550.

Mittal, A., Kumar, P.V., Banerjee, S., Rao, A.R., Kumar, A., 1999. Modulatory potential of *Spirulina fusiformis* on carcinogen metabolizing enzymes in Swiss albino mice. Phytother Res. 13, 111—114.

Nemoto-Kawamura, C., Hirahashi, T., Nagai, T., Yamada, H., Katoh, T., Hayashi, O., 2004. Phycocyanin enhances secretary IgA antibody response and suppresses allergic IgE antibody response in mice immunized with antigen-entrapped biodegradable microparticles. J. Nutr. Sci. Vitaminol. 50, 129—136.

Pasquet, V., Chérouvrier, J.R., Farhat, F., Thiéry, V., Piot, J.M., Bérard, J.B., Kaas, R., Serive, B., Patrice, T., Cadoret, J.P., Picot, L., 2011. Study on the microalgal pigments extraction process: performance of microwave assisted extraction. Process. Biochem. 46, 59—67.

Phang, S.M., Miah, M.S., Chu, W.L., Hashim, M., 2000. *Spirulina* culture in digested sago starch factory waste water. J. Appl. Phycol. 12, 395—400.

Ponce-Canchihuaman, J.C., Perez-Mendez, O., Hernandez-Munoz, R., Torres-Duran, P.V., Juarez-Oropeza, M.A., 2010. Protective effects of *Spirulina maxima* on hyperlipidaemia and oxidative-stress induced by lead acetate in the liver and kidney. Lipids Health Dis. 9, 35.

Price, J.A., Sanny, C., Shevlin, D., 2002. Inhibition of mast cells by algae. J. Med. Food 5, 205—210.

Qishen, P., Baojiang, G., Kolman, A., 1989. Radioprotective effect of extract from *Spirulina platensis* in mouse bone marrow cells studied by using the micronucleus test. Toxicol. Lett. 48, 165—169.

Reddy, C.M., Bhat, V.B., Kiranmai, G., Reddy, M.N., Reddanna, P., Madyastha, K.M., 2000a. Selective inhibition of cyclooxygenase-2 by C-phycocyanin, a biliprotein from *Spirulina platensis*. Biochem. Biophys. Res. Commun. 277, 599–603.

Reddy, C.M., Bhat, V.B., Kiranmai, G., Reddy, M.N., Reddanna, P., Madyastha, K.M., 2000b. Selective inhibition of cyclooxygenase-2 by C-phycocyanin, a biliprotein from *Spirulina platensis*. Biochem. Biophys. Res. Commun. 3, 599–603.

Remirez, D., González, A., Merino, N., González, R., Ancheta, O., Romay, C., Rodríguez, S., 1999. Effect of phycocyanin in zymosan-induced arthritis in mice-phycocyanin as an antiarthritic compound. Drug Develop. Res. 48, 70–75.

Remirez, D., Ledón, N., González, R., 2002. Role of histamine in the inhibitory effects of phycocyanin in experimental models of allergic inflammatory response. Mediat. Inflamm. 11, 81–85.

Richmond, A.E., 1986. Microalgae. CRC Critical Reviews in Biotechnology. Vol. 4, Issue 4. CRC Press, Boca Raton, FL, pp. 349–438.

Romay, C., Armesto, J., Remirez, D., Gonzalez, R., Ledon, L., Garcia, I., 1998a. Antioxidant and anti-inflammatory properties of c-phycocyanin from blue-green algae. Inflamm. Res. 47, 36–41.

Romay, C., Delgado, R., Remirez, D., González, R., Rojas, A., 2001. Effects of phycocyanin extract on tumor necrosis factor-alpha and nitrite levels in serum of mice treated with endotoxin. Arzneimittelforschung 51, 733–736.

Romay, C., Ledón, N., González, R., 1998b. Further studies on anti-inflammatory activity of phycocyanin in some animal models of inflammation. Inflamm. Res. 47, 334–338.

Romay, C., Ledón, N., González, R., 1999. Phycocyanin extract reduces leukotriene B4 levels in arachidonic acid-induced mouse-ear inflammation test. J. Pharm. Pharmacol. 51, 641–642.

Romay, C., Ledón, N., González, R., 2000. Effects of phycocyanin extract on prostaglandin E2 levels in mouse ear inflammation test. Arzneimittelforschung 50, 1106–1109.

Sánchez, M., Castillo, B.J., Rozo, C., Rodríguez, I., 2003. *Spirulina* (*Arthrospira*): an edible microorganism: a review. Universitas Scientiarum 8, 1–16.

Sarada, D.V.L., Kumar, C.S., Rengasamy, R., 2011. Purified C-phycocyanin from *Spirulina platensis* (Nordstedt) Geitler: a novel and potent agent against drug resistant bacteria. World J. Microb. Biot. 27, 779–783.

Sasson, A., 1997. Micro Biotechnologies: Recent Developments and Prospects for Developing Countries. BIOTEC Publication 1/2542. United Nations Educational, Scientific and Cultural Organization (UNESCO), Place de Fontenoy, Paris, France, pp. 11–31.

Schwartz, J., Shklar, G., 1987. Regression of experimental hamster cancer by beta carotene and algae extracts. J. Oral Maxillofac. Surg. 45, 510–515.

Schwartz, J., Shklar, G., Reid, S., Trickler, D., 1988. Prevention of experimental oral cancer by extracts of *Spirulina-Dunaliella* algae. Nutr. Cancer 11, 127–134.

Shastri, D., Kumar, M.M.M., Kumar, A., 1999. Modulation of lead toxicity by *Spirulina fusiformis*. Phytother. Res. 13, 258–260.

Shklar, G., Schwartz, J., 1988. Tumor necrosis factor in experimental cancer regression with alphatocopherol, beta-carotene, canthaxanthin and algae extract. Eur. J. Cancer Clin. Oncol. 24, 839–850.

Subhashini, J., Mahipal, S.V., Reddy, M.C., Mallikarjuna Reddy, M., Rachamallu, A., Reddanna, P., 2004. Molecular mechanisms in C-phycocyanin induced apoptosis in human chronic myeloid leukemia cell line-K562. Biochem. Pharmacol. 68, 453–462.

Takai, Y., Hossayamada, Y., Kato, T., 1991. Effect of water soluble and water insoluble fractions of *Spirulina* over serum lipids and glucose resistance of rats. J. Jpn Soc. Nutr. Food Sci. 44, 273–277.

Thaakur, S.R., Jyothi, B., 2007. Effect of *Spirulina maxima* on the haloperidol induced tardive dyskinesia and oxidative stress in rats. J. Neural Transm. 114, 1217–1225.

Tobon-Velasco, J.C., Palafox-Sanchez, V., Mendieta, L., García, E., Santamaría, A., Chamorro-Cevallos, G., Limón, I.D., 2013. Antioxidant effect of *Spirulina* (*Arthrospira*) *maxima* in a neurotoxic model caused by 6-OHDA in the rat striatum. J. Neural Transm. 120, 1179–1189.

Vadiraja, B., Gaikwad, N., Madyastha, K., 1998. Hepatoprotective effect of C-phycocyanin: protection for carbon tetrachloride and R-(+)-pulegone-mediated hepatotoxicity in rats. Biochem. Biophys. Res. Commun. 249, 428–431.

Vonshak, A., 2002. Use of *Spirulina* biomass. In: Vonshak, A. (Ed.), *Spirulina platensis* (*Arthrospira*) Physiology Cell Biology and Biotechnology. Taylor & Francis, London, ISBN 0-203-48396-0, pp. 159–173.

Wikfors, H.G., Ohno, M., 2001. Impact of algal research in aquaculture. J. Phycol. 37, 968–974.

Yang, H.N., Lee, E.H., Kim, H.M., 1997. *Spirulina platensis* inhibits anaphylactic reaction. Life Sci. 61, 1237–1244.

Zhang, H., Chen, T., Jiang, J., Wong, Y.S., Yang, F., Zheng, W., 2011. Selenium-containing allophycocyanin purified from selenium-enriched *Spirulina platensis* attenuates AAPH-induced oxidative stress in human erythrocytes through inhibition of ROS generation. J. Agric. Food Chem. 59, 8683–8690.

Applications of Microalgae-Derived Active Ingredients as Cosmeceuticals

BoMi Ryu[1], S.W.A. Himaya[2], Se-Kwon Kim[3,4]

[1]School of Pharmacy, The University of Queensland, Brisbane, QLD, Australia;
[2]The Institute for Molecular Bioscience, The University of Queensland, St Lucia, QLD, Australia;
[3]Marine Bioprocess Research Center, Pukyong National University, Busan, Republic of Korea;
[4]Specialized Graduate School Science and Technology Convergence, Pukyong National University,
Department of Marine-Bio Convergence Science, Busan, Republic of Korea

1. INTRODUCTION

Microalgae use by humans was reported approximately 2000 years ago in China, when people used *Nostoc* to survive during a period of famine; edible *Spirulina* and *Aphanizomenon* species also have been used for many centuries in Africa and Mexico (Jensen et al., 2001; Olaizola, 2003; Spolaore et al., 2006). Attempts to develop commercial products from microalgae were adopted in the late 1800s and early 1900s. In 1910, Allen and Nelson in Berlin, Germany, proposed a source of lipids that could be used as food for aquaculture purposes; Harder and von Witsch proposed their use as fuel; and the commercialization of *Chlorella* production for nutrition succeeded in Japan, Taiwan, and the United States in the 1960s (de la Noue and de Pauw, 1988; Harder and von Witsch, 1942; Preisig and Andersen, 2005).

With the commercial production of microalgae, microalgae research grew and diversified significantly (Kawaguchi, 1980). Interest in the potential of microalgae has been concentrated on their wide range of compositions and metabolites, such as proteins, lipids, carbohydrates, and pigments for health products, food and feed additives, cosmetics, and energy sources (Arad and Yaron, 1992; Christaki et al., 2011; Guedes et al., 2011; Markou and Nerantzis, 2013; Skjanes et al., 2013; Stolz and Obermayer, 2005).

Microalgae (photosynthetic planktons) are able to regenerate and protect themselves from stress in severe conditions. They have protective mechanisms to prevent the accumulation of free radicals and reactive oxygen species, thus counteracting cell-damaging activities. Thus, much interest in microalgae has resulted because they possess valuable characteristics for the cosmeceutical field. The properties of microalgae are rapidly gaining importance as a natural substitute for synthetic components in cosmetics. Microalgae can be used as a cosmetic ingredient

Handbook of Marine Microalgae
http://dx.doi.org/10.1016/B978-0-12-800776-1.00020-0

in two ways: as an excipient in the formulation, such as a stabilizer or emulsifier, or as the pharmaceutical agent itself.

This chapter provides a broad review of the application of microalgae as cosmetics or cosmeceutical sources, as well as the high-value molecules extracted from these microorganisms.

2. COSMECEUTICAL VALUE OF MICROALGAE

Although microalgae blooms mainly occur in aquatic habitats, they also could inhabit severe conditions, including high or low temperatures, anaerobiosis, high salinity, photo-oxidation, high osmotic pressure, and ultraviolet radiation. As a result of microalgae's response and adaptation to environmental change, they have received much attention due to their more efficient utilization of their abundant availability that cannot be found in other organisms (Chu et al., 2005; Kaya et al., 2005; Teoh et al., 2004; Wong et al., 2007).

2.1 Commercialized Cosmetics and Cosmeceuticals

Many commercialized products using microalgae as cosmetics and cosmeceuticals are manufactured based on the activity of the microalgae itself or its ingredients; however, research to find the potential sources has not always resulted in commercial applications. A few sources of microalgae that are commercialized in the skin care market include antiaging creams, refreshing/regenerating care products, emollients and anti-irritants in peelers, sunscreen cream, and hair care products (Table 1). These cosmetics and cosmeceuticals include the algae in an extract formulation or bioactive components derived from microalgae. The products could offer promising and innovative alternatives to existing cosmetics and drives the development of new functions for cosmetic products (Table 2).

By the early 2000s, numerous companies in Europe and the United States started to launch cosmetics that were made by extraction of microalgae such as *Spirulina*, *Chlorella*, *Arthrospira*, *Anacystis*, *Halymenia*, *Nannochloro*, and *Dunaliella*. These microalgae act on the epidermis to erase vascular imperfections, boost collagen synthesis, and possibly prevent wrinkle formation eventually (Stolz and Obermayer, 2005).

Phytomer (France) used *Chlorella vulgaris* extraction, which is claimed to neutralize inflammation and improve the skin's natural protection. Skinicer, produced by Ocean Pharma in Germany, included the extraction of *Arthrospira maxima* to strengthen the skin's natural protection and regeneration, thus allowing the preservation of the youthful characteristics of the skin. Photosomes and XCELL-30 from Estee Lauder and Greentech, respectively, in the United States also used microalgae to strengthen skin's immunity. Pentapharm in Switzerland produced Pepha-Tight using extraction of *Nannochloropsis oculata*, which is reported to have anti aging properties via skin contraction and regeneration.

Microalgae are known as potential sources for high-value products, including polyunsaturated fatty acids (PUFAs; e.g., γ-linoleic acid or alguronic acid), polysaccharides, carotenoids, and phycobiliproteins (e.g., phycocyanin or phycoerythrin), which have protective activity against stress. Therefore, microalgae are cultured on a large scale for the production of these chemicals (Table 1). Efforts to obtain the active chemicals have led to the development of advanced cosmetics, which act on a specific target on the skin.

Pepha-Ctive by Pentapharm (Switzerland) and Dermochlorella by Codif Recherche and Nature (France) contain rich carotenoids from *Dunaliella salina* and *C. vulgaris*, respectively, that are known to have protective activity against ultraviolet light and oxidative damage,

TABLE 1   Microalgae in Commercialized Cosmetics

Species	Product	Ingredient	Activity claimed	Company and references
Spirulina spp.	*Spirulina firming algae mask*	Extract	Improves moisture balance of skin and strengthens skin's immunity	Optimum Derma Aciditate (ODA, Great Britain) Chakdar et al. (2012)
Spirulina spp.	*Spirulina whitening facial mask*	Extract	Improves skin complexion, reduces wrinkles	Ferenes Cosmetics Chakdar et al. (2012)
Chlorella Vulgaris	Phytomer	Extract	Neutralizes inflammation and improves skin's natural protection	Phytomer (France)
Arthrospira maxima	Skinicer	Extract	Strengthens skin's natural protection and regeneration	Ocean Pharma (Germany)
Anacystis nidulans	Photosomes	Extract	Protects the skin against sun damage and strengthens skin's immunity	Estee Lauder (USA)
Halymenia durvillei	XCELL-30	Extract	Improves radiance and skin luminosity	Greentech (USA)
Nannochloropsis oculata	Pepha-Tight	Extract	Excellent skin-tightening properties (short- and long-term effects)	Pentapharm LTD (Switzerland) Stolz and Obermayer (2005)
Dunaliella salina	Pepta- Ctive	β-Carotene	Stimulates cell proliferation and turnover and positively influences the energy metabolism of skin	Pentapharm LTD (Switzerland) Stolz and Obermayer (2005)
Spirulina platensis	Protulines	γ-Linoleic acid	Help combat early skin aging, exerting a tightening effect and preventing wrinkle formation	Exsymol S.A.M (Monaco) Stolz and Obermayer (2005), Sanatur GmbH (Germany)
Anacystis nidulans	Algenist	Alguronic acid	Strengthens skin's immunity	Solazyme (France) Harwati (2013)
Porphyridium spp.	Alguard	Sulfated polysaccharide	Protects the skin against sun damage and microorganisms	Frutarom (Israel) Harwati (2013)

Continued

TABLE 1   Microalgae in Commercialized Cosmetics—cont'd

Species	Product	Ingredient	Activity claimed	Company and references
C. vulgaris	Dermochlorella	Cartenoids	Stimulate collagen synthesis in skin supporting tissue regeneration and wrinkle reduction	Codif Recherche and Nature (France) Stolz and Obermayer (2005) Microlife S.R.L. (Italy)
Spirulina spp.	Lina-Blue	Phycocyanin	Cosmetics (eye shadow)	Dainippon Ink and Chemicals Inc.(Japan) Arad et al. (1997), Arad and Yaron (1992), and Hirata et al. (2000)
Porphyra spp., Porphyridium spp.	–	Phycoerythrin	Cosmetics (face powder cake, eye shadow)	Dainippon Ink and Chemicals Inc.(Japan) Arad et al. (1997)

which can lead to premature aging and other disorders. Carotenoids are a diverse class of naturally occurring pigments in photosynthesis that has been found to have a defensive role in the protection of cells and tissues from oxidative stress. The antioxidant activities of carotenoids are mostly based on their lipophilicity, which is likely to accumulate in lipophilic compartments such as membranes or lipoproteins (Stahl and Sies, 2003; Tapiero et al., 2004).

In addition to carotenoids, PUFAs (which have high total lipid proportions) are used as a lipophilic source, with significant antioxidant activity (Sahu et al., 2013; Zhukova and Aizdaicher, 1995). Protulines (Exsymol S.A.M, Monaco, Sanatur GmbH, Germany) and Algenist (Solazyme, France) contain PUFAs, γ-linoleic acid, and alguronic acid (Harwati, 2013; Stolz and Obermayer, 2005). Sulfated polysaccharides from red microalgae Porphyridium spp. are used in Alguard (Frutarom, Israel), which has protective activity from sun damage and microorganisms.

Phycobiliproteins (mainly phycocyanin and phycoerythrin) are photosynthetic pigments used as cosmetic dyes. For example, Linablue, which is used in products such as face powder and eye shadow, is produced using Spirulina, Porphyra, or Porphyridium by Dainippon Ink and Chemicals (Japan) (Arad et al., 1997).

2.2 Valuable Sources of Microalgae as Cosmetics and Cosmeceuticals

Research on microalgae's activities as cosmetics and cosmeceuticals has led to the development of a multitude of products. Cosmetics contain active components derived from microalgae, such as PUFAs, carotenoids, polysaccharides, and phycobiliproteins, which are incorporated into cream, lotions, and ointments, as discussed previously (Table 1). These sources are considered to have functional cosmeceutical effects or drug-like effects on skin that affect its appearance. Some components derived from microalgae are claimed to have potential as cosmetics and cosmeceuticals, but they have not yet been commercialized. Some examples are examined in this section.

Mycosporine-like amino acids (MAAs) from Spirulina, Chlorella, and Dunaliella are known to act as sunscreens to reduce ultraviolet light-induced damage (Atkin et al., 2006; Balskus and Walsh, 2010; Dionisio-Se Se, 2010; Garciapichel et al., 1993; Priyadarshani and Rath, 2012). MAAs are small secondary metabolites

TABLE 2 Potential Application of Microalgae as Cosmetics

Species	Group	Components	Claimed activity	References
Spirulina spp. *Gloeocapsa* spp.	Cyanobacteria	Mycosporine-like amino acids (MAAs)	Protect against sun damage	Balskus and Walsh (2010) and Garciapichel et al. (1993)
Spirulina platensis, Dunaliella salina, Chlorella vulgaris, and *Nannochloropsis oculata*	Cyanobacteria, Chlorophyta, Heterokontophyta	Sporopollenin, Scytonemin, mycosporine-like amino acids (MAAs)	Protect against sun damage	Atkin et al. (2006), Dionisio-Se Se (2010), and Priyadarshani and Rath (2012)
Odontella aurita	Heterokontophyta	Polyunsaturated fatty acids (PUFAs)	Prevents oxidative stress on skin	Pulz and Gross (2004)
Traustochytrium, Botryococcus braunii and *Schizochytrium mangrovei*	Chlorophyta, *Thraustochytrium*	Squalene	Improves skin elasticity and moisture retention, prevents age spots and hyperpigmentation	Achitouv et al. (2004), Fan et al. (2010), and Spanova and Daum (2011)
Haematococcus pluvialis	Chlorophyta	Astaxanthin	Protects against sun damage	Guerin et al. (2003) and Tominaga et al. (2012)
Dunaliella spp.	Chlorophyta	Lutein, Zeaxanthin, and Canthaxanthin	Enhances a tanned appearance and/or protects against sun damage	Gierhart and Fox (2013)
Dunaliella bardawil	Chlorophyta	Carotenoids, β-carotene	Improves bioavailability and antioxidant properties on skin	Walker et al. (2005)
Porphyridium spp.	Rhodophyta	Polysaccharides, Phycoerythrin	Improves antioxidant properties on skin; an additive for cosmetics	Pulz and Gross (2004) and Spolaore et al. (2006)

produced by marine organisms that live in environments with high volumes of sunlight; they are assumed to have scavenging activity against free radicals and reactive oxygen species, which cause sun damage. Their metabolites generally absorb at 310 or 320 nm and consist of a cyclohexenone ring conjugated with a nitrogen substituent of an amino acid or amino alcohol (Bandaranayake, 1998; Conde et al., 2000).

Squalene, which is a natural triterpene and an important intermediate in the endogenous synthesis of sterol, is reported to have several beneficial properties, including being a natural antioxidant that hydrates skin and prevents age spots and hyperpigmentation (Achitouv et al., 2004; Fan et al., 2010; Spanova and Daum, 2011). Microalgae do not accumulate as much squalene in shark liver, which is the traditional source, but their advantage is fast and massive growth. Squalene isolation from *Traustochytrium*, *Botryococcus braunii*, and *Schizochytrium mangrovei* has been reported (Banerjee et al., 2002; Bhattacharjee et al., 2001; Jiang et al., 2004; Yue and Jiang, 2009). Also, PUFAs from *Odontella aurita* are claimed to have antioxidant activity that maintains youthful skin (Pulz and Gross, 2004).

Carotenoids, such as astaxanthin, lutein, zeaxanthin, and canthaxanthin, which are rich in *Haematococcus* and *Dunaliella*, act as a protectant against sun damage; they are also reported to have antioxidant activity (Gierhart and Fox, 2013; Guerin et al., 2003; Tominaga et al., 2012; Walker et al., 2005). According to Palombo et al. (2007), lutein and zeaxanthin significantly enhanced the elasticity of skin when applied orally, topically, or both; they also had a synergistic effect on cutaneous hydration when applied as a combined treatment rather than alone (Palombo et al., 2007).

Polysaccharides and phycoerythrin, which is a red protein-pigment complex from the light-harvesting phycobiliprotein family from *Porphyridium*, have been reported to enhance antioxidant properties on skin and could be used as a dye in cosmetics (Pulz and Gross, 2004; Spolaore et al., 2006).

3. CONCLUSION

Cosmetics and cosmeceuticals are increasingly using active ingredients from natural sources. Microalgae are an emerging source of these ingredients, representing an exciting new frontier for cosmeceutical manufacturers. Microalgae are readily available and able to grow quickly, producing the desired components in mass quantities. With the increasing demand and interest with regard to finding new ingredients, microalgae's multibiological effects and better delivery systems make them potential sources for cosmetic and cosmeceutical ingredients. Therefore, research into microalgae use for cosmetics and cosmeceuticals requires a comprehensive approach for overall skin care.

References

Achitouv, E., Metzger, P., Rager, M.N., Largeau, C., 2004. C-31-C-34 methylated squalenes from a Bolivian strain of *Botryococcus braunii*. Phytochemistry 65, 3159–3165.

Arad, S., Cohen, E., Yaron, A., 1997. Coloring materials. In: Google Patents.

Arad, S.M., Yaron, A., 1992. Natural pigments from red microalgae for use in foods and cosmetics. Trends Food Sci. Tech. 3, 92–97.

Atkin, S.L., Backett, S.T., Mackenzie, G., 2006. Topical formulations containing sporopollenin. In: Google Patents.

Balskus, E.P., Walsh, C.T., 2010. The genetic and molecular basis for sunscreen biosynthesis in cyanobacteria. Science 329, 1653–1656.

Bandaranayake, W.M., 1998. Mycosporines: are they nature's sunscreens? Nat. Prod. Rep. 15, 159–172.

Banerjee, A., Sharma, R., Chisti, Y., Banerjee, U.C., 2002. *Botryococcus braunii*: a renewable source of hydrocarbons and other chemicals. Crit. Rev. Biotechnol. 22, 245–279.

Bhattacharjee, P., Shukla, V.B., Singhal, R.S., Kulkarni, P.R., 2001. Studies on fermentative production of squalene. World J. Microb. Biotechnol. 17, 811–816.

Chakdar, H., Jadhav, S.D., Dhar, D.W., Pabbi, S., 2012. Potential applications of blue green algae. J. Sci. Ind. Res. 71, 13–20.

Christaki, E., Florou-Paneri, P., Bonos, E., 2011. Microalgae: a novel ingredient in nutrition. Int. J. Food Sci. Nutr. 62, 794–799.

Chu, W., Wong, C., Teoh, M., Phang, S., 2005. Response and Adaptation of Algae to the Changing Global Environment. Algal Culture Collections and the Environment. Tokai University Press, Kanagawa pp. 177–195.

Conde, F.R., Churio, M.S., Previtali, C.M., 2000. The photoprotector mechanism of mycosporine-like amino acids. Excited-state properties and photostability of porphyra-334 in aqueous solution. J. Photochem. Photobiol. B Biol. 56, 139–144.

Dionisio-Se Se, M.L., 2010. Aquatic microalgae as potential sources of UV-screening compounds. Philipp. J. Sci. 139, 5–19.

Fan, K.W., Aki, T., Chen, F., Jiang, Y., 2010. Enhanced production of squalene in the thraustochytrid Aurantiochytrium mangrovei by medium optimization and treatment with terbinafine. World J. Microb. Biotechnol. 26, 1303–1309.

Garciapichel, F., Wingard, C.E., Castenholz, R.W., 1993. Evidence regarding the UV sunscreen role of a mycosporine-like compound in the Cyanobacterium Gloeocapsa Sp. Appl. Environ. Microbiol. 59, 170–176.

Gierhart, D.L., Fox, J.A., 2013. Protection against sunburn and skin problems with orally-ingested high-dosage zeaxanthin. In: Google Patents.

Guedes, A.C., Amaro, H.M., Malcata, F.X., 2011. Microalgae as sources of carotenoids. Mar. Drugs 9, 625–644.

Guerin, M., Huntley, M.E., Olaizola, M., 2003. Haematococcus astaxanthin: applications for human health and nutrition. Trends Biotechnol. 21, 210–216.

Harder, R., von Witsch, H., 1942. Bericht über Versuche zur Fettsynthese mittels autotropher Mikroorganismen. Forschungsdienst Sonderheft 16, 270–275.

Harwati, T.U., 2013. Cultivation of Microalgae: Lipid Production, Evaluation of Antioxidant Capacity and Modeling of Growth and Lipid Production. PhD Dissertation, Carolo-Wilhelmina Technical University, Braunschweig.

Hirata, T., Tanaka, M., Ooike, M., Tsunomura, T., Sakaguchi, M., 2000. Antioxidant activities of phycocyanobilin prepared from Spirulina platensis. J. Appl. Phycol. 12, 435–439.

Jensen, G.S., Ginsberg, D.I., Drapeau, C., 2001. Blue-green algae as an immuno-enhancer and biomodulator. J. Amer. Nutraceut. Assoc. 3, 24–30.

Jiang, Y., Fan, K.W., Wong, R.D.Y., Chen, F., 2004. Fatty acid composition and squalene content of the marine micro-alga Schizochytrium mangrovei. J. Agric. Food Chem. 52, 1196–1200.

Kawaguchi, K., 1980. In: Shelef, G., Soeder, C.J. (Eds.), Micro-algae Production Systems in Asia. Algae Biomass: Production and Use/[Sponsored by the National Council for Research and Development, Israel and the Gesellschaft fur Strahlen-und Umweltforschung (GSF), Munich, Germany].

Kaya, K., Watanabe, M.M., Kasai, F., 2005. Algal Culture Collections and the Environment. Tokai University Press.

Markou, G., Nerantzis, E., 2013. Microalgae for high-value compounds and biofuels production: a review with focus on cultivation under stress conditions. Biotechnol. Adv. 31, 1532–1542.

de la Noue, J., de Pauw, N., 1988. The potential of microalgal biotechnology: a review of production and uses of microalgae. Biotechnol. Adv. 6, 725–770.

Olaizola, M., 2003. Commercial development of microalgal biotechnology: from the test tube to the marketplace. Biomol. Eng. 20, 459–466.

Palombo, P., Fabrizi, G., Ruocco, V., Ruocco, E., Fluhr, J., Roberts, R., Morganti, P., 2007. Beneficial long-term effects of combined oral/topical antioxidant treatment with the carotenoids lutein and zeaxanthin on human skin: a double-blind, placebo-controlled study. Skin Pharmacol. Physiol. 20, 199–210.

Preisig, H.R., Andersen, R.A., 2005. Historical review of algal culturing techniques. In: Andersen, R.A. (Ed.), Algal Culturing Techniques. Academic Press, NY, pp. 1–12.

Priyadarshani, I., Rath, B., 2012. Commercial and industrial applications of micro algae—a review. J. Algal Biomass Utln. 3, 89–100.

Pulz, O., Gross, W., 2004. Valuable products from biotechnology of microalgae. Appl. Microbiol. Biotechnol. 65, 635–648.

Sahu, A., Pancha, I., Jain, D., Paliwal, C., Ghosh, T., Patidar, S., Bhattacharya, S., Mishra, S., 2013. Fatty acids as biomarkers of microalgae. Phytochemistry 89, 53–58.

Skjanes, K., Rebours, C., Lindblad, P., 2013. Potential for green microalgae to produce hydrogen, pharmaceuticals and other high value products in a combined process. Crit. Rev. Biotechnol. 33, 172–215.

Spanova, M., Daum, G., 2011. Squalene—biochemistry, molecular biology, process biotechnology, and applications. Eur. J. Lipid Sci. Technol. 113, 1299–1320.

Spolaore, P., Joannis-Cassan, C., Duran, E., Isambert, A., 2006. Commercial applications of microalgae. J. Biosci. Bioeng. 101, 87–96.

Stahl, W., Sies, H., 2003. Antioxidant activity of carotenoids. Mol. Aspects Med. 24, 345–351.

Stolz, P., Obermayer, B., 2005. Manufacturing microalgae for skin care. Cosmetics Toiletries 120, 99–106.

Tapiero, H., Townsend, D.M., Tew, K.D., 2004. The role of carotenoids in the prevention of human pathologies. Biomed. Pharmacother. 58, 100–110.

Teoh, M.-L., Chu, W.-L., Marchant, H., Phang, S.-M., 2004. Influence of culture temperature on the growth, biochemical composition and fatty acid profiles of six Antarctic microalgae. J. Appl. Phycol. 16, 421–430.

Tominaga, K., Hongo, N., Karato, M., Yamashita, E., 2012. Cosmetic benefits of astaxanthin on humans subjects. Acta Biochim. Pol. 59, 43–47.

Walker, T.L., Purton, S., Becker, D.K., Collet, C., 2005. Microalgae as bioreactors. Plant Cell Rep. 24, 629–641.

Wong, C., Chu, W., Marchant, H., Phang, S., 2007. Comparing the response of Antarctic, tropical and temperate microalgae to ultraviolet radiation (UVR) stress. J. Appl. Phycol. 19, 689–699.

Yue, C.J., Jiang, Y., 2009. Impact of methyl jasmonate on squalene biosynthesis in microalga *Schizochytrium mangrovei*. Proc. Biochem. 44, 923–927.

Zhukova, N.V., Aizdaicher, N.A., 1995. Fatty-acid composition of 15 species of marine microalgae. Phytochemistry 39, 351–356.

Supercritical Fluid Extraction of Microalgae (*Chlorella vulagaris*) Biomass

Ali Bahadar[1,2], *M. Bilal Khan*[3], *M.A. Asim K. Jalwana*[4]

[1]Department of Chemical and Materials Engineering, King Abdulaziz University, Rabigh, Saudi Arabia; [2]School of Chemical and Materials Engineering (SCME), National University of Sciences and Technology (NUST), Islamabad, Pakistan; [3]Center for Advanced Studies in Energy (CAE), National University of Sciences and Technology (NUST), Islamabad, Pakistan; [4]School of Electrical Engineering and Computer Sciences (SEECS), National University of Sciences and Technology (NUST), Islamabad, Pakistan

1. INTRODUCTION

Among the various biomasses that may be used for biodiesel production, microalgae have great potential to produce renewable liquid fuels. Algae grow in open ponds, raceway ponds, and closed systems (Mata et al., 2010). Naturally, they are grown in open ponds, but they can be cultivated in specially fabricated raceway ponds and in closed photobioreactor systems (Gouveia and Oliveira, 2009). One of the major advantages of algae is that they do not require land for cultivation. Also, they have a higher lipid content than any terrestrial plants (Chen et al., 2011; Li et al., 2008). The comparative oil yields of various feedstocks are presented in Table 1.

For efficient and selective component extraction, supercritical fluid extraction (SCFE) has been used extensively as a separation method in the food processing and pharmaceutical industries. Supercritical carbon dioxide (SC-CO_2) fluid extraction has emerged as a greener technology than conventional petroleum-based

solvent extraction techniques. SC-CO_2 is a widely and commonly used solvent in SCFE. The unique characteristics of SC-CO_2 have made it attractive for separating essential oils, neutral lipids, flavors, fragrances, antioxidants, and pigments from both terrestrial and nonterrestrial biomasses. SC-CO_2 is lipophilic and is able to separate most nonpolar solutes. The separation of CO_2 is

TABLE 1   Oil Yield Comparisons of Various Biofuel Crops

Biofuel crops	Oil yield (L/ha/y)
Corn	174
Soybean	446
Jatropha	2200
Coconut	2889
Oil palm	6050
Microalgae (70% oil)	146,900
Microalgae (30% oil)	70,000

Handbook of Marine Microalgae
http://dx.doi.org/10.1016/B978-0-12-800776-1.00021-2

easy, simple, and leaves no residues in the extract as compared with petroleum-based solvent extraction (Rizvi et al., 1986; Tonthubthimthong et al., 2001). The solvation capability of SC-CO_2 can be tuned by changing the pressure and temperature of the system, making it desirable to separate specific compounds from plant and animal resources. CO_2 is nontoxic, inexpensive, and abundantly available, with a critical temperature of 31 °C and a pressure of 1072 psi (Eggers, 1985; Johnson et al., 1998).

In the supercritical phase, the liquid-like density increases the interactions between the substrate. The solvent- and gas-like diffusivity properties allow excellent mass transfer. In addition, low viscosities like gas and zero surface tension allow the SC-CO_2 to penetrate easily into a microporous matrix to extract the desired compounds (Amajuoyi, 2001). Thus, the collective actions of density, diffusivity, surface tension, and viscosity, together with the tenability of temperature and pressure dependence, make SCFE an attractive green separation technique for desired products (Meireles, 2003).

In the study presented in this chapter, we separated oil from a promising biofuel feedstock, *Chlorella vulgaris* (microalgae), to produce biodiesel.

2. THEORETICAL MODELS FOR SUPERCRITICAL FLUID EXTRACTION

Advances in green chemistry and biorefinery concepts have led to the extensive use of SCFE for liquids and solids in recent years. Conventional solvent-based extraction has certain drawbacks, such as the presence of traces of solvent in the final product and the possibility of causing thermal degradation (Reverchon and De Marco, 2006; Pourmortazavi and Hajimirsadeghi, 2007). SC-CO_2 extraction is a green technology that promises to replace organic solvent extraction. When the temperature and pressure of a fluid increase above its critical point, the fluid behaves as both a liquid

and a gas. This method is very efficient for extracting lipids for several reasons: (1) The crude lipid products are solvent free. (2) The solvent rapidly penetrates the algal cells, giving a higher lipid yield. (3) Solvent power is a function of fluid density, which can be tuned by adjusting the temperature and pressure to get neutral lipids (acylglycerols). (4) Supercritical fluids are noncorrosive, nontoxic, nonflammable, and inert. (5) No degumming is required because SC-CO_2 does not solubilize polar phospholipids (Sahena et al., 2009; Mendes et al., 2003; Herrero et al., 2006).

Hence, it is essential to provide suitable modeling of the SCFE to improve operating conditions and simulate the global process. In this chapter, some most significant and physically sound models published in the literature for SCFE of solid matrices and liquid, such as the linear driving force (LDF), shrinking core, broken and intact cells, and a combination of Broken Plus Intact Cells (BIC) and shrinking core models, are discussed. The basic equations used for these models are presented in Table 2 (Huang et al., 2012; Pawliszyn, 1993).

1. Linear Driving Force Model

The LDF model assumes that the mass transfer flux is proportional to the difference between an average solute concentration in the particle and the solute concentration in equilibrium with the fluid phase. This model demonstrates that when the solute concentration in the particle follows a parabolic profile, the LDF model is exact (Liaw et al., 1979). Do and Rice (1986) have shown that these assumptions are valid, except at the beginning of the extraction of each particle in the column, when the concentration profile is still very sharp.

2. The Shrinking Core Model

The shrinking core model assumes that there is a sharp boundary between the extracted and nonextracted portions of the particle. As extraction is carried out, this boundary is reduced until it reaches the center of the particle (with all solute being exhausted).

TABLE 2   Different Kinetic Models used for SCFE and Their Respective Equations

S/N	Model name	Equations
1	Linear driving force Model	$q_i^* = q_i^*(C_i)$
		$t_{min} = \frac{0.4 d_p^2}{D_e}$
		$\frac{\partial \bar{q}_i}{\partial t} = k_{LDF}(\bar{q}_i - q_i^*)$
		$k_{LDF} = \frac{15 D_c}{R_p^2}$
		$q_i^* = K_i C_i$
		$J_f a_p = \rho_S k_{LDF}(\bar{q}_i - q_i^*)$
		$\frac{\in \partial \overline{C}_{s,i}}{\partial t} + (1 - \in_p)\rho_s \frac{\partial \bar{q}_i}{\partial t} = K_{LDF}(\overline{C}_{s,i} - C_i)$
		$\frac{\partial \bar{q}_i}{\partial t} = K_{LDF}(\bar{q}_i - q_i^*)$
		$\frac{1}{K_{LDF}} = \frac{1}{k_f a_p} + \frac{R_p}{5 D_e a_p} \leftrightarrow K_{LDF} = \frac{5 k_f a_p}{5 + \frac{k_f R_p}{D_e}}$
		$q_i^* = K_i \overline{C}_{s,i}$
		$J_f a_p = K_{LDF}(\overline{C}_{s,i} - C_i)$
		$t = 0 \leftrightarrow \overline{C}_{s,i} = \overline{C}_{s,i,0}, \bar{q} = q_{i,0}$
2	The shrinking core Model	$\frac{\partial \bar{q}_i}{\partial t} = k_f a_p(C_{s,i}\big|_{r=R_p} - C_i)$
		$\frac{D_c}{r^2} \frac{\partial}{\partial r}\left(r^2 \frac{\partial C_{s,i}}{\partial r}\right) = 0$
		$\bar{q}_i = q_{i,0}\left(\frac{r_c}{R_p}\right)^3$
		$J_f = k_f(C_{s,i}\big|_{r=R_p} - C_i)$
		$\forall_r, t = 0 \; q_i = q_{i,0}$
		$\forall_t, r = r_c \; C_{s,i} = C_{s,i}^*$
		$\forall_t, r = R_p - \frac{D_e \partial C_{s,i}}{\partial r} = k_f(C_{s,i} - C_i)$
3	Broken plus intact cells model	$\frac{\partial C_{s,i}}{\partial t} = -J_f a_p \begin{cases} k_f(C_i^* - C_i)C_{s,i} \geq solubility \\ \frac{k_s}{k_i}(C_i^* - C_i)C_{s,i} < solubility \end{cases}$
		$\frac{\zeta \partial C_{broken,i}}{\partial t} = -J_f a_f + J_S a_s$
		$\frac{(1-\zeta)\partial C_{intact,i}}{\partial t} = -J_s a_s$
		$J_f = k_f(C_i^* - C_i)$
		$J_s = k_S(C_{intact,i} - C_{broken,i})$
		$\frac{\zeta \partial C_{broken,i}}{\partial t} = -J_{broken} a_p$
		$\frac{(1-\zeta)\partial C_{intact,i}}{\partial t} = -J_{intact} a_p$
		$J_{broken} = k_f(C_i^*\big|_{broken} - C_i)$
		$J_{intact} = k_s(C_i^*\big|_{intact} - C_i)$
		$J_f = J_{intact} + J_{broken}$
		$t = 0 \begin{cases} C_{broken,i} = C_{broken,i,0} \\ C_{intact,i} = C_{intact,i,0} \end{cases}$

Continued

TABLE 2   Different Kinetic Models used for SCFE and Their Respective Equations—cont'd

S/N	Model name	Equations
4	Broken plus intact cells + shrinking model	$\frac{\partial \bar{q}_{s,i}}{\partial t} = -J_f a_p$ $J_f = k'(C_i^* - C_i)$ $t = 0 \quad q_i = q_{i,0}$

3. Broken Plus Intact Cells Model

This model shows how, after grinding, the extraction is done. A few cells break down during the grinding process and a few remain intact. This model takes both types into account.

4. Broken Plus Intact Cells and Shrinking Core Combined Model

This model combines both the broken plus intact cells and shrinking core models. The model therefore adds the properties of both. For example, extraction of oil from milled grape seed, due to milling there is cell breakdown while the layered structure suites to shrinking.

2.1 Response Surface Methodology

Response surface methodology (RSM) is a statistical method that uses quantitative data from appropriate experimental designs to determine and simultaneously solve multivariate equations. These equations can be graphically represented as response surfaces, which can be used in three ways: (1) to describe how the test variables affect the response; (2) to determine the interrelationships among the test variables; and (3) to describe the combined effects of all test variables on the response (Raymond and Montgomery, 2002; Özkal et al., 2005).

2.2 Materials and Methods

2.2.1 Microalgae (C. vulgaris)

Samples of the locally grown freshwater algal species *C. vulgaris* were taken from the Pakistan

Agriculture Research Council. The sample was inoculated in a glass column by providing specified media, light, and air for its initial growth. Bold Basal Medium and Modified Bold Basal Medium were used as the stock solutions. Microalgae were cultivated in a closed solarized air lift tubular photobioreactor, as shown in Figure 1. Microalgae were dried by naturally available sunlight for 4 h to avoid decomposition, as shown in Figure 2.

2.2.2 Soxhlet Solvent Extractions

The initial oil content in microalgae species (*C. vulgaris*) was found by conventional *n*-hexane solvent extraction using a Soxhlet extractor for 14 h. The total lipid content for three replicates was found to be in the range of $18 \pm 1.2\%$ w/w.

2.2.3 Supercritical Fluid Extraction of Microalgae (C. vulgaris)

The SC-CO_2 fluid extraction of microalgae biomass was performed using an SFT-150 bench scale SC-CO_2 fluid extraction unit (Supercritical Fluid Extraction Fluids, Inc.). The system consisted

FIGURE 1   In-house designed tubular photobioreactor at CES, NUST.

(a) **(b)**

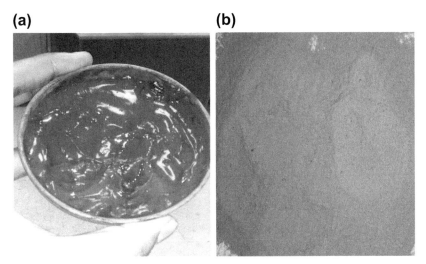

FIGURE 2   (a) Wet microalgae, (b) dried microalgae.

of a 1000-mL pressure vessel and could be pressurized up to 10,000 psi. Bone-dry CO_2 contained in dip tube cylinders was purchased from Linde Pakistan Ltd. The process flow diagram is illustrated in Figure 3.

A 30-g algal biomass was used in each experiment. In each experiment, the biomass was loaded into the pressure vessel; if needed, glass spheres were added to make a completely packed bed vessel. First, the desired temperature was achieved in the vessel before pressurizing. The diaphragm pump was used to pressurize the vessel; to ensure the liquid feed of CO_2, the CO_2 was fed into a chiller ($-5\,^{\circ}C$) directly from cylinders. Then, the liquid CO_2 was pumped and discharged into the pressure vessel from the bottom and the desired operating pressure was achieved. The biomass was soaked each time for 10 min, then dynamic valve was opened for 10 min at 0.375 L/min. Typically, a 1000-mL vessel exchanges 7.5 volumes; here, we used 8.5 volumes to get all of the oil out of the pressure vessel, ensuring that all of the oil would flow out of the pressure vessel (Chrastil, 1982). A restrictor valve was used to regulate the purged CO_2 flow rate. As a large pressure drop

occurs from inside the pressure vessel to atmospheric pressure, the restrictor valve is heated up to 80 $^{\circ}C$ to avoid freezing. The extracted oil was collected in the glass collection vessel during the dynamic cycle and CO_2 was purged to the outside. Again, the biomass was soaked for 10 min. The static/dynamic cycling continued for 3 h. The oil in the collection was weighed and the extraction yield was calculated by the following equation:

$$\% \text{ Mass collected } = (\text{gram of oil collected}/ \\ \text{gram of biomass loaded})* \\ 100$$

$$(1)$$

Experimental runs were carried out for a wide range of operating conditions. The temperatures ranged from 40 to 50 $^{\circ}C$, while pressure ranges from 4000 to 9000 psi were used for both of the biomasses.

2.2.4 Experimental Designs for RSM

For RSM, a central composite design model consisting of 30 runs with six replicas at the central points was used to optimize the process

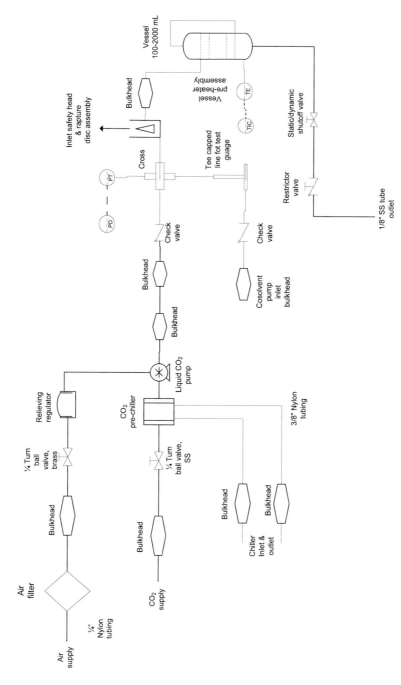

FIGURE 3   Process flow diagram for sft-150 supercritical fluid extraction unit.

variables (temperature, pressure, flow rate, and static/dynamic cycling time). Optimization and experimental data analysis using RSM were performed using Design Expert 8.0.7.1 statistical software. Data was fitted into a polynomial equation that shows all of the possible interactions of the SCFE parameters (X_1, X_2, X_3, and X_4) and their effects on the response (the yield). For RSM, a preliminary experimental study was conducted. Based on that study, the independent variables used in the Central Composite Designs (CCD) model are presented in Table 3. To analyze the response (Y), the experimental data were fitted with the polynomial regression equation of the following form:

$$Y = \beta_0 + \sum_{i=1}^{k} \beta_i X_i$$
$$+ \sum_{i=1}^{k} \beta_{ii} X_i^2 + \sum_{i=1}^{k} \sum_{j>i}^{k}$$
$$\beta_{ij} X_i X_j + \varepsilon i = 1, 2, \ldots, k; j = 2, \ldots, k; i \neq j$$
$$(2)$$

Here, Y is the response (percent of oil yield), β_0 is a constant, and β_i, β_{ii}, β_{ij} are the linear, quadratic, and interaction terms, respectively. X_i and X_j are the levels of the independent variables.

The analysis of variance (ANOVA) was also used to evaluate the quality of the fitted model for microalgae oil extraction in RSM. The statistical difference was based on the total error with a confidence level of 95%.

TABLE 3   Experimental Ranges used in the CCD for Oil Yield for *Chlorella vulgaris* Biomass

Factors	Codes	Levels		
		−1	0	+1
Temperature (°C)	X_1	40	50	60
Pressure (MPa)	X_2	4000	5500	7000
Flow rate (L/min)	X_3	1	1.5	3
Extraction time (h)	X_4	1	1.5	3

3. RESULTS AND DISCUSSION

3.1 Supercritical Fluid Extraction of Microalgae (C. *vulgaris*)

SC-CO_2 was also used to extract lipids from *C. vulgaris* in the temperature range of 40−80 °C and pressure of 4000−9000 psi. Light green oil was obtained in all the experimental runs. The effects of temperature and pressure on SCFE of microalgae (*C. vulgaris*) are shown in Figures 4 and 5. Because the supercritical extraction of seeds is strongly dependent on temperature and pressure, *C. vulgaris* oil extraction also shows its dependence on these two variables.

Figure 4 shows the yield of algal oil against temperature at constant pressure. The maximum yield was found at the highest temperature (80 °C) and highest pressure (9000 psi), and the lowest yield was obtained at the lowest pressure (4000 psi) and highest temperature (80 °C). At the lowest pressure (4000 psi), the highest yield of algae oil was obtained at 40 °C and the lowest at 80 °C. Subsequently, at the highest pressures (9000 psi), an inversion effect was observed as compared to 4000 psi. At 6000−6500 psi, the effect of temperature was less and yielded the same value in the range of 14.5−17.7%. The inversion effect above and below at intermediate pressures of 5500−6500 psi was observed, as shown in Figure 5.

This effect was also reported in previous SCFE studies of different microalgae strains to extract neutral lipids and also for specific components of interest. This type of behavior is typical of a solid-fluid binary system, termed the *cross-over phenomenon*. Below this pressure, an increase in temperature will decrease the solubility of microalgae biomass in SC-CO_2. Above the cross-over pressure, an isobaric increase in temperature will increase the microalgae biomass solubility in SC-CO_2, thus increasing the overall yield. The cross-over effect mainly occurs due to two factors: density and volatility.

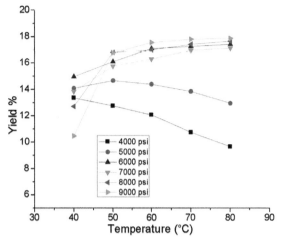

FIGURE 4   Effect of temperature on oil yield of *Chlorella vulgaris*.

At higher densities, the SC-CO$_2$ is exposed to more surface area, translating to a higher concentration that allows more penetration; thus, more oil dissolves from the biomass and more oil is extracted. The effect of density is more

prominent below the cross-over pressure, which increases the solubility and subsequently the yield at lower temperatures. Above the cross-over pressure, the volatility effect gives way to higher yields at higher temperatures.

Figure 6 shows the scanning electron microscopy (SEM) of unextracted microalgae cells and ruptured cells of algae after extraction when SC-CO$_2$ is used.

3.2 Response Surface Analysis

3.2.1 Response Surface Analysis for Microalgae (C. vulgaris)

Four extraction process parameters—temperature, pressure, flow rate of CO$_2$, and extraction time—were selected to study the effects on SC-CO$_2$ extraction of microalgae (*C. vulgaris*) using CCD design in RSM. The extraction yields of algal oil obtained under 30 different process conditions are presented in Table 4.

The results of the ANOVA, *F*-test, and probability are shown in Table 5. The oil yield (*Y*) was

FIGURE 5   Effect of pressure on oil yield of *Chlorella vulgaris*.

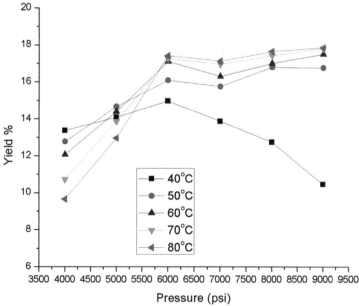

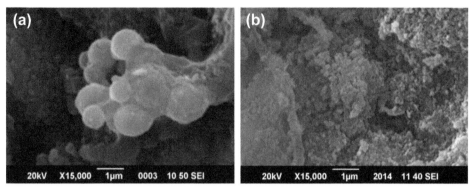

FIGURE 6 (a) SEM showing oil pouches of algae (b) SEM showing ruptured oil pouches of algae after oil extraction using SCFE.

estimated by the following quadratic multiple regression equation:

actual and predicted yields of C. vulgaris oil is shown in Figure 7 and is in close agreement.

$$\text{Extraction Yield} = +14.82 + 0.20 * \text{A} + 1.43 * \text{B} + 0.15 * \text{C} + 0.34 * \text{D} + 1.24 * \text{A} * \text{B} + 0.30 * \text{A} * \text{C}$$
$$- 0.17 * \text{A} * \text{D} - 0.51 * \text{B} * \text{C} + 0.14 * \text{B} * \text{D} + 0.22 * \text{C} * \text{D} + 0.46 * \text{A}^2 - 0.92$$
$$* \text{B}^2 + 1.20 * \text{C}^2 - 1.32 * \text{D}^2$$

(3)

Here, the extraction yield is the yield of microalgae (C. vulgaris) oil (%), A is the extraction temperature (°C), B is the extraction pressure (psi), C is the flow rate of CO_2 (g/min), and D is the extraction time (h).

The F-values and the probability values in Table 5 indicate that the model is significant. This reflects that the model and actual runs of SCFE of C. vulgaris were best fitted. The lack of fit value was not significant, implying that the model was fit for SC-CO_2 extraction of C. vulgaris oil. R^2 was 0.962, which shows accurate fitness of the model. The regression quadratic equation (Eqn. (3)) explains all the response surfaces and is accurate for predicting the yield of the microalgae oil. Table 5 represents the significant terms ($p > 0.001$) that effect the extraction yield and their interactions with each other. The plot of

3.2.2 Analysis of Response Surface for Microalgae (C. vulgaris)

A three-dimensional (3D) plot is used to represent the significant ($p < 0.05$) statistical interaction of variables in surface response for algal oil extraction, which is shown in Figure 8(a–f). Four key SCFE parameters—temperature (A), pressure (B), flow rate (C), and time (D)—were selected to study the SC-CO_2 extraction process by CCD design in RSM on the orthogonal test using actual experiment runs.

RSM was employed using CCD design on SCFE of C. vulgaris to investigate the effects of SCFE parameters on algal oil yield within experimental space runs. Response surfaces and 3D contour plots from the model under investigation were obtained. Pressures of 4000–7000 psi (higher pressures were not

TABLE 4   CCD Matrix of Factors and the Responses of Oil Yield for Microalgae (*Chlorella vulgaris*)

Run	A Temperature (°C)	B Pressure (psi)	C Flowrate (g/min)	D Time (h)	Actual yield (%)	Predicted yield (%)
1	50	5500	2	2	14.37	14.66
2	60	7000	1	3	17.19	17.22
3	40	4000	1	1	13.32	13.35
4	40	4000	3	3	14.58	14.60
5	50	5500	2	2	15.1	14.66
6	50	5500	2	2	14.64	14.66
7	40	4000	1	3	13.54	13.72
8	50	5500	2	2	14.47	14.66
9	40	4000	3	1	13.51	13.46
10	40	7000	1	3	15.82	15.48
11	60	4000	3	3	13.08	12.84
12	60	7000	3	1	16.96	16.75
13	50	5500	2	2	15.14	14.60
14	50	4500	2	2	13.79	13.24
15	50	5500	2	2	14.96	14.66
16	50	5500	3	2	15.78	15.94
17	50	8500	2	2	13.73	13.83
18	60	7000	3	3	17.7	17.65
19	50	5500	2	2	14.35	14.66
20	50	5500	2	3	13.64	13.80
21	60	5500	2	2	14.98	15.42
22	50	5500	2	2	14.32	14.66
23	60	4000	1	3	10.36	10.47
24	40	7000	1	1	14.29	14.51
25	40	7000	3	1	12.8	12.68
26	70	5500	2	2	17.36	17.29
27	60	4000	1	1	11.1	10.93
28	40	7000	3	3	14.26	14.41
29	60	7000	1	1	17.13	17.09
30	60	4000	3	1	12.18	12.52

TABLE 5   ANOVA for Response Surface Quadratic Model of *Chlorella vulgaris* oil Extraction

Source	Sum of squares	df	Mean square	F value	p-value prob > F	
Model	85.60441	14	6.114601	47.73043	<0.0001	Significant
A-Temperature	0.685538	1	0.685538	5.351296	0.0353	
B-pressure	37.64491	1	37.64491	293.8553	<0.0001	
C-flow rate	0.298882	1	0.298882	2.333069	0.1475	
D-Time	1.652791	1	1.652791	12.90165	0.0027	
AB	25.1001	1	25.1001	195.9308	<0.0001	
AC	2.2201	1	2.2201	17.33005	0.0008	
AD	0.6889	1	0.6889	5.377538	0.0349	
BC	3.744225	1	3.744225	29.22733	<0.0001	
BD	0.342225	1	0.342225	2.6714	0.1230	
CD	0.600625	1	0.600625	4.688465	0.0469	
A^2	3.913424	1	3.913424	30.5481	<0.0001	
B^2	11.57596	1	11.57596	90.36167	<0.0001	
C^2	2.233732	1	2.233732	17.43646	0.0008	
D^2	2.357389	1	2.357389	18.40172	0.0006	
Residual	1.921604	15	0.128107			
Lack of fit	1.075917	8	0.13449	1.113209	0.4503	Not significant
Pure error	0.845688	7	0.120813			
Cor total	87.52602	29				

considered due to negative cost implications on the SCFE process) and temperatures of 40–80 °C, with a flow rate of 1–3 g/min, were used in this RSM. In Figure 8, these responses and plots show the effects of temperature and pressure on algal oil yield, where the CO_2 flow rate was 2.0 g/min and extraction time was 2 h. Here, temperature and pressure are both affecting the yield and are significant terms, according to Table 5.

It was observed in this contour that when the pressure decreased from 7000 psi to 4000 psi, the effect of the extraction temperature decreased from 40 °C to 80 °C; low yield was observed at 4000 psi. At a high pressure of 7000 psi and an increasing temperature from 40 °C to 80 °C, the algal oil yield increased from 14.2 wt% to 17.2 wt%. At 4000–4800 psi, the increase in this temperature did not affect the yield. This was because 4800 psi is near the cross-over pressure for *C. vulgaris* yield (the cross-over phenomenon occurs between 5000 psi and 5500 psi), as discussed in Section 3 and Figure 4. The solubility of *C. vulgaris* oil increases with an increase in temperature above the cross-over pressure. As the pressure decreases to cross-over pressure, the solubility decreases due to the combined effects of solvent density and solute volatility (Bulley et al., 1984; Norulaini et al., 2009). Also in Figure 8(a), it can be noted that at higher

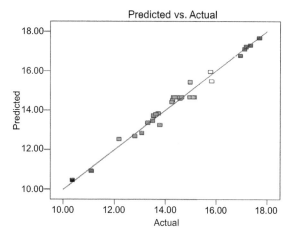

FIGURE 7   Predicted vs. actual yields of microalgae oil extraction.

temperatures the effect of pressure was more prominent than at lower temperatures. Increasing pressure from 4000 psi to 7000 psi at 40 °C increased the oil yield of *C. vulgaris* from 13.1 wt% to 14.8 wt%; at the same pressure change at 80 °C, the yield increased from 12.1 wt% to 16.8 wt%. This was due to the higher effects of pressure changes at high temperatures compared to the low temperatures on solubility.

Figure 8(b) and (c) show the interaction of flow rate with temperature and pressure. The oil yield of microalgae increased with the flow rate of CO_2 and temperature at constant pressure (Figure 8(b)). This was due to decrease in density of the solvent; consequently, the vapor pressure of the solute increases, which leads to an increase in the solubility of microalgae in $SC-CO_2$ and its mass transfer rate (Wei et al., 2009; Wang et al., 2008). Figure 8(c) visualizes the possible interaction of pressure and flow rate at constant temperature. With an increase in pressure from 4000 psi to 7000 psi and flow rate from 1 m/min to 3 m/min, the oil yield increases. With an increase in pressure at constant temperatures, the solubility increases and an increase in flow rate facilitates the mass transfer rate, hence decreasing the mass transfer resistance due to

an increase in convection and decrease in film thickness. The effects of time on the other three SCFE parameters with regard to the yield of microalgae are also presented in Figure 8(d—f).

4. CONCLUSIONS

1. Supercritical CO_2 fluid extraction was employed on microalgae (*C. vulgaris*) for biofuel production. SCFE was demonstrated to be a valuable green technology as compared to petroleum-based solvent extraction techniques in terms of the environment, extraction time, and extraction efficiency. The results using SCFE displayed in this chapter were in close agreement with the Soxhlet petroleum-based extraction. Both experimental and simulated results were discussed.

2. The RSM using CCD was employed on microalgae. A quadratic regression model was fitted well and was sufficient to explain and predict the optimized process conditions for microalgae oil yield using $SC-CO_2$ extraction within the experimental ranges. A response surface plot represents the linear and quadratic terms of pressure, temperature, flow rate, and extraction time. The predicted and actual oil yields for *C. vulgaris* were in close agreement.

3. The optimum yield for microalgae was 17.7 wt% at a pressure of 7000 psi, temperature of 60 °C, CO_2 flow rate of 3 g/min, and extraction time of 3 h, which was in close agreement with the *n*-hexane-based solvent extraction. Thus, the model using CCD of RSM can be used to find the optimal SCFE conditions for microalgae.

4. The actual experimental results were also discussed and the yield of algal oil at wide ranges of temperatures (40—80 °C) and pressure (4000—9000 psi) was presented. The maximum yield was found at the highest temperature (80 °C) and highest pressure

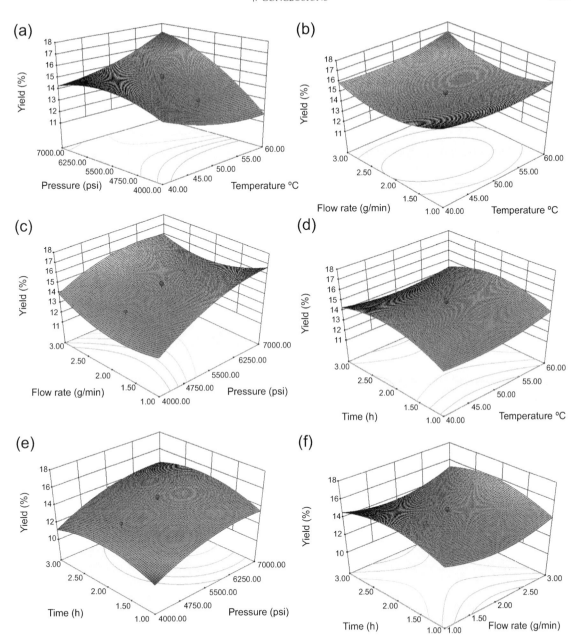

FIGURE 8 (a–f). Plot of response surfaces for microalgae (*Chlorella vulgaris*) oil yield. (a) Effect of temperature and pressure on yield at a fixed flow rate of 2.0 g/min and extraction time of 2.0 h. (b) Effect of temperature and flow rate on yield at a fixed pressure of 5500 psi and an extraction time of 2.0 h. (c) Effect of pressure and flow rate on yield at a fixed temperature of 50 °C and an extraction time of 2.0 h. (d) Effect of temperature and time on yield at a fixed pressure of 5500 psi and an extraction time of 2.0 h. (e) Effect of pressure and time on yield at a fixed temperature of 50 °C and an extraction time of 2.0 h. (f) Effect of flow rate and time on yield at a fixed pressure of 5500 psi and a temperature of 50 °C.

(9000 psi), whereas the lowest yield was obtained at the lowest pressure (4000 psi) and highest temperature (80 °C). At the lowest pressure (4000 psi), the highest yield of algae oil was obtained at 40 °C and the lowest at 80 °C. Subsequently, at the highest pressure (9000 psi), an inversion effect was observed as compared to 4000 psi. At 6000–6500 psi, the effect of temperature was less and yielded the same value in the range of 14.5–17.7%. The oil obtained from the feedstock was used for biodiesel production.

References

Li, Y., et al., 2008. Biofuels from microalgae. Biotechnol. prog. 24 (4), 815–820.

Amajuoyi, K.I. (2001). Behavior and Elimination of Pesticide Residues During Supercritical Carbon Dioxide Extraction of Essential Oils of Spice Plants and Analysis of Pesticides in High-Lipid-Content Plant Extracts (Doctoral dissertation, Technische Universität München, Universitätsbibliothek).

Do, D.D., Rice, R.G., 1986. Validity of the parabolic profile assumption in adsorption studies. AIChE J. 32, 149–154.

Bulley, N.R., Fattori, M., Meisen, A., Moyls, L., 1984. Supercritical fluid extraction of vegetable oil seeds. JAOCS 61 (8), 1362–1365.

Chen, C.-Y., et al., 2011. Cultivation, photobioreactor design and harvesting of microalgae for biodiesel production: a critical review. Bioresour. Technol. 102 (1), 71–81.

Chrastil, J., 1982. Solubility of solids and liquids in supercritical gases. J. Phys.Chem. 86 (15), 3016–3021.

Eggers, R., 1985. High pressure extraction of oil seed. JAOCS 62, 1222–1230.

Gouveia, L., Oliveira, A.C., 2009. Microalgae as a raw material for biofuels production. J. ind. microbiol. biotechnol. 36 (2), 269–274.

Herrero, M., Cifuentes, A., Ibanez, E., 2006. Sub-and supercritical fluid extraction of functional ingredients from different natural sources: plants, food-by-products, algae and microalgae: a review. Food chem. 98, 136–148.

Huang, Z., Shi, X-han, Jiang, W-juan, 2012. Theoretical models for supercritical fluid extraction. J. Chromatogr. A 1250, 2–26.

Johnson, L.A., 1998. Recovery, refining, converting, and stabilizing edible fats and oils. In: Akoh, C.C., Min, D.B. (Eds.), Food Lipids: Chemistry, Nutrition, and Biotechnology. Marcel Dekker, Inc., New York, NY, pp. 181–228.

Liaw, C.H., Wang, J.S.P., Greenkorn, R.A., Chao, K.C., 1979. Kinetics of fixed-bed adsorption: a new solution. AIChE J. 25, 376–381.

Mata, T.M., Martins, A.A., Caetano, N.S., 2010. Microalgae for biodiesel production and other applications: a review. Renewable Sustainable Energy Rev. 14 (1), 217–232.

Meireles, M.A.A., 2003. Supercritical extraction from solid: process design data (2001–2003). Curr. Opin. Solid State Mat. Sci. 7, 321–330.

Mendes, R.L., Nobre, B.P., Cardoso, M.T., Pereira, A.P., Palavra, A.F., 2003. Supercritical carbon dioxide extraction of compounds with pharmaceutical importance from microalgae. Inorg. Chim. Acta 356, 328–334.

Norulaini, N.A.N., Setianto, W.B., Zaidul, I.S.M., Nawi, A.H., Azizi, C.Y.M., Mohd Omar, A.K., 2009. Effects of supercritical carbon dioxide extraction parameters on virgin coconut oil yield and medium-chain triglyceride content. Food Chem. 116, 193–197.

Özkal, S.G., Yener, M.E., Bayındırlı, L., 2005. Response surfaces of apricot kernel oil yield in supercritical carbon dioxide. LWT-Food Sci. Technol. 38 (6), 611–616.

Pourmortazavi, S.M., Hajimirsadeghi, S.S., 2007. Supercritical fluid extraction in plant essential and volatile oil analysis. J. Chromatogr. A 1163, 2–24.

Pawliszyn, J., 1993. Kinetic model of supercritical fluid extraction. J. chromatogr. sci. 31 (1), 31–37.

Rizvi, S.S.H., Daniels, J.A., Benado, A.L., Zollweg, J.A., 1986. Supercritical fluid extraction: operating principles and food applications. Food Technol. 40, 57–64.

Reverchon, E., De Marco, I., 2006. Supercritical fluid extraction and fractionation of natural matter. J. Supercrit. Fluid 38, 146–166.

Raymond, H.M., Montgomery, D.C., 2002. Response Surface Methodology: Process and Product Optimization Using Designed Experiment, second ed. A Wiley-Interscience Publication, USA.

Sahena, F., Zaidul, I., Jinap, S., Karim, A., Abbas, K., Norulaini, N., et al., 2009. Application of supercritical CO_2 in lipid extraction—A review. J. Food Eng. 95, 240–253.

Tonthubthimthong, P., Chuaprasert, S., Douglas, P., Luewisutthichat, W., 2001. Supercritical CO_2 extraction of nimbin from neem seed—an experimental study. J. Food Eng. 47, 289–293.

Wei, Z.-J., Liao, A.-M., Zhang, H.-X., Liu, J., Jiang, S.-T., 2009. Optimization of supercritical carbon dioxide extraction of silkworm pupal oil applying the response surface methodology. Biores. Technol. 100, 4214–4219.

Wang, L., Weller, C.L., Schlegel, V.L., Carr, T.P., Cuppett, S.L., 2008. Supercritical CO_2 extraction of lipids from grain sorghum dried distillers grains with soluble. Biores. Technol. 99, 1337–1382.

Exploiting the Molecular Genetics of Microalgae: From Strain Development Pipelines to the Uncharted Waters of Mass Production

Julian N. Rosenberg[1,2], Victor H. Oh[1], Geng Yu[1],
Bernardo J. Guzman[1], George A. Oyler[1,2], Michael J. Betenbaugh[1]

[1]Johns Hopkins University, Department of Chemical & Biomolecular Engineering, Baltimore, MD, USA;
[2]Synaptic Research, Baltimore, MD, USA

1. INTRODUCTION

Cyanobacteria and microalgae comprise countless species and strains that are adept at surviving in diverse habitats (Tirichine and Bowler, 2011). As a result, these microbes have evolved versatile metabolic functions, which give rise to diverse biomass compositions. Since the early 1960s, various classes of metabolites produced by algae in response to environmental cues have been used as animal-free nutritional additives and high-value pigments for cosmetics and biotechnology (Pulz and Gross, 2004). Beyond natural products, marine and freshwater algae are emerging as systems for the tailored biosynthesis of energy-dense hydrocarbons for biofuels, antioxidant pigments, and free fatty acids for health care and consumer products, and recombinant protein products as therapeutics and nutraceuticals (Rasala and Mayfield, 2014).

Currently, no standard expression system exists for the production of recombinant proteins or specialty biomolecules. While most recombinant pharmaceuticals are produced by bacteria and yeast, the choice of hosts is largely dictated by the characteristics of the end product. While bacteria and yeast are considered economic platforms because of their relatively low operating costs and inexpensive media components, yeast and eukaryotic algae may be more suitable hosts for the production of complex therapeutics due to their ability to perform posttranslational modifications necessary for functional protein activity (Fletcher et al., 2007; Walker et al., 2005). In addition to algae, other plant cells have been developed for suspension culture and are becoming more prevalent in biomanufacturing (e.g., tobacco, carrot) (Manuell et al., 2007; Shaaltiel et al., 2007; Sun et al., 2011; Xu et al., 2011). However, cost projections for antibodies

Handbook of Marine Microalgae
http://dx.doi.org/10.1016/B978-0-12-800776-1.00022-4

produced by algae are only 0.002 USD per gram, compared to 150 and 0.05 USD per gram in mammalian and plant expression systems, respectively (Mayfield et al., 2003).

As for commodity-scale production of biochemicals or biofuels, the choice between terrestrial, microbial, and aquatic crops also depends strongly on the target compounds and balance between natural metabolic traits of the host and tools available for genetic improvement. Microalgae are an attractive alternative to current platforms and possess many of the hallmarks of a sustainable feedstock. With simple growth requirements of carbon dioxide and sunlight, photosynthetic microbes exhibit fast growth rates and produce biomass that is generally regarded as safe (GRAS) for human consumption. As such, natural products derived from algae have contributed to markets including infant formula, fish feed, human nutritional supplements, and cosmetic pigments, as well as specialty chemicals for research purposes (Rosenberg et al., 2008). Furthermore, established genetic transformation methods exist for a number of species (Rosales-Mendoza et al., 2012; Surzycki et al., 2009), which has enabled the use of algae as model photosynthetic organisms for biosensors (Marshall et al., 2012; Bachman, 2003), biological fuel cells (Powell et al., 2014), and biocomputing (Harrop et al., 2014; Zhang et al., 2011). For biofuel and agrochemical production, the scale of algal cultivation will require farms that are many orders of magnitude larger than most current operations. These immense projects are coupled with entirely different technical hurdles, ranging from the logistics of water handling to regulatory control and potential for varying degrees of public acceptance.

Although outdoor evaluation of algal cultivation has begun in raceway ponds and other photobioreactor systems (Table 1), microalgae and cyanobacteria have yet to reach production maturity at commercial scales (tens of thousands of acres). Algae biomass is still relatively expensive to produce due to high capital and operating costs and is currently only economical for high-value products. However, with improved biomass yields and more favorable biomass compositions, the technoeconomics and overall sustainability have the potential to become viable (Rogers et al., 2014). One approach to increased yields stems from genetic enhancement of algal species, which has only recently become readily available. Although the ability to perform genetic manipulation and engineer metabolism is still limited to a select subset of algal species, the molecular genetic framework that has been successful with these strains will undoubtedly be applied to other species. As such, genetically modified (GM) algae have now entered strain development pipelines and are currently being validated for commercial-scale cultivation. This chapter addresses technological advancements in molecular genetic tools for algae, highlights examples of biosynthetic pathway design and selection of industrial fitness, and discusses the regulatory aspects of growing GM algae outdoors.

2. ADVANCES IN MOLECULAR TOOLS FOR CHLOROPLAST AND NUCLEAR GENE EXPRESSION

Because algae can be leveraged for food, feed, fuel, and high-value recombinant proteins, a multitude of different natural microalgal and cyanobacterial species have been characterized for specific production objectives (Sheehan et al., 1998; Rodolfi et al., 2009; Griffiths and Harrison, 2009; Li et al., 2011; Liu et al., 2011; Vazhappilly and Chen, 1998). However, some posit that we have reached the limit of naturally occurring traits and further bioprospecting efforts will not yield meaningful results in a timely manner (Gressel, 2008). Others even claim that natural variation and recombination of traits by means of cross-breeding will be exhausted by 2050 (Sayre, 2014). In direct response to this bottleneck, the ability to manipulate genomes with

TABLE 1   Companies and Academic Algae Programs with Advanced Research and Development Projects in the United States

Program	Location	Technology	Regulatory status
COMPANIES			
Hydromentia	Ocala, FL	Benthic algal turf scrubber, bioremediation	
Sapphire Energy	Columbus, NM	Open raceway ponds for green crude	TERA approved
Triton Algae Innovations	San Diego, CA	Strain engineering for recombinant proteins	
Joule Unlimited Technologies	Hobbs, NM	Photosynthetic biofuel secretion in PBR	MCAN reviewed
Bioprocess Algae	Shenandoah, IA	Solid-phase growth, integrated biorefinery	
Algenol	Fort Myers, FL	Photosynthetic ethanol secretion in PBR	MCAN filed
Aurora Algae	Hayward, CA	Strain development for food, feed, and fuel	
Phycal	Highland Heights, OH	Biofuels from auto/ hetero algae, cassava	
Algae to Omega	Oakland Park, FL	Closed bioreactors for nutritional products	
Heliae	Gilbert, AZ	Modular ponds for mixotrophic growth	
Solazyme	Peoria, IL	Heterotrophic algae for nutritional oils	MCAN filed
Cellana	Kona, HI	Hybrid system for biofuel and fish feed	
UNIVERSITIES			
Arizona State University	Mesa, AZ	Outdoor raceway ponds and PBRs	
California Polytechnic	San Luis Obispo, CA	Algae biofuel from wastewater	
New Mexico State University	Las Cruces, NM	Geothermal greenhouse testbed	
UC-San Diego/ Sapphire	La Jolla, CA	Outdoor raceway ponds and PBRs	TERA approved

This list of companies with large-scale facilities showcases some groups that have received regulatory approval from the US Environmental Protection Agency for outdoor field testing and commercial production. Universities with demonstration-scale test beds are also examining alternative pathways to algal biomass production in parallel.

MCAN, Microbial commercial activity notice; PBR, photobioreactor; TERA, Toxic Substances Control Act Environmental Release Application.

facile molecular genetic tools may open the door to previously unattainable traits.

In developing various microalgal species as platforms for genetic engineering, the few strains that can be stably genetically transformed are limited to members of *Chlamydomonas, Chlorella, Volvox, Haematococcus,* and *Dunaliella* genera (Rosenberg et al., 2008). While the molecular physiology, genetics, and growth characteristics of *Chlamydomonas reinhardtii* have been extensively documented, it is not an ideal species for commercial production (Raja et al., 2008). Even in this model organism, the commercial application of recombinant protein expression has been hindered by low expression levels.

2.1 Genetic Tools to Improve Transgene Expression and Recombinant Protein Yield

Although there are many obstacles that prevent microalgae from being a viable expression system on a commercial scale, significant advancements in the development of genetic engineering tools have ushered in tremendous efforts to optimize the "upstream" side of process design. While recombinant protein expression at commercially viable levels has only been reported from the chloroplast genome, both nuclear and mitochondrial genomes have achieved successful recombinant protein expression (Zhang and Potvin, 2010). The disparity between expression levels from chloroplast and nuclear transformants can be attributed to the way transgenes are incorporated into the respective genomes. Genetic transformation of the chloroplast is almost exclusively executed through homologous recombination. Thus, the transgene inserted into the plastid genome can be site-specific, leading to precise targeting of "silent sites," which are regions void of any known coding sequences (Tam and Lefebvre, 1993; Goldschmidt-Clermont, 1991). Conversely, expression levels of barely 1% of total soluble protein (TSP) are considered high for nuclear expression as a result of random transgene integration.

Compared to 5—10% TSP in the chloroplast, the nuclear genome does provide other advantages, such as posttranslational modifications and localization of protein products to organelles (Zhang and Potvin, 2010).

2.1.1 Constitutively Active Promoters and Untranslated Regions

To achieve high-level transcription of transgenes, promoters positioned directly upstream of genes regulate transcription, while untranslated regions (UTRs) mediate mRNA stability and initiation/termination of translation/transcription (Cadoret et al., 2012). In *C. reinhardtii,* the most widely used endogenous promoters for transgene expression from the nuclear genome have had success in expressing transgenes individually, but chimeric promoters, such as the rubisco (RbcS2) promoter fused to heat shock protein (Hsp70A) regulatory elements, have been shown to be more effective by many orders of magnitude (Cadoret et al., 2012). In the chloroplast, the most effective promoters have been taken from photosynthetically associated genes (*e.g., psbA*) and those likened to energy generation (*e.g., atpA*), along with their respective 5′/3′ UTRs (Rosales-Mendoza et al., 2012; Surzycki et al., 2009).

Although the PsbD promoter and RbcL promoters have also been used to successfully express recombinant proteins in the chloroplast, the highest reported expressions were only ~0.25 and ~0.7% TSP, respectively (Rosales-Mendoza et al., 2012). A head-to-head comparison of the most widely used nuclear promoters in *C. reinhardtii* concluded that the PsaD (photosystem I complex) promoter gave the highest transgene (luciferase) expression levels when paired with its terminator sequence. Using this PsaD promoter/terminator expression vector, the authors were able to achieve accumulation of a human butyrylcholinesterase (huBuChE) protein fused to luciferase up to 0.4% TSP (Kumar et al., 2013). A summarized list of widely used promoter and UTR pairs is shown in Table 2.

TABLE 2 Summary of Useful Genetic Elements for Bioengineering in Various Microalgal Species

Genome	Category	Element	Gene source	References
Nuclear	Promoter	β-tub	Beta-tubulin	Blankenship and Kindle (1992)
Nuclear	Promoter	TubA1	Alpha-tubulin	Kozminski, Diener, and Rosenbaum (1993)
Nuclear	Promoter	Pcy1	Plastocyanin	Quinn and Merchant (1995)
Nuclear	Promoter	AtpC	Chloroplast ATPase gamma-subunit	Quinn and Merchant (1995)
Nuclear	Promoter	RbcS2	Ribulose biphosphate carboxylase small subunit	Kindle (1998)
Nuclear	Promoter	Hsp70A	Heat shock protein 70A	Schroda, Blocker, and Becka (2000)
Nuclear	Promoter	PsaD	Photosystem I complex	Fischer and Rochaix (2001)
Nuclear	Expression vector	ble-2A-**X**	Foot-and-mouth-disease-virus 2A peptide	Rasala et al. (2012)
Nuclear	Gene-stacking strategy	DNA assembly method	Constructing multigene pathway in *Saccharomyces cerevisiae* and transfer to algae	Noor-Mohammadi et al. (2013)
Nuclear	Nuclear-targeting vector	ble-2A-**X**-2×SV40	Tandem localization sequence from SV40	Rasala et al. (2014)
Nuclear	Mitochondria-targeting vector	ble-2A-atpA-**X**	Mitochondrial ATP synthase alpha-subunit	Rasala et al. (2014)
Nuclear	ER-targeting vector	ble-2A-bip1-**X**-HDEL	Hsp70 superfamily chaperone	Rasala et al. (2014)
Nuclear	ER-targeting vector	ble-2A-ars1-**X**-HDEL	Arylsulfatase secretion peptide	Rasala et al. (2014)
Nuclear	Chloroplast-targeting vector	ble-2A-psaD-**X**	Photosystem I protein	Rasala et al. (2014)
Nuclear	Multigene expression vector	ble-E2A-**X**-F2A-**Y**	Equine-rhinitis-A-virus 2A peptide, FMDV 2A peptide	Rasala et al. (2014)
Nuclear	Gene-stacking strategy	Mating	Cross-breeding transgene lines and sorting progeny	Rasala et al. (2014)
Nuclear	Selectable marker	GAT	Glyphosate acetyltransferase	Bruggeman et al. (2014)

Continued

TABLE 2   Summary of Useful Genetic Elements for Bioengineering in Various Microalgal Species—cont'd

Genome	Category	Element	Gene source	References
Nuclear	Selectable marker	protox rs-3	Protoporphyrinogen oxidase with *rs-3* mutation	Bruggeman et al. (2014)
Nuclear	Selectable marker	PDS (R268T)	Modified *Chlamydomonas reinhardtii* phytoene desaturase gene to encode Thr at position 268 in place of Arg	Bruggeman et al. (2014)
Chloroplast	Promoter	atpA	Adenosine triphosphatase alpha subunit	Sun et al. (2003)
Chloroplast	Promoter	psbD	Photosystem II D1 subunit	Manuell et al. (2007)
Chloroplast	Promoter	rbcL	Ribulose biphosphate carboxylase large subunit	Dreesen, Charpin-El Hamri, and Fussenegger (2010)
Chloroplast	Promoter	psbA	Photosystem II psbA	Rasala et al. (2010)
Chloroplast	Vector	Gateway® system	Lambda-phage *att*	Oey et al. (2014)

X and Y represent hypothetical transgenes of interest.

2.1.2 Codon Optimization Using Gene Synthesis

In designing transgene expression vectors, codon dependency is another important factor affecting the expression of recombinant proteins in microalgae. This causes shifts in tRNA abundance that potentially lead to translational stalling, premature translation termination, translation frame shifting, and amino acid misincorporation (Kurland and Gallant, 1996). For example, *C. reinhardtii* has an extremely high GC content, around 61%, and its chloroplast genome preferentially uses codons with adenine or thiamine nucleotides in the third position rather than guanine or cytosine (Leon-Banares et al., 2004; Nakamura et al., 1999). Codon optimization of a transgene for a particular microalgal genome can lead to a significant improvement in recombinant protein expression, as shown in a 2002 study by Franklin et al., where an 80-fold increase in green fluorescent protein (GFP) accumulation was observed when *C. reinhardtii* was transformed with codon-optimized GFP (Franklin et al., 2002). The same phenomenon has also been observed as a result of codon optimization in diatoms (Corbeil et al., 2012). Given that the price of gene synthesis has dropped dramatically to less than 1 USD per base pair, codon optimization is now a more widely accessible tool to ensure high-level expression in various species of microalgae.

2.1.3 Endogenous Introns

Although the precise mechanism of how introns upregulate transgene expression is still unclear, it has been shown in other organisms, including microalgae, that inserting introns from native genes in heterologous sequences results in increased protein yields (Rosales-Mendoza et al., 2012). A 2009 study by Eichler-Stahlberg, Weisheit, Ruecker, and Heitzer reported a 4-fold increase in a *Renilla*-luciferase gene used as a reporter, with the random insertion of at least one intron into the *rbsc2* gene. Interestingly, none of the three *rbsc2* introns alone increased nuclear expression, but integration in their physiological order and number had a positive effect (Specht et al., 2010; Rosales-Mendoza et al., 2012).

2.2 Significant Advances in Genetic Engineering of Microalgae

The prospects of microalgae as a recombinant protein expression system has attracted incredible efforts to prepare microalgae for mainstream implementation as a standard commercial platform. For nuclear expression, a major obstacle has been low transgene expression and recombinant protein yields. This is largely due to the high level of transgene silencing that takes place in the nuclear genome. In the past, work has been done using ultraviolet mutagenesis to develop strains with enhanced transgene expression with underwhelming results (Zhang and Potvin, 2010).

In recent years, development of new genetic tools for both the nuclear and chloroplast genomes of model algal organisms represents major advancements in this discipline. Furthermore, with the development of low-cost and high-throughput nucleotide sequencing, a wealth of algae genomes have become available for public and private use. These advancements are described in detail here and summarized in Table 3.

2.2.1 Coupling Selectable Markers with Transgenes for High-Level Nuclear Expression

Several advancements have been developed to overcome poor transgene expression in the nucleus. In 2012, Rasala et al. developed a nuclear expression strategy to overcome transgene silencing using the foot-and-mouth-disease-virus (FMDV) 2A self-cleaving sequence to transcriptionally link the transgene of interest to an antibiotic selection marker in *C. reinhardtii*. In this study, robust accumulation of monomeric, cytosolic GFP and xylanase (*xyl1*) was achieved by linking their expression to bleomycin antibiotic resistance (*ble*) via the 2A peptide. The study further demonstrated an ~100-fold increase in xylanase expression using the construct coupled with *ble*-2A. Additionally, by fusing an endogenous secretion signal between *ble*-2A and *xyn1*, significant accumulation of active xylanase was detected in the culture media (Rasala et al., 2012).

The FMDV 2A peptide encodes an approximately 20 amino acid sequence that self-cleaves during translation elongation of the 2A sequence. When the 2A links two genes in a single open reading frame, the product is two discrete proteins, with the short 2A sequence fused to the C-terminus of the first protein product. While the FMDV 2A and similar self-cleavage sequences have been used for heterologous gene expression in many eukaryotic systems, this study showed that the 2A peptide is fully functional in the algal cytoplasm. The bleomycin resistance is a key factor in the high expression of the transgene of interest. The Ble protein functions by antibiotic sequestration rather than enzymatic inactivation, binding to bleomycin (or Zeocin™) in stoichiometric equivalence. Therefore, high levels of expression are necessary to survive antibiotic selection (Rasala et al., 2012).

The development of this FMDV 2A nuclear expression strategy is a significant achievement, enabling the *C. reinhardtii* nucleus to achieve protein accumulation levels comparable to the high expression levels previously only seen in the chloroplast. In 2013, Rasala et al. further applied this technique to demonstrate the first successful expression of an extensive set of six fluorescent proteins (blue mTagBFP, cyan mCerulean, green CrGFP, yellow Venus, orange tdTomato, and red mCherry) from the *C. reinhardtii* nuclear genome. All fluorescent proteins were localized throughout the cytoplasm, well-expressed, and easily detectable in whole cells. Additionally,

TABLE 3   Genome Information for Green Microalgae, Diatoms, and Cyanobacteria Species

Status	Species and strain number	Genome size	Reference
Sequenced and annotated	*Chlamydomonas reinhardtii* CC-503	120 Mbps	Merchant et al. (2007)
	Cyanidioschyzon merolae 10D	16 Mbps	Matsuzaki et al. (2004)
	Cyanophora paradoxa LB555	140 Mbps	Price et al. (2012)
	Thalassiosira pseudonana CCMP 1335	34 Mbps	Armbrust et al. (2004)
	Thalassiosira oceanica CCMP 1005	81.6 Mbps	Lommer et al. (2012)
	Phaeodactylum tricornutum CCMP 2561	27.4 Mbps	Bowler et al. (2008)
	Ectocarpus siliculosus Ec32	214 Mbps	Cock et al. (2010)
	Aureococcus anophagefferens	56 Mbps	Gobler et al. (2011)
	Nannochloropsis gaditana CCMP 526	29 Mbps	Radakovits et al. (2012)
	Nannochloropsis oceanica CCMP 1779	28.7 Mbps	Vieler et al. (2012)
	Synechocystis sp. PCC 6803	3.6 Mbps	Kaneko et al. (1996)
Sequenced, unannotated	*Picochlorum* sp. SENEW3	13.5 Mbps	NAABB
	Auxenochlorella protothecoides UTEX 25	21.4 Mbps	Unpublished data
	Chrysochromulina tobin	75.9 Mbps	NAABB
	Dunaliella salina CCAP 19/18	To be announced	JGI
	Chlorella sorokiniana UTEX 1230	To be announced	Rosenberg et al. (2014)
	Chlorella vulgaris UTEX 395	To be announced	Rosenberg et al. (2014)
	Botryococcus braunii UTEX 572 (race A)	166 Mbps	JGI
	Coccomyxa subellipsoidea sp. C-169	49 Mbps	Blanc et al. (2012)
	All six *Nannochloropsis* species	25.3–32.1 Mbps	Wang et al. (2014)
In progress	*Tetraselmis* sp. LANL1001	220 Mbps	NAABB
	Chlorococcum sp. DOE0101	120 Mbps	NAABB
	Chlorella sp. DOE1412 aka NAABB 2412	55 Mbps	NAABB
	C. sorokiniana Phycal 1228 (UTEX 1230)	55 Mbps	Unpublished data
	Haematococcus pluvialis	To be announced	Unpublished data
	Chlorella zofingiensis UTEX 32	To be announced	Unpublished data
	Chlorella minutissima UTEX 2341	To be announced	NCBI: PRJNA245012
	Chlorella emersonii	To be announced	NCBI: PRJNA243839
	Pelagophyceae sp. CCMP2097	To be announced	JGI

This compilation of industrial and academic sequencing projects represents the current status of genomes for photosynthetic microbes. Comparative genomics underway with many of these organisms, particularly *Chlorella* species relevant to biofuel production.

the researchers showed that they could fluorescently label an endogenous protein by expressing a mCerulean-α-tubulin fusion protein through the ble-2A expression vector. The fusion protein localized to the cytoskeleton and flagella and the cells exhibited normal cellular function (Rasala et al., 2013).

In 2014, Rasala et al. constructed a set of transformation vectors that enabled protein targeting to four distinct subcellular locations: the nucleus, mitochondria, endoplasmic reticulum (ER), and chloroplast. Each organelle-targeting vector was generated by inserting a respective transit sequence between the ble-2A gene and a gene encoding a different fluorescent protein (Rasala et al., 2014). Enabling specified localization of the expressed recombinant protein is a significant step in minimizing proteolytic degradation (Rosales-Mendoza et al., 2012).

2.2.2 Transformation Vectors and Selectable Markers

In 2013, Life Technologies Corporation released the first commercially available genetic modification and expression systems for *C. reinhardtii* and *Synechococcus elongatus*. The new GeneArt® *Chlamydomonas* Protein Expression Kits and GeneArt® *Synechococcus* Protein Expression Kits contain vectors claimed to drive expression levels up to 10% TSP and new proprietary transformation reagents that improve the transformation efficiency up to as much as 1000-fold over current techniques (LifeTech Inc., 2013).

While the Life Technologies vector expresses single genes, multigene integration has also been developed in *C. reinhardtii* using either genetic linker elements to express multiple genes from the same transcript or cross-breeding of genetically modified (GM) strains. In 2013, Noor-Mohammadi et al. developed a new method for multiple transgene expression in the nucleus of *C. reinhardtii* by first assembling the cassettes in *Saccharomyces cerevisiae* by *in vivo* homologous recombination. The assembled plasmid was then isolated from *S. cerevisiae* and

transformed into the nuclear genome of *C. reinhardtii*. Using this method, they demonstrated a proof of concept by expressing up to three reporter proteins (Ble, AphVIII, and GFP) simultaneously from the nuclear genome of *C. reinhardtii* (Noor-Mohammadi et al., 2013). These new tools for multi-gene engineering may enable the introduction of more sophisticated biochemical pathways into the nuclear genome of *C. reinhardtii*.

More recently, complementary methods for robust expression of multiple nuclear-encoded transgenes within a single cell of *C. reinhardtii* have been developed. The first method involves a modification of their original ble-2A expression strategy by including a second 2A peptide from equine rhinitis A virus (E2A) followed by a third protein coding sequence. When properly integrated, the multi-cistronic transgene cassette successfully expressed two different fluorescent proteins localized to separate organelles with high levels of accumulation. The second method involves a gene-stacking strategy using successive rounds of cross-breeding to generate transgenic algae strains that express up to four fluorescent proteins with subcellular localization, shown in Figure 1 (Rasala et al., 2014). In addition to multiple fluorescent genes, many novel selectable markers are being bioprospected. Bruggeman et al. developed and tested the use of three genes conferring resistance to the herbicides glyphosate, oxyfluorfen, and norflurzon. Separate transgenic *C. reinhardtii* cell lines expressing a synthetic glyphosate acetyltransferase (GAT), a mutated *Chlamydomonas* protoporphyrinogen oxidase (protox, PPO), and modified *Chlamydomonas* phytoene desaturase (PDS) genes exhibited increased tolerance to their respective herbicides by 2.7-, 136-, and 40-fold, respectively. Additionally, the low concentration requirements for both norflurazon and oxyfluorfen suggest that these two herbicides may be effective crop protection tools for commercial-scale algal production facilities (Bruggeman et al., 2014).

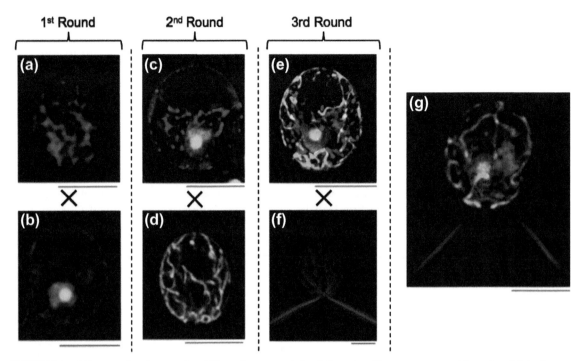

FIGURE 1 **Micrographs of successive *Chlamydomonas reinhardtii* mating illustrate gene stacking of multiple fluores-cent proteins.** After cross breeding a mating-type plus (mt⁺) strain expressing *mCherry* targeted to the ER (a) with a mating-type minus (mt⁻) strain expressing *mCerulean* targeted to the nucleus (b), progeny were screened for transgenic cells that expressed both fluorescent proteins (c). These cells were crossed again with transgenic cells expressing *Venus* targeted to the mitochondria (d). Progeny expressing three distinct fluorescent proteins (e) underwent a third round of mating with a strain stably expressing α-*tubulin-mTagBFP* (f) to obtain progeny successfully expressing four different fluorescent proteins, all local-ized to four unique subcellular organelles with high levels of expression (g). *Adapted from Rasala et al. (2014) "Enhanced Genetic Tools for Engineering Multigene Traits into Green Algae" and licensed under CC BY 4.0 (creativecommons.org/license/by/4.0).*

Although there is a long history of genetic manipulation of the *C. reinhardtii* chloroplast genome, Oey et al. introduced the lambda-phage based Gateway® system to microalgae as an alternative to traditional cloning, which can be complex and laborious (Oey et al., 2014). The Gateway® system uses recombination sequences that facilitate the transfer of DNA fragments between vectors (Hartley et al., 2000). Oey et al. used GFP to demonstrate how the flexible, versatile, interchangeable vector sys-tem allows rapid insertion of different expres-sion cassettes into destination vectors of choice for high-throughput cloning (Oey et al., 2014).

2.3 Sequencing and Functional Annotation of Microalgal Genomes

While targeted molecular genetic tools have allowed researchers to achieve even more pre-cise engineering design goals, fundamental insight into microalgal metabolism can be reached through comparative genomics and, in turn, leveraged for industrial production of desired metabolites. Viewed from phylogenetic relationship with land plants, the closest algal relatives of land plants are Rodophyta (red algae), Chlorophyta (green algae), Glaucophyta and Charophyta. In Rodophyta, five species' genomes have been sequenced and annotated, with *Cyanidoschyzon merolae* being the first (Matsuzaki et al., 2004). Many more Chlorophyta genomes are publicly available, with *C. reinhardtii* (Merchant et al., 2007) being the most well-characterized representative. More recently, the first genome in the phylum Glaucophyta became available with the sequencing of *Cyanophora paradoxa* (Price et al., 2012). Unfortunately, no genomic sequence information is available for Charophyta organisms yet.

Eukaryotic algae that are distantly related to land plants are classified as Stramenopiles (or Heterokonts), including Bacillariophyta (diatoms), Phaeophyta (brown algae), Pelagophyte (golden-brown algae), and Eustigmatophyceae (*Nannochloropsis*). Diatoms with completed genome sequences are *Thalassiosira pseudonana* (Armbrust et al., 2004), *Thalassiosira oceanica* (Lommer et al., 2012), and *Phaeodactylum tricornutum* (Bowler et al., 2008). *Ectocarpus siliculosus* (Cock et al., 2010) in Phaeophyta and *Aureococcus anophagefferens* (Gobler et al., 2011) in Pelagophytes have also been sequenced. In Eustigmatophyceae, *Nannochloropsis gadaiana* (Radakovits et al., 2012) and *Nannochloropsis oceanica* (Vieler et al., 2012) genomes were initially released, although the genomes of all six *Nannochloropsis* species are now available (Wang et al., 2014). Compared to the eukaryotic algal genomes, more Cyanophyta (cyanobacteria or blue-green algae) have been sequenced. *Synechocystis* sp. PCC 6803 (Kaneko et al., 1996) was the first cyanobacteria to be fully sequenced. Since then, more than 90 cyanobacterial genomes have been released and more remain as ongoing sequencing projects (Nakao et al., 2010; Shih et al., 2013).

Several industrially relevant algae species are also in the process of having their genomes sequenced, including *Dunaliella salina*, *Haematococcus pluvialis*, and *Chlorella zofingiensis*. *D. salina* is cultured to produce β-carotene, while *H. pluvialis* and *C. zofingiensis* are cultured for astaxanthin. After screening 2200 algae isolates for biofuel production, the National Alliance for Advanced Biofuels and Bioproducts (NAABB) has finished the sequencing and assembly of eight distantly related strains with potential for biomass and lipid accumulation under large-scale cultivation conditions, including *Picochlorum* sp., *Chrysochromulina tobin*, *Nannochloropsis salina*, *Tetraselmis* sp., and a number of *Chlorella* isolates (Table 3). Besides the *Chlorella* strains chosen by NAABB, other fast-growing oleaginous *Chlorella* strains have also been successfully sequenced, such as *Chlorella sorokiniana* UTEX 1230 and *Chlorella vulgaris* (UTEX 259 and UTEX 395) (Rosenberg et al., 2014). *Botryococcus braunii* is a slow-growing algae, but it was also chosen for genome sequencing because 90% of its biomass can be converted to drop-in fuels. *Pelagophyceae* sp. and *Cyclotella cryptica* are next on the list for sequencing due to their potential as biofuel sources. For more about the ongoing algal genome sequencing projects, readers are referred to the Joint Genome Institute's project list (JGI, 2014).

2.4 Computational Tools for Genome-Scale Bioinformatics and Metabolic Network Interrogation

To focus bioengineering efforts toward the fine-tuning of algal metabolism, computation biology provides a number of avenues for metabolic and genetic elucidation. With the exponentially expanding algae genome sequence information, functional annotation of algal genes has become a challenge. The Algal Functional Annotation Tool is the first algae-focused gene list annotation tool (Lopez et al., 2011), which integrates several independent databases to annotate large gene lists.

Gene functional annotation may be further improved by incorporating protein-oriented subcellular localization prediction tools, such as PredAlgo (Tardif et al., 2012). The knowledge obtained from gene annotation can be further used to construct genome metabolic models and the first photosynthetic algal model has recently been built (Chang et al., 2011). This first green algae genome metabolic model has offered insights into the effects of light quality on algal metabolism. Besides genome annotation and modeling, the gene expression profiles of algae grown under various conditions generated by next-generation sequencing have been integrated in a freely online web portal AlgaePath for transcript abundance analysis and co-expression analysis (Zheng et al., 2014).

A complementary experimental approach to these computational tools, called metabolic flux analysis, allows for the elucidation of rates of metabolite generation and consumption in a biological system, which provides valuable information on cellular metabolite utilization (Zamboni et al., 2009). By cultivating algal organisms using ^{13}C-labeled organic substrates, considerable pathway information can be deduced based on the final distribution of labeled-carbon tracer in the biomass by gas chromatograph mass spectrometry (GC–MS). For example, this method has been applied to understanding carbon metabolism in the heterotrophic microalga *Chlorella sorokiniana* (Xiong et al., 2010) and photosynthetic processes in the cyanobacterium *Synechocystis* sp. PCC 6803 (Young et al., 2011). Applying this technique to photosynthetic organisms has led to an increased study of proteomics and metabolomics in microalgae (Guarnieri and Pienkos, 2014) and has offered a means to improve and validate the results from computational tools.

3. APPLICATION OF METABOLIC ENGINEERING FOR BIOSYNTHESIS AND SYNTHETIC BIOLOGY

The accumulation of endogenous compounds and overexpression of transgenic products can benefit from metabolic engineering either through enhanced product generation with increased throughput in desirable pathways or by inhibiting alternative pathways. While

genetic modification of cyanobacteria can be more technically feasible than microalgae, the integration of transgene fragments can lead to novel functions in algae (Heidorn et al., 2011). In recent years, microalgae have been genetically engineered for improved photosynthetic efficiency by redirecting electron flow within the photosystems (Veyel et al., 2014; Lassen et al., 2013). Moreover, certain strains can secrete byproducts under conditions of continuous growth (Mendez et al., 2009; Rasala et al., 2012). Increasing product yield is also possible through the addition of matrix attachment region binding proteins or by modulating gene silencing/expression (Wang et al., 2010; Allen et al., 2000). Microalgae have also been used as a host for human antibody production, demonstrating their capabilities in recombinant protein and vaccine development (Hempel et al., 2011; Surzycki et al., 2009; Mayfield and Franklin, 2005; Gregory et al., 2013; Jones et al., 2013). In the interest of renewable oil production, carbohydrate biosynthetic pathways have been redirected toward lipid production (Li et al., 2010a; Ramazanov and Ramazanov 2006, Lohr et al., 2012). These and other accomplishments of metabolic engineering are compiled in Table 4. This compendium of examples combining metabolic engineering with synthetic biology demonstrates that product biosynthesis can be enhanced without hindering the cell's ability to grow.

Commercial endeavors are also pioneering strain development pipelines based on newly bioprospected host species that exhibit robust baseline growth aptitudes and are also amenable to genetic transformation (Figure 2). After superior strains are identified, transgene expression systems can be developed by adapting known genetic elements and selectable markers to the strains' unique genome. Various classical methods of genetic transformation, including agrobacterium vectors, plant promoters, and other basic methods of gene incorporation (e.g., microparticle bombardment, electroporation)

TABLE 4   Summary of Metabolic Engineering Accomplishments in Microalgae

Target modification	Product	Host species	Genome	Reference
Light-harvesting complex size	Photosynthesis efficiency	*Chlamydomonas reinhardtii*	Chloroplast	Beckmann et al. (2009)
Carbon fixation	Organic carbon	*Synechococcus elongatus*	Nucleus	Shih et al. (2014)
Deletion of mating and recombination genes	Strains cannot spread genetic information	All algae species	Nucleus	Henley et al. (2013)
Rearrange essential gene under inducible promoter	Strain cannot spread genetic information	All algae species	Nucleus	Allutt et al. (2013)
Mutagenesis of gene silencing mechanism	Enhanced gene expression	*C. reinhardtii*	Nucleus	Neupert et al. (2009)
Stm6 mutation	Hydrogen	*C. reinhardtii*	Nucleus	Kruse et al. (2005)
PDS L516F overexpression	Astaxanthin	*Chlorella zofigiensis*	Nucleus	Liu et al. (2014)
Knockout of starch biosynthesis	Lipids	*C. reinhardtii*	Nucleus	Li et al. (2010b)
Knockdown of lipid catabolism	Lipids	*Thalassiosira pseudonana*	Nucleus	Trentacoste et al. (2013)
Expression of exogenous genes	Lipids	*Phaeodactylum tricornutum*	Nucleus	Radakovits et al. (2010)
	Ribose sensor	*T. pseudonana*	Nucleus	Marshall et al. (2012)
	Human antibodies	*P. tricornutum*	Nucleus	Hempel et al. (2011)
	Enzyme (xylanase)	*C. reinhardtii*	Nucleus	Rasala et al. (2012)
	Immunotoxin	*C. reinhardtii*	Chloroplast	Tran et al. (2012)

Further characterization of photosynthetic microbes and their genomes has led to significant advances in the bioengineering of microalgae. Applications range from renewable energy and high-value products for human health to biocontainment and biosensing.

can then be used to accomplish transgene integration (Rosenberg et al., 2008). Concurrently, gene targets that may confer industrial fitness, expanded metabolic capabilities, or increased rates of biomass production can be determined in a somewhat independent manner.

Trait discovery campaigns may follow a few different approaches to achieve lead transgene candidates, including (1) panning large cDNA libraries generated from different organisms grown under different conditions; (2) pursuing directed evolution of proteins for variants with enhanced enzymatic or regulatory characteristics; and (3) executing hypothesis-based selection of previously characterized genes with desired function. Once gene targets are identified, they can be cloned into expression vectors and transformed in an algal platform ready to enter the GM strain pipeline for assessment and validation. For example, competition-driven selection of industrial fitness can be accomplished using turbidostat experiments, in which a mixed culture of wild-type algal cells are grown with a minority population of engineered organisms.

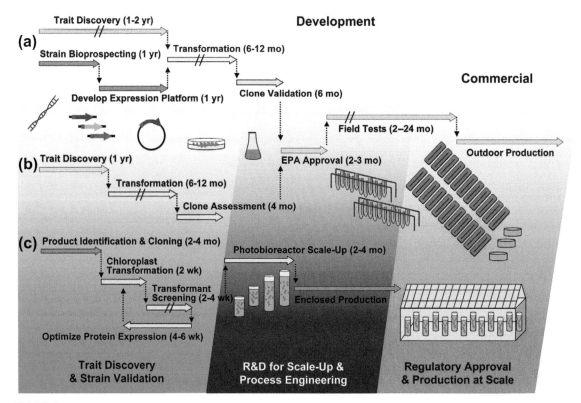

FIGURE 2   **Approximate timelines for commercial development of transgenic microalgae.** This figure depicts strain development milestones for three production scenarios: (a) commodity-scale products derived from GM algae biomass (e.g., biofuel), including an initial bioprospecting stage for a novel strain platform; (b) similar commodity-scale GM algae production using an existing transgenic strain platform; and (c) recombinant expression of high-value proteins in the C. *reinhardtii* chloroplast. While EPA approval is required for open-air cultivation of transgenic algae, strains grown in closed photobioreactors under greenhouse facilities may likely evade such regulatory regimes (Johanningmeier and Fischer, 2010).

While keeping the culture's cell density constant, GM cells with particular survival advantages will dominate the population over the growth period and can be recovered for further analysis and characterization.

3.1 Biocontainment: Proactive Measures for Maintenance of Culture Integrity

As a result of improved molecular genetic tools, the pace of microalgal strain development and the range of industrially relevant traits have increased dramatically (Stephens et al., 2013). Both commercial and academic endeavors are now poised to reach advanced stages of metabolic engineering, process development, and research-scale production of recombinant cell lines in both contained and open-air environments (Table 1). Open-air cultivation of microalgae and cyanobacteria poses uniquely different risks with potentially greater probability of unintended mutations and nonnegligible release to the environment by means of accidental spills (Gressel et al., 2013a, 2013b). As such, it is important to proactively address concerns with preventive measures so that these organisms only grown in their intended bioreactors.

During commercial cultivation, microalgae encounter abiotic and biotic stress factors that can limit cell growth (Affenzeller et al., 2009; Radakovits et al., 2010; Vuttipongchaikij, 2012). Because cellular survival during environmental stresses relies on robust stress response networks (Vinocur and Altman, 2005), engineering these pathways can be an important trait for industrial fitness, but it may also increase the organisms' ability to survive outside of contained cultivation (Torres et al., 2003) One method to prevent these microorganisms from surviving in the wild can be conferred by inserting a "suicide" gene that will only allow the organism to grow under specific conditions. Through the addition of a lethal gene that remains inactive in the presence of a specific compound, the culture will grow. When the organism encounters an environment where the compound is not present, the lethal gene will be activated, leading to self-destruction. This can be accomplished in conjunction with an essential gene linked to an inducible promoter, where the promoter is activated by another compound only found in the artificial environment (Henley et al., 2013; Allnutt et al., 2013). Alternatively, the use of gene deletion as a method to reduce an alga's fitness is an alternative approach to prevent it from thriving outside of a raceway pond or photobioreactor. By removing genes used for mating or recombination, the microorganism will not be able to pass specific traits onto wild-type organisms (Henley et al., 2013).

In terms of the ecological effects of inadvertent escape of GM algae, the resilience and response to GM algae release have been modeled and reviewed extensively (Henley et al., 2013; Flynn, Greenwell, Lovitt, & Shields, 2010; Shurin et al., 2013). Nearly all findings agree that transgenic strains do not possess the ability to out-compete naturally occurring species in the wild. Invasion studies and bioburden testing in which GM organisms are purposefully released into natural bodies of water, soil, and air support the conclusion that GM algae survival rates are many of orders of magnitude less than wild-type strains. Furthermore, molecular diagnostic approaches to tracking both culture integrity as well as the presence of potential contaminant organisms holds potential for future tracking of GM algae populations (Fulbright, Dean, Wardle, Lammers, Chisholm, 2014).

4. IMPLICATIONS FOR REGULATORY CONTROL OVER GM ALGAE

For nearly two decades, genetic improvement of microbial hosts and agricultural crops has impacted areas of human life ranging from advanced medicines to our daily meals. With recent debate over the role genetically modified organisms (GMOs) in our food supply, its labeling, and a growing hesitance toward the "unnatural" in general, it appears that the benefits of these biotechnological advancements greatly outweigh any potential negative effects. Some advantages of genetically enhanced plants include: (1) lower cost of production, reduced energy inputs, and less acreage required for cultivation (all due to increased productivity); (2) improved biochemical composition of the biomass by expanding the plants metabolic capabilities for nutrition, fuel, and chemical applications; (3) stress tolerance to enable cultivation in non-ideal climates (e.g., drought tolerance for arid climates); (4) and the potential for pharmaceutical production using a host that is devoid of human pathogens.

In recent years, GMO development has explored new ground with the first cases of GM insects and fish to combat disease transmission and improve the sustainability and profitability of aquaculture (Franz et al., 2014; aquabounty.com). Theses two particular cases are receiving attention as landmark accomplishments of biotechnology, but also controversial applications of genetic engineering as the tides of public concern for commercial cultivation of transgenic organisms continue to rise. While public sentiments disagree on the use of GMOs in foods, immediate concern for chemical and fuel production using transgenic crops is less severe, but not absent. Despite the fact that GM algae may not apply extensively to food products, algae oils (initially developed for biofuels) are being included in personal products (e.g., soaps) (Strom, 2014). As such, the potential for concerns by the public about GM algae and discordance between policies governing their distribution exists. Ecological impacts of GM microbes cultivated in open systems will also necessitate relevant regulatory regimes and adequate control measures for this emerging industry (Snow & Smith, 2012).

Initial regulatory oversight for large-scale cultivation of microalgae in the United States was considered by the US Department of Agriculture's Animal and Plant Health Inspection Service (USDA APHIS), the Environmental Protection Agency (EPA), and possibly the US Food and Drug Administration (FDA). Ultimately, transgenic algae for renewable fuel production fell under the EPA's existing mechanisms of assessing biotechnological use of microorganisms is appropriate; however, future applications of algal cultivation for agricultural use may be regulated by the USDA. In 1998, the EPA established two routes for regulatory review of transgenic microbes, focusing on associated environmental safety and human pathogen control. Photosynthetic microbes and their byproducts are regulated under the EPA's Toxic Substances Control Act (TSCA), which requires a "TSCA Environmental Release Application" (TERA) to be filed before pre-commercial field-testing can be conducted in open systems. For commercial-scale operations, a "Microbial

Commercial Activity Notice" (MCAN) must be filed with the EPA to rule out the possibility that any recombinant products or non-essential transgenes may carry toxic or allergenic effects. Historically, MCANs and TERAs have sought approval for the production of biochemicals, such as industrial enzymes (e.g., amylase, phytase), cellulosic ethanol, and soil additives using recombinant heterotrophs. These host organisms were predominantly bacteria and fungi including *Escherichia coli*, *Trichoderma*, *Pseudomonas*, *Bacillus*, *Zymomonas*, *Saccharomyces*, *Bradyrhizobium*, and *Pichia* species cultivated in closed bioreactors with contained effluent. Before new embodiments of GM microalgae can even be considered for EPA approval, a lengthy pipeline of trait selection and design of a suitable expression system must be developed for each application. This timeline can span many months to multiple years prior to filing TERA or MCAN requests (Figure 2). For example, native strains of algae and cyanobacteria are often bioprospected as platforms for biofuel production with certain selection criteria, such as robust growth, hardiness during conditions of environmental stress, resilient population dynamics in response to intruding microbes, and resistance to predation.

With a wide range of efforts focused on enabling the transition of these organisms from bench-top laboratory experiments to field trials, two algal biofuel companies have already received regulatory approval for the controlled outdoor cultivation of proprietary transgenic algae and cyanobacteria, which is truly an industry first (Joule Unlimited Technologies, 2012; Sapphire Energy, 2013). Two additional companies currently have EPA TSCA applications under review (Table 1). Aside from Sapphire Energy, all of these applications involve contained growth systems, but still represent significant steps toward commercialization of GM algae in outdoor environments. One of roughly five national testbeds for open-air algae cultivation is the University of California—San Diego

Biology Field Station, which is the designated experimental site for Sapphire's TERA. Scale-up of open cultures at this facility using polyethylene hanging bags and air-lift driven raceway ponds has been described recently by Schoepp et al. (2014) with anticipated transition to a 300-acre commercial site in Columbus, New Mexico. With a nearly exponential increase in MCAN filings between 2010 and 2013, it is anticipated that GM microbes will become even more prevalent in industrial biochemical production.

5. CONCLUSIONS AND PERSPECTIVES

The recent development of new molecular tools has enabled significant advancements in harnessing the full potential of microalgae as a platform for producing nutraceuticals, specialty biomolecules, renewable biofuels, and bulk chemicals. For the first time, strategies for expressing multiple genes from a single cell were successfully developed for *C. reinhardtii* allowing for the introduction of more complex biochemical pathways into the nuclear genome of microalgae. Through methods of coupling transgene expression with selectable markers, protein accumulation levels previously only seen from the chloroplast were achieved from the nuclear genome. The availability of new transformation vectors, genetic engineering kits, and selectable markers has simplified the traditional cloning process allowing for more rapid strain development.

In light of the advances in molecular genetic tools for microalgae and cyanobacteria that have occurred over the past 5 years, it is clear that metabolic engineering has improved upon the framework of the traditional microalgal cell. With this commercial-scale growth, conscientious efforts toward biocontainment and regulatory jurisdictions are also being taken to prevent unintended consequences of GM algae in natural ecosystems. However, algal communities grown in open-air systems cannot be kept axenic, and it is well understood that biomass composition is strongly affected by intruding organisms. While conventional mechanisms of crop protection and biological control have proven effective at large scales (McBride et al., 2013), lipid accumulation phenotypes certainly vary depending on trophic conditions with interdependence between competing phototrophs and predators (Letcher et al., 2013). Therefore, future developments in this field may build upon synthetic biology tools for individual organisms by engineering mutualism or symbiosis between two or more organisms with a unique opportunity to access an expanded complexity of metabolism within microbial consortia (Zengler and Pallson, 2012). By examining monocultures in crop rotation and year-round polycultures, we may eventually reach a better understanding of "synthetic ecology" within industrial algal production systems (Kazamia et al., 2012; Hamilton, 2014; Carney et al., 2014).

Acknowledgments

This work was supported in part by funds from the National Science Foundation under grant number NSF-EFRI-1332344 to MJB and a fellowship to JNR from the Johns Hopkins Environment, Energy, Sustainability & Health Institute (E²SHI). This material is also based upon work supported by the U.S. Department of Energy, Office of Science, Biological Systems Science Division, under Award Number DE-SC0012658.

References

Affenzeller, M.J., Darehshouri, A., Andosch, A., Lütz, C., Lütz-Meindl, U., 2009. Salt stress-induced cell death in the unicellular green alga *Micrasterias denticulata*. J. Exp. Bot. 60 (3), 939–954.

Allen, G.C., Spiker, S., Thompson, W.F., 2000. Use of matrix attachment regions (MARs) to minimize transgene silencing. Plant Mol. Biol. 43, 361–376.

Allnutt, F.C.T., Postier, B., Sayre, R., Coury, D., Kumar, A., Swanson, A., Abad, M., Perrine, Z. 2013. Biosecure genetically modified algae. U.S. Patent Application. Pub. No. US 2013/0109098 A1.

Armbrust, F.V., Berges, J.A., Bowler, C., Green, B.R., Martinez, D., Putnam, N.H., et al., 2004. The genome of the diatom *Thalassiosira pseudonana*: ecology, evolution, and metabolism. Science 306, 79–86.

Bachman, T., 2003. Transforming cyanobacteria into bio-reporters of biological relevance. Trends Biotechnol. 20, 247–249.

Beckmann, J., Lehr, F., Finazzi, G., Hankamer, B., Posten, C., Wobbe, L., Kruse, O., 2009. Improvement of light to biomass conversion by de-regulation of light-harvesting protein translation in *Chlamydomonas reinhardtii*. J. Biotechnol. 142, 70–77.

Blanc, G., Agarkova, I., Grimwood, J., Kuo, A., Brueggeman, A., et al., 2012. The genome of the polar eukaryotic microalga *Coccomyxa subellipsoidea* reveals traits of cold adaptation. Genome Biol. 13, R39.

Blankenship, J.E., Kindle, K.L., 1992. Expression of chimeric genes by light-regulatable cabll-1 promoter in *Chlamydomonas reinhardtii*. A cabll-1/nit1 gene functions as a dominant selectable marker in a nit1- nit2-strain. Mol. Cell Biol. 12, 5268–5279.

Bowler, C., Allen, A.E., Badger, J.H., Grimwood, J., Jabbari, K., Kuo, A., et al., 2008. The *Phaeodactylum* genome reveals the evolutionary history of diatom genomes. Nature 456, 239–244.

Bruggeman, A.J., Kuehler, D., Weeks, D.P., 2014. Evaluation of three herbicide resistance genetic transformations and for potential crop protection in algae production. Plant Biotechnol. J. 1–9. http://dx.doi.org/10.1111/pbi.12192.

Cadoret, J.P., Matthieu, G., Bruno, S.J., 2012. Microalgae, functional genomics and biotechnology. Adv. Bot. Res. 64, 285–341.

Carney, L.T., Reinsch, S.S., Lane, P.D., Solberg, O.D., Jansen, L.S., Williams, K.P., Trent, J.D., Lane, T.W., 2014. Microbiome analysis of a microalgal mass culture growing in municipal wastewater in a prototype OMEGA photobioreactor. Algal Res. 4, 52–61.

Chang, R.L., Ghamsari, L., Manichaikul, A., Hom, E.F., Balaji, S., Fu, W., Shen, Y., Hao, T., Palsson, B.O., Salehi-Ashtiani, K., et al., 2011. Metabolic network reconstruction of *Chlamydomonas* offers insight into light-driven algal metabolism. Mol. Syst. Biol. 7, 518.

Cock, J.M., Sterck, L., Rouze, P., Scornet, D., Allen, A.E., Amoutzias, G., et al., 2010. The *Ectocarpus* genome and the independent evolution of multicellularity in brown algae. Nature 465, 617–621.

Corbeil, L.B., Hildebrand, M., Shrestha, R., Davis, A., Schrier, R., Oyler, G.A., et al., 2012. Diatom-Based Vaccines. U.S. Application No. 14/353,721.

Dreesen, I.A., Charpin-El Hamri, G., Fussenegger, M., 2010. Heat-stable oral alga-based vaccine protects mice from *Staphylococcus aureus* infection. J. Biotechnol. 145, 273–280.

Eichler-Stahlberg, A., Weisheit, W., Ruecker, O., Heitzer, M., 2009. Strategies to facilitate transgene expression in *Chlamydomonas reinhardtii*. Planta 229, 873–883.

Fischer, N., Rochaix, J.D., 2001. The flanking regions of PsaD drive efficient gene expression in the nucleus of the green alga *Chlamydomonas reinhardtii*. Mol. Genet. Genomics 265, 888–894.

Fletcher, S.P., Muto, M., Mayfield, S., 2007. Optimization of recombinant protein expression in the chloroplasts of green algae. Adv. Exp. Med. Biol. 616, 90–98.

Flynn, K.J., Greenwell, H.C., Lovitt, R.W., Shields, R.J., 2010. Selection for fitness at the individual or population levels: Modelling effects of genetic modifications in microalgae on productivity and environmental safety. J. Theor. Biol. 263, 269–280.

Franklin, S., Ngo, B., Efuet, E., Mayfield, S.P., 2002. Development of a GFP reporter gene for *Chlamydomonas reinhardtii* chloroplast. Plant J. 30, 733–744.

Franz, A.W.E., Clem, R.J., Passarelli, A.L., 2014. Novel genetic and molecular tools for the investigation and control of dengue virus transmission by mosquitoes. Curr. Trop. Med. Rep. 1, 21–31.

Fulbright, S.P., Dean, M.K., Wardle, G., Lammers, P.J., Chisholm, S., 2014. Molecular diagnostics for monitoring contaminants in algal cultivation. Algal Res. 4, 41–51.

Gobler, C.J., Berry, D.L., Dyhrman, S.T., Wilhelm, S.W., Salamov, A., et al., 2011. Niche of harmful alga *Aureococcus anophagefferens* revealed through ecogenomics. Proc. Natl. Acad. Sci. USA 108, 4352–4357.

Goldschmidt-Clermont, M., 1991. Transgenic expression of aminoglycoside adenine trasnferase in the chloroplast: a selectable marker for site-directed transformation of *Chlamydomonas*. Nucleic Acids Res. 19, 4083–4089.

Gregory, J.A., Topol, A.B., Doerner, D.Z., Mayfield, S., 2013. Alga-produced cholera toxin-Pfs25 fusion proteins as oral vaccines. Appl. Environ. Microbiol. 79 (13), 3917–3925.

Gressel, J., 2008. Genetic Glass Ceilings: Transgenics for Crop Biodiversity. ISBN: 0801887194, Johns Hopkins University Press: Baltimore, MD.

Gressel, J., van der Vlugt, C.J.B., Bergmans, H.E.N., 2013a. Cultivated microalgae spills: hard to predict/easier to mitigate risks. Trends Biotechnol. 32, 65–69.

Gressel, J., van der Vlugt, C.J.B., Bergmans, H.E.N., 2013b. Environmental risks of large scale cultivation of microalgae: mitigation of spills. Algal Res. 2, 286–298.

Griffiths, M., Harrison, S.L., 2009. Lipid productivity as a key characteristic for choosing algal species for biodiesel production. J. Appl. Phycol. 21, 493–507.

Guarnieri, M.T., Pienkos, P.T., March 2014. Algal omics: unlocking bioproduct diversity in algae cell factories. Photosynth. Res. 1–9.

Hamilton, C., 2014. Exploring the utilization of complex algal communities to address algal pond crash and increase annual biomass production for algal biofuels. U.S. Dep. Energy. BETO Report DOE/EE-1059.

Harrop, S.J., Wilk, K.E., Dinshaw, R., Collini, E., Mirkovic, T., et al., 2014. Single-residue insertion switches the quaternary structure and exciton states of cryptophyte light-harvesting proteins. Proc. Natl. Acad. Sci. USA 111, E2666–E2675.

Hartley, J.L., Temple, G.F., Brasch, M.A., 2000. DNA cloning using in vitro site-specific recombination. Genome Res. 10, 1788–1795.

Heidorn, T., Camsund, D., Huang, H., Lindberg, P., Oliveira, P., Stensjö, K., Lindblad, P., 2011. Synthetic biology in cyanobacteria engineering and analyzing novel functions. Methods Enzymol. 497, 539–579.

Hempel, F., Lau, J., Klingl, A., Maier, U.G., 2011. Algae as protein factories: expression of a human antibody and the respective antigen in the diatom *Phaeodactylum tricornutum*. PLoS One 6 (12), 1–7.

Henley, W.J., Litaker, R.W., Novoveská, L., Duke, C.S., Quemada, H.D., Sayre, R.T., 2013. Initial risk assessment of genetically modified (GM) microalgae for commodity-scale biofuel cultivation. Algal Res. 2, 66–77.

JGI, 2014. Joint Genome Institute: Project List (accessed 01.07.14.) Available online: http://genome.jgi.doe.gov/genome-projects/pages/projects.jsf.

Johanningmeier, U., Fischer, D., 2010. Perspective for the use of genetic transformants in order to enhance the synthesis of the desired metabolites: engineering chloroplasts of microalgae for the production of bioactive compounds. Adv. Exp. Med. Biol. 698, 144–151.

Jones, C.S., Luong, T., Hannon, M., Tran, M., Gregory, J.A., Shen, Z., Briggs, S.P., Mayfield, S.P., 2013. Heterologous expression of the C-terminal antigenic domain of the malaria vaccine candidate Pfs48/45 in the green algae *Chlamydomonas reinhardtii*. Appl. Microbiol. Biotechnol. 97, 1987–1995.

Joule Unlimited Technologies, 2012. Contained Use of Modified *Synechococcus* Strain JPS1 for the Production of Ethanol. Joule Unlimited Technologies, Inc. EPA MCAN Case Number: TS-JUT001. (accessed 08.04.14.) Available online: http://www.epa.gov/biotech_rule/pubs/submiss.htm

Kaneko, T., Sato, S., Kotani, H., Tanaka, A., Asamizu, E., Nakamura, Y., et al., 1996. Sequence analysis of the genome of the unicellular cyanobacterium *Synechocystis* sp. strain PCC6803. II. Sequence determination of the entire genome and assignment of potential protein-coding regions. DNA Res. 3, 109–136.

Kazamia, E., Aldridge, D.C., Smith, A.G., 2012. Synthetic ecology—a way forward for sustainable algal biofuel production? J. Biotechnol. 162, 163–169.

Kindle, K.L., 1998. Nuclear transformation: technology and applications. In: Rochaix, J.D., Goldschimdt-Clermont, M., Merchant, S. (Eds.), The Molecular biology of chloroplast and mitochondria in *Chlamydomonas*. Kluwer Academic Publishers Dordrecht, The Netherlands, pp. 41–61.

Kozminski, K.G., Diener, D.R., Rosenbaum, J.L., 1993. High level expression of non-acetylatable alpha tubulin in *Chlamydomonas reinhardtii*. Cell Motil. Cytoskeleton 25, 158–170.

Kruse, O., Rupprecht, J., Bader, K.P., Thomas-Hall, S., Schenk, P.M., Finazzi, G., Hankamer, B., 2005. Improved photobiological H_2 production in engineered green algal cells. J. Biol. Chem. 280, 34170–34177.

Kumar, A., Falcao, V.R., Sayre, R.T., 2013. Evaluating nuclear transgene expression systems in *Chlamydomonas reinhardtii*. Algal Res. 2, 321–332.

Kurland, C., Gallant, J., 1996. Errors of heterologous protein expression. Curr. Opin. Biotechnol. 7 (5), 489–493.

Lassen, L.M., Nielsen, A.Z., Ziersen, B., Gnanasekaran, T., Møller, B.L., Jensen, P.E., 2013. Redirecting photosynthetic electron flow into light-driven synthesis of alternative products including high-value bioactive natural compounds. ACS Synth. Biol. 2, 308–315.

Leon-Banares, R., Gonzalez-Ballester, D., Galvan, A., Fernandez, E., 2004. Transgenic microalgae as green cell-factories. Trends Biotechnol. 22 (1), 45–52.

Letcher, P.M., Lopez, S., Schmieder, R., Lee, P.A., Behnke, C., Powell, M.J., McBride, R.C., 2013. Characterization of *Amoeboaphelidium protococcarum*, an algal parasite new to the cryptomycota isolated from an outdoor algal pond used for the production biofuel. PLoS One 8 (2), e56232.

Li, Y., Han, D., Hu, G., Dauvillee, D., Sommerfeld, M., Ball, S., Hu, Q., 2010. *Chlamydomonas* starchless mutant defective in ADP-glucose pyrophosphorylase hyper-accumulates triacylglycerol. Metab. Eng. 12, 387–391.

Li, Y., Han, D., Hu, G., Sommerfeld, M., Hu, Q., 2010. Inhibition of starch synthesis results in overproduction of lipids in *Chlamydomonas reinhardtii*. Biotechnol. Bioeng. 107 (2), 258–268.

Li, J., Zhu, D., Niu, J., Shen, S., Wang, G., 2011. An economic assessment of astaxanthin production by large scale cultivation of *Haematococcus pluvialis*. Biotechnol. Adv. 29, 568–574.

LifeTech Inc., 2013. Algae Expression & Engineering Products to Enhance Your Results (accessed 01.07.14.). Available online: http://www.lifetechnologies.com/us/en/home/life-science/protein-expression-and-analysis/protein-expression/algae-engineering-kits.html.

Liu, J., Huang, J., Sun, Z., Zhong, Y., Jiang, Y., Chen, F., 2011. Differential lipid and fatty acid profiles of photoautotrophic and heterotrophic *Chlorella zofingiensis*: assessment of algal oils for biodiesel production. Bioresour. Technol. 102, 106–110.

Liu, J., Sun, Z., Gerken, H., Huang, J., Jiang, Y., Chen, F., 2014. Genetic engineering of the green alga *Chlorella zofingiensis*: a modified norflurazon-resistant phytoene desaturase gene as a dominant selectable marker. Appl. Microbiol. Biotechnol. 98, 5069–5079.

Lohr, M., Schwender, J., Polle, J.E.W., 2012. Isoprenoid biosynthesis in eukaryotic phototrophs: a spotlight on algae. Plant Sci. 185–186, 9–22.

Lommer, M., Specht, M., Roy, A.S., Kraemer, L., Andreson, R., Gutowska, M.A., et al., 2012. Genome and low-iron response of an oceanic diatom adapted to chronic iron limitation. Genome Biol. 13, R66.

Lopez, D., Casero, D., Cokus, S.J., Merchant, S.S., Pellegrini, M., 2011. Algal functional annotation tool: a web-based analysis suite to functionally interpret large gene lists using integrated annotation and expression data. BMC Bioinf. 12, 282.

Manuell, A.L., Beligni, M.V., Elder, J.H., Siefker, D.T., Tran, M., Weber, A., McDonald, T.L., Mayfield, S.P., 2007. Robust expression of a bioactive mammalian protein in *Chlamydomonas* chloroplast. Biotechnol. J. 5, 402–412.

Marshall, K.E., Robinson, E.W., Hengel, S.M., Paša-Tolić, L., Roesijadi, G., 2012. FRET imaging of diatoms expressing a biosilica-localized ribose sensor. PLoS One 7 (3), e33771.

Matsuzaki, M., Misumi, O., Shin, I.T., Maruyama, S., Takahara, M., et al., 2004. Genome sequence of the ultrasmall unicellular red alga *Cyanidioschyzon merolae* 10D. Nature 428, 653–657.

Mayfield, S.P., Franklin, S.E., 2005. Expression of human antibodies in eukaryotic micro-algae. Vaccine 23 (15), 1828–1832.

Mayfield, S.P., Franklin, S.E., Lerner, R.A., 2003. Expression and assembly of a fully active antibody in algae. Proc. Natl Acad. Sci. USA 100 (2), 438–442.

McBride, R., Behnke, C., Botsch, K., Heaps, N., Meenach, C., 2013. Use of Fungicides in Liquid Systems. PCT/US2012/060120, WO/2013056166.

Mendez, M., Mayfield, S., O'Neill, B., Poon, Y., Lee, P., Behnke, C.A., Fang, S., 2009. Molecule Production by Photosynthetic Organisms. U.S. Patent Application Pub. No. 2009/0280545 A1.

Merchant, S.S., Prochnik, S.E., Vallon, O., Harris, E.H., Karpowicz, S.J., et al., 2007. The *chlamydomonas* genome reveals the evolution of key animal and plant functions. Science 318, 245–250.

Nakamura, Y., Gojobori, T., Ikemura, T., 1999. Codon usage tabulated from the international DNA sequence databases. Nucleic Acids Res. 27, 292–298.

Nakao, M., Okamoto, S., Kohara, M., Fujishiro, T., Fujisawa, T., Sato, S., Tabata, S., Kaneko, T., Nakamura, Y., 2010. CyanoBase: the cyanobacteria genome database update 2010. Nucleic Acids Res. 38, D379–D381.

Neupert, J., Karcher, D., Bock, R., 2009. Generation of *Chlamydomonas* strains that efficiently express nuclear transgenes. Plant J. 57, 1140–1150.

Noor-Mohammadi, S., Pourmir, A., Johannes, T.W., 2013. Method for assembling and expressing multiple genes in the nucleus of microalgae. Biotechnol. Lett. 36 (3), 561–566.

Oey, M., Ross, I.L., Hankamer, B., 2014. Gateway-assisted vector construction to facilitate expression of foreign proteins in the chloroplast of single celled algae. PLoS One 9 (2), e86841.

Powell, R.J., White, R., Hill, R.T., 2014. Merging metabolism and power: development of a novel photobioelectric device driven by photosynthesis and respiration. PLoS One 9 (1), e86518.

Price, D.C., Chan, C.X., Yoon, H.S., Yang, E.C., Qiu, H., Weber, A.P.M., Schwacke, R., et al., 2012. *Cyanophora paradoxa* genome elucidates origin of photosynthesis in algae and plants. Science 335, 843–847.

Pulz, O., Gross, W., 2004. Valuable products from biotechnology of microalgae. Appl. Microbiol. Biotechnol. 65, 635–648.

Quinn, J.M., Merchant, S., 1995. Two copper-responsive elements associated with the *Chlamydomonas Cyc6* gene function as target for transcriptional activator. Plant Cell 7, 623–638.

Radakovits, R., Eduafo, P.M., Posewitz, M.C., 2011. Genetic engineering of fatty acid chain length in *Phaeodactylum tricornutum*. Metab. Eng. 13, 89–95.

Radakovits, R., Jinkerson, R.E., Darzins, A., Posewitz, M.C., 2010. Genetic engineering of algae for enhanced biofuel production. Eukaryotic Cell 9 (4), 486–501.

Radakovits, R., Jinkerson, R.E., Fuerstenberg, S.I., Tae, H., Settlage, R.E., Boore, J.L., Posewitz, M.C., 2012. Draft genome sequence and genetic transformation of the oleaginous alga *Nannochloropis gaditana*. Nat. Commun. 3, 686.

Raja, R., Hemaiswarya, S., Kumar, N.A., Sridhar, S., Rengasamy, R., 2008. A perspective on the biotechnological potential of microalgae. Crit. Rev. Microbiol. 34, 77–88.

Ramazanov, A., Ramazanov, Z., 2006. Isolation and characterization of a starchless mutant of *Chlorella pyrenoidosa* STL-PI with a high growth rate, and high protein and polyunsaturated fatty acid content. Phycol. Res. 54, 255–259.

Rasala, B.A., Barrera, D.J., Ng, J., Plucinak, T.M., Rosenberg, J.N., Weeks, D.P., Oyler, G.A., Peterson, T.C., Haerizade, F., Mayfield, S.P., 2013. Expanding the spectral palette of fluorescent proteins for the green microalgae *Chlamydomonas reinhardtii*. Plant J. 74, 545–556.

Rasala, B.A., Chao, S.S., Pier, M., Barrera, D.J., Mayfield, S.P., 2014. Enhanced genetic tools for engineering multigene traits into green algae. PLoS One 9 (4), e94028.

Rasala, B.A., Lee, P.A., Shen, Z., Briggs, S.P., Mendez, M., Mayfield, S.P., 2012. Robust expression and secretion of xylanase1 in *Chlamydomonas reinhardtii* by fusion to a selection gene and processing with the FMDV 2A peptide. PLoS One 7 (8), e43349.

Rasala, B., Mayfield, S.P., 2014. Photosynthetic bio-manufacturing in green algae; production of recombinant proteins for industrial, nutritional, and medical uses. Photosynth. Res. http://dx.doi.org/10.1007/s11120-014-9994-7. published ahead of print March 22, 2014.

Rasala, B.A., Muto, M., Lee, P.A., Jager, M., Cardoso, R.M., Behnke, C.A., et al., 2010. Production of therapeutic proteins in algae, analysis of expression of seven human proteins in the chloroplast of *Chlamydomonas reinhardtii*. Plant Biotechnol. J. 8, 719–733.

Rodolfi, L., Chini Zittelli, G., Bassi, N., Padovani, G., Biondi, N., Bonini, G., Tredici, M.R., 2009. Microalgae for oil: strain selection, induction of lipid synthesis and outdoor mass cultivation in a low-cost photobioreactor. Biotechnol. Bioeng. 102, 100–112.

Rogers, J.N., Rosenberg, J.N., Guzman, B.J., Oh, V.H., Mimbela, L.E., et al., 2014. A critical analysis of paddlewheel-driven raceway ponds for algal biofuel production at commercial scales. Algal Res. 4, 76–88.

Rosales-Mendoza, S., Paz-Maldonado, L.M.T., Soria-Guerra, R.E., 2012. *Chlamydomonas reinhardtii* as a viable platform for the production of recombinant proteins: current status and perspectives. Plant Cell Rep. 31, 479–494.

Rosenberg, J.N., Kobayashi, N., Barnes, A., Noel, E.A., Betenbaugh, M.J., Oyler, G.A., 2014. Comparative analyses of three *Chlorella* species in response to light and sugar reveal distinctive lipid accumulation patterns in the microalga *C. sorokiniana*. PLoS One 9, e92460.

Rosenberg, J.N., Oyler, G.A., Wilkinson, L., Betenbaugh, M.J., 2008. A green light for engineered algae: redirecting metabolism to fuel a biotechnology revolution. Curr. Opin. Biotechnol. 19, 430–436.

Sapphire Energy, 2013. Evaluation of Genetically-modified *Scenedesmus dimorphus* in Open Ponds for the Production of Green Crude. Sapphire Energy, Inc. EPA TERA Case Numbers: R13–0003 through R13-0007. (accessed 08.04.14.). Available online: http://www.epa.gov/biotech_rule/pubs/submiss.htm.

Sayre, R., 2014. Genetically modified algae: a risk benefit analysis. U.S. Dep. Energy Webinar (accessed 22.05.14.) Available online: http://youtu.be/q8ElBFI8VHs.

Schoepp, N.G., Stewart, R.L., Sun, V., Quigley, A.J., Mendola, D., Mayfield, S.P., Burkart, M.D., 2014. System and method for research-scale outdoor production of microalgae and cyanobacteria. Bioresour. Technol. 166, 273-281

Schroda, M., Blocker, D., Beck, C.F., 2000. The HSP70A promoter as a tool for the improved expression of transgenes in *Chlamydomonas*. Plant J. 21, 121–131.

Shaaltiel, Y., Bartfeld, D., Hashmueli, S., Baum, G., Brill-Almon, E., Galili, G., Dym, O., Boldin-Adamsky, S.A., Silman, I., Sussman, J.L., et al., 2007. Production of glucocerebrosidase with terminal mannose glycans for enzyme replacement therapy of Gaucher's disease using a plant cell system. Plant Biotechnol. J. 5, 579–590.

Sheehan, J., Dunahay, T., Benemann, J., Roessler, P., 1998. A Look Back at the U.S. Department of Energy's Aquatic Species Program: Biodiesel from Algae Golden, Colorado. TP-580–24190, National Renewable Energy Laboratory.

Shih, P.M., Wu, D., Latifi, A., Axen, S.D., Fewer, D.P., Talla, E., Calteau, A., et al., 2013. Improving the coverage of the cyanobacterial phylum using diversity-driven genome sequencing. Proc. Natl. Acad. Sci. USA 110, 1053–1058.

Shih, P.M., Zarzycki, J., Niyogi, K.K., Kerfeld, C.A., 2014. Introduction of a synthetic CO_2-fixing photorespiratory bypass into a cyanobacterium. J. Biol. Chem. 289, 9493–9500.

Shurin, J.B., Abbott, R.L., Deal, M.S., Kwan, G.T., Litchman, E., McBride, R.C., et al., 2013. Industrial-strength ecology: trade-offs and opportunities in algal biofuel production. Ecol. Lett. 16, 1393–1404.

Snow, A.A., Smith, V.H., 2012. Genetically engineered algae for biofuels: a key role for ecologists. BioSci. 62, 765–768.

Specht, E., Miyake-Stoner, S., Mayfield, S.P., 2010. Microalgae come of age as a platform for recombinant protein production. Biotechnol. Lett. 32 (10), 1373–1383.

Stephens, E., Ross, I.L., Hankamer, B., 2013. Expanding the microalgal industry—continuing controversy or compelling case? Curr. Opin. Chem. Biol. 17, 1–9.

Strom, S., 2014. Companies Quietly Apply Biofuel Tools to Household Products. The New York Times. Available online: http://www.nytimes.com/2014/05/31/business/biofuel-tools-applied-to-household-soaps.html (accessed 01.06.14.).

Sun, M., Qian, K., Su, N., Chang, H., Liu, J., Shen, G., 2003. Foot-and-mouth disease virus VP1 protein fused with cholera toxin B subunit expressed in *Chlamydomonas reinhardtii* chloroplast. Biotechnol. Lett. 25, 1087–1092.

Sun, Q.-Y., Ding, L.-W., Lomonossoff, G.P., Sun, Y.-B., Luo, M., Li, C.-Q., Jiang, L., Xu, Z.-F., 2011. Improved expression and purification of recombinant human serum albumin from transgenic tobacco suspension culture. J. Biotechnol. 155, 164–172.

Surzycki, R., Greenham, K., Kitayama, K., Dibal, F., Wagner, R., Rochaix, J.D., Ajam, T., Surzycki, S., 2009. Factors effecting expression of vaccines in microalgae. Biologicals 37, 133–138.

Tam, L.W., Lefebvre, P.A., 1993. Cloning of flagellar genes in *Chlamydomonas reinhardtii* by DNA insertional mutagenesis. Genetics 135, 375–384.

Tardif, M., Atteia, A., Specht, M., Cogne, G., Rolland, N., Brugiere, S., Hippler, M., Ferro, M., Bruley, C., Peltier, G., et al., 2012. PredAlgo: a new subcellular localization prediction tool dedicated to green algae. Mol. Biol. Evol. 29 (12), 3625–3639.

Tirichine, L., Bowler, C., 2011. Decoding algal genomes: tracing back the history of photosynthetic life on Earth. Plant J. 66, 45–57.

Torres, B.1, Jaenecke, S., Timmis, K.N., García, J.L., Díaz, E., 2003. A dual lethal system to enhance containment of recombinant micro-organisms. Microbiology 149, 3595–3601.

Tran, M., Van, C., Barrera, D.J., Pettersson, P.L., Peinado, C.D., Bui, J., Mayfield, S.P., 2012. Production of unique immunotoxin cancer therapeutics in algal chloroplasts. Proc. Natl. Acad. Sci. USA 110, E15–E22.

Trentacoste, E.M., Shrestha, R.P., Smith, S.R., Glé, C., Hartmann, A.C., Hildebrand, M., Gerwick, W.H., 2013. Metabolic engineering of lipid catabolism increases microalgal lipid accumulation without compromising growth. Proc. Natl. Acad. Sci. USA 110, 19748–19753.

Vazhappilly, R., Chen, F., 1998. Eicosapentaenoic acid and docosahexaenoic acid production potential of microalgae and their heterotrophic growth. J. Am. Oil Chem. Soc. 75, 393–397.

Veyel, D., Erban, A., Fehrle, I., Kopka, J., Schroda, M., 2014. Rationales and approaches for studying metabolism in eukaryotic microalgae. Metabolites 4, 184–217.

Vieler, A., Wu, G., Tsai, C.H., Bullard, B., Cornish, A.J., Harvey, C., Reca, I.B., et al., 2012. Genome, functional gene annotation, and nuclear transformation of the heterokont oleaginous alga *Nannochloropsis oceanica* CCMP1779. PLoS Genet. 8, e1003064.

Vinocur, B., Altman, A., 2005. Recent advances in engineering plant tolerance to abiotic stress: achievements and limitations. Curr. Opin. Biotechnol. 16 (2), 123–132.

Vuttipongchaikij, S., 2012. Genetic manipulation of microalgae for improvement of biodiesel production. Thai J. Genet. 5 (2), 130–148.

Walker, T.L., Purton, S., Becker, D.K., Collet, C., 2005. Microalgae as bioreactors. Plant Cell Rep. 24, 629–641.

Wang, T.Y., Han, Z.M., Chai, Y.R., Zhang, J.H., 2010. A mini review of MAR-binding proteins. Mol. Biol. Rep. 37, 3553–3560.

Wang, D., Ning, K., Li, J., Hu, J., Han, D., Wang, H., Zeng, X., Jing, X., et al., 2014. *Nannochloropsis* genomes reveal evolution of microalgal oleaginous traits. PLoS Genet. 10, e1004094.

Xiong, W., Liu, L., Wu, C., Yang, C., Wu, Q., 2010. 13C-tracer and gas chromatography-mass spectrometry analyses reveal metabolic flux distribution in the oleaginous microalga *Chlorella protothecoides*. Plant Physiol. 154 (2), 1001–1011.

Xu, J., Ge, X., Mc, D., 2011. Towards high-yield production of pharmaceutical proteins with plant cell suspension cultures. Biotechnol. Adv. 29, 278–299.

Young, J.D., Shastri, A.A., Stephanopoulos, G., Morgan, J.A., 2011. Mapping photoautotrophic metabolism with isotopically nonstationary (13)C flux analysis. Metab. Eng. 13 (6), 656–665.

Zamboni, N., Fendt, S.M., Ruhl, M., Sauer, U., 2009. (13)C-based metabolic flux analysis. Nat. Protoc. 4 (6), 878–892.

Zengler, K., Palsson, B.O. 2012. A road map for the development of community systems (CoSy) biology. Nat. Rev. Microbiol. 10, 366–372.

Zhang, F., Vierock, J., Yizhar, O., Fenno, L.E., Tsunoda, S., Kianianmomeni, A., Prigge, M., Berndt, A., Cushman, J., Polle, J., Magnuson, J., Hegemann, P., Deisseroth, K., 2011. The microbial opsin family of optogenetic tools. Cell 147 (7), 1446–1457.

Zhang, Z., Potvin, G., 2010. Strategies for high-level recombinant protein expression in transgenic microalgae: a review. Biotechnol. Adv. 28, 910–918.

Zheng, H.-Q., Chiang-Hsieh, Y.-F., Chien, C.-H., Hsu, B.-K., Liu, T.-L., Chen, C.-N., Chang, W.-C., 2014. AlgaePath: comprehensive analysis of metabolic pathways using transcript abundance data from next-generation sequencing in green algae. BMC Genomics 15 (1), 196.

Microalgal Systems Biology Through Genome-Scale Metabolic Reconstructions for Industrial Applications

Seong-Joo Hong, Choul-Gyun Lee

Marine Bioenergy Research Center, Department of Biological Engineering, Inha University, Korea

1. INTRODUCTION

Since the publication of the complete genome sequence of *Synechocystis* sp. PCC6803, an enormous amount of genomic data for microalgae has been made available on the web. As of November 2013, genome information for 39 cyanobacteria and 24 eukaryotic microalgae is publicly accessible. Additionally, the development of high-throughput omics technologies has driven the accumulation of information on microalgal metabolism. With the evolution of omics technologies, systems biology has become a tool for processing massive amounts of omics data; it has revealed global microalgal metabolic networks and enhanced our understanding of cellular physiology and regulation. Using in silico models of microalgae generated by systems biology, researchers can investigate the photosynthetic pathways that convert inorganic carbon to organic carbon and cellular responses to environmental changes. Such studies will provide opportunities for enhancing microalgal biomass and productivity.

Microalgae are a major natural source of many valuable compounds. In industry, microalgae have been used in aquaculture as live feeds, in food production as color additives, in cosmetics as vitamin supplements, and in food as a source of astaxanthin and polyunsaturated fatty acids. Recently, owing to oil shock and atmospheric pollution, microalgae have attracted attention as precursors for biodiesel production that consume carbon dioxide. These industrial applications originated from the diversity of microalgae and their photosynthetic ability. Photosynthesis provides sustainability because it converts carbon dioxide and light energy into biomass, and diverse microalgae function as "cell factories" to produce industrially important metabolites (Burja et al., 2003).

Despite the various industrial applications for microalgae, there are significant problems in biomass production. Light energy is a major requirement for photosynthesis, and although the sun produces a vast amount of energy, it is a limiting factor for microalgal growth. Light energy for photosynthesis is lost due to several causes, from the reflection of radiation to the formation of biomass (Zhu et al., 2008). If the problems in the photosynthetic process, from photochemical inefficiency to respiration, are improved, the theoretical photosynthetic efficiency could be increased by 80%. Another limitation for algal biotechnology is that the metabolic pathways of microalgae species differ greatly because of symbiosis and evolutionary selection (Facchinelli and Weber, 2011). Therefore, their overall cellular metabolic networks need to be analyzed through a systematic approach.

The purpose of this chapter is to present the current state of microalgal omics technologies and metabolic network modeling of cyanobacteria and eukaryotic microalgae. This chapter also provides information on photosynthesis and lipid metabolism, which are discussed from the perspective of the microalgal industry for integrating the fields of systems biology and microalgal biotechnology.

2. BIOLOGICAL FEATURES: PHOTOSYNTHESIS AND LIPID METABOLISM

2.1 Photosynthesis

Photosynthesis occurs via two main reactions: light-dependent and light-independent reactions. In the light-dependent reaction, adenosine triphosphate (ATP) and nicotinamide adenine dinucleotide phosphate (NADPH) are generated via proton motive force and the electron transport chain (Figure 1). The production of ATP via photosynthesis is called photophosphorylation. Generally, photophosphorylation is divided into two types: noncyclic and cyclic phosphorylation. In cyclic phosphorylation, electrons move within a closed loop. The electrons from light energy begin in photosystem I and pass from plastoquinone to the cytochrome bc-type complex. Then, they are transferred to the cytochrome c-type complex before returning to chlorophyll or bacteriophyll. This electron chain generates the proton motive force required for ATP synthesis; however, with the exception of the system in purple bacteria, NADPH is not produced. Noncyclic phosphorylation can

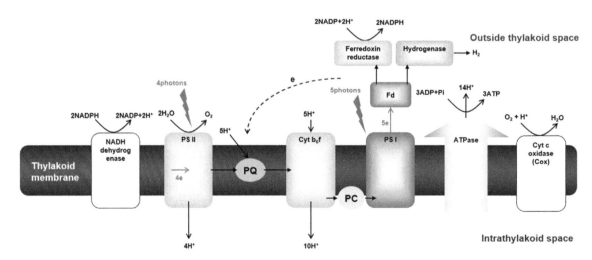

FIGURE 1   Schematic of light-dependent photosynthesis.

produce both ATP and NADPH with photosystems I and II.

The ATP:NADPH ratio in the Calvin cycle is 3:2 because 18 ATP and 12 NADPH are required to synthesize one hexose molecule from six CO_2 molecules. However, ATP synthase in chloroplasts has a 14-fold, not a 12-fold, rotational proportion (Seelert et al., 2000). Thus, the ATP:NADPH ratio for noncyclic phosphorylation is $3 \times (12/14):2$. Noncyclic phosphorylation combined with cyclic phosphorylation is the solution to fulfill the proton deficiency for ATP synthesis (Allen, 2003). To reconstruct a stoichiometric photosynthetic model, it is assumed that the H^+ ions for ATP synthesis are transferred by three mechanisms (Hong and Lee, 2007):

1. The role of plastoquinone: Plastoquinone is reduced to plastoquinol by four electrons generated when a water molecule is divided into O_2 and H^+ in photosystem II.
2. Metabolic consumption of H^+: Proton ions from photosystem II move to the lumen via the thylakoid membrane and play an important role as electron donors in photosystem I (Hervas et al., 2003).
3. Proton ion pumping: H^+ ions move via conformational changes in protein pumps, such as cytochrome b_6f.

Based on the following four previously published assumptions, the protons can be pumped across the thylakoid membrane:

1. Spinach ATP synthase has 14-fold rotational symmetry (Seelert et al., 2000).
2. The structure of the cytochrome b_6f complex was determined by isomorphous replacement, and the cytochrome b_6f complex transfers $2H^+$ to the lumen (Kurisu et al., 2003; Nelson and Yocum, 2006).
3. The ratio of photosystem I to photosystem II reaction centers is between 1.1 and 1.2. Therefore, photosystem I must recycle one of every five electrons (Allen, 2002).
4. A novel thylakoid membrane protein (*ssr2016*, PGR5) transfers the recycled electron

from ferredoxin to plastoquinone (Yeremenko et al., 2005).

Additionally, the hydrogenase in the thylakoid membrane functions as an electron valve to maintain redox stability under the overloaded electron transfer caused by excess light energy (Appel et al., 2000). For example, *Synechocystis* sp. PCC 6803 possesses a bidirectional hydrogenase that is able to take up and generate hydrogen (Tamagnini et al., 2002). This enzyme consists of five components, HoxE, HoxF, HoxH, HoxU, and HoxY. It is known to interact with NADH:plastoquinone oxidoreductase and transport electrons from photosystem I during photosynthesis (Schmitz et al., 2002). Studies of hydrogen production systems have been accelerated using this hydrogenase (Khetkorn et al., 2013; McNeely et al., 2014; Navarro et al., 2009; Ortega-Ramos et al., 2014; Pinto et al., 2012; Schmitz et al., 2002). Moreover, the green microalgae *Chlamydomonas reinhardtii* has also garnered attention as the best eukaryotic hydrogen producer due to its transformation ability (Chochois et al., 2009).

Organic energy molecules (ATP and NADPH) from the light-dependent reaction are consumed by the synthesis of glucose from carbon dioxide in the Calvin cycle. The equation for the light-independent reaction is as follows:

$$6CO_2 + 12NADPH + 18ATP \rightarrow C_6H_{12}O_6 + 12NADP + +18ADP + 18Pi$$

This biosynthetic process, which converts CO_2 to energy-rich compounds powered by the energy generated by cleavage of the P-bond of ATP, is called CO_2 fixation.

2.2 Lipid Metabolism

Lipid composition and metabolism in cyanobacteria differ from those in bacteria and plants. Cyanobacterial cells contain four major glycerolipids: monogalactosyldiacylgycerol (MGDG), digalactosyldiacylglycerol (DGDG), sulfoquinovosyldiacylglycerol (SQDG), and phosphatidylglycerol (PG). Other glycerolipids, such as phosphatidyl choline, phosphatidylinositol, phosphatidylethanolamine, diphosphatidylglycerol (cardiolipin), and phosphatidylserine, are not found in cyanobacteria (Wada and Murata, 1998). Lipid biosynthesis is divided into the Gro3P (glycerol-3-phosphate) pathway and the GrnP (dihydroxyacetone phosphate) pathway.

In the case of bacteria and plants, the enzymatic activities of the GrnP pathway are deficient (Athenstaedt and Daum, 1999). Accordingly, they use the Gro3P pathway to generate glycerolipids using glycerol 3-phosphate. In particular, cyanobacteria require

acyl carrier protein as a cofactor not only to elongate fatty acids but also to generate glycerolipids. As shown in Figure 2, glycerol 3-phosphate passes through two acylation steps leading to the formation of phosphatidic acid. Phosphatidic acid is an essential intermediate of glycerolipid metabolism because the major glycerolipids are generated from phosphatidic acid. Diacylglycerol is converted into glucosyldiacylglycerol by glucosyldiacylglycerol synthase, which is encoded by *sll1377* (Awai et al., 2006). MGDG is produced from diacylglycerol by stereochemical isomerization at the C-4 atom of the glucose unit. DGDG is generated by transfer of galactose (Sato and Murata, 1982), and SQDG is synthesized by sulfolipid

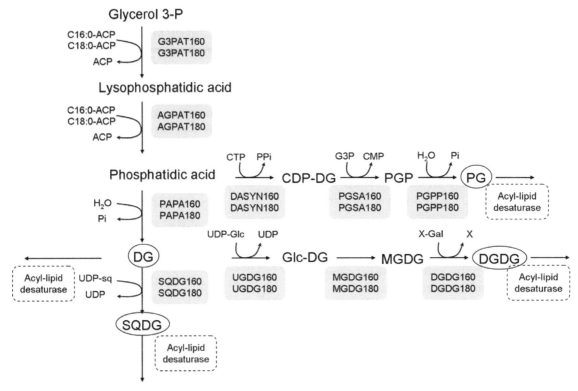

FIGURE 2   Biosynthesis of glycerolipids in cyanobacteria. DG, diacylglycerol; SQDG, sulfoquinovosyldiacylglycerol; ACP, acyl carrier protein; CDP-DG, PGP, phosphatidylglycerolphosphate; PG, phosphatidylglycerol; DGDG, digalactosyldiacylglycerol; MGDG, monogalactosyldiacylgycerol.

biosynthase, which is encoded by *sqdX* (Guler et al., 2000). These major glycerolipids contain only hexadecanoate, octadecanoate, and their polyunsaturated fatty acids because phosphatidic acid, which is a precursor of glycerolipids in cyanobacteria, consists of hexadecanoate (C16:0) and octadecanoate (C18:0). Desaturation of the fatty acids in glycerolipids is catalyzed by acyl-lipid desaturase (Murata and Wada, 1995). *Synechocystis* sp. PCC6803 has three types of acyl-lipid desaturases, and these desaturases generate polyunsaturated fatty acids at the $\Delta 6$, $\Delta 9$, and $\Delta 12$ positions of C16 and C18 fatty acids (Gombos et al., 1996).

The de novo fatty acid synthesis in eukaryotic microalgae strongly resembles that in cyanobacteria due to their endosymbiosis. The elongation of fatty acids occurs until C16:0, C16:3, C18:0, and C18:1 are synthetized in chloroplasts. These fatty acids are released to the endoplasmic reticulum for chain elongation and desaturation (Chi et al., 2008). They are converted to triacylglycerol (TAG) and exported into the cytosol as lipid droplets (Scott et al., 2010). De novo TAG synthesis in the chloroplasts of *Chlamydomonas reinhardtii* was reported by Fan et al. (Fan et al., 2011). TAG is formed and stored in both the chloroplast and the cytosol under nitrogen starvation conditions.

3. SYSTEMS BIOLOGY RESOURCES FOR MICROALGAE: MULTI-OMICS DATA

The repository of diverse high-throughput experimental data produced by using omics technologies, such as genomics, transcriptomics, proteomics, and metabolomics, has expanded the availability of bioinformatics tools and facilitated analysis of the complexity in cellular processes and the relationships between genotypes and phenotypes with systems biology.

3.1 Genomics

The early published genome sequence of *Synechocystis* became a driving force for accumulating a large quantity of genome sequence information in public databases (Table 1). The CyanoBase of the Kazusa institute in Japan, which was set up in 1995, is a notable database for *Synechocystis* sp. PCC6803 that is continuously updated along with the user interface. The last update (2010) included a circular chromosome map, gene list, function classifications, and search capability to improve data accessibility. In this update, the data presentation was improved by the addition of genome context, a table view, protein domains, and a word cloud (Nakao et al., 2010). The CyanoBase also makes comparisons between cyanobacteria strains more efficient because this database provides the complete genome sequences and genomic information for 39 cyanobacterial strains.

Due to the large size of their chromosomes, the complete genome of a eukaryotic microalga was published about 10 years later than the first cyanobacterial genome. The *Chlamydomonas reinhardtii* nuclear genome was the starting point for whole genome analyses of eukaryotic microalgae (Merchant et al., 2007). The complete genome sequences of several eukaryotic microalgae, including *Phaeodactylum tricornutum*, *Ostreococcus lucimarinus*, *Ostreococcus tauri*, *Micromonas pusilla*, and *Coccomyxa subellipsoidea*, have been determined by whole genome shotgun sequencing (Blanc et al., 2012; Derelle et al., 2006; Palenik et al., 2007; Worden et al., 2009). Since the development of next-generation sequencing (NGS) technology for high-throughput low-cost genotyping, genomic analyses of microalgae is now relatively straightforward compared to analyses using previous methods. NGS technology was used to obtain complete genome information for *Nannochloropsis gaditana*, *Porphyridium purpureum*, and *Monoraphidium neglectum* (Battchikova et al., 2010; Bogen et al., 2013; Radakovits et al., 2012). In

TABLE 1   Genome and Pathway Databases for Cyanobacteria and Eukaryotic Microalgae

Database	Content	Website
SPECIALIZED DATABASES FOR CYANOBACTERIA		
CyanoBase	Genomes	http://genome.kazusa.or.jp/
CyanoMutant	Mutants	
Cyano2Dbase	Proteins	
CYORF	Gene annotation	http://cyano.genome.ad.jp/
PCC	Description and properties of cyanobacterial strains	www.pasteur.fr
Cyanosite	Experimental protocols	www.cyanosite.bio.purdue.edu/
SynechoNET	Protein–protein interactions	bioportal.kobic.re.kr/SynechoNET/
CyanoCluster	Homologous proteins in cyanobacteria and plastids	http://cyanoclust.c.u-tokyo.ac.jp/
SPECIALIZED DATABASES FOR EUKARYOTIC MICROALGAE		
Chlamydomonas connection	*Chlamydomonas* genomes	www.chlamy.org/
Nannochloropsis genome portal	*Nannochloropsis* genomes	www.nanchloropsis.org
Porphyridium purpureum genome project	*Porphyridium* genome	http://cyanophora.rutgers.edu/porphyridium
Ostreococcus tauri		
DATABASES CONTAINING OMICS INFORMATION		
NCBI	Genes, proteins, and compounds	www.ncbi.nlm.nih.gov/
JGI	Genomes	www.jgi.doe.gov/
JCVI	Genomes	www.jcvi.org/
BRENDA	Enzyme classification information	www.brenda-enzymes.info/
UniProt	Proteins	www.uniprot.org/
KEGG	Genes, proteins, and pathways	www.kegg.com/
TCDB	Transporters	www.tcdb.org

addition, a large number of genome projects supported by the Joint Genome Institute of the US. Department of Energy (http://genome.jgi.doe.gov/genome-projects/) are ongoing.

3.2 Transcriptomics

Transcriptomic data can be easily obtained owing to the availability of the first commercial microalgal microarray called the IntelliGene™ CyanoCHIP (Takara Bio Inc., Japan). The term transcriptome refers to the total set of RNA molecules, including mRNAs, rRNAs, tRNAs, and non-coding RNAs, produced by a genome. The transcriptome can be controlled by internal and external factors under various conditions. It serves as a dynamic link between an organism's genome and its phenotype (Velculescu et al., 1997). Therefore, transcriptomic studies on *Synechocystis* sp. PCC6803 using microarrays have mainly focused on gene expression levels and regulation under certain conditions. Studies using microarray analyses include gene regulation under various conditions, including high-light intensity (Hihara et al., 2001; Singh et al., 2008; Tu et al., 2004), phosphate limitation (Suzuki et al., 2004), nitrogen depletion (Osanai et al., 2006), iron deficiency and reconstitution (Singh et al., 2003), inorganic carbon limitation (Wang et al., 2004), salt and osmotic stress (Kanesaki et al., 2002; Mikami et al., 2002), light—dark transition (Gill et al., 2002; Kucho et al., 2005; Schmitt and Stephanopoulos, 2003), and low temperature shock (Inaba et al., 2003), the effect of redox state using inhibitors of the photosynthetic electron chain (Hihara et al., 2003), oxidative stress induction by methyl viologen (Kobayashi et al., 2004), red and far-red light for phytochrome mutants (Hubschmann et al., 2005), heat shock (Singh et al., 2006; Suzuki et al., 2005, 2006), acid stress (Ohta et al., 2005), growth mode shift from light to glucose (Lee et al., 2007), alkaline conditions (Summerfield and Sherman, 2008), sulfur starvation (Zhang

et al., 2008), and the effect of light quality (Singh et al., 2009). NGS technology can overcome the limited ability of microarray systems since the full catalog of genes can be analyzed and diverse RNA molecules can be quantified (Ozsolak and Milos, 2011). Georg et al. attempted to screen and identify the function of antisense RNAs (asRNAs) using northern blotting, 5'-RACE, and microarrays. They found 73 candidate *cis*-asRNAs and 60 free-standing genes among the putative ncRNAs (Georg et al., 2009). In comparison, Mitschke et al. discovered many more candidate noncoding RNAs using pyrosequencing, including 1112 aTSS (annotated RNAs) producing asRNAs, and 1165 gTSS (upstream of annotated genes) giving rise to mRNAs, 821 iTSS (annotated genes by inverse origin) for internal sense transcripts, and 429 nTSS (in intergenic spacers) (Mitschke et al., 2011).

NGS technology has even impacted the transcriptomic analysis of eukaryotic microalgae without whole genome sequences. Lipid and starch metabolism in *Dunaliella tertiolecta* and *Botryococcus braunii* were reconstructed via de novo transcriptomic studies for biofuel production (Molnar et al., 2012; Rismani-Yazdi et al., 2011). Comparative transcriptomic analyses of *Chlorella variabilis*, *Nannochloropsis gaditana*, and *C. reinhardtii* under nitrogen depleted and nitrogen replete conditions revealed differential expression of lipid metabolism genes in these strains (Boyle et al., 2012; Carpinelli et al., 2014; Guarnieri et al., 2011). Therefore, transcriptomic analyses are essential for systems biology because de novo and comparative transcriptomic analyses can provide a frame work for metabolic reconstruction instead of genomics and metabolic network analysis combined with mathematical simulations.

3.3 Proteomics

Proteomics is the analysis of all the proteins expressed by a genome. For metabolic reconstruction, proteomics can provide additional information that cannot be observed in mRNA expression analyses. The tools typically used for protein analysis are two-dimensional gel electrophoresis (2DE) techniques. Under defined conditions, specific spots are sorted and extracted for protein identification. After that, the amino acid composition of the extracted proteins is determined by mass estimation. The common method used for mass estimation is matrix-assisted laser desorption/ionization time-of-flight mass spectrometry (MALDI-TOF MS). Recently, increased implementation of ESI/MS has provided more accurate mass measurements.

Proteomic analyses of *Synechocystis* sp. PCC6803 began in 1999, with the construction of a 2DE database. The Kazusa institute isolated 234 soluble protein spots and analyzed them along with their genome data. In 2000, Wang et al. identified peripheral proteins from the thylakoid membrane. They detected 200 spots by 2DE, and 116 proteins were extracted and analyzed by MALDI-TOF MS. Among them, 78 proteins were expressed from 58 genes, and their metabolic categories were classified as photosynthesis and the carotenoid pathway (Wang et al., 2000). In the same year, Fluda et al. reported the identification of proteins expressed under osmotic shock conditions via 2DE and MALDI-TOF MS analysis (Fulda et al., 2002). Huang et al. isolated proteins in the plasma and outer membranes by sucrose density centrifugation and aqueous two-phase partitioning. They analyzed proteins related to transport systems (Huang et al., 2004, 2002). Herranen et al. also analyzed proteins in the thylakoid membrane. They isolated 20 spots and predicted photosynthetic electron flow through the isolated proteins and confirmed their predictions by comparison of wild type to a ΔpsbA1 mutant under various growth modes. In addition, they clearly determined the structure and function of the type I-NADH dehydrogenase in *Synechocystis* sp. PCC6803

(Herranen et al., 2004). After *Synechocystis* was shown to have three compartments by analysis of signal peptides and N-terminal segments (Rajalahti et al., 2007), Pisareva et al. attempted to analyze proteins purified from the plasma membrane, and Wang et al. isolated and identified proteins extracted from the outer, plasma, and thylakoid membranes (Pisareva et al., 2007; Wang et al., 2009). The function and regulation of proteins from these membranes were inspected continuously under different stress conditions, such as high pH and copper- and iron-depleted conditions (Castielli et al., 2009; Zhang et al., 2009). Proteins binding to the membrane are thought to have significant roles in photosynthesis and respiration. Therefore, proteomic studies of membrane proteins will continue with an aim to determine the complete photosynthetic mechanism. A different approach for proteomic analysis is to examine the expression changes of proteins in carbon metabolism under various trophic conditions. Proteins expressed under heterotrophic conditions were reported by Kurian et al. Differential expression of proteins in the citric acid cycle, pentose phosphate pathway, and CO_2 fixation was examined under photoautotrophic and heterotrophic conditions (Kurian et al., 2006). Battchikova et al. identified 19% of the cyanobacterial proteome and quantified the expression changes in 17% of the theoretical ORFs under CO_2 limitation by using isobaric tagging for relative and absolute quantification (iTRAQ) technique (Battchikova et al., 2010).

Thus far, proteomic analyses of microalgae have focused on lipid accumulation for biofuel production, and nitrogen starvation is a major factor that induces lipid accumulation in microalgae. In a study by Nguyen et al. (Nguyen et al., 2012), 33 *C. reinhardtii* proteins regulated in lipid metabolism were identified among a group of 248 proteins. The global proteomic changes in *Nannochloropsis oceanica* IMET-1 observed under long-term nitrogen starvation included global downregulation of protein expression, maintained expression of proteins involved in glycolysis and fatty acid synthesis, and upregulation of proteins involved in nitrogen scavenging and protein turnover (Dong et al., 2013). Under heterotrophic nitrogen deprivation, *Chlorella protothecoides* was shown to upregulate 13 proteins, including proteins involved in photosynthesis, protein synthesis/folding, gene regulation, and β-oxidation of fatty acids. Under these conditions, 15 proteins in carbohydrate metabolism, stress response and defense, amino acid biosynthesis, and secondary metabolite biosynthesis were downregulated (Li et al., 2013). *P. tricornutum* was shown to redirect the metabolic network from carbon flux toward lipid accumulation (Yang et al., 2014). Investigation of the *Synechocystis* sp. proteome under diverse environmental stresses, including nitrogen deficiency, revealed that a common stress response is activation of atypical pathways for the acquisition of carbon and nitrogen from urea and arginine (Wegener et al., 2010). Quia et al. attempted to analyze the overall protein expression changes in *Synechocystis* sp. in the resistance mechanism upon exposure to ethanol, hexane, and butanol (Qiao et al., 2012; Tian et al., 2013; Wegener et al., 2010; Zhu et al., 2013). These proteomic studies provided an overall view of complicated metabolic networks and serve as useful references for metabolic reconstructions.

3.4 Metabolomics

Metabolomics is the global study of all the metabolites in a cellular system (Rochfort, 2005). Metabolites are the intermediate compounds used as substrates and generated as products by enzymes translated from the genome. Metabolomics plays an important role in metabolic reconstruction, which was improved by the addition of missing metabolic pathways through qualitative analyses and is used to construct biomass object functions (BOFs) by quantitative analysis.

Cyanobacterial metabolomic studies have mainly focused on carbon footprinting to screen bioactive natural products (Burja et al., 2001; Tan, 2007). An early study of intracellular metabolites was reported by Pearce et al. in 1969. They traced carbon flow in the citric acid cycle with [^{14}C]-acetate. Consequently, cyanobacteria were found to have an incomplete citric acid cycle by measuring metabolites and assaying enzymes (Pearce et al., 1969). The intracellular metabolites in central metabolism were measured by capillary electrophoresis—mass spectrometry under photoautotrophic and mixotrophic conditions (Takahashi et al., 2008). The analytical methods used for fingerprinting of *Synechocystis* sp. PCC6803 were fast filtering and centrifugation combined with gas chromatography-mass spectrometry (GC—MS) (Krall et al., 2009).

Metabolomics analyses of eukaryotic microalgae have also focused on the differences in lipid characteristics between species and the changes in lipid composition with growth. Inoculum size affected the cell growth, lipid accumulation, and metabolic changes of *Chlorella sorokiniana* (Lu et al., 2012). Metabolic changes in lipids were observed during different growth stages of *Nitzschia closterium f. minutissima* (Su et al., 2013). Nitrogen-deficient conditions also induced changes in the lipid profile of *C. reinhardtii*, *Nannochloropsis oceanica*, and *Dunaliella tertiolecta* (Courant et al., 2013; Kim et al., 2013; Xiao et al., 2013).

Compared with transcriptomics and proteomics, there are currently a limited number of published metabolomics studies. However, the amount of metabolomic data for microalgae will likely increase and contribute to the development of in silico models as a useful reference to fill the metabolic gaps and construct BOFs with the development of advanced analytical instruments for liquid chromatography and mass spectrometry.

4. IN SILICO ANALYSES: GENOME-SCALE METABOLIC RECONSTRUCTION AND SIMULATION

4.1 Metabolic Reconstruction and Simulation

An in silico model of microalgae for gene-protein-reaction associations (GPR associations) was generated based on complete genome sequences (Thiele and Palsson, 2010). An ORF database that includes genome annotation was downloaded from the genome database Website (Table 2). The layout of the metabolic reconstruction was obtained using the KEGG pathway map, as well as the ChlamyCyc (http://pmn.lantcyc.org) and AraCyc (www.arabidopsis.org) metabolic pathway databases. Translated metabolic proteins were identified using genome annotation tools such as CYORF and CyanoBase, and were assigned to biochemical reactions with

TABLE 2   Network Properties of Genome-Scale Metabolic Reconstructions for *Synechocystis* sp. PCC6803 and Eukaryotic Microalgae

Strain	Year	N_m	N_r	Status	References
IN SILICO MODELS OF SYNECHOCYSTIS SP. PCC6803					
Synechocystis sp. PCC6803	2002	—	29	C	Yang et al. (2002)
	2005	—	72	C	Shastri and Morgan (2005)
	2007			C	Hong and Lee (2007)
	2008	704	831 510[a]	W	Pengcheng (2009)
	2009	29	43	C	Navarro et al. (2009)
	2010	291	380	W	Knoop et al. (2010)
	2010	790	882	W	Montagud et al. (2010)
	2011	911	956	W	Montagud et al. (2011)
	2012	795	863	W	Nogales et al. (2012)
IN SILICO MODELS OF EUKARYOTIC MICROALGAE					
Chlamydomonas reinhardtii	2009	458	484	C	Boyle and Morgan (2009)
	2009	259	467	C	Manichaikul et al. (2009)
	2011	278	280	C	Cogne et al. (2011)
	2011	1068	2090	W	Chang et al. (2011)
	2011	1869	1725	W	Dal'Molin et al. (2011)
Ostreococcus lucimarinus	2012	1100	964	W	Krumholz et al. (2012)
Ostreococcus tauri	2012	1014	871	W	Krumholz et al. (2012)

N_g, number of genes; N_m, number of metabolites; N_r, number of reactions.
C, in silico model of central metabolism; W, whole genome-scale metabolic reconstruction.
[a] *Recounted data because the published reactions overlapped.*

the EC number on the BRENDA Website. To create the biochemical reactions, the catalytic activity and cofactors were validated by referring to the universal protein resource UniProt. The formula and charge of the compounds in each biochemical reaction were calculated using the pK_a value of their functional group. The reversibility of reactions was determined according to thermodynamic correlations, such as the specificity of enzymes related to high-energy metabolites (ATP, GTP, etc.). The metabolic pathways were built using GPR association to identify the biochemical reactions regulated by gene expression.

The BOF for the simulation includes biomass synthesis and maintenance requirements. The most important ATP maintenance requirements were cellular molarity, maintaining cellular osmolality, macromolecule turnover, and maintenance of transmembrane gradients. Through the biochemical reactions in the in silico model, a stoichiometric matrix, $S(m \times n)$, was generated to analyze the metabolic characterization of Synechocystis sp. PCC6803. This matrix could be solved by flux balance analysis based on linear programming (Varma and Palsson, 1993a,b). For the computational analysis of cell growth, BOF was defined as an object function to be maximized in simulation software, such as COBRA Toolbox.

4.2 The Current Status of the in Silico Model

With the development of modern omics technologies, it has become possible to build whole-genome metabolic reconstructions and to characterize them more accurately. Systems biology aims at a system-level understanding of the cellular systems by integrating the transcriptome, proteome, metabolome, and genome (Kitano, 2002). The first step in systems biology is to form a cellular network using various omics data. The first metabolic model of Synechocystis sp. PCC6803 was published by Yang et al.

They reconstructed glycolysis, the citric acid cycle, Calvin cycle, and pentose phosphate pathway, but not photosynthesis. Using a carbon isotope labeling technique, the flux distribution of the reconstructed pathways was investigated under heterotrophic and mixotrophic conditions (Yang et al., 2002). An in silico model of the complete central metabolism was reconstructed by Shastri and Morgan. This model included two rounds of ATP and NADPH generation by photon absorption and BOF in three growth modes obtained by calculating cellular composition (Shastri and Morgan, 2005).

After the oil crisis in 2008, Synechocystis sp. PCC6803 has attracted great attention as a bioenergy producer. Because this microalga was not only already fully sequenced but also known as an efficient transformant with a pilus system, Pengcheng attempted to reconstruct a whole-genome model to predict cell growth and ethanol production following insertion of ethanol production genes (Pengcheng, 2009). Navarro et al. analyzed hydrogen production in the central metabolism of Synechocystis (Navarro et al., 2009). Knoop et al. attempted to build a primary metabolic network to understand the regulation of Synechocystis photosynthesis (Knoop et al., 2010). A large-scale genome network of Synechocystis was published by Montagud et al. (Montagud et al., 2010, 2011). They simulated cell growth under three physiologically relevant growth conditions by flux balance analysis and flux coupling analysis. Through integration of transcriptomic data, they found metabolic hot spots during light-shifting growth regimes. In 2012, a manually curated in silico model of Synechocystis sp. PCC6803 was published by Nogales et al. (Nogales et al., 2012). This model was the first in microalgae to reconstruct metabolic reactions stoichiometrically and thermodynamically. This study showed that the electron flow pathways in photosynthesis are simulated under carbon-limited and light-limited conditions, and metabolic flux was analyzed under various trophic conditions by flux balance analysis (FBA).

A metabolic reconstruction of central metabolism in the eukaryotic microalga *Chlorella pyrenoidosa* was proposed by Yang et al. (Yang et al., 2000). This model included 61 metabolites and 67 metabolic reactions in two compartments. The influence of carbon and energy metabolism under various trophic modes was explained by the use of metabolic flux analysis, even though this model was not based on genomic information. After the complete genome sequence of *C. reinhardtii* was determined in 2007, the first genome-based model of central metabolism was published by Boyle et al. (Boyle and Morgan, 2009). The reconstructed network was built with 484 metabolic reactions and 458 intracellular metabolites in three compartments, and was used to predict metabolic fluxes under three growth conditions: autotrophic, heterotrophic, and mixotrophic growth, by FBA. The second genome-based model (of *C. reinhardtii*) was reconstructed with verification using bioinformatics and experimental tools to fill the metabolic gaps due to the incomplete genome sequence (Manichaikul et al., 2009). This model contained 259 reactions and 467 metabolites in five compartments. To validate the model, they compared the results of FBA simulation with the literature on growth yield and photosynthetic oxygen exchange. Another constraint-based model of *C. reinhardtii* was developed to investigate the differences in thermodynamic and energetic processes between respiration and photosynthesis (Cogne et al., 2011). Their modeling showed that respiration interacted with photosynthesis under autotrophic conditions through regulation of redox state during photosynthesis and maintenance of the ATP supply. A genome-scale metabolic reconstruction has been built for *C. reinhardtii*, including 1068 metabolites and 2190 metabolic reactions (Chang et al., 2011). This model was introduced to determine the photon absorption at various wavelengths for quantitative growth prediction

according to light source efficiency, and comprehensively expanded our understanding of lipid metabolism to engineer industrial strains for biofuel production. Validation of the model focused on gene knockout simulations and various growth simulations under light regulated and trophic conditions. At the same time, AlgaGEM, which is another genome-scale metabolic reconstruction of *C. reinhardtii*, contains 1725 unique reactions and 1869 metabolites in four compartments (Dal'Molin et al., 2011). The reconstruction methodology for AlgaGEM was based on the GEM of *Mus musculus*, *Arabidopsis* (AraGEM), maize sorghum, and sugarcane (C4GEM) using MATLAB and the COBRA Toolbox. AlgaGEM predicted phosphoglycolate recycling in photorespiration and the metabolic changes for enhancement of H_2 production yield under mixotrophic conditions. In addition to *C. reinhardtii*, the metabolic networks of the picoalgae *O. lucimarinus* and *Ostreococcus tauri* were reconstructed, and they included 1100 metabolites and 964 reactions, and 1014 metabolites and 871 reactions, respectively. The models of the *Ostreococcus* strains focused on thermodynamically constrained, elementally balanced methodologies for the metabolic reconstruction, and were functionally evaluated with the KEGG database and the gap-filling approaches between phylogenetic distance and sequence similarity.

Notwithstanding the efforts of the latest research, the phenotypes of microalgae under various culture conditions have not been fully covered by in silico model analyses. For example, the flux distribution of the *Synechocystis* sp. model could not explain the published growth patterns, such as light-activated heterotrophic growth (Anderson and McIntosh, 1991). Additionally, the difficulty in the genetic transformation of microalgae due to epigenetic silencing causes a big problem for in silico model validation by knockout simulation (van Dijk et al., 2006). To improve the accuracy of in silico models, the high-throughput experimental data produced by omics technologies will be used for the integration, and additional genetic tools for microalgae will be developed for validation.

5. INDUSTRIAL APPLICATIONS OF SYSTEMS BIOLOGY: ENGINEERING METABOLIC NETWORKS

Microalgae contain numerous bioactive compounds that can be of commercial use. They have been used as additives in infant formula, supplements in aquaculture, natural dyes, cosmeceuticals, and pharmaceuticals. Nowadays, biofuel precursors produced by microalgae are important for environmental crises due to their sustainability. To improve the productivity of target compounds in large-scale systems, it is necessary to optimize both the upstream and downstream bioprocesses. Systems biology approaches are useful tools for strain improvement in terms of the upstream bioprocesses.

In Figure 3, the photosynthetic carbon flux shows that various carbon metabolites of microalgae are synthesized in the chloroplast. Antioxidant compounds, such as carotenoids and tocopherols, are produced as part of a protective mechanism against oxidative stress. The pigments, such as chlorophyll and phycobiliproteins, are involved in photosynthesis and are synthesized in the chloroplast. Interestingly, de novo synthesis of fatty acids occurs in the

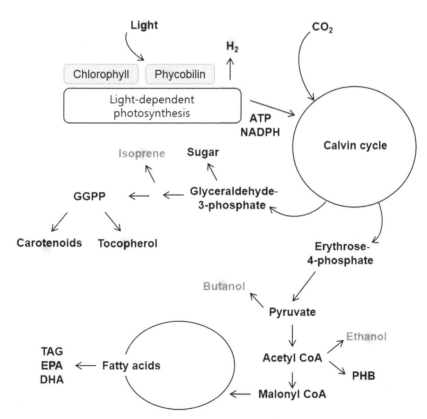

FIGURE 3   Carbon flux pathways and valuable metabolites in photosynthetic microorganisms, including cyanobacteria and the chloroplasts in eukaryotic microalgae.

chloroplast, and 16- or 18-carbon fatty acids are produced, just like in cyanobacteria. Therefore, carbon partitioning between fatty acids, carbohydrates, and proteins in the chloroplast will be a key factor for optimizing single-cell productivity. To enhance the productivity of the target metabolites produced in the chloroplast, organic connectivity must be considered. Genome-scale metabolic network reconstruction allows an in-depth understanding of the overall connectivity among the biomass components in the networks of microalgae.

Microalgal in silico models were not directly used to develop industrial strains because models based on genome information were first released in 2012. However, *Synechocystis* was used to develop algal biofuel systems by photosynthetic carbon flux analysis (Liu et al., 2010). The rate-controlling enzyme, acetyl-CoA carboxylase, was overexpressed, and other pathways competing with fatty acid synthesis were blocked to overproduce free fatty acids. The engineered *Synechocystis* sp. produced free fatty acids up to 133 ± 12 mg L^{-1} day^{-1}, which corresponds to 50% of the biomass, despite the long lag phase.

Furthermore, understanding the overall metabolic network in target microalgae is very important for generating engineered strains for bioindustry. Metabolic reconstruction will aid in systems level analysis and cell behavior prediction via genotype–phenotype relationships.

6. SUMMARY

Genome-scale metabolic reconstructions of microalgae for bioindustrial generation of valuable metabolites are still in the early stages. In this chapter, photosynthesis and lipid metabolism have been reviewed with an aim to increase our understanding of metabolic specificity in microalgae. Omics studies of microalgae have become the basis for systems biology. Through integration of multi-omics data,

systems biology of microalgae can provide a deep understanding of cellular mechanisms. Furthermore, these powerful tools can be used to describe metabolic pathways and modify target genes to produce valuable compounds.

References

Allen, J.F., 2002. Photosynthesis of ATP—electrons, proton pumps, rotors, and poise. Cell 110 (3), 273–276.

Allen, J.F., 2003. Cyclic, pseudocyclic and noncyclic photophosphorylation: new links in the chain. Trends Plant Sci. 8 (1), 15–19.

Anderson, S.L., McIntosh, L., 1991. Light-activated heterotrophic growth of the cyanobacterium *Synechocystis* sp. strain PCC6803: a blue-light-requiring process. J. Bacteriol. 173 (9), 2761–2767.

Appel, J., Phunpruch, S., Steinmuller, K., Schulz, R., 2000. The bidirectional hydrogenase of *Synechocystis* sp. PCC6803 works as an electron valve during photosynthesis. Arch. Microbiol. 173 (5–6), 333–338.

Athenstaedt, K., Daum, G., 1999. Phosphatidic acid, a key intermediate in lipid metabolism. Eur. J. Biochem. 266 (1), 1–16.

Awai, K., Kakimoto, T., Awai, C., et al., 2006. Comparative genomic analysis revealed a gene for monoglucosyldiacylglycerol synthase, an enzyme for photosynthetic membrane lipid synthesis in cyanobacteria. Plant Physiol. 141 (3), 1120–1127.

Battchikova, N., Vainonen, J.P., Vorontsova, N., et al., 2010. Dynamic changes in the proteome of *Synechocystis* 6803 in response to CO_2 limitation revealed by quantitative proteomics. J. Proteome Res. 9 (11), 5896–5912.

Blanc, G., Agarkova, I., Grimwood, J., et al., 2012. The genome of the polar eukaryotic microalga *Coccomyxa subellipsoidea* reveals traits of cold adaptation. Genome Biol. 13 (5).

Bogen, C., Al-Dilaimi, A., Albersmeier, A., et al., 2013. Reconstruction of the lipid metabolism for the microalga *Monoraphidium neglectum* from its genome sequence reveals characteristics suitable for biofuel production. BMC Genomics 14.

Boyle, N.R., Morgan, J.A., 2009. Flux balance analysis of primary metabolism in *Chlamydomonas reinhardtii*. BMC Systems Biol. 3.

Boyle, N.R., Page, M.D., Liu, B.S., et al., 2012. Three acyltransferases and nitrogen-responsive regulator are implicated in nitrogen starvation-induced triacylglycerol accumulation in Chlamydomonas. J. Biol. Chem. 287 (19), 15811–15825.

Burja, A.M., Banaigs, B., Abou-Mansour, E., Burgess, J.G., Wright, P.C., 2001. Marine cyanobacteria—a prolific source of natural products. Tetrahedron 57 (46), 9347–9377.

Burja, A.M., Dhamwichukorn, S., Wright, P.C., 2003. Cyanobacterial postgenomic research and systems biology. Trends in Biotechnol. 21 (11), 504–511.

Carpinelli, E.C., Telatin, A., Vitulo, N., et al., 2014. Chromosome scale genome assembly and transcriptome profiling of *Nannochloropsis gaditana* in nitrogen depletion. Mol. Plant 7 (2), 323–335.

Castielli, O., De la Cerda, B., Navarro, J.A., Hervas, M., De la Rosa, M.A., 2009. Proteomic analyses of the response of cyanobacteria to different stress conditions. Febs Lett. 583 (11), 1753–1758.

Chang, R.L., Ghamsari, L., Manichaikul, A., et al., 2011. Metabolic network reconstruction of *Chlamydomonas* offers insight into light-driven algal metabolism. Mol. Systems Biol. 7.

Chi, X.Y., Zhang, X.W., Guan, X.Y., et al., 2008. Fatty acid biosynthesis in eukaryotic photosynthetic microalgae: Identification of a microsomal delta 12 desaturase in *Chlamydomonas reinhardtii*. J. Microbiol. 46 (2), 189–201.

Chochois, V., Dauvillee, D., Beyly, A., et al., 2009. Hydrogen production in *chlamydomonas*: photosystem II-dependent and -independent pathways differ in their requirement for starch metabolism. Plant Physiol. 151 (2), 631–640.

Cogne, G., Rugen, M., Bockmayr, A., et al., 2011. A model-based method for investigating bioenergetic processes in autotrophically growing eukaryotic microalgae: application to the green algae *Chlamydomonas reinhardtii*. Biotechnol. Progress 27 (3), 631–640.

Courant, F., Martzolff, A., Rabin, G., et al., 2013. How metabolomics can contribute to bio-processes: a proof of concept study for biomarkers discovery in the context of nitrogen-starved microalgae grown in photobioreactors. Metabolomics 9 (6), 1286–1300.

Dal'Molin, C.G.D., Quek, L.E., Palfreyman, R.W., Nielsen, L.K., 2011. AlgaGEM—a genome-scale metabolic reconstruction of algae based on the *Chlamydomonas reinhardtii* genome. Bmc Genomics 12.

Derelle, E., Ferraz, C., Rombauts, S., et al., 2006. Genome analysis of the smallest free-living eukaryote *Ostreococcus tauri* unveils many unique features. Proc. Natl Acad. Sci. USA 103 (31), 11647–11652.

Dong, H.P., Williams, E., Wang, D.Z., et al., 2013. Responses of *Nannochloropsis oceanica* IMET1 to long-term nitrogen starvation and recovery. Plant Physiol. 162 (2), 1110–1126.

Facchinelli, F., Weber, A.P., 2011. The metabolite transporters of the plastid envelope: an update. Front Plant Sci. 2, 50.

Fan, J.L., Andre, C., Xu, C.C., 2011. A chloroplast pathway for the de novo biosynthesis of triacylglycerol in *Chlamydomonas reinhardtii*. Febs Lett. 585 (12), 1985–1991.

Fulda, S., Norling, B., Schoor, A., Hagemann, M., 2002. The Slr0924 protein of *Synechocystis* sp. strain PCC6803 resembles a subunit of the chloroplast protein import complex and is mainly localized in the thylakoid lumen. Plant Mol. Biol. 49 (1), 107–118.

Georg, J., Voss, B., Scholz, I., et al., 2009. Evidence for a major role of antisense RNAs in cyanobacterial gene regulation. Mol. Systems Biol. 5.

Gill, R.T., Katsoulakis, E., Schmitt, W., et al., 2002. Genome-wide dynamic transcriptional profiling of the light-to-dark transition in *Synechocystis* sp. strain PCC6803. J. Bacteriol. 184 (13), 3671–3681.

Gombos, Z., Wada, H., Varkonyi, Z., Los, D.A., Murata, N., 1996. Characterization of the Fad12 mutant of *Synechocystis* that is defective in delta12 acyl-lipid desaturase activity. Biochimica Et Biophysica Acta—Lipids and Lipid Metabolism 1299 (1), 117–123.

Guarnieri, M.T., Nag, A., Smolinski, S.L., et al., 2011. Examination of triacylglycerol biosynthetic pathways via *de novo* transcriptomic and proteomic analyses in an unsequenced microalga. PLoS One 6 (10).

Guler, S., Essigmann, B., Benning, C., 2000. A cyanobacterial gene, sqdX, required for biosynthesis of the sulfolipid sulfoquinovosyldiacylglyerol. J. Bacteriol. 182 (2), 543–545.

Herranen, M., Battchikova, N., Zhang, P.P., et al., 2004. Towards functional proteomics of membrane protein complexes in *Synechocystis* sp. PCC6803. Plant Physiol. 134 (1), 470–481.

Hervas, M., Navarro, J.A., De La Rosa, M.A., 2003. Electron transfer between membrane complexes and soluble proteins in photosynthesis. Acc. Chem. Res. 36 (10), 798–805.

Hihara, Y., Kamei, A., Kanehisa, M., Kaplan, A., Ikeuchi, M., 2001. DNA microarray analysis of cyanobacterial gene expression during acclimation to high light. Plant Cell 13 (4), 793–806.

Hihara, Y., Sonoike, K., Kanehisa, M., Ikeuchi, M., 2003. DNA microarray analysis of redox-responsive genes in the genome of the cyanobacterium *Synechocystis* sp. strain PCC6803. J. Bacteriol. 185 (5), 1719–1725.

Hong, S.J., Lee, C.G., 2007. Evaluation of central metabolism based on a genomic database of *Synechocystis* PCC6803. Biotechnol. Bioprocess Engin. 12 (2), 165–173.

Huang, F., Hedman, E., Funk, C., et al., 2004. Isolation of outer membrane of *Synechocystis* sp. PCC6803 and its proteomic characterization. Mol. Cell. Proteomics 3 (6), 586–595.

Huang, F., Parmryd, I., Nilsson, F., et al., 2002. Proteomics of *Synechocystis* sp. strain PCC6803-Identification of plasma membrane proteins. Mol. Cell. Proteomics 1 (12), 956–966.

Hubschmann, T., Yamamoto, H., Gieler, T., Murata, N., Borner, T., 2005. Red and far-red light alter the transcript profile in the cyanobacterium *Synechocystis* sp. PCC6803: impact of cyanobacterial phytochromes. FEBS Lett. 579 (7), 1613–1618.

Inaba, M., Suzuki, I., Szalontai, B., et al., 2003. Gene-engineered rigidification of membrane lipids enhances the cold inducibility of gene expression in Synechocystis. J. Biol. Chem. 278 (14), 12191–12198.

Kanesaki, Y., Suzuki, I., Allakhverdiev, S.I., Mikami, K., Murata, N., 2002. Salt stress and hyperosmotic stress regulate the expression of different sets of genes in *Synechocystis* sp. PCC6803. Biochem. Biophysical Res. Communications 290 (1), 339–348.

Khetkorn, W., Khanna, N., Incharoensakdi, A., Lindblad, P., 2013. Metabolic and genetic engineering of cyanobacteria for enhanced hydrogen production. Biofuels 4 (5), 535–561.

Kim, S.H., Liu, K.H., Lee, S.Y., et al., 2013. Effects of light intensity and nitrogen starvation on glycerolipid, glycerophospholipid, and carotenoid composition in *Dunaliella tertiolecta* culture. PLoS One 8 (9).

Kitano, H., 2002. Systems biology: a brief overview. Science 295 (5560), 1662–1664.

Knoop, H., Zilliges, Y., Lockau, W., Steuer, R., 2010. The metabolic network of *Synechocystis* sp. PCC6803: systemic properties of autotrophic growth. Plant Physiol. 154 (1), 410–422.

Kobayashi, M., Ishizuka, T., Katayama, M., et al., 2004. Response to oxidative stress involves a novel peroxiredoxin gene in the unicellular cyanobacterium *Synechocystis* sp. PCC6803. Plant Cell Physiol. 45 (3), 290–299.

Krall, L., Huege, J., Catchpole, G., Steinhauser, D., Willmitzer, L., 2009. Assessment of sampling strategies for gas chromatography-mass spectrometry (GC-MS) based metabolomics of cyanobacteria. J. Chromatog. B-Analyt. Technol. Biomed. Life Sci. 877 (27), 2952–2960.

Krumholz, E.W., Yang, H., Weisenhorn, P., Henry, C.S., Libourel, I.G., 2012. Genome-wide metabolic network reconstruction of the picoalga *Ostreococcus*. J. Exp. Bot. 63 (6), 2353–2362.

Kucho, K., Okamoto, K., Tsuchiya, Y., et al., 2005. Global analysis of circadian expression in the cyanobacterium *Synechocystis* sp. strain PCC6803. J. Bacteriol. 187 (6), 2190–2199.

Kurian, D., Jansen, T., Maenpaa, P., 2006. Proteomic analysis of heterotrophy in *Synechocystis* sp. PCC6803. Proteomics 6 (5), 1483–1494.

Kurisu, G., Zhang, H.M., Smith, J.L., Cramer, W.A., 2003. Structure of the cytochrome b_6f complex of oxygenic photosynthesis: tuning the cavity. Science 302 (5647), 1009–1014.

Lee, S., Ryu, J.Y., Kim, S.Y., et al., 2007. Transcriptional regulation of the respiratory genes in the cyanobacterium *Synechocystis* sp. PCC6803 during the early response to glucose feeding. Plant Physiol. 145 (3), 1018–1030.

Li, Y.Q., Yuan, Z.Q., Mu, J.X., Chen, D., Feng, B., 2013. Proteomic analysis of lipid accumulation in *Chlorella protothecoides* cells by heterotrophic N deprivation coupling cultivation. Energy and Fuels 27 (7), 4031–4040.

Liu, Y.R., Ma, W.M., Mi, H.L., 2010. Increase of the expression and activity of ferredoxin-NADP(+) oxidoreductase in the cells adapted to low CO_2 in the cyanobacterium *Synechocystis* 6803. Photosynthetica 48 (3), 417–420.

Lu, S.H., Wang, J.X., Niu, Y.H., et al., 2012. Metabolic profiling reveals growth related FAME productivity and quality of *Chlorella sorokiniana* with different inoculum sizes. Biotechnol. Bioeng. 109 (7), 1651–1662.

Manichaikul, A., Ghamsari, L., Hom, E.F.Y., et al., 2009. Metabolic network analysis integrated with transcript verification for sequenced genomes. Nat. Methods 6 (8), 589–592.

McNeely, K., Kumaraswamy, G.K., Guerra, T., et al., 2014. Metabolic switching of central carbon metabolism in response to nitrate: application to autofermentative hydrogen production in cyanobacteria. J. Biotechnol. 182–183 (1), 83–91.

Merchant, S.S., Prochnik, S.E., Vallon, O., et al., 2007. The *Chlamydomonas* genome reveals the evolution of key animal and plant functions. Science 318 (5848), 245–251.

Mikami, K., Kanesaki, Y., Suzuki, I., Murata, N., 2002. The histidine kinase Hik33 perceives osmotic stress and cold stress in Synechocystis sp. PCC6803. Mol. Microbiol. 46 (4), 905–915.

Mitschke, J., Georg, J., Scholz, I., et al., 2011. An experimentally anchored map of transcriptional start sites in the model cyanobacterium *Synechocystis* sp. PCC6803. Proc. Nat.l Acad. Sci. the USA 108 (5), 2124–2129.

Molnar, I., Lopez, D., Wisecaver, J.H., et al., 2012. Bio-crude transcriptomics: gene discovery and metabolic network reconstruction for the biosynthesis of the terpenome of the hydrocarbon oil-producing green alga, Botryococcus braunii race B (Showa). Bmc Genomics 13.

Montagud, A., Navarro, E., de Cordoba, P.F., Urchueguia, J.F., Patil, K.R., 2010. Reconstruction and analysis of genome-scale metabolic model of a photosynthetic bacterium. Bmc Systems Biol. 4.

Montagud, A., Zelezniak, A., Navarro, E., et al., 2011. Flux coupling and transcriptional regulation within the metabolic network of the photosynthetic bacterium *Synechocystis* sp. PCC6803. Biotechnol. J. 6 (3), 330–342.

Murata, N., Wada, H., 1995. Acyl-Lipid desaturases and their importance in the tolerance and acclimatization to cold of cyanobacteria. Biochem. J. 308, 1–8.

Nakao, M., Okamoto, S., Kohara, M., et al., 2010. CyanoBase: the cyanobacteria genome database update 2010. Nucleic Acids Res. 38, D379–D381.

Navarro, E., Montagud, A., de Cordoba, P.F., Urchueguia, J.F., 2009. Metabolic flux analysis of the hydrogen production potential in *Synechocystis* sp. PCC6803. Int. J. Hydrogen Energy 34 (21), 8828–8838.

Nelson, N., Yocum, C.F., 2006. Structure and function of photosystems I and II. Annu. Rev. Plant Biol. 57, 521–565.

Nguyen, H.M., Baudet, M., Cuiné, S., et al., 2012. Proteomic profiling of oil bodies isolated from the unicellular green microalga *Chlamydomonas reinhardtii*: with focus on proteins involved in lipid metabolism. Proteomics 11 (21), 4266–4273.

Nogales, J., Gudmundsson, S., Knight, E.M., Palsson, B.O., Thiele, I., 2012. Detailing the optimality of photosynthesis in cyanobacteria through systems biology analysis. Proc. Natl Acad. Sci. USA 109 (7), 2678–2683.

Ohta, H., Shibata, Y., Haseyama, Y., et al., 2005. Identification of genes expressed in response to acid stress in *Synechocystis* sp. PCC6803 using DNA microarrays. Photosynth. Res. 84 (1–3), 225–230.

Ortega-Ramos, M., Jittawuttipoka, T., Saenkham, P., et al., 2014. Engineering *synechocystis* PCC6803 for hydrogen production: influence on the tolerance to oxidative and sugar stresses. PLoS One 9 (2).

Osanai, T., Imamura, S., Asayama, M., et al., 2006. Nitrogen induction of sugar catabolic gene expression in *Synechocystis* sp. PCC6803. DNA Res. 13 (5), 185–195.

Ozsolak, F., Milos, P.M., 2011. RNA sequencing: advances, challenges and opportunities. Nat. Rev. Genetics 12 (2), 87–98.

Palenik, B., Grimwood, J., Aerts, A., et al., 2007. The tiny eukaryote *Ostreococcus* provides genomic insights into the paradox of plankton speciation. Proc. Natl Acad Sci. USA 104 (18), 7705–7710.

Pearce, J., Leach, C.K., Carr, N.G., 1969. The incomplete tricarboxylic acid cycle in the blue-green alga *Anabaena variabilis*. J. General Microbiol. 55 (3), 371–378.

Pengcheng, F., 2009. Genome-scale modeling of *Synechocystis* sp. PCC6803 and prediction of pathway insertion. J. Chem. Technol. Biotechnol. 84 (4), 473–483.

Pinto, F., van Elburg, K.A., Pacheco, C.C., et al., 2012. Construction of a chassis for hydrogen production: physiological and molecular characterization of a *Synechocystis* sp. PCC6803 mutant lacking a functional bidirectional hydrogenase. Microbiol.-Sgm 158, 448–464.

Pisareva, T., Shumskaya, M., Maddalo, G., Ilag, L., Norling, B., 2007. Proteomics of *Synechocystis* sp. PCC6803-Identification of novel integral plasma membrane proteins. Febs J. 274 (3), 791–804.

Qiao, J.J., Wang, J.X., Chen, L., et al., 2012. Quantitative iTRAQ LC-MS/MS proteomics reveals metabolic responses to biofuel ethanol in cyanobacterial *Synechocystis* sp. PCC6803. J. Proteome Res. 11 (11), 5286–5300.

Radakovits, R., Jinkerson, R.E., Fuerstenberg, S.I., et al., 2012. Draft genome sequence and genetic transformation of the oleaginous alga *Nannochloropis gaditana*. Nat. Communications 3.

Rajalahti, T., Huang, F., Klement, M.R., et al., 2007. Proteins in different *Synechocystis* compartments have distinguishing N-terminal features: a combined proteomics and multivariate sequence analysis. J. Proteome Res. 6 (7), 2420–2434.

Rismani-Yazdi, H., Haznedaroglu, B.Z., Bibby, K., Peccia, J., 2011. Transcriptome sequencing and annotation of the microalgae *Dunaliella tertiolecta*: pathway description and gene discovery for production of next-generation biofuels. Bmc Genomics 12.

Rochfort, S., 2005. Metabolomics reviewed: a new "Omics" platform technology for systems biology and implications for natural products research. J. Nat. Prod. 68 (12), 1813–1820.

Sato, N., Murata, N., 1982. Lipid biosynthesis in the blue-green alga, *Anabaena variabilis*: I. Lipid classes. Biochimica et Biophysica Acta (BBA) - Lipids and Lipid Metabolism 710 (3), 271–278.

Schmitt, W.A., Stephanopoulos, G., 2003. Prediction of transcriptional profiles of *Synechocystis* PCC6803 by dynamic autoregressive modeling of DNA microarray data. Biotechnol. Bioeng. 84 (7), 855–863.

Schmitz, O., Boison, G., Salzmann, H., et al., 2002. HoxE - a subunit specific for the pentameric bidirectional hydrogenase complex (HoxEFUYH) of cyanobacteria. Biochimica Et Biophysica Acta-Bioenergetics 1554 (1–2), 66–74.

Scott, S.A., Davey, M.P., Dennis, J.S., et al., 2010. Biodiesel from algae: challenges and prospects. Curr. Opin. Biotechnol. 21 (3), 277–286.

Seelert, H., Poetsch, A., Dencher, N.A., et al., 2000. Structural biology. Proton-powered turbine of a plant motor. Nature 405 (6785), 418–419.

Shastri, A.A., Morgan, J.A., 2005. Flux balance analysis of photoautotrophic metabolism. Biotechnol. Progress 21 (6), 1617–1626.

Singh, A.K., Bhattacharyya-Pakrasi, M., Elvitigala, T., et al., 2009. A systems-level analysis of the effects of light quality on the metabolism of a cyanobacterium. Plant Physiol. 151 (3), 1596–1608.

Singh, A.K., McIntyre, L.M., Sherman, L.A., 2003. Microarray analysis of the genome-wide response to iron deficiency and iron reconstitution in the cyanobacterium *Synechocystis* sp. PCC6803. Plant Physiol. 132 (4), 1825–1839.

Singh, A.K., Summerfield, T.C., Li, H., Sherman, L.A., 2006. The heat shock response in the cyanobacterium *Synechocystis* sp. strain PCC6803 and regulation of gene expression by HrcA and SigB. Arch. Microbiol. 186 (4), 273–286.

Singh, M., Satoh, K., Yamamoto, Y., Kanervo, E., Aro, E.M., 2008. In vivo quality control of photosystem II in cyanobacteria, *Synechocystis* sp. PCC6803: D1 protein degradation and repair under the influence of light, heat and darkness. Indian J. Biochem. Biophysics 45 (4), 237–243.

Su, X.L., Xu, J.L., Yan, X.J., et al., 2013. Lipidomic changes during different growth stages of *Nitzschia closterium* f. minutissima. Metabolomics 9 (2), 300–310.

Summerfield, T.C., Sherman, L.A., 2008. Global transcriptional response of the alkali-tolerant cyanobacterium *Synechocystis* sp. strain PCC6803 to a pH 10 environment. Appl. Environ. Microbiol. 74 (17), 5276–5284.

Suzuki, I., Kanesaki, Y., Hayashi, H., et al., 2005. The histidine kinase Hik34 is involved in thermotolerance by regulating the expression of heat shock genes *in Synechocystis*. Plant Physiol. 138 (3), 1409−1421.

Suzuki, I., Simon, W.J., Slabas, A.R., 2006. The heat shock response of *Synechocystis* sp. PCC6803 analysed by transcriptomics and proteomics. J. Exp. Botany 57 (7), 1573−1578.

Suzuki, S., Ferjani, A., Suzuki, I., Murata, N., 2004. The SphS-SphR two component system is the exclusive sensor for the induction of gene expression in response to phosphate limitation in Synechocystis. J. Biological Chem. 279 (13), 13234−13240.

Takahashi, H., Uchimiya, H., Hihara, Y., 2008. Difference in metabolite levels between photoautotrophic and photomixotrophic cultures of *Synechocystis* sp. PCC6803 examined by capillary electrophoresis electrospray ionization mass spectrometry. J. Exp. Botany 59 (11), 3009−3018.

Tamagnini, P., Axelsson, R., Lindberg, P., et al., 2002. Hydrogenases and hydrogen metabolism of cyanobacteria. Microbiol. Mol. Biol. Rev. 66 (1).

Tan, L.T., 2007. Bioactive natural products from marine cyanobacteria for drug discovery. Phytochemistry 68 (7), 954−979.

Thiele, I., Palsson, B.O., 2010. A protocol for generating a high-quality genome-scale metabolic reconstruction. Nat. Protocols 5 (1), 93−121.

Tian, X.X., Chen, L., Wang, J.X., Qiao, J.J., Zhang, W.W., 2013. Quantitative proteomics reveals dynamic responses of *Synechocystis* sp. PCC6803 to next-generation biofuel butanol. J. Proteomics 78, 326−345.

Tu, C.J., Shrager, J., Burnap, R.L., Postier, B.L., Grossman, A.R., 2004. Consequences of a deletion in dspA on transcript accumulation in *Synechocystis* sp. strain PCC6803. J. Bacteriol. 186 (12), 3889−3902.

van Dijk, K., Xu, H., Cerutti, H., 2006. Epigenetic silencing of transposons in the green alga *Chlamydomonas reinhardtii*. In: Small RNAs. Springer, pp. 159−178.

Varma, A., Palsson, B.O., 1993a. Metabolic capabilities of *Escherichia coli*. 1. Synthesis of biosynthetic precursors and cofactors. J. Theoretical Biol. 165 (4), 477−502.

Varma, A., Palsson, B.O., 1993b. Metabolic capabilities of *Escherichia coli*. 2. Optimal-growth patterns. J. Theoretical Biol. 165 (4), 503−522.

Velculescu, V.E., Zhang, L., Zhou, W., et al., 1997. Characterization of the yeast transcriptome. Cell 88 (2), 243−251.

Wada, H., Murata, N., 1998. Membrane lipids in cyanobacteria. In: Siegenthaler, P., Murata, N. (Eds.), Lipids in Photosynthesis: Structure, Function and Genetics. Kluwer Academic Publishers, pp. 65−81.

Wang, H.L., Postier, B.L., Burnap, R.L., 2004. Alterations in global patterns of gene expression in *Synechocystis* sp. PCC6803 in response to inorganic carbon limitation and the inactivation of ndhR, a LysR family regulator. J. Biological Chem. 279 (7), 5739−5751.

Wang, Y.C., Sun, J., Chitnis, P.R., 2000. Proteomic study of the peripheral proteins from thylakoid membranes of the cyanobacterium *Synechocystis* sp. PCC6803. Electrophoresis 21 (9), 1746−1754.

Wang, Y.C., Xu, W., Chitnis, P.R., 2009. Identification and bioinformatic analysis of the membrane proteins of *Synechocystis* sp. PCC6803. Proteome Sci. 7.

Wegener, K.M., Singh, A.K., Jacobs, J.M., et al., 2010. Global proteomics reveal an atypical strategy for carbon/nitrogen assimilation by a cyanobacterium under diverse environmental perturbations. Mol. Cell. Proteomics 9 (12), 2678−2689.

Worden, A.Z., Lee, J.-H., Mock, T., et al., 2009. Green Evolution and Dynamic Adaptations Revealed by Genomes of the Marine Picoeukaryotes *Micromonas* 324 (5924), 268−272.

Xiao, Y., Zhang, J.T., Cui, J.T., Feng, Y.G., Cui, Q., 2013. Metabolic profiles of *Nannochloropsis oceanica* IMET1 under nitrogen-deficiency stress. Bioresource Technol. 130, 731−738.

Yang, C., Hua, Q., Shimizu, K., 2000. Energetics and carbon metabolism during growth of microalgal cells under photoautotrophic, mixotrophic and cyclic light-autotrophic/dark-heterotrophic conditions. Biochem. Engin. J. 6 (2), 87−102.

Yang, C., Hua, Q., Shimizu, K., 2002. Metabolic flux analysis in *Synechocystis* using isotope distribution from C-13-labeled glucose. Metabolic Engin. 4 (3), 202−216.

Yang, Z.K., Ma, Y.H., Zheng, J.W., et al., 2014. Proteomics to reveal metabolic network shifts towards lipid accumulation following nitrogen deprivation in the diatom *Phaeodactylum tricornutum*. J. Appl. Phycol. 26 (1), 73−82.

Yeremenko, N., Jeanjean, R., Prommeenate, P., et al., 2005. Open reading frame *ssr2016* is required for antimycin A-sensitive photosystem I-driven cyclic electron flow in the cyanobacterium *Synechocystis* sp. PCC6803. Plant and Cell Physiol. 46 (8), 1433−1436.

Zhang, L.F., Yang, H.M., Cui, S.X., et al., 2009. Proteomic analysis of plasma membranes of cyanobacterium *Synechocystis* sp. strain PCC6803 in response to high pH stress. J. Proteome Res. 8 (6), 2892−2902.

Zhang, Z., Pendse, N.D., Phillips, K.N., Cotner, J.B., Khodursky, A., 2008. Gene expression patterns of sulfur starvation in *Synechocystis* sp. PCC6803. BMC Genomics 9.

Zhu, H.J., Ren, X.Y., Wang, J.X., et al., 2013. Integrated OMICS guided engineering of biofuel butanol-tolerance in photosynthetic *Synechocystis* sp. PCC6803. Biotechnol. Biofuels 6.

Zhu, X.G., Long, S.P., Ort, D.R., 2008. What is the maximum efficiency with which photosynthesis can convert solar energy into biomass? Curr. Opin. Biotechnol. 19 (2), 153−159.

Genetic Engineering of Marine Microalgae to Optimize Bioenergy Production

Pavan P. Jutur, Asha A. Nesamma

DBT-ICGEB Centre for Advanced Bioenergy Research, New Delhi, India

1. INTRODUCTION

Marine algae, which include cyanobacteria, eukaryotic microalgae, and seaweed, are responsible for approximately 40−50% of the photosynthesis that occurs on earth each year (Falkowski et al., 1998). They are seen as increasingly attractive options for bioenergy production to improve fuel security and reduce CO_2 emissions because they are renewable, potentially carbon-neutral, and scalable alternative reserves (Larkum et al., 2012). Furthermore, emerging microalgal technologies contain bait sources for biofuels, including oils (e.g., triglycerides), polysaccharides (e.g., algin, agar), and pigments (e.g., phycobiliproteins, carotenoids). They also have potential as new pharmaceuticals, nutraceuticals, therapeutics, industrial chemicals, and protein-rich animal feeds. This has shifted the research paradigm significantly toward the optimization of low-cost bioenergy constituents, which will be further advanced by genetic engineering and enhanced algae-based sustainable valuables (Apt and Behrens, 1999; Beer et al., 2009; Chisti, 2007; Georgianna and Mayfield, 2012; Gimpel et al., 2013; Hannon et al., 2010; Jones and Mayfield, 2012; Jutur and Asha, 2015; Lin et al., 2011; Pulz and Gross, 2004; Radakovits et al., 2010; Raja et al., 2008; Rasala et al., 2014; Rosenberg et al., 2008; Specht et al., 2010; Specht and Mayfield, 2014; Spolaore et al., 2006; Walker et al., 2005; Wijffels et al., 2013).

However, little is known regarding the commercial potential of these marine microalgae due to their complexity, their unique genetic and evolutionary patterns, the lack of available molecular tools, and overall poor expression of heterologous genes from the nuclear genome in microalgal species, with the major concern being at least partially due to rapid gene silencing; (Cerutti et al., 1997; Fuhrmann et al., 1999; Neupert et al., 2009; Rasala et al., 2012). A set of validated vectors for targeting transgene products to specific subcellular locations generally does not exist, nor does a vector to allow the expression of multiple nuclear-encoded genes within a single cell.

Genetic engineering of marine microalgae as cell factories and marine bioreactors has been investigated (Khan et al., 2009; Larkum et al., 2012; León-Bañares et al., 2004; Liu and Benning, 2013; Mata et al., 2010; Qin et al., 2012; Tabatabaei et al., 2011; Wijffels et al., 2013). Because of the rapidly growing demand for the development of bioenergy from marine microalgae and the profitable market for cosmetics and pharmaceuticals from algal natural products, genetic engineering of marine algae has been attracting an increasing amount of attention as a crucial systemic technology to overcome the biomass problem in industrial applications (John et al., 2011), modify the metabolic pathway for high-value products (Schmidt et al., 2010), engineer biobricks, and design an artificial photoautotroph for the rising and promising field of synthetic biology (Heidorn et al., 2011; Keasling and Venter, 2013; Muers, 2011; Wang et al., 2012). Nevertheless, to date, only a few marine algae species have been genetically manipulated successfully. The development of this technology is in its early stages and much remains to be done to optimize the business models and lifecycle analysis, as well as in testing pilot- and demonstration-scale systems (Frank et al., 2013; Han et al., 2014; Kumar et al., 2010; Pfromm et al., 2011; Resurreccion et al., 2012; Shirvani et al., 2011; Sills et al., 2013).

Investigations are underway to determine optimal microalgae species, select strains, and implement subsequent genetic/metabolic engineering. Modifications involve target improvement of cellular activities by the manipulation of enzymatic, transport, and regulatory functions of the photosynthetic cells using biological modulators and/or engineering methods to refine sustainable fuel production systems (Franz et al., 2013; Gong and Jiang, 2011; Han et al., 2014; Larkum, 2010; Larkum et al., 2012; Lu et al., 2011; Radakovits et al., 2010). This chapter reviews the research progress and potential approaches in the use of marine microalgae to optimize bioenergy production through genetic engineering.

2. TOOLS AND TECHNIQUES FOR GENETIC ENGINEERING IN MICROALGAE

Marine microalgal genomes provide leads for understanding fundamental genetic manipulation and later followed by optimization through genetic engineering will not only identify the gene loci and the distribution patterns of metabolic pathways and enzymes, but also aid in the identification of elements including *cis*-acting elements, *trans*-acting factors, and other regulatory elements that can improve the production of bioenergy precursors (Qin et al., 2012). The critical and fundamental issue of marine algal biotechnology is to obtain "seed" culture in commercial large-scale cultivation to produce valuable products for algal economic feasibility. Several genera of marine algae have been successfully cultivated for commercial exploitations, such as *Dunaliella, Laminaria, Undaria, Porphyridium, Nannochloropsis, Porphyra*, and *Gracilaria* (Hallmann, 2007; Lamers et al., 2008; Raja et al., 2008; San Pedro et al., 2013). For an effective expression system to be successful in marine algae, several scenarios play a crucial role in the production of commercial bioenergy such as economical target construction, technological methods at low cost, engineering design for better instrumentation, and certainly aspects related to biosafety issues (Lin et al., 2011).

Bioprospecting of microalgae can be carried out from versatile environments using different methodologies, including physical extraction from crude marine water samples using micromanipulation (Kacka and Donmez, 2008; Moreno-Garrido, 2008; Subburamu, 2013), isolation of individual cells by dilution procedures, selection by antibiotics, and enrichment of cultures using specific selection parameters (e.g., photoautotrophic conditions)

(Franz et al., 2013; Larkum et al., 2012; San Pedro et al., 2013). Individual microalgal strains can be isolated based on traditional colony selection or high-throughput fluorescence-activated cell sorting approaches, before these axenic cultures are cryopreserved to prevent any genetic drift (Subburamu, 2013). Screening of microalgae is categorized based on their optimal growth and/or the production of specific metabolites of relevance. Commonly used methods for initial screens are response surface modeling and principle component analysis to identify conditions and the key variables controlling these metabolic changes among microalgae (Mohamed et al., 2013; Shen et al., 2014).

Few technical criteria where microalgae have significant advantages are shorter life cycles (either in hours or in days instead of seasonal cycles), when compared to other bioenergy crops systems. Their unicellular nature assisting in the miniaturization of breeding systems, which greatly reduces cost and the ability to replicate both sexually and asexually, accelerates the generation of genetic diversity. The bioprospecting of microalgae for particular phenotypes with the application of UV and chemical mutagenesis with specific selection parameters can greatly enhance the rate of strain development (Bonente et al., 2011; Bougaran et al., 2012; Cagnon et al., 2013), and further screening can be done by either using flow cytometry or other high-throughput methodologies (Larkum et al., 2012; Subburamu, 2013; Xie et al., 2014a).

The analyses of microalgal growth parameters are essential under controlled conditions of light intensity, pH, CO_2 supply, and mixing, that can be directly deployed in the development of different types of photobioreactors such as flat-plate, tubular, helical, and other more complex geometric designs; thus these systems are advancing rapidly in the diversity of their design and will make a significant contribution to futuristic solutions in algal bioenergy production (Posten, 2009; Rodolfi et al., 2009; Tredici, 2007) (Figure 1).

Construction and application of engineered microalgal mutants require comprehensive knowledge of the microalgal genomes and access to molecular and gene manipulation tools, including selectable markers, vectors and techniques for systematic insertion in screening libraries (Larkum et al., 2012). Advanced technologies such as RNAi-targeted gene up- and down-regulation approaches and other facets of transcriptional regulation will enhance the research progress, where limited knowledge is available only for few microalgae species that fulfill even some of these crucial requirements toward sustainable bioenergy production (Cerutti et al., 2011; Molnar et al., 2009). Extensively, some model systems like *Chlamydomonas reinhardtii* (Pratheesh et al., 2014), *Phaeodactylum tricornutum* (Bowler et al., 2008; Xie et al., 2014b; Zhang and Hu, 2013), *Chlorella kessleri* (El-Sheekh, 1999; Rathod et al., 2013), *Porphyridium* (Lapidot et al., 2002), *Nannochloropsis* (Kilian et al., 2011; Radakovits et al., 2012), and *Dunaliella salina* (Feng et al., 2009) have been successfully transformed, but new tools will be developed as genomes and the factors regulating them are increasingly understood in detail.

For any successful transformation in marine microalgae, certain hurdles are associated with gene efficiency, integration, or stability which are essential to overcome by implementing measures where suitable vectors for successful incorporation, reporter genes, and homologous promoters need to be exploited along with new DNA incorporation methods such as optimized homologous recombination strategies and databases for modeling species-specific codon usage between donor and host organism would be more beneficial (Hallmann, 2007; Kilian et al., 2011). Reverse genetics approaches a new tool for direct isolation of mutants to produce targeted knockouts where reports and/or patents are in pipeline for few species such as *Chlamydomonas*, *Nannochloropsis*, and *Dunaliella*, capable of performing efficient homologous gene recombination in the nucleus with desirable phenotype

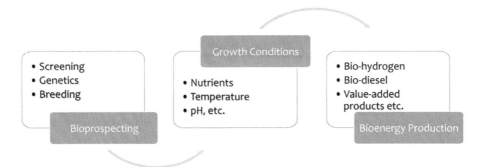

FIGURE 1   Schematic representation of bioenergy production in microalgae.

and/or genotype characteristics (Cha et al., 2011; Gonzalez-Ballester et al., 2011; Pazour and Witman, 2000).

Certainly, one of the major obstacles in microalgal biofuel production is maximization of the light capture efficiency by photosynthesis because this is the first limiting step in all biofuel production processes, wherein out of ~43% of the energy in the solar spectrum that can be captured via photosynthesis only ~4—8% is converted by wild-type strains to chemical energy in the form of biomass (Blankenship et al., 2011; Larkum, 2010). Furthermore, understanding both biology and engineering advances to maximize the light reactions—a complex phenomenon (i.e., chemical reduction of CO_2, the complex interplay between spectral range, light capture mechanism) plays a crucial role in balancing of CO_2 and O_2 supply to rubisco, as it catalyzes the competing reactions of CO_2 fixation and oxygenation (photorespiration) (Larkum et al., 2012).

3. INTEGRATED "OMICS" - SOLUTIONS FOR PATHWAY ENGINEERING

The commercial relevance of microalgae including cyanobacteria as potential feedstock for renewable biofuels gained importance due to diminishing oil reserves and demand for sustainable system capable of converting atmospheric CO_2 to substantial biomass and valuable biofuels along with low- or/and high-value products. A broad range of industrially relevant products include biofuels, nutraceuticals, therapeutics, industrial chemicals, and animal feeds that have shifted research paradigm significantly toward the production of bioenergy constituents under low-cost platform, and further improvements through genetic engineering will enable and enhance microalgal-based bioproducts (Rasala et al., 2014). However, the transgenic marine microalgae are yet to be exploited for optimized bioenergy production. The major concern to optimize microalgae for genetic engineering is lack of molecular tools and overall poor expression of heterologous genes from the nuclear genome of many microalgae species, and also may be at least partially due to their rapid silencing (Cerutti et al., 1997; Neupert et al., 2009; Rasala et al., 2012). Tools such as promoters, transcriptional terminators, ribosome binding sites, and other regulatory factors, etc. for synthetic biology and genetic engineering in marine microalgae are yet to be developed or still in their infancy (Wang et al., 2012). A novel scenario has been pipelined to understand the behavior of biological systems as a whole by integration of "Omics" research, where the metabolic pathways are often highly regulated and

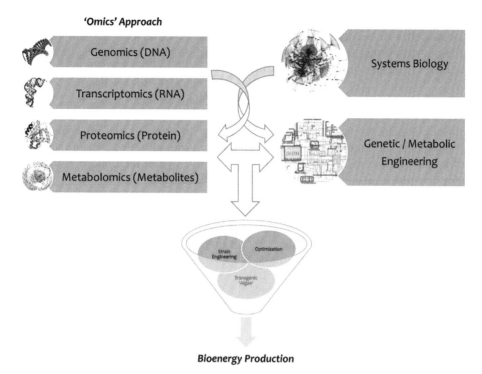

'Omics' Approach

Genomics (DNA)

Transcriptomics (RNA)

Proteomics (Protein)

Metabolomics (Metabolites)

Systems Biology

Genetic / Metabolic Engineering

Strain Engineering

Optimization

Transgenic 'Algae'

Bioenergy Production

FIGURE 2 Conceptual illustration of Integrative "Omics" research with systems biology and genetic engineering approach for optimization of microalgae for bioenergy production.

connected with a number of both feed-forward and feed-back mechanisms that can act positively and/or negatively ultimately affecting the systems output (Jutur and Asha, 2015). Understanding the marine microalgal system through integrated "Omics" research will lead to the identification of relevant enzyme-encoding genes, and reconstruct the metabolic pathways involved in the biosynthesis and degradation of precursor molecules that may have potential for biofuel production, aiming toward the vision of tomorrow's bioenergy needs (Figure 2).

The integrated "Omics" biology is foreseen as foundation for hypothesis-driven research predicting the role among cellular molecules which includes transcriptome, proteome, and metabolome and their interactions in the cells, which would lead to a systematic quantified model of cellular metabolism at a genome scale (Ishii and

Tomita, 2009; Zhang et al., 2010), which will eventually provide new insights into marine microalgal cellular metabolism. Liu et al. (2013) defines systems biology as an inter-disciplinary science that studies the complex interactions and cumulative behavior of a cell or an organism. From "Omics" to systems biology, the primary goal is to develop conceptual knowledge of a biological system by evaluating its behavior and interaction between its individual components (Jamers et al., 2009). One of the key steps in this process involves modeling, where the structure of the system is unraveled by mathematical algorithms allowing its dynamics, but also allow the prediction of the system's response to perturbations. A framework of studies in systems biology initially understand the structure and identify key elements in the system, such as gene networks, protein interactions, and metabolic pathways

(Ideker et al., 2001), thus constructing a primary model of systems behavior. Secondly, the system is perturbed either genetically or using environmental stimuli, and corresponding responses are measured using high-throughput measurement tools, wherein the data generated at different levels of biological organization are integrated with each other and with the current model of the system. Obviously, the model is adapted in such a way that the experimentally observed phenomena correspond best with the model's predictions. These steps are continually repeated, thereby expanding and refining the model until the model's predictions reflect biological reality (Liu et al., 2013). For microalgae, few studies on systems biology with extensive computational modeling efforts have been reported so far to our knowledge (Hildebrand et al., 2013; Veyel et al., 2014). Studies on metabolic, genomic, and transcriptomic data in *C. reinhardtii* provide genome-wide insights into the regulation of the metabolic networks under anaerobic conditions associated with H_2 production (Mus et al., 2007). Metabolic network reconstruction in model alga *C. reinhardtii* encompasses organism's metabolism and genome annotation, providing a platform for omics data analysis and phenotype prediction. The morphological characteristics and metabolome profiles of the oil-rich alga *P. ellipsoidea* exposed to $+N$ and $-N$ conditions were analyzed to determine how lipids synthesize and the mechanisms in which they accumulate in $-N$ conditions. This study revealed few hypothetical metabolisms where advanced systems biology approaches such as metabolic flux analysis, turnover analysis, and pulse-chase experiments are required for more effective understanding of the physiological phenomenon occurring among these microalgal strains (Ito et al., 2013).

Integrating biological and optical data, reconstructing a genome-scale metabolic network offers insight into algal metabolism and potential for genetic engineering and efficient light source design, a pioneering resource for studying light-driven metabolism and quantitative systems biology (Chang et al., 2011). Systems biology will also make available synthetic biology tools more reliable, enabling the precise control of transcription and translation regardless of the under-controlled gene (Mutalik et al., 2013). Similarly, simplified understanding of genetic systems created in synthetic biology will provide systems biology with new insights into the fundamentals of native gene regulation among microalgae, thus allowing simultaneous integration of "Omics" data at global scales in living cells, forms a direct networking between systems and synthetic biologists that will hasten rapid progress in areas pertaining to genomics, transcriptomics, proteomics, and metabolomics (Liu et al., 2013).

4. METABOLIC ENGINEERING TO OPTIMIZE BIOENERGY PRODUCTION

Understanding of metabolic network in marine microalgae is highly complex and intricate, and the characterization of the metabolic pathways is required before metabolic engineering to optimize bioenergy production. Genetic engineering plays a key role in transforming these systems into the desired cell factories and higher productivity based on the crucial enzymes that can be targeted for various metabolic pathways (Radakovits et al., 2010; Williams, 2010). Such "wonder" strains also known as "transgenic" generated through genetic engineering approach should have a combination of characteristics such as faster growth, high productivity, broader resistance, and suitable for large-scale cultivation in different environments. Another upcoming technique is gene silencing by RNA interference (RNAi) that occurs either through repression of transcription, termed post-transcriptional gene silencing, or through mRNA degradation (Angaji et al., 2010; Xu et al., 2010), due to its high targeted specificity and efficiency used for analyzing the biological function of the targeted gene and manipulating the metabolic process by sequence-specific knockdown (Cerutti et al., 2011; Dafny-Yelin et al., 2007). A wide range of RNAi components, which promotes stable gene repression and the transient gene silencing, were identified in the algae, such as red alga *P. yezoensis* (Liang et al., 2010), diatom *P. tricornutum* (De Riso et al., 2009), *T. pseudonana* (Armbrust et al., 2004), green alga *D. salina* (Jia et al., 2009), and brown alga *Ectocarpus siliculosus* (Cock et al., 2010). Metabolic engineering of microalgae for increased biofuel precursors is still in its infancy due to the fact that sequenced genomes, transformation techniques, promoters to drive gene overexpression, selection markers and many more are still under progress, and it remains questionable if a single-gene overexpression/deletion will be sufficient enough to redirect the whole metabolic flux in cells, and/or whether a multi-gene manipulation of a regulatory gene approach is more beneficial (Courchesne et al., 2009; Schuhmann et al., 2011).

A deeper understanding of carbon flux throughout the full range of cellular processes is necessary to optimize both growth and lipid accumulation in marine microalgae cultures to facilitate bioenergy production. Engineering metabolic pathways which are unique and practical for increasing biofuel yields from eukaryotic microalgae without compromising growth are essential for optimizing the vision of tomorrow's bioenergy needs.

5. CONCLUDING REMARKS

Genetic engineering of marine microalgae will be a challenging task to optimize bioenergy production where the primary constraint is strain development involving improved methods for carbon concentrating and partitioning, manipulation of metabolic pathways for production of sustainable biofuel precursors. The availability of favorable growth environmental platforms to promote these optimized engineered algal strains are important in terms of economic competitiveness and environmental impact. Despite the numerous hurdles, these organisms still represent one of the best options available as a source of renewable bioenergy that has rationale to be economically viable for production of biofuels from marine microalgae. The interdisciplinary approach and integrated "Omics" research will provide leads to optimize the growth parameters, harvesting techniques, and low-cost processing of these microalgal, systems, thus creating an extremely versatile and efficient production platform for bioenergy.

References

Angaji, S.A., Hedayati, S.S., Poor, R.H., Madani, S., Poor, S.S., Panahi, S., 2010. Application of RNA interference in treating human diseases. J. Genet. 89, 527–537.

Apt, K.E., Behrens, P.W., 1999. Commercial developments in microalgal biotechnology. J. Phycol. 35, 215–226.

Armbrust, E.V., Berges, J.A., Bowler, C., Green, B.R., Martinez, D., Putnam, N.H., Zhou, S., Allen, A.E., Apt, K.E., Bechner, M., Brzezinski, M.A., Chaal, B.K., Chiovitti, A., Davis, A.K., Demarest, M.S., Detter, J.C., Glavina, T., Goodstein, D., Hadi, M.Z., Hellsten, U., Hildebrand, M., Jenkins, B.D., Jurka, J., Kapitonov, V.V., Kroger, N., Lau, W.W., Lane, T.W., Larimer, F.W., Lippmeier, J.C., Lucas, S., Medina, M., Montsant, A., Obornik, M., Parker, M.S., Palenik, B., Pazour, G.J., Richardson, P.M., Rynearson, T.A., Saito, M.A., Schwartz, D.C., Thamatrakoln, K., Valentin, K., Vardi, A., Wilkerson, F.P., Rokhsar, D.S., 2004. The genome of the diatom *Thalassiosira pseudonana*: ecology, evolution, and metabolism. Science 306, 79–86.

Beer, L.L., Boyd, E.S., Peters, J.W., Posewitz, M.C., 2009. Engineering algae for biohydrogen and biofuel production. Curr. Opin. Biotechnol. 20, 264–271.

Blankenship, R.E., Tiede, D.M., Barber, J., Brudvig, G.W., Fleming, G., Ghirardi, M., Gunner, M.R., Junge, W., Kramer, D.M., Melis, A., Moore, T.A., Moser, C.C., Nocera, D.G., Nozik, A.J., Ort, D.R., Parson, W.W., Prince, R.C., Sayre, R.T., 2011. Comparing photosynthetic and photovoltaic efficiencies and recognizing the potential for improvement. Science 332, 805–809.

Bonente, G., Formighieri, C., Mantelli, M., Catalanotti, C., Giuliano, G., Morosinotto, T., Bassi, R., 2011. Mutagenesis and phenotypic selection as a strategy toward domestication of *Chlamydomonas reinhardtii* strains for improved performance in photobioreactors. Photosynth. Res. 108, 107–120.

Bougaran, G., Rouxel, C., Dubois, N., Kaas, R., Grouas, S., Lukomska, E., Le Coz, J.-R., Cadoret, J.-P., 2012. Enhancement of neutral lipid productivity in the microalga *Isochrysis* affinis *Galbana* (T-Iso) by a mutation-selection procedure. Biotechnol. Bioeng. 109, 2737–2745.

Bowler, C., Allen, A.E., Badger, J.H., Grimwood, J., Jabbari, K., Kuo, A., Maheswari, U., Martens, C., Maumus, F., Otillar, R.P., Rayko, E., Salamov, A., Vandepoele, K., Beszteri, B., Gruber, A., Heijde, M., Katinka, M., Mock, T., Valentin, K., Verret, F., Berges, J.A., Brownlee, C., Cadoret, J.P., Chiovitti, A., Choi, C.J., Coesel, S., De Martino, A., Detter, J.C., Durkin, C., Falciatore, A., Fournet, J., Haruta, M., Huysman, M.J., Jenkins, B.D., Jiroutova, K., Jorgensen, R.E., Joubert, Y., Kaplan, A., Kroger, N., Kroth, P.G., La Roche, J., Lindquist, E., Lommer, M., Martin-Jezequel, V., Lopez, P.J., Lucas, S., Mangogna, M., McGinnis, K., Medlin, L.K., Montsant, A., Oudot-Le Secq, M.P., Napoli, C., Obornik, M., Parker, M.S., Petit, J.L., Porcel, B.M., Poulsen, N., Robison, M., Rychlewski, L., Rynearson, T.A., Schmutz, J., Shapiro, H., Siaut, M., Stanley, M., Sussman, M.R., Taylor, A.R., Vardi, A., von Dassow, P., Vyverman, W., Willis, A., Wyrwicz, L.S., Rokhsar, D.S., Weissenbach, J., Armbrust, E.V., Green, B.R., Van de Peer, Y., Grigoriev, I.V., 2008. The *Phaeodactylum* genome reveals the evolutionary history of diatom genomes. Nature 456, 239–244.

Cagnon, C., Mirabella, B., Nguyen, H.M., Beyly-Adriano, A., Bouvet, S., Cuine, S., Beisson, F., Peltier, G., Li-Beisson, Y., 2013. Development of a forward genetic screen to isolate oil mutants in the green microalga *Chlamydomonas reinhardtii*. Biotechnol. Biofuels 6, 178.

Cerutti, H., Johnson, A.M., Gillham, N.W., Boynton, J.E., 1997. Epigenetic silencing of a foreign gene in nuclear transformants of *Chlamydomonas*. Plant Cell 9, 925–945.

Cerutti, H., Ma, X., Msanne, J., Repas, T., 2011. RNA-mediated silencing in algae: biological roles and tools for analysis of gene function. Eukaryot. Cell 10, 1164–1172.

Cha, T.S., Chen, C.F., Yee, W., Aziz, A., Loh, S.H., 2011. Cinnamic acid, coumarin and vanillin: alternative phenolic compounds for efficient Agrobacterium-mediated transformation of the unicellular green alga, *Nannochloropsis* sp. J. Microbiol. Methods 84, 430–434.

Chang, R.L., Ghamsari, L., Manichaikul, A., Hom, E.F., Balaji, S., Fu, W., Shen, Y., Hao, T., Palsson, B.O., Salehi-Ashtiani, K., Papin, J.A., 2011. Metabolic network reconstruction of *Chlamydomonas* offers insight into light-driven algal metabolism. Mol. Syst. Biol. 7, 518.

Chisti, Y., 2007. Biodiesel from microalgae. Biotechnol. Adv. 25, 294–306.

Cock, J.M., Sterck, L., Rouze, P., Scornet, D., Allen, A.E., Amoutzias, G., Anthouard, V., Artiguenave, F., Aury, J.M., Badger, J.H., Beszteri, B., Billiau, K., Bonnet, E., Bothwell, J.H., Bowler, C., Boyen, C., Brownlee, C., Carrano, C.J., Charrier, B., Cho, G.Y., Coelho, S.M., Collen, J., Corre, E., Da Silva, C., Delage, L., Delaroque, N., Dittami, S.M., Doulbeau, S., Elias, M., Farnham, G., Gachon, C.M., Gschloessl, B., Heesch, S., Jabbari, K., Jubin, C., Kawai, H., Kimura, K., Kloareg, B., Kupper, F.C., Lang, D., Le Bail, A., Leblanc, C., Lerouge, P., Lohr, M., Lopez, P.J., Martens, C., Maumus, F., Michel, G., Miranda-Saavedra, D., Morales, J., Moreau, H., Motomura, T., Nagasato, C., Napoli, C.A., Nelson, D.R., Nyvall-Collen, P., Peters, A.F., Pommier, C., Potin, P., Poulain, J., Quesneville, H., Read, B., Rensing, S.A., Ritter, A., Rousvoal, S., Samanta, M., Samson, G., Schroeder, D.C., Segurens, B., Strittmatter, M., Tonon, T., Tregear, J.W., Valentin, K., von Dassow, P., Yamagishi, T., Van de Peer, Y., Wincker, P., 2010. The *Ectocarpus* genome and the independent evolution of multicellularity in brown algae. Nature 465, 617–621.

Courchesne, N.M., Parisien, A., Wang, B., Lan, C.Q., 2009. Enhancement of lipid production using biochemical, genetic and transcription factor engineering approaches. J. Biotechnol. 141, 31–41.

Dafny-Yelin, M., Chung, S.M., Frankman, E.L., Tzfira, T., 2007. pSAT RNA interference vectors: a modular series for multiple gene down-regulation in plants. Plant Physiol. 145, 1272–1281.

De Riso, V., Raniello, R., Maumus, F., Rogato, A., Bowler, C., Falciatore, A., 2009. Gene silencing in the marine diatom *Phaeodactylum tricornutum*. Nucleic Acids Res. 37, e96.

El-Sheekh, M.M., 1999. Stable transformation of the intact cells of *Chlorella kessleri* with high velocity microprojectiles. Biol. Plant 42, 209–216.

Falkowski, P.G., Barber, R.T., Smetacek, V., 1998. Biogeochemical controls and feedbacks on ocean primary production. Science 281, 200–206.

Feng, S., Xue, L., Liu, H., Lu, P., 2009. Improvement of efficiency of genetic transformation for *Dunaliella salina* by glass beads method. Mol. Biol. Rep. 36, 1433–1439.

Frank, E.D., Elgowainy, A., Han, J., Wang, Z., 2013. Life cycle comparison of hydrothermal liquefaction and lipid extraction pathways to renewable diesel from algae. Mitig Adapt. Strateg. Glob. Change 18, 137–158.

Franz, A.K., Danielewicz, M.A., Wong, D.M., Anderson, L.A., Boothe, J.R., 2013. Phenotypic screening with oleaginous microalgae reveals modulators of lipid productivity. ACS Chem. Biol. 8, 1053–1062.

Fuhrmann, M., Oertel, W., Hegemann, P., 1999. A synthetic gene coding for the green fluorescent protein (GFP) is a versatile reporter in *Chlamydomonas reinhardtii*. Plant J. 19, 353–361.

Georgianna, D.R., Mayfield, S.P., 2012. Exploiting diversity and synthetic biology for the production of algal biofuels. Nature 488, 329−335.

Gimpel, J.A., Specht, E.A., Georgianna, D.R., Mayfield, S.P., 2013. Advances in microalgae engineering and synthetic biology applications for biofuel production. Curr. Opin. Chem. Biol. 17, 489−495.

Gong, Y., Jiang, M., 2011. Biodiesel production with microalgae as feedstock: from strains to biodiesel. Biotechnol. Lett. 33, 1269−1284.

Gonzalez-Ballester, D., Pootakham, W., Mus, F., Yang, W., Catalanotti, C., Magneschi, L., de Montaigu, A., Higuera, J., Prior, M., Galvan, A., Fernandez, E., Grossman, A., 2011. Reverse genetics in *Chlamydomonas*: a platform for isolating insertional mutants. Plant Methods 7, 24.

Hallmann, A., 2007. Algal transgenics and biotechnology. Transgenic Plant J. 81−98.

Han, S.F., Jin, W.B., Tu, R.J., Wu, W.M., 2014. Biofuel production from microalgae as feedstock: current status and potential. Crit. Rev. Biotechnol. Early online, 1−14. http://dx.doi.org/10.3109/07388551.2013.835301.

Hannon, M., Gimpel, J., Tran, M., Rasala, B., Mayfield, S., 2010. Biofuels from algae: challenges and potential. Biofuels 1, 763−784.

Heidorn, T., Camsund, D., Huang, H.H., Lindberg, P., Oliveira, P., Stensjo, K., Lindblad, P., 2011. Synthetic biology in cyanobacteria engineering and analyzing novel functions. Methods Enzymol. 497, 539−579.

Hildebrand, M., Abbriano, R.M., Polle, J.E., Traller, J.C., Trentacoste, E.M., Smith, S.R., Davis, A.K., 2013. Metabolic and cellular organization in evolutionarily diverse microalgae as related to biofuels production. Curr. Opin. Chem. Biol. 17, 506−514.

Ideker, T., Galitski, T., Hood, L., 2001. A new approach to decoding life: systems biology. Annu. Rev. Genomics Hum. Genet. 2, 343−372.

Ishii, N., Tomita, M., 2009. Multi-omics data-driven systems biology of *E. coli*. In: Lee, S. (Ed.), Systems Biology and Biotechnology of *Escherichia coli*. Springer, Netherlands, pp. 41−57.

Ito, T., Tanaka, M., Shinkawa, H., Nakada, T., Ano, Y., Kurano, N., Soga, T., Tomita, M., 2013. Metabolic and morphological changes of an oil accumulating trebouxiophycean alga in nitrogen-deficient conditions. Metabolomics 9, 178−187.

Jamers, A., Blust, R., De Coen, W., 2009. Omics in algae: paving the way for a systems biological understanding of algal stress phenomena? Aquat. Toxicol. 92, 114−121.

Jia, Y., Xue, L., Liu, H., Li, J., 2009. Characterization of the glyceraldehyde-3-phosphate dehydrogenase (GAPDH) gene from the halotolerant alga *Dunaliella salina* and inhibition of its expression by RNAi. Curr. Microbiol. 58, 426−431.

John, R.P., Anisha, G.S., Nampoothiri, K.M., Pandey, A., 2011. Micro and macroalgal biomass: a renewable source for bioethanol. Bioresour. Technol. 102, 186−193.

Jones, C.S., Mayfield, S.P., 2012. Algae biofuels: versatility for the future of bioenergy. Curr. Opin. Biotechnol. 23, 346−351.

Jutur, P.P., Asha, A.N., 2015. Marine microalgae: Exploring the systems through an omics approach for biofuel production. In: Kim, S-K., Lee, C-G. (Eds.), Marine Bioenergy: Trends and Developments. Taylor & Francis Group, pp. 149−162.

Kacka, A., Donmez, G., 2008. Isolation of *Dunaliella* spp. from a hypersaline lake and their ability to accumulate glycerol. Bioresour. Technol. 99, 8348−8352.

Keasling, J.D., Venter, J.C., 2013. Applications of synthetic biology to enhance life. Bridge 43, 47−58.

Khan, S.A., Rashmi, Hussain, M.Z., Prasad, S., Banerjee, U.C., 2009. Prospects of biodiesel production from microalgae in India. Renewable Sustainable Energy Rev. 13, 2361−2372.

Kilian, O., Benemann, C.S.E., Niyogi, K.K., Vick, B., 2011. High-efficiency homologous recombination in the oil-producing alga *Nannochloropsis* sp. Proc. Natl. Acad. Sci. USA 108, 21265−21269.

Kumar, A., Ergas, S., Yuan, X., Sahu, A., Zhang, Q., Dewulf, J., Malcata, F.X., van Langenhove, H., 2010. Enhanced CO_2 fixation and biofuel production via microalgae: recent developments and future directions. Trends Biotechnol. 28, 371−380.

Lamers, P.P., Janssen, M., De Vos, R.C., Bino, R.J., Wijffels, R.H., 2008. Exploring and exploiting carotenoid accumulation in *Dunaliella salina* for cell-factory applications. Trends Biotechnol. 26, 631−638.

Lapidot, M., Raveh, D., Sivan, A., Arad, S.M., Shapira, M., 2002. Stable chloroplast transformation of the unicellular red alga *Porphyridium* species. Plant Physiol. 129, 7−12.

Larkum, A.W., 2010. Limitations and prospects of natural photosynthesis for bioenergy production. Curr. Opin. Biotechnol. 21, 271−276.

Larkum, A.W.D., Ross, I.L., Kruse, O., Hankamer, B., 2012. Selection, breeding and engineering of microalgae for bioenergy and biofuel production. Trends Biotechnol. 30, 198−205.

León-Bañares, R., González-Ballester, D., Galván, A., Fernández, E., 2004. Transgenic microalgae as green cell-factories. Trends Biotechnol. 22, 45−52.

Liang, C., Zhang, X., Zou, J., Xu, D., Su, F., Ye, N., 2010. Identification of miRNA from *Porphyra yezoensis* by high-throughput sequencing and bioinformatics analysis. PLoS One 5, e10698.

Lin, H., Qin, S., Jiang, P., 2011. Biotechnology of seaweeds: Facing the coming decade. In: Kim, S-K. (Ed.), Handbook of Marine Macroalgae: Biotechnology and Applied Phycology. John Wiley & Sons Ltd, pp. 424–430.

Liu, B., Benning, C., 2013. Lipid metabolism in microalgae distinguishes itself. Curr. Opin. Biotechnol. 24, 300–309.

Liu, D., Hoynes-O'Connor, A., Zhang, F., 2013. Bridging the gap between systems biology and synthetic biology. Front. Microbiol. 4, 211.

Lu, J., Sheahan, C., Fu, P., 2011. Metabolic engineering of algae for fourth generation biofuels production. Energy Environ. Sci. 4, 2451–2466.

Mata, T.M., Martins, A.A., Caetano, N.S., 2010. Microalgae for biodiesel production and other applications: a review. Renewable Sustainable Energy Rev. 14, 217–232.

Mohamed, M.S., Tan, J.S., Mohamad, R., Mokhtar, M.N., Ariff, A.B., 2013. Comparative analyses of response surface methodology and artificial neural network on medium optimization for Tetraselmis sp. FTC209 grown under mixotrophic condition. Scientific World J. 2013, 948940.

Molnar, A., Bassett, A., Thuenemann, E., Schwach, F., Karkare, S., Ossowski, S., Weigel, D., Baulcombe, D., 2009. Highly specific gene silencing by artificial microRNAs in the unicellular alga Chlamydomonas reinhardtii. Plant J. 58, 165–174.

Moreno-Garrido, I., 2008. Microalgae immobilization: current techniques and uses. Bioresour. Technol. 99, 3949–3964.

Muers, M., 2011. Synthetic biology: simplifying design. Nat. Rev. Genet. 13, 72.

Mus, F., Dubini, A., Seibert, M., Posewitz, M.C., Grossman, A.R., 2007. Anaerobic acclimation in Chlamydomonas reinhardtii: anoxic gene expression, hydrogenase induction, and metabolic pathways. J. Biol. Chem. 282, 25475–25486.

Mutalik, V.K., Guimaraes, J.C., Cambray, G., Lam, C., Christoffersen, M.J., Mai, Q.A., Tran, A.B., Paull, M., Keasling, J.D., Arkin, A.P., Endy, D., 2013. Precise and reliable gene expression via standard transcription and translation initiation elements. Nat. Methods 10, 354–360.

Neupert, J., Karcher, D., Bock, R., 2009. Generation of Chlamydomonas strains that efficiently express nuclear transgenes. Plant J. 57, 1140–1150.

Pazour, G.J., Witman, G.B., 2000. Forward and reverse genetic analysis of microtubule motors in Chlamydomonas. Methods 22, 285–298.

Pfromm, P.H., Amanor-Boadu, V., Nelson, R., 2011. Sustainability of algae derived biodiesel: a mass balance approach. Bioresour. Technol. 102, 1185–1193.

Posten, C., 2009. Design principles of photo-bioreactors for cultivation of microalgae. Eng. Life Sci. 9, 165–177.

Pratheesh, P.T., Vineetha, M., Kurup, G.M., 2014. An efficient protocol for the agrobacterium-mediated genetic transformation of microalga Chlamydomonas reinhardtii. Mol. Biotechnol. 56, 507–515.

Pulz, O., Gross, W., 2004. Valuable products from biotechnology of microalgae. Appl. Microbiol. Biotechnol. 65, 635–648.

Qin, S., Lin, H., Jiang, P., 2012. Advances in genetic engineering of marine algae. Biotechnol. Adv. 30, 1602–1613.

Radakovits, R., Jinkerson, R.E., Darzins, A., Posewitz, M.C., 2010. Genetic engineering of algae for enhanced biofuel production. Eukaryot. Cell 9, 486–501.

Radakovits, R., Jinkerson, R.E., Fuerstenberg, S.I., Tae, H., Settlage, R.E., Boore, J.L., Posewitz, M.C., 2012. Draft genome sequence and genetic transformation of the oleaginous alga Nannochloropsis gaditana. Nat. Commun. 3, 686.

Raja, R., Hemaiswarya, S., Kumar, N.A., Sridhar, S., Rengasamy, R., 2008. A perspective on the biotechnological potential of microalgae. Crit. Rev. Microbiol. 34, 77–88.

Rasala, B.A., Chao, S.S., Pier, M., Barrera, D.J., Mayfield, S.P., 2014. Enhanced genetic tools for engineering multigene traits into green algae. PLoS One 9, e94028.

Rasala, B.A., Lee, P.A., Shen, Z., Briggs, S.P., Mendez, M., Mayfield, S.P., 2012. Robust expression and secretion of Xylanase1 in Chlamydomonas reinhardtii by fusion to a selection gene and processing with the FMDV 2A peptide. PLoS One 7, e43349.

Rathod, J.P., Prakash, G., Pandit, R., Lali, A.M., 2013. Agrobacterium-mediated transformation of promising oil-bearing marine algae Parachlorella kessleri. Photosynth. Res. 118, 141–146.

Resurreccion, E.P., Colosi, L.M., White, M.A., Clarens, A.F., 2012. Comparison of algae cultivation methods for bioenergy production using a combined life cycle assessment and life cycle costing approach. Bioresour. Technol. 126, 298–306.

Rodolfi, L., Chini Zittelli, G., Bassi, N., Padovani, G., Biondi, N., Bonini, G., Tredici, M.R., 2009. Microalgae for oil: strain selection, induction of lipid synthesis and outdoor mass cultivation in a low-cost photobioreactor. Biotechnol. Bioeng. 102, 100–112.

Rosenberg, J.N., Oyler, G.A., Wilkinson, L., Betenbaugh, M.J., 2008. A green light for engineered algae: redirecting metabolism to fuel a biotechnology revolution. Curr. Opin. Biotechnol. 19, 430–436.

San Pedro, A., Gonzalez-Lopez, C.V., Acien, F.G., Molina-Grima, E., 2013. Marine microalgae selection and culture conditions optimization for biodiesel production. Bioresour. Technol. 134, 353–361.

Schmidt, B.J., Lin-Schmidt, X., Chamberlin, A., Salehi-Ashtiani, K., Papin, J.A., 2010. Metabolic systems analysis to advance algal biotechnology. Biotechnol. J. 5, 660–670.

Schuhmann, H., Lim, D.K.Y., Schenk, P.M., 2011. Perspectives on metabolic engineering for increased lipid contents in microalgae. Biofuels 3, 71–86.

Shen, Y., Xu, X., Zhao, Y., Lin, X., 2014. Influence of algae species, substrata and culture conditions on attached microalgal culture. Bioprocess. Biosyst. Eng. 37, 441–450.

Shirvani, T., Yan, X., Inderwildi, O.R., Edwards, P.P., King, D.A., 2011. Life cycle energy and greenhouse gas analysis for algae-derived biodiesel. Energy Environ. Sci. 4, 3773–3778.

Sills, D.L., Paramita, V., Franke, M.J., Johnson, M.C., Akabas, T.M., Greene, C.H., Tester, J.W., 2013. Quantitative uncertainty analysis of life cycle assessment for algal biofuel production. Environ. Sci. Technol. 47, 687–694.

Specht, E., Miyake-Stoner, S., Mayfield, S., 2010. Micro-algae come of age as a platform for recombinant protein production. Biotechnol. Lett. 32, 1373–1383.

Specht, E.A., Mayfield, S.P., 2014. Algae-based oral recombinant vaccines. Front. Microbiol. 5, 60.

Spolaore, P., Joannis-Cassan, C., Duran, E., Isambert, A., 2006. Commercial applications of microalgae. J. Biosci. Bioeng. 101, 87–96.

Subburamu, K., 2013. Strain selection for biodiesel production. In: Bux, F. (Ed.), Biotechnological Applications of Microalgae. CRC Press, pp. 17–44.

Tabatabaei, M., Tohidfar, M., Jouzani, G.S., Safarnejad, M., Pazouki, M., 2011. Biodiesel production from genetically engineered microalgae: future of bioenergy in Iran. Renewable Sustainable Energ Rev. 15, 1918–1927.

Tredici, M.R., 2007. Mass Production of Microalgae: Photobioreactors. In: Richmond, A., Hu, R. (Eds.), Handbook of Microalgal Culture. Blackwell Publishing Ltd, pp. 178–214.

Veyel, D., Erban, A., Fehrle, I., Kopka, J., Schroda, M., 2014. Rationales and approaches for studying metabolism in eukaryotic microalgae. Metabolites 4, 184–217.

Walker, T.L., Purton, S., Becker, D.K., Collet, C., 2005. Microalgae as bioreactors. Plant Cell Rep. 24, 629–641.

Wang, B., Wang, J., Zhang, W., Meldrum, D.R., 2012. Application of synthetic biology in cyanobacteria and algae. Front. Microbiol. 3, 344.

Wijffels, R.H., Kruse, O., Hellingwerf, K.J., 2013. Potential of industrial biotechnology with cyanobacteria and eukaryotic microalgae. Curr. Opin. Biotechnol. 24, 405–413.

Williams, N., 2010. New biofuel questions. Curr. Biol. 20, R219–R220.

Xie, B., Stessman, D., Hart, J.H., Dong, H., Wang, Y., Wright, D.A., Nikolau, B.J., Spalding, M.H., Halverson, L.J., 2014a. High-throughput fluorescence-activated cell sorting for lipid hyperaccumulating *Chlamydomonas reinhardtii* mutants. Plant Biotechnol. J. 12, 872–882.

Xie, W.H., Zhu, C.C., Zhang, N.S., Li, D.W., Yang, W.D., Liu, J.S., Sathishkumar, R., Li, H.Y., 2014b. Construction of novel chloroplast expression vector and development of an efficient transformation system for the diatom *Phaeodactylum tricornutum*. Mar Biotechnol. 16, 538–546.

Xu, G., Sui, N., Tang, Y., Xie, K., Lai, Y., Liu, Y., 2010. One-step, zero-background ligation-independent cloning intron-containing hairpin RNA constructs for RNAi in plants. New Phytol. 187, 240–250.

Zhang, C., Hu, H., 2013. High-efficiency nuclear transformation of the diatom *Phaeodactylum tricornutum* by electroporation. Mar Genomics 16, 63–66.

Zhang, W., Li, F., Nie, L., 2010. Integrating multiple "omics" analysis for microbial biology: application and methodologies. Microbiology 156, 287–301.

Genetic Optimization of Microalgae for Biohydrogen Production

Suphi S. Oncel[1], Ayse Kose[1], Cecilia Faraloni[2]

[1]Ege University, Engineering Faculty, Department of Bioengineering, Izmir, Turkey;
[2]CNR, Istituto per lo Studio degli Ecosistemi Sezione di Firenze, Firenze, ITALY

1. INTRODUCTION

Hydrogen produced from biological sources—known as biohydrogen—is renewable, sustainable, and environmentally friendly. Theoretically, biohydrogen is a good candidate for a reliable energy source; however, in reality, the scenario changes because of major bottlenecks such as low productivity and storage issues (Melis et al., 2000; Ghirardi et al., 2006; Melis and Happe, 2001; Kosourov et al., 2007; Oncel and Sabankay, 2012; Oncel, 2013; Torzillo et al., 2014). Its feasibility and contribution to the global hydrogen economy are other challenges to be addressed, but the main drawbacks come from the lack of an adequate production strategy and knowledge. Attempts to eliminate the challenges associated with biohydrogen and introduce a well-established production strategy are becoming more positively viewed by society. However, these challenges will likely remain for a long period of time.

Biohydrogen can be produced by diverse taxonomical clusters of microorganisms with certain metabolisms (Figure 1), generally named as bacteria (Srirangan et al., 2011), microalgae, and cyanobacteria (Lindblad et al., 2002; Allakhverdiev et al., 2010; Ghirardi and Mohanty, 2010; McKinlay and Harwood, 2010; Oncel, 2013). The focus of this chapter is on microalgal biohydrogen production strategies, but it will also discuss the essential information about cyanobacterial biohydrogen.

Microalgae and cyanobacteria are photosynthetic microorganisms that fix CO_2 under natural light (sunlight) or artificial light into complex macromolecules and building blocks. From a metabolic perspective, biohydrogen can be produced under certain illumination and anaerobic conditions (Benemann, 1997). This metabolic evolution has provided new insights on the research of biohydrogen production after the first attempt by Gaffron and Rubin in the 1940s (Gaffron and Rubin, 1942). In 2000, Melis and coworkers introduced a new method that separated production into aerobic and anaerobic phases (Melis et al., 2000). Sulfur deprivation in the anaerobic stage is key for prolonged biohydrogen production because of the amino acid sequence of the D1 protein in the photosystem II (PSII) repair system (Faraloni and Torzillo et al., 2010; Scoma et al., 2012; Kose

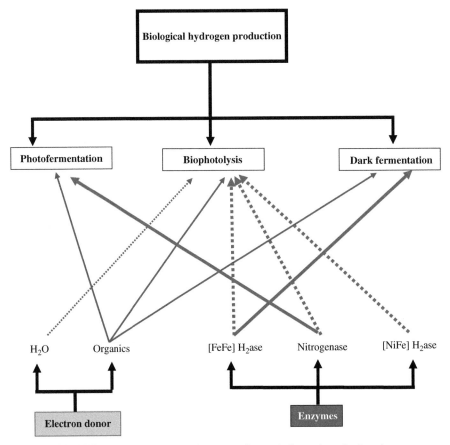

FIGURE 1   Biohydrogen production pathways in bacteria and microalgae.

and Oncel, 2014). Currently, light—dark cycles (Oncel and Sukan, 2009), purging inert gas to remove oxygen, and inhibitor chemicals are used to prolong biohydrogen production (Melis and Happe, 2001). Laboratory-scale and pilot-scale photobioreactors have also been constructed in an effort to scale up production (Gianelli et al., 2009; Oncel and Sabankay, 2012; Oncel and Kose, 2014). Some studies on photobological hydrogen production have focused on screening potential biohydrogen producer species (Melis et al., 2000; Antal et al., 2003; Oncel, 2013). Immobilization techniques have been adapted for biohydrogen production to use cell captured beads in aerobic/anaerobic

cycles (Hahn et al., 2007; Kosourov and Seibert, 2009; Laurinavichene et al., 2008).

A current trend in biohydrogen production from either microalgae or cyanobacteria is to adapt genetic engineering tools to develop sustainable biohydrogen-producing machinery or cell factories (Mathews and Wang, 2009; Kruse and Hankamer, 2010; Srirangan et al., 2011). Numerous studies have been conducted to screen and sequence new genomes that may have the potential for biohydrogen production and complete whole genome sequencing (Beer et al., 2009; Radokovits et al., 2010), investigate protein engineering, and identify the protein machinery involved in hydrogen production

(Meyer, 2007), enhance the oxygen tolerance of hydrogenase enzymes (Meuser et al., 2012), investigate engineering in starch catabolism (Kruse et al., 2005), examine trophic conversion (Doebbe et al., 2007; Mathews and Wang, 2009), and investigate engineering of direct biophotolysis mechanisms and D1 protein activity (Faraloni and Torzillo, 2010; Kose and Oncel, 2014). This chapter highlights the concept of microalgal hydrogen production from a metabolic perspective. The model microalgae *Chlamydomonas reinhardtii* is used to examine the genetic engineering aspects.

2. STATE OF THE ART: PHOTOBIOHYDROGEN PRODUCTION BY MICROALGAE

As unicellular, photosynthetic, and oxygen-evolving microorganisms, microalgae have remarkable theoretical potential to produce biohydrogen (Melis et al., 2000). The initial step for both photosynthesis and biohydrogen production is light capture at PSII in the light-harvesting complex II (LHCII). Water splits into electrons, protons, and O_2. Biophotolysis is essential to microalgal or cyanobacterial biohydrogen evolution (Srirangan et al., 2011). It is classified as direct or indirect biophotolysis based on the use of an electron donor or water/organic substrates, respectively (Kim and Kim, 2011).

After the first step, electrons are transferred in electron transport chains of chloroplasts (the photosynthetic chain in the cytoplasm of cyanobacteria); under aerobic conditions, CO_2 is fixed into organic complex molecules. The Plastoquinone (PQ) pool plays a critical role in the transportation of e^- and H^+ through the electron transport chain, either for photosynthesis or for biohydrogen production (Wykoff et al., 1998). Because the substrates of hydrogenase enzymes are e^- and H^+, the effective supply of these components directly affects the enzyme activity and

hydrogen yield. Thus, the PQ pool plays a crucial role in the processes for both direct and indirect biophotolysis. Apart from photosynthesis, when the metabolism shifts to anoxia, certain genes are activated to synthesize enzymes called hydrogenases, which catalyze biohydrogen evolution (Meyer, 2007).

Hydrogen evolution in aerobic conditions may be viewed as proof that microalgae are capable of catalyzing biohydrogen; however, under aerobic conditions, hydrogenase enzymes are deactivated by the O_2 that is produced or exists in the production chamber (Schulz, 1996; Melis, 2002). Researchers determined that microalgae are capable of producing visible amounts of hydrogen in anaerobic dark conditions, which encouraged further studies (Melis, 2002). The reaction is thought to be the same in dark fermentative bacteria; however, later studies showed that sulfur deprivation is key for biohydrogen production under anaerobic and illuminated conditions (Melis et al., 2000). Before the introduction of a two-stage protocol, Wykoff et al. (1998) showed that photosynthetic activity under sulfur-deprived conditions in aerobic cultures decreases due to impairment of the D1 protein in the PSII system, which has sulfur-containing amino acids as building blocks. Also, light increases the supply of H^+ and e^- to the PQ pool, and the total amount of transported H^+ and e^- to hydrogenases results in a dramatic increase of biohydrogen evolution when compared to dark conditions. Light acts as a reaction accelerator, which demonstrates that the photosynthetic electron transport chain is essential for microalgal hydrogen production (Figure 2).

It is known that microalgae and cyanobacteria are good candidates when photobiological hydrogen production is the target. *Spirulina*, *Synechococcus*, *Anabaene*, *Chlorella* sp., and *Chlamydomonas* sp. are model strains in terms of microalgal studies (Oncel, 2013). Among them, *Synechococcus* (Antal and Lindblad, 2005) and

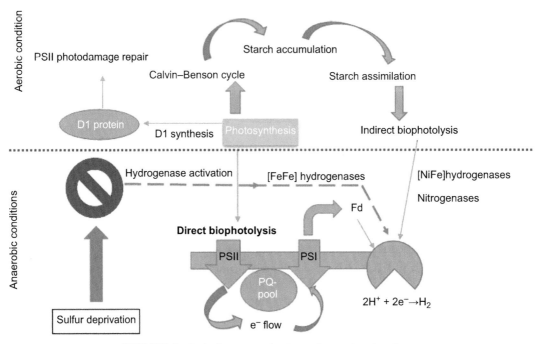

FIGURE 2   Biohydrogen production pathways in microalgae.

C. reinhardtii are thought to be the best hydrogen producers, according to current knowledge (Torzillo et al., 2014). Metabolic engineering is an efficient tool to enhance yield in several ways (Mathews and Wang, 2009), but limitations come from a lack of knowledge in the whole genome sequencing of certain important species (Radakovits et al., 2010). *C. reinhardtii* has been one of the most studied species because of its fully sequenced genome, high photosynthetic activity, ease of cultivation, and high biohydrogen production capacity (Oncel, 2013); however, cyanobacteria are also thought to be promising candidates because of their various biohydrogen production pathways and hydrogen-producing enzyme diversity (Carrieri et al., 2011). The ultimate goal of genetic engineering tools for biohydrogen production is to establish oxygen-tolerant microalgae species that enable single-step biohydrogen production.

3. HYDROGEN-CATALYZING ENZYMES IN MICROALGAL METABOLISM

The enzymes responsible for biological hydrogen production are generally referred as hydrogenases, but their enzyme maturation, enzyme activity, and structure diversity vary by microbial source. In N_2-fixing cyanobacteria, nitrogenases are also able to generate significant amounts of hydrogen under anaerobic and nitrogen-deficient conditions (Carrieri et al., 2011; Kim and Kim, 2011; Oh et al., 2011). Both enzymes play a major role in the reduction of H^+ ions to H_2 gas. The catalyzed reaction

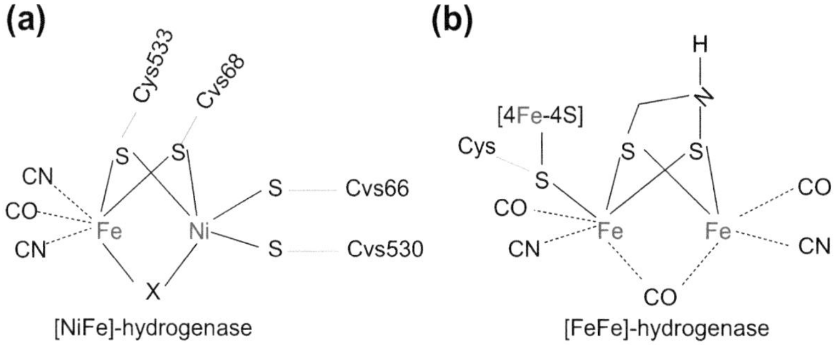

(a)

(b)

[NiFe]-hydrogenase

[FeFe]-hydrogenase

FIGURE 3 Hydrogenase enzymes in microalgae (Kim and Kim, 2011).

$(2H^+ + 2e^- \rightarrow H_2)$ seems to be a very basic, single-step reaction equation, but the mechanism is far more complex and yet to be elucidated.

3.1 Hydrogenases

Hydrogenases are noteworthy enzymes in evolutionary studies. They are found in most microorganisms, from Archaea to eukaryotes (Meyer, 2007). The genetic diversity of hydrogenases is the reason for the different hydrogen production capacities among species. The hydrogenases can be classified into three main groups: [NiFe] hydrogenases, [Fe] hydrogenases, and [FeFe] hydrogenases (Meyer, 2007). [Fe] hydrogenases are a small group of enzymes whose active site consists of Fe–S bonds (Meyer, 2007; Kim and Kim, 2011; Oh et al., 2011).

3.1.1 [FeFe] Hydrogenases

[FeFe] hydrogenases in microalgae are known to be coded in the nucleus; however, after maturation, the enzyme is localized in the chloroplast (Stephenson and Stickland, 1931; Meyer, 2007). It is the best hydrogen-catalyzing enzyme in terms of catalytic activity (Florin et al., 2001; Vignais et al., 2001; Meyer, 2007). It has a monomeric composition and a molecular weight of around 45–50 kDa (Meyer, 2007). Bacteria and microalgae have the same cluster of hydrogenases. The enzyme activity, channels in which O_2 is diffused, active sites, and molecular weight vary by species; microalgal hydrogenases are the smallest enzymes (Srirangan et al., 2011).

The active site, called the H cluster, is composed of [FeFe] bonds with sulfur bridges and also 4Fe–4S residue (Nicolet et al., 2001; Meyer, 2007), with non-proteinous ligands at CO and CN junctions (Figure 3(a)). [FeFe] hydrogenases can catalyze either hydrogen evolution or proton release from hydrogen (Meyer, 2007). The active site of the enzyme, where electrons are transferred from ferrodoxin, is highly sensitive to even trace amounts of oxygen, causing reversible activity losses; this is also the reason why biohydrogen cannot be accumulated naturally under oxygenic conditions.

[FeInstitute] hydrogenases are encoded by *hyd*A (*hyd*A1, *hyd*A2) genes (Happe and Kaminski, 2002; Forestier et al., 2003; Gadaux et al., 2013). Studies have shown that *hyd*A1 is much more active than *hyd*A2 (Meuser et al., 2012). Maturation of the active enzyme is controlled by a set of genes *hyd*E, *hyd*F, *hyd*G (Boyer et al., 2008). To set a mature hydrogenase enzyme, maturation proteins are required to be synthesized. Structural studies of maturation enzymes and sequence analysis of maturation genes prove that other than mature enzyme maturation system is also highly sensitive to oxygen (Posewitz et al., 2005; Böck et al., 2006; Meyer, 2007). Thus, strict anaerobiosis is required for maturation genes to be activated and translated in the cytoplasm (Skjanes et al., 2008). If hydrogenase enzymes are that sensitive to oxygen, what is needed for this pathway to be used for biotechnological purposes? Researchers are struggling with this question.

3.1.2 [NiFe] Hydrogenases

[NiFe] hydrogenases constitute the most distributed class of hydrogenases. They are divided into four major groups (i.e., Groups I–IV) according to enzyme activity and taxonomy (Kim and Kim, 2011). Cyanobacteria have [NiFe] hydrogenases as bidirectional and uptake hydrogenase enzymes. The basic structure of the enzyme consists of two compartments. The active site is the larger one ($\sim$60 kDa), which has [NiFe] bonds, whereas the small subunit ($\sim$30 kDa) only has an Fe–S (4Fe–4S or 3Fe–4S) cluster (Figure 3(b)). Four cysteine residues are bonded to metallogenic compartments with sulfur bonds. The small subunit transfers electrons to the active site and protons are reduced to hydrogen. [NiFe] hydrogenases can also work as uptake hydrogenases, oxidizing hydrogen gas to form electrons and protons that are later used to reduce NADPH. Sequence studies show that [FeFe] and [NiFe] hydrogenases are not related, but they both have a dimetallic compartment, CO and CN ligands attached to iron, and cysteine residues (Kim and Kim, 2011).

3.2 Nitrogenases

The role of nitrogenases in cyanobacteria is different from hydrogenases. To have an active nitrogenase enzyme in terms of catalyzing biohydrogen production, the culture environment must be strictly anaerobic. The evolutionary nature of nitrogenase enzymes fixes nitrogen into ammonium ions (Carrieri et al., 2011). Genetic regulation and synthesis of nitrogenases only occurs in cyanobacteria when nitrogen sources such as ammonium, urea, and nitrate are depleted. Therefore, the use of nitrogenase would require nitrogen-starved cultures in order to produce hydrogen catalyzed by nitrogenase enzymes via indirect biophotolysis.

Although microalgae only have [FeFe] hydrogenases to catalyze hydrogen gas production, cyanobacteria can use [NiFe] hydrogenases (bidirectional or uptake) in vegetative forms; heterocyst forms can use nitrogenases known as MoFe nitrogenases (Shestakov and Mikheeva, 2006; Srirangan et al., 2011). Under N_2-containing environments, the nitrogenase enzyme uses 2 mol of adenosine triphosphate (ATP) to transfer a single electron. When 1 mol of N_2 is converted into ammonia, 16 mol of ATP is consumed and 1 mol of H_2 is generated (Srirangan et al., 2011). In the absence of N_2, only 4 mol of ATP is required and 4 mol of H_2 is generated (Kim and Kim, 2011). Thus, nitrogenase-catalyzed biohydrogen production is an energy-expensive process, and the turnover rate of nitrogenase is lower than [NiFe] and [FeFe] hydrogenases (Meyer, 2007; Oh et al., 2011; Srirangan et al., 2011).

Because hydrogenase and nitrogenase enzymes are catalytic motors for biohydrogen production, establishing a route around enzyme activity is another research objective. The bottleneck of the enzyme machinery is oxygen sensitivity; thus, oxygen-tolerant production should be introduced to obtain realistic numbers. Other than environmental studies, genetic engineering and transcriptome studies may be more effective in designing a permanent strategy (Nguyen et al., 2008), which is aiming toward one-step production (Srirangan et al., 2011). The problem is the structural diversity of the enzymes; the isolation of both maturation proteins and mature enzymes is another issue. Microalgal enzyme machinery studies have been conducted with bacterial systems and adapted to algal systems (Nicolet et al., 2001; Meyer, 2007).

4. PHOTOSYNTHESIS AND BIOPHOTOLYSIS

Microalgal metabolism is essential for photosynthesis when an adaptation to heterotrophic cultivation has not been introduced. Photosynthesis can be defined as the conversion of solar energy into biochemical substances such as proteins, carbohydrates, and lipids, with O_2

as a byproduct. Photosynthesis is a complex redox reaction starting in the thylakoid membranes of the chloroplast, when light is captured by the PSII reaction center. In cyanobacteria, photosynthesis occurs in the photosynthetic membrane (Masojidek et al., 2013) in the cytoplasm. The photosynthetic apparatus transfers electrons from PSII to PQ pool, Cytb6f, Photosystem I (PSI), and ferrodoxin, respectively. Finally, CO_2 is fixed in the form of starch (in microalgae) or glycogen (in cyanobacteria) in the Calvin—Benson cycle (Srirangan et al., 2011).

Photobiological hydrogen is generated via a pathway similar to photosynthesis. However, because of strict anaerobiosis, the final ferrodoxin donates electrons to hydrogenase enzymes. Photobiological hydrogen production faces its very first challenge in the initial phase of photosynthesis (Benemann, 1997): light conversion efficiency is very low in natural conditions. A significant enhancement can be done under controlled laboratory environments, although these results are still far less than the theoretical limit of 13.4% (Torzillo and Seibert, 2013).

Biophotolysis is the initial step of microalgal biohydrogen production. The very basic definition of biophotolysis is the process of light-dependent water splitting at photosynthetic reaction centers. This process is most promising in terms of sustainability because the required components are only water and light, which are thought to be convenient, accessible, and cost-effective sources (Carrieri et al., 2011; Kim and Kim, 2011; Oh et al., 2011). Biophotolysis is divided into two major groups according to the electron donor: direct biophotolysis and indirect biophotolysis.

4.1 Direct Production of Biohydrogen (Direct Biophotolysis)

Direct biophotolysis is thought to be the most promising pathway for sustainable hydrogen production. The reaction itself is independent from any organic compounds, only requiring water and light to supply electrons and protons in the hydrogen production pathway (Hallenback et al., 2012). Light is the reaction initiator; when it reaches PSII, water splits into its protons and electrons and oxygen. Step-by-step oxidation and reduction reactions occur until the electrons reach hydrogen-catalyzing enzymes. Electrons are transferred from PSII to the PQ pool, with a larger PQ pool resulting in higher electron flow, then PSI and finally ferrodoxin as the final electron donor to hydrogenases. This flow occurs in microalgal [FeFe] hydrogenase-mediated reactions; cyanobacteria also use nitrogenase enzymes or [NiFe] hydrogenases. Direct biophotolysis is limited by oxygen sensitivity; also, nitrogenases require higher amounts of ATP, with a reduction of NAD or NADP required in the case of [NiFe] hydrogenases (Melis and Happe, 2001; Mathews and Wang, 2009; Antal et al., 2011; Srirangan et al., 2011; Masojidek et al., 2013).

4.2 Indirect Production of Biohydrogen

Indirect biophotolysis requires starch or glycogen reserves for hydrogen production. The advantage of indirect biophotolysis is the ability to use other carbon sources to provide protons and electrons for hydrogen generation (Mathews and Wang, 2009). Endogenous sources are used via anaerobic fermentation, resulting in CO_2 production as a byproduct rather than O_2 as in direct biophotolysis. Indirect biophotolysis mechanism is studied mostly in cyanobacteria (Carrieri et al., 2011). Indirect biophotolysis is promising in terms of avoiding oxygen, but it is still a challenging pathway for microalgae.

4.2.1 Two-Stage Hydrogen Production: Sulfur Deprivation Approach for Sustainable Biohydrogen Production

Two-stage hydrogen production in green microalgae is a breakthrough in the development of sustainable hydrogen production methods

(Melis et al., 2000), the very first step in the road toward commercialization. The method covers both aerobic and anaerobic production and also points out some certain metabolic essentials related to microalgal biohydrogen production, which is a highly conserved metabolic pathway among biohydrogen-producing species.

Before the introduction of two-stage protocol, Wykoff et al. (1998) found that when sulfur is deprived in culture, the photosynthetic activity is also decreased. The amino acid sequences of the D1 protein in the PSII photodamage repair system consist of sulfur-containing amino acids; sulfur-deprived cultures also showed less D1 synthesis. Melis et al. (2000) separated aerobic and anaerobic stages to produce biomass and biohydrogen, respectively. Cells were cultivated in a sulfur-containing medium in aerobic conditions but transferred into sulfur-deprived media, and sulfur-containing substances were changed with counterparts. When sulfur is deprived under anaerobic conditions, photosynthetic activity falls behind respiration. When complete anaerobiosis is achieved, hydrogenase enzymes are also activated. The solution comes with a problem: water splitting provides protons and electrons for hydrogenases, but oxygen is also generated as a byproduct. The accumulation of metabolic oxygen blocks the hydrogenase activity. In addition, when D1 protein synthesis is decreased, the PSII reaction center cannot be fixed as it is in photosynthesis. The activity of PSII is also decreased, along with the decreased proton and electron flow to hydrogenases (Faraloni and Torzillo, 2010).

A two-stage protocol is a way to sustain biohydrogen and pave the way for further studies and developments; however, there are some challenges that cannot be avoided physically. Metabolic bottlenecks include the oxygen sensitivity of hydrogenases, low light conversion efficiency, electron loss at cyclic electron transfer around PSI, low hydrogen production efficiency, and a limited number of species with hydrogen-producing nature (Melis and Happe, 2001; Antal

and Lindblad, 2005; Kosourov et al., 2007). In addition, process-related problems include expensive photobiorector design, inadequate know-how and technology to produce and storage hydrogen, and scale-up issues with the two-stage protocol (Srirangan et al., 2011; Gianelli and Torzillo, 2012; Oncel and Sabankay, 2012; Oncel, 2013). The metabolic issues may be overcome with genetic engineering tools. Genome studies have attempted to modify the D1 activity of wild-type strains (Faraloni and Torzillo, 2010; Kose and Oncel, 2014). These studies showed that mutations in the D1 protein have a positive effect of obtaining higher yields of hydrogen for a longer period of time, but oxygen sensitivity remains a problem.

5. METABOLIC PATHWAY ENGINEERING AS A TOOL TO DEVELOP HYDROGEN-PRODUCING CELL FACTORIES: IMPROVEMENT OF H_2 PRODUCTION BY GENETIC OPTIMIZATION OF MICROALGAE

Metabolic engineering has been used mainly to enhance the desired properties of a certain component or target product, introduce new features to an organism, or adapt new techniques for further strain development (Leon-Banares et al., 2004; Gimpel et al., 2013). In the case of biohydrogen production, an array of metabolic engineering techniques are being studied to avoid the main challenges of biohydrogen (Wu et al., 2011; Posewitz et al., 2005; Godaux et al., 2013).

Green microalgae are suitable platforms for genetic engineering, mutations, and metabolic pathway engineering. Their diverse genetic material (nucleus, mitochondria, and chloroplast) offers a reliable base for those who wants to control the genes or gene products (Rosenberg et al., 2008). Microalgae have various applications in genetic engineering, with perhaps the most sophisticated one being the production of vaccines using microalgae as a host organism (Mayfield and Franklin, 2005).

New tools (Martens and Liese, 2004) and genome sequencing for certain microalgal or cyanobacterial strains have been introduced in genetic engineering studies. Systems biology and omics (genomics, transcriptomics, proteomics, metabolomics) may be used to study the integration of essential pathways, regulation metabolism, diversity, and new characteristics (Matthew et al., 2009; Nguyen et al., 2008; Rupprecth, 2009). To develop an efficient metabolism to sustain microalgal biohydrogen production, challenges must be addressed. A detailed list of genetic studies is presented in Table 1.

5.1 *Chlamydomonas reinhardtii* as a Model Organism for H_2 Production

Microalgal hydrogen production has been studied for many years (Gaffron and Rubin, 1942; Melis, 2007; Oncel, 2013). In recent years, it gained more attention when prolonged H_2 production was achieved with comparison to theoritical efficiency; 0.1% light conversion efficiency to hydrogen has been achieved under sulfur deprived conditions (Melis et al., 2000). *C. reinhardtii* was used for this very promising study, and it also became the model for biohydrogen studies. *C. reinhardtii* has been known for long time with regard to its photoheterotrophic and heterotrophic growth characteristics. It has been used in several genomic studies (Mayfield and Franklin, 2005; Rosenberg et al., 2008). Its chloroplast, mitochondrial, and nuclear genomes have been sequenced (Radokovits et al., 2010). Moreover, both its chloroplast and nuclear DNA are easily transformed, enabling genetic modifications to obtain strains with enhanced H_2 production capabilities (Kruse and Hankamer, 2010).

5.1.1 Metabolism of C. reinhardtii during Sulfur Deprivation

Nutrient deprivation, particularly of sulfur, is known to promote the downregulation of photosynthetic O_2 production, establishing anaerobic conditions in the cultures (Wykoff et al., 1998). In *C. reinhardtii*, sulfur starvation may induce a decrease in photosynthetic activity to 75% of the initial value within 24 h and create anaerobic conditions. The reason is that under sulfur starvation and light exposure, the repair cycle is blocked due to lack of sulfur; the cells are not able to resynthesize proteins, particularly the D1 protein, which is associated with PSII and is important for its functionality. The role of the D1 protein in biohydrogen production is critical. The D1 protein synthesis is related to oxygen evolution, which is the inhibitor for hydrogenase enzymes. Recent investigations are related to the role of D1 protein activity with a special approach: obtaining D1 mutant strains to see the effect of mutations on biohydrogen production (Faraloni and Torzillo, 2010). Will it be helpful to sustain biohydrogen production and level up?

During the phase in which the O_2 evolution is present (aerobic phase), starch accumulation in *C. reinhardtii* can be detected, which represents a storage of excess reducing power that is degraded in parallel to H_2 production (Melis et al., 2000; Melis, 2007). The starch metabolism plays a central role in H_2 production, as its catabolism sustains the electron transport chain to hydrogenase. In particular, during the aerobic phase, starch is accumulated and subsequently degraded under anaerobic conditions, sustaining the H_2 production (Melis et al., 2000; Torzillo and Siebert, 2013). It has been proposed that starch catabolism sustains respiration in mitochondria, maintains anoxic environment for the PSII-dependent direct pathway (Melis, 2007), and supplies electrons to the chlororespiratory pathway for the activation of hydrogenase enzymes by the electron transport chain in chloroplasts via an indirect pathway using PSI (Fouchard et al., 2005; Mus et al., 2005; Melis, 2007).

Chochois et al. (2009) concluded that starch breakdown feeds electrons to the PQ pool. Acetate in the culture medium is used by microalgae as a carbon source during the aerobic phase and results in the biomass accumulation. There is an

TABLE 1 Key Studies in Genetic Engineering for Microalgal Biohydrogen Production

Aim	Method	Results	Contributions	References
CARBON METABOLISM				
Trophic conversion	Gene insertion *to* C. *reinhardtii* stm6 strain HUP1 gene from *Chlorella kessleri*	• Stm6Glc4 mutant cell growth observed under dark conditions supplemented with glucose • Exogenous glucose utilization to generate biohydrogen (150% increase)	• Various carbon sources can be used as substrates for biohydrogen production after successful transformation • Biohydrogen production under dark conditions • A new approach for sustainable and feasible hydrogen production	Doebbe et al. (2007)
Increase the amount of the storage material	Random mutagenesis	• Starch over accumulation, increase in the cellular respiration rate, inhibition of cyclic electron transfer around PSI, increase in the hydrogen production • The mutant is named as stm6 (state transition mutant)	• Role of endogenous substrate on biohydrogen production	Kruse et al. (2005)
Action of starch assimilation related to biohydrogen production	Two-stage culture	• Sta mutants produced less hydrogen than the wild type • Sta6 starchless mutant produced hydrogen • Addition of DCMU had no effect on light activation of hydrogenase • Sta mutant with DCMU addition had no hydrogen production	• Direct and indirect biophotolysis is the electron source for biohydrogen production • Biohydrogen can be produced via direct biophotolysis in the absence of starch • PSII is a must to sustain biohydrogen production	Chochois et al. (2009)
	Amino acid substitute mutant strains (Y67A, Y68A, Y72A) cultivated according to the two-stage protocol	• Increased O_2 evolution (Y68A, highest) • Increase in Rubisco followed by a rapid decrease • Y67A showed 15- to 16-fold greater hydrogen production than wild-type	• The effects of Rubisco on hydrogen production and the relation with sulfur nutrition are introduced • Rubisco mutation could be a new approach to observe biohydrogen production without nutrient alteration	Esquível et al. (2006), Pinto et al. (2013)

PHOTOSYNTHESIS AND BIOPHOTOLYSIS

Increase the photosynthetic efficiency with regulating LHC size	• DNA insertion mutagenesis	• Smaller size in the LHC-I and LHC-II complexes • Increase in the Chl a/Chl b ratio (from 2.6 to 6) • Decrease in the total chlorophyll/cell	• Increase in the light utilization efficiency could have a significant role in the PSII activity • Ability to work under high-light conditions (outdoor experiments)	Tetali et al. (2007)
Understand the electron transfer and its role in microalgal biohydrogen production metabolism	• PSII-deficient strain (FuD7) and Rubisco deficient strain (CC-2803) cultivated according to two-stage protocol	• FuD7 activated *HydA1* genes but no further H$_2$ production occurred • Strain CC-2803 showed decreased photosynthetic efficiency but hydrogen is produced after a fast anaerobiosis	• Effect of the electron supply sources for biohydrogen production	Hemschemeier et al. (2008)
Decrease the D1 protein activity	• Amino acid deletions in D1 protein	• Reduced amount of chlorophyll content/dry weight, duration of aerobic phase • Increase in the respiration/photosynthesis ratio, starch accumulation, light to hydrogen conversion efficiency (3.2%) total hydrogen produced • Prolonged H$_2$ production • Effective mutants are D239–40, D240–41, D240	• Role of D1 protein amino acid sequences in biohydrogen production is highlighted • Introduction of new mutant strains with elevated biohydrogen production capacity	Faraloni and Torzillo (2010) and Oncel and Kose (2014)
Decrease the PSII repair activity of D1 protein	• Antisense transformation of sulfate uptake genes *SulP*/*SulP2*	• Decrease in the *SulP* synthesis • Impaired sulfate uptake • Decrease the O$_2$ evolution rate • Hydrogen production under anaerobic sulfur-containing conditions	• Sulfate is the key nutrient for D1 synthesis • Decrease in the *SulP* synthesis could be a key for nutritional modification of biohydrogen production from microalgae	Schroda et al. (1999), Melis and Chen (2005) and Chen et al. (2005)
Drop PSII activity below respiration	• Chloroplast expression system (copper sensitive cyc6 promotor)	• Inhibition in the PSII synthesis • Strict anaerobiosis is required for the expression system; the results were not satisfied	• The role of PSII and the importance of anaerobic environment have been established	Surzycki et al. (2007)

Continued

TABLE 1 Key Studies in Genetic Engineering for Microalgal Biohydrogen Production—cont'd

Aim	Method	Results	Contributions	References
Increasing the photochemical utilization of light	• RNAi approach to down regulate LHC gene family	• *Stm3LR3* had significantly reduced levels of LHCI and LHCII mRNAs and proteins • Reduced levels of fluorescence, sensitivity to photoinhibition • Higher photosynthetic quantum yield • Increased efficiency of cell cultivation	• Mutations in LHC gene family can be used to cultivate cells under high-light conditions • A route for outdoor cultures • Improved light penetration properties and potentially reduced risk of oxidative photodamage of PSII for enhanced biohydrogen production	Mussgnug et al. (2007)
HYDROGENASE ENZYMES				
To overcome oxygen sensitivity	• Rational mutagenesis (in silico mutagenesis and volumetric oxygen accessibility maps) • Gene shuffling, DNA extraction	• Mutant strain L283W 29% increased in hydrogen production; hydrogen production is decreased 90% in other strains • Redox properties of H cluster is sequenced. Fe—S bonding sequence is highlighted. Conserved genetic structure is observed	• Oxygen diffusion channels and their effect on the globular hydrogenase enzyme structure are highlighted • Introduction of recombinant hydrogenase library to obtain O₂-tolerant enzymes	Ghirardi et al. (2006), Nagy et al. (2007), Boyd et al. (2009) and Meyer (2007)
Identify the role of hydrogenase coding genes in hydrogen production	• Artificial mRNA silencing	• Downregulation of *HydA1* (four fold lower activity), *HydA2* and Hydrogenase-like protein (*Hyd3*) • *HydA2* and *Hyd3* mutations have no significant effect on hydrogen production	The regulation mechanism of hydrogenase genes under hypoxia	Godman et al. (2010)
Role of Hyd genes in photobiological biohydrogen production	• Insertional mutagenesis of psL72 vector to *HydA2* and *HydA1* genes	• Hyd enzymes capable of fermentative hydrogen production • *HydA1* is the most abundant gene responsible for hydrogen production	*HydA1* is the dominant gene in terms of hydrogen production	Meuser et al. (2012)

obvious pH increase in the culture during the aerobic phase; a shift to anaerobiosis results in a decrease in the pH value (increase from 7.4 to 8.5 in aerobic phase and decrease to 7.9 in anaerobiosis). The changes in the pH value demonstrate the metabolic pathway of the organism itself: during the aerobic phase, cells use acetate, whereas during anaerobiosis, respiration releases CO_2 to the culture media.

When the O_2 evolution balances with the respiration rate (Melis et al., 2000; Melis, 2007), this condition is fundamental for the synthesis and activity of hydrogenase in *C. reinhardtii*. Under these conditions, it can produce hydrogen for some days, reversibly. When anaerobiosis is attained in the cultures, a lag time is needed to induce H_2 production (i.e., lag phase), which is the time necessary to synthesize and activate the hydrogenase enzyme. Hydrogenase enzymes catalyze a reduction reaction (H^+ ions to H_2); thus, the redox potential changes from positive to negative, up to -550 mV. This phenomenon may be related to the fact that the hydrogen production pathway uses required components from a pool that is established during the aerobic phase, making it possible to partially decrease the degree of reduction of the plastoquinone.

In the following H_2-production phase, both the PSII-dependent (due to the residual PSII activity) and PSII-independent pathways act as major sources of electrons. It has been demonstrated that the PSII contribution to the process is dominant with respect to starch mobilization (Ghirardi et al., 2000), providing up to 80% of the supplied reducing power (Kruse et al., 2005; Scoma et al., 2012; Volgusheva et al., 2013). After a period of 2–3 days, depending on the strain, the damage on PSII becomes consistent, no photosynthetic activity can be detected, the starch is degraded, and the H_2 production stops.

5.1.2 Screening of C. reinhardtii Mutant Strains for H_2 Production

An important approach to increase overall H_2 production has focused on *C. reinhardtii* mutant strains with upgraded hydrogen production properties. Increased H_2 production has been reported in D1 protein mutants (Posewitz et al., 2005; Kruse et al., 2005; Mathews and Wang, 2009; Torzillo et al., 2009; Faraloni and Torzillo, 2010). D1 protein is a crucial component of the photosynthetic electron transport chain (Tolletter et al., 2011), acting at the reducing side of photosystem II (Edelma and Mattoo, 2008). The changes in D1 protein activity and functionality, and their implication in the H_2 production process, in *C. reinhardtii* have attracted much attention after the introduction of sulfur starvation and the downregulation of PSII activity, which cause the reduction in the oxygen evolution rate (Wykoff et al., 1998; Takahashi et al., 2001). However, it is not adequate to establish a compatible amount of biohydrogen in comparison of even among other biofuel sources. Biohydrogen as a direct metabolic product should be enhanced via genetic engineering.

The D1 protein is a light-dependent protein with the highest turnover rate (Edelman et al., 1984). In particular, the Q_B binding site of the D1 reaction center protein is thought to be essential to electron flow through the photosynthetic chain, mainly related to PSII (Vermaas and Ikeuchi, 1991); also, it is the mechanism responsible for photodamage (Ohad et al., 1994).

This site is located within a stromal hydrophilic loop between transmembrane helices IV and V. Known as the D–E loop, it contains an amino acid sequence that is conserved among cyanobacteria, algae, and higher plants (Trebst, 1987; Sobolev and Edelman, 1995). Moreover, this amino acid region is involved in the rapid turnover of the D1 protein (Kettunen et al.,

1996); thus, this photobiological hydrogen production site is thought to be a critical observation in the PSII-dependent pathway that microalgae use (Kless et al., 1994; Nixon et al., 1995; Mäenpää et al., 1995).

For the study of H_2 production by D1 protein mutant strains, mutants were obtained from wild-type (11/32b) genetic manipulation, as previously described (Johanningmeier and Heiss, 1993). D1 protein mutation of different regions of D1 protein involved various functions of the D1, promoting diverse changes in the phenotypic characteristics and photosynthetic parameters of the strains. The mutations influenced the yield of the hydrogen production in a different manner (Faraloni and Torzillo, 2010). In most of the strains, the mutated D1 protein conferred the capability for a higher H_2 production than in the wild-type strain. Only few exceptions that did not produce any hydrogen were observed.

During H_2 production, several biochemical pathways operate together; although the total H_2 outputs resulted in a wide range, some of the most productive strains showed certain common features. In particular, they showed a reduced amount of chlorophyll/dry weight and chlorophyll/cell ratios, high photosynthesis and respiration rates, high starch accumulation, and high synthesis of xanthophyll cycle pigments. Moreover, the induction of the synthesis of lutein was observed to be higher in these strains than in the wild type. Some examples of the most productive strains are reported in Table 2.

In particular, phenotypic characteristics of two amino-acid deleted mutants, D240 and D239-40, and a mutant with two amino-acid substitutions, L159I/N230Y, are summarized in Table 3 and compared with the wild type. The deletion of the D240 and D239-40 is located in the D1 protein region involved in the binding of Q_B and D1 degradation, while the mutation of the other mutant strain involves a region implicated in the electron donor capacity to the oxygen evolving complex (OEC).

These strains exhibited phenotypic characteristics that are considered to be important for the improvement of the H_2 production process. In particular, all the mutants showed an amount of chlorophyll per dry weight biomass that was 36–56% lower than in the wild type, which is a useful requisite because the cultures can grow under relatively high cell densities, improving photosynthesis efficiency by reducing the chlorophyll antenna size or the number of light-harvesting complexes (Polle et al., 2002; Beckmann et al., 2009; Donald

TABLE 2   Some Phenotypic Characteristics of the Most Productive D1 Protein Mutant Strains, Compared to the Wild Type[1]

Strain	D1 region affected by mutation	Chl % of dry weight	Chl/cell (µg 10^{-6})	Respiration (R) (µmol O_2 mg^{-1} Chl h^{-1})	P_{max} (µmol O_2 mg^{-1} Chl h^{-1})	R/P ratio %
Wild-type	—	3.63 (±0.51)	3.57 (±0.12)	72 (±0.25)	260 (±2.8)	27.70 (±0.21)
D240	D1 degradation	1.60 (±0.18)	1.56 (±0.05)	95 (±5.8)	220 (±25.0)	43.18 (±2.04)
D239-40	D1 degradation	2.33 (±0.32)	2.08 (±0.09)	78 (±1.75)	119 (±2.0)	65.55 (±0.36)
L159I/N230Y	Q_B interaction	1.80 (±0.20)	3.32 (±0.14)	190 (±2.5)	487 (±36.0)	39.01 (±2.20)

[1] Chlorophyll content and cellular parameters in Chlamydomonas reinhardtii wild-type and mutants D240 and D239-40. Maximum photosynthesis and respiration rates measured during the logarithmic phase of growth. P_{max}: maximum rate (oxygen evolution plus dark respiration) of light saturation photosynthetic oxygen evolution; R/P: respiration rate versus oxygen evolution rate.

TABLE 3 H_2 Production Parameters and Changes in Pigment Composition after the H_2 Production of the Most Productive D1 Protein Mutant Strains, Compared to the Wild Type[1]

Strain	Aerobic phase (h)	Lag phase (h)	Production time (h)	H_2 total volume (ml l^{-1})	H_2 production rate)(ml l^{-1} h^{-1}	Pigment composition at the end of H_2 production			
						V	A	Z	L
Wild-type	34 (±1)	16 (±5)	55 (±4)	26 (±5)	0.47 (±0.11)	13.06 (±0.22)	8.87 (±0.73)	77.20 (±3.63)	191.41 (±1.47)
D240	2 (±1)	26 (±4)	207 (±39)	318 (±23)	1.54 (±0.31)	16.12 (±8.03)	4.62 (±2.26)	108.91 (±13.06)	323.64 (±5.31)
D239-40	3 (±2)	30 (±1)	183 (±70)	475 (±50)	2.60 (±0.18)	23.73 (±0.93)	12.66 (±1.92)	103.31 (±14.46)	288.84 (±12.45)

[1] The pigment content is reported as mmol mol^{-1} Chl a.; V: violaxanthin; A: antheraxanthin; Z: zeaxanthin; L: lutein.

et al., 2011). With these cells under sulfur starvation, it is possible to work with denser cultures, which can reach anaerobiosis faster.

Quantification of the photosynthetic parameters revealed that the D1 protein mutation conferred different properties, as deleted mutants exhibited lower oxygen evolution rates and higher respiration rates than those for the wild type. By contrast, double amino acid substitution mutant L159I-N230Y led to increased oxygen evolution and respiration rates, which were 87% and 164% higher than in the wild type, respectively. An increase in the respiration-to-photosynthesis ratio is a very useful peculiarity in *C. reinhardtii* strains for the production of hydrogen. This helps the culture to shift to anaerobiosis in a shorter period of time, which means that the lag phase is decreased. In the cultures of the best producing mutant strains, H_2 production was 12—19 times greater than in the wild type; these strains exhibited better performance in overall biohydrogen production, as well as increased biohydrogen production rates.

A very interesting strain was found to be the D239-40 mutant. Among D1 protein mutant strains, D239-40 has a higher amount of starch, which is required for survival. However, the hydrogen production is not only related to starch catabolism; PSII activity is another critical effect.

The results indicated that D239-40 has higher fluorescence activity, which can be counted as an increase in the number of electrons from light-dependent water splitting being transferred to certain hydrogenase enzymes. Another important feature of the D1 mutants is the relatively high level of fluorescence yield (0.300) compared to the wild type (below 0.1) during the H_2 production phase. This may be related to a higher xanthophyll cycle pool found in these strains, which allows greater protection of PSII from photo-inactivation (Faraloni and Torzillo, 2010).

The findings suggest that genetic engineering is a reliable tool to achieve sustainable and prolonged biohydrogen production, which may help to introduce technoeconomic and applicable production. The diversity of metabolic pathways indeed contributes to the development of new insights. In particular, the D1 protein mutations could provide different phenotypes to the mutated strains. Altering the photosynthetic performance may alter the rate of oxygen evolution, thereby ameliorating the performance in terms of hydrogen production. It has been shown that mutations at level of the Q_B binding and the OEC interaction site of the D1 protein are important in improving H_2 productivity. Moreover, acting at the level of metabolism of carbohydrates is a crucial step, allowing greater accumulation of starches from the medium and changing the metabolism.

Apart from the route of biohydrogen in microalgal metabolism, the main problem is the light conversion efficiency in biohydrogen production, which is still less than 1% under controlled laboratory conditions. Using mutants in properly designed photobioreactors (PBRs) (3.22% of Photosynthetic Active Radiation (PAR), Scoma et al., 2012a), immobilizing algae (1.53% of PAR, Kosourov and Seibert, 2009), or controlling partial gas pressure within the culture chamber (1.65% of PAR, Kosourov et al., 2012) may enhance the light-to-fuel efficiency. However, these methods are applied in laboratory conditions; results in outdoor environments will be far below these conditions (0.055% of solar light, Scoma et al., 2012b).

5.1.3 Chlorophyll Fluorescence Changes during Sulfur Starvation

The chlorophyll fluorescence measurement is considered to be a useful tool to monitor photosynthetic activity. It is commonly used to evaluate the effect of different kinds of stresses on the photosynthetic apparatus. During sulfur starvation, PSII activity decreases, resulting in protein degradation and the inability to resynthesize new protein; hence, chlorophyll fluorescence can be used to measure the changes in PSII maximum and effective quantum yields, both during the aerobic phase under sulfur starvation and in the occurrence of anaerobiosis and hydrogen production.

A typical example of the kinetics of chlorophyll fluorescence during sulfur starvation has been reported for some mutant strains and the wild type, as shown in Figure 4. As the oxygen content decreases, a reduction in chlorophyll fluorescence, measured as effective PSII quantum yield, $\Delta F/F'_m$, can be detected. $\Delta F/F'_m$ drops from 0.6 to 0.2. The rapid drop in fluorescence is followed by hydrogen production, which was first reported by Antal et al. (2003).

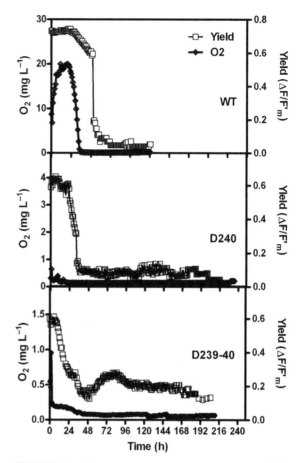

FIGURE 4   Fluorescence and oxygen changes of D1 mutant strains D239-40 and D240 with time. Measurements of the effective quantum yield of PSII ($\Delta F/F'_m$) and dissolved oxygen (pO$_2$) are shown. *Chlamydomonas reinhardtii* wild-type and mutant strains D240 and D239-40 are compared during incubation under sulfur starvation for hydrogen production.

The transition from state 1 to state 2 has photoprotective properties, such as the partial migration of the LHCII from PSII to PSI, which is known to occur in *C. reinhardtii* with the migration of more than 85% of LHCII (Endo and Asada, 1996; Finazzi et al., 1999; Cournac et al., 2002; Cardol et al., 2003).

As soon as the anaerobic condition is attained, hydrogenase is expressed and becomes active. This may induce a partial reincrease in DF/F'_m, which is very accentuated in the culture of the strain D239-40 as a consequence of the hydrogenase activity. This partially oxidizes photosynthetic carriers, including the PQ pool, thereby establishing the linear electron transfer from PSII and hence the primary photochemical yield. The measurement of chlorophyll induction kinetics is a complementary technique to investigate nonphotochemical reduction of the PQ pool. It is generally accepted that chlorophyll (Chl) a fluorescence kinetics reflect the progressive reduction of the photosynthetic electron transport chain (Tóth et al., 2007)—that is, the reduction of Q_A to Q_A^-, and the reduction of the PQ pool.

The effect of sulfur starvation on the PSII reaction centers has been studied. Wykoff et al. (1998) reported the formation of PSII Q_B nonreducing centers and consequently a limitation of the rapid electron transfer from Q_A^- to Q_B. This retardation may induce further PSII damage, with a reduction in the number of functional PSII reaction centers. The measurement of the rate of closure of the PSII reaction centers has been determined by changes in the Mo parameter, which is a measurement of the slope of the transient curve (Wykoff et al., 1998; Antal et al., 2006, 2007).

The use of *C. reinhardtii* strains with a reduced constitutive capability to transfer electrons further than Q_A has been found to be important for improving H_2 production under sulfur starvation (Faraloni and Torzillo, 2010). It has been shown that D1 protein mutant strains, which exhibited transient curves with increased Mo and V_J parameters with respect to the wild-type strain, were able to achieve a higher H_2 production performance (Figure 5). In particular, these phenotypic characteristics were translated, under sulfur starvation, into an earlier decrease of photosynthetic activity than in the wild-type. This induced anaerobiosis in a very short time (2 h), whereas the wild-type reached anaerobic conditions in 34 h (Faraloni and Torzillo, 2010).

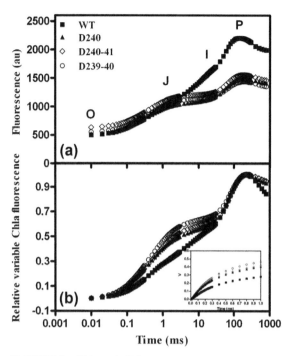

FIGURE 5 Chlorophyll fluorescence changes of D1 protein mutant strains with time. Shown is a comparison of the kinetic rise in Chl a fluorescence (a) and relative variable Chl a fluorescence (b). $V_t = (F_t - F_0)/(F_m - F_0)$ in *C. reinhardtii*: WT (■), D240 (▲), D239-40 (▼), and D240-41, D1 (○) protein mutant strains, measured in photomixotrophically grown cultures. The inset shows the initial rise in V_t (from 50 μs to 1 ms) of the curves on a linear scale.

6. CONCLUSIONS

Microalgal biohydrogen production has a sophisticated metabolism, regarding both the regulation of the photosynthesis pathway as a result of anaerobiosis and the enzymes that catalyze the metabolic hydrogen evaluation. Thus, biohydrogen production is a trending topic in the urgent area of renewable fuel sources. Using a two-stage protocol, microalgal biohydrogen production can be studied in laboratory conditions, with an aim to reach commercial scales.

After the two-step strategy, studies investigated environmental conditions. The main challenge for the future is to obtain a direct H_2-conversion process. This goal may be reached by providing mutants with O_2-tolerant hydrogenase, a truncated antenna size, and inhibited or reduced pathways of competitors for electron sources (Marín-Navarro et al., 2010). The utilization of engineered strains may improve biohydrogen production in outdoor cultures, with the direct utilization of solar light and CO_2 to improve the efficiency of the process.

The ultimate goal of microalgal biohydrogen production is to establish one-step biohydrogen production with an oxygen-tolerant species. However, the taxonomic studies of microalgal hydrogenase are known to be really conserved among the species; therefore, finding a naturally oxygen-tolerant hydrogen evolving species will be challenging. The sophisticated tools of biotechnology and bioengineering could combine with genomics and metabolism to discover the desired species. Genetic manipulations may be helpful in designing hydrogen-producing machinery concepts that highlight the metabolic essentials.

References

Allakhverdiev, S.I., Thavasi, V., Kreslavski, V.D., Zharmukhamedov, S.K., Klimov, V.V., Ramakrishna, S., Los, D.A., Mimuro, M., Nishihara, H., Carpentier, R., 2010. Photosynthetic hydrogen production. J. Photochem. Photobiol. C 11, 101–113.

Antal, T., Lindblad, P., 2005. Production of H_2 by sulphur-deprived cells of the unicellular cyanobacteria *Gloeocapsa alpicola* and *Synechocystis* sp. PCC 6803 during dark incubation with methane or at various extracellular pH. J. Appl. Microbiol. 98, 114–120.

Antal, T.K., Krendeleva, T.E., Laurinavichene, T.V., Makarova, V.V., Ghirardi, M.L., Rubin, A.B., Tsygangov, A.A., Seibert, M., 2003. The dependence of algal H_2 production on photosystem II and O_2 consumption in sulfur-deprived *Chlamydomonas reinhardtii* cells. BBA 1607, 153–160.

Antal, T.K., Krendeleva, T.E., Rubin, A.B., 2007. Study of photosystem 2 heterogeneity in the sulfur-deficient green alga *Chlamydomonas reinhardtii*. Photosynth. Res. 94 (1), 13–22.

Antal, T.K., Krendeleva, T.E., Rubin, A.B., 2011. Acclimation of green algae to sulfur deficiency: underlying mechanisms and application for hydrogen production. Appl. Microbiol. Biotechnol. 89, 3–15.

Antal, T.K., Volgusheva, A.A., Kukarskikh, G.P., Krendeleva, T.E., Tusov, V.B., Rubin, A.B., 2006. Examination of chlorophyll fluorescence in sulfur-deprived cells of *Chlamydomonas reinhardtii*. Biofizika 51 (2), 292–298.

Beckmann, J., Lehr, F., Finazzi, G., Hankamer, B., Posten, C., Wobbe, L., Kruse, O., 2009. Improvement of light to biomass conversion by de-regulation of light-harvesting protein translation in *Chlamydomonas reinhardtii*. J. Biotechnol. 142, 70–77.

Beer, L.L., Boyd, E.S., Peters, J.W., Posewitz, M.C., 2009. Engineering algae for biohydrogen and biofuel production. Curr. Opin. Biotechnol. 20, 264–271.

Benemann, J.R., 1997. Feasibility analysis of photobiological hydrogen production. Int. J. Hydrogen Energ. 22 (10/11), 979–987.

Böck, A., King, P.W., Blokesch, M., Posewitz, M.C., 2006. Maturation of hydrogenases. Adv. Microb. Physiol. 51, 1–72.

Boyd, E.S., Spear, R.J., Peters, J.W., 2009. [FeFe] hydrogenase genetic diversity provides insight into molecular adaptation in a saline microbial mat community. Appl. Environ. Microbiol. 75 (13), 4620–4623.

Boyer, M.E., Stapleton, J.A., Kuchenreuther, J.M., Wang, C., Swartz, J.R., 2008. Cell-free synthesis and maturation of [FeFe] hydrogenases. Biotechnology and Bioengineering 99 (1), 59–67.

Cardol, P., Gloire, G., Havaux, M., Remacle, C., Matagne, R., Franck, F., 2003. Photosynthesis and state transitions in mitochondrial mutants of *Chlamydomonas reinhardtii* affected in respiration. Plant Physiol. 133, 2010–2020.

Carrieri, D., Wawrousek, K., Eckert, C., Yu, J., Maness, P.-J., 2011. The role of bidirectional hydrogenases in cyanobacteria. Bioresour. Technol. 102, 8368–8377.

Chen, H.-C., Newton, J.A., Melis, A., 2005. Role of SulP, a nuclear-encoded chloroplast sulfate permease, in sulfate transport and H_2 evolution in *Chlamydomonas reinhardtii*. Photosynth. Res. 84, 289–296.

Chochois, V., Dauvillée, D., Beyly, A., Tolleter, D., Cuiné, S., Timpano, H., Ball, S., Cournac, L., Peltier, G., 2009. Hydrogen Production in *Chlamydomonas*, Photosystem II-dependent and-independent pathways differ in their requirement for starch metabolism. Plant Physiol. 151, 631–640.

Cournac, L., Latouche, G., Cerovic, Z., Redding, K., Ravenel, J., Peltier, G., 2002. In vivo interactions between photosynthesis, mitorespiration, and chlororespiration in *Chlamydomonas reinhardtii*. Plant Physiol. 129, 1921–1928.

Doebbe, A., Rupprecht, J., Beckmann, J., Mussgnug, J.H., Hallmann, A., Hankamer, B., Kruse, O., 2007. Functional integration of the HUP1 hexose symporter gene into the genome of *C. reinhardtii*, impacts on biological H_2 production. J. Biotechnol. 131 (1), 27−33.

Donald, R.O., Xinguang, Z., Melis, A., 2011. Optimizing antenna size to maximize photosynthetic efficiency. Plant Physiol. 155 (1), 79−85.

Edelman, M., Mattoo, A.K., 2008. D1-protein dynamics in photosystem II, the lingering enigma. Photosynth. Res. 98, 609−620.

Edelman, M., Mattoo, A.K., Marder, J.B., 1984. Three hats of the rapidly metabolized 32 kD protein thylakoids. In: Ellis, R.T. (Ed.), Chloroplast Biogenesis. Cambridge University Press, Cambridge, pp. 283−302.

Endo, T., Asada, K., 1996. Dark induction of the non-photochemical quenching of chlorophyll fluorescence by acetate in *Chlamydomonas reinhardtii*. Plant Cell Physiol. 37 (4), 551−555.

Esquivel, M.G., Pinto, T.S., Matin-Navarro, J., Moreno, J., 2006. Substitution of Tyrosine Residues at the Aromatic Cluster around the ßA-ßB Loop of Rubisco Small Subunit Affects the Structural Stability of the Enzyme and the in Vivo Degradation under Stress Conditions. Biochemistry 45, 5745−5753.

Faraloni, C., Torzillo, G., 2010. Phenotypic characterization and hydrogen production in *Chlamydomonas reinhardtii* Q_B binding D1 protein mutants under sulfur starvation, changes in chlorophyll fluorescence and pigment composition. J. Phycol. 46, 788−799.

Finazzi, G., Furia, A., Barbagallo, R.M., Forti, G., 1999. State transitions, cyclic and linear transport and photophorylation in *Chlamydomonas reinhardtii*. BBA 1413, 117−129.

Florin, L., Tsokoglou, A., Happe, T., 2001. A novel type of iron hydrogenase in the green alga *Scenedesmus obliquus* is linked to the photosynthetic electron transport chain. J. Biol. Chem. 276, 6125−6132.

Forestier, M., King, P., Zhang, L., Posewitz, M., Schwarzer, S., Happe, T., 2003. Expression of two [Fe]-hydrogenases in *Chlamydomonas reinhardtii* under anaerobic conditions. Eur. J. Biochem. 270, 2750−2758.

Fouchard, S., Hemschemeier, A., Caruana, A., Pruvost, J., Legrand, J., Happe, T., Peltier, G., Cournac, L., 2005. Autotrophic and mixotrophic hydrogen photoproduction in sulfur-deprived *Chlamydomonas* cells. Appl. Env. Microbiol. 71 (10), 6199−6205.

Gaffron, H., Rubin, J., 1942. Fermentative and photochemical production of hydrogen in algae. J. Gen. Physiol. 26, 219−240.

Ghirardi, M., Mohanty, P., 2010. Oxygenic hydrogen production-current status of the technology. Curr. Sci. India 98, 499−507.

Ghirardi, M.L., Cohen, J., King, P., Schulten, K., Kim, K., Seibert, M., 2006. [FeFe]-hydrogenases and photobiological hydrogen production. SPIE 6340, U257−U262.

Ghirardi, M.L., Zhang, L., Lee, J.W., Flynn, T., Seibert, M., Greenbaum, E., Melis, A., 2000. Microalgae, a green source of renewable H_2. TIBTECH 18, 506−511.

Giannelli, L., Torzillo, G., 2012. Hydrogen production with the microalga Chlamydomonas reinhardtii grown in a compact tubular photobioreactor immersed in a scattering light nanoparticle suspension. Int. J. Hydrogen Energ. 37, 16951−16961.

Giannelli, L., Scoma, A., Torzillo, G., 2009. Interplay between light intensity, chlorophyll concentration and culture mixing on the hydrogen production in sulfur-deprived *Chlamydomonas reinhardtii* cultures grown in laboratory photobioreactors. Biotechnol. Bioeng. 104, 76−90.

Gimpel, J.A., Specht, E.A., Georgianna, D.R., Mayfield, S.P., 2013. Advances in microalgae engineering and synthetic biology applications for biofuel production. Curr. Opin. Biotech. 17, 489−495.

Godaux, D., Emonds-Alta, B., Berne, N., Ghysels, B., Alric, J., Remacle, C., Cardol, P., 2013. A novel screening method for hydrogenase-deficient mutants in *Chlamydomonas reinhardtii* based on in vivo chlorophyll fluorescence and photosystem II quantum yield. Int. J. Hydrogen Energ. 38, 1826−1836.

Godman, J.E., Molnar, A., Baulcombe, D.C., Bakl, J., 2010. RNA silencing of hydrogenase(-like) genes and investigation of their physiological roles in the green alga *Chlamydomonas reinhardtii*. BioChem. J. 431, 345−351.

Hahn, J.J., Ghirardi, M.L., Jacoby, W.A., 2007. Immobilized algal cells used for hydrogen production. J. Biol. Chem. 37, 75−79.

Hallenback, P.C., Abo-Hashesh, M., Ghosh, D., 2012. Strategies for improving biological hydrogen production. Bioresour. Technol. 110, 1−9.

Happe, T., Kaminski, A., 2002. Differential regulation of the Fe hydrogenase during anaerobic adaptation in the green alga *Chlamydomonas reinhardtii*. Eur. J. Biochem. 269, 1022−1032.

Hemschemeier, A., Fouchard, S., Cournac, L., Peltier, G., Happe, T., 2008. Hydrogen production by *Chlamydomonas reinhardtii*: an elaborate interplay of electron sources and sinks. Planta 227, 397−407.

Johanningmeier, U., Heiss, S., 1993. Construction of a *Chlamydomonas reinhardtii* mutant with an intronless *psbA* gene. Plant Mol. Biol. 22 (1), 91−99.

Kettunen, R., Tyystjärvi, E., Aro, E.M., 1996. Degradation pattern of photosystem II reaction center protein D1 in intact leaves. Plant Physiol. 111, 1183−1190.

Kim, D.-H., Kim, M.-S., 2011. Hydrogenases for biohydrogen production. Bioresour. Technol. 102, 8423−8431.

Kless, H., Oren-Shamir, M., Malkin, S., McIntosh, L., Edelman, M., 1994. The D−E region of the D1 protein is involved in multiple quinine and herbicide interaction in photosystem II. Biochemistry 33 (34), 10501–10507.

Kose, A., Oncel, S., 2014. Biohydrogen production from engineered microalgae *Chlamydomonas reinhardtii*. Adv. Energ. Res. 2 (1), 1–9.

Kosourov, S.N., Seibert, M., 2009. Hydrogen photoproduction by nutrient-deprived *Chlamydomonas reinhardtii* cells immobilized within thin alginate films under aerobic and anaerobic conditions. Biotechnol. Bioeng. 102, 50–58.

Kosourov, S.N., Batyrova, K.A., Petushkova, E.P., Tsygankov, A.A., Ghirardi, M.L., Seibert, M., 2012. Maximizing the hydrogen photoproduction yields in Chlamydomonas reinhardtii cultures: the effect of the h2 partial pressure. Int. J. Hydro. Energ. 37, 8850–8858.

Kosourov, S., Patrusheva, E., Ghirardi, M.L., Seibert, M., Tsygankov, A., 2007. A comparison of hydrogen photoproduction by sulfur-deprived Chlamydomonas reinhardtii under different growth conditions. J. Biotechnol. 128, 776–787.

Kruse, O., Hankamer, B., 2010. Microalgal hydrogen production. Curr. Opin. Biotechnol. 21, 238–243.

Kruse, O., Rupprecht, J., Bader, K.P., Thomas-Hall, S., Schenk, P.M., Finazzi, G., Hankamer, B., 2005a. Improved photobiological H$_2$ production in engineered green algal cells. J. Biol. Chem. 280 (40), 34170–34177.

Kruse, O., Rupprecht, J., Bader, K.P., Thomas-Hall, S., Schenk, P.M., Finazzi, G., Hankamer, B., 2005b. Improved photobiological H$_2$ production in engineered green algal cells. J. Biol. Chem. 280, 34170–34177.

Laurinavichene, T.V., Kosourov, S.N., Ghirardi, M.L., Seibert, M., Tsygankov, A.A., 2008. Prolongation of H$_2$ photoproduction by immobilized,sulfur-limited *Chlamydomonas reinhardtii* cultures. J. Biotechnol. 134, 275–277.

Leon-Banares, R., Gonza, D., Galvan, A., Fernandez, E., 2004. Transgenic microalgae as green cell-factories. Trends Biotechnol. 22 (1), 45–52.

Lindblad, P., Christensson, K., Lindberg, P., Fedorov, A., Pinto, F., Tsygankov, A., 2002. Photoproduction of H$_2$ by wild type *Anabaena* PCC7120 and a hydrogen uptake deficient mutant: from laboratory experiments to outdoor culture. Int. J. Hydrogen Energ. 27, 1271–1281.

Mäenpää, P., Miranda, T., Tyystjarvi, E., Tyystjarvi, T., Govindjee, Ducret, J.M., Etienne, A.L., Kirilovsky, D., 1995. A mutation in the D-de loop of D1 modifies the stability of the S2QA-and S2BB-state in photosystem II. Plant Physiol. 107 (1), 187–197.

Marin-Navarro, J., Esquivel, M.G., Moreno, J., 2010. Hydrogen production by Chlamydomonas reinhardtii revisited: rubisco as biotechnological target. World J. Microbiol. Biotechnol. 26, 1785–1793.

Martens, R., Liese, A., 2004. Biotechnological applications of hydrogenases. Curr. Opin. Biotechnol. 15, 343–348.

Masojidek, J., Torzillo, G., Koblizek, M., 2013. Photosynthesis in microalgae. In: Richmond, A., Hu, Q. (Eds.), Handbook of Microalgal Culture: Applied Phycology and Biotechnology, Second ed. Wiley, pp. 21–36.

Mathews, J., Wang, G., 2009. Metabolic pathway engineering for enhanced biohydrogen production. Int. J. Hydrogen Energ. 34, 7404–7416.

Matthew, T., Zhou, W., Rupprecht, J., Lim, L., Thomas-Hall, S.R., Doebbe, A., Kruse, O., Hankamer, B., Marx, U.C., Smith, S.M., Schenk, P.M., 2009. The metabolome of *Chlamydomonas reinhardtii* following induction of anaerobic H$_2$ production by sulfur depletion. Curr. Opin. Biotechnol. 284, 23415–23425.

Mayfield, S.P., Franklin, S.E., 2005. Expression of human antibodies in eukaryotic micro-algae. Vaccine 23, 1828–1832.

Mckinlay, J.B., Harwood, C.S., 2010. Photobiological production of hydrogen gas as a biofuel. Curr. Opin. Biotechnol. 21, 244–251.

Melis, A., 2007. Photosynthetic H$_2$ metabolism in *Chlamydomonas reinhardtii* (unicellular green algae). Planta. 226, 1075–1086.

Melis, A., 2002. Green alga hydrogen production: progress, challenges and prospects. Int. J. Hydrogen Energ. 27, 1217–1228.

Melis, A., Chen, H.-C., 2005. Chloroplast sulfate transport in green algae—genes, proteins and effects. Photosynth. Res. 86, 99–307.

Melis, A., Happe, T., 2001. Hydrogen production: green algae as a source of energy. Plant Physiol. 127, 740–748.

Melis, A., Zhang, L., Forestier, M., Ghirardi, M., Seibert, M., 2000. Sustained photobiological hydrogen gas production upon reversible inactivation of oxygen evolution in the green alga *Chlamydomonas reinhardtii*. Plant Physiol. 122, 127–135.

Meuser, J.E., D'Adamo, S., Jinkerson, R.E., Mus, F., Yang, W., Ghirardi, M.L., Seibert, M., Grossman, A.R., Posewitz, M.C., 2012. Genetic disruption of both Chlamydomonas reinhardtii [FeFe]-hydrogenases: Insight into the role of HYDA2 in H2 production. Biochem. Bioph. Res. Co. 417 (2), 704–709.

Meyer, J., 2007. [FeFe] hydrogenases and their evolution: a genomic perspective. Cell. Mol. Life Sci. 64, 1063–1084.

Mus, F., Cournac, L., Cardettini, V., Caruana, A., Peltier, G., 2005. Inhibitor studies on non-photochemical PQ reduction and H$_2$ photoproduction in *Chlamydomonas reinhardtii*. Biochim. Biophys. Acta 1708, 322–332.

Mussgnug, J.H., Thomas-Hall, S., Rupprecht, J., Foo, A., Klassen, V., McDowall, A., Schenk, P.M., Kruse, O., Hankamer, B., 2007. Engineering photosynthetic light capture: impacts on improved solar energy to biomass conversion. Plant Biotechnol. J. 5 (6), 802–814.

Nagy, L.E., Meuser, J.E., Plummer, S., Seibert, M., Ghirardi, M.L., King, P.W., Ahmann, D., Posewitz, M.C., 2007. Application of gene shuffling for the rapid generation of novel [FeFe]-hydrogenase libraries. Biotechnol. Lett. 29, 421–430.

Nguyen, A.V., Thomas-Hall, S.R., Malnoe, A., Timmins, M., Mussgnug, J.H., Rupprecht, J., Kruse, O., Hankamer, O., Schenk, P.M., 2008. Transcriptome for photobiological hydrogen production induced by sulfur deprivation in the green alga *Chlamydomonas reinhardtii*. Eukaryot. Cell. 7 (11), 1965–1979.

Nicolet, Y., de Lacey, A.L., Vernede, X., Fernandez, V.M., Hatchikian, E.,C., Fontecilla-Camps, J.C., 2001. Crystallographic and FTIR spectroscopic evidence of changes in Fe coordination upon reduction of the active site of the Fe-only hydrogenase from *Desulfovibrio desulfuricans*. J. Am. Chem. Soc. 123, 1596–1601.

Nixon, P.J., Komenda, J., Barber, J., Deak, Z., Vass, I., Diner, B.A., 1995. Deletion of the PEST-like region of photosystem two modifies the QB-binding pocket but does not prevent rapid turnover of D1. J. Biol. Chem. 270, 14919–14927.

Oh, Y.-K., Raj, S.M., Jung, G.Y., Park, S., 2011. Current status of the metabolic engineering of microorganisms for biohydrogen production. Bioresour. Technol. 102, 8357–8367.

Ohad, I., Kren, N., Zer, H., Gong, H., Mor, T.S., Gal, A., Tal, S., Domovich, Y., 1994. Light-induced degradation of the photosystem II reaction centre D1 protein in vivo, an integrative approach. In: Backer, N.R., Bowyer, J.R. (Eds.), Photoinhibition of Photosynthesis, from Molecular Mechanisms to the Field. Bios Scientific Publishers, Oxford, pp. 161–178.

Oncel, S., Sukan, F.V., 2009. Photo-bioproduction of hydrogen by *Chlamydomonas reinhardtii* using a semi-continuous process regime. J. Hydrogen Energ. 34, 7592–7602.

Oncel, S., 2013. Microalgae for a macroenergy world. Renew. Sust. Energy Rev. 26, 241–264.

Oncel, S., Kose, A., 2014. Comparison of tubular and panel type photobioreactors for biohydrogen production utilizing *Chlamydomonas reinhardtii* considering mixing time and light intensity. Bioresour. Technol. 151, 265–270.

Oncel, S., Sabankay, M., 2012. Microalgal biohydrogen production considering light energy and mixing time as the two key features for scale-up. Bioresour. Technol. 121, 228–234.

Pinto, T.S., Malcata, F.X., Arrabaça, J.D., Silva, J.M., Spreitzer, R.J., Esquivel, M.G., 2013. Rubisco mutants of *Chlamydomonas reinhardtii* enhance photosynthetic hydrogen production. Appl. Microbiol. Biotechnol. 97, 5635–5643.

Polle, J.E.W., Kanakagiri, S., Jin, E.S., Masuda, T., Melis, A., 2002. Truncated chlorophyll antenna size of the photosystems — a practical method to improve microalgal productivity and hydrogen production in mass culture. Int. J. Hydrogen Energ. 27, 1257–1264.

Posewitz, M.C., King, P.W., Smolinski, S.L., Smith, R.D., Ginley, A.R., Ghirardi, M.L., 2005. Identification of genes required for hydrogenase activity in *Chlamydomonas reinhardtii*. Biochem. Soc. Trans. 33, 102–104.

Radakovits, R., Jinkerson, R.E., Darzins, A., Posewitz, M.C., 2010. Genetic engineering of algae for enhanced biofuel production. Eukaryot. Cell. 9 (4), 486.

Rosenberg, J.N., Oyler, G.A., Wilkinson, L., Betenbaughü, M.J., 2008. A green light for engineered algae: redirecting metabolism to fuel a biotechnology revolution. Curr. Opin. Biotechnol. 19, 430–436.

Rupprecht, J., 2009. From system biology to fuel-*Chlamydomonas reinhardtii* as a model for a systems biology approach to improve biohydrogen production. J. Biotechnol. 142, 10–20.

Schroda, M., Vallon, O., Wollman, F.A., Beck, C.F., 1999. A chloroplast-targeted heat shock protein 70 (HSP70) contributes to the photoprotection and repair of PSII during and after photoinhibition. Plant Cell 11, 1165–1178.

Schulz, R., 1996. Hydrogenases and hydrogen production in eukaryotic organisms and cyanobacteria. J. Mar. Biotechnol. 4, 16–22.

Scoma, A., Gianelli, L., Faraloni, C., Torzillo, G., 2012a. Outdoor H_2 production in a 50-L tubular photobioreactor by means of a sulfur-deprived culture of the microalga *Chlamydomonas reinhardtii*. J. Biotechnol. 157, 620–627.

Scoma, A., Krawietz, D., Faraloni, C., Giannelli, L., Happe, T., Torzillo, G., 2012b. Sustained H_2 production in a *Chalmydomonas reinhardtii* D1 protein mutant. J. Biotechnol. 157, 613–619.

Shestakov, S.V., Mikheeva, L.E., 2006. Genetic control of hydrogen metabolism in cyanobacteria. Russian Journal of Genetics 42 (11), 1272–1284.

Skjanes, K., Knutsen, G., Kallqvist, T., Lindblad, P., 2008. H_2 production from marine and freshwater species of green algae during sulfur deprivation and considerations for bioreactor design. Int. J. Hydrogen Energ. 33, 511–521.

Sobolev, V., Edelman, M., 1995. Modeling the Quinone-B binding site of the photosystem-II-reaction center using notions of complementary and contact-surface between atoms. Proteins. Struct. Funct. Genet. 21, 214–225.

Srirangan, K., Pyne, M.E., Chou, C.P., 2011. Biochemical and genetic engineering strategies to enhance hydrogen production in photosynthetic algae and cyanobacteria. Bioresour. Technol. 102, 8559–8604.

Stephenson, M., Stickland, S.H., 1931. Hydrogenase: a bacterial enzyme activating molecular hydrogen. I. The properties of hydrogenases. Biochem. J. 25, 205–214.

Surzycki, R., Cournac, L., Peltier, G., et al., 2007. Potential for hydrogen production with inducible chloroplast gene expression in *Chlamydomonas*. Proc. Natl. Acad. Sci. 104, 17548–17553.

Takahashi, H., Braby, C.E., Grossman, A.R., 2001. Sulfur economy and cell wall biosynthesis during sulfur limitation of *Chlamydomonas reinhardtii*. Plant Physiol. 127, 665–673.

Tetali, S.D., Mitra, M., Melis, A., 2007. Development of the light-harvesting chlorophyll antenna in the green alga *Chlamydomonas reinhardtii* is regulated by the novel Tla1 gene. Planta 225 (4), 813–829.

Tolletter, D., Ghysels, B., Alric, J., Petroutsos, D., Tolstygina, et al., 2011. Control of hydrogen photoproduction by the proton gradient generated by cyclic electron flow in *Chlamydomonas reinhardtii*. Plant Cell 23 (7), 2619–2630.

Torzillo, G., Scoma, A., Faraloni, C., Ena, A., Johanningmeier, U., 2009. Increased hydrogen photoproduction by means of a sulfur-deprived *Chlamydomonas reinhardtii* D1 protein mutant. Int. J. Hydrogen Energ. 34, 4529–4536.

Torzillo, G., Scoma, A., Faraloni, C., Giannelli, L., 2014. Advances in the biotechnology of hydrogen production with the microalga *Chlamydomonas reinhardtii*. Crit. Rev. Biotechnol. 1549–7801.

Torzillo, G., Seibert, M., 2013. Hydrogen production by microalgae. In: Richmond, A., Hu, Q. (Eds.), Handbook of Microalgal Culture: Applied Phycology and Biotechnology, Second ed. Wiley, pp. 417–444.

Tóth, S.Z., Schansker, G., Garab, G., Strasser, R.J., 2007. Photosynthetic electron transport activity in heat-treated barley leaves, the role of internal alternative electron donors to photosystem II. BBA Bioenerg. 1767, 295–305.

Trebst, A., 1987. The three-dimensional structure of the herbicide binding niche on the reaction center polypeptides of photosystem II. Z. Naturforsch. 42, 742.

Vermaas, W.F.J., Ikeuchi, M., 1991. Photosystem II. In: Bogorad, L., Vasil, I.K. (Eds.), The Photosynthetic Apparatus, Molecular Biology and Operation. Cell Culture and Somatic Cell Genetics of Plants, vol. 7B. Academic Press, San Diego, pp. 25–111.

Vignais, P.M., Billoud, B., Meyer, J., 2001. Classification and phylogeny of hydrogenases. FEMS Microbiol. Rev. 54, 455–501.

Volgusheva, A., Styring, S., Mamedov, F., 2013. Increased photosystem II stability promotes H_2 production in sulfur-deprived *Chlamydomonas reinhardtii*. PNAS 110 (18), 7223–7228.

Wu, S., Xu, L., Wang, R., Liu, X., Wang, Q., 2011. A high yield mutant of *Chlamydomonas reinhardtii* for photoproduction of hydrogen. Int. J. Hydrogen Energ. 36, 14134–14140.

Wykoff, D.D., Davies, J.P., Melis, A., Grossman, A.R., 1998. The regulation of photosynthetic electron transport during nutrient deprivation in *Chlamydomonas reinhardtii*. Plant Physiol. 117, 129–139.

Genetic Engineering of Microalgae for Production of Value-added Ingredients

Asha A. Nesamma, Kashif M. Shaikh, Pavan P. Jutur

DBT-ICGEB Centre for Advanced Bioenergy Research, New Delhi, India

1. INTRODUCTION

Increasing demand for energy, the depletion of fossil-based resources, global climate change, and environmental issues are emerging as the main challenges that have motivated the search for alternative "clean" energy sources, along with an impetus for a transition to a bio-based economy with a low carbon footprint (Hariskos and Posten, 2014; Wichuk et al., 2014). Due to their unique properties, microalgae including cyanobacteria represent an extremely diverse group of organisms, considered as promising feedstocks for applications in food and feed production, bioactive pharmaceuticals, nutraceuticals, functional foods, and biofuels (de la Noue and de Pauw, 1988; Lee et al., 2010; Pangestuti and Kim, 2011; Pulz and Gross, 2004; Vanthoor-Koopmans et al., 2013; Wijffels and Barbosa, 2010). The genetic diversity corresponds to a broad range of habitats almost found in all environments, although knowledge of their organismal and biochemical diversity is not yet fully known. The major advantages that render microalgal species as a potential new generation of feedstock for biofuel production and high value-added renewables are their high growth rates and higher photosynthetic efficiencies when grown under specific environments (Klein-Marcuschamer et al., 2013); they might be cultivated on nonarable land and/or even offshore, consequently they do not compete for arable land for food production (Brennan and Owende, 2013; Markou and Nerantzis, 2013; Singh et al., 2011b).

Some salient features of microalgae as a source of renewable fuels are:

- the ability to utilize sunlight as their only source of energy that can be converted into energy-rich biomolecules;
- their efficient capture and utilization of atmospheric CO_2, thus reducing greenhouse gas emissions and environmental issues;
- the ability of some microalgal species to grow in diverse environments such as marine, hypersaline, wastewater, or other sources of water, which are not feasible for any domestic purposes; and
- their most promising feature is their faster growth rate, thus facilitating high rates of biomass production.

Murray et al. (2013) proposed the scenario of sustainable production of biologically active

Handbook of Marine Microalgae
http://dx.doi.org/10.1016/B978-0-12-800776-1.00026-1

molecules of marine-based origin to develop a harmonious relationship with the marine environments by applying sustainable natural practices to the valorization of high value-added biomolecules (HVABs) from marine life (Murray et al., 2013; Urreta et al., 2014); such an approach will permit a conscientious means to maximize both human and economic benefits from the marine environments, an excellent solution toward meeting the soaring demands for tomorrow's bioenergy needs.

Microalgae under stress conditions have the ability to alter their biomass composition and accumulate lipids and/or carbohydrates, the major precursors for biofuel production. Meanwhile, some species cultivated under stress conditions accumulate—along with the lipids and carbohydrates—specific secondary metabolites that are HVABs, such as pigments, vitamins, carotenoids, etc., which are relevant to the cosmetic, food, or pharmaceutical industries (Skjånes et al., 2012). Therefore, improvements in economic feasibility through the concept of biorefinery can be achieved with advanced genetic engineering and modeling, where these microalgae will have the ability to simultaneously produce specific HVABs and biofuels under specific conditions (Campenni et al., 2013; Carriquiry et al., 2011; Durmaz, 2007; Nobre et al., 2013; Singh et al., 2011a).

2. HIGH VALUE-ADDED BIOMOLECULES IN MICROALGAE

Microalgae are an excellent source of HVABs, having a high content of antioxidants and pigments, that is, carotenoids such as fucoxanthin, β-carotene, astaxanthin, lutein, phycobiliproteins, long-chain polyunsaturated fatty acids (LC-PUFAs), polysaccharides, and proteins (Balavigneswaran et al., 2013; Gouveia, 2014; Kawee-ai et al., 2013; Kim and Mendis, 2006; Reyes et al., 2014; Rodriguez-Garcia and Guil-Guerrero, 2008; Samarakoon and Jeon,

2012; Spolaore et al., 2006; Sun et al., 2014). Co-extraction of other HVABs such as eicosapentaenoic acid (EPA), docosahexaenoic acid (DHA), vitamin E (α-tocopherol), and arachidonic acid (AA) may also further enhance the nutritional and/or nutraceutical value of these microalgae (Dewapriya and Kim, 2014; Durmaz, 2007; Ryan and Symington, 2014; Ryckebosch et al., 2014).

Carotenoids are hydrophobic pigments with 40-carbon structures and mainly classified into two groups: the carotenes (nonoxygenated molecules) and the xanthophylls (oxygenated molecules) (Markou and Nerantzis, 2013). In microalgae, the significant roles played by the carotenoids are in the process of photosynthesis such as light harvesting, photoprotection, free radical scavenging, excess energy dissipation, and structure stabilization (Frank and Cogdell, 1996). Marine microalgae such as *Dunaliella salina* (Fu et al., 2014), *Tetraselmis suecica* (Jo et al., 2012), *Isochrysis galbana*, (Custódio et al., 2014), and *Pavlova salina* (Zhou et al., 2007) are further exploited for commercial production (Ahmed et al., 2014) as carotenoids are associated with various health benefits, for example, helping in the prevention of age-related macular degeneration and cataract formation (Snodderly, 1995; Weikel et al., 2012), cancers (Gerster, 1993; Lupulescu, 1994; Willett, 1994), rheumatoid arthritis, muscular dystrophy, and cardiovascular diseases (Giordano et al., 2012; Kohlmeier and Hastings, 1995), and they may also have an effect on the immune system and influence chronic diseases (Meydani et al., 1995; Park et al., 2010).

Diverse pharmaceutical applications and their commercial relevance led to screening of new HVABs with biological activity from marine microalgae that have the potential to prevent or reduce the impact of several lifestyle-related diseases such as antiviral (including anti-HIV), antitumor, antibiotic, cytotoxic, enzyme inhibitory agents, and other therapeutic applications along with antimicrobial (antibacterial, antifungal, antiprotozoal) as well as biomodulatory effects such as immunosuppressive and anti-inflammatory issues (Burja et al., 2001; de Jesus Raposo et al., 2013; Guedes et al., 2013; Shanab et al., 2012; Shibata et al., 2007, 2003). Furthermore, microalgae HVABs are effective in the reduction of cardiocirculatory and coronary diseases, wounds, gastric ulcers, constipation, anemia, hypertension, atherosclerosis, and diabetes (Lee and Jeon, 2013; Nuño et al., 2013; Yamaguchi, 1996).

These microalgae can also synthesize polysaccharides that can be used as emulsion stabilizers or as bioflocculants, isoprene molecules for synthetic rubber, adhesives, and surgical gloves, etc., and polyhydroxyalkanoate for bioplastics (Matos et al., 2013). Biofertilizers are the growth-promoting substances produced from these microalgae that will be eco-friendly organic agro-input and are more cost-effective than current chemical fertilizers (Painter, 1993). Finally, the microalgae biomass leftovers after the extraction of added-value compounds will be used for the production of liquid biofuels (bioethanol, biodiesel, biobutanol, and bio-oil) (Gouveia and Oliveira, 2009; Miranda et al., 2012) and/or gaseous biofuels (biomethane, biohydrogen, syngas, etc.) (Ferreira et al., 2013; Marques et al., 2011).

Unfortunately, the economic viability of microalgae-based biofuels and HVABs is still not feasible and sustainable (Clarens et al., 2010; Norsker et al., 2011; Razon and Tan, 2011; Soratana and Landis, 2011). The only solution is the coproduction of HVABs; simultaneously, the environmental issues could eventually offset the high production costs of mass microalgae cultivation and would support a microalgae-based bio-economy such as food, feed, energy, pharmaceutical, cosmetic, and chemical industries (Gouveia, 2014). The main bottleneck of the microalgal biorefinery approach is the availability of optimized separation techniques to overcome preliminary hurdles (Vanthoor-Koopmans et al., 2013; Wijffels et al.,

2010), and these should also be commonly applicable for a variety of end products (HVABs) (Brennan and Owende, 2010; da Silva et al., 2014).

Major pathways of carotenoid metabolism are discussed below in some species such as *Chlorella vulgaris*, *D. salina*, *Haematococcus pluvialis*, and *Phaeodactylum tricornutum*. Carotenoids play an essential role in the light-harvesting complex of microalgae and higher plants. Carotenoid biosynthesis is complex, as it is coordinated with the biogenesis of chlorophylls, the photosynthetic apparatus, and electron transport (Bohne and Linden, 2002; Cardol et al., 2011). Some carotenoids, such as diadinoxanthin, diatoxanthin, and fucoxanthin, are only present in diatoms, while others, such as δ-carotene, ε-carotene, α-carotene, lutein, and astaxanthin, are only produced in green algae (Hildebrand et al., 2012). *D. salina* can overproduce β-carotene and lutein under stress conditions, and *H. pluvialis* is a good producer of astaxanthin (Katsuda et al., 2004). As these microalgae are able to synthesize very diverse carotenoid biomolecules, characterization of metabolic pathways is an important step prior to engineering microalgal strains for industrial applications.

Biorefinery is an integration of various processes and equipment to produce biofuels, bioenergy, and HVABs from biomass, aiming to provide sustainable processing of biomass feedstock and thus maximizing its value into a range of commercial products and energy (Demirbas, 2009; Nobre et al., 2013; Subhadra, 2010; Vanthoor-Koopmans et al., 2013). The microalgal-based biorefinery concept (Figure 1) includes: cultivation of microalgae, biomass harvesting, cell disruption and compound extraction, fractionation, and purification. The main aim of the biorefinery processes is to separate and recover the desired compounds from the same biomass batch in a reproducible manner as intact commercial HVABs. The fractionation step is considered to be one of the major

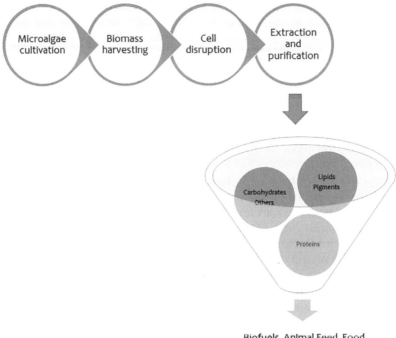

FIGURE 1   Schematic representation of microalgal biorefinery concept.

bottlenecks because of the difficulty of separating the various HVABs without causing damage to other relevant fractions (Vanthoor-Koopmans et al., 2013).

Microalgal proteins might also be of special interest because of their nutrition quality, due to their good profile and proportion of the amino acids (Becker, 2007), but their poor stability and their denaturation patterns under acid, alkali, or highly thermal conditions while being extracted and separated make them challenging for commercial production (Zeng et al., 2013). After the extraction of the desired HVABs, the remaining unexploited biomass could be utilized by one of the biomass energy conversion technologies, like anaerobic digestion, anaerobic fermentation, and liquefaction, to produce biofuels (Alzate et al., 2014; Ehimen et al., 2011; Nobre et al., 2013).

3. PATHWAY ENGINEERING OF MICROALGAE FOR ENHANCED HIGH VALUE-ADDED BIOMOLECULES

These microalgae are photosynthetic eukaryotes capable of fixing CO_2 into biomass with a higher efficiency of photosynthesis than vascular plants (Dismukes et al., 2008); they are also highly diverse in their evolutionary patterns (Armbrust, 2009). The availability of some advanced genetic modification tools (Radakovits et al., 2010) makes it possible to engineer microalgae for the efficient production of relevant biofuels and bioproducts. In addition, microalgae are a rich source of natural value-added products, such as carotenoids and unsaturated fatty acids. There is a high demand in global markets (Cordero et al., 2011) for naturally synthesized carotenoids such as β-carotene and lutein. However, the productivity of carotenoids in microalgae has been low, and the economic viability of algal biotechnology is limited by processing costs and photosynthetic efficiency.

The combination of iterative metabolic engineering and lipidomics can help drive forward both our understanding of acyl metabolism in diatoms and also the establishment of *P. tricornutum* as an algal synthetic biology chassis to produce high value fatty acids such as omega-3 LC-PUFAs or medium-chain fatty acids for biofuels (Hamilton et al., 2014; Radakovits et al., 2011). The metabolic relationship between the n-6 and n-3 fatty acid series in the marine flagellate *Pavlova lutheri*, which is a microalga known to be rich in LC-PUFAs and able to produce large amounts of n-3 fatty acids, such as EPA, 20:5n-3, and DHA, 22:6n-3 (Guihéneuf et al., 2013), was demonstrated by studying microalgal LC-PUFA biosynthesis pathways with reference to desaturase and elongase activities *in vivo*, using externally radiolabeled fatty acid precursors as substrates.

The expression of exogenous carotenogenic genes in *C. reinhardtii* (León et al., 2007) showed that there are limitations to obtain high-yield productions of new carotenoids, which are not feasible in native strains; these constraints are due to our limited knowledge of the mechanisms and signals that control carotenoids biosynthesis, modification, and storage. Nevertheless, the main challenge here is to optimize the strains for producing higher levels of carotenoids through metabolic engineering of the carotenoid biosynthetic pathway.

Engineering of microalgae *N. oceanica* CY2 revealed that medium composition, nitrogen source concentration, and light sources all play crucial roles in affecting the microalgae growth and EPA accumulation (Chen et al., 2013). The most efficient EPA production with lower electricity consumption occurred when the microalgae were grown in modified BG11 medium at a $NaNO_3$ concentration of 1.50 g/L using LED-blue as the light source, showing an optimal EPA content (5.57%) and EPA productivity (12.29 mg/l/day), indicating the great potential of using this system for the commercial production of EPA from microalgae (Chen et al., 2013).

The relationship between abiotic environmental stresses and lutein biosynthesis in the green microalga *D. salina* (Fu et al., 2014) demonstrated that experimentation using the rational design of adaptive evolution can be an effective and promising approach for optimizing the production of carotenoids. However, there are a few limitations that need to be addressed to understand carotenoid metabolism and characterize the relevant rate-limiting steps prior to the rational design. Such approaches to characterizing metabolic pathways can also pave the way for further optimization of microalgal cell factories through metabolic and/or pathway engineering.

4. INDUSTRIAL/COMMERCIAL RELEVANCE

Various species of microalgae accumulate high concentrations of specific pigments. Red, orange, and yellow carotenoid pigments such as β-carotene from *Dunaliella* (Fu et al., 2014), astaxanthin from *Haematococcus* (Reyes et al., 2014), and canthaxanthin (from various green algae) have wide applications as antioxidants, nutritional supplements, food-coloring agents, color-enhancement agents in fish and shellfish, and sunscreens (Wichuk et al., 2014). Squalene, a carotenoid precursor that occurs in all microalgae, is a widely used cosmetic and skin care preparation (Huang et al., 2009). Many rapidly growing eukaryotic microalgae, such as species of *Porphyridium*, *Nannochloropsis*, *Phaeodactylum*, and *Nitzschia*, can accumulate high concentrations of neutral lipids that have long-chain

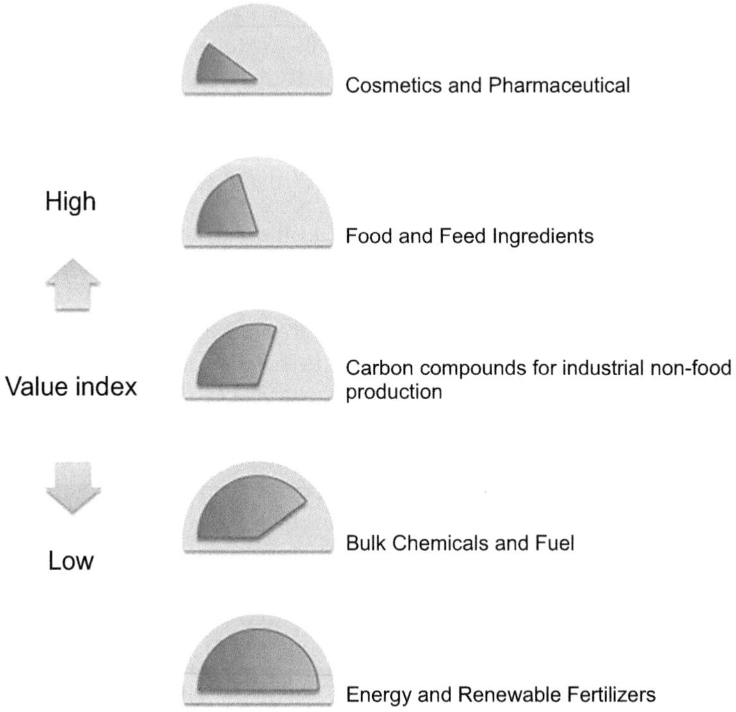

FIGURE 2   Production of low and high value-added biomolecules (HVABs) among microalgae through genetic engineering.

High

Value index

Low

Cosmetics and Pharmaceutical

Food and Feed Ingredients

Carbon compounds for industrial non-food production

Bulk Chemicals and Fuel

Energy and Renewable Fertilizers

omega-3 polyunsaturated fatty acids, such as EPA, DHA, and AA, are of particular interest since they are essential components of human and animal diets (Dewapriya and Kim, 2014; Durmaz, 2007; Ryan and Symington, 2014; Ryckebosch et al., 2014). *Botryococcus braunii* is a green algae that often forms extensive blooms and which, under appropriate conditions, can synthesize and secrete hydrocarbons that accumulate up to 86% of the alga dry weight. Although technical limitations have thus far precluded commercial production of this alga, studies have shown that certain strains (mainly within "A Race" *Botryococcus*) can produce substantial quantities of hydrocarbons during exponential growth (Hirose et al., 2013).

Microalgae are a rich source of novel low and high-value bioactive compounds that may have diverse applications in human and animal medicine, as well as in agriculture (Figure 2).

Detailed screening during the last 20 years has revealed a whole new range of molecules with antibiotic, antiviral, and anticancer activities as well as anti-inflammatory, hypocholesterolemic, enzyme inhibition, and various other pharmacological activities (Burja et al., 2001; de Jesus Raposo et al., 2013; Guedes et al., 2013; Shanab et al., 2012; Shibata et al., 2007, 2003). Some marine genera include species of *Dunaliella*, *Nanochloropsis*, *Isochrysis*, and *Pavlova* (Hallmann, 2007) that are used in farmed fish and shellfish industries, representing a huge and expanding market for fresh, frozen, or dried biomass from algae that accumulate specific compounds of HVABs. With the exception of some microalgal compounds that are already produced at a commercial level, the majority of the microalgal HVABs are either not established in the market or still not commercialized. It seems clear that there is a future wider scope for production of sustainable environmentally conscientious marine microalgal high-value biomolecules (Borowitzka, 2013).

5. CONCLUSIONS

Microalgae (including the cyanobacteria) are commercial resources of HVABs such as β-carotene, astaxanthin, docosahexaenoic acid, EPA, phycobilin pigments, and algal extracts extensively used in cosmeceuticals, nutraceuticals, and functional foods. In the last few years, there has been renewed interest in microalgae as commercial sources of these and other HVABs, driven in part by the attempts to develop commercially viable biofuels from microalgae. The potential HVABs that can be derived from microalgae through genetic engineering will provide leads for their commercial development and optimize product quality requirements and assurance, and along with the legal and regulatory environment, ultimately provide an efficient microalgae-based bio-refinery process.

References

Ahmed, F., Fanning, K., Netzel, M., Turner, W., Li, Y., Schenk, P.M., 2014. Profiling of carotenoids and antioxidant capacity of microalgae from subtropical coastal and brackish waters. Food Chem. 165, 300–306.

Alzate, M.E., Muñoz, R., Rogalla, F., Fdz-Polanco, F., Pérez-Elvira, S.I., 2014. Biochemical methane potential of microalgae biomass after lipid extraction. Chem. Eng. J. 243, 405–410.

Armbrust, E.V., 2009. The life of diatoms in the world's oceans. Nature 459, 185–192.

Balavigneswaran, C.K., Sujin Jeba Kumar, T., Moses Packiaraj, R., Veeraraj, A., Prakash, S., 2013. Anti-oxidant activity of polysaccharides extracted from *Isocrysis galbana* using RSM optimized conditions. Int. J. Biol. Macromol. 60, 100–108.

Becker, E.W., 2007. Micro-algae as a source of protein. Biotechnol. Adv. 25, 207–210.

Bohne, F., Linden, H., 2002. Regulation of carotenoid biosynthesis genes in response to light in *Chlamydomonas reinhardtii*. BBA-Gene Struct. Expr. 1579, 26–34.

Borowitzka, M.A., 2013. High-value products from microalgae—their development and commercialisation. J. Appl. Phycol. 25, 743–756.

Brennan, L., Owende, P., 2010. Biofuels from microalgae-A review of technologies for production, processing, and extractions of biofuels and co-products. Renew. Sust. Energ. Rev. 14, 557–577.

Brennan, L., Owende, P., 2013. Biofuels from microalgae: towards meeting advanced fuel standards. In: Lee, J.W. (Ed.), Advanced Biofuels and Bioproducts. Springer, New York, pp. 553–599.

Burja, A.M., Banaigs, B., Abou-Mansour, E., Grant Burgess, J., Wright, P.C., 2001. Marine cyanobacteria—a prolific source of natural products. Tetrahedron 57, 9347–9377.

Campenni, L., Nobre, B.P., Santos, C.A., Oliveira, A.C., Aires-Barros, M.R., Palavra, A.M., Gouveia, L., 2013. Carotenoid and lipid production by the autotrophic microalga *Chlorella protothecoides* under nutritional, salinity, and luminosity stress conditions. Appl. Microbiol. Biotechnol. 97, 1383–1393.

Cardol, P., Forti, G., Finazzi, G., 2011. Regulation of electron transport in microalgae. BBA - Bioenerg. 1807, 912–918.

Carriquiry, M.A., Du, X., Timilsina, G.R., 2011. Second generation biofuels: economics and policies. Energ. Pol. 39, 4222–4234.

Chen, C.-Y., Chen, Y.-C., Huang, H.-C., Huang, C.-C., Lee, W.-L., Chang, J.-S., 2013. Engineering strategies for enhancing the production of eicosapentaenoic acid (EPA) from an isolated microalga *Nannochloropsis oceanica* CY2. Bioresour. Technol. 147, 160–167.

Clarens, A.F., Resurreccion, E.P., White, M.A., Colosi, L.M., 2010. Environmental life cycle comparison of algae to other bioenergy feedstocks. Environ. Sci. Technol. 44, 1813–1819.

Cordero, B.F., Obraztsova, I., Couso, I., Leon, R., Vargas, M.A., Rodriguez, H., 2011. Enhancement of lutein production in *Chlorella sorokiniana* (chorophyta) by improvement of culture conditions and random mutagenesis. Mar. Drugs 9, 1607–1624.

Custódio, L., Soares, F., Pereira, H., Barreira, L., Vizetto-Duarte, C., Rodrigues, M., Rauter, A., Albério, F., Varela, J., 2014. Fatty acid composition and biological activities of *Isochrysis galbana* T-ISO, *Tetraselmis* sp. and *Scenedesmus* sp.: possible application in the pharmaceutical and functional food industries. J. Appl. Phycol. 26, 151–161.

Demirbas, A., 2009. Biorefineries: current activities and future developments. Energ. Convers. 50, 2782–2801.

Dewapriya, P., Kim, S.-K., 2014. Marine microorganisms: an emerging avenue in modern nutraceuticals and functional foods. Food Res. Int. 56, 115–125.

Dismukes, G.C., Carrieri, D., Bennette, N., Ananyev, G.M., Posewitz, M.C., 2008. Aquatic phototrophs: efficient alternatives to land-based crops for biofuels. Curr. Opin. Biotechnol. 19, 235–240.

Durmaz, Y., 2007. Vitamin E (α-tocopherol) production by the marine microalgae *Nannochloropsis oculata* (Eustigmatophyceae) in nitrogen limitation. Aquaculture 272, 717–722.

Ehimen, E.A., Sun, Z.F., Carrington, C.G., Birch, E.J., Eaton-Rye, J.J., 2011. Anaerobic digestion of microalgae residues resulting from the biodiesel production process. Appl. Energ. 88, 3454–3463.

Ferreira, A.F., Ortigueira, J., Alves, L., Gouveia, L., Moura, P., Silva, C.M., 2013. Energy requirement and CO_2 emissions of bioH_2 production from microalgal biomass. Biomass Bioenerg. 49, 249–259.

Frank, H.A., Cogdell, R.J., 1996. Carotenoids in photosynthesis. Photochem. Photobiol. 63, 257–264.

Fu, W., Paglia, G., Magnusdottir, M., Steinarsdottir, E., Gudmundsson, S., Palsson, B., Andresson, O., Brynjolfsson, S., 2014. Effects of abiotic stressors on lutein production in the green microalga *Dunaliella salina*. Microb. Cell Fact. 13, 3.

Gerster, H., 1993. Anticarcinogenic effect of common carotenoids. Int. J. Vitam. Nutr. Res. 63, 93–121.

Giordano, P., Scicchitano, P., Locorotondo, M., Mandurino, C., Ricci, G., Carbonara, S., Gesualdo, M., Zito, A., Dachille, A., Caputo, P., Riccardi, R., Frasso, G., Lassandro, G., Di Mauro, A., Ciccone, M.M., 2012. Carotenoids and cardiovascular risk. Curr. Pharm. Des. 18, 5577–5589.

Gouveia, L., 2014. From tiny microalgae to huge biorefineries. Oceanography 2, 2332–2632.

Gouveia, L., Oliveira, A.C., 2009. Microalgae as a raw material for biofuels production. J. Ind. Microbiol. Biotechnol. 36, 269–274.

Guedes, A., Gião, M., Seabra, R., Ferreira, A., Tamagnini, P., Moradas-Ferreira, P., Malcata, F., 2013. Evaluation of the antioxidant activity of cell extracts from microalgae. Mar. Drugs 11, 1256–1270.

Guihéneuf, F., Ulmann, L., Mimouni, V., Tremblin, G., 2013. Use of radiolabeled substrates to determine the desaturase and elongase activities involved in eicosapentaenoic acid and docosahexaenoic acid biosynthesis in the marine microalga *Pavlova lutheri*. Phytochemistry 90, 43–49.

Hallmann, A., 2007. Algal transgenics and biotechnology. Transgenic Plant J. 81–98.

Hamilton, M.L., Haslam, R.P., Napier, J.A., Sayanova, O., 2014. Metabolic engineering of *Phaeodactylum tricornutum* for the enhanced accumulation of omega-3 long chain polyunsaturated fatty acids. Metab. Eng. 22, 3–9.

Hariskos, I., Posten, C., 2014. Biorefinery of microalgae—opportunities and constraints for different production scenarios. Biotechnol. J. 9, 739–752.

Hildebrand, M., Davis, A.K., Smith, S.R., Traller, J.C., Abbriano, R., 2012. The place of diatoms in the biofuels industry. Biofuels 3, 221–240.

Hirose, M., Mukaida, F., Okada, S., Noguchi, T., 2013. Active hydrocarbon biosynthesis and accumulation in a Green alga, *Botryococcus braunii* (Race a). Eukaryot. Cell 12, 1132–1141.

Huang, Z.-R., Lin, Y.-K., Fang, J.-Y., 2009. Biological and pharmacological activities of squalene and related compounds: potential uses in cosmetic dermatology. Molecules 14, 540–554.

de Jesus Raposo, M.F., de Morais, R.M.S.C., de Morais, A.M.M.B., 2013. Health applications of bioactive compounds from marine microalgae. Life Sci. 93, 479–486.

Jo, W.S., Yang, K.M., Park, H.S., Kim, G.Y., Nam, B.H., Jeong, M.H., Choi, Y.J., 2012. Effect of microalgal extracts of Tetraselmis suecica against UVB-induced Photoaging in human skin Fibroblasts. Toxicol. Res. 28, 241–248.

Katsuda, T., Lababpour, A., Shimahara, K., Katoh, S., 2004. Astaxanthin production by *Haematococcus pluvialis* under illumination with LEDs. Enzym. Microb. Tech. 35, 81–86.

Kawee-ai, A., Kuntiya, A., Kim, S.M., 2013. Anticholinesterase and antioxidant activities of fucoxanthin purified from the microalga *Phaeodactylum tricornutum*. Nat. Prod. Commun. 8, 1381–1386.

Kim, S.-K., Mendis, E., 2006. Bioactive compounds from marine processing byproducts—a review. Food Res. Int. 39, 383–393.

Klein-Marcuschamer, D., Chisti, Y., Benemann, J.R., Lewis, D., 2013. A matter of detail: assessing the true potential of microalgal biofuels. Biotechnol. Bioeng. 110, 2317–2322.

Kohlmeier, L., Hastings, S.B., 1995. Epidemiologic evidence of a role of carotenoids in cardiovascular disease prevention. Am. J. Clin. Nutr. 62, 1370s–1376s.

Lee, S.-H., Jeon, Y.-J., 2013. Anti-diabetic effects of brown algae derived phlorotannins, marine polyphenols through diverse mechanisms. Fitoterapia 86, 129–136.

Lee, S.H., Kang, H.J., Lee, H.-J., Kang, M.-H., Park, Y.K., 2010. Six-week supplementation with *Chlorella* has favorable impact on antioxidant status in Korean male smokers. Nutrition (Burbank, Los Angeles County, California) 26, 175–183.

León, R., Couso, I., Fernández, E., 2007. Metabolic engineering of ketocarotenoids biosynthesis in the unicelullar microalga *Chlamydomonas reinhardtii*. J. Biotechnol. 130, 143–152.

Lupulescu, A., 1994. The role of vitamins A, beta-carotene, E and C in cancer cell biology. Int. J. Vitam. Nutr. Res. 64, 3–14.

Markou, G., Nerantzis, E., 2013. Microalgae for high-value compounds and biofuels production: a review with focus on cultivation under stress conditions. Biotechnol. Adv. 31, 1532–1542.

Marques, A.E., Barbosa, A.T., Jotta, J., Coelho, M.C., Tamagnini, P., Gouveia, L., 2011. Biohydrogen production by *Anabaena* sp. PCC 7120 wild-type and mutants under different conditions: light, nickel, propane, carbon dioxide and nitrogen. Biomass Bioenerg. 35, 4426−4434.

Matos, C.T., Gouveia, L., Morais, A.R.C., Reis, A., Bogel-Lukasik, R., 2013. Green metrics evaluation of isoprene production by microalgae and bacteria. Green Chem. 15, 2854−2864.

Meydani, S.N., Wu, D., Santos, M.S., Hayek, M.G., 1995. Antioxidants and immune response in aged persons: overview of present evidence. Am. J. Clin. Nutr. 62, 1462, 1462s−1476s.

Miranda, J.R., Passarinho, P.C., Gouveia, L., 2012. Bioethanol production from *Scenedesmus obliquus* sugars: the influence of photobioreactors and culture conditions on biomass production. Appl. Microbiol. Biotechnol. 96, 555−564.

Murray, P.M., Moane, S., Collins, C., Beletskaya, T., Thomas, O.P., Duarte, A.W., Nobre, F.S., Owoyemi, I.O., Pagnocca, F.C., Sette, L.D., McHugh, E., Causse, E., Perez-Lopez, P., Feijoo, G., Moreira, M.T., Rubiolo, J., Leiros, M., Botana, L.M., Pinteus, S., Alves, C., Horta, A., Pedrosa, R., Jeffryes, C., Agathos, S.N., Allewaert, C., Verween, A., Vyverman, W., Laptev, I., Sineoky, S., Bisio, A., Manconi, R., Ledda, F., Marchi, M., Pronzato, R., Walsh, D.J., 2013. Sustainable production of biologically active molecules of marine based origin. N. Biotechnol. 30, 839−850.

de la Noue, J., de Pauw, N., 1988. The potential of microalgal biotechnology: a review of production and uses of microalgae. Biotechnol. Adv. 6, 725−770.

Nobre, B.P., Villalobos, F., Barragan, B.E., Oliveira, A.C., Batista, A.P., Marques, P.A., Mendes, R.L., Sovova, H., Palavra, A.F., Gouveia, L., 2013. A biorefinery from *Nannochloropsis* sp. microalga-extraction of oils and pigments. Production of biohydrogen from the leftover biomass. Bioresour. Technol. 135, 128−136.

Norsker, N.-H., Barbosa, M.J., Vermuë, M.H., Wijffels, R.H., 2011. Microalgal production—a close look at the economics. Biotechnol. Adv. 29, 24−27.

Nuño, K., Villarruel-López, A., Puebla-Pérez, A.M., Romero-Velarde, E., Puebla-Mora, A.G., Ascencio, F., 2013. Effects of the marine microalgae *Isochrysis galbana* and *Nannochloropsis oculata* in diabetic rats. J. Func. Foods 5, 106−115.

Painter, T.J., 1993. Carbohydrate polymers in desert reclamation: the potential of microalgal biofertilizers. Carbohyd. Polym. 20, 77−86.

Pangestuti, R., Kim, S.-K., 2011. Biological activities and health benefit effects of natural pigments derived from marine algae. J. Func. Foods 3, 255−266.

Park, J.S., Chyun, J.H., Kim, Y.K., Line, L.L., Chew, B.P., 2010. Astaxanthin decreased oxidative stress and inflammation and enhanced immune response in humans. Nutr. Metab. (Lond.) 7, 18.

Pulz, O., Gross, W., 2004. Valuable products from biotechnology of microalgae. Appl. Microbiol. Biotechnol. 65, 635−648.

Radakovits, R., Eduafo, P.M., Posewitz, M.C., 2011. Genetic engineering of fatty acid chain length in *Phaeodactylum tricornutum*. Metab. Eng. 13, 89−95.

Radakovits, R., Jinkerson, R.E., Darzins, A., Posewitz, M.C., 2010. Genetic engineering of algae for enhanced biofuel production. Eukaryot. Cell 9, 486−501.

Razon, L.F., Tan, R.R., 2011. Net energy analysis of the production of biodiesel and biogas from the microalgae: *Haematococcus pluvialis* and *Nannochloropsis*. Appl. Energ. 88, 3507−3514.

Reyes, F.A., Mendiola, J.A., Ibañez, E., del Valle, J.M., 2014. Astaxanthin extraction from *Haematococcus pluvialis* using CO_2-expanded ethanol. J. Supercrit. Fluids 92, 75−83.

Rodriguez-Garcia, I., Guil-Guerrero, J.L., 2008. Evaluation of the antioxidant activity of three microalgal species for use as dietary supplements and in the preservation of foods. Food Chem. 108, 1023−1026.

Ryan, L., Symington, A.M., 2014. Algal-oil supplements are a viable alternative to fish-oil supplements in terms of docosahexaenoic acid (22:6n−3; DHA). J. Func. Foods . http://dx.doi.org/10.1016/j.jff.2014.06.023.

Ryckebosch, E., Bruneel, C., Termote-Verhalle, R., Goiris, K., Muylaert, K., Foubert, I., 2014. Nutritional evaluation of microalgae oils rich in omega-3 long chain polyunsaturated fatty acids as an alternative for fish oil. Food Chem. 160, 393−400.

Samarakoon, K., Jeon, Y.-J., 2012. Bio-functionalities of proteins derived from marine algae—a review. Food Res. Int. 48, 948−960.

Shanab, S.M.M., Mostafa, S.S.M., Shalaby, E.A., Mahmoud, G.I., 2012. Aqueous extracts of microalgae exhibit antioxidant and anticancer activities. Asian Pac. J. Trop. Biomed. 2, 608−615.

Shibata, S., Hayakawa, K., Egashira, Y., Sanada, H., 2007. Hypocholesterolemic mechanism of *chlorella*: Chlorella and its indigestible fraction enhance hepatic cholesterol catabolism through up-regulation of cholesterol 7α-hydroxylase in rats. Biosci. Biotechnol. Biochem. 71, 916−925.

Shibata, S., Natori, Y., Nishihara, T., Tomisaka, K., Matsumoto, K., Sansawa, H., Nguyen, V.C., 2003. Antioxidant and anti-cataract effects of *Chlorella* on rats with streptozotocin-induced diabetes. J. Nutr. Sci. Vitaminol. (Tokyo) 49, 334−339.

da Silva, T.L., Gouveia, L., Reis, A., 2014. Integrated microbial processes for biofuels and high value-added products: the way to improve the cost effectiveness of biofuel production. Appl. Microbiol. Biotechnol. 98, 1043–1053.

Singh, A., Nigam, P.S., Murphy, J.D., 2011a. Mechanism and challenges in commercialisation of algal biofuels. Bioresour. Technol. 102, 26–34.

Singh, A., Nigam, P.S., Murphy, J.D., 2011b. Renewable fuels from algae: an answer to debatable land based fuels. Bioresour. Technol. 102, 10–16.

Skjånes, K., Rebours, C., Lindblad, P., 2012. Potential for green microalgae to produce hydrogen, pharmaceuticals and other high value products in a combined process. Crit. Rev. Biotechnol. 33, 172–215.

Snodderly, D.M., 1995. Evidence for protection against age-related macular degeneration by carotenoids and antioxidant vitamins. Am. J. Clin. Nutr. 62, 1448s–1461s.

Soratana, K., Landis, A.E., 2011. Evaluating industrial symbiosis and algae cultivation from a life cycle perspective. Bioresour. Technol. 102, 6892–6901.

Spolaore, P., Joannis-Cassan, C., Duran, E., Isambert, A., 2006. Commercial applications of microalgae. J. Biosci. Bioeng. 101, 87–96.

Subhadra, B.G., 2010. Sustainability of algal biofuel production using integrated renewable energy park (IREP) and algal biorefinery approach. Energ. Pol. 38, 5892–5901.

Sun, L., Wang, L., Li, J., Liu, H., 2014. Characterization and antioxidant activities of degraded polysaccharides from two marine Chrysophyta. Food Chem. 160, 1–7.

Urreta, I., Ikaran, Z., Janices, I., Ibañez, E., Castro-Puyana, M., Castañón, S., Suárez-Alvarez, S., 2014. Revalorization of *Neochloris oleoabundans* biomass as source of biodiesel by concurrent production of lipids and carotenoids. Algal Res. 5, 16–22.

Vanthoor-Koopmans, M., Wijffels, R.H., Barbosa, M.J., Eppink, M.H.M., 2013. Biorefinery of microalgae for food and fuel. Bioresour. Technol. 135, 142–149.

Weikel, K.A., Chiu, C.J., Taylor, A., 2012. Nutritional modulation of age-related macular degeneration. Mol. Aspects Med. 33, 318–375.

Wichuk, K., Brynjólfsson, S., Fu, W., 2014. Biotechnological production of value-added carotenoids from microalgae: emerging technology and prospects. Bioengineered 5, 204–208.

Wijffels, R.H., Barbosa, M.J., 2010. An outlook on microalgal biofuels. Science 329, 796–799.

Wijffels, R.H., Barbosa, M.J., Eppink, M.H.M., 2010. Microalgae for the production of bulk chemicals and biofuels. Biofuel Bioprod. Bior. 4, 287–295.

Willett, W.C., 1994. Micronutrients and cancer risk. Am. J. Clin. Nut. 59, 1162S–1165S.

Yamaguchi, K., 1996. Recent advances in microalgal bioscience in Japan, with special reference to utilization of biomass and metabolites: a review. J. Appl. Phycol. 8, 487–502.

Zeng, Q., Wang, Y., Li, N., Huang, X., Ding, X., Lin, X., Huang, S., Liu, X., 2013. Extraction of proteins with ionic liquid aqueous two-phase system based on guanidine ionic liquid. Talanta 116, 409–416.

Zhou, X.R., Robert, S.S., Petrie, J.R., Frampton, D.M., Mansour, M.P., Blackburn, S.I., Nichols, P.D., Green, A.G., Singh, S.P., 2007. Isolation and characterization of genes from the marine microalga *Pavlova salina* encoding three front-end desaturases involved in docosahexaenoic acid biosynthesis. Phytochemistry 68, 785–796.

Genetic Engineering of Microalgae for Production of Therapeutic Proteins

P.T. Pratheesh[1], M. Vineetha[2]

[1]School of Biosciences, Mahatma Gandhi University, Kottayam, Kerala, India; [2]Department of Microbiology, Government Arts and Science College, Kozhinjampara, Palakkad, Kerala, India

1. INTRODUCTION

Developments in the field of recombinant DNA technology have revolutionized the pharmaceutical industry in recent years, creating a global multibillion dollar industry. GBI Research analysis in 2012 reported that the global therapeutic proteins' market value was estimated to be $93 billion in 2010 and expected to reach $141.5 billion in 2017. To meet rising demands, bioreactors are becoming increasingly more important for the production of large amounts of recombinant proteins, particularly in cases where traditional sources are limited due to cost and/or availability.

1.1 Expression Systems Used in Therapeutic Protein Production

There are a number of expression systems available for the large-scale production of therapeutic proteins; prokaryotic and eukaryotic systems are the two general categories of expression systems. Though different expression systems are currently available for heterologous protein expression, each of these systems has

its own merits and demerits in terms of protein yield, ease of manipulation, cost of production, and cost of operation (Dove, 2002). The rapid growth of bacteria, high production rates, and the ease of genetic manipulation-prokaryotic expression systems are often the methods of choice for small proteins and peptides. However, they lack machinery for expressing eukaryotic proteins, which undergo a variety of posttranslational modifications like proper folding, glycosylation, phosphorylation, formation of disulfide bridges, etc.

As unicellular eukaryotes, yeasts combine the benefits of the bacterial system with eukaryotic characteristics of gene expression, but posttranslational modifications are often performed incorrectly, for example, resulting in misfolded and hyperglycosylated products. Mammalian cell cultures for the expression of recombinant proteins exhibit a number of advantages over the microbial systems; however, animal cell cultures are often difficult to handle, particularly in large-scale production, and costs for media and cultivation are considerably high, leading to a price range of $150–1000 per gram protein before purification. A major disadvantage of mammalian cell cultures is the risk of contamination of the isolated protein with potentially pathogenic agents such as viruses and prions. Methods using transgenic animals delivering recombinant proteins in their milk have gained much attention (Maga, 2005). But the establishment of transgenic animals is still extremely time-consuming and very expensive.

Plants offer an attractive system for expression of recombinant proteins and perhaps the best economic alternative for the expression of multimeric proteins such as antibodies. Proteins purified from plants should be free from toxins and viral agents that may be present in preparations from bacteria or mammalian cell culture. While plants afford an economy of scale unprecedented in the biotechnology industry, there are several inherent drawbacks to this approach. The length of time required from the initial transformation event to having usable quantities of product can take up to 2 years for crops such as tobacco and over 3 years for species such as corn. Another concern surrounds the expression of human therapeutics in food plants, with the potential for gene flow (via pollen) to surrounding crops.

Eukaryotic microalgae as an alternative platform for recombinant protein production has been gaining a lot of attention in recent years. Protein production in transgenic algae could presumably offer many of the same advantages as transgenic plants, including cost, safety, and rapid scalability. This chapter discusses different factors involved in therapeutic protein production using microalgae and its current and future perspectives.

2. ADVANTAGES OF THERAPEUTIC PROTEIN PRODUCTION IN MICROALGAE

Although transgenic microalgal technology is still in its infancy, microalgae may represent the "best of both worlds" by combining simple and inexpensive growth requirements and capabilities for posttranscriptional and posttranslational processing of plants, with the rapid growth rate and potential for high-density culture of microorganisms (Walker et al., 2005). Unicellular photosynthetic green algae are most commonly used for protein production as they only require inexpensive salt-based media, carbon dioxide, and light for growth. Most green algae are also classified as generally regarded as safe, making purification and processing of expressed products much less onerous for many targeted applications. Contrary to transgenic plants, which must be strictly contained to avoid the transfer of transgenic material to surrounding wild-type flora by airborne vectors, microalgae can be cultivated in open facilities as no such transfer can occur. On the economics side, based on recombinant antibody production studies, the cost of production per gram of functional antibody is $150, $0.05, and $0.002 in mammalian, plant, and microalgal bioreactor systems, respectively, making the latter system very economically attractive (Mayfield et al., 2003). A comparison of different recombinant protein expression systems is shown in Table 1.

Despite the recent surge of interest and successful transformation of a myriad of microalgal species, transgenic strains belonging to the *Chlamydomonas*, *Chlorella*, *Volvox*, *Haematococcus*, and *Dunaliella* genera remain the most widely used and studied (Griesbeck et al., 2006; Raja et al., 2008; Rosenberg et al., 2008); many obstacles remain to be overcome before microalgae can be considered standard expression systems. The green microalga *Chlamydomonas reinhardtii* has been the center of attention for a number of human therapeutic protein developments. Since all three genomes (chloroplast, mitochondrial, and nuclear) have been completely sequenced, they provide valuable information for genetic engineering studies. Each of these genomes can be transformed in *Chlamydomonas*, and have distinct transcriptional, translational, and posttranslational properties that make them distinct.

TABLE 1   Comparison of Different Recombinant Protein Expression Systems

| Expression systems | Expression system characteristics | | | | | | | |
| | Molecular | | | | Operational | | | |
	Glycosylation	Gene size	Sensitivity to shear stress	Recombinant product yield	Production time	Cost of cultivation	Scale-up cost	Cost of storage
Bacteria	None	Unknown	Medium	Medium	Short	Medium	High	Low (−20 °C)
Yeast	Incorrect	Unknown	Medium	High	Medium	Medium	High	Low (−20 °C)
Insect	Correct, but depends on strain and product	Limited	High	Medium to high	Long	High	High	High (liquid N_2)
Mammalian cells	Correct	Limited	High	Medium to high	Long	High	High	High (liquid N_2)
Plant cells	Correct	Unlimited	N/A	High	Long	Low	Very low	Low (room temperature)
Unicellular microalgae	Correct	Unlimited	Low	Generally low	Short	Very low	Low	Low (room temperature)

These factors make *Chlamydomonas* the ideal microalgal species for therapeutic protein development studies.

3. DEVELOPMENTS IN MICROALGAL TRANSFORMATION TECHNOLOGY

Choice of suitable selectable marker genes, reporter genes, and efficient promoters are key factors governing transformation. These molecular tools can greatly enhance, reduce, or even silence transgene expression. The studies conducted so far in this regard have facilitated the improvement of microalgal transformation studies in recent years.

3.1 Selectable Marker Genes

An important aspect of transformation is the choice of an appropriate selective agent and its corresponding resistance gene. Selectable marker genes may encode an enzyme capable of either detoxifying a phytotoxic compound (negative selection) or metabolizing a substrate (e.g., a carbon source) that wild-type cells cannot utilize (positive selection). In microalgal transformation studies, a number of homologous and heterologous selectable marker genes have been used for both nuclear (Table 2) and chloroplast (Table 3) transformation.

3.2 Promoters

One important aspect in the development of a transformation system has been the choice of a promoter to drive expression of the transgene. One of the most widely used promoters in plant molecular biology is the cauliflower mosaic virus 35S (CaMV 35S) promoter. Although the CaMV 35S promoter drives strong and constitutive expression in most dicotyledonous and some monocotyledonous plants (Benfey et al., 1990),

TABLE 2   Homologous and Heterologous Selectable Marker Genes Used for Nuclear Transformation of Microalgae

Gene	Gene product
HETEROLOGOUS GENES	
aadA	Aminoglycoside 3'-adenyl transferase from *Escherichia coli*
Als	*Acetolactate synthase gene mutant*
aphA	Aminoglycoside 3-phosphotransferase from *Streptomyces rimosus* Kanamycin
aph7	Aminoglycoside 3-phosphotransferase from *Streptomyces hygroscopicus*
aphVIII	Aminoglycoside 3-phosphotransferase from *Streptomyces rimosus*
ble	Bleomycin-binding protein of *Streptoalloteichus hindustanus*
hpt	Hygromycin phosphotransferase of *E. coli*
nptII	Neomycin phosphotransferase II of *E. coli*
Oee-1	Oxygen evolving enhancer protein
HOMOLOGOUS GENES	
AC29	AC29 gene product (albino-3 homolog)
Als	Acetolactate synthase
Arg7	Argininosuccinate lyase
Cry1	Mutant cytosolic ribosomal protein S14
Nic7	Guinolinate synthetane A
Nit1	Nitrate reductase
Ppx1	Protoporphyrinogen oxidase
Thi10	Hydroxyethyl thiazole kinase

it has not proven to be a useful promoter in most algal species. Algal transformation has been most successful using promoters derived from highly expressed algal genes (Table 4). A widely used promoter for *Chlamydomonas* transformation is derived from the 5' untranslated region of the *C. reinhardtii* ribulose bisphosphate carboxylase/oxygenase small subunit (*RbcS2*) (Stevens and Purton, 1997). It was also shown that transformation frequency was significantly increased when *Chlamydomonas* introns (particularly the first intron of *RbcS2*) were introduced into the coding region of the *ble* selectable marker gene (Lumbreras et al., 1998). This intron appears to contain a transcriptional enhancer element as it can act in an orientation-independent manner and is effective when placed either upstream or downstream of the promoter. Synthetic promoters have also been developed by fusing the promoter from the *Chlamydomonas Hsp70A* (heat shock protein 70A) gene to other *Chlamydomonas* promoters. The *Hsp70A* promoter serves as a transcriptional enhancer of promoters *RbcS2*, β2-tubulin, and *Hsp70B* leading to high-level expression under inducing conditions (Schroda et al., 2000).

TABLE 3 Homologous and Heterologous Selectable Marker Genes Used for Chloroplast Transformation of Microalgae

Gene	Gene product
HETEROLOGOUS GENES	
aadA	Aminoglycoside 3'-adenyl transferase from Escherichia coli
aphA6	Aminoglycoside 3-phosphotransferase from Streptomyces rimosus
Cat	Chloramphenicol acetyltransferase
nptII	Neomycin phosphotransferase gene from E. coli
nit1	Nitrate reductase
HOMOLOGOUS GENES	
atpA,B,E	α-, β-, and ε-subunits of the CF1 ATP synthase complex
chlL/chlN	Protochlorophyllide reductase
petA & D	Subunits I and IV of cytochrome b_6/f complex
psaA	Photosystem I reaction center protein
psaB	Photosystem I reaction center protein
psaC	Iron sulfur protein of photosystem I
psbA	Photosystem II reaction center protein D1
psbC	P6 photosystem II core protein
psbD	D2 protein of photosystem II
rbcL	Rubisco large subunit
rps4 & rps12	Ribosomal protein S4 & S12
tscA	Photosystem 1 complex
16S rDNA	16S rRNA
23S rDNA	23S rRNA

TABLE 4 Commonly Used Promoters for Microalgal Transformation Studies

Promoter	Derivation
NUCLEAR	
Amt	Chlorella virus adenine methyltransferase
Ca1/Ca2	Carbonic anhydrase
CabII-1	Chlorophyll a/b-binding protein of photosystem I
CaMV 35S	Cauliflower mosaic virus 35S
Hsp70A	Heat shock protein 70A
Hsp70B	Heat shock protein 70B
Nia1 (Nit1)	NAD(P)H nitrate reductase
Nos	Agrobacterium tumefaciens nopaline synthase
PsaD	Photosystem II subunit D
RbcS2	Small subunit of ribulose-1,5 bisphosphate carboxylase/oxygenase
TubB2	β2-tubulin
SV40	Animal adenovirus SV40
CHLOROPLAST	
atpA	α-subunit of the ATP synthase CF1 complex
atpB	β-subunit of the ATP synthase CF1 complex
chlL	UV resistance
petD	Subunit IV of cytochrome $b6/f$ complex
rbcL	Large subunit of ribulose
rrn16	Bisphosphate carboxylase 16S rRNA

3.3 Reporter Genes

Transformation systems are generally developed using an efficient reporter gene, which encodes a protein that can be easily detected and quantified in transgenic lines. The most commonly used reporter genes in higher plants are *uidA*, which encodes β-glucuronidase (GUS), and the green fluorescent protein gene (*gfp*) from the bioluminescent jellyfish *Aequorea victoria* (Chalfie et al., 1994). There has been considerable difficulty, to date, in the expression of heterologous reporter genes in the nuclear genome of *Chlamydomonas*. The *uidA* gene has not been a successful reporter gene for nuclear transformation of *C. reinhardtii* even when under

TABLE 5   Commonly Used Reporter Genes for Transformation Studies in Microalgae

Reporter gene	Gene product	Assay method
NUCLEAR		
Ars	Arylsulfatase	Chromogenic detection using X-SO$_4$ as a substrate
GFP	Green fluorescent protein of *Aequorea victoria*	Luminescence or antibody based
PC1	NADPH: protochlorophyllide oxidoreductase	Assay protochlorophyllide reductase mRNA or enzyme activity
Cgluc	Luciferase of *Gaussia princeps*	Luminescence
CHLOROPLAST		
aadA	Aminoglycoside 3′-adenyl transferase from *Escherichia coli*	AAD assay
GFP	Green fluorescent protein of *Aequorea victoria*	Fluorescence or antibody based
LUC	Luciferase from *Renilla reniformis*	Luminescence or antibody based
uidA	β-Glucuronidase of *E. coli*	Histochemical or fluorometric

the control of a native promoter (Blankenship and Kindle, 1992). Commonly used reporter genes for transformation studies in microalgae are shown in Table 5.

The use of GFP as a reporter gene in both the nuclear and chloroplast genome of *Chlamydomonas* has been developed (Franklin et al., 2002). Unfortunately, the unmodified GFP gene under the control of heterologous promoters gave poor expression when used as a reporter for nuclear transformation of *Chlamydomonas* (Fuhrmann et al., 1999). Reasons for poor expression may include the predominance of A/T rich codons in the native gene and the possible presence or absence of sequence motifs that regulate transcription or the processing and targeting of the transcript. To overcome these limitations, a modified *gfp* gene has been synthesized utilizing the codon preference of *C. reinhardtii* nuclear genes (Fuhrmann et al., 1999). Additional modifications known to modify the spectral properties of GFP protein for fluorescence imaging were also introduced. This modified gene was fused in-frame to the *ble* gene and expressed

under the control of the *RbcS2* promoter. GFP fluorescence was observed in the nucleus, demonstrating nuclear accumulation of the GFP-BLE fusion protein (Fuhrmann et al., 1999). GFP fusions have proven to be a useful tool for the in vivo study of dynamic processes such as flagellar and centriole assembly and cell cycle events in *Chlamydomonas* (Ruiz-Binder et al., 2002).

4. METHODS USED IN MICROALGAL TRANSFORMATION

Appropriate transformation method used for the delivery of foreign DNA to target microalgal genome greatly influences transformation efficiency and expression. Research over the years has developed a number of methods for genetic manipulation of *C. reinhardtii*, making it a top candidate for biotechnological applications (Fuhrmann, 2002). Although some of these methods may not have significantly changed since their initial development, they are still being applied and studied. A number of recent studies have presented methods that seem to offer advantages over earlier techniques.

4.1 Cell Wall-Deficient Strains

The use of cell wall-deficient strains, or the removal of the cell walls from wild-type strains, greatly increases the number of transformants recovered following transformation. Protocols for cell wall removal have been developed that facilitate the study of microalgae. These protocols involve the mating of mating type plus (mt+) and mating type minus (mt−) gametes of *C. reinhardtii*. The specific cell−cell recognition resulting from flagellar interaction leads to the release of enzymes, autolysin or lysin, that cause cell wall degradation. These enzymes can be purified and used as a pretreatment to transformation. A detailed protocol for production and purification of these enzymes is given by Buchanan and Snell (1988), and a detailed study of the mating process was reported by Hoffmann and Beck (2005).

4.2 Particle Bombardment

Bombardment of target cells with DNA-coated metallic particles is a widespread, simple, effective, and highly reproducible transformation method. This method has been successfully employed for the transformation of most standard cellular expression systems, and it is therefore not surprising that it is also useful for the study of microalgae. The main drawback of the particle bombardment method is the cost of the required specialized equipment. Although the number of transformants recovered following particle bombardment can be low, it remains the most effective method for the transformation of chloroplasts, as it allows for the delivery of multiple copies of recombinant DNA through both the cellular and chloroplast membranes, increasing the chance for a successful integration event to occur (Boynton and Gillham, 1993).

This method has been shown to be effective for the stable nuclear (Mayfield and Kindle, 1990) and chloroplast (Ramesh et al., 2011) transformation of *C. reinhardtii*, the transformation of *Volvox carteri* (Schiedlmeier et al., 1994), *Chlorella sorokiana* (Dawson et al., 1997), *Chlorella ellipsoidea* (Chen et al., 1998), and *Chlorella kessleri* (El-Sheekh, 1999) species, transient transformation of *Haematococcus pluvialis* (Teng et al., 2002), and the stable nuclear transformation of the diatom *Phaeodactylum tricornutum* (Apt et al., 1996). Studies have shown that the particle bombardment method is also effective for the transformation of more complex algal species, such as the multicellular *Gonium pectorale* (Lerche and Hallmann, 2009).

4.3 Glass Beads Method

A simple and effective transformation method consists of agitating cell wall−deficient microalgal cells with recombinant DNA, polyethylene glycol, and glass beads, which greatly increases transformation efficiency. Despite the drop in cell viability to 25% following agitation with the beads, a nuclear transformation efficiency of about 103 transformants/μg DNA was achieved using this method, and an efficiency of 50 transformants/μg DNA was achieved for the transformation of *C. reinhardtii* chloroplasts (Kindle, 1990). Compared to the particle bombardment method, the glass beads method is simpler, more efficient for nuclear transformations, and much less expensive as it does not require specialized equipment. Studies have shown that glass beads method is more efficient than particle bombardment for the transformation of microalga *Dunaliella salina* (Feng et al., 2009).

4.4 Silicon Carbide Whisker Method

A similar protocol, using silicon carbon whiskers instead of glass beads to pierce cells, has also been used successfully (Dunahay, 1993;

Wang et al., 1995). Cells are transformed by mixing with the SiC whiskers and DNA, and vortexing briefly. Although the exact mechanism for whisker-mediated transformation is unknown, it is known that fractured SiC crystals readily form sharp cutting edges. The surface of SiC whiskers is negatively charged, which probably results in there being little affinity between whiskers and DNA in a neutral pH medium. It has been suggested that the whiskers do not carry the DNA into the treated cells but function as needles that facilitate DNA delivery by cell perforation and abrasion during mixing (Wang et al., 1995). The cell viability following agitation is much improved, but due to low transformation efficiencies, high cost of materials, and health concerns associated with the handling of the whiskers, the glass beads are generally preferred. However, SiC whiskers have since been reported to be extremely hazardous to humans, and therefore are rarely used.

4.5 Electroporation

Electroporation or electropermeabilization is a transformation technique that uses induction of macromolecular uptake by exposing cell walls to high-intensity electrical field pulses. The effectiveness of microalgal electroporation was first reported by Brown et al. (1991). Electroporation specifically disrupts lipid bilayers, leading to efficient molecular transport across the plasma membrane. Efficient electroporation-mediated transformation was achieved in both wild-type and cell wall−deficient *Chlamydomonas* cells (Brown et al., 1991). The transformation efficiency of electroporation is two orders of magnitude higher than the glass beads method, and only requires relatively simple equipment. Important parameters affecting the effectiveness of electroporation include field strength, pulse length, medium composition, temperature, and membrane characteristics, as well as the concentration of DNA (Wang et al., 2007). Electroporation was successfully used for the transformation

microalga *D. salina* (Sun et al., 2008), *Dunaliella viridis* (Sun et al., 2006) and *Dunaliella tertiolecta* (Walker et al., 2005), *C. reinhardtii* (Kovar et al., 2002; Ladygin, 2004), *Chlorella* sp. (Wang et al., 2007), and *Nannochloropsis oculata* (Chen et al., 2008; Li and Tsai, 2009).

4.6 *Agrobacterium tumefaciens*-Mediated Transformation

The prospects of *A. tumefaciens*-mediated transformation were first reported in microalgae by Kumar et al. (2004). The microalga *C. reinhardtii* was successfully transformed with *uidA* (β-glucuronidase), *gfp*, and *hpt* (hygromycin phosphotransferase) genes in the presence of acetosyringone (AS). Later, Kumar and Rajam reported a successful transformation of *C. reinhardtii* even in the absence of AS (Kumar and Rajam, 2007), with about 50-fold increase in transformants compared to the commonly used glass beads method for genetic transformation in *C. reinhardtii*.

The major difficulty associated with the nuclear transformation of *C. reinhardtii* was gene silencing, as in many other organisms (Manuell and Mayfield, 2006). Integration of transgenes into the nuclear genome of *C. reinhardtii* and their expression without any silencing was obtained in the studies conducted by Kumar et al. (2004) and Kumar and Rajam (2007). This can be attributed to the possible advantages of *Agrobacterium* T-DNA to be targeted to and integrated at potentially transcribable regions of the genome (Hiei et al., 1994). Similar promising results were observed in the *Agrobacterium*-mediated genetic transformation of other microalgal species, *H. pluvialis* (Kathiresan and Sarada, 2009), *Dunaliella bardawil* (Anila et al., 2011), *Nannochloropsis* sp. (Cha et al., 2011), *Schizochytrium* (Cheng et al., 2012), and *Chlorella vulgaris* (Cha et al., 2012). The *Agrobacterium*-mediated genetic transformation technique is a simple, stable, and efficient method for transgene experiment studies in microalgae.

5. CURRENT STATUS OF ALGAL EXPRESSION SYSTEM

Since the first demonstration of mammalian protein expression in the chloroplast (Mayfield et al., 2003), therapeutic protein expression in microalgae has progressed fairly rapidly. Rasala et al. (2010) attempted the expression of a set of seven recombinant proteins in the chloroplast of *C. reinhardtii* and met with very good success. This work demonstrates that recombinant protein expression in algal chloroplasts is on par with any other expression platform, and shows that expression of complex mammalian proteins is as likely to be achieved in algae as it is in any eukaryotic system. In a recent study, Hepatitis B virus capsid antigen was expressed in the nuclear genome of *C. reinhardtii* (Soria-Guerra et al., 2014). This was the first algal vaccine to be expressed from the nuclear genome, but it only accumulated to 0.05% TSP. More than 25 different proteins of significant therapeutic value have been expressed in microalgae (Table 6).

Algal-produced human vaccine production platform is expected to develop as an alternative for very expensive vaccines like human papillomavirus or for novel vaccines against diseases

TABLE 6   Therapeutic Proteins Expressed in Microalgae

Gene	Function	Source
HSV8-lsc	First mammalian protein-expressed antibody	Mayfield et al. (2003)
CTB-VP1	Cholera toxin B subunit fused to foot and mouth disease VP1	Sun et al. (2003)
HSV8-scFv	Classic single-chain antibody	Mayfield and Franklin (2005)
hMT-2	Human metallothionine-2	Zhang et al. (2006)
hTRAIL	Human tumor necrosis factor-related apoptosis-inducing ligand (TRAIL)	Yang et al. (2006)
M-SAA	Bovine mammary-associated serum amyloid	Manuell et al. (2007)
CSFV-E2	Swine fever virus E2 viral protein	He et al. (2007)
hGAD65	Diabetes-associated autoantigen human glutamic acid decarboxylase 65	Wang et al. (2008)
ARS2-crEpo-his6	Human erythropoietin fused to ARS2 export sequence w/6xhis tag	Eichler-Stahlberg et al. (2009)
83K7C	Full-length IgG1 human monoclonal antibody against anthrax protective antigen 83	Tran et al. (2009)
IgG1	Murine and human antibodies (LC and HC)	Tran et al. (2009)
VP28	White spot syndrome virus protein 28	Surzycki et al. (2009)
CTB-D2	D2 fibronectin-binding domain of *Staphylococcus aureus* fused with the cholera toxin B subunit	Dreesen et al. (2010)
10NF3, 14FN3	Domains 10 and 14 of human fibronectin, potential antibody mimics	Rasala et al. (2010)

Continued

TABLE 6   Therapeutic Proteins Expressed in Microalgae—cont'd

Gene	Function	Source
M-SAA-Interferon β1	Multiple sclerosis treatment fused to M-SAA	Rasala et al. (2010)
Proinsulin	Blood sugar level—regulating hormone, type I diabetes treatment	Rasala et al. (2010)
VEGF	Human vascular endothelial growth factor isoform 121	Rasala et al. (2010)
HMGB1	High-mobility group protein B1	Rasala et al. (2010)
HBsAg	Hepatitis B surface antigen	
*Pfs*25 and *Pfs*28	*Plasmodium falciparum* surface proteins	Gregory et al. (2012)
*Pfs*25 -CTB	*P. falciparum* surface protein *Pfs*25 fused to cholera toxin B subunit	Gregory et al. (2013)
αCD22PE40	Gene encoding a single-chain antibody (scFv) genetically fused to domains II and III of Exotoxin A (PE40) from *Pseudomonas aeruginosa*	Tran et al. (2013)
αCD22CH23PE40	Gene that contained the hinge and CH2 and CH3 domains of a human IgG1 placed between the αCD22 scFv antibody and PE40	Tran et al. (2013)
HBcAg	HBcAg-based antigen carrying angiotensin II	Soria-Guerra et al. (2014)
E7GGG	Human papilloma virus type 16 E7 protein, attenuated mutant	Demurtas et al. (2013)

for which no alternative currently exists (Martinez et al., 2012). Studies of algal-produced vaccine antigens verified that dried algae can be stored at room temperature for long periods of time without losing antigenic properties (Dreesen et al., 2010; Gregory et al., 2013). Microalgae have the potential to revolutionize the way subunit vaccines are made and delivered—from current costly parenteral administration of purified protein to an inexpensive oral algae tablet with effective mucosal and systemic immune reactivity (Specht and Mayfield, 2014).

A significant obstacle for algal protein expression systems is the lack of production systems optimized for large-scale growth and harvesting of algae under photoautotrophic conditions. The main limitations arise from limited gas exchange and light penetration in large cultures, especially at the high cell densities required to keep costs low (Ugwu et al., 2008). Recombinant protein production in algal chloroplast also has some limitations. Proteins expressed in the algal chloroplast cannot be secreted, which means that the cells must be harvested and lysed in order to purify the desired recombinant protein. In addition, like bacteria, the algal chloroplast is not capable of protein glycosylation. However, both these limitations can be overcome if the recombinant gene is expressed from the nuclear genome of microalgae. Nuclear transformation of *C. reinhardtii* is well established, and many heterologous genes have been expressed from the nuclear genome (Leon-Banares et al., 2004). But the disadvantage of recombinant protein expression from the microalgal nuclear genome is product yield. To date, the yield achieved through nuclear expression tends to be much lower compared to chloroplast expression. Because the algal expression system is still in its infancy, there are many potential improvement strategies that still need to be tested to increase recombinant protein accumulation. As each of these improvements is introduced, we can expect production to increase and costs to decrease, potentially making algal recombinant protein production the most cost-effective system for a variety of recombinant proteins.

6. CONCLUSIONS

The research data available till date undoubtedly establish the capacity of microalgae to produce complex therapeutic proteins. Extensive research on optimal transformation constructs and gene optimization have greatly increased yields of recombinant protein, though further work is still needed to address nuclear gene silencing, plastid auto-attenuation, and to optimize reactor design for large-scale use. The use of microalgae for the production of therapeutic proteins has enormous economic and biotechnology promise for the future, as they have the potential to be the alternative to current mammalian, yeast, or bacterial recombinant protein production systems.

References

Anila, N., Chandrashekar, A., Ravishankar, G.A., Sarada, R., 2011. Establishment of *Agrobacterium tumefaciens*-mediated genetic transformation in *Dunaliella bardawil*. Eur. J. Phycol. 46, 36–44.

Apt, K.E., Kroth-Pancic, P.G., Grossman, A.R., 1996. Stable nuclear transformation of the diatom *Phaeodactylum tricornutum*. Mol. Gen. Genet. 252, 572–579.

Benfey, P.N., Ren, L., Chua, N.H., 1990. Tissue-specific expression from CaMV 35S enhancer subdomains in early stages of plant development. EMBO J. 9, 1677–1684.

Blankenship, J.E., Kindle, K.L., 1992. Expression of chimeric genes by the light-regulated *cabII*-1 promoter in *Chlamydomonas reinhardtii*: a *cabII*-1/*nit1* gene functions as a dominant selectable marker in a *nit1- nit2-* strain. Mol. Cell. Biol. 12, 5268–5279.

Boynton, J.E., Gillham, N.W., 1993. Chloroplast transformation in *Chlamydomonas*. Methods Enzymol. 217, 510–536.

Brown, L.E., Sprecher, S.L., Keller, L.R., 1991. Introduction of exogenous DNA into *Chlamydomonas reinhardtii* by electroporation. Mol. Cell. Biol. 11, 2328–2332.

Buchanan, M.J., Snell, W.J., 1988. Biochemical studies on lysin, a cell wall degrading enzyme released during fertilization in *Chlamydomonas*. Exp. Cell. Res. 179, 181–193.

Cha, T.S., Chen, C.F., Yee, W., Aziz, A., Loh, S.H., 2011. Cinnamic acid, coumarin and vanillin: alternative phenolic compounds for efficient *Agrobacterium*-mediated transformation of the unicellular green alga, *Nannochloropsis* sp. J. Microbiol. Methods 84, 430–434.

Cha, T.S., Yee, W., Aziz, A., 2012. Assessment of factors affecting *Agrobacterium* mediated genetic transformation of the unicellular green alga, *Chlorella vulgaris*. World J. Microbiol. Biotechnol. 28, 1771–1779.

Chalfie, M., Tu, Y., Euskirchen, G., Ward, W.W., Prasher, D.C., 1994. Green fluorescent protein as a marker for gene expression. Science 263, 802–805.

Chen, H.L., Li, S.S., Huang, R., Tsai, H.J., 2008. Conditional production of a functional fish growth hormone in the transgenic line of *Nannochloropsis oculata* (Eustigmatophyceae). J. Phycol. 44, 768–776.

Chen, Y., Li, W.B., Bai, Q.H., Sun, Y.R., 1998. Study on transient expression of GUS gene in *Chlorella ellipsoidea* (Chlorophyta), by using biolistic particle delivery system. Chin. J. Oceanol. Limnol. 47, 9–16.

Cheng, R., Ma, R., Li, K., Rong, H., Lin, X., Wang, Z., Yang, S., Ma, Y., 2012. *Agrobacterium tumefaciens* mediated transformation of marine microalgae *Schizochytrium*. Microbiol. Res. 167, 179–186.

Dawson, H.N., Burlingame, R., Cannons, A.C., 1997. Stable transformation of *Chlorella*: rescue of nitrate reductase-deficient mutants with the nitrate reductase gene. Curr. Microbiol. 35, 356–362.

Demurtas, O.C., Massa, S., Ferrante, P., Venuti, A., Franconi, R., Giuliano, G., 2013. A *Chlamydomonas*-derived human papillomavirus 16 E7 vaccine induces specific tumor protection. PLoS One 8, e61473. http://dx.doi.org/10.1371/journal.pone.0061473.

Dove, A., 2002. Uncorking the biomanufacturing bottleneck. Nat. Biotechnol. 20, 777–779.

Dreesen, I.A., Charpin-ElHamri, G., Fussenegger, M., 2010. Heat stable oral alga-based vaccine protects mice from *Staphylococcus aureus* infection. J. Biotechnol. 145, 273–280.

Dunahay, T.G., 1993. Transformation of *Chlamydomonas reinhardtii* with silicon carbide whiskers. Biotechniques 15, 452–455, 7–8, 60.

Eichler-Stahlberg, A., Weisheit, W., Ruecker, O., et al., 2009. Strategies to facilitate transgene expression in *Chlamydomonas reinhardtii*. Planta 229, 873–883.

El-Sheekh, M.M., 1999. Stable transformation of the intact cells of *Chlorella kessleri* with high velocity microprojectiles. Biol. Plant. 42, 209–216.

Feng, S., Xue, L., Liu, H., Lu, P., 2009. Improvement of efficiency of genetic transformation for *Dunaliella salina* by glass beads method. Mol. Biol. Rep. 36, 1433–1439.

Franklin, S., Ngo, B., Efuet, E., Mayfield, S.P., 2002. Development of a GFP reporter gene for *Chlamydomonas reinhardtii* chloroplast. Plant J. 30, 733–744.

Fuhrmann, M., 2002. Expanding the molecular toolkit for *Chlamydomonas reinhardtii*: from history to new frontiers. Protist. 15, 357–364.

Fuhrmann, M., Oertel, W., Hegemann, P., 1999. A synthetic gene coding for the green fluorescent protein (GFP) is a versatile reporter in *Chlamydomonas reinhardtii*. Plant J. 19, 353–361.

Gregory, J.A., Li, F., Tomosada, L.M., Cox, C.J., Topol, A.B., Vinetz, J.M., Mayfield, S.P., 2012. Algae produced Pfs25 elicits antibodies that inhibit malaria transmission. PLoS One 7, e37179. http://dx.doi.org/10.1371/journal.pone. 0037179.

Gregory, J.A., Topol, A.B., Doerner, D.Z., Mayfield, S., 2013. Alga produced cholera toxin pfs25 fusion proteins as oral vaccines. Appl. Environ. Microbiol. 79, 3917–3925. http://dx.doi.org/10.1128/AEM.00714-13.

Griesbeck, C., Kobl, I., Heitzer, M., 2006. *Chlamydomonas reinhardtii*. Mol. Biotechnol. 34, 213–223.

He, D.M., Qian, K.X., Shen, G.F., Zang, Z.F., Li, Y.N., Su, Z.L., Shao, H.B., 2007. Recombination and expression of classical swine fever virus (CSFV) structural protein E2 gene in *Chlamydomonas reinhardtii* chloroplasts. Colloids Surf. B Biointerfaces. 55, 26–30.

Hiei, Y., Ohta, S., Komari, T., Komashiro, T., 1994. Efficient transformation of rice (*Oryza sativa* L.) mediated by *Agrobacterium* and sequence analysis of the boundaries of the T-DNA. Plant J. 6, 271–282.

Hoffmann, X., Beck, C.F., 2005. Mating-induced shedding of cell walls, removal of walls from vegetative cells, and osmotic stress induce presumed cell wall genes in Chlamydomonas. Plant Physiol. 139, 999–1014.

Kathiresan, S., Sarada, R., 2009. Towards genetic improvement of commercially important microalga *Haematococcus pluvialis* for biotech applications. J. Appl. Phycol. 21, 553–558.

Kindle, K.L., 1990. High-frequency nuclear transformation of *Chlamydomonas reinhardtii*. Proc. Natl Acad. Sci. USA 87, 1228–1232.

Kovar, J.L., Zhang, J., Funke, R.P., Weeks, D.P., 2002. Molecular analysis of the acetolactate synthase gene of *Chlamydomonas reinhardtii* and development of a genetically engineered gene as a dominant selectable marker for genetic transformation. Plant J. 29, 109–117.

Kumar, S.V., Rajam, M.V., 2007. Induction of *Agrobacterium tumefaciens vir* genes by the green alga—*Chlamydomonas reinhardtii*. Curr. Sci. 92, 1727–1729.

Kumar, S.V., Misquitta, R.W., Reddy, V.S., Rao, B.J., Rajam, M.V., 2004. Genetic transformation of the green alga—*Chlamydomonas reinhardtii* by *Agrobacterium tumefaciens*. Plant Sci. 166, 731–738.

Ladygin, V.G., 2004. Efficient transformation of mutant cells of *Chlamydomonas reinhardtii* by electroporation. Process Biochem. 39, 1685–1691.

Leon-Banares, R., Gonzalez-Ballester, D., Galvan, A., Fernandez, E., 2004. Transgenic microalgae as green cell factories. Trends Biotechnol. 22, 45–52.

Lerche, K., Hallmann, A., 2009. Stable nuclear transformation of *Gonium pectorale*. BMC Biotechnol. 9, 64.

Li, S.S., Tsai, H.J., 2009. Transgenic microalgae as a non-antibiotic bactericide producer to defend against bacterial pathogen infection in the fish digestive tract. Fish Shellfish Immunol. 26, 316–325.

Lumbreras, V., Stevens, D.R., Purton, S., 1998. Efficient foreign gene expression in *Chlamydomonas reinhardtii* mediated by an endogenous intron. Plant J. 14, 441–447.

Maga, E.A., 2005. Genetically engineered livestock: closer than we think? Trends Biotechnol. 23, 533–535.

Manuell, A.L., Mayfield, S.P., 2006. A bright future for *Chlamydomonas*. Genome Biol. 7, 327.

Manuell, A.L., Beligni, M.V., Elder, J.H., Siefker, D.T., Tran, M., Weber, A., McDonald, T.L., Mayfield, S.P., 2007. Robust expression of a bioactive mammalian protein in *Chlamydomonas* chloroplast. Plant Biotechnol. J. 5, 402–412.

Martinez, C.A., Giulietti, A.M., Talou, J.R., 2012. Research advances in plant-made flavivirus antigens. Biotechnol. Adv. 30, 1493–1505.

Mayfield, S.P., Kindle, K.L., 1990. Stable nuclear transformation of *Chlamydomonas reinhardtii* by using a *C. reinhardtii* gene as the selectable marker. Proc. Natl Acad. Sci. USA 87, 2087–2091.

Mayfield, S.P., Franklin, S.E., Lerner, R.A., 2003. Expression and assembly of a fully active antibody in algae. Proc. Natl Acad. Sci. USA 100, 438–442.

Mayfield, S.P., Franklin, S.E., 2005. Expression of human antibodies in eukaryotic micro-algae. Vaccine 23, 1828–1832.

Raja, R., Hemaiswarya, S., Kumar, N.A., Sridhar, S., Rengasamy, R., 2008. A perspective on the biotechnological potential of microalgae. Crit. Rev. Microbiol. 34, 77–88.

Ramesh, V.M., Bingham, S.E., Andrew, N.W., 2011. A Simple Method for Chloroplast Transformation in *Chlamydomonas reinhardtii*. Photosynthesis Research Protocols. Humana Press, NY, pp. 313–320.

Rasala, B.A., Muto, M., Lee, P.A., Jager, M., Cardoso, R.M., Behnke, C.A., Kirk, P., Hokanson, C.A., Crea, R., Mendez, M., Mayfield, S.P., 2010. Production of therapeutic proteins in algae, analysis of expression of seven human proteins in the chloroplast of *Chlamydomonas reinhardtii*. Plant Biotechnol. J. 8, 719—733.

Rosenberg, J.N., Oyler, G.A., Wilkinson, L., Betenbaugh, M.J., 2008. A green light for engineered algae: redirecting metabolism to fuel a biotechnology revolution. Curr. Opin. Biotechnol. 19, 430—436.

Ruiz-Binder, N.E., Geimer, S., Melkonian, M., 2002. In vivo localisation of centrin in the green alga *Chlamydomonas reinhardtii*. Cell Motil. Cytoskeleton 52, 43—55.

Schiedlmeier, B., Schmitt, R., Müller, W., Kirk, M.M., Gruber, H., Mages, W., Kirk, D.L., 1994. Nuclear transformation of *Volvox carteri*. Proc. Natl Acad. Sci. USA 91, 5080—5084.

Schroda, M., Blocker, D., Beck, C.F., 2000. The HSP70A promoter as a tool for the improved expression of transgenes in *Chlamydomonas*. Plant J. 21, 121—131.

Soria-Guerra, R.E., Ramírez-Alonso, J.I., Ibáñez-Salazar, A., Govea-Alonso, D.O., Paz-Maldonado, L.M.T., Bañuelos-Hernández, B., Korban, S.S., Rosales-Mendoza, S., 2014. Expression of an HBcAg-based antigen carrying angiotensin II in *Chlamydomonas reinhardtii* as a candidate hypertension vaccine. Plant Cell Tissue Organ Cult. 116, 133—139.

Specht, E.A., Mayfield, S.P., 2014. Algal-based oral recombinant vaccines. Front. Microbiol. 5, 1—7.

Stevens, D.R., Purton, S., 1997. Genetic engineering of eukaryotic algae: progress and prospects. J. Phycol. 33, 713—722.

Sun, G.H., Zhang, X.C., Sui, Z.H., Mao, Y.X., 2008. Inhibition of pds gene expression via the RNA interference approach in *Dunaliella salina* (Chlorophyta). Mar. Biotechnol. 10, 219—226.

Sun, M., Qian, K.X., Su, N., Chang, H., Liu, J., Chen, G., 2003. Foot-and-mouth disease virus VP1 protein fused with cholera toxin B subunit expressed in *Chlamydomonas reinhardtii* chloroplast. Biotechnol. Lett. 25, 1087—1092.

Sun, Y., Gao, X.S., Li, Q.Y., Zhang, Q.Q., Xu, Z.K., 2006. Functional complementation of a nitrate reductase defective mutant of a green alga *Dunaliella viridis* by introducing the nitrate reductase gene. Gene 377, 140—149.

Surzycki, R., Greenham, K., Kitayama, K., Dibal, F., Wagner, R., Rochaix, J.D., Ajam, T., Surzyki, S., 2009. Factors effecting expression of vaccines in microalgae. Biologicals 37, 133—138.

Teng, C., Qin, S., Liu, J., Yu, D., Liang, C., Tseng, C., 2002. Transient expression of lacZ in bombarded unicellular alga *Haematococcus pluvialis*. J. Appl. Phyco. 14, 495—500.

Tran, M., Van, C., Barrera, D.J., Pettersson, P.L., Peinado, C.D., Bui, J., Mayfield, S.P., 2013. Production of unique immunotoxin cancer therapeutics in algal chloroplasts. Proc. Natl Acad. Sci. USA 110, 15—22.

Tran, M., Zhou, B., Pettersson, P.L., Gonzalez, M.J., Mayfield, S.P., 2009. Synthesis and assembly of a full-length human monoclonal antibody in algal chloroplasts. Biotechnol. Bioeng. 104, 663—673.

Ugwu, C.U., Aoyagi, H., Uchiyama, H., 2008. Photobioreactors for mass cultivation of algae. Bioresour. Technol. 99, 4021—4028.

Walker, T.L., Becker, D.K., Dale, J.L., Collet, C., 2005. Towards the development of a nuclear transformation system for *Dunaliella tertiolecta*. J. Appl. Phycol. 17, 363—368.

Wang, K., Drayton, P., Frame, B., Dunwell, J., Thompson, J., 1995. Whisker mediated plant transformation—an alternative technology. In Vitro Cell. Dev. Biol. 31, 101—104.

Wang, T.Y., Xue, L.X., Hou, W.H., Yang, B.S., Chai, Y.R., Ji, X., Wang, Y.F., 2007. Increased expression of transgene in stably transformed cells of *Dunaliella salina* by matrix attachment regions. Appl. Microbiol. Biotechnol. 76, 651—657.

Wang, X.F., Brandsma, M., Tremblay, R., Maxwell, D., Jevnikar, A.M., Huner, N., Ma, S., 2008. A novel expression platform for the production of diabetes-associated autoantigen human glutamic acid decarboxylase (hGAD65). BMC Biotechnol. 8, 87.

Yang, Z., Li, y, Chen, F., Li, D., Zang, Z., Liu, Y., Zeng, D., Wang, Y., Shen, G., 2006. Expression of human soluble TRAIL in *Chlamydomonas reinhardtii* chloroplast. Chin. Sci. Bull. 51, 1703—1709.

Zhang, Y.K., Shen, G.F., Ru, B.G., 2006. Survival of human metallothionein-2 transplastomic *Chlamydomonas reinhardtii* to ultraviolet B exposure. Acta Biochim. Biophys. Sin. 38, 187—193.

An Expressed Sequence Tag Database Analysis of Fatty Acid Genes in *Stichococcus bacillaris* Strain Siva2011

Keat H. Teoh, Ganapathy Sivakumar

Arkansas State University, Arkansas Biosciences Institute and College of Agriculture and Technology, Jonesboro, AR, USA

1. INTRODUCTION

Algal biomass can be used in the pharmaceutical or energy sectors by producing useful lipids, hydrocarbons, and valuable small molecules (Sivakumar et al., 2012). Indeed, algae provide a carbon-neutral feedstock to these industries and could help minimize the negative effects of CO_2 in the environment (Sivakumar et al., 2010). The *Stichococcus bacillaris* strain siva2011 (UTEX 3000) is a green microalga that produces significant lipid (over 30%) and biomass (Sivakumar et al., 2014b). In addition, it reproduces quickly and has efficient photosynthetic and lipid biosynthetic mechanisms as well as can tolerate a wide range of environments. The *S. bacillaris* strain siva2011 has shown special commercial potential for several reasons: (1) it biosynthesizes a high degree of unsaturated fatty acids: methyl hexadecatrienoic acid (C16:3), oleic acid (C18:1), linoleic acid (C18:2), and linolenic acid (C18:3); the predominant saturated fatty acid is palmitic acid (C16:0) (Figure 1); (2) it also biosynthesizes

three hydrocarbons such as *n*-nonadecane ($C_{19}H_{40}$), nonacosane ($C_{29}H_{60}$), and heptadecane ($C_{17}H_{36}$), as well as two free fatty acids C16:0 and C18:3 (Sivakumar et al., 2014b); and (3) it has the ability to biosynthesize bioactive vitamin E, *RRR*-α-tocopherol (Sivakumar et al., 2014a). The essential fatty acids and natural antioxidant *RRR*-α-tocopherol can be used in the pharmaceutical sector as an alternative to fish and vegetable oils. The other fatty acids and hydrocarbons can be used in liquid fuels such as biodiesel and jet fuel.

Nevertheless, the *S. bacillaris* strain siva2011 is a new species; the genomic information for the lipid metabolism is unavailable. In order to improve fatty acids or triacylglycerols (TAGs) overproduction in *S. bacillaris* strain siva2011 via metabolic engineering, the fatty acid and TAG pathways genes need to be screened. The lipid metabolism has been extensively studied in microorganisms, plants, and mammals. However, the biosynthesis and regulation of fatty acids and TAGs in microalgae

Handbook of Marine Microalgae
http://dx.doi.org/10.1016/B978-0-12-800776-1.00028-5

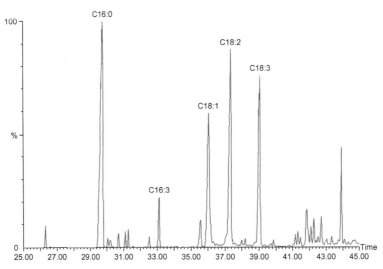

FIGURE 1   Gas chromatography mass spectrometry profile of fatty acid methyl esters from *Stichococcus bacillaris* strain siva2001, biomass from 50-mL flask culture on day 4. Methyl palmitate (C16:0), methyl hexadecatrienoate (C16:3), methyl oleate (C18:1), methyl linoleate (C18:2), and methyl linolenate (C18:3).

remain unclear. Although genome sequence information is available for several microalgae (Khozin-Goldberg and Cohen, 2011; Wang et al., 2014), much of the current understanding of fatty acid and TAG biosynthesis in microalgae is based on knowledge gained in plants and animals. The fatty acid pathway is the key metabolism for TAG or hydrocarbon biosynthesis in *S. bacillaris* strain siva2011, and the manipulation of fatty acid biosynthesis requires a complete understanding of pathway genes and corresponding enzyme functions. Thus, the objective of this research was to identify the likely fatty acid pathway genes that control the fatty acid biosynthesis in *S. bacillaris* strain siva2011. Expressed sequence tag (EST) databases have been established for a number of algal species (Radakovits et al., 2010; Palenik et al., 2007; Grossman, 2005; Armbrust et al., 2010). Access to the sequence information of these genes will greatly facilitate genetic manipulation for high fatty acid production. Identifying *S. bacillaris* strain siva2011 genes that

encode for fatty acids could allow comprehensive characterization of this pathway and knowledge-based modification of lipid metabolism. This chapter reviews current efforts toward screening the fatty acid pathway genes in *S. bacillaris* strain siva2011 using the cDNA library.

2. FATTY ACID, TAG, AND HYDROCARBON BIOSYNTHESIS

The fatty acid and TAG biosynthesis in microalgae occurs in two separate organelles: the fatty acid is synthesized in the plastid and TAG assembly is in the endoplasmic reticulum (ER) (Jillian et al., 2013; Wang et al., 2014). Fatty acid synthesis begins with the carboxylation of plastidial acetyl-CoA to form malonyl-CoA by acetyl-CoA carboxylase. Then malonyl CoA-acyl carrier protein (ACP) acyltransferase catalyzes the transfer of a malonyl group from malonyl-CoA to the malonyl-ACP. The encoded protein may be part of a fatty acid synthase; fatty acids are biosynthesized by a type II fatty acid synthase in an NADPH-dependent reaction. Thus, consecutively adding two carbons following each elongation cycle does the following: (1) condenses acetyl-CoA with malonyl-ACP to form a β-ketoacyl-ACP by action of 3-ketoacyl-ACP synthase; (2) the β-keto group is reduced by the NADPH-dependent β-ketoacyl-ACP reductase to form a β-hydroxy intermediate; (3) the β-hydroxy group is dehydrated by the β-hydroxy acyl-ACP dehydratase to form an enoyl-ACP; and (4) the reduction of the enoyl chain by enoyl-ACP reductase produces an acyl-ACP with an elongated acyl chain. This cycle repeats several times until C14−C20 fatty acids are on ACP; a thioesterase catalyzes hydrolysis of the long-chain fatty acid and release from ACP. Some of these fatty acids are precursors of glycolipids and polar lipids, which are building blocks of the semipermeable bilayer membranes. The fatty acids esterified to acyl-CoA can be channeled through the cytosol into specialized domains of the ER for neutral TAG assembly. In the ER, these fatty acids can be further modified by a variety of membrane-bound desaturases, thus producing mono- and polyunsaturated fatty acids. The esterified free fatty acids are introduced into the different positions of the glycerol-3-phosphate backbone with sequential transfer of acyl groups from the acyl-CoA pool by acyltransferases of the Kennedy pathway to generate the final product, a TAG (Kennedy, 1961). The hydrocarbon biosynthesis from fatty acid is by the reduction of acyl-ACP to fatty aldehydes, which is catalyzed by acyl-ACP reductase. In other words, an acyl-ACP reductase cleaves the thioester bond and uses NADPH to reduce the fatty acid to an aldehyde (Lennen and Pfleger, 2013). The aldehyde decarbonylase catalyzes aldehydes to alkanes via decarbonylation, a cytochrome P450 enzyme catalyze decarboxylative oxidation of fatty acids to matured hydrocarbons (Peralta-Yahya et al., 2012).

3. ESTABLISHMENT OF cDNA LIBRARY

3.1 RNA Isolation

The *S. bacillaris* strain siva2011 cells were cultured according to Sivakumar et al. (2014b). Exponential growth and significant fatty acid biosynthesis were noticed on the fourth day. Thus the full-length cDNA library was constructed from 4-day-old *S. bacillaris* strain siva2011 cells. The total RNA was isolated from *S. bacillaris* strain siva2011 cells following the protocol described in the aurum™ total RNA fatty and fibrous tissue kit (BioRad, Hercules, CA). A DNase I digestion step was included in the extraction to remove any contamination by DNA. Concentrations and integrity of the total RNA were assessed using the experion™ automated electrophoresis system (BioRad). The mRNA was purified from the total RNA using the Dynabeads® mRNA purification kit (Invitrogen, Carlsbad, CA) following the manufacturer's protocol.

3.2 Full-Length cDNA Library Construction

The full-length cDNA of *S. bacillaris* strain siva2011 was constructed using the Super-Script® full-length cDNA library construction kit ll (Invitrogen). The purified mRNA served as a template for the synthesis of first-strand cDNA using an oligo dT primer that included a biotinylated Gateway® recombination site *att*B2. The reverse transcription was carried out with SuperScript® lll RT (Invitrogen). Truncated cDNA/RNA moieties that resulted from the first-strand cDNA synthesis were removed with RNAse I digestion, and full-length cDNAs were selected using a Cap-antibody that recognized the Cap structure at the 5′ end of full-length mRNA. An adapter that included a Gateway® *att*B1 recombination site was ligated to the 5′ end of the cDNA, and the second strand was synthesized using 3′ and 5′ primers that included the attB2 and attB1 sites, respectively. The cDNA library containing both the 3′ and 5′ adapters was cloned into a Gateway® vector, pDNOR™ 222 through Gateway® BP recombination. The cloned cDNAs were transformed into ElectroMAX™ DH10B™ T1 phage-resistant competent cells (Invitrogen) using the Eppendorf electroporator 2510 (Eppendorf, Hamburg, Germany) with the settings at 2.2 kV, 200 Ω, and 25 μF. The cDNA library was plated on Laura Broth (LB) supplemented with 50 μg mL^{-1} kanamycin and grown overnight at 37 °C. The transformed clones were analyzed with restriction digestion as well as colony PCR. The restriction digestion was performed with restriction enzyme BsrG I, and the colony PCR was carried out with primer M13 forward and a primer specific to the 3′ adapter. A total of 2880 individual clones of the cDNA library were selected from solid LB plus kanamycin plates. The clones were grown on 96-well plates containing 200 μL of freezing medium consisting of liquid LB with 50 μg mL^{-1} kanamycin and 8% glycerol. The plates were incubated overnight in a 37 °C shaker platform shaking at 200 rpm.

3.3 Sanger Sequencing

DNA from a saturated culture, transformed with a plasmid from a DNA library and grown in 96-well plates, was amplified by rolling cycle amplification according to the manufacturer's protocol (GE Healthcare Bio-Sciences). Sequencing reactions were performed from only the 5′ end of the clones using ABI prism BigDye terminator cycle sequencing protocols developed by applied biosystems (Perkin-Elmer Corp., Foster City, CA). Cycle sequencing reactions were performed in 10 μL reaction volume by adding 10−20 ng of the purified PCR product, 5 pmol primer, 1 μL BigDye™ terminator, and 2 μL sequencing reaction buffer,

according to the cyclic profile recommended by the manufacturer. Excess dye-labeled terminators were removed by ethanol precipitation. Purified extension products were dried in a SpeedVac (ThermoSavant, Holbrook, NY) and then suspended in Hi-Di formamide. Sequencing reactions were then analyzed on applied biosystems 3730 genetic analyzers using POP-7 sieving matrix and 1X capillary buffer.

3.4 Sequence Assembly and Analysis

Initial assembly of the sequences was performed with Paracel Transcript Assembler (PTA) version 3.0.0 (Paracel Inc, Pasadena, CA). In PTA, all sequences were masked for universal and species-specific vector sequences, adapters, and PCR primers used in cDNA libraries. Escherichia coli contamination as well as mitochondrial and ribosomal RNA genes of S. bacillaris strain siva2011 were identified and removed from input sequences using default settings to ascertain the novelty of the sequences. The poly (A/T) tails and intrinsic repeats, such as simple sequence repeats and short interspersed elements, were annotated prior to clustering and assembly. Low base-call quality data were trimmed from the ends of individual sequences, and the sequences with length <50 bp were excluded from consideration during initial pair-wise comparison. After cleanup, sequences were passed to the PTA clustering module for pair-wise comparison and then to the CAP3-based PTA module for assembly. The large-scale homology searches of the consensus sequences resulting from the PTA assemblies against the NCBI's NR and NT databases using BLAST were conducted using a computational pipeline. To obtain a more complete description of gene function, for each query sequence, the top 100 BLAST hits were retrieved, and the best scoring BLAST hit and the tentative GO classification with e-value ≤ 1e-4 were annotated to query sequences. These GO term assignments were organized around GO hierarchies that were divided into biological processes, cellular components, and molecular functions. Clones with e-values higher than 1e-4 are not statistically significant and therefore are not included in searching for GO terms.

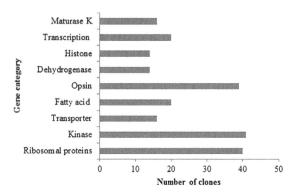

FIGURE 2 Top gene category in *Stichococcus bacillaris* cDNA library.

4. ESTs ANALYSIS OF FATTY ACID PATHWAY GENES IN *S. BACILLARIS* STRAIN SIVA2011

The library consisted of 2880 ESTs. The raw ESTs were subjected to a filtering process that removed 27 sequences that did not meet the filtering criteria. The remaining 2853 sequences were assembled into overlapping contiguous sequence (contigs) and singlets. The assembly resulted in a total of 380 contigs from 1394 sequences and 1459 singlets. These assembled ESTs represent unique transcripts that can be mined for targeted gene discovery. The median length of the assembled ESTs falls within 601–700 nucleotides, and the number of ESTs per contig ranges from 2 to 49.

A homology search of the 1839 assembled sequences against the nonredundant (NR) database at the NCBI using BLASTX yielded 1024 (55.6%) queries with good hits and 815 queries without hits based on the e-value cutoff of less than or equal to 1e-4. Of the 1024 queries with good hits, 63% have at least a match with sequences from fungi, 8% with sequences from plants, 1% with sequences from algae, and the remainder 28% with sequences from nonplant organisms. The top nine categories of proteins represented in the cDNA library are opsin, kinase, ribosomal, transcription, maturase K, fatty acid, histone, dehydrogenase, and transporter (Figure 2). A few of the clones in the *S. bacillaris* strain siva2011 cDNA library showed high-sequence homology (e < 1e-34) to genes associated with fatty acid biosynthesis, modification and metabolism such as acetyl-CoA

carboxylase, ACP, 3-ketoacyl-ACP synthase, β-ketoacyl-ACP reductase, fatty acid-binding proteins, and palmitoyl thioesterase (Table 1). The identified likely genes and corresponding clones were as follows.

4.1 Acetyl-CoA Carboxylase

In green algae, fatty acids are mainly biosynthesized from acetyl-CoA in plastids. Acetyl-CoA carboxylase (ACCase) catalyzes the first committed step in the de novo synthesis of fatty acids by converting acetyl-CoA to malonyl-CoA (Liu and Benning, 2013). Malonyl-CoA is an important intermediate that serves as a carbon donor for fatty acid synthesis. There are two physically distinct types of ACCases: heteromeric and homomeric. The heteromeric type of ACCase is a multifunctional enzyme complex consisting of multiple subunits. The heteromeric type ACCase is found in prokaryotes and in the plastids of plants. The homomeric ACCase consists of a single large polypeptide found in the cytosol of plants and other eukaryotes (Sasaki and Nagono, 2004). ACCases isolated from two unicellular algae, the diatom *Cyclotella cryptica* (Roessler, 1990) and *Isochrysis galbana* (Livne and Sukenik, 1990), were very similar to ACCases in the plastids of higher plants in that

TABLE 1 Putative Fatty Acid Genes Represented in the cDNA Library of *Stichococcus bacillaris* Strain siva2011

Clones	Amino acid sequences	Putative gene
SB1H07	GSSRCHLFPTIRFNPVVPRHGCKAMGTHIPRLHAYLPMCTLCLGGLLIANADPR FLPTAI	β-subunit of acetyl-CoA carboxylase (ACCase)
SB3G08	MMSLNCRCTAPTAVRTSFRSGAPLAARRVAVSHRAARLVTRAAVDKDAVLEDVR GIISEQLGKSKDEVTANAKFVDLGADSLDTVEIMMALEEKFGLELDEEGAEKIATV QEAADLISSQVADK	Acyl carrier protein (ACP)
SB4A01	MPSKQRGPAYVLGVGMTKFIKPRGKVDYTELGFEAGVKALLDAQINYDDVDQGV ACYCYGDSTCGQRVFYQFGMTQIPIYNVNNNCSTGSTGLAMARTLVGSGAADCV MVVGFEKMMPGSLQSFFNDRENPTGTTVKMMAETRGVTNSPGAAQMFGNAGR EYMEKYGARPRTLLRSPA	3-Ketoacyl-ACP synthase (KAS III)
SB16C12[a]	MADITIYPSLRDKVVLITGGADGIGAAAVELFCRQNARVVFLDIDTDRANALIDRL KSELPSVPVPQFKHCDVTNLDRLKTCAEKTLAAHEGRVDVLVNNAFGSSPKTKAP TSEITPESFDFDINA**S**LRHQFFLTQYIVPSMRSRGSGSIVNMG**S**TWRIPATDVPVY**S**T A**K**AAVLGMTRVHAREFGVDGVRVNSVMPGSTATRRQRELVLTQEYEAMTMETQ A**I**RVVEPIEVARVILFLASDDASAVTGGSHVVDGGWVGDT	β-ketoacyl-ACP reductase
Contig 100	MALKDDKFPSSWAFDSISSALEGEQERKDAIKQGKAIFGFTLKNTEGETASWHIDL KTTGTVGKGLGEKPTVTLILSDADFGELIAGKANAQKMFMSGRMKIKGDVMKAT KLEPILKKAQTKSKL	Fatty acid binding protein
SB6C06	LWHITTSGEEKLTHNTLTMRSITNSVAAVLLLAGQTAYAQSDLRAYDESDDTPLPL VIWHGLGDSFDGDGIQQVGRLAEEIHPGTFVYTVQLAADGNGDRSATFFGNVTQ QIESVCEALAEHPILSTAPAIDAIGFSQGGQFLRGYVEQCNYPPIRSLVTFGSQHNGI IEFRACGTTDWLCRSAMALLRFNTWSSFVQNRLVPAQYYRDPSTPETYETYLKNSN FLADANNERILKNTKYADNIAQLKNFV	Palmitoyl thioesterase

Amino acid residues in bold and underlined are conserved residues of the active site.

[a] *Amino acid residues underlined represent conserved NADP binding motif in SDR proteins.*

they consist of multiple subunits. SB1H07 is a partial clone in the *S. bacillaris* strain siva2011 cDNA library, which consists of 265 nucleotides (Table 1). The clone contains a translated protein of 60 amino acid residues that was used to search the protein database for homologs. At the top of the hits were two proteins identified as the β-subunit of ACCase of the plants, *Silene furticosa* and *Nothofagus nitida*. There is 39% identity between the amino acid sequences of the top two hits and the translated protein of the clone.

4.2 Acyl Carrier Protein

ACP is a carrier of acyl intermediates during fatty acid synthesis. It is a small and highly conserved protein present in all organisms. It is a monomeric protein closely associated with the other elements of the fatty acid synthase complex. The clone SB3G08 for ACP in the *S. bacillaris* strain siva2011 cDNA library is 788 nucleotides long and includes an open reading frame (ORF) of 375 nucleotides (Table 1). The translated ORF, which has 124 amino acid residues, shows highest homology to the plastid ACP of microalgae *Coccomyxa subellipsoidea* and *Heaematcoccus pluvialis* with 66% and 58% identity, respectively. A conserved acpP domain is identified in the translated sequence (Nguyen et al., 2014). This clone is a member of the PP-binding superfamily, which includes ACPs. Members of the PP-binding family have an attachment site for a 4'-phosphopantetheine prosthetic group through a serine residue. This prosthetic group acts as a "swinging arm" for the attachment of activated fatty acid through a thiol ester bond at the distal thiol of the phosphopantetheine moiety.

4.3 Ketoacyl-ACP Synthase

The clone SB4A01 in the *S. bacillaris* strain siva2011 cDNA library showed a very strong

homology at the amino acid level with lipid transfer proteins (Table 1). The amino acid identity with the lipid transfer proteins of plants and fungi is greater than 80%, and the e-value is <1e-100. Further analysis of the amino acid sequence revealed the presence of a sterol carrier protein (SCP)-x thiolase domain and one of the three cysteine residues that make up the active site of this domain. The thiolase domain is associated with SCP-x isoforms, which might be a peroxisome-associated thiolase. The SCP-x proteins are members of the condensing enzyme superfamily. Ketoacyl-ACP synthase (KAS III) is one of the three condensing enzymes used in fatty acid biosynthesis. KAS III initiates the fatty acid elongation in the type II fatty acid synthase system found in bacteria and plants. The elongation begins with the condensation reaction between acetyl-CoA and malonyl-ACP. KAS III directs this process by specifically using acetyl-CoA over acyl-CoA. The resulting product is a 4-carbon acetoacetyl-ACP that serves as a substrate for the next four sequential reactions involving the addition of two-carbon units (Jillian et al., 2013).

4.4 β-Ketoacyl-ACP Reductase

The *S. bacillaris* strain siva2011 cDNA of clone SB16C12 is 1006 nucleotides long and includes an ORF consisting of 795 nucleotides (Table 1). The translated protein derived from the ORF has 264 amino acid residues. A sequence homology search based on the amino acid sequences produces strong matches with two β-ketoacyl-ACP reductases of *Metarhizium acridum* at 66% and 65% amino acid identity, respectively. Further analysis of the amino acid sequence of clone SB16C12 shows the presence of a short chain dehydrogenase/reductase (SDR) domain, the conserved NADP-binding motif, TGXXX [AG]XG, and all the four conserved amino acid residues of the active site. It is very likely that clone SB16C12 belongs to the family of SDR proteins. SDRs are a functionally diverse family of oxidoreductases that catalyze oxidation and reduction reactions and often require cofactors such as NADP or NADPH. The NADPH-dependent reduction of β-ketoacyl-ACP to β-hydroxyacyl-ACP is catalyzed by β-ketoacyl reductase, which is the first reductive step in the elongation cycle of fatty acid biosynthesis (Price et al., 2001).

4.5 Fatty Acid Binding Protein

Three clones in the *S. bacillaris* strain siva2011 cDNA library have sequences that overlap each other to form contig100 (Table 1). This contig is 663 nucleotides long and includes an ORF that has 378 nucleotides. The protein translated from the sequence of the ORF has 125 amino acid residues. Results from the homology search with the translated protein showed that contig100 suggests a sequence very similar to the fatty acid binding protein of various fungi. The highest similarity at 76% and 75% identity is with the fatty acid binding proteins of *Gaeumannomyces graminis* and *Neurospora crassa*, respectively. A search for the conserved domain within the translated sequence identified a sterol carrier protein 2 (SCP2) domain. This domain is involved in the binding of sterols. The fatty acid binding proteins are also known as lipid transfer proteins; they are a family of carrier proteins for fatty acids or other lipophilic substances. These proteins were proposed to facilitate the transfer of fatty acids across the plastid membranes (Weisiger, 2002).

4.6 Palmitoyl Thioesterase

The *S. bacillaris* strain siva2011 cDNA clone SB6C06 is 755 nucleotides long (Table 1). The translated protein shares the highest identity, 75% and 73%, with palmitoyl protein thioesterases from two fungi, *Glomerella graminicola* and *Colletotrichum higginsianum*, respectively. This clone harbors a conserved palm thioest

domain. Proteins with this domain are part of the palmitoyl thioesterase superfamily. The thioesterases play a role in the termination of fatty acid synthesis. Thioesterases are a class of enzymes that hydrolyze the thioester bond between the acyl chain and the ACP moiety. This hydrolysis reaction releases the unesterified fatty acid, allowing the fatty acid to be transported outside the plastid for further modification in the ER. Palmitoyl thioesterases from plants have been shown to have a preference for C16:0 fatty acids (Jones et al., 1995). This clone does not appear to contain the complete protein. It has only about 75% of the amino acid residues reported for palmitoyl thioesterases in the NCBI protein database. All the missing amino acids are in the C-terminal of the protein.

Lu et al. (2011) reported that in transgenic oilseeds the expression level of key enzymes for biosynthesis of fatty acids was usually low. To enhance production, additional genes from natural species are required. Since *S. bacillaris* strain siva2011 biosynthesizes significant amounts of fatty acids, the strain could be used as a potential candidate. To clone genes encoding enzymes for fatty acids in *S. bacillaris* strain siva2011, a reference transcriptome is necessary. A wide array of next-generation sequencing efforts and transcriptome annotation was established in microalgae to screen the fatty acids pathway genes (Lv et al., 2013; Zheng et al., 2013). In addition, complete or semicomplete genome sequences of *Chlamydomonas reinhardtii* (Merchant et al., 2007), *Phaeodactylum tricornutum* (Bowler et al., 2008), etc. are available, which could help comparative genomic analysis of the algal fatty acid pathway. Time-course transcriptome analyses of *S. bacillaris* strain siva2011 could facilitate identification of the upregulated fatty acids pathway genes and their expression profiles. Once identified, these genes could enhance the fatty acids production via metabolic engineering, which might be a robust way to produce natural fatty acids for pharmaceutical and energy industries.

5. CONCLUSIONS

A total of six clones in the *S. bacillaris* strain siva2011 cDNA library were identified as potential fatty acid genes based on their high-sequence homology to fatty acid genes in the NR database of NCBI. Four clones have complete ORFs that can be cloned and expressed in the *E. coli* system. The functions of their proteins need to be validated and characterized through biochemical assays with substrates appropriate for each protein. Full-length EST can be obtained for the two clones that have incomplete ORF through the 5′ and 3′ rapid amplification of cDNA end reactions. The information gained from the cDNA library can add to the growing understanding of fatty acid biosynthesis in *S. bacillaris* strain siva2011. The TAGs pathway conversion genes from the ER were not represented in this library. To identify these genes in *S. bacillaris* strain siva2011, an assembled reference transcriptome is necessary. The established *S. bacillaris* strain siva2011 cDNA library and other available microalgae EST libraries could serve as valuable resources for identifying lipid biosynthesis genes as well as genes regulating the lipid pathway. The primary sequence data from these ESTs can be used to modify the fatty acids metabolic pathway of *S. bacillaris* strain siva2011 to improve lipid production.

Acknowledgments

This research was funded by the Arkansas Biosciences Institute grant (# 262178 and 200109).

References

Armbrust, E.V., Berges, J.A., Bowler, C., Green, B.R., Martinez, D., Putnam, N.H., Zhou, S., Allen, A.E., Apt, K.E., Bechner, M., Brzezinski, M.A., Chaal, B.K., Chiovitti, A., Davis, A.K., Demarest, M.S., Detter, J.C., Glavina, T., Goodstein, D., Hadi, M.Z., Hellsten, U., Hildebrand, M., Jenkins, B.D., Jurka, J., Kapitonov, V.V., Kröger, N., Lau, W.W., Lane, T.W., Larimer, F.W., Lippmeier, J.C., Lucas, S., Medina, M., Montsant, A., Obornik, M., Parker, M.S., Palenik, B., Pazour, G.J., Richardson, P.M., Rynearson, T.A., Saito, M.A., Schwartz, D.C., Thamatrakoln, K., Valentin, K., Vardi, A., Wilkerson, F.P., Rokhsar, D.S., 2010. The genome of the diatom *Thalassiosira pseudonana*: ecology, evolution, and metabolism. Science 306, 79—86.

Bowler, C., Allen, A.E., Badger, J.H., Grimwood, J., Jabbari, K., Kuo, A., Maheswari, U., Martens, C., Maumus, F., Otillar, R.P., 2008. The *Phaeodactylum* genome reveals the evolutionary history of diatom genomes. Nature 456, 239—244.

Grossman, P., 2005. Path toward algal genomics. Plant Physiol. 137, 410—427.

Jillian, L., Blatti, J.L., Michaud, J., Burkart, M.D., 2013. Engineering fatty acid biosynthesis in microalgae for sustainable biodiesel. Curr. Opin. Chem. Biol. 17, 496—505.

Jones, A., Davies, H.M., Voelker, T.A., 1995. Palmitoyl-acyl carrier protein (ACP) thioesterase and the evolutionary-origin of plant acyl-ACP thioesterases. Plant Cell 7, 359—371.

Kennedy, E.P., 1961. Biosynthesis of complex lipids. Fed. Proc. Am. Soc. Exp. Biol. 20, 934—940.

Khozin-Goldberg, I., Cohen, Z., 2011. Unraveling algal lipid metabolism: recent advances in gene identification. Biochimie 93, 91—100.

Lennen, R.M., Pfleger, B.F., 2013. Microbial production of fatty acid-derived fuels and chemicals. Curr. Opin. Biotechnol. 24, 1044—1053.

Liu, B., Benning, C., 2013. Lipid metabolism in microalgae distinguishes itself. Curr. Opin. Biotechnol. 24, 300—309.

Livne, A., Sukenik, A., 1990. Acetyl-coenzyme a carboxylase from the marine Prymnesiophyte *Isochrysis galbana*. Plant Cell Physiol. 31, 851—858.

Lu, C., Wallis, J.G., Browse, J., 2011. Construction of a full-length cDNA library from castor endosperm for high-throughput functional screening. Methods Mol. Biol. 729, 37—52.

Lv, H., Qu, G., Qi, X., Lu, L., Tian, C., Ma, Y., 2013. Transcriptome analysis of *Chlamydomonas reinhardtii* during the process of lipid accumulation. Genomics 101, 229—237.

Merchant, S.S., Prochnik, S.E., Vallon, O., Harris, E.H., Karpowicz, S.J., Witman, G.B., Terry, A., Salamov, A., Fritz-Laylin, L.K., Maréchal-Drouard, L., 2007. The *Chlamydomonas* genome reveals the evolution of key animal and plant functions. Science 318, 245.

Nguyen, C., Haushalter, R.W., Lee, D.J., Markwick, P.R., Bruegger, J., Caldara-Festin, G., Finzel, K., Jackson, D.R., Ishikawa, F., O'Dowd, B., McCammon, J.A., Opella, S.J., Tsai, S.C., Burkart, M.D., 2014. Trapping the dynamic acyl carrier protein in fatty acid biosynthesis. Nature 505, 427—431.

Palenik, B., Grimwood, J., Aerts, A., Rouze, P., Salamov, A., Putnam, N., Dupont, C., Jorgensen, R., Derelle, E., Rombauts, S., Zhou, K., Otillar, R., Merchant, S.S., Podell, S., Gaasterland, T., Napoli, C., Gendler, K., Manuell, A., Tai, V., Vallon, O., Piganeau, G., Jancek, S., Heijde, M., Jabbari, K., Bowler, C., Lohr, M., Robbens, S., Werner, G., Dubchak, I., Pazour, Q., Ren, I., Paulsen, C., Delwiche, J., Schmutz, D., Rokhsar, Y., Van de Peer, H., Moreau, I.V., Grigoriev, G.J., 2007. The tiny eukaryote *Ostreococcus* provides genomic insights into the paradox of plankton speciation. Proc. Natl. Acad. Sci. USA 104, 7705–7710.

Peralta-Yahya, P.P., Zhang, F., del Cardayre, S.B., Keasling, J.D., 2012. Microbial engineering for the production of advanced biofuels. Nature 488, 320–328.

Price, A.C., Zhang, Y.M., Rock, C.O., White, S.W., 2001. Structure of β-ketoacyl-[acyl carrier protein] reductase from *Escherichia coli*: negative cooperativity and its structural basis. Biochemistry 40, 12772–12781.

Radakovits, R., Jinkerson, R.E., Darzins, A., Posewitz, M.C., 2010. Genetic engineering of algae for enhanced biofuel production. Eukaryot. Cell. 4, 486–501.

Roessler, P.G., 1990. Purification and characterization of acetyl-coA carboxylase from the diatom *Cyclotella cryptica*. Plant Physiol. 92, 73–78.

Sasaki, Y., Nagono, Y., 2004. Plant acetyl-coA carboxylase: structure, biosynthesis, regulation, and gene manipulation for plant breeding. Biosci. Biotechnol. Biochem. 68, 1175–1184.

Sivakumar, G., Vail, D.R., Xu, J., Burner, D.M., Lay, J.O., Ge, X., Weathers, P.J., 2010. Bioethanol and biodiesel: alternative liquid fuels for future generations. Eng. Life Sci. 10, 8–18.

Sivakumar, G., Xu, J., Thompson, R.W., Yang, Y., Randol-Smith, P., Weathers, P.J., 2012. Integrated green algal technology for bioremediation and biofuel. Bioresour. Technol. 107, 1–9.

Sivakumar, G., Jeong, K., Lay, J.O., 2014a. Biomass and *RRR-α*-tocopherol production in *Stichococcus bacillaris* strain siva2011 in a balloon bioreactor. Microb. Cell Fact. 13, 79.

Sivakumar, G., Jeong, K., Lay, J.O., 2014b. Bioprocessing of *Stichococcus bacillaris* strain siva2011. Biotechnol. Biofuels. 7, 62.

Wang, D., Ning, K., Li, J., Hu, J., Han, D., Wang, H., Zeng, X., Jing, X., Zhou, Q., Su, X., Chang, X., Wang, A., Wang, W., Jia, J., Wei, L., Xin, Y., Qiao, Y., Huang, R., Chen, J., Han, B., Yoon, K., Hill, R.T., Zohar, Y., Chen, F., Hu, Q., Xu, J., 2014. *Nannochloropsis* genomes reveal evolution of microalgal oleaginous traits. PLOS Genet. 10, e1004094.

Weisiger, R.A., 2002. Cytosolic fatty acid binding proteins catalyze two distinct steps in intracellular transport of their ligands. Mol. Cell Biol. 239, 35–43.

Zheng, M., Tian, J., Yang, G., Zheng, L., Chen, G., Chen, J., Wang, B., 2013. Transcriptome sequencing, annotation and expression analysis of Nannochloropsis sp. at different growth phases. Gene 523, 117–121.

Microalgae-based Wastewater Treatment

Cynthia Alcántara[1], Esther Posadas[1], Benoit Guieysse[2], Raúl Muñoz[1]

[1]Valladolid University, Department of Chemical Engineering and Environmental Technology, Valladolid, Spain; [2]Massey University, School of Engineering and Advanced Technology, Palmerston North, New Zealand

1. INTRODUCTION

Maturation and facultative ponds provide simple and cost-effective wastewater treatment (WWT) capabilities that are used by thousands of communities worldwide (Shilton, 2005). However, while microalgae contribute to pollutant removal in these systems (Rittmann and McCarty, 2001), neither facultative nor maturation ponds are specifically designed and operated to optimize algal biomass productivity and recovery (Craggs et al., 2012). These systems are therefore not considered as representative of microalgae-based WWT in this chapter where an emphasis is instead given to WWT in high-rate algae ponds (raceway ponds). This technology has been shown to support efficient treatment for a broad range of influents (García et al., 2000; De Godos et al., 2009a; Tarlan et al., 2002; Park et al., 2011a), including at full scale in 5-ha ponds (Craggs et al., 2012). As witness of the vitality of this field, several 10-ha facilities were being commissioned across the world at the time of writing.

Table 1 describes key wastewater pollutants and the typical mechanisms used to remove them during conventional and microalgae-based treatment. In the latter, Microalgae photosynthesis takes place 'parallel' to normal biological reactions occurring during conventional aerobic biological treatment (Table 1), thereby boosting nutrient assimilation while providing oxygen to aerobic heterotrophs. A key driver behind the current popularity of microalgae-based WWT is that this technology offers synergetic advantages for algal biofuel feedstock production and biological WWT: Indeed, *in situ* photosynthetic aeration reduces the costs and environmental impacts associated with conventional mechanical aeration, while photosynthetic productivity improves the recovery of energy and/or nutrients via biomass anaerobic digestion or biomass use as slow-release fertilizer (Mulbry et al., 2005; Alcántara et al., 2013). Finally, the high pH and dissolved O_2 concentrations induced by microalgal photosynthesis can also enhance nutrient and heavy metal removal and trigger pathogen deactivation, as discussed below.

TABLE 1 Summary of the Main Pollutants Found in Wastewater and Conventional versus Microalgae-based Removal Mechanisms

Pollutants (associated risks)	Conventional treatment mechanisms	Microalgae-based WWT
Suspended solids (reduce water clarity, clog waterways, many solids are also biodegradable organic pollutants).	Gravity settling prior to or during biological treatment.	Microalgae-based WWT systems are well mixed and not designed for the removal of inert (nonbiodegradable) solids. Prior primary settling of suspended solids may often be needed to avoid operational issues and improve light penetration in the influent.
Biodegradable organic pollutants[a] (uncontrolled biodegradation triggers oxygen depletion in wastewater discharge recipients, some organics are also hazardous).	Aerobic or anaerobic biodegradation by heterotrophs, the former often preferred for treating low-strength effluents such as domestic wastewater and inhibitory and/or recalcitrant influents such as industrial wastewaters.	Aerobic biodegradation by heterotrophs, microalgae provide additional metabolic abilities and support aerobic heterotrophic degraders via exchange of substrates (Section 2.).
Nutrients[b] (trigger uncontrolled algae growth in recipient ecosystems).	Microbial assimilation, nitrification/denitrification processes (see Section 2.2), biological phosphate removal via accumulation of poly-P in bacterial cells, chemical phosphate precipitation.	Microalgae photosynthesis enhances nutrient assimilation by boosting biomass productivity. Environmental conditions also favor N volatilization and P precipitation (Section 2.2). Certain algae are capable of intracellular P accumulation as stable poly-P.
Pathogens (active vectors of human and animal diseases).	Disinfection in maturation ponds, chemical disinfection (e.g., UV irradiation, chlorine, ozone).	Microalgae photosynthesis can promote pathogen disinfection by contributing to raise pH and dissolved oxygen concentration. The high-illuminated surface/volume ratio also favors disinfection via UV radiation (Section 2.3).
Heavy metals (acute and chronic toxicity, bioaccumulation).	Chemical precipitation (lime), electrochemical precipitation, ion exchange, adsorption onto GAC[c] or various biomaterials (e.g., waste wood chips, bacteria or yeasts from industrial fermentations).	Microalgae-based WWT has the potential to generate significant amounts of cost-effective biosorbent (Section 2.4). Algae photosynthesis can also favor heavy metal precipitation at high pH.

[a] Typically expressed as bCOD (biodegradable chemical oxygen demand, $g\ m^{-3}$), which represents the amount of biodegradable organic pollutants expressed as the amount of oxygen required for their oxidation to CO_2, or BOD_5 (biochemical oxygen demand), which represents the amount of oxygen required by aerobic heterotrophs to convert biodegradable organic pollutants into biomass and CO_2 in 5 days. Typically, $bCOD = 1.6\ BOD_5$ in domestic wastewater.

[b] In the context of WWT, nutrients include nitrogen (mainly found as NH_4^+ in domestic wastewater, minor forms including NO_3^-, NO_2^- and organic-N) and phosphorus (mainly found as phosphate and organic phosphate).

[c] Granulated activated carbon.

2. MECHANISMS OF POLLUTANT REMOVAL

Microalgae productivity is dictated by numerous biological (e.g., photosynthetic response), environmental (e.g., temperature), design, and operational factors during WWT. Photosynthetic productivity can be reasonably estimated based on the average light intensity reaching a given specific location. For example, assuming a photosynthetic efficiency (PE) of 2.5% of the total light irradiance and a biomass heat value of $21 \, kJ \, g^{-1}$, a yearly average productivity of around $14-21 \, g$ biomass m^{-2} per day can be expected depending on location (Table 2).

Given that heterotrophic growth would typically yield $\sim 3 \, g$ biomass m^{-2} per day during WWT in high-rate algal ponds (HRAPs) (Box 1), a total biomass productivity of $19-26 \, g \, m^{-2}$ per day would be expected, which is consistent with experimental rates reviewed by Park et al. (2011a). These productivity estimates can be used to predict photosynthetic oxygenation and nutrient assimilation rates as detailed below. It should be noted that these productivities can only be achieved if inorganic carbon supply does not limit photosynthesis, a condition that often requires external CO_2 supply during episodes of strong photosynthetic activity (Park and Craggs, 2010).

2.1 Organic Pollutant Removal

Numerous microalgae are capable of growing photoautotrophically using inorganic carbon and sunlight as carbon and energy sources, respectively, heterotrophically using organic compounds as carbon and energy sources, and mixotrophically by combining phototrophic and heterotrophic metabolisms (Barsanti and Gualteri, 2006). In addition, microalgae synergistically interact with aerobic heterotrophs and autotrophs via exchange of substrates (Figure 1). These metabolic abilities and interactions support the biodegradation of biodegradable organic pollutants (i.e., biodegradable Chemical Oxygen Demand (bCOD)), reduce energy use and its associated impacts (see Box 1 and Section 6), and reduce the risks associated with hazardous pollutants and pathogens (Muñoz et al., 2004).

Considering an HRAP with a typical depth of 0.3 m and microalgae productivity of 20 g algae m^{-2} per day, the HRT theoretically required for stabilizing 200 g O_2 m^{-3} of BOD would be approximately 4 days (i.e., a surface loading of 0.075 m^3 wastewater m^{-2} land per day) when process oxygenation is only provided by O_2 diffusion from the atmosphere. However, this HRT decreases to 1.3 days (0.23 m^3 $(m^2)^{-1}$ per day) when active photosynthetic oxygenation supplies oxygen demand (typically 1.5 g of O_2

TABLE 2 Predicted Yearly Average Photosynthetic Algal Biomass, Oxygenation, and Carbon Fixation Productivities at Various Locations (Guieysse et al., 2013a)

Climate	Solar irradiance (GJ m^{-2} per year)	Average productivity (g m^{-2} per day)		
		Algal biomass	O_2	C
Arid	7.73	21.4	33.2	11.21
Mediterranean	6.59	18.3	28.4	9.59
Subtropical	6.41	17.8	27.6	9.33
Tropical	5.01	13.9	21.5	7.28
Temperate	5.44	15.1	23.4	7.91

PE = 2.5%, biomass heat value = 21 kJ g^{-1}.

BOX 1

CASE STUDY OF PRIMARY-SETTLED DOMESTIC WASTEWATER TREATMENT IN HRAPS

Input data and assumptions: The average flow rate of 0.325 m^3 per capita per day and primary-settled influent loading factors of 68 g bCOD per capita per day, 13.3 g N per capita per day, and 3.28 g P per capita per day were based on the typical values for US individual residences (Metcalf and Eddy, 2003), assuming 50% bCOD removal efficiency during primary treatment and a bCOD/BOD_5 ratio of 1.6. The HRAP was designed based on a depth of 0.3 m and an HRT of 6 days. Heterotrophic growth was modeled using standard kinetics for aerobic wastewater treatment with activate sludge at 20 °C (Metcalf and Eddy, 2003) assuming well-mixed steady-state conditions. Algal productivity was estimated to be 20 $g\,m^{-2}$ per day based on simulations shown in Table 2. The N and P contents of the heterotrophic and phototrophic biomasses were estimated to be 9% and 1%, respectively.

Results: The net yield of heterotrophic growth at 20 °C was estimated to be 0.30 g biomass dry weight g^{-1} bCOD loaded (computation not shown), yielding a productivity of 20.5 g biomass per capita per day and N and P assimilatory removals of 13.9% and 6.25%, respectively. The area required for treatment, 6.5 m^2 per capita per day, supported a high phototrophic productivity (130 g per capita per day) and thereby significantly improved N (100%) and P assimilation (46%). Photosynthetic aeration (195 g per capita per day) exceeded heterotrophic oxygen demand

(42.9 g per capita per day) and boosted potential energy recovery from biomass more than six-fold. An energy input for mixing of 0.023 kWh m^{-3} wastewater treated (0.16 $W\,m^{-3}$ reactor or 0.0074 kWh per capita per day) was calculated based on Borowitzka (2005) assuming a pond channel width of 6 m, a mean fluid velocity of 0.15 m^{-1}, a paddle wheel efficiency of 0.17, and a Manning coefficient of 0.012 (smooth plastic on granular earth). Based on an electricity generation carbon footprint of 362 (EU) - 610 (USA) g $CO_2\,kWh^{-1}$ (IPCC, 2014), this energy input would generate a CO_2 footprint of 8.3–14 g $CO_2\,m^{-3}$ wastewater treated (2.7–4.6 g CO_2 per capita per day). In comparison, the same influent treated using the activated sludge process (energy input 0.33–0.62 kWh m^{-3}, Plappally and Lienhard, 2012) would contribute to 119–378 g $CO_2\,m^{-3}$ wastewater treated (38.7–123 g CO_2 per capita per day). Anaerobic digestion of the biomass generated could generate 48 L CH_4 per capita per day (0.736 kWh per capita per day) during microalgae-based WWT against 6.56 L CH_4 per capita per day (0.100 kWh per capita per day) during conventional treatment, thus yielding energy ratios of nearly 100 against and 0.5–1 (assuming 80% digestion efficiency, methane yield of 0.4 $L\,g^{-1}$ biomass loaded, methane high heat value of 55 $kJ\,g^{-1}$, methane density of 0.67 $kg\,m^{-3}$).

per g alga cultivated when ammonium is the only N source) (Figure 2). HRAP can therefore entail a considerable reduction in HRT/land use compared to stabilization ponds ($\approx$ 15–30 days) (Kivaisi, 2001). Although HRAP are typically

operated at high HRT in comparison to activated sludge processes (Metcalf and Eddy, 2003), photosynthetic oxygenation significantly lowers energy demand, which represents an economic and environmental advantage (40–60% of the

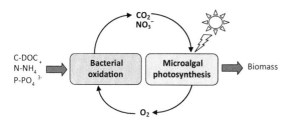

FIGURE 1 Principle of photosynthetic oxygenation (C-DOC = dissolved organic carbon).

total economic cost in a WWTP using activated sludge as secondary treatment is associated with the mechanical O_2 supply, see Box 1) (Chae and Kang, 2013).

2.2 Nitrogen and Phosphorus Removal

Several mechanisms cause nutrient removal during microalgae-based WWT. Under typical operational conditions in HRAPs, assimilation and abiotic mechanisms can support N and P removal efficiencies as high as 90—98% and 90—95% (De Godos et al., 2010; Posadas et al., 2014b).

2.2.1 Assimilatory Nutrient Removal

Nutrient removal via assimilation is limited during conventional biodegradable chemical oxygen demand (bCOD) removal because aerobic heterotrophs must sacrifice a considerable amount of organic substrate for energy generation. Consequently, there is not enough organic carbon available in primary-settled domestic

wastewater to ensure efficient N and P removal via assimilation only (Box 1). This limitation is irrelevant during photosynthetic growth as nutrient assimilation is mainly a function of light and inorganic carbon availability. The contents of N and P in microalgae range from 6.6% to 9.3% (Chisti, 2007; Oswald, 1988) and from 0.2% to 3.9% (Powell et al., 2009), respectively. The large variability in P content is due to luxury phosphorus uptake by some microalgae species, when phosphorus is accumulated as polyphosphate. Luxury uptake depends on the dissolved phosphate concentration, light intensity, and temperature (Powell et al., 2008, 2009). Assuming a microalgal nitrogen and phosphorous content of 9% and 1%, respectively, a HRT of 7.5 days would be required to completely remove via assimilation the concentrations of N and P typical in medium-strength domestic wastewater (Figure 3). Therefore, microalgae-based WWT allows for simultaneous N and P removal at relatively short HRT, which represents an advantage in comparison with conventional stabilization ponds (higher HRT than in HRAPs).

2.2.2 Abiotic Nutrient Removal

Microalgal photosynthesis can cause the wastewater pH to increase when the rate of photosynthesis is carbon limited, a situation that occurs frequently in algae raceway ponds at peak sun hours. This rise in pH shifts the NH_4^+/NH_3 equilibrium toward NH_3 formation, which then increases the rate of N removal via

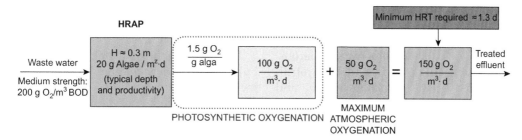

FIGURE 2 HRT required for stabilizing medium-strength domestic wastewater in a photosynthetically oxygenated HRAP.

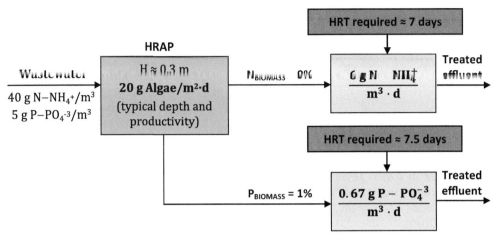

FIGURE 3   Estimated HRTs required for the treatment of typical N and P concentrations in medium-strength domestic wastewaters.

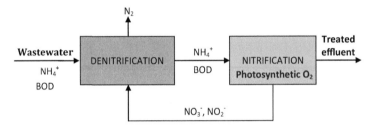

FIGURE 4   Schematic of a denitrification–nitrification process implemented in algal–bacterial photobioreactors.

ammonia volatilization. It also favors phosphate precipitation with Ca^{2+} as follows:

$$3HPO_4^{2-} + 5Ca^{2+} + 4OH^- \rightarrow Ca_5(OH)(PO_4)_3 \\ + 3H_2O$$

(1)

2.2.3 Dissimilatory Nutrient Removal

Conventional biological nitrogen removal typically involves the oxidation of NH_4^+ into NO_2^- and NO_3^- by chemolithotrophic aerobic bacteria and archaea, a process known as nitrification, followed by the reduction of NO_2^- and NO_3^- into N_2 under anoxic conditions by heterotrophic bacteria, a process known as denitrification (Rittmann and McCarty, 2001). During microalgae-based WWT, these processes may

occur simultaneously due to the occurrence of diffusional gradients between the inner part of the algal–bacterial flocs or biofilms and the culture broth (De Godos et al., 2009b). A recent study in our laboratory successfully implemented two-stage and single-stage denitrification–nitrification processes based on a photosynthetically oxygenated nitrification (De Godos et al., 2014) (Figure 4).

2.3 Pathogen Removal

Microalgae photosynthetic activity can contribute to the deactivation of pathogens by increasing wastewater pH, temperature, and dissolved oxygen concentration (Muñoz and Guieysse, 2006). For example, Schumacher et al. (2003) found a reduction in the concentration of total coliforms and *Escherichia coli* of

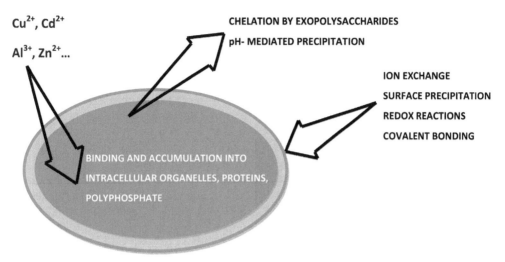

FIGURE 5   Passive and active mechanisms underlying heavy metal removal from wastewaters in microalgae biosorbents.

four and six orders of magnitude concomitantly with a pH increase from 8.4 to 10.5, respectively. The key role of alkaline pH in *E. coli* removal was also confirmed by Heubeck et al. (2007), who observed significantly higher *E. coli* removals at pH 9.5 ($\approx 100\%$) than at pH 8 ($\approx 50\%$) in an HRAP treating domestic wastewater. Similarly, El Hamouri et al. (1994) reported total coliform removal rates of 92% in an HRAP treating domestic wastewater at a pH of 9.4 and dissolved oxygen concentrations of approximately $20 \text{ mg O}_2 \text{ L}^{-1}$. In addition to the effect of pH, solar irradiation is also suspected to contribute to disinfection during microalgae-based WWT (Oswald, 2003). For example, Craggs et al. (2004) reported that sunlight irradiation caused an increase in the *E. coli* removal by 75% attributed in a 0.2-m-depth HRAP treating digested dairy farm effluent.

2.4 Heavy Metal Removal

Microalgae support heavy metal removal through a combination of active and passive mechanisms (Figure 5). These mechanisms include:

- Physical adsorption, ion exchange (with protons from R—COOH, R—OH, and R—NH functional groups), and chemisorptions (Chojnacka et al., 2005) are passive mechanisms that occur at a cell surface level. These fast (0–10 h) and reversible initial mechanisms are followed by irreversible heavy metal covalent bonding or surface precipitation (Wilde and Benemann, 1993; Ozturk et al., 2014).
- Exposure to high metal concentrations causes microalgae to synthesize and excrete metal-chelating exopolysaccharides that decrease the toxicity associated with dissolved heavy metals such as chromium and cadmium (Ozturk et al., 2014).
- The active ATP-driven transport of heavy metals inside algae cells and their subsequent binding to Class III metallothioneins or polyphosphates can significantly contribute to wastewater detoxification (Yu and Wang, 2004; Perales-Vela et al., 2006; Pereira et al., 2013).
- Heavy metal precipitation is also mediated when pH increases during carbon-limited photosynthesis.

Overall, the biosorption capacity of microalgae for a specific heavy metal depends on cell biochemistry (a function of the algae genotype and phenotype), pH, and the synergistic (increased membrane permeability) or antagonistic (competition for active sites) interactions within other heavy metals (Chojnacka et al., 2005). Eukaryotic microalgae genera such as *Chlorella*, *Chlamydomonas*, or *Scenedesmus* exhibit typical bioaccumulation capacities of 30–200 mg metal g^{-1} microalgae for Cu, Zn, and Pb (Akhtar et al., 2003; Maznah et al., 2012), and accumulations as high as 240–420 mg metal g^{-1} microalgae have been reported for the cyanobacterium *Spirulina* (Chojnacka et al., 2005). These capacities are suitable for full-scale use as biosorbent based on criteria set for similar materials (Muñoz et al., 2006). While the high cost of axenic microalgae biomass prohibits the use of this biosorbent, the increasing implementation of microalgae-based WWT may contribute to generate significant amounts of low-cost residual algal–bacterial suitable for heavy metal removal.

3. ENERGY RECOVERY AND CARBON SEQUESTRATION

Microalgae typically consume 1.8 kg of CO_2 per kg of biomass photosynthesized (Alcántara et al., 2013). This ability to convert CO_2 into organic biomass can be used to mitigate climate change following biomass conversion into biochar (Lehmann et al., 2006) or biofuel following fermentation (bioethanol), lipid transesterification (biodiesel), or anaerobic digestion (biomethane), the latter option currently being the most economical with CH_4 yields ranging from 0.15 to 0.4 m^3 per kg of microalgae (Alzate et al., 2012). Given that microalgae photosynthesis is often carbon-limited in HRAPs, WWT can be combined with CO_2 capture from point-source emissions. Despite the inhibitory CO_2 concentration thresholds in microalgae that are strain specific, tolerances to CO_2 concentrations of up to 50% have been reported in *Scenedesmus Obliquus* strains (Lam et al., 2012; Arbib et al., 2014).

4. PROCESS DESIGN AND OPERATION

Microalgae-based WWT systems are designed and operated to optimize light supply (high surface/volume ratio) and provide adequate mixing, carbon supply, and degassing under low hydrodynamic stress, while minimizing construction and operation costs (Muñoz and Guieysse, 2006; Tredici, 2004). Numerous configurations have been tested for WWT and these can be broadly categorized as either suspended-growth or biofilm-growth systems (Figure 6).

4.1 Suspended-Growth Systems

4.1.1 HRAPs

HRAPs, also known as raceway ponds, are open shallow ponds equipped with a paddle wheel that continuously mixes and promotes the access of microalgae to light and nutrients (Figure 7(a), Table 3). These systems are relatively easy to construct and operate, and therefore overwhelmingly preferred for microalgae production (De Godos et al., 2009a). The use of HRAPs for WWT started in the early 1950s (Oswald et al., 1957) and is broadly considered as the most (only) cost and resource efficient means to mass-produce microalgae biomass for biofuel generation. Although these systems have been tested at a meaningful scale for WWT, there is still a lack of long-term field experience for providing precise design and operational guidelines.

4.1.2 HRAP Design Guidelines

The design and operation of any WWT unit depend on influent properties, removal targets, how the particular unit is integrated within the complete plant, and local conditions (e.g., discharge ecosystem sensitivity, associated nuisances and hazards, temperature, access to energy, etc.). Local climatic conditions are of

FIGURE 6 Suspended-growth raceway (a) and attached-growth algal turf scrubber (b) treating domestic wastewater at the Department of Chemical Engineering and Environmental Technology, Valladolid University (Spain). *Photograph by the author.*

paramount importance during microalgae-based WWT because on-site solar irradiance largely dictates photosynthetic activity (Table 2), evaporation losses (Guieysse et al., 2013a), and temperature changes (Bechet et al., 2011) in these systems. The targeted efficiency is also critical because, as illustrated in Box 1, complete P-removal will normally require a larger treatment area than N and bCOD removal. In the case of bCOD removal, HRAPs for secondary WWT are typically designed by matching the photosynthetic oxygenation rate with the amount of oxygen required by aerobic heterotrophs to oxidize the biodegradable organic matter present in wastewater (Rittmann and McCarty, 2001). This normally requires an estimation of light penetration and biomass concentration in relation to biomass productivity; an

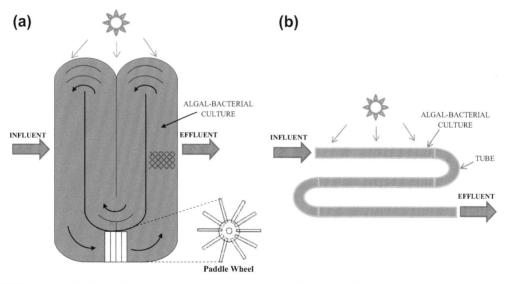

FIGURE 7 Schematic of a (a) high-rate algal pond photobioreactor and (b) enclosed tubular algal–bacterial photobioreactor.

TABLE 3 Key Design and Operation Parameters of HRAPs (Oswald, 1988; Molina-Grima, 1999)

Parameter	Typical range
Investment costs	$\approx 10 \in m^{-2}$
Typical size of single unit	1500–5000 m^2
Depth	0.15–0.40 m
Length–width ratio (L:W)	40:1
Recirculation rate	0.15–0.30 m s^{-1}
Engine rotation rate	5–20 rpm
Productivity	10–25 g m^{-2} per d
Power consumption	0.12–1 W m$^{-3}_{reactor}$

TABLE 4 Design and Operation Parameters of Tubular Photobioreactors (Janssen et al., 2003)

Parameter	Typical value
Investment costs	100 $\in$ m^{-2}
Tube length	10–100 m
Tube diameter	2–12 cm
Frequency of light/dark cycles	>1 s^{-1}
Culture rate	1 m s^{-1}
Productivity	15–27 g m^{-2} per day
Power consumption	200–1000 W m$^{-3}_{reactor}$

Janssen, M., Tramper, J., Mur, L.R., Wijffels, R.H. (2003). Enclosed Outdoor Photobioreactors: Light Regime, Photosynthetic Efficiency, Scale-up, and Future Prospects. Biotechnol. Bioeng. 81, 193–210.

empirical protocol for this purpose was established by Oswald and coworkers based on 30 years of experience on full-scale domestic WWT in HRAPs in California (Oswald, 1988; Molina-Grima, 1999). A similar approach is described by Rittmann and McCarty (2001) where (1) photosynthetic oxygenation is calculated at a given location and season; (2) a target HRT is determined with consideration of microalgae growth kinetics and safety factors of 3–10; and (3) pond area is determined from the HRT by fixing depth. Regardless of the method considered, the pilot and full-scale HRAP described in the literature are typically designed with an HRT of 2–8 d and depth of 0.2–0.5 m (Shilton, 2005). More advanced design can account for specific heterotroph kinetics, gas transfer, hydraulics, and a more complex model for photosynthesis and broth temperature (Bechet et al., 2013; Buhr and Miller, 1983; Shilton, 2005).

4.1.3 Enclosed Suspended-Growth Photobioreactors

Enclosed photobioreactors (Figure 7(b)) maintain microalgae cultivation broth protected from the environment, which allows the maintenance of monoalgal and even axenic cultures in industrial microalgae mass cultivation. Tubular photobioreactors (TPBRs) constitute the most commonly implemented enclosed microalgae culture technology (Table 4).

The efficient control of the operational conditions and the higher ratio of illuminated surface/volume in these photobioreactors compared to open ponds allow operating with a high biomass concentration (Arbib et al., 2013). The main disadvantages of enclosed photobioreactors are high cost and operational issues such as temperature control (Bechet et al., 2010) and fouling. Arbib et al. (2013) thus reported that productivity (initially 35 g biomass m^{-2} per day) and nutrient removal (initially 98% for N and P) of a 0.350 m^3 TPBR treating wastewater collapsed after 30 days of operation due to internal biofouling reducing light supply.

4.2 Biofilm Photobioreactors

Biofilm photobioreactors (BPBRs) are based on the attachment of microalgae and heterotrophs onto the photobioreactor walls (Boelee et al., 2011; Muñoz et al., 2009). BPBRs potentially allow for the simultaneous recovery of carbon and nutrients from wastewaters in the form of easily harvestable biofilm particles, and the production of a biomass-free effluent (Christenson and Sims, 2011a). Other potential advantages for WWT include efficient nutrient removal rates at low HRTs, low risks of total biomass washout, and microalgae protection against pollutant toxicity as a result of the diffusional gradients (Hoffman, 1998). On the negative side, continuous exposure of attached microalgae to high irradiances is expected to induce photosaturation or photoinhibition under outdoor conditions (Cuaresma et al., 2011). Different BPBR configurations have been evaluated for the treatment of a large variety of influents (Table 5), although feasibility at full scale is still uncertain (Christenson and Sims, 2011a). Biomass harvesting in biofilm photobioreactors is also a critical issue during photobioreactor maintenance.

Algal turf scrubber photobioreactors (ATSs) consist of an inclined plastic mesh (solid support) for microalgae (mainly filamentous) and benthic bacteria immobilization with intermittent wave surges operated with a downstream wastewater flow (Figure 8(a)). The main advantages of ATSs are their simple design and construction and their high biomass productivities ($15-27\,\mathrm{g\,m^{-2}}$ per day) (Adey et al., 1993). ATSs have been successfully applied for the treatment of aquacultural, agricultural, domestic, and industrial wastewaters (Adey and Loveland, 1991; Mulbry et al., 2008; Craggs et al., 1996; Adey et al., 1996). WWT in ATSs is performed by the periphyton (a mixture of microalgae and heterotrophic bacteria). Biomass harvesting in ATSs is normally carried out by periodic vacuuming of the growth surface, which removes inhibitory metabolites or predators accumulated in the ecosystem and enables a continuous biomass production and nutrient removal (Craggs et al., 1996). ATSs exhibit high

TABLE 5   Different BPBR Configurations for WWT

Photobioreactor	Wastewater	Total cultivation surface (m²)	Scale	References
ATS	Secondary domestic	1021	Large	Craggs et al. (1996)
ATS	Raw and anaerobically digested dairy manure	1	Laboratory	Mulbry et al. (2008)
ATS	Raw domestic	0.5	Laboratory	Posadas et al. (2013)
EBPR	Raw domestic	1	Laboratory	Zamalloa et al. (2013)
EBPR	Swine manure	1.8	Laboratory	De Godos et al. (2009b)
EBPR	Raw domestic	0.5	Laboratory	Posadas et al. (2014a)
RBAP	Synthetic petroleum hydrocarbon-rich	0.83	Laboratory	Chavan and Mukherji (2010)
RBAP	Secondary domestic	4.3	Laboratory	Christenson and Sims (2011b)
Flow-cell algal biofilm	Synthetic/real domestic	0.02	Laboratory	Boelee et al. (2011)
Periphyton-fish	Fish farm	48	Large	Rectenwald and Drenner (2000)

ATS: algal-turf scrubber; EBPR: enclosed algal−bacterial biofilm photobioreactor; RBAP: rotating biological algal-bacterial photobioreactor.
Rectenwald, L.L., and Drenner, R. (2000). Nutrient Removal from Wastewater Effluent Using an Ecological Water Treatment System. Environ. Sci. Technol. 34, 522−526.

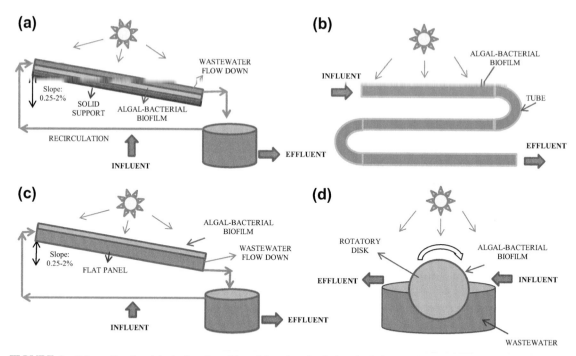

FIGURE 8   Schematic of an (a) algal turf scrubber; (b) enclosed tubular algal–bacterial photobioreactor; (c) enclosed flat panel algal–bacterial photobioreactor; (d) rotating biological algal–bacterial photobioreactor.

evaporative losses (1–7 L m^{-2} per day at laboratory scale) and therefore high carbon and N–NH$_4^+$ removal rates by stripping, which causes a deterioration in the quality of the treated wastewater and a decrease in both biomass productivity and nutrients recycling potential, respectively (Murphy and Berberoglu, 2012; Posadas et al., 2013).

Enclosed algal–bacterial biofilm photobioreactors (EBPRs) were devised in order to overcome the disadvantages of open ATSs during the treatment of agricultural effluents, domestic, and industrial wastewaters (González et al., 2008; De Godos et al., 2009b; Posadas et al., 2014a; Zamalloa et al., 2013; Muñoz et al., 2009). Although the establishment of phototrophic biofilms reduces light penetration and should therefore impact carbon and nutrient removal, the few lab-scale trials conducted with EBPR have shown promising results, with N and P removal

efficiencies of 94–100% and 70–90%, respectively, at influent loadings of 656 ± 37 mg N–NH$_4^+$ L^{-1} and   117 ± 19 mg P–PO$_4^{3-}$ L^{-1} (De Godos et al., 2009b). Tubular (Figure 8(b)) (González et al., 2008; De Godos et al., 2009b; Muñoz et al., 2009; Posadas et al., 2014a) and flat panel (Figure 8(c)) (Zamalloa et al., 2013; Ruiz et al., 2013) photobioreactors are the most common configurations used during laboratory-scale WWT, although these systems have not yet been tested outdoors at a meaningful scale.

Rotating biological algal-bacterial photobioreactors (RBAPs) consist of a series of rotating discs partially submerged in wastewater. The algal-bacterial biofilm develops onto the surface of the rotating discs (Christenson and Sims, 2011a) (Figure 8(d)). To date, RBAPs have been exclusively applied for tertiary WWT at lab scale (Torpey et al., 1971; Christenson and Sims, 2011b).

5. MICROALGAE HARVESTING

Efficient harvesting of the algal–bacterial biomass generated during microalgae-based WWT processes is critical to environmental performance and economic viability of biomass valorization (Posten, 2009; Acién et al., 2012). However, and despite significant research ongoing in this area, the only economical alternative for WWT remains gravity settling. Fortunately, experimental evidence suggests HRAP generates algal–bacterial flocks that quickly settled under quiescent conditions (Shilton, 2005). Moreover, recent research on full-scale HRAPs treating real effluents has demonstrated that biomass recycling considerably stabilizes ecology and favors fast-settling algae species (Park et al., 2011b, 2013; De Godos et al., 2014).

6. ENVIRONMENTAL IMPACT OF MICROALGAE-BASED WWT

6.1 Land Use

As can be seen in Box 1, WWT in HRAPs requires a significant amount of land, for example, 6.5 m^2 per capita per day in the case study, although a lower HRT may be applicable if only bCOD removal is targeted. Yet, even a conservative estimate of 5–6 m^2 per capita (high PE, external CO_2 supply) could represent a considerable burden for domestic WWT as this would normally require the availability of flat land near dwellings to reduce sewage collection costs and risks. It should, however, be noted that because the cost of microalgae-based WWT is largely driven by the cost of land, expenditure is not lost as the land value would remain, and even increase, and is thus reclaimable. In comparison, the majority of the expenditure incurred during "low-footprint intensive" WWT is related to mechanical/electrical/civil equipment and structures that depreciate in value.

6.2 Climate Change

Due to photosynthetic oxygenation, microalgae-based WWT has a considerably lower carbon footprint than conventional alternatives using mechanical aeration (Box 1). Atmospheric carbon fixation can even support climate change mitigation if the biomass photosynthesized is used for bioenergy production (see below) or converted into biochar for long-term carbon storage. On the negative side, various studies have reported N_2O emissions from algal cultures caused by either associated microorganisms or microalgae (Fagerstone et al., 2011; Guieysse et al., 2013b). While the carbon footprint associated with the emissions rates hitherto reported is minor, further research is needed in this area.

6.3 Energy Use

During conventional biological WWT, biomass productivity is limited by the amount of biodegradable organic material found in the influent. The positive impact of photosynthetic biomass production on potential energy recovery is evident in the case study shown in Box 1 where productivity increased more than sevenfold at a considerably lower energy input, yielding energy recovery ratios potentially 100–200 fold higher than during conventional secondary biological treatment.

6.4 Water Footprint

Although algae-based WWT improves water quality, significant amounts of water may be lost from local ecosystems due to free-surface evaporation. Guieysse et al. (2013a) predicted evaporation up to 2.275 m^3 (m^2)$^{-1}$ per year in Arizona in 0.25-m algae raceway ponds, which represented 15% of the amount of water treated. Although this may sound "affordable," the ecological and economical value of water depends on local conditions. In Arizona, an evaporation rate of $2.275 \, m^3 \, (m^2)^{-1}$ per year represents more than 40 years of rainfall equivalent, so caution must be taken before implementing HRAPs in water-stressed areas.

7. IMPLEMENTATION

If long-term full-scale experience is critically lacking to assess the true 'application niche' of microalgae-based WWT, this technology is likely only feasible for small-scale secondary and tertiary treatment in HRAPs at the present stage of development. The vigorous research ongoing in the field, however, bares testimony to its unique potential for energy and nutrient recovery and recent critical achievements, such as demonstrations of long-term species control and economical biomass harvesting, strongly advocate for further implementation.

Acknowledgments

Valladolid University, the regional government of Castilla y León, the European Social Fund and Spanish Ministry of Economy and Competitiveness are gratefully acknowledged for funding doctoral contracts of Cynthia Alcántara and Esther Posadas (Contract No. E-47-2011-0053564 and project Ref. VA024U14/GR76/RTA2013-00056-C03-02/Red Novedar.

References

Acién, F.G., Fernandez, J.M., Magán, J.J., Molina, E., 2012. Production cost of a real microalgae production plant and strategies to reduce it. Biotechnol. Adv. 30 (6), 1344–1353.

Adey, W.H., Loveland, K., 1991. Dynamic Aquaria: Building Living Ecosystems. Academic Press, NewYork.

Adey, W.H., Luckett, C., Jensen, K., 1993. Phosphorus removal from natural waters using controlled algal production. Restor. Ecol. 1, 29–39.

Adey, W.H., Luckett, C., Smith, M., 1996. Purification of industrially contaminated ground waters using controlled ecosystems. Ecol. Eng. 3, 191–212.

Akhtar, N., Saeed, A., Iqbal, M., 2003. Chlorella sorokiniana immobilized on the biomatrix of vegetable sponge of Luffa cylindrica: a new system to remove cadmium from contaminated aqueous medium. Bioresour. Technol. 88, 163–165.

Alcántara, C., Garcia-Encina, P.A., Munoz, R., 2013. Evaluation of mass and energy balances in the integrated microalgae growth-anaerobic digestion process. Chem. Eng. J. 221, 238–246.

Alzate, M.E., Muñoz, R., Rogalla, F., Fdz-Polanco, R., Perez, S.I., 2012. Biochemical methane potential of microalgae: influence of substrate to inoculum ratio, biomass concentration and pretreatment. Bioresour. Technol. 123, 488–494.

Arbib, Z., Ruiz, J., Álvarez-Díaz, P., Garrido-Pérez, C., Barragan, J., Perales, J.A., 2013. Effect of pH control by means of flue gas addition on three different photo-bioreactors treating urban wastewater in long-term operation. Ecol. Eng. 57, 226–235.

Arbib, Z., Ruiz, J., Álvarez-Díaz, P., Garrido-Pérez, C., Perales, J.A., 2014. Capability of different microalgae species for phytoremediation processes: wastewater tertiary treatment, CO_2 bio-fixation and low cost biofuels production. Water Res. 49, 465–474.

Barsanti, L., Gualteri, P., 2006. Algae: anatomy, biochemistry and biotechnology. In: General Overview (Chapter 1: 1–6); Photosynthesis (Chapter 3: 135–150) and Algal Culturing (Chapter 6:209–228). Taylor and Francis group, Boca Ratón (USA).

Bechet, Q., Shilton, A., Fringer, O.B., Muñoz, R., Guieysse, B., 2010. Mechanistic modeling of broth temperature in outdoor photobioreactors. Environ. Sci. Technol. 44, 2197–2203.

Bechet, Q., Shilton, A., Park, J.B.K., Craggs, R.J., Guieysse, B., 2011. Universal temperature model for shallow algal ponds provides improved Accuracy. Environ. Sci. Technol. 45, 3702–3709.

Bechet, Q., Shilton, A., Guieysse, B., 2013. Modeling the effects of light and temperature on algae growth: state of the art and critical assessment for productivity prediction during outdoor cultivation. Biotechnol. Adv. 31, 1648–1663.

Boelee, N.C., Temmink, H., Janssen, M., Buisman, C.J.N., Wijffels, R.H., 2011. Nitrogen and phosphorus removal from municipal wastewater effluent using microalgal biofilms. Water Res. 45, 5925–5933.

Borowitzka, M.A., 2005. Culturing microalgae in outdoor ponds. In: Andersen, I.R.A. (Ed.), Algal Culturing Techniques. Elsevier, Academic Press, New York, pp. 205–218.

Buhr, H.O., Miller, S.B., 1983. A dynamic model of the high-rate algal-bacterial wastewater treatment pond. Water Res. 17, 29–37.

Chae, K.Y., Kang, J., 2013. Estimating the energy independence of a municipal wastewater treatment plant incorporating green energy resources. Energ. Convers. Manage. 72, 664–672.

Chavan, A., Mukherji, S., 2010. Effect of co-contaminant phenol on performance of a laboratory-scale RBC with algal-bacterial biofilm treating petroleum hydrocarbon-rich wastewater. J. Chem. Technol. Biot. 85 (6), 851–859.

Chisti, Y., 2007. Biodiesel from microalgae. Biotechnol. Adv. 25, 294–306.

Chojnacka, K., Chojnacki, A., Gorecka, H., 2005. Biosorption of Cr3+, Cd2+ and Cu2+ ions by blue-green algae Spirulina sp.: kinetics, equilibrium and the mechanism of the process. Chemosphere 59, 75–84.

Christenson, L., Sims, R., 2011a. Production and harvesting of microalgae for wastewater treatment, biofuels and bioproducts. Biotechnol. Adv. 29, 686–702.

Christenson, L., Sims, R., 2011b. Rotating algal biofilm reactor and spool harvester for wastewater treatment with biofuels by-products. Biotechnol. Bioeng. 109, 1674–1684.

Craggs, R.J., Adey, W., Jessup, B.K., Oswald, W.J., 1996. A controlled stream mesocosm for tertiary treatment of sewage. Ecol. Eng. 6, 149–169.

Craggs, R.J., Zwart, A., Nagels, J.W., Davies-Colley, R.J., 2004. Modelling sunlight disinfection in a high rate pond. Ecol. Eng. 22, 113–122.

Craggs, R., Sutherland, D., Campbell, H., 2012. Hectare-scale demonstration of high rate algal ponds for enhanced wastewater treatment and biofuel production. J. Appl. Phycol. 24, 329–337.

Cuaresma, M., Janssen, M., Vílchez, C., Wijffels, R.H., 2011. Horizontal or vertical photobioreactors? How to improve microalgae photosynthetic efficiency. Bioresour. Technol. 102, 5129–5137.

Fagerstone, K.D., Quinn, J.C., Bradley, T.H., De Long, S.K., Marchese, A.J., 2011. Quantitative measurement of direct nitrous oxide emissions from microalgae cultivation. Environ. Sci. Technol. 45 (21), 9449–9456.

De Godos, I., Blanco, S., García-Encina, P., Becares, E., Muñoz, R., 2009a. Long term operation of high rate algae ponds for the bioremediation of Piggery wastewaters at high loading rates. Bioresour. Technol. 100 (19), 4332–4339.

De Godos, I., González, C., Becares, E., García-Encina, P.A., Muñoz, R., 2009b. Simultaneous nutrients and carbon removal during pretreated swine slurry degradation in a tubular biofilm photobioreactor. Appl. Microbiol. Biotechnol. 82, 187–194.

De Godos, I., Blanco, S., García-Encina, P.A., Becares, E., Muñoz, R., 2010. Influence of flue gas sparging on the performance of high rate algae ponds treating agro-industrial wastewaters. J. Hazard. Mater. 179, 1049–1054.

De Godos, I., Vargas, V., Guzmán, H., Soto, R., García, B., García-Encina, P., Muñoz, R., 2014. Assessing carbon and nitrogen removal in a novel anoxic-aerobic algal-bacterial photobioreactor configuration with enhanced biomass sedimentation. Water Res. 61, 77–85.

García, J., Mujeriego, R., Hernández-Marine, M., 2000. High rate algal pond operating strategies for urban wastewater nitrogen removal. J. Appl. Phycol. 12, 331–339.

González, C., Marciniak, J., Villaverde, S., García-Encina, P.A., Muñoz, R., 2008. Efficient nutrient removal from swine manure in a tubular biofilm photobioreactor using algae-bacteria consortia. Wat. Sci. Technol. 58, 95–102.

Guieysse, B., Béchet, Q., Shilton, A., 2013a. Variability and uncertainty in water demand and water footprint assessments of fresh algae cultivation based on case studies from five climatic regions. Bioresour. Technol. 128, 317–323.

Guieysse, B., Plouviez, M., Coilhac, M., Cazali, L., 2013b. Nitrous oxide (N_2O) production in axenic Chlorella vulgaris microalgae cultures: evidence, putative pathways, and potential environmental impacts. Biogeosciences 10 (10), 6737–6746.

El Hamouri, B., Khallayoune, K., Bouzoubaa, K., Rhallabi, N., Chalabi, M., 1994. High-rate algal pond performances in faecal coliformes and helminth egg removals. Water Res. 28 (1), 171–174.

Heubeck, S., Craggs, R.J., Shilton, A., 2007. Influence of CO_2 scrubbing from biogas on the treatment performance of a high rate algal pond. Water Sci. Technol. 55 (11), 193–200.

Hoffman, J.P., 1998. Wastewater treatment with suspended and non suspended algae. J. Phycol. 34, 757–763.

Intergovermental Panel on Climate Change (IPCC). (2014). https://www.ipcc.ch/pdf/special-reports/sroc/Tables/t0305.pdf.

Kivaisi, A.K., 2001. The potential for constructed wetlands for wastewater treatment and reuse in developing countries: a review. Ecol. Eng. 16, 545–560.

Lam, M.K., Lee, K.T., Mohamed, A.R., 2012. Current status and challenges on microalgae-based carbon capture. Int. J. Greenhouse Gas Control 10, 456–469.

Lehmann, J., Gaunt, J., Rondon, M., 2006. Bio-char sequestration in terrestrial ecosystems- A review. Mitigation Adaptation Strateg. Glob. Change 11, 403–427.

Maznah, W.O.W., Al-Fawwaz, A.T., Surif, M., 2012. Biosorption of copper and zinc by immobilized and free algal biomass, and the effects of metal biosorption on the growth and cellular structure of Chlorella sp. and Chlamydomonas sp. isolated from rivers in Penang, Malaysia. J. Environ. Sci. 24 (8), 1386–1393.

Metcalf, Eddy (Eds.), 2003. Wastewater Engineering: Treatment and Reuse, fourth ed. Mc Graw Hill.

Molina-Grima, E., 1999. Microalgae mass culture methods. In: Flickinger, M.C., Drew, S.W. (Eds.), Encyclopedia of Bioprocess Technology: Fermentation, Biocatalysis and Bioseparation. John Wiley & Sons.

Mulbry, W., Kebede-Westhead, E., Pizarro, C., Sikora, L., 2005. Recycling of manure nutrients: use of algal biomass from dairy manure treatment as a slow release fertilizer. Bioresour. Technol. 96, 451–458.

Mulbry, W., Kondrad, S., Buyer, J., 2008. Treatment of dairy and swine manure effluents using freshwater algae: fatty acid content and composition of algal biomass at different manure loading rates. J. Appl. Phycol. 20, 1079–1085.

Muñoz, R., Köllner, C., Guieysse, B., Mattiasson, B., 2004. Photosynthetically oxygenated salicylate biodegradation in a continuous stirred tank photobioreactor. Biotechnol. Bioeng. 87 (6), 797–803.

Muñoz, R., Guieysse, B., 2006. Algal-bacterial processes for the treatment of hazardous contaminants: a review. Water Res. 40, 2799–2815.

Muñoz, R., Alvarez, T., Muñoz, A., Terrazas, E., Guieysse, B., Mattiasson, B., 2006. Sequential removal of heavy metals ions and organic pollutants using an algal-bacterial consortium. Chemosphere 63, 903–911.

Muñoz, R., Köllner, C., Guieysse, B., 2009. Biofilm photobioreactors for the treatment of industrial wastewaters. J. Hazard. Mater. 161, 29–34.

Murphy, T.E., Berberoglu, H., 2012. Temperature fluctuation and evaporative loss rate in an algae biofilm photobioreactor. J. Sol. Energy Eng. 134 (011002), 1–9.

Oswald, W.J., Gotaas, H.B., Golueke, C.G., Kellen, W.R., 1957. Algae in waste treatment. Sewage Ind. Wastes 29, 437–455.

Oswald, W.J., 1988. Large-scale algal culture systems (engineering aspects). In: Borowitzka, Borowitzka, L.J. (Eds.), Micro-algal Biotechnology. Cambridge University Press, Cambridge, pp. 357–394.

Oswald, W.J., 2003. My sixty years in applied algology. J. Appl. Phycol. 15, 99–106.

Ozturk, S., Aslim, B., Suludere, Z., Tan, S., 2014. Metal removal of cyanobacterial exopolysaccharides by uronic acid content and monosaccharide composition. Carbohyd. Polym. 101, 265–271.

Park, J.B.K., Craggs, R.J., 2010. Wastewater treatment and algal production in high rate algal ponds with carbon dioxide addition. Water Sci. Technol. 61, 633–639.

Park, J.B.K., Craggs, R.J., Shilton, A.N., 2011a. Wastewater treatment high rate algal ponds for biofuel production. Bioresour. Technol. 102 (1), 35–42.

Park, J.B.K., Craggs, R.J., Shilton, A.N., 2011b. Recycling algae to improve species control and harvest efficiency from a high rate algal pond. Water Res. 45 (20), 6637–6649.

Park, J.B.K., Craggs, R.J., Shilton, A.N., 2013. Investigating why recycling gravity harvested algae increases

harvestability and productivity in high rate algal ponds. Water Res. 47 (14), 4904–4917.

Perales-Vela, H.V., Peña-Castro, J.M., Cañizares-Villanueva, R.O., 2006. Heavy metal detoxification in eukaryotic microalgae. Chemosphere 64, 1–10.

Pereira, M., Bartolomé, M.C., Sánchez-Fortum, S.S., 2013. Bioadsorption and bioaccumulation of chromium trivalent in Cr (III)- tolerant microalgae: a mechanism for chromium resistance. Chemosphere 93, 1057–1063.

Plappally, A.K., Lienhard, J.H., 2012. Energy requirements for water production, treatment, end use, reclamation, and disposal. Renew. Sustain. Energy Rev. 16, 4818–4848.

Posadas, E., García-Encina, P.A., Soltau, A., Domínguez, A., Díaz, I., Muñoz, R., 2013. Carbon and nutrient removal from centrates and domestic waste-water using algal–bacterial biofilm bioreactors. Bioresour. Technol. 139, 50–58.

Posadas, E., García-Encina, P.A., Domínguez, A., Díaz, I., Becares, E., Blanco, S., Muñoz, R., 2014a. Enclosed tubular and open algal–bacterial biofilm photobioreactorsfor carbon and nutrient removal from domestic wastewater. Ecol. Eng. 67, 156–164.

Posadas, E., Muñoz, A., García-González, M.C., Muñoz, R., García-Encina, P.A., 2014b. A case study of a pilot high rate algal pond for the treatment of fish farm and domestic wastewaters. J. Chem. Technol. Biot. http://dx.doi.org/10.1002/jctb.4417.

Posten, C., 2009. Design principles of photo-bioreactors for cultivation of microalgae. Eng. Life Sci. 9, 165–177.

Powell, N., Shilton, A.N., Pratt, S., Chisti, Y., 2008. Factors Influencing luxury uptake of phosphorus by microalgae in waste stabilization ponds. Environ. Sci. Technol. 42 (16), 5958–5962.

Powell, N., Shilton, A., Chisti, Y., Pratt, S., 2009. Towards a luxury uptake process via microalgae-defining the polyphosphate dynamics. Water Res. 43 (17), 4207–4213.

Rectenwald, L.L., Drenner, R., 2000. Nutrient Removal from Wastewater Effluent Using an Ecological Water Treatment System. Environ. Sci. Technol. 34, 522–526.

Rittmann, B.E., McCarty, P.L., 2001. Environmental Biotechnology: Principles and Applications. McGraw-Hill Book Co, New York.

Ruiz, J., Álvarez-Díaz, P.D., Arbiba, Z., Garrido-Pérez, Barragána, J., Perales, J.A., 2013. Performance of a flat panel reactor in the continuous culture of microalgae in urban wastewater: prediction from a batch experiment. Bioresour. Technol. 127, 456–463.

Schumacher, G., Blume, T., Sekoulov, I., 2003. Bacteria reduction and nutrient removal in small wastewater treatment plants by an algal biofilm. Water Sci. Technol. 47, 195–202.

Shilton, A. (Ed.), 2005. Pond Treatment Technology. IWA Publishing, London, United Kingdom.

Tarlan, E., Dilek, F.B., Yetis, U., 2002. Effectiveness of algae in the treatment of a wood-based pulp and paper industry wastewater. Bioresour. Technol. 84, 1–5.

Torpey, W.N., Heukelekian, H., Kaplovsky, A.J., Epstein, R., 1971. Rotating disks with biological growth prepares wastewater for disposal or reuse. J. Water Pollut. Control Fed. 43, 2181–2188.

Tredici, M., 2004. Mass production of microalgae: photobioreactors. In: Richmond, A. (Ed.), Handbook of Microalgal Culture: Biotechnology and Applied Phycology. páginas, Oxford, pp. 178–213.

Wilde, E.W., Benemann, J.R., 1993. Bioremoval of heavy metals by the use of microalgae. Biotechnol. Adv. 11, 781–812.

Yu, R.-Q., Wang, W.-X., 2004. Biokinetics of cadmium, selenium, and zinc in freshwater alga *Scenedesmus obliquus* under different phosphorus and nitrogen conditions and metal transfer to *Daphnia magna*. Environ. Pollut. 129, 443–456.

Zamalloa, C., Boon, N., Verstraete, W., 2013. Decentralized two-stage sewage treatment by chemical–biological flocculation combined with microalgae biofilm for nutrient immobilization in a roof installed parallel plate reactor. Bioresour. Technol. 130, 152–160.

Bioremediation of Heavy Metals by Microalgae

Laura Bulgariu[1], *Maria Gavrilescu*[1,2]

[1]"Gheorghe Asachi" Technical University of Iaşi, Faculty of Chemical Engineering and Environmental Protection, Department of Environmental Engineering and Management, Iaşi, Romania; [2]Academy of Romanian Scientists, Bucharest, Romania

1. INTRODUCTION

The rapid growth of industrialization and urban populations has affected the quality of surface waters, mainly due to the excessive input of hazardous substances that are discharged daily. Of various types of water pollutants, the heavy metals are among the most dangerous due to their toxicity, persistence, resistance to biological degradation, and long-term accumulation in the food chain (Mendoza et al., 1998), thus causing serious health problems for many forms of life. Even as they have a serious negative impact on the environment, heavy metals continue to largely be used in various industries, such as mining, plating, smelting, plastics, textiles, painting etc. (Abdel-Ghani and El-Chaghaby, 2014), as important material in various technological processes. Consequently, industrial and municipal wastewaters contain, in many cases, large amounts of heavy metals, which represent one of the major sources of natural water pollution. As a result, it is beneficial to eliminate the heavy metals from industrial

waste streams, so as to reduce their impact on the environment. This could be also essential economically, due to the high value of such metals.

Conventional remediation methods for the removal of heavy metals from wastewaters include chemical precipitation, coagulation, ion exchange, membrane processing, electrochemical techniques, adsorption on activated carbon etc. (Dabrowski et al., 2004; Wan Ngah and Hanafiah, 2008; Llanos et al., 2010; Gantam et al., 2014). The selection of a particular remediation technique depends on various factors, like type and concentration of heavy metals, effluent heterogeneity, required level of cleanup, as well as cost of decontamination process (Gavrilescu, 2004). Unfortunately, most of these methods are disadvantageous since they (1) involve high operation costs; (2) are inefficient especially when large volumes of wastewater with low heavy metal contents (1–100 dissolved metal/L) have to be treated (Montazer-Rahmati et al., 2011); (3) are not environmentally friendly because large inputs of chemical reagents are required; and (4) could generate secondary wastes that require disposal or further treatments. Due to progressively stricter governmental legislations, the cost-effective remediation methods have become more interesting.

In recent years, biosorption developed into a promising alternative method for environmental remediation and has proved to be very effective in the removal and recovery of numerous toxic heavy metals from aqueous effluents. The advantages of this method include: high efficiency, cost-effectiveness (especially when the biosorbent can be recycled and heavy metals recovered for reuse), wide adaptability, strong metal binding capacity, high efficiency in dilute effluents, and environmental friendliness.

The success of the biosorption process is largely determined by the selection of biosorbent material and the design of the biosorption process. Thus, various types of biological materials, including agricultural wastes, peat, algae, fungi, bacteria, yeasts, cellular products, etc. (Gavrilescu, 2004; Demirbas, 2008; Lupea et al., 2012), have been tested in various experimental conditions in order to establish their performance in the heavy metals uptake processes. In most cases, these biological materials have proven to have the economic potential to be used as biosorbents for the reduction of environmental pollution.

The use of microalgae for the removal of heavy metals from aqueous effluents has been well studied over the past 40 years (Fu and Wang, 2011; Hubbe et al., 2011). Since the 1980s, many studies in the literature have been devoted to the utilization of microalgae as biosorbents for the remediation of contaminated environmental compartments, especially with heavy metals (Hubbe et al., 2011). The microalgae are particularly attractive because they are available in large quantities in many regions of the world, can grow both in fresh and saltwater in various climatic conditions, are low cost to prepare, and have an excellent retention capacity for most toxic heavy metals (Inthorn et al., 2002; Munoz et al., 2006).

Generally, the microalgae used in the remediation processes have the ability to accumulate toxic heavy metals from aqueous effluents both by passive uptake and via the metabolic activity of living organisms. Even if the living microalgae have shown promising capabilities in the removal processes of heavy metals from different types of wastewater, their use is limited by different factors that affect their growth (pH of wastewater, contents of heavy metals, etc.), which can influence the efficiency of the remediation process in various degrees. From this perspective, the nonliving microalgae are more profitable for industrial applications, mainly because (1) dead microalgae biomass can be stored at room temperature; (2) they can be used as biosorbents for a long period of time without losing their biosorptive characteristics; (3) they are not affected by the toxicity of heavy metals from polluted effluents; and (4) the nonliving microalgae have a comparable or even higher biosorption capacity than living microalgae, and this characteristic can be significantly improved by various simple chemical treatments (Gautan et al., 2014).

In this chapter we discuss the potential use of microalgae for bioremediation of aqueous effluents that contain heavy metal ions. A detailed description of the factors that influence the heavy metal biosorption process is outlined, along with new updates on biosorption process modeling and some recent advances in the elucidation of the retention mechanism. The analysis indicates that the microalgae have the potential to become an effective and economical biosorbent for the removal of heavy metals from industrial waste effluents.

2. MICROALGAE CHARACTERISTICS

Microalgae are considered microscopic photosynthetic organisms that can be found in all aquatic environments (salty to fresh water) or

TABLE 1 Chemical Composition of Some Microalgae Expressed on a Dry Mass Basis (http://www.oilgae.com; Rojan et al., 2011)

Microalgae	Starch	Protein	Carbohydrates	Lipids	Nucleic acid
Chlorella vulgaris	12–17	51–58	12–17	14–22	4–5
Chlorella pyrenoidosa	21.5	51	26	2	–
Scenedesmus obliquus	23.7	50–56	10–17	12–14	3–6
Spirogyra sp.	43.3	6–20	33–64	11–21	–
Dunaliella salina		57	32	6	–
Dunaliella bioculata		49	4	8	–
Spirulina maxima	32.2	60–71	13–16	6–7	3–4.5
Porphyridium cruentum	28.0	28–39	40–57	9–14	–

can be cultivated in either indoor or outdoor conditions. They are unicellular species that can exist individually or in chains, while their size can vary from a few micrometers to a few hundreds of micrometers, depending on the species.

Different criteria can be applied for the classification of microalgae, including their pigmentation, life cycle, or basic cellular structure. If their abundance is considered, the most important classes of microalgae are diatoms (*Bacillariophyceae*), green microalgae (*Chlorophyceae*), and golden microalgae (*Chrysophyceae*) (Demirbas, 2010). The differences among these types of marine algae consist essentially in the structure of cell walls, where the retention of heavy metal ions takes place.

The cell walls of microalgae generally contain important quantities of starch and glycogen, but also cellulose, hemicellulose, and polysaccharides (Arief et al., 2008; Wang and Chen, 2008). These constituents contain numerous reactive functional groups (e.g., amino, hydroxyl, carboxyl, sulfate, etc.) that can be involved in chemical binding with metal ions and are responsible for the excellent biosorption potential of microalgae. Table 1 summarizes the main constituents present in the structure of some microalgae.

Although the constituents of the cell walls provide an extensive array of ligands with various functional groups capable of binding metal ions, their role in the biosorption process depends on several factors. Thus, the number of active sites on the biosorbent material surface, the accessibility and chemical state, and the affinity of metal ions for a given functional group will directly influence the efficiency of the biosorption process.

3. MECHANISMS OF HEAVY METAL UPTAKE BY ALGAE

Elucidation of the biosorption mechanism involved in the metal uptake process is essential for successful utilization of this method in practice and for biosorbent regeneration in multiple re-useable cycles. Usually, the biosorption processes are quite complex, and the overall metal uptake mechanism results from the combination of different elementary mechanisms, such as electrostatic interactions, ion exchange, complexation, chelation adsorption, microprecipitation, etc. which occur concomitantly or successively (Volesky, 1987; Gavrilescu, 2004).

As a function of elementary processes involved in the uptake process, the biosorption mechanism can be classified as: *chemical biosorption*, which involves a chemical reaction, and *physical biosorption*, where the retention of metal ions occurs via van der Waals or electrostatic interaction. The difference between these two types of biosorption mechanisms is given by the magnitude of enthalpy change (ΔH) for a given biosorption process, calculated from biosorption isotherms obtained at different temperatures (Senthilkumaret al., 2006). Thus, it was considered that an enthalpy change ranging from 0.5 to 5 kcal mol^{-1} (2.1−20.9 kJ mol^{-1}) indicates a physical biosorption mechanism, while a value of enthalpy change between 5 and 100 kcal mol^{-1} (20.9−418.4 kJ mol^{-1}) shows that the chemical interactions are predominant in the biosorption process (Deng et al., 2007). In general, the biosorption process on nonliving microalgae follows a chemical mechanism, while the main important factors that determine the nature of elementary processes are (1) type of functional groups present on the microalgae surface; (2) nature of heavy metal species from aqueous solution; and (3) characteristics of aqueous solution (pH, ionic strength, presence of competing ions etc.).

1. *Ion exchange* − is considered to be the dominant mechanism of heavy metals biosorption on microalgae (Davis et al., 2003; Herrero et al., 2006), where the elementary interactions may range from physical (electrostatic or van der Waals forces) to chemical (ionic or covalent).

In general, the microalgae contain in their structure mobile metal ions such as K^+, Na^+, Ca^{+2}, and Mg^{+2}, which are bound to the acid functional group of the microalgae. In the biosorption process, these anions are exchanged with heavy metals, according to the reaction:

$$R^- - X^+ + M^+ \rightleftarrows R^- - M^+ + X^+ \qquad (1)$$

where R^- is the functional group from the microalgae surface; X^+ is a mobile ion (e.g., Na^+, K^+, Ca^{2+}, Mg^{2+}, etc.); and M^+ is the heavy metal ion present in aqueous solution.

The microalgae can retain heavy metal ions from aqueous solution and release such mobile ions in solution.

The biosorption of various heavy metals (such as Pb(II), Cd(II), Cu(II), Zn(II), etc.) using different types of microalgae (*Scenedesmus obliquus*, *Chlorella pyrenoidosa*) occurs predominantly by ion-exchange interaction (Zhou et al., 2012; Mirghaffari et al., 2014). The experimental studies have shown that the amount of light metal ions is higher at the end of biosorption process.

2. *Complexation* − is another possible mechanism that can appear in the biosorption process, which involves the formation of a complex on the cell surface, between heavy metal ions from aqueous solution and functional groups of microalgae. For example, Aksu et al. (1992) have shown that the biosorption of Cu(II) ions onto *Chlorella vulgaris* occurs via a complexation mechanism that involves the formation of coordination bonds between metal ions and amino and carboxyl groups of the polysaccharides from the microalgae cell wall. It should be noted that in the complexation mechanism, both electrostatic interactions and covalent and/or coordinative interactions are involved, and, in comparison with an ion-exchange mechanism, the resulting superficial complexes are more stable. Because of this, the regeneration of such biosorbents requires the utilization of strong complexing agents, like ethylenediaminetetraacetic acid.

Nevertheless, the complexation interactions have been evidenced as elementary interactions in many biosorption processes on various types of microalgae, especially at a high initial concentration of heavy metal ions.

3. *Microprecipitation* — can take place when the solution pH drastically increases during biosorption and/or even when concentrations of metal ions in aqueous effluents increase up to their saturation. In this case, the heavy metals from the solution can precipitate, and the obtained microprecipitates are deposited on the biomass surface. Microprecipitation can take place dependent or not on the nature of microalgae, and can produce a distortion of biosorption results and hinder the determination of the amount of metal ions uptake.

4. FACTORS INFLUENCING THE HEAVY METALS BIOSORPTION ON MICROALGAE

The design of an efficient heavy metals bioremediation process for aqueous effluents by biosorption on microalgae mainly supposes the identification and optimization of the most important factors that can influence the uptake capacity of the biosorbent. This happens because it is well known that the removal of metal ions from aqueous solution by biosorption takes place with maximum efficiency only in well-defined experimental conditions.

In the case of microalgae, the most important factors that affect their biosorptive performance can be divided into two categories (Han et al., 2007): (1) biomass factors, such as growth medium, specific surface properties of microalgae, pretreatment of cells, etc.; and (2) process factors, such as initial pH of aqueous solution, biosorbent concentration, contact time, temperature, experimental methodology, bed height, solution flow rate, heavy metal concentration, etc.

1. *Biomass factors*

Growth and development conditions can influence the biosorptive performance of microalgae. Even if the data presented in the literature

(http://www.oilgae.com) indicated that the microalgae growth in saline medium have a higher content of polysaccharides than those developed in fresh water (see Table 1), their efficiency in biosorption process varied in a wide interval. However, it was demonstrated that the microalgae having a large number of available functional groups on their surface will exhibit better biosorption characteristics, but this depends on both the nature of microalgae and the pretreatment of biomass cells before utilization as biosorbent.

Usually, the microalgae biomass is centrifuged at different speed and time intervals, in order to obtain the raw biomass. This biomass is then pretreated, in most cases by drying, and thus the biosorbent results, which is more easily conserved, and can be used for a long period of time. This pretreatment step is most often carried out at a temperature interval of 50–60 °C during 12–24 h, because under these conditions the biomass is not degraded and the superficial functional groups are not altered.

2. *Process factors*

Depending on the methodology, the main process factors that significantly influence the biosorptive performance of microalgae and that should be optimized are (1) *for batch systems*—solution pH, biosorbent dose, contact time, temperature; (2) *for continuous systems*—height of biosorbent bed, flow rate of aqueous solution containing heavy metals, initial concentration of heavy metal ions.

Solution pH it is one of the most important experimental parameters that influences not only the speciation and solubility of heavy metal ions but also the dissociation degree of functional groups from the biosorbent surface (e.g., hydroxyl, carboxyl, carbonyl, amino, etc.), considered as biosorption sites (Gao and Wang, 2007). From this point of view, many studies have shown that pH values at the interval of 3 and 6.5 lead to increased biosorption capacity of the most heavy metals species on microalgae

TABLE 2   Optimal Experimental Conditions used in the Case of Heavy Metals Biosorption on Some Microalgae

Microalgae	Heavy metal	pH	Biosorbent dose, g L^{-1}	Contact time, min	Temperature, °C	References
Scenedesmus obliquus	Zn(II) Cu(II)	6.0–7.0 5.0–7.0	0.02 0.03	90 60	25 Room temperature	Monteiro et al. (2011) Kumar et al. (2014)
Scenedesmus quadricauda	Cd(II) Pb(II)	5.0	0.20	60	Room temperature	Mirghaffari et al. (2014)
Chlamydomonas reinhardtii	Cr(VI) Pb(II), Cd(II), Hg(II)	2.0 5.0–6.0	0.10 0.20	120 60	Room temperature 25	Arica et al. (2005) Bayramoglu et al. (2006)
Chlorella vulgaris	Cd(II), Ni(II)	4.0	–	120	25	Aksu and Donmez (2006)
Chlorella sorokiniana	Cr(III)	4.0–5.0	0.1	60	25	Nasreen et al. (2008)
Oedogonium hatei	Cr(VI) Ni(II)	2.0 5.0	0.8 0.7	110 80	45 25	Gupta and Rastogi (2009) Gupta et al. (2010)

(Table 2). This is in agreement with the chemistry of metal solution, where heavy metals possess a high solubility and are present in solution as simple ionic species, having the most toxic effect and higher bioavailability (Stumm and Morgan, 1996). Consequently, heavy metal speciation and solubility as a function of pH variations is one of the most important criteria in selecting the type of microalgae biosorbent. At pH values lower than 3, the uptake capacity of microalgae is lower, mainly because of the competition between protons and heavy metal ions for the binding sites of biosorbent (Priyadarshani et al., 2011; Zhao et al., 2013). Above pH 6.5, the heavy metal ions tend to precipitate as hydroxides, and only a low amount of heavy metals remain in solution and can be retained by interactions with superficial groups of microalgae.

The *biosorbent dose* is another parameter that should be optimized in order to ensure the economic and environmental feasibility of the bioremediation process. Thus, the utilization of large amounts of biosorbent not only will increase the cost of biosorption processes but also will generate large amounts of waste loaded with heavy metals, which will have a negative impact on the environment. On the other hand, a too small amount of microalgae will significantly affect the biosorption efficiency, and the bioremediation processes will become inadequate for large-scale applications. Therefore, in order to design an economically viable bioremediation system for heavy metal ion biosorption, the optimal biosorbent dose should be selected after a careful comparison of the experimental results. Table 2 summarizes the values of optimal biosorbent dose used for the removal of heavy metal ions from aqueous environments, in different types of microalgae.

The *contact time* also has an important role in ensuring the efficiency of the biosorption process. Unsatisfactory value of this parameter could drastically limit the practical use of a given biosorption process, even if its efficiency in heavy metal ions removal is high. In a large number of studies from the literature, the efficiency of heavy metal ion biosorption on microalgae increases with increasing the contact time, and the biosorption process generally reaches equilibrium within a time interval of 180 min (Table 2). In most studies, the experimental observations have shown that no further significant biosorption is observed after 3 h in the case of microalgae biosorbents (Wang and Chen, 2008).

The influence of *temperature* is, in the case of microalgae biosorbents, more important for the thermodynamic description of biosorption process than for the increase of heavy metals uptake efficiency. In many studies from the literature has been shown that a drastically increased temperature (up to 40 °C) determined a slight increase or decrease of microalgae biosorption capacity (with $5-15\,mg\,g^{-1}$) (Aksu, 2001; Febrianto et al., 2009). Considering these results, it is recommended that for large-scale application, the biosorption of heavy metals from aqueous solution on microalgae should be performed at ambient temperature (see Table 2), because the cost of operation will be kept low under these conditions.

Unfortunately, batch systems are usually limited to the bioremediation of small volumes of aqueous effluents. The use of continuous systems seems to be more adequate for the large-scale application of bioremediation processes using microalgae biosorbents, where the biosorbent can be used in multiple biosorption—desorption cycles (Febrianto et al., 2009). However, it should be mentioned that the utilization of microalgae in continuous systems has an important disadvantage—easy clogging of the column, due to the small size of biosorbent particles. Therefore, to ensure an adequate flow rate of the aqueous solution through the column, many studies from the literature have proposed the immobilization of microalgae in various matrices, which improves the mechanical strength, particle size, and resistance to chemicals that could be present in the aqueous effluent. The main immobilization method of microalgae involves the entrapment into polymers or the natural retention onto inert and porous materials, such as silica, agar, polyacrylamide, alginates, cellulose, and different cross-linking agents (Valdman and Leite, 2000).

The biosorption of heavy metal ions on microalgae in continuous systems is mainly influenced by three key experimental parameters: bed height, flow rate of aqueous solution, and initial concentration of heavy metal ions. According to the results found in the literature, the metal uptake capacity and the breakthrough time increase with the increase of the bed height, which means an increased total surface area of biosorbent. Also, the increase of flow rate directly affects the efficiency of the bioremediation process and the treated effluent volume by decreasing breakthrough and exhaustion time. The uptake capacity of metal during biosorption decreases with increased initial concentration of metal in solution, because the biosorbent saturates faster at high concentration. For this reason, the optimal values that ensure the efficient delivery of the bioremediation process have to be established for each of these parameters, for practical applications.

The use of microalgae for heavy metals biosorption in continuous systems facilitates the treatment of large volumes of aqueous effluents, but research is still underway in other areas related to biosorption in continuous systems.

5. EQUILIBRIUM ISOTHERM MODELS AND EVALUATION OF BIOSORPTIVE PERFORMANCE

The performance of certain biosorbent materials can be described by equilibrium biosorption isotherms, which are characterized by definite parameters directly correlated with the surface properties and affinity of this for different heavy metal ions. Equilibrium isotherm parameters were successfully used to predict practical biosorption capacity and optimize biosorption systems design (Pelhivan and Arslan, 2007; Bulgariu et al., 2013).

TABLE 3　Langmuir Isotherm Parameters for the Biosorption of Heavy Metals on Some Microalgae

Microalgae	Heavy metal	R^2	q_{max}, mg g^{-1}	K_L, g L^{-1}	References
Chlorella vulgaris	Cr(VI)	0.960	163.93	0.036	Gokhale et al. (2008)
	Cd(II)	0.999	80.00	0.020	Aksu and Donmez (2006)
	Ni(II)	0.999	70.90	0.030	
Chlorella sorokiniana	Cr(III)	0.991	56.56	0.110	Nasreen et al. (2008)
Scenedesmus obliquus	Zn(II)	–	836.50	–	Monteiro et al. (2011)
Microcystis novacekii	Pb(II)		70.00		Ribeiro et al. (2010)
Scenedesmus quadricauda	Cd(II)	0.996	135.13	0.07	Mirghaffari et al. (2014)
	Pb(II)	0.983	333.33	0.03	
Oedogonium hatei	Cr(VI)	0.993	31.02	0.034	Gupta and Rastogi (2009)
	Ni(II)	0.9991	40.90	0.026	Gupta et al. (2010)

Many mathematical models were developed to describe biosorption equilibrium, but the *Langmuir* and *Freundlich isotherm models* remain the most commonly used, essentially because these models can be applied for the mathematical description of the biosorption process, both for single component and multicomponent systems (Chong and Volesky, 1995; Bulgariu et al., 2007).

The *Langmuir model* was frequently found to be the best-fit model for the experimental data obtained during the biosorption of different heavy metals on various types of microalgae. This model is based on three assumptions (Febrianto et al., 2009): (1) the biosorption process occurs until heavy metal ions form a monolayer coverage on the outer surface of microalgae biosorbent; (2) the surface of microalgae biosorbent can be considered homogeneous, and all binding sites are alike; and (3) each heavy metal from aqueous solution will interact with a corresponding binding site from the microalgae surface, while this process is independent of the degree of sites occupation.

Starting from the mathematical expression of this model (Eqn. (2)) (Chong and Volesky, 1995; Bulgariu et al., 2007):

$$\frac{c}{q} = \frac{1}{q_{max} \cdot K_L} + \frac{c}{q_{max}} \qquad (2)$$

where q_{max} is the maximum sorption capacity upon complete saturation of sorbent surface (mg g^{-1}) and K_L is the Langmuir constant (L mg^{-1}), related to the sorption/desorption energy; the parameters q_{max} and K_L can be evaluated from the intercepts and the slopes of linear plots c/q versus c. Table 3 summarizes the values of Langmuir isotherm parameters obtained in the case of heavy metal biosorption onto various microalgae.

The *Freundlich isotherm* model is an empirical model, which can be used for the description of biosorption process on a heterogeneous surface or a surface supporting sites of different affinities, assuming that the stronger binding sites are occupied first, while the binding strength decreases with increasing the degree of sites occupation (Farooq et al., 2010). This model is used to estimate the biosorption intensity of heavy metal ions toward a given microalgae biosorbent and can be expressed in linear forms with the relation:

$$\lg q = \lg K_F + \frac{1}{n}\lg c \qquad (3)$$

where K_F and n are constants of the Freundlich isotherm model, which can be determined from the slopes and intercepts of lg q versus lg c dependences. A favorable biosorption process

TABLE 4 Freundlich Isotherm Parameters for the Biosorption of Heavy Metals on Some Microalgae

Microalgae	Heavy metal	R^2	n	K_F, mg g^{-1}	References
Chlorella vulgaris	Cr(VI)	0.980	0.42	18.90	Gokhale et al. (2008)
	Cd(II)	1.000	2.13	2.92	Aksu and Donmez (2006)
	Ni(II)	0.999	1.81	3.68	
Chlorella sorokiniana	Cr(III)	0.877	2.85	10.67	Nasreen et al. (2008)
Scenedesmus quadricauda	Cd(II)	0.807	2.01	10.06	Mirghaffari et al. (2014)
	Pb(II)	0.840	1.42	10.01	

tends to have Freundlich constant n between 1 and 10 (Febrianto et al., 2009). Larger values of n (smaller values for $1/n$) indicate strong interactions between biosorbent and heavy metal ions, while a value of n equal to 1 suggests a linear biosorption process, leading to identical biosorption energy for all sites. The values of *Freundlich isotherm* parameters obtained for the biosorption of some heavy metal ions on microalgae biosorbents are summarized in Table 4.

It should be noted that the *Freundlich isotherm* model cannot be used to predict the biosorption equilibrium data at extreme concentrations of heavy metal ions. This is because the mathematical equation is not linear at very low concentrations and has no limited expression at very high concentrations (Febrianto et al., 2009). Fortunately, this is not a problem, because a moderate concentration range is frequently used in most biosorption studies.

6. KINETICS MODELING OF HEAVY METALS BIOSORPTION ON MICROALGAE

Mathematical models that can describe the kinetics of batch biosorption process operated under different experimental conditions are very useful for scale-up studies or process optimization. Various kinetic models can be used to investigate the mechanism of the metal ion biosorption from aqueous solution onto various microalgae and to explain the transport of metal ions to the surface of biosorbent. The most frequently used are *pseudo-first-order* and *pseudo-second-order* models.

The *pseudo-first-order* kinetics model considers that the rate of occupation of biosorption sites is proportional to the number of unoccupied sites (Gerente et al., 2007) and can be written in its linear form as given by Eqn (5):

$$\log(q_e - q_t) = \log q_e - \frac{k_1}{2.303}t \qquad (5)$$

where q_e and q_t are the amounts of metal ions retained on the microalgae biosorbent at equilibrium and at time t, respectively (mg g^{-1}), and k_1 is the rate constant of the pseudo-first order kinetic model (1 min^{-1}).

The *pseudo-second-order* kinetic model is based on the fact that metal ions displace alkaline-earth ions from the algae biosorption sites and, therefore, with respect to the biosorption sites, the metal ions sorption can be considered to be a *pseudo-second-order* reaction (Ho and McKay, 1999; Gerente et al., 2007). The linear expression of this model is given by Eqn (6):

$$\frac{t}{q_t} = \frac{1}{k_2 \cdot q_e^2} + \frac{t}{q_e} \qquad (6)$$

where k_2 is the rate constant of *pseudo-second order* kinetic model (g mg^{-1} min)

The parameters of the *pseudo-first order* and the *pseudo-second order* kinetics models, calculated from the linear plots of $\log(q_e - q_t)$ versus t in the case of *pseudo-first-order* model, and of t/q_t versus t in the case of *pseudo-second-order*

TABLE 5   Kinetic Parameters for the Biosorption of Heavy Metals on Some Microalgae

| Microalgae | Heavy metal | Pseudo-first-order kinetics model | | | Pseudo-second-order kinetics mode | | | References |
		R^2	q_e, mg g^{-1}	k_1, 1 min^{-1}	R^2	q_e, mg g^{-1}	k_2, g mg^{-1} min	
Scenedesmus	Cd(II)	0.638	7.54	0.08	0.999	35.84	0.09	Mirghaffari
quadricauda	Pb(II)	0.929	8.11	0.08	0.999	32.36	0.07	et al. (2014)
Oedogonium	Pb(II)	0.903	30.22	0.018	0.996	63.29	0.01	Gupta and Rastogi (2008)
Oedogonium hatei	Ni(II)	0.982	25.37	0.004	0.982	28.11	0.007	Gupta et al. (2010)
Ulothrix cylindricum	As(III)	0.862	1.4	0.039	0.999	4.2	0.151	Tuzen et al. (2009)

kinetics model, for the biosorption of some heavy metals onto various types of microalgae, are summarized in Table 5.

In most cases, the heavy metals binding on microalgae complies with the *pseudo-second order kinetic* model, which suggests that the rate controlling steps in the biosorption process are the chemical interactions involving ion exchange and/or sharing of electrons between heavy metal ions from aqueous solution and superficial functional groups of biosorbent. Similar behaviors have been reported for various types of low-cost biomasses used as sorbents (Donmez et al., 1999; Febrianto et al., 2009; Montazer-Rahmati et al., 2011). In addition, the high values of rate constants (k_2) obtained in the case of these biosorbents indicate that the rate of the biosorption process is limited by the availability of heavy metal ions and the accessibility of functional groups from biomass surface to interact. When the availability of the superficial functional groups is higher, as is the case of salted water microalgae, the rate of biosorption process is also higher.

7. CONCLUSIONS

Bioremediation of aqueous effluents that contain various heavy metals by biosorption on microalgae is a simple method that has received attention in recent years since it has good efficiency, minimizes secondary (chemical or biological) wastes, and uses low-cost materials. Data presented in studies from the literature have shown that these biosorbents can be successfully used for the removal of heavy metals from large volumes of aqueous effluents, with relatively low metal ion concentrations (10–100 mg L^{-1}).

Microalgae are available in large quantities in many regions that can grow both in salted and fresh water, and their utilization as biosorbent is mainly determined by the variety of functional groups and relatively uniform distribution of these on the biosorbent surface, as well as by their reduced preference for alkali and alkali-earth metal ions, in comparison with heavy metals. In biosorption processes, the microalgae act as a chemical substrate of biologic origin, with a resistant structure, where the functional groups from biomass skeleton represent the binding sites from heavy metals from aqueous solution.

When designing bioremediation processes for the removal of heavy metals by biosorption on microalgae, the following have to be considered:

(1) The efficiency of biosorption process depends on several factors related to both the microalgae characteristics (particle size, growth and development conditions, pretreatment of cells, etc.), and the retention process (solution

pH, biosorbent dose, contact time, temperature, work methodology, height of biosorbent bed, flow rate of aqueous solution, heavy metals concentration, etc.), which must be optimized;

(2) The biosorption process can be easily modeled using several equilibrium and kinetics models that are very well known and that provide useful information about the mechanism of heavy metals biosorption on microalgae. The experiments must be started in batch conditions, to obtain some basic information, and then the possible applicability of biosorption systems on a large scale should be tested under dynamic continuous-flow conditions. The performance assessment of biosorption systems and the prediction based on both equilibrium and dynamic studies eventually lead to sizing of the equipment.

Research to date, as described in this chapter, clearly indicates that microalgae have the potential to become an effective and economical biosorbent for the bioremediation of aqueous effluents containing heavy metals.

References

Abdel-Ghani, N.T., El-Chaghaby, G.A., 2014. Biosorption for metal ions removal from aqueous solutions: a review of recent studies. Int. J. Latest. Res. Sci. Technol. 3, 24–42.

Aksu, Z., Sag, Y., Kutsal, T., 1992. The biosorption of copper (II) by *Chlorella vulgaris* and Zoogloea ramigera. Environ. Technol. 13, 579–586.

Aksu, Z., 2001. Equilibrium and kinetic modelling of cadmium(II) biosorption by *C. vulgaris* in a batch system: effect of temperature. Sep. Purif. Technol. 21, 285–294.

Aksu, Z., Donmez, G., 2006. Binary biosorption of cadmium(II) and nickel(II) onto dried *Chlorella vulgaris*: Co-ion effect on mono-component isotherm parameters. Process. Biochem. 41, 860–868.

Arica, Y.M., Tuzun, I., Yalcin, E., Ince, O., Bayramoglu, G., 2005. Utilisation of native, heat and acid-treated microalgae *Chlamydomonas reinhardtii* preparations for biosorption of Cr(VI) ions. Process Biochem. 40, 2351–2358.

Arief, V.O., Trilestari, K., Sunarso, J., Indraswati, N., Ismadji, S., 2008. Recent progress on biosorption of heavy metals from liquids using low cost biosorbents: characterization, biosorption parameters and mechanism studies. Clean Soil Air Water 36, 937–962.

Bayramoglu, G., Tuzun, I., Celik, G., Yilmaz, M., Arica, M.Y., 2006. Biosorption of mercury(II), cadmium(II) and lead(II) ions from aqueous system by microalgae *Chlamydomonas reinhardtii* immobilized in alginate beads. Int. J. Mineral Process. 81, 35–43.

Bulgariu, L., Cojocaru, C., Robu, B., Macoveanu, M., 2007. Equilibrium isotherms studies for the sorption of lead ions from aqua solutions using Romanian peat sorbent. Environ. Eng. Manag. J. 6, 425–430.

Bulgariu, L., Lupea, M., Bulgariu, D., Rusu, C., Macoveanu, M., 2013. Equilibrium study of Pb(II) and Cd(II) biosorption from aqueous solution on marine green algae biomass. Environ. Eng. Manag. J. 12, 183–190.

Chong, K.H., Volesky, B., 1995. Description of two-metal biosorption equilibria by Langmuir-type models. Biotechnol. Bioeng. 47, 451–460.

Dabrowski, A., Hubicki, Z., Podkoscielny, P., Robens, E., 2004. Selective removal of the heavy metal ions from waters and industrial wastewaters by ion-exchange method. Chemosphere 56, 91–106.

Davis, T.A., Volesky, B., Mucci, A., 2003. A review of the biochemistry of heavy metal biosorption by brown algae. Water Res. 37, 4311–4330.

Demirbas, A., 2008. Heavy metal adsorption onto agro-based waste materials: a review. J. Hazard. Mater. 157, 220–229.

Demirbas, A., 2010. Use of algae as biofuel sources. Energy Convers. Manag. 51, 2738–2749.

Deng, L., Su, Y., Su, H., Wang, X., Zhu, X., 2007. Sorption and desorption of lead (II) from wastewater by green algae *Cladophora fascicularis*. J. Hazard. Mater. 143, 135–146.

Donmez, G., Aksu, Z., Ozturk, A., Kutsal, T., 1999. A comparative study on heavy metal biosorption characteristics of some algae. Process. Biochem. 34, 885–892.

Farooq, U., Kozinski, J.A., Khan, M.A., Athar, M., 2010. Biosorption of heavy metal ions using wheat based biosorbents—a review of the recent literature. Bioresour. Technol. 101, 5043–5053.

Febrianto, J., Kosasih, A.N., Sunarso, J., Ju, Y.H., Indrawati, N., Ismadji, S., 2009. Equilibrium and kinetic studies in adsorption of heavy metals using biosorbent: a summary of recent studies. J. Hazard. Mater. 162, 616–645.

Fu, F., Wang, Q., 2011. Removal of heavy metal ions from wastewaters: a review. J. Environ. Manag. 92, 407–418.

Gao, R., Wang, J., 2007. Effects of pH and temperature on isotherm parameters of chlorophenols biosorption to anaerobic granular sludge. J. Hazard. Mater. 145, 398–403.

Gavrilescu, M., 2004. Removal of heavy metals from the environment by biosorption. Eng. Life Sci. 3, 219–232.

Gautam, R.K., Mudhoo, A., Lofrano, G., Chattopadhyaya, M.C., 2014. Biomass-derived biosorbents for metal ions sequestration: adsorbent modification and activation methods and adsorbent regeneration. J. Environ. Chem. Eng. 2, 239–259.

Gerente, C., Lee, V.K.C., Lee, P., McKay, G., 2007. Application of chitosan for the removal of metals from wastewaters by adsorption—mechanisms and models review. Crit. Rev. Environ. Sci. Technol. 37, 41−127.

Gokhale, S.V., Jyoti, K.K., Lele, S.S., 2008. Kinetic and equilibrium modeling of chromium (VI) biosorption on fresh and spent *Spirulina platensis/Chlorella vulgaris* biomass. Bioresour. Technol. 99, 3600−3608.

Gupta, V.K., Rastogi, A., 2008. Biosorption of lead(II) from aqueous solutions by non-living algal biomass *Oedogonium* and *Nostoc* sp. A comparative study. Coll. Surf. B 64, 170−178.

Gupta, V.K., Rastogi, A., 2009. Biosorption of hexavalent chromium by raw and acid-treated green alga *Oedogonium hatei* from aqueous solutions. J. Hazard. Mater. 163, 396−402.

Gupta, V.K., Rastogi, A., Nayak, A., 2010. Biosorption of nickel onto treated alga (*Oedogonium hatei*): application of isotherm and kinetic models. J. Coll. Interf. Sci. 32, 533−539.

Han, X., Wong, Y.S., Wong, M.H., Tam, N.F.Y., 2007. Biosorption and bioreduction of Cr(VI) by a microalgal isolate. *Chlorella miniata*. J. Hazard. Mater. 146, 65−72.

Herrero, R., Cordero, B., Lodeiro, P., Rey-Castro, C., SastreDeVicente, M.E., 2006. Interactions of cadmium(II) and protons with dead biomass of marine algae Fucus sp. Marine Chem. 99 (1−4), 106−116.

Ho, Y.S., McKay, G., 1999. Pseudo-second-order model for sorption processes. Process. Biochem. 34, 451−465. http://www.oilgae.com/algae/comp/comp.html#sthash.4s8c0svH.dpuf.

Hubbe, M.A., Hasan, S.H., Ducoste, J.J., 2011. Cellulosic substrates for removal of pollutants from aqueous systems: a review. 1. Metals Biores. 6, 2161−2914.

Inthorn, D., Sidtitoon, N., Silapanuntakul, S., Incharoensakdi, A., 2002. Sorption of mercury, cadmium and lead by microalgae. Sci. Asia 28, 253−261.

Kumar, R., Singh, K., Sarkar, S., Sethi, L.N., 2014. Accumulation of Cu by microalgae *Scenedesmus obliquus* and *Synechocystis* sp. PCC 6803. J. Environ. Sci. Toxicol. Food Technol. 8, 64−68.

Llanos, J., Williams, P.M., Cheng, S., Rogers, D., Wright, C., Perez, A., Canizares, P., 2010. Characterization of a ceramic ultrafiltration membrane in different operational states after its use in a heavy-metal ion removal process. Water Res. 44, 3522−3530.

Lupea, M., Bulgariu, L., Macoveanu, M., 2012. Biosorption of Cd(II) from aqueous solutions on marine algae biomass. Environ. Eng. Manag. J. 11, 607−615.

Mendoza, C.A., Cortes, G., Munoz, D., 1998. Heavy metal pollution in soils and sediments of rural developing district 063, Mexico. Environ. Toxicol. Water Qual. 11, 327−333.

Mirghaffari, N., Moeini, E., Farhadian, O., 2014. Biosorption of Cd and Pb ions from aqueous solutions by biomass of the green microalga, *Scenedesmus quadricauda*. J. Appl. Phycol. http://dx.doi.org/10.1007/s10811-014-0345-z.

Montazer-Rahmati, M.M., Rabbari, P., Abdulali, A., Keshtkar, A.R., 2011. Kinetics and equilibrium studies on biosorption of cadmium, lead, and nickel ions from aqueous solutions by intact and chemically modified brown algae. J. Hazard. Mater. 185, 401−407.

Monteiro, C.M., Castro, P.M.L., Malcata, F.X., 2011. Biosorption of zinc ions from aqueous solution by the microalga. *Scenedesmus obliquus*. Environ. Chem. Lett. 9, 169−176.

Munoz, R., Alvarez, M.T., Munoz, A., Terrazas, E., Guieysse, B., Mattiasson, B., 2006. Sequential removal of heavy metal ions and organic pollutants using an algal-bacterial consortium. Chemosphere 63, 903−911.

Nasreen, K., Muhammad, I., Iqbal, Z.S., Javed, I., 2008. Biosorption characteristics of unicellular green alga *Chlorella sorokiniana* immobilized in loofa sponge for removal of Cr(III). J. Environ. Sci. 20, 231−239.

Pelhivan, E., Arslan, G., 2007. Removal of metal ions using lignite in aqueous solution—low cost biosorbents. Fuel Proces. Technol. 88, 99−106.

Priyadarshani, I., Sahu, D., Rath, B., 2011. Microalgal bioremediation: current practices and perspectives. J. Biochem. Technol. 3, 299−304.

Ribeiro, R.F.L., Magalhães, S.M.S., Barbosa, F.A.R., Nascentes, C.C., Campos, I.C., Moraes, D.C., 2010. Evaluation of the potential of microalgae *Microcystis novacekii* in the removal of Pb^{2+} from an aqueous medium. J. Hazard. Mater. 179, 947−953.

Rojan, P.J., Anisha, G.S., Madhavan Nampoothiri, K., Pandey, A., 2011. Micro and macroalgal biomass: a renewable source for bioethanol. Biores. Technol. 102, 186−193.

Senthilkumar, R., Vijayaraghavan, K., Thilakavathi, M., Iyer, P.V.R., Velan, M., 2006. Seaweeds for the remediation of wastewaters contaminated with zinc(II) ions. J. Hazard. Mater. 136, 791−799.

Stumm, W., Morgan, J.J., 1996. Aquatic Chemistry: Chemical Equilibria and Rates in Natural Waters. John Wiley & Sons Inc., New York.

Tuzen, M., Sari, A., Mendil, D., Uluozlu, O.D., Soylak, M., Dogan, M., 2009. Characterization of biosorption process of As(III) on green algae ulothrix cylindricum. J. Hazard. Mater. 566−572.

Valdman, E., Leite, S.G.F., 2000. Biosorption of Cd, Zn and Cu by *Sargassum* sp. waste biomass. Bioprocess. Eng. 22, 171−173.

Volesky, B., 1987. Biosorbents for metal recovery. Trends Biotechnol. 5, 96−101.

Wan Ngah, W.S., Hanafiah, M.K.M., 2008. Removal of heavy metal ions from wastewater by chemically modified plant wastes as adsorbents: a review. Biores. Technol. 99, 3935–3948.

Wang, J., Chen, C., 2008. Biosorbents for heavy metals removal and their future. Biotechnol. Adv. 27, 195–226.

Zhou, G.J., Peng, F.Q., Zhang, L.J., Ying, G.G., 2012. Biosorption of zinc and copper from aqueous solutions by two freshwater green microalgae *Chlorella pyrenoidosa* and *Scenedesmus obliquus*. Environ. Sci. Pollut. Res. 19, 2918–2929.

Zhao, Y., Wang, B., Liu, C., Wu, Y., 2013. Biosorption of trace metals from aqueous multimetal solutions by green microalgae. Chim. J. Geochem. 32, 385–391.

Bioremediation with Microalgae: Toward Sustainable Production of Biofuels

J. Paniagua-Michel

Laboratory for Bioactive Compounds and Bioremediation, Department of Marine Biotechnology, Centro de Investigación Científica y de Educación Superior de Ensenada (CICESE), Ensenada, BC, México

1. INTRODUCTION

The increased global shortages of fossil fuels, and their respective contributions to pollution and global warming, have negatively impacted almost all the primary activities of all societies and industries. New options for renewable and sustainable energy sources have become a priority because of the increase in population growth, which has impacted the human condition and life quality at the global level. Under these circumstances, the development of sustainable and eco-friendly sources of energy has become one of the most important challenges in this century, and new sources of renewable feedstocks that would produce sustainable replacement fuels have become a priority. Under this scenario, bioprospecting for renewable and cost-effective sources of energy for the future is mandatory for the world's societies and governments (International Energy Agency IEA, 2013). Recent reports have indicated that microalgae can be utilized for low-cost and environmentally friendly wastewater treatment and bioremediation when compared to other common treatment technologies (Pittman et al.,

2011; Rahman et al., 2012). This chapter presents the roles played by the bioremediation on wastewater effluents in selected species of microalgae that can be used as feedstock biomass for the production of liquid biofuels. Coupling bioremediation to the carbon neutral production of biofuels can be considered as a promising low-cost alternative to reduce the cost per yield of liquid biofuels affordable for a demanding sector.

2. BIOREMEDIATION AND REMOVAL OF NUTRIENT IONS WITH MICROALGAE

Municipal wastewater effluents are one of the largest sources of pollution, by volume, being discharged to surface water bodies in aquatic environments (Zamora-Castro et al., 2008). These effluents from municipal treatment plants (WWTPs) are heavily charged with nutrients (nitrogen and phosphorus) affecting both freshwater and sea life, causing eutrophication and degradation of the marine ecosystem (Sriram and Seenivasan, 2012). Advanced wastewater treatments exhibit several advantages when integrated to the microalgae biomass production (Christenson & Sims, 2011; Pittman et al., 2011). Recently, wastewater bioremediation has been envisioned as one of the most viable alternatives for cost-effective liquid biofuels production and renewable energy (Schneider et al., 2013). A wide number of microalgae have been used for the successful removal of pollutants and bioremediation of wastewater effluents, such as *Chlorococcum* (Zamora-Castro et al., 2008; Rosales and Paniagua-Michel, 2014), *Arthrospira* (*Spirulina*), *Scenedesmus*, *Chlorella*, *Chlamydomonas*, *Phormidium*, and *Botryococcus* (Fathi et al., 2013; Kong et al., 2010; Pittman et al., 2010; Stephens et al., 2010). In general terms, photoautotrophic biological assimilation of wastewater nutrients can be less expensive, more efficient, and ecologically safer than physical/chemical removal processes (Christenson and Sims, 2012; Oswald, 2003). Microalgae can be placed as efficient bioremediators of wastewater and organic wastes and at the same time a safe producer of biomass for recycling (Paniagua-Michel et al., 1987) in other bioproductive industrial applications, viz, biofuels. It is expected that the combination of bioremediation of domestic wastewater, biomass production, and industrial energy generation could help overcome the present obstacles of unsustainability.

3. MICROALGAE AND BIOFUELS PRODUCTION

In principle, microalgae grown under normal and induced conditions are able to accumulate 20—75% of lipids as part of their dry biomass (Sivakumar et al., 2012). Photosynthetic algae use energy from the sun to sustain their fast growth rate and to accumulate a high quantity of lipids and reserve of oil, carbohydrates, polysaccharides, proteins, pigments, vitamins, minerals, and other cellular substances susceptible to be biotechnologically exploited (Schneider et al., 2013; Paniagua-Michel et al., 2012).

The biochemical composition of microalgae has been considered an attribute for an alternative feedstock source for biofuel production, besides their several advantages over traditional land crops to produce higher biomass productivity on the basis of land area (Fathi et al., 2013; McGinn et al., 2011). The high productivity exhibited by microalgae, up to 100,000 L he^{-1}, is an excellent outcome when compared to productive crops (palm, yield 5959 L he^{-1} (Schneider et al., 2013). In general terms, practically, microalgae can double their biomass within 24 h and accumulate an average oil content in most of them up to 30—80% by weight of dry biomass (Cho et al., 2012; Chisti, 2007). For biofuel production, algae need to have a lipid content exceeding 20% (Dalrymple et al., 2013), and some researchers even suggest 40%. Usable lipids were assumed to be 20% and 50% of the algae dry weight for moderate-strength wastewater and low-strength pond water, respectively (Dalrymple et al., 2013). In these calculations, an algal oil-to-biofuel conversion efficiency of 80% was used, which is similar to that obtained for vegetable oil. Microalgae oil production per unit area of land far exceeds other oil crops such as corn, soybean, coconut, and oil palm by as much as two to three orders of magnitude. Specific microalgae, like *Botryococcus braunii* and *Nannochloropsis* sp., are well suited for this alternative (Sydney et al., 2011). Furthermore, they do not compete for arable land, can be produced year-round in suitable climates, and are likely to recover more quickly from adverse effects. The biofuel potential for the various algae is shown in Table 1. The total potential volume of biofuel obtained is approximately 269,545, which can, on average, fuel 450 cars per year (assuming 15,000 miles year^{-1} with an average of 25 miles per gallon) (Dalrymple et al., 2013). Concerning the production values of algae under lab-controlled conditions, a rough estimate yields from 3000 to 14, 000 gallons of oil/acre/year, which is a 130-fold increase over soybean, the leading feedstock for biodiesel production

TABLE 1   Biomass and Lipid Productivities of Selected Microalgae Cultured on Wastewater of Different Origin

Wastewater	Microalgae	Biomass (DW) (mg L^{-1} day^{-1})	Lipids (mg L^{-1} day^{-1})	References
Municipal (centrate)	Chlamydomonas reinhardtii	2000	505	Kong et al. (2010)
Municipal (ST)	Scenedesmus obliquus	26[a]	8[e]	[f]
Municipal (ST)	Botryococcus brauni	345.6[b]	62	[g]
Agricultural (piggery manure)	B. braunii	700[c]	69	[h]
Agricultural (dairy wastewater)	Chlorella sp. Micractinium sp. Actinastrum sp.	59[d]	17	[i]
Industrial carpet mill	Dunaliella tertiolecta	28	4.3	[j]

DW, Dry weight.
ST, Secondary treatment.
[a] Estimated from the biomass value of 1.1 mg L^{-1} h^{-1}.
[b] Estimated from the biomass value of 14.4 mg L^{-1} h^{-1}.
[c] Estimated from the biomass value of 7 g L^{-1} after 10 days.
[d] Estimated from lipid productivity and lipid content value.
[e] Fatty acid content and productivity determined rather than total lipid.
[f] Martinez, M. E., Sanchez, S., Jimenez, J. M., Yousfi, F. E., Munoz, L. (2000). Nitrogen and phosphorus removal from urban wastewater by the microalga Scenedesmus obliquus. Bioresour. Technol., 73, 263–272.
[g] Orpez, R., Martinez, M. E., Hodaifa, G., Yousfi, F. E., Jbari, N., Sanchez, S. (2009) Growth of the microalga Botryococcus braunii in secondarily treated sewage. Desalination, 246, 625–630.
[h] An, J. Y., Sim, S. J., Lee, J. S., Kim, B. W. (2003). Hydrocarbon production from secondarily treated piggery wastewater by the green alga Botryococcus braunii. J. Appl. Phycol., 15, 185–191.
[i] Woertz, I., Feffer, A., Lundquist, T. and Nelson, Y. (2009). Algae Grown on Dairy and Municipal Wastewater for Simultaneous Nutrient Removal and Lipid Production for Biofuel. Feedstock. J. Environ. Eng., 135, 1115–1122.
[j] Chinnasamy, S., Bhatnagar, A., Hunt, R. W., Das, K. C. (2010). Microalgae cultivation in a wastewater dominated by carpet mill effluents for biofuel applications. Bioresour. Technol., 101, 3097–3105.
Modified after Pittman et al. (2011); Rawat et al. (2011).

(Sturm et al., 2011). In Figure 1, these microalgal properties and uniqueness have been integrated with respective strategies for the enhancement of the feasibility of algae-based biofuels.

3.1 Transesterification and Biodiesel Production from Microalgae

Biodiesel is currently recognized as a green and alternative renewable fuel because it is nontoxic, biodegradable, and has lower emission of GHG when burned (Lam et al., 2012; Demirbas et al., 2011). From microalgae, this biofuel is made from triglycerides and of three chains of fatty acids linked by a glycerol molecule (Hu et al.,

2008). Moreover, once the biodiesel process is developed, after hydrolysis, the residual biomass can potentially be used for bioethanol production (Schneider et al., 2013). During the process of reproduction and biomass accumulation, microalgae accumulate a large quantity of triacylglycerides (>50%) of their dry mass and free fatty acids that can be used to produce biodiesel (Lam et al., 2012; Chisti, 2007). These long-chain hydrocarbons can be converted to biodiesel and jet fuel or cracked like petroleum for liquid fuel (Sivakumar et al., 2010; Niehaus et al., 2011). Algal biodiesel production essentially involves two main steps: (1) extraction of oils from the biomass, and (2) conversion (transesterification) of oils

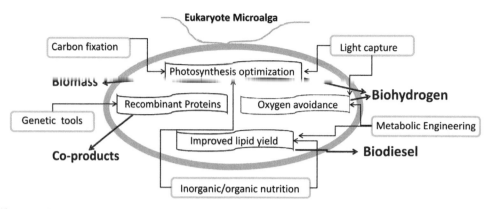

FIGURE 1   A schematic representation of the integration of strategies for improving the feasibility of algae-based biofuels. *Modified after Gimpel et al. (2013).*

(fatty acids) to biodiesel (alkyl esters) (Schenk et al., 2008). To date, biodiesel production from algae biomass has been generally performed by one of the following methods: (1) A two-step protocol in which algae oil is extracted with organic solvent and then converted to biodiesel using a catalyst, such as an acid, a base, or an enzyme; (2) direct production of biodiesel from algae biomass using an acid catalyst at atmospheric pressure and ambient temperature; and (3) one-step, single-pot conversion to biodiesel at high pressure and high temperature in the absence of a catalyst (Martinez-Guerra et al., 2014; Chen et al., 2012). *Chlorella* sp. exhibited a maximum lipid to FAME conversion of around 88% after a reaction time of 2 h, using 0.04 mol of sulfuric acid, 500:1 mol of methanol, and a temperature of 90 °C (Martinez-Guerra et al., 2014). The in situ transesterification process is applied on microalgae for biodiesel production (Ehimen et al., 2010). Lipids are extracted from the biomass by methanol and catalyzed by the acid, which concurrently transesterifies the extracted lipids to produce fatty acid methyl esters (Figure 2). Recent studies on in situ transesterification using an acid catalyst for the production of biodiesel were developed using *Arthrospira platensis*. The total lipid content of *A. Platensis* was 0.1095 g g^{-1} biomass (El-shimi et al., 2013). The optimum level of fatty acid methyl ester (84.7%) was

obtained at 100% (wt./wt.oil) catalyst concentration. Velasquez-Orta et al. (2012) developed a successful process of alkaline in situ transesterification in *Chlorella vulgaris*; their results were comparable to those obtained with an acid catalyst (Miao and Wu, 2006). The algae biomass-to-methanol ratio ($\sim$1:12) is considered a clever issue to optimize lipid recovery ($\sim$18%) and FAEE conversion (96.1%) (Martinez-Guerra et al., 2014).

3.2 Bioethanol from Microalgae: A Potential Biotechnology

Bioethanol, or perhaps fuel ethanol, is a biomass-derived, biodegradable, and environmental friendly fuel produced from different feedstocks such as cellulosic biomass. Bioethanol is produced from biomass by the fermentation of available carbohydrates, usually simple sugars, into bioethanol and carbon dioxide, via the following chemical process:

$$C_nH_{2n}O_n(\text{Sugar}) \rightarrow n/3\ C_2H_5OH + n/3(CO_2)$$
$$+ \text{Heat}$$

$$(1)$$

Most carbohydrate molecules (($CH_2O)_n$) have the potential to produce bioethanol; however, the primary sources for current bioethanol production are sugar, starch, cellulose, and

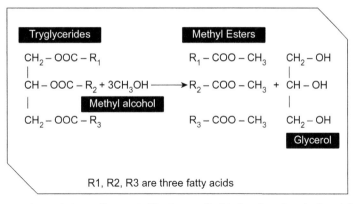

FIGURE 2 The main reaction and steps of transesterification applicable in microalgae feedstock for biodiesel production.

hemicellulose (Harun et al., 2014; Kim et al., 2004). During microbial hydrolysis, these carbohydrate polymers are broken into simple sugars, followed by fermentation to yield bioethanol (Kumar et al., 2013). Microalga genera such as *Chlorella, Dunaliella, Chlamydomonas, Scenedesmus, Nannochloropsis,* and *Spirulina* are some of the most abundant organisms on earth, existing in salt or fresh water. Studies in marine microalgae reported dark fermentation in the marine green algae *Chlorococcum littorale,* which produced 450 μmol ethanol g^{-1} at 30 °C (Ueno et al., 1998). Despite the real potential of microalgae biofuel, its feasibility for commercialization is still questionable. Recent calculations have increased the expectation to have lower cost per yield, a condition associated with their ability to reduce greenhouse emissions through the replacement of fossil fuels (Fathi et al., 2013; Brune et al., 2009) concomitantly with the use of low-cost nutrients.

3.3 Hydrogen Photoproduction by Microalgae and Wastewater Bioremediation

Hydrogen is one of the ideal alternative energy sources to replace fossil fuel (Li, and Yu, 2011; Hallenbeck and Ghosh, 2009). Hydrogen can be obtained from one of the following sources: fossil fuel, water, and biomass (Beer et al., 2009). Photosynthetic eukaryotic microalgae and some cyanobacteria have the ability to simultaneously photoproduce molecular hydrogen and oxygen, using water as the only electron donor, a process called water biophotolysis (Ghirardi et al., 2007). Photosynthesis as the clever photosynthetic water-splitting agent, in coordination with the activity of hydrogenase enzymes, is functionally linked to H$_2$ production. A remarkable feature is that green microalgae and cyanobacteria contain only one of two major types of hydrogenases, [FeFe] or [NiFe] enzymes (Ghirardi et al., 2007). In spite of the fact that biohydrogen production from wastewater still is not developed on an industrial scale, great progress in the development of such technologies is underway. Expression of algal cells [FeFe]-hydrogenase enzymes that catalyze the following ferredoxin (Fd)-linked reaction occurs under anaerobic conditions (Kosourov et al., 2012):

$$2H + 2Fred \quad H_2 + 2Fdox \qquad (2)$$

Studies by Batyrova et al. (2012) have indicated that sustained H$_2$ photoproduction can be induced by subjecting *Chlamydomonas reinhardtii* cultures to sulfur and phosphorus deprivation, respectively. Studies on the maximum H$_2$ output have registered ∼70 ml L^{-1} in cultures with Chl content of ~1 mg/l (Batyrova et al., 2012). *C. reinhardtii cultures* deprived of inorganic sulfur are capable of prolonged H_2 photoproduction, not only in fresh water but also in marine strains, but obstacles due to the high concentrations of sulfates in seawater make sulfur deprivation a hard task for hydrogen gas photoproduction by microalgae. Moreover, the photoproduction in nutrient-deprived algae cannot be assured due to the high degradation of the RuBisCo enzyme at the moment that H_2 photoproduction begins (Kosourov et al., 2012). The possibility that nutrient-deprived algae utilize H_2 gas by the chlororespiration pathway from [FeFe]-hydrogenase(s) to O_2 through Fd, NADPþ/NADPH and the PQ pool can be considered (Kosourov et al., 2012). Under these circumstances, bioremediation plays an important role considering the limitations in nutrients for the induction to the photoproduction of hydrogen, viz, phosphate and sulfate from wastewater/seawater. Bioremediation with microalgae offers the possibility of control, counterbalance and regulated nutrient sufficient, enriching and/or exhausting conditions in ionic components of wastewater and seawater with their different technological alternatives and circuits.

4. BIOREMEDIATION: TROPHIC PATHWAY TOWARD SUSTAINABLE BIOFUELS

Bioremediation is a biotechnology that involves the use of organisms to remove, break down, and control or neutralize pollutants nutrients from a contaminated site into less toxic or nontoxic substances (Paniagua-Michel et al., 2005). Particularly, microorganisms and microalgae under certain circumstances have the ability to generate different levels of ions nutrients. The ability of bioremediation to accomplish different ratios of micro- and macronutrients, carbon:nitrogen:phosphorus (C:N:P), allow supply of the required nutrient condition according to the

microalgae requirement. In spite of the efficiency of certain microalgae to taking up nitrogen and phosphate from the surrounding media, efficient strategies for nutrient recycling need to be developed if algae are going to be used to produce significant amounts of fuel. In either case, the potential to expand the variety of nutrients utilized by algae can be harnessed, and therefore improve the overall yield of products. Microalgae can be grown photoautotrophically or heterotrophically. The heterotrophic or mixotrophic algal culture on a reduced carbon source exhibits several advantages, viz, the culture conditions are highly controlled and reproducible and higher cell densities can be achieved, which decreases harvesting costs (Rasala et al., 2013). A great number of microalgae species, however, are strict autotrophs, which can be exploited more practically in bioremediation programs. The enrichment of glucose in the growth media generates a remarkable production of 150% more hydrogen by transformed *C. reinhardtii* cells than the parental strain (Doebbe et al., 2007). Despite the advantages of heterotrophic culture for biofuel production, the enrichment by a carbon source produces an undesirable increase in cost when compared to phototrophic systems, because the special conditions of the high risk of contamination require indoor enclosed bioreactors.

It is recognized that the development and production of economically viable biofuels, mainly biodiesel of algae, depend on the yield and quality of production of oil. Different strategies aiming to increase oil production and lipid quality have been developed. In *C. vulgaris*, values of $0.2 \, kg \, m^{-3} \, d^{-1}$ oil were reported when first grown under nutrient-sufficient conditions followed by nutrient deprivation, but higher lipid contents were reported when the cells depleted their N naturally (Scott et al., 2010). Many algae can grow heterotrophically with an external carbon source, rather than photosynthetically. In this condition the N:C ratio is altered, and the lipid production is increased as the effect produced when growing algae with low N levels under autotrophic conditions. For example, an enhancement in the lipid levels was seen when sugars were added to photosynthetically grown cells of *C. prototothecoides*, resulting in lipid levels of up to 11.8 kg m^{-3} day^{-1} and 58% of oil content per dry weight (Xiong et al., 2010).

5. ADVANCES IN GENETIC AND METABOLIC ENGINEERING OF MICROALGAE BIOFUELS

Nowadays, genetic and in silico developments have advanced in the characterization of microalgal genomes aiming to harness biofuel production (Gimpel et al., 2013). Sequences of microalgal genomes of academic or industrial interest are actually accessed from the different bank of sequences, which has greatly facilitated genetic manipulation (Radakovits et al., 2010). Introns, which initially were conceived of as selfish genes, are widespread in microalgal genomes. The presence or absence of introns has been used for identification of open reading frames in genomic DNA, and has recently been used to resolve identification and differentiation between closely related species of carotenogenic microalgae (Olmos et al., 2012). The current advances in microalgal genetics for biofuel production include genetic transformation and stable heterologous gene expression. At present, the advent of genome sequences for algae opens the possibility to apply metabolic engineering as a significant mechanism for increasing yields of TAGs or engineering pathways for novel algal biofuel molecules (Scott et al., 2010; Yoon et al., 2011). In the case of *C. reinhardtii*, all three genomes—nuclear, chloroplast, and mitochondria—have been successfully sequenced (Rasala et al., 2013). It is well known that the mechanisms that control lipid production are basic steps to enable genetic manipulation of microalgae for enhancing growth and sustainable photoproduction of biofuels (Christenson and Sims, 2011). For instance, cellular lipid yield can be increased by blocking the metabolic pathways involved in the biosynthesis of starch or by diminishing lipid catabolism (Radakovits et al., 2010; Radakovits et al., 2011).

The cellular oil content of microalgae is a key factor to achieve an efficient yield of biodiesel. *C. vulgaris* grown under nutrient-sufficient conditions accumulated a lipid content between 14% and 30% of dry weight (Illman et al., 2000), a condition related to the insufficient N for protein production necessary for growth, while excess carbon from photosynthesis is deviated into storage molecules (TAGs or starch). In cultures with N-depleted media, the total fatty acid content on a per cell basis increased by 2.4-fold, after 72 h, and the total fatty acids (higher than 60%) are esterified to TAG in oil bodies, a condition indicating that TAG formation depends on de novo synthesis of fatty acids. The presence of a major lipid droplet protein (MDLP) was shown by proteomic analysis, which was particularly abundant in the microalgal lipid bodies. A direct correlation was confirmed between the MLDP mRNA transcript abundance, with a corresponding increase in lipid droplets after N depletion in *C. reinhardtii*, corroborating that this protein regulates the lipid droplet size in this microalgae. This process also provides information on gene transcript, protein, and metabolic activities during lipid accumulation in algae (Scott et al., 2010). In the case of starch-deficient *C. reinhardtii* strains, increased accumulation levels of TAGs during nitrogen deprivation were identified as a response of interference in its ADP pyrophosphorylase or isoamylase genes, respectively (Sivankumar et al., 2012; Wang et al., 2009). The engineering of this Chlorophyte occurred in photobioreactors that were exposed to 200−400 µmol m^{-2} s^{-1} (~10−20%) (Sivakumar et al., 2012; Gordon and Polle, 2007), which increased biomass yield by downregulation expression of its light-harvesting antenna complexes (Beckman et al., 2009).

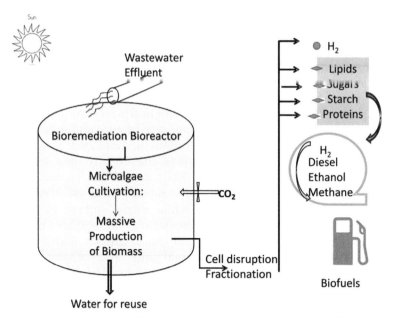

FIGURE 3   General scheme of the integration of bioremediation of secondary wastewater effluents to the microalgal culture and respective potential for development of a biorefinery and biofuels.

6. COUPLING BIOREMEDIATION AND MICROALGAL BIOFUEL: SUSTAINABLE APPROACHES

Successful photoautotrophically produced microalgae in municipal wastewater can be grown once the control of parameters such as concentrations of dissolved inorganic nutrients, primarily ammonium and phosphates, as well as light limitation and fluctuations in temperature are achieved (Park et al., 2012; Apt and Behrens, 1999; Lee, 2001). Anaerobically digested agricultural waste materials and catfish processing waste have been used to feed algal cultures, which were used as feedstock to produce microalgal oil and methane energy, as well as other by-products, for example, fertilizer (Sivakumar et al., 2012; Prajapati et al., 2013). Since algae are known to grow in wastewater, a possible synergistic solution is to co-locate and integrate algal production with nutrients obtained from rich wastewater and utilization of CO_2 from power

plant to produce flue gas (Dalrymple et al., 2013). This approach essentially reduces the cost of algal production and consequently leads to the sustainable production of biofuels, while preventing eutrophication and mitigating CO_2 emissions (Chynoweth et al., 1993).

The lipid production for algae grown on wastewater with nitrogen levels of ~ 30 mg L^{-1} was carried out by Woertz et al. (2009), obtaining 3 g dry wt m^{-2} day^{-1} and 30% lipids by dry weight, respectively. Integrating microalgal biomass production with municipal wastewater treatment and industrial CO_2 emissions are alternative to achieve cost-effective microalgal production yields. Effective removal of N and P from municipal wastewater is limited by the processing capacity of available microalgal cultivation systems. The selection of the biofuel of interest (biodiesel, bioethanol, or biohydrogen) and the microalgal strain will determine the bioprocess and treatment methods of biomass to coupling bioremediation and biofuel production, as exemplified in Figure 3. Nevertheless, some remaining challenges on the production of biofuels and bio-products using algal biomass have been handicapped by an inability to find a reliable and cost-effective method of producing and harvesting large quantities of algae feedstock (Christenson and Sims, 2011; Chisti et al., 2008). The respective integration of microalgae with light and mineral ion utilization, carbon flow, and engineering of biofuels' substrates will contribute to the reduction in the cost per yield and the achievement of carbon-neutral conditions, key factors to achieve sustainable biofuels production.

References

Apt, K.E., Behrens, P.W., 1999. Commercial developments in microalgal biotechnology. Rev. J. Phycol. 35, 215–216.

Batyrova, K.A., Tsygankov, A.A., Kosourov, S.N., 2012. Sustained hydrogen photoproduction by phosphorous-deprived Chlamydomonas reinhardtii cultures. Intl. J. Hydrogen Energy 37, 8834–8839.

Beckmann, J., Lehr, F., Finazzi, G., Hankamer, B., Posten, C., Wobbe, L., Krusea, O., 2009. Improvement of light to biomass conversion by de-regulation of light-harvesting protein translation in Chlamydomonas reinhardtii. J. Biotechnol. 142, 70–77.

Beer, L.L., Boyd, E.S., Peters, J.W., Posewitz, M.C., 2009. Engineering algae for biohydrogen and biofuel production. Curr. Opin. Biotechnol. 20, 264–271.

Brune, D.E., Lundquist, T.J., Benemann, J.R., 2009. Microalgal biomass for greenhouse gas reductions: potential for replacement of fossil fuels and animal feeds. J. Environ. Eng. 135, 1136–1144.

Chen, L., Liu, T., Zhang, W., Chen, X., Wang, J., 2012. Biodiesel production from algae oil high in free fatty acids by two-step catalytic conversion. Bioresour. Technol. 111, 208–214.

Chisti, Y., 2007. Biodiesel from microalgae. Biotechnol. Adv. 25, 294–306.

Chisti, Y., 2008. Biodiesel from microalgae beats bioethanol. Trends Biotechnol. 26, 126–131.

Cho, S., Luong, T.T., Lee, D., Oh, Y.K., Lee, T., 2012. Reuse of effluent water from a municipal wastewater treatment plant in microalgae cultivation for biofuel production. Bioresour. Technol. 1, 1–7.

Christenson, L., Sims, R., 2011. Production and harvesting of microalgae for wastewater treatment, biofuels, and bioproducts. Biotechnol. Adv. 29, 686–702.

Christenson, L.B., Sims, R.C., 2012. Rotating algal biofilm reactor and spool harvester for wastewater treatment with biofuels by-products. Biotechnol. Bioeng. 109, 1674–1684.

Chynoweth, D.P., Turick, C.E., Owens, J.M., Jerger, D.E., Peck, M.W., 1993. Biochemical methane potential of biomass and waste feedstocks. Biomass Bioenergy 5, 95–111.

Dalrymple, O.K., Halfhide, T., Udom, I., Gilles, B., Wolan, J., Zhang, Q., Ergas, S., 2013. Wastewater use in algae production for generation of renewable resources: a review and preliminary results. Aquatic Biosystems 5, 2.

Demirbas, M.F., 2011. Biofuels from algae for sustainable development. Appl. Energy 88, 3473–3480.

Doebbe, A., Rupprecht, J., Beckmann, J., Mussgnug, J.H., Hallmann, A., Hankamer, B., Kruse, O., 2007. Functional integration of the HUP1 hexose symporter gene into the genome of C. reinhardtii: impacts on biological H(2) production. J. Biotechnol. 131, 27–33.

Ehimen, E.A., Sun, Z.F., Carrington, C.G., 2010. Variables affecting the in situ transesterification of microalgae lipids. Fuel 89, 677–684.

El-Shimi, H.I., Attia, N.K., El-Sheltawy, S.T., El-Diwani, G.I., 2013. Biodiesel production from Spirulina-platensis microalgae by in-situ transesterification process. J. Sustainable Bioenergy Systems 3, 224–233.

Fathi, A.M., Azooz, M.M., Al-Fredan, M.A., 2013. Phycoremediation and the potential of sustainable algal biofuel production using wastewater. Am. J. Appl. Sci. 10, 189–194.

Ghirardi, M.L., Posewitz, M.C., Maness, P.C., Dubini, A., Yu, J., Seibert, M., 2007. Hydrogenases and hydrogen photoproduction in oxygenic photosynthetic organisms. Annu. Rev. Plant Biol. 58, 71–91.

Gimpel, A.J., Specht, A.E., Georgianna, R.D., Mayfield, P.S., 2013. Advances in microalgae engineering and synthetic biology applications for biofuel production. Curr. Opin. Chem. Biol. 17, 489–495.

Gordon, J.M., Polle, J.E.W., 2007. Ultrahigh bioproductivity from algae. App. Microbiol. Biotechnol. 76, 969–975.

Hallenbeck, P.C., Ghosh, D., 2009. Advances in fermentative biohydrogen production: the way forward? Trends Biotechnol. 27, 287–297.

Harun, R., Yip, J.W.S., Thiruvenkadam, S., Ghani, W.A.W.A.K., Cherrington, T., Danquah, M.K., 2014. Algal biomass conversion to bioethanol—a step-by-step assessment. Biotechnol. J. 9, 73–86.

Hu, Q., Sommerfeld, M., Jarvis, E., Ghirardi, M., Posewitz, M., Seibert, M., 2008. Microalgal triacylglycerols as feedstocks for biofuel production: perspectives and advances. Plant J. 54, 621–639.

International Energy Agency IEA, 2013. Key World Statistics. IEA.

Kim, S., Dale, B.E., 2004. Global potential bioethanol production from wasted crops and crop residues. Biomass Bioenergy 26, 361–375.

Kong, Q.X., Li, L., Martinez, B., Chen, P., Ruan, R., 2010. Culture of microalgae Chlamydomonas reinhardtii in wastewater for biomass feedstock production. Appl. Biochem. Biotechnol. 160, 9–18.

Kosourov, S.N., Batyrova, K.A., Petushkova, E.P., Tsygankov, A.A., Ghirardi, M.L., Seibert, M., 2012. Maximizing the hydrogen photoproduction yields in *Chlamydomonas reinhardtii* cultures: the effect of the H₂ partial pressure. Int. J. Hydrogen Energy 37, 8850–8858.

Kumar, S., Gupta, R., Kumar, G., Sahoo, D., Kuhad, R.C., 2013. Bioethanol production from *Gracilaria verrucosa*, a red alga, in a biorefinery approach. Bioresour. Technol. 135, 150–156.

Lam, M.K., Lee, K.T., 2012. Microalgae biofuels: a critical review of issues, problems and the way forward. Biotechnol. Adv. 30, 673–690.

Lee, Y.K., 2001. Microalgal mass culture systems and methods: their limitation and potential. J. Appl. Phycol. 13, 307–315.

Li, W.W., Yu, H.Q., 2011. From wastewater to bioenergy and biochemicals via two-stage bioconversion processes: a future paradigm. Biotechnol. Adv. 29, 972–982.

Illman, A.M., Scrangg, A.H., Shales, S.W., 2000. Increase in Chlorella strains calorific values when grown in low nitrogen medium. Enzyme Microb. Technol. 27, 631–635.

Martinez-Guerra, E., Gnaneswar Gude, V., Mondala, A., Holmes, W., Hernandez, R., 2014. Extractive-transesterification of algal lipids under microwave irradiation with hexane as solvent. Bioresour. Technol. 156, 240–247.

McGinn, P.J., Dickinson, K.E., Bhatti, S., Frigon, J.C., Guiot, S.R., O'Leary, S.J., 2011. Integration of microalgae cultivation with industrial waste remediation for biofuel and bioenergy production: opportunities and limitations. Photosynth. Res. 109, 231–247.

Miao, X., Wu, Q., 2006. Biodiesel production from heterotrophic microalgal oil. Bioresour. Technol. 97, 841–846.

Niehaus, T.D., Okada, S., Devarenne, T.P., Watt, D.S., Sviripa, V., Chappell, J., 2011. Identification of unique mechanisms for triterpene biosynthesis in *Botryococcus braunii*. Proc. Natl Acad. Sci. USA 108, 12260–12265.

Olmos-Soto, J., Paniagua-Michel, J., Contreras, R., Ochoa, L., 2012. DNA fingerprinting intron-sizing method to accomplish a specific, rapid, and sensitive identification of carotenogenic *Dunaliella* species. Methods in Mol. Biol. 892, 269–281.

Oswald, W.T., 2003. My sixty years in applied algology. J. Appl. Phycol. 23, 259–270.

Paniagua-Michel, J., Farfan, B.C., Buckle-Ramirez, L.F., 1987. Culture of marine microalgae with natural biodigested resources. Aquaculture 64, 249–256.

Paniagua-Michel, J., Franco-Rivera, A., Cantera, J.J.L., Stein, L.Y., 2005. Activity of nitrifying biofilms constructed in low-density polyester enhances bioremediation of a coastal wastewater effluent. World J. Microbiol. Biotechnol. 21, 1371–1377.

Paniagua-Michel, J., Olmos-Soto, J., Acosta-Ruiz, M., 2012. Pathways of carotenoid biosynthesis in bacteria and microalgae. In: Barredo, J.-L. (Ed.), Microbial Carotenoids from Bacteria and Microalgae: Methods and Protocols, Methods in Molecular Biology, Vol. 892. Springer, pp. 1–12.

Park, K.C., Whitney, C., McNichol, J.C., Dickinson, K.E., MacQuarrie, S., Skrupski, B.P., Zou, J., Wilson, K.E., O'Leary, S.J.B., McGinn, P.J., 2012. Mixotrophic and photoautotrophic cultivation of 14 microalgae isolates from Saskatchewan, Canada: potential applications for wastewater remediation for biofuel production. J. App. Phycol. 24, 339–348.

Pittman, J.K., Dean, A.P., Osundeko, O., 2011. The potential of sustainable algal biofuel production using wastewater resources. Bioresour. Technol. 102, 17–25.

Prajapati, S.K., Kaushik, P., Malik, A., Vijay, V.K., 2013. Phycoremediation coupled production of algal biomass, harvesting and anaerobic digestion: possibilities and challenges. Biotechnol. Adv. 31, 1408–1425.

Radakovits, R., Jinkerson, R.E., Darzins, A., Posewitz, M.C., 2010. Genetic engineering of algae for enhanced biofuel production. Eukaryot Cell 9, 486–501.

Radakovits, R., Eduafo, P.M., Posewitz, M.C., 2011. Genetic engineering of fatty acid chain length in *Phaeodactylum tricornutum*. Metabolic Eng. 13, 89–95.

Rahman, A., Ellis, J.T., Miller, C.D., 2012. Bioremediation of domestic wastewater and production of bioproducts from microalgae using waste stabilization ponds. J. Biorem. Biodegrad. 3, e113.

Rasala, B.A., Gimpel, J.A., Tran, M., Hannon, M.J., Miyake-Stoner, S.J., Specht, E.A., Mayfield, S.P., 2013. Genetic engineering to improve algal biofuels production. Algae for biofuels and Energy. Dev. App. Phycol. 5, 99–113.

Rawat, I., Ranjith, K., Mutanda, T., Bux, F., 2011. Dual role of microalgae: Phycoremediation of domestic wastewater and biomass production for sustainable biofuels production. Appl. Energy 88, 3411–3424.

Rosales-Morales, A., Paniagua-Michel, J., 2014. Bioremediation of hexadecane and diesel oil is enhanced by photosynthetically produced marine Biosurfactants. J. Biorem. Biodegrad. 34, 1–5.

Schenk, P., Thomas-Hall, S., Stephens, E., Marx, U., Mussgnug, J., Posten, C., Kruse, O., Hankamer, B., 2008. Second generation biofuels: high-efficiency microalgae for biodiesel production. Bioenergy Res. 1, 20–43.

Schneider, R.C.S., Bjerk, T.R., Gressler, P.D., Souza1, M.P., Corbellini, V.A., Lobo, E.A., 2013. Potential production of biofuel from microalgae biomass produced in wastewater. http://dx.doi.org/10.5772/52439.

Scott, S.A., Davey, M.P., Dennis, J.S., Horst, I., Howe, C.J., Lea-Smith, D.J., Smith, A.G., 2010. Biodiesel from algae: challenges and prospects. Curr. Opin. Biotechnol. 21, 277–286.

Sivakumar, G., Vail, D.R., Xu, J., Burner, D.M., Lay, J.O., Ge, X., Weathers, P.J., 2010. Bioethanol and biodiesel: alternative liquid fuels for future generations. Eng Life Sci. 10, 8–18.

Sivakumar, G., Xu, J., Thompson, R.W., Yang, Y., Randol-Smith, P., Weathers, P.J., 2012. Weathers. Integrated green algal technology for bioremediation and biofuel. Bioresour. Technol. 107, 1–9.

Sriram, S., Seenivasan, R., 2012. Microalgae cultivation in wastewater for nutrient removal. J Algal Biomass Utilization 3, 9–13.

Stephens, E., Ross, I.L., King, Z., Mussgnug, J.H., Kruse, O., Posten, C., Borowitzka, M.A., Hankamer, B., 2010. An economic and technical evaluation of microalgal biofuels. Nat. Biotechnol. 28, 126–128.

Sturm, B.S.M., Lamer, S.L., 2011. An energy evaluation of coupling nutrient removal from wastewater with algal biomass production. App. Energy 88 (10), 3499–3506.

Sydney, E.B., da Silva, T.E., Tokarski, A., Novak, A.C., Carvalho, D.J.C., Woiciecohwski, A.L., Larroche, C., Soccol, C.R., 2011. Screening of microalgae with potential for biodiesel production and nutrient removal from treated domestic sewage. Appl. Energy 88, 3291–3294.

Ueno, Y., Kurano, N., Miyachi, S., 1998. Ethanol production by dark fermentation in the marine green alga, chlorococcum littorale. J. Ferment. Bioeng. 86, 38–43.

Velasquez-Orta, S.B., Lee, J.G.M., Harvey, A., 2012. Alkaline in situ transesterification of *Chlorella vulgaris*. Fuel 94, 544–550.

Wang, Z.T., Ullrich, N., Joo, S., Waffenschmidt, S., Goodenough, U., 2009. Algal lipid bodies: stress induction, purification, and biochemical characterization in wild-type and starchless *Chlamydomonas reinhardtii*. Eukaryot Cell 8, 1856–1868.

Woertz, I., Feffer, A., Lundquist, T., Nelson, Y., 2009. Algae grown on dairy and municipal wastewater for simultaneous nutrient removal and lipid production for biofuel feedstock. J. Environ. Eng. 135, 1115–1122.

Xiong, W., Gao, C., Yan, D., Wu, C., Wu, Q., 2010. Double CO(2) fixation in photosynthesis-fermentation model enhances algal lipid synthesis for biodiesel production. Bioresour. Technol. 101, 2287–2293.

Yoon, S.M., Kim, S.Y., Li, K.F., Yoon, B.H., Choe, S., Kuo, M.M.C., 2011. Transgenic microalgae expressing *Escherichia coli* AppA phytase as feed additive to reduce phytate excretion in the manure of young broiler chicks. Appl. Microbiol. Biotechnol. 91, 553–563.

Zamora-Castro, J., Paniagua-Michel, J., Lezama-Cervantes, C., 2008. A novel approach for bioremediation of a coastal marine wastewater effluent based on artificial microbial mats. Marine Biotechnol. 10, 181–189.

32

Phycoremediation-Coupled Biomethanation of Microalgal Biomass

Poonam Choudhary, Arghya Bhattacharya, Sanjeev K. Prajapati, Prachi Kaushik, Anushree Malik

Applied Microbiology Laboratory, Centre for Rural Development and Technology, Indian Institute of Technology (IIT) Delhi, Hauz Khas, New Delhi, India

1. INTRODUCTION

Microalgae have been considered as potential feedstock for renewable energy to reduce global warming and treat wastewater. Being photosynthetic, microalgae use CO_2 from the atmosphere or waste gases, uptake nutrients from wastewater, and grow into a biomass rich in carbohydrates, proteins, and lipids. However, microalgal technology still requires innovative scientific and technological breakthroughs to make it feasible. The main challenges identified include low productivity for large-scale processes, as well as the high costs associated with the recovery and processing of biomass to convert biofuels and other products. If the approach is coupled with other low-input co-processes, such as wastewater treatment and CO_2 sequestration from flue gases, it would greatly aid in the economics of the entire process, as well as contribute to greenhouse gas (GHG) abatement by not producing additional fossil fuel as is generally required for product formation.

Various reports have targeted microalgae cultivation in wastewater for biofuel production (Prajapati et al., 2013a). The advantage of using neat wastewater for algae cultivation lies in the possibility of enhanced biomass production attributed to the synergistic growth of algae with native bacteria in the selected wastewater (Rawat et al., 2011). Moreover, algal biomass production could be enhanced further by providing elevated CO_2 levels using waste gas streams, such as flue gas (Kumar et al., 2010). Anaerobic digestion (AD) is another attractive technology for bioconversion of produced biomass into methane. However, the biomass composition—and hence the methane potential—of wastewater-grown algae might differ from the algae grown in standard nutrient medium. Another attractive feature of AD is the resulting nutrient flux in the form of digestate, which could be used as a nutrient supplement for algae growth.

In the case of algal biomethane production, it is possible to construct a closed-loop process for its coupling with phycoremediation and nutrient recycling. Lifecycle assessment (LCA) provides an excellent platform for assessing the environmental impact and feasibility of such a closed-loop process. This chapter reviews the current state of the art of algal-related work on phycoremediation-coupled biomass production along with CO_2 sequestration, AD of algal biomass, and possible closed-loop processes. LCA-based assessment is also discussed to predict the feasibility of such processes.

2. WASTEWATER AS A POTENTIAL NUTRIENT SOURCE FOR MICROALGAE CULTIVATION

Microalgae are photosynthetic organisms. Under natural environmental conditions, algae grow by absorbing sunlight, assimilating CO_2 from the atmosphere, and deriving nutrients from the aquatic habitat. In addition to the phototrophic method (i.e., sunlight as energy and CO_2 as carbon source), microalgae can be grown through heterotrophic (organic carbon as the energy and carbon source), mixotrophic (growth under the phototrophic and/or heterotrophic mode), and photoheterotrophic (light as energy and organic carbon as carbon source) cultivation methods. Depending upon the application of microalgae and the available resources, a suitable cultivation mode can be adopted. For example, heterotrophic cultivation is a favored mode when high lipid productivity is desired. Large-scale production of microalgae has a lot of potential in wastewater treatment as well. Wastewater that has an excess of nitrogen and phosphorus content when discharged into the environment leads to eutrophication (Christenson and Sims, 2011). Microalgae production can use different types of nutrient-rich wastewater, thereby reducing the water and fertilizer demand considerably, as well as improving the water quality to dischargeable limits (Prajapati et al., 2013b).

The minimal nutritional requirements of a microalga can be estimated using the approximate molecular formula of microalgal biomass: $CO_{0.48}H_{1.83}N_{0.11}P_{0.01}$ (Chisti, 2007). Any growth medium that is utilized for the production of microalgae must provide inorganic elements, such as nitrogen and phosphorus, in excess to constitute the algal cell. Microalgal biomass also contains approximately 50% carbon (on a dry-weight basis), which is usually derived from carbon dioxide. The artificial cultivation of microalgae in wastewater under natural conditions is demonstrated in Figure 1.

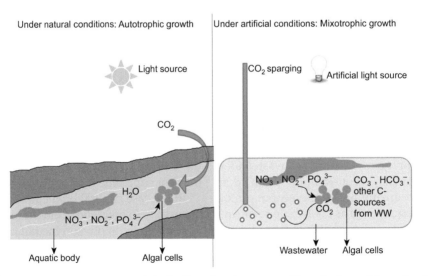

FIGURE 1   Schematic representation of algal cells undergoing autotrophic and mixotrophic modes of growth.

In mixotrophic growth, the simultaneous occurrence of phototrophic and heterotrophic modes may occur. The CO_2 requirements of microalgae may be met by either direct sparging of the gas into the wastewater or through the carbon-rich sources (carbonates, bicarbonates) present in the wastewater.

2.1 Types of Potential Wastewater and Nutrient Availability

Researchers have used different types of wastewater that are rich in nitrogen and phosphorus for the cultivation of microalgae with subsequent treatment of the wastewater. Some of the studies have been tabulated in Table 1. Data indicate the utilization of different microalgae for the treatment of wastewater, ranging from highly specific livestock wastewater to a more variable urban wastewater. Microalgae remove nutrients such as nitrogen and phosphorus from wastewater through direct uptake into the algal cells. It is important to note that effective removal and utilization of nitrogen and phosphorus take place at a suitable nitrogen-to-phosphorus ratio of the wastewater, which varies by microalgae species.

2.2 Microalgae Cultivation Systems Utilizing Wastewater

Different configurations for the cultivation methods have been proposed by researchers for the production of microalgae using wastewater as the cultivation medium. These configurations can be broadly classified into the following three groups: open ponds, closed reactors, and immobilized systems (Christenson and Sims, 2011). An open pond used for commercial microalgae production can either be a circular pond (Figure 2(a)) or a raceway pond; in this closed-loop recirculation channel, mixing and circulation are achieved using a paddlewheel and the flow is guided through bends and baffles, which are intermittently placed in the flow channel (Figure 2(b)). Although open-pond systems for microalgae cultivation are relatively inexpensive to build and operate, they suffer from a major limitation: low productivity, which is due to contamination, poor mixing, and evaporative water loss leading to inefficient use of CO_2 (Chisti, 2007).

Photobioreactors (PBRs) are closed-culture tubular systems made of transparent material (glass/plastic) and are based on single-specie microalgae culture. The configuration of tubular PBRs may use a vertical (Figure 2(c)), horizontal (Figure 2(d)), or helical design (Figure 2(e)). Limitations posed by open-pond systems related to contamination, mixing, evaporative loss, and pH and temperature control can be overcome with the use of PBRs, leading to higher cell densities (Prajapati et al., 2013a).

Open-pond and closed-culture bioreactors are based on suspended microalgae growth systems, which are often limited by the problems faced during harvesting of the culture. Thus, immobilized/attached-growth microalgae culture systems (Figure 2(f)) have been developed; these systems have an added advantage of high biomass density and require less water and land (Christenson and Sims, 2011). Different materials, such as polystyrene foam, cardboard, polyethylene fabric, and loofah sponge, have been used to fabricate attached-growth systems for microalgae (Johnson and Wen, 2010).

If a comparison is made between the nutrient removal efficiency and biomass production potential of different cultivation methods, higher biomass production ($20-45 \, \mathrm{g \, m^{-2} \, day^{-1}}$) has been observed with tubular PBRs, compared with the $10-20 \, \mathrm{g \, m^{-2} \, day^{-1}}$ biomass production obtained with raceway ponds. However, greater phosphorus removal has been observed in raceway-pond configurations (96% removal compared with 86% in tubular PBRs). Another cultivation method, the rotating algal biofilm reactor (RABR), was tested by Christenson and Sims (2012); they achieved average removal rates of 2.1 and $14.1 \, \mathrm{g \, m^{-2} \, day^{-1}}$ for total dissolved

TABLE 1 Wastewater Treatment and Algal Biomass Production

Wastewater	Algae species	Nutrient concentration (mg L^{-1})	Algal productivity	Nutrient removal	References
Wastewater from metro plant	*Chlamydomonas reinhardtii*	TAN: 67; TDP: 120	2.00 g L^{-1} day^{-1}	TAN: 55.8 mg L^{-1} day^{-1}; TDP: 17.4 mg L^{-1} day^{-1}	Kong et al. (2010).
Artificial urban wastewater	*Scenedesmus obliqus*	TAN: 32.5; NO$_3$–N: 2.0; TDP: 2.5	GR: 0.401 day^{-1}	TAN: 100%; TDP: 83.3%	Ruiz-Marin et al. (2010).
	Chlorella vulgaris		GR:0.377 day^{-1}	TAN: 60.1%; TDP: 80.3%	
Municipal wastewater	*Chlorella* sp.	COD: 2250; TAN: 71.8; TDP: 201.5	GR: 0.948 day^{-1}	COD removal: 83.0%; TAN: 78.3%; TDP: 85.6%	Wang et al. (2010).
Concentrated municipal wastewater	*Chlorella* sp.	COD: 2389; TAN: 85.9; TKN:132.3; TDP: 215	0.92 g L^{-1} day^{-1}	COD removal: 90.8%; TAN: 93.9%; total nitrogen removal: 89.1%; TDP: 80.9%	Li et al. (2011).
Concentrated municipal wastewater	*Auxenochlorella protothecoides*	COD: 2324; TAN: 91; TKN:134; TDP: 212	1.1 g L^{-1}	COD removal: 79.10%; TAN: 100%; nitrogen removal: 90.60%; TDP: 98.48%	Zhou et al. (2012).
Drain wastewater	*Chroococcus* sp.1	COD: 310.46; NO$_3$–N: 9.8; TAN: 10; TDP: 26.89	1.05 g L^{-1}	COD removal: 70.0%; TAN: 100%	Prajapati et al. (2013).
Urban wastewater	*Desmodesmus communis*	NO$_3$–N: 0.60; TAN: 33.62; TDP: 1.54	0.138–0.227 g L^{-1} day^{-1}	TAN: $\approx$ 100%; TDP: $\approx$ 100%	Samori et al. (2013).
Livestock wastewater	*Chroococcus* sp. 1	COD: 2965; NO$_3$–N: 74.67; TAN: 160.67; TDP: 201.67	4.44 g L^{-1}	COD removal: 76.6%; NO$_3$–N removal: 83.7%; TAN: 98.0%; TDP: 84.5%	Prajapati et al. (2014b).

COD: chemical oxygen demand; GR: growth rate; TAN: total ammoniacal nitrogen; TDP: total dissolved phosphorus.

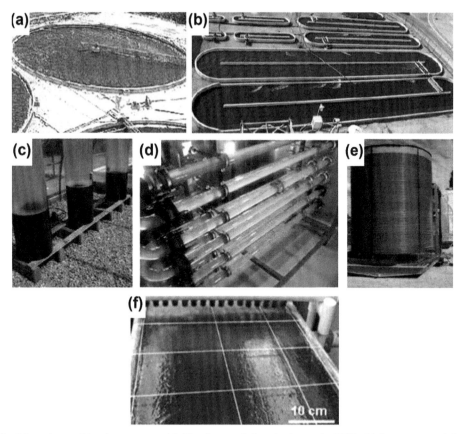

FIGURE 2   Microalgae cultivation systems: (a) Circular pond (Lundquist et al., 2010). (b) Raceway pond (http://algae-energy.co.uk/biofuel_production/cultivation/). (c) Vertical tank (Chinnasamy et al., 2010). (d) A 1000-L helical tubular photo-bioreactor (Chisti, 2007). (e) Horizontal tubular photobioreactor (Bitog et al., 2011). (f) Attached-growth system (Ozkan et al., 2012).

phosphorus and total dissolved nitrogen, respectively. Biomass production in RABR systems ranged from 5.5 g m^{-2} day^{-1} at bench scale to as high as 31 g m^{-2} day^{-1} at pilot scale, which is better than raceway ponds but less than what has been achieved with tubular PBRs.

2.3 Biomass Production Potential in Wastewater

The growth of microalgae in wastewater is the most important parameter for the simultaneous removal of nutrients from wastewater and uptake by the algal cells. Therefore, the reactor configurations and cultivation methods, as well as the algal species and its acclimatization to the wastewater, play important roles in nutrient removal. Microalgae belonging to the *Chlorella*, *Spirulina*, and *Scendesmus* genera have been widely employed for the removal of nutrients from wastewater. Prajapati et al. (2013b) compared the performance of aquatic and terrestrial counterparts of microalgae isolated from drain wastewater and surrounding soil, respectively, in removing nutrients from wastewater and biomass production. Their results indicated that native isolates (from wastewater) produce higher biomass compared with those isolated from soil when grown in drainage wastewater. Moreover, the better synergy of the isolated strains with native microbial populations in removing nutrients from unsterilized wastewater demonstrates the superiority of such strains in wastewater treatment processes.

Because the wastewater characteristics vary largely depending upon the source, site-specific algal–wastewater combinations need to be worked out to avoid failures. Furthermore, certain other modifications, such as nutrient supplementation, CO_2 sequestration, etc., can be carried out to increase the biomass production potential of wastewater.

3. COUPLING CO$_2$ SEQUESTRATION WITH MICROALGAE CULTIVATION IN WASTEWATER

3.1 Microalgal Carbon Dioxide Fixation

Increasing carbon dioxide concentrations in the atmosphere are responsible for more than 50% of the global warming potential of all GHGs; because of the likelihood of further increases, there is an urgent need to address carbon sequestration technology more rationally and effectively (Singh and Ahluwalia, 2012). The existing carbon capture and storage technologies (absorption, adsorption, cryogenic distillation, gas-separation membranes) are considered to be short-term solutions with no environmental sustainability (Kumar et al., 2010). A promising technology could be the photosynthesis-driven microalgal fixation of CO_2 due to its simple nutritional requirements, unmatched CO_2 fixation rate over higher plants, and lack of requirement for the further disposal of the trapped CO_2. The ability of microalgae to use low-quality water, such as municipal, industrial, or agricultural wastewater, as a source of nitrogen, phosphorus, and minor nutrients can be rendered more sustainable by coupling microalgal fixation of CO_2 generated by agricultural or industrial processes (Figure 3). Hence, the combination of CO_2 fixation from flue gas and nutrient utilization from wastewater may provide economic and environmental benefits as a result of the decreased cost of water and chemicals required for microalgae growth, while providing a pathway for wastewater treatment, reduction of overall carbon emissions, and the

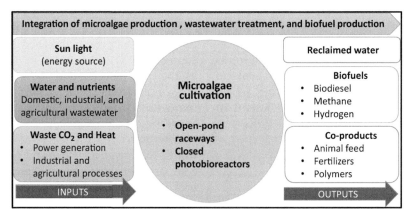

FIGURE 3   An integrated system for coupling microalgae-mediated CO_2 sequestration with wastewater treatment. *Modified from Kumar et al. (2010).*

subsequent utilization of the generated biomass for energy production.

3.2 Mechanisms and Key Parameters of Microalgal CO_2 Fixation

Microalgal photosynthesis converts carbon dioxide into organic compounds using light energy and releases molecular oxygen by the CO_2 concentrating mechanism (CCM). CCM directly results in an increase of the photosynthetic rate by the enhancement of CO_2 levels at the active site of ribulose bisphosphate carboxylase-oxygenase (Rubisco) by transporting inorganic carbon or carbon dioxide into the cell and ultimately decreasing photorespiration. The main enzyme involved in the process called carbonic anhydrase (CA) assures the availability, conversion, and equilibrium of dissolved inorganic carbon forms inside microalgal cells.

Because microalgal CO_2 fixation performance is mainly controlled by CA enzymes, the factors influencing the enzyme activity are key factors in controlling CO_2 sequestration technology. Apart from suitable CO_2-fixing microalgae, the factors influencing microalgal—CO_2 fixation usually include physicochemical parameters (e.g., CO_2 concentration, sulfur and nitrogen oxides in combustion flue gas, initial inoculation density, culture temperature, light, nutrients, pH) and hydrodynamic parameters (e.g., aeration, mixing, mass transfer). The physicochemical parameters of extremely high or low values of CO_2 concentration, pH, and temperature directly affect the activity of CA and Rubisco (Rawat et al., 2011). Therefore, maintaining an optimal range or value for these parameters is important for efficient CO_2 fixation.

Unlike the physicochemical parameters, hydrodynamic factors affect microalgal photosynthesis or CO_2 uptake indirectly. For instance, the appropriate flow and mixing in gas—liquid—solid (CO_2—medium—microalgae) are necessary to enhance mass transfer, ensure efficient light and temperature distribution,

ensure nutrient availability, and prevent microalgal sedimentation (Zhao and Su, 2014). These parameters can be addressed by designing suitable PBRs to maintain an optimum flow rate and mixing and increase mass transfer (Jajesniak et al., 2014).

3.3 Process Kinetics and Critical Parameters Affecting CO_2 Sequestration

After selecting a suitable microalgal strain, the next step is the selection of an appropriate cultivation system or conditions. On an industrial scale, two systems that have been extensively used are open-pond and closed PBR technologies (Brennan and Owende, 2010). However, it is still unclear whether an open-pond or closed PBR system is better for CO_2 sequestration. Although raceway ponds are the most commonly used artificial growth systems because of their cost-effectiveness, they are limited by large surface area requirements, high cultivation costs, low biomass productivity, and significant CO_2 losses to the atmosphere. Closed PBR configurations, including vertical column reactors (bubble columns or air-lift), tubular reactors, and flat-plate reactors, utilize CO_2 much more efficiently than open systems because of their efficient light distribution and higher mass transfer (Pires et al., 2012).

A comparative analysis of two different cultivation modes for growth and CO_2 biofixation of a *Chlorella* sp. showed that closed cultivation substantially enhanced microalgal performance, with 1.78- and 5.39-fold greater growth rate and carbon biofixation, respectively (Zhao et al., 2011). Most of the research on microalgal-CO_2 fixation technology has been done on a laboratory scale, with limited reports on pilot-scale implementation. The performance of microalgal-CO_2 fixation and biomass production heavily depend on the culture process conditions; for example, microalgal growth under flue gas conditions is usually more complex than under atmospheric conditions.

Waste gases from industries typically contain 3–30% CO_2 as compared to atmospheric CO_2 (approximately 0.036%) and can be an effective supplement to stimulate microalgal growth (Kumar et al., 2010). Kao et al. (2014) investigated the growth of a mutant *Chlorella* sp., aerated with flue gases from a coke oven, hot stove, and power plant of a steel plant; they found two-fold higher growth rate than those obtained using air or 25% CO_2 aeration. The reason could be the tolerance of mutant species to high CO_2 levels. However, at high concentrations of CO_2 (>5%), growth for intolerant microalgae can be suppressed. This is often attributed to acidification of the cellular content, which eventually hinders growth (Kumar et al., 2010). Considering CO_2 concentrations in flue gases (3–30% CO_2), it is important to identify strains capable of growing under very high CO_2 concentrations. Additionally, microalgal growth can be maintained at a 100% CO_2 concentration by controlling changes in pH and only releasing CO_2 to the microalgae on demand (Olaizola, 2003).

Besides CO_2, flue gases also contain NO_x, SO_x, and heavy metals such as nickel, vanadium, and mercury. According to researchers, the presence of NO_x in flue gases poses little or no problem for microalgal growth; difficulty arises in the presence of SO_x, which decreases the pH due to the formation of sulfurous acid (Kumar et al., 2010). Also, particular concentrations of nickel (>1 ppm) and vanadium (>0.1 ppm) decrease microalgal productivity, whereas mercury can be remediated by certain microalgal species (Van Den Hende et al., 2012). To avoid these limitations, pretreatment by denitrification and desulfurization, along with cooling, dedusting, and the selection of CO_2-tolerant microalgae (*Dunaliella tertiolecta*, *Tetraselmis* sp., and *Chlorella* sp. T-1) could be suitable options. Apart from toxic components, the temperatures of flue gases are extremely high (around 120 °C) and could have an adverse effect on cells if incorporated directly (Kumar et al., 2010). Therefore, the feasibility of sequestering CO_2 from flue gas would

either depend on using thermophilic microalgal species (e.g., *Synechococcus elongates*) or installing a heat-exchange system.

Previous research initiatives (Nakajima and Ueda, 2000) suggest that practical CO_2 utilization from flue gases using microalgae still requires innovative scientific and technological breakthroughs to make this a feasible technology. Unless coupled with other technologies, microalgal technology is unlikely to make a considerable contribution to solving the CO_2 problem globally. Potential couplings include wastewater treatment, production of useful metabolites, and biofuels. Supplementation with CO_2 in wastewater is expected to increase algal biomass productivity and could be a viable approach because process requirements and objectives overlap significantly. However, CO_2 uptake efficiency under wastewater conditions become more complex because of high turbidity and suspended matter. For instance, *Chlorella vulgaris* was cultivated in wastewater discharged by steel-making plants to develop an economically feasible system to simultaneously remove ammonia from wastewater and CO_2 (15% v/v) from flue gas (Yun et al., 1997). The CO_2 fixation and ammonia removal rates were 26.0 and 0.92 g m^{-3} h^{-1}, respectively. In another study, Chinnasamy et al. (2010) cultivated native mixotrophic algal strains (*Chlamydomonas globosa*, *Chlorella minutissima*, and *Scenedesmus bijuga*) in untreated wastewater from the carpet industry using raceways, vertical reactors, and polybags aerated with 5–6% of CO_2. The highest biomass productivity was obtained in polybags (21.1 g m^{-2} day^{-1}) followed by vertical tank reactors and raceways (5.9 g m^{-2} day^{-1}).

3.4 Challenges and Process Advances

Research on microalgal CO_2 fixation has expanded greatly in recent years, both in scope and diversity. However, several challenges for microalgal CO_2 sequestration remain unaddressed and needs intensive research. Significant advances have been made (Jajesniak et al., 2014) in targeting all stages of large-scale algal cultivation, from CO_2 uptake to product extraction (Table 2). Apart from the advances in PBR engineering, future research should consider open systems for the widespread use of biological CO_2 mitigation. Among these, technologies that supply adequate and continuous CO_2, nutrients and light to microalgal cells, consume minimal energy, and release less CO_2 into the atmosphere are in demand. Enhanced CO_2 levels needed for efficient microalgal growth and metabolism have been provided by existing sources, such as CO_2 from ammonia plants or flue gases from power stations. However, due to the presence of toxic sulfur and nitrogen oxides, flue gas may hamper algal applications in the medical and food markets. Hence, CO_2 from existing manure plants or novel algal-based biogas plants could be incorporated into algal PBRs, resulting in enhanced biomass production with subsequent utilization into biogas production, making whole process economically and environmentally sustainable.

TABLE 2 Recent Advances for Process Improvement in Microalgal-Based CO_2 Sequestration Technologies

Process parameters	Recent advances	References
CO_2 supply and mass transfer	Using hollow fiber membrane (HFM)	Kalontarov et al. (2014)
Photobio-reactor	Design of an airlift-driven raceway reactor	Ketheesan and Nirmalakhandan (2012)
	External carbonation column	Putt et al. (2011)
	Design of a photobioreactor using Taylor vortex flow	Kong, Shanks, and Vigil (2013)
Biomass recovery	Dissolved air flotation	Singh and Olsen (2011)
	pH-mediated algae flocculation	Liu et al. (2013)
Light supply	Solar-tracked PBR (photobioreactor)	Hindersin et al. (2014)
Microalgal strain	High-throughput screening method for rapid identification of microalgae with high CO_2 affinity or tolerance	Liu et al. (2013)

4. BIOMETHANE POTENTIAL OF WASTEWATER-GROWN MICROALGAL BIOMASS

As discussed in the previous sections, microalgae have unmatched potential for wastewater treatment-coupled biomass production. Also, the biomass production of microalgae can be further increased by using waste CO_2 as a carbon source. Once the algal biomass is available, the next step is to convert the biomass into biofuels. There are several routes available for algal biomass to biofuel conversion, including biodiesel, bioethanol, direct combustion, and gaseous fuels such as biomethane. However, drying/dewatering is an essential step in most biofuel production processes, including for biodiesel. On the other hand, the process of biomethane production through AD uses algal slurry instead of dry biomass. Furthermore, the superiority of algal biomass over other AD substrates could be attributed to its relatively balanced biochemical composition and availability of essential micronutrients for AD microbes (Prajapati et al., 2013a,b). Additionally, during algal AD, nitrogen, phosphorous, and other micronutrients stored in algal biomass are released into the digestate, which could serve as a concentrated nutrient source for further algal growth (Prajapati et al., 2014a). However, such nutrient recycling is not feasible in other biofuel production routes. Because of these advantages, microalgae have emerged as a potential feedstock for biomethane production. However, industrial explorations of algal biomethanation have been limited. The different aspects of algal biomethane generation are discussed below.

4.1 Biomass Composition and Theoretical Methane Potential

One of the basic criteria for selecting an algal biomass for AD is its biomass composition. The major components of algal biomass are divided into three major groups: lipids, carbohydrates, and proteins. Furthermore, as reported by Prajapati et al. (2013a), if the biochemical composition of any substrate including microalgae is known, the theoretical biomethane potential (TMP) can easily be estimated. Alternatively, stoichiometric methane potential (SMP) instead of TMP can also be used to describe the possible methane potential of any given algal biomass. The values of SMP can be estimated using the equation by Symons and Buswell and the empirical formula of algal biomass (Prajapati et al., 2014b). Although TMP and SMP give rough estimates of the methane potential of algal biomass, they could easily augment the time-consuming and tedious laboratory-based approach for practical methane potential estimation.

4.2 Effect of Cultivation Conditions on Biomass Composition and Methane Potential

As discussed previously, the TMP/SMP solely depends on the biomass composition. This, in turn, depends on the cultivation conditions. For instance, microalgae accumulate lipids under nitrogen starvation conditions (Menon et al., 2013) and carbohydrates under phosphorous limitations (Markou et al., 2013). Hence, the biomass composition of the same alga may differ when grown in different types of wastewater, depending on the nutrient profile

of the wastewater (Prajapati et al., 2014c). On a mass scale, microalgae cultivation for biofuel applications is generally carried out using wastewaters, so it is necessary to consider the variations in algal biomass due to changes in the nutrient profile of wastewater. Furthermore, the light conditions (e.g., dark:light cycle, light intensity, color) also affect the algal growth and biomass composition (Monlau et al., 2014). Interestingly, sometimes the harvesting method can also affect the biomethane potential. For instance, during fungal-assisted algal harvesting, significant enhancements in TMP were observed due to the additional carbon contributed by the biomass of the fungal strain used (Prajapati et al., 2014c).

Hence, along with the biomass composition, the methane potential of algal biomass can be greatly affected by the cultivation conditions and growth medium. An algal strain with high lipid content would have high TMP because the specific methane yield for lipids is significantly higher than for proteins and carbohydrates, whereas microalgae with high-protein contents would have low TMPs. In contrast, microalgae with higher lipid contents will have poor digestibility compared with microalgae that have high carbohydrate and protein contents. On the other hand, if microalgae have significantly higher carbohydrate content, it is possible that its AD may fail with the accumulation of volatile fatty acids due to the relatively quick digestion of carbohydrates. Similarly, AD of high-protein microalgae may suffer from ammonia accumulation and inhibition. Also, the low carbon-to-nitrogen (C/N) ratio of microalgae with high protein can reduce the activity of anaerobic microflora. Hence, microalgae with a balanced composition of lipids, carbohydrates, and proteins may be the most suitable choice for fermentative biomethane production through AD (Prajapati et al., 2013a). *Chroococcus* sp. (Prajapati et al., 2013b, 2014d) is an excellent example of an algal biomass with a balanced composition and hence good methane potential and digestibility. However, the effects of cultivation conditions and growth medium/wastewater need to be considered in the determination of the actual biomethane potential of microalgae.

4.3 Anaerobic Digestion of Microalgal Biomass and Process Hurdles

Because of their great potential for biomethane and relatively favorable composition, microalgae have been subjected to extensive research for AD. The various studies on algal AD have been summarized in our review (Prajapati et al., 2013a). Among the reviewed strains, *Chlamydomonas reinhardtii* was found to have the highest biogas production (up to $0.587 \text{ m}^3 \text{ kg}^{-1} \text{ VS}_{fed}$ (VS, Volatile solids)) with a digestion time of 30 days. Although microalgae have good composition and potential for biomethane potential, there are certain limitations in this process. The identified hurdles in the algal AD process are two-fold: the algal cell-wall resistance to AD and the poor activity of anaerobic microflora due to low C/N ratios. The optimal C/N ratio of AD lies between 20 and 30, whereas algal biomass usually have a C/N ratio of 6–9 (Prajapati et al., 2014e). Furthermore, most unicellular microalgae possess cell walls composed of mainly cellulose, with some amount of hemicellulose/xylan and chitin, which makes the algal cell wall recalcitrant for AD.

A popular approach for overcoming the digestibility problem is the pretreatment of algal biomass before AD. A range of physiochemical methods are available for algal biomass pretreatment. However, high costs and energy inputs in the pretreatment stage make these processes nonviable on an industrial scale. Furthermore, the formation of furanic and phenolic compounds during the thermochemical pretreatment of algal biomass could reduce the biomethane production by inhibiting growth and activities of AD microflora (Monlau et al., 2014).

Some reports have indicated the suitability of biological methods for algal biomass pretreatment. Muñoz et al. (2014) investigated the use of cellulolytic marine bacteria for enzymatic pretreatment of algal biomass and observed methane yields that increased by 140–159%. Similarly, Miao et al. (2013) used natural storage as a pretreatment and recorded a 37% improvement in biomethane yield. Such biological methods have been reported to show excellent performance in laboratory-scale studies and have great potential to improve the feasibility of algal-based biomethanation processes. However, further scale-up and validation are still required to successfully apply these approaches on an industrial scale.

The C/N ratio of algal feed can be improved through its codigestion with carbon-rich waste/substrate. In our study, it was observed that the methane yield from *Chroococcus* sp. biomass was enhanced by 44% under its codigestion with cattle dung at a 1:1 ratio (on a VS basis). This could be attributed to the improved C/N ratio (13.0) of feed under codigestion compared with microalgae alone, which have a C/N ratio of 9.26 (Prajapati et al., 2014e). Similarly, Zhao and Ruan (2013) observed good results for the codigestion of algal biomass with kitchen waste. The diverse strategies that can be adopted to improve the feasibility of algal AD, including coupling with phycoremediation, are depicted in Figure 4.

5. THE CLOSED-LOOP PROCESS

Microalgae is an attractive option for the bioremediation of wastewater and simultaneous energy recovery. Using sunlight and CO_2 available from the atmosphere, microalgae can utilize the nitrogen and phosphorus present in the wastewater to produce biomass. The wastewater-grown microalgae can then be used to produce energy via a variety of routes, such as biomethane, biodiesel, bioethanol, and

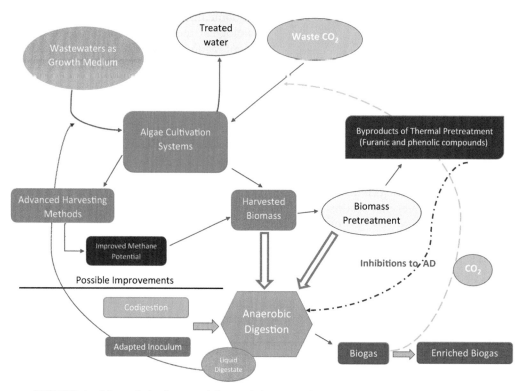

FIGURE 4   Schematic for the central theme of phycoremediation-coupled biomethane production.

biohydrogen, using processing technologies such as AD, pyrolysis, gasification, catalytic cracking, and enzymatic or chemical transesterification.

The basic idea behind a closed-loop process is to reuse the waste from these processes for the cultivation of microalgae. The digestate obtained from AD is rich in nutrients, such as nitrogen and phosphorus. Similarly, the waste from biodiesel production consists of a solid that is rich in fatty acids and a nutrient-rich aqueous phase. Also, the CO_2 produced during these processes can be coupled with microalgal biomass cultivation to earn carbon credits. Nutrient recycling of the waste from algal biofuel production seems to be a feasible option for mitigating the environmental impacts and offset a part of the nutrient requirement for

cultivation. Also, recycling the supernatant after algal harvest reduces the overall water demand for the process.

The first use of a closed-loop concept dates back to 1959, when Golueke and Oswald used the algal-digested slurry directly for algal growth. Studies by Iyovo et al. (2010) demonstrated a closed-loop process where poultry manure, paper pulp, and algal waste sludge were mixed in varied proportions for codigestion to produce biomethane. The solid part of the digestate was used as a fertilizer and the liquid part was used as a nutrient source for microalgae cultivation. Prajapati et al. (2014a) showed the feasibility of such a closed-loop process, using different concentrations of digestate for the cultivation of *Chroococcus* sp. Uggetti et al. (2014) demonstrated that the digestate

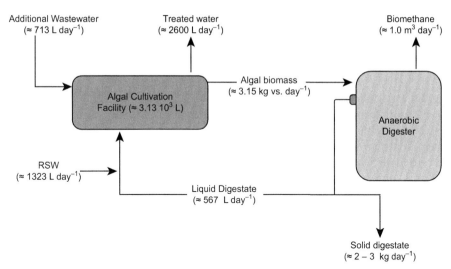

FIGURE 5   Schematic diagram of a closed-loop process (biomethane production capacity of $1.0\,m^3\,day^{-1}$). *Adapted from Prajapati et al. (2014a).*

slurry was an effective substrate for microalgal growth, promoting biomass production up to $2.6\,gTSS\,L^{-1}$. Figures 5 and 6 illustrate the concept of a closed-loop process for the production of biogas.

5.1 Assessing the Feasibility of a Closed-Loop Process Using LCA

Due the rapid exhaustion of conventional energy sources, the focus is very much on the production of algal biofuels. LCA of algal biofuels might be an effective tool to compare alternative energy routes in terms of environmental impact and indirect natural resource costs (Pittman et al., 2011). LCA is often referred to as a "cradle-to-grave" analysis, which consists of the following steps:

1. *Goal and scope definitions*: This defines the purpose of the LCA study and the basis of the functional unit that is the quantitative reference for the study. For example, for the production of biodiesel, the functional unit can be chosen as 1 ton of biodiesel.

2. *Inventory analysis*: Inventory analysis collects all the data of the processes within a system and relates them to the functional unit of the study. Inventory analysis results, which

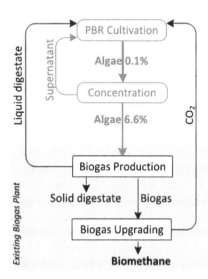

FIGURE 6   A closed-loop process involving production of biomethane from microalgae integrated with an existing biogas plant. *Adapted from Wang et al. (2013).*

consist of a flowchart relating various processes to the functional unit, are called the lifecycle inventory table.

3. *Impact assessment*: Results from the previous step are analyzed to understand the environmental impacts of the process. This aids in taking mitigating measures for minimizing the possible impact.

4. *Interpretation*: This step aims to evaluate the results from inventory analysis and impact assessment and compare them with the goal of the study, as defined in the first step.

5.2 LCA Analysis of a Closed-Loop Process of Microalgal Biomethane Production

There are few reports on LCA of microalgal biomethanation (Dave et al., 2013, Zhang et al., 2014). One of the relevant studies by Wang et al. (2013), which reported an LCA analysis of microalgal biomethane production integrated with an existing biogas plant in Sweden, is discussed here.

The waste from a biogas plant in Växtkraft in Västerås, Sweden (production capacity of 54,000 GJ biomethane and 1979 t CO_2 annually) is integrated with microalgal cultivation. Due to the cold climate for the majority of the year, the cultivation of microalgae was done in a PBR during the warmer months. The liquid digestate and CO_2 from the biogas plant were used for microalgal cultivation. The cultivated microalgae were then concentrated by the methods of natural settling and centrifugation, then used for biomethane production. The supernatant was recycled for microalgae cultivation. Because the production of energy was a concern, the functional unit was chosen as 1 GJ of biomethane produced. The system boundaries included feedstock production to biomethane sold at the gate of the biogas plant, including the stages of cultivation, concentration, biogas production, conditioning, and transportation. The study mainly focused on the total energy

utilized and the GHG emissions from the entire process. The energy efficiency was evaluated by the net energy ratio (NER):

$$\text{NER} = E_\text{p}/E_\text{LC} \qquad (1)$$

where E_p is the energy of the product and E_LC is the energy of the lifecycle processes. If a process is to be feasible, the NER value should be greater than 1. For the production of 1 GJ of energy (1000 MJ), an input of 650.42 MJ of energy was required. Hence, the NER came to be around 1.54, indicating that it is a favorable process. PBR infrastructure only takes up 4.7% of the total energy process, but the sparging and CO_2 injection increases the energy cost. Also, due to the self-shading of a vertical PBR, its height cannot be increased without limitations. Hence, cultivating microalgae in open raceway ponds where possible may increase the NER value, but it will also increase the footprint of the process. Thus, for regions where there is a tropical climate, this type of a closed-loop system is a feasible option for increasing biomethane production.

5.3 Suitable Approach for a Closed-Loop Process

The closed-loop process appears to be an interesting approach for offsetting the nutrient requirements for microalgal growth and proper disposal of the wastes generated from algal biofuel production. However, this process has yet to be used on a commercial scale. Although the wastes from algal biofuel production processes are nutrient rich, they cannot be directly used for the mass cultivation of microalgae. Prajapati et al. (2014a) showed that the algal digestate from AD needs to be diluted first with water for use in microalgae cultivation. A concentration of more than 30% of slurry inhibited the growth of microalgae. This was due to the fact that the dark color of the slurry provided a shading effect, which hindered the photosynthetic activity of the microalgae. Also, the ammonia concentration was high due to high nitrogen content, which could also be toxic to the microalgal cells. The slurry obtained after biofuel production process has to be processed in order to be reused. Because it is a colloidal solution, separating the solid and liquid parts adds to the cost. The solid and liquid parts need to be separated because the solid particles shield the sunlight and interfere with the photosynthetic activities of the microalgal cells. Hence, to make the closed-loop approach a commercially feasible and acceptable process, these issues need to be addressed carefully.

6. CONCLUSION AND PERSPECTIVES

Phycoremediation-coupled biomethanation is a promising technology because it leads to the production of biofuels (or other industrially important products) together with a potential reduction of GHG emissions and the treatment of wastewaters. However, many challenges encountered during CO_2 sequestration still need to be addressed, such as improvements in microalgal cultivation systems to capture CO_2 with simultaneous consumption of nutrients from wastewater. Also, the costs associated with different processes should be reduced by coupling two or more processes.

To maximize the overall efficiency of the technology, geographical considerations must be taken into account to reduce temperature and solar radiation fluctuations. Open-pond microalgae cultivation in tropical areas, located as close as possible to point sources of flue gases (as a CO_2 source) and waste streams (as a nutrient source), would be an economical and sustainable approach. The coupling can be made more economical when untreated wet microalgal biomass is incorporated directly into biogas plants to produce algal-based biogas ("algas").

Ongoing research activities have shown higher methane potential for microalgae compared with cow dung, but this needs more realistic implementation. The biogas production from microalgae creates methane as the main product and nutrient-rich digestate (i.e., slurry) as a byproduct. Recycling the nutrients from the slurry and assimilating them into algal biomass can result in further feedstock for the process without additional costs for nutrient media, while simultaneously performing remediation of the waste stream.

To assess the economic feasibility and environmental sustainability of this coupled process, LCA is an essential tool. LCA can identify problems during different stages of the process and can highlight the environmental performance and technological innovation opportunities.

Acknowledgment

This study was supported by the Ministry of New and Renewable Energy, Government of India. The authors gratefully acknowledge the Senior Research Fellowship and Research Associateship granted by Council of Scientific and Industrial Research, Government of India.

References

Brennan, L., Owende, P., 2010. Biofuels from microalgae—A review of technologies for production, processing, and extractions of biofuels and co-products. Renewable Sustainable Energy Rev. 14, 557–577.

Bitog, J., Lee, I.-B., Lee, C.-G., Kim, K.-S., Hwang, H.-S., Hong, S.-W., et al., 2011. Application of computational fluid dynamics for modeling and designing photobioreactors for microalgae production: a review. Comput. Electron. Agric. 76, 131–147.

Christenson, L., Sims, R., 2011. Production and harvesting of microalgae for wastewater treatment, biofuels, and bioproducts. Biotechnol. Adv. 29, 686–702.

Chisti, Y., 2007. Biodiesel from microalgae. Biotechnol. Adv. 25, 294–306.

Christenson, L.B., Sims, R.C., 2012. Rotating algal biofilm reactor and spool harvester for wastewater treatment with biofuels by-products. Biotechnol. Bioeng. 109, 1674–1684.

Chinnasamy, S., Bhatnagar, A., Hunt, R.W., Das, K.C., 2010. Microalgae cultivation in a wastewater dominated by carpet mill effluents for biofuel applications. Bioresour. Technol. 101, 3097–3105.

Van Den Hende, S., Vervaeren, H., Boon, N., 2012. Flue gas compounds and microalgae:(Bio-) chemical interactions leading to biotechnological opportunities. Biotechnol. Adv. 30, 1405–1424.

Dave, A., Huang, Y., Rezvani, S., McIlveen-Wright, D., Novaes, M., Hewitt, N., 2013. Techno-economic assessment of biofuel development by anaerobic digestion of European marine cold-water seaweeds. Bioresour. Technol. 135, 120–127.

Li, Y., Chen, Y.-F., Chen, P., Min, M., Zhou, W., Martinez, B., et al., 2011. Characterization of a microalga Chlorella sp. well adapted to highly concentrated municipal wastewater for nutrient removal and biodiesel production. Bioresour. Technol. 102, 5138–5144.

Golueke, C.G., Oswald, W.J., 1959. Biological conversion of light energy to the chemical energy of methane. Appl. Microbiol. 7, 219–227.

Hindersin, S., Leupold, M., Kerner, M., Hanelt, D., 2014. Key parameters for outdoor biomass production of Scenedesmus obliquus in solar tracked photobioreactors. J. Appl. Phycol. 1–11.

Iyovo, G.D., Du, G., Chen, J., 2010. Sustainable bioenergy bioprocessing: biomethane production, digestate as biofertilizer and as supplemental feed in algae cultivation to promote algae biofuel commercialization. J. Microb. Biochem. Technol. 2, 100–106.

Johnson, M.B., Wen, Z., 2010. Development of an attached microalgal growth system for biofuel production. Appl. Microbiol. Biotechnol. 85, 525–534.

Jajesniak, P., Omar Ali, H., Wong, T., 2014. Carbon dioxide capture and utilization using biological systems: opportunities and challenges. J. Bioprocess Biotech. 4, 2.

Kumar, A., Ergas, S., Yuan, X., Sahu, A., Zhang, Q., Dewulf, J., et al., 2010. Enhanced CO(2) fixation and biofuel production via microalgae: recent developments and future directions. Trends Biotechnol. 28, 371–380.

Kao, C.Y., Chen, T.Y., Chang, Y.B., Chiu, T.W., Lin, H.Y., Chen, C.D., et al., 2014. Utilization of carbon dioxide in industrial flue gases for the cultivation of microalga Chlorella sp. Bioresour. Technol. 166C, 485–493.

Kong, Q.X., Li, L., Martinez, B., Chen, P., Ruan, R., 2010. Culture of microalgae Chlamydomonas reinhardtii in wastewater for biomass feedstock production. Appl. Biochem. Biotechnol. 160, 9–18.

Kalontarov, M., Doud, D.F.R., Jung, E.E., Angenent, L.T., Erickson, D., 2014. Hollow fibre membrane arrays for CO_2 delivery in microalgae photobioreactors. RSC Adv. 4, 1460.

Ketheesan, B., Nirmalakhandan, N., 2012. Feasibility of microalgal cultivation in a pilot-scale airlift-driven raceway reactor. Bioresour. Technol. 108, 196–202.

Kong, B., Shanks, J.V., Vigil, R.D., 2013. Enhanced algal growth rate in a Taylor vortex reactor. Biotechnol. Bioeng. 110, 2140–2149. http://dx.doi.org/10.1002/bit.24886.

Liu, Z., Zhang, F., Chen, F., 2013. High throughput screening of CO_2-tolerating microalgae using GasPak bags. Aquat. Biosyst. 9, 23.

Lundquist, T.J., Woertz, I.C., Quinn, N., Benemann, J.R., 2010. A realistic technology and engineering assessment of algae biofuel production. Energy Biosci. Inst. 1.

Menon, K.R., Balan, R., Suraishkumar, G.K., 2013. Stress induced lipid production in Chlorella vulgaris: relationship with specific intracellular reactive species levels. Biotechnol. Bioeng. 110, 1627–1636.

Markou, G., Angelidaki, I., Georgakakis, D., 2013. Carbohydrate-enriched cyanobacterial biomass as feedstock for bio-methane production through anaerobic digestion. Fuel 111, 872–879.

Miao, H., Lu, M., Zhao, M., Huang, Z., Ren, H., Yan, Q., et al., 2013. Enhancement of Taihu blue algae anaerobic digestion efficiency by natural storage. Biores. Technol. 149, 359–366.

Monlau, F., Sambusiti, C., Barakat, A., Quéméneur, M., Trably, E., Steyer, J.P., et al., 2014. Do furanic and phenolic compounds of lignocellulosic and algae biomass hydrolyzate inhibit anaerobic mixed cultures? A comprehensive review. Biotechnol. Adv. 32 (5), 934–951. http://dx.doi.org/10.1016/j.biotechadv.2014.04.007.

Muñoz, C., Hidalgo, C., Zapata, M., Jeison, D., Riquelme, C., Rivas, M., 2014. Use of cellulolytic marine bacteria for enzymatic pretreatment in microalgal biogas production. Appl. Environ. Microbiol. 80 (14), 4199–4206. http://dx.doi.org/10.1128/aem.00827-14.

Nakajima, Y., Ueda, R., 2000. The effect of reducing light-harvesting pigment on marine microalgal productivity. J. Appl. Phycol. 12, 285–290.

Olaizola, M., 2003. Microalgal removal of CO_2 from flue gases: changes in medium pH and flue gas composition do not appear to affect the photochemical yield of microalgal cultures. Biotechnol. Bioprocess Eng. 8, 360–367.

Ozkan, A., Kinney, K., Katz, L., Berberoglu, H., 2012. Reduction of water and energy requirement of algae cultivation using an algae biofilm photobioreactor. Bioresour. Technol. 114, 542–548.

Prajapati, S.K., Kaushik, P., Malik, A., Vijay, V.K., 2013a. Phycoremediation coupled production of algal biomass, harvesting and anaerobic digestion: possibilities and challenges. Biotechnol. Adv. 31, 1408–1425.

Prajapati, S.K., Kaushik, P., Malik, A., Vijay, V.K., 2013b. Phycoremediation and biogas potential of native algal isolates from soil and wastewater. Bioresour. Technol. 135, 232–238.

Pires, J.C.M., Alvim-Ferraz, M.C.M., Martins, F.G., Simões, M., 2012. Carbon dioxide capture from flue gases using microalgae: engineering aspects and biorefinery concept. Renewable Sustainable Energy Rev. 16, 3043–3053.

Prajapati, S.K., Kumar, P., Malik, A., Vijay, V.K., 2014a. Bioconversion of algae to methane and subsequent utilization of digestate for algae cultivation: a closed loop bioenergy generation process. Bioresour. Technol. 158, 174–180.

Prajapati, S.K., Malik, A., Vijay, V.K., 2014b. Comparative evaluation of biomass production and bioenergy generation potential of *Chlorella* spp. through anaerobic digestion. Appl. Energy 114, 790–797.

Prajapati, S., Kumar, P., Malik, A., Choudhary, P., 2014c. Exploring pellet forming filamentous fungi as tool for harvesting non-flocculating unicellular microalgae. BioEnergy Res. 7 (4), 1430–1440. http://dx.doi.org/10.1007/s12155-014-9481-1.

Prajapati, S.K., Kumar, P., Malik, A., Vijay, V.K., 2014d. Bioconversion of algae to methane and subsequent utilization of digestate for algae cultivation: a closed loop bioenergy generation process. Bioresour. Technol. 158, 174–180.

Prajapati, S.K., Choudhary, P., Malik, A., Vijay, V.K., 2014e. Algae mediated treatment and bioenergy generation process for handling liquid and solid waste from dairy cattle farm. Bioresour. Technol. 167, 260–268.

Pittman, J.K., Dean, A.P., Osundeko, O., 2011. The potential of sustainable algal biofuel production using wastewater resources. Bioresour. Technol. 102, 17–25.

Putt, R., Singh, M., Chinnasamy, S., Das, K.C., 2011. An efficient system for carbonation of high-rate algae pond water to enhance CO_2 mass transfer. Bioresour. Technol. 102, 3240–3245.

Rawat, I., Kumar, R.R., Mutanda, T., Bux, F., 2011. Dual role of microalgae: phycoremediation of domestic wastewater and biomass production for sustainable biofuels production. Appl. Energy 88, 3411–3424.

Ruiz-Marin, A., Mendoza-Espinosa, L.G., Stephenson, T., 2010. Growth and nutrient removal in free and immobilized green algae in batch and semi-continuous cultures treating real wastewater. Bioresour. Technol. 101, 58–64.

Singh, U.B., Ahluwalia, A.S., 2012. Microalgae: a promising tool for carbon sequestration. Mitig. Adapt. Strateg. Glob. Change 18, 73–95.

Samorì, G., Samorì, C., Guerrini, F., Pistocchi, R., 2013. Growth and nitrogen removal capacity of *Desmodesmus communis* and of a natural microalgae consortium in a batch culture system in view of urban wastewater treatment: part I. Water Res. 47, 791–801.

Singh, A., Olsen, S.I., 2011. A critical review of biochemical conversion, sustainability and life cycle assessment of algal biofuels. Appl. Energy 88, 3548–3555.

Uggetti, E., Sialve, B., Latrille, E., Steyer, J.P., 2014. Anaerobic digestate as substrate for microalgae culture: the role of ammonium concentration on the microalgae productivity. Bioresour. Technol. 152, 437–443.

Wang, X., Nordlander, E., Thorin, E., Yan, J., 2013. Microalgal biomethane production integrated with an existing biogas plant: a case study in Sweden. Appl. Energy 112, 478–484.

Wang, L., Min, M., Li, Y., Chen, P., Chen, Y., Liu, Y., et al., 2010. Cultivation of green algae *Chlorella* sp. in different wastewaters from municipal wastewater treatment plant. Appl. Biochem. Biotechnol. 162, 1174–1186.

Yun, Y.S., Lee, S.B., Park, J.M., Lee, C.I., Yang, J.W., 1997. Carbon dioxide fixation by algal cultivation using wastewater nutrients. J. Chem. Technol. Biotechnol. 69, 451–455.

Zhao, B., Su, Y., 2014. Process effect of microalgal-carbon dioxide fixation and biomass production: a review. Renewable Sustainable Energy Rev. 31, 121–132.

Zhao, B., Zhang, Y., Xiong, K., Zhang, Z., Hao, X., Liu, T., 2011. Effect of cultivation mode on microalgal growth and CO_2 fixation. Chem. Eng. Res. Des. 89, 1758–1762.

Zhao, M.-X., Ruan, W.-Q., 2013. Biogas performance from co-digestion of Taihu algae and kitchen wastes. Energy Convers. Manage. 75, 21–24.

Zhang, Y., Kendall, A., Yuan, J., 2014. A comparison of on-site nutrient and energy recycling technologies in algal oil production. Resour. Conserv. Recycl. 88, 13–20.

Zhou, W., Min, M., Li, Y., Hu, B., Ma, X., Cheng, Y., et al., 2012. A hetero-photoautotrophic two-stage cultivation process to improve wastewater nutrient removal and enhance algal lipid accumulation. Bioresour. Technol. 110, 448–455.

N$_2$-Fixing Cyanobacteria: Ecology and Biotechnological Applications

Kirsten Heimann[1,2,3], *Samuel Cirés*[1,2,3]

[1]James Cook University, College of Marine and Environmental Sciences, Townsville, QLD, Australia;
[2]James Cook University, Centre for Sustainable Fisheries and Aquaculture, Townsville, QLD, Australia;
[3]James Cook University, Comparative Genomics Centre, Townsville, QLD, Australia

1. PHYSIOLOGY AND GENETICS OF NITROGEN FIXATION

1.1 Introduction

As an essential component of biomolecules, including amino acids and nucleic acids, nitrogen (N) is one of the key elements for life. Together with phosphorus, N is considered to be the main limiting factor in aquatic ecosystems. Photoautotrophic prokaryotic cyanobacteria assimilate N from combined nitrogen sources (ammonium, urea, nitrite, and nitrate) and, in diazotrophic species, from atmospheric nitrogen gas (N$_2$). Ammonium and urea are readily incorporated into organic compounds as NH_4^+, which is the preferred intracellular form of N, whereas intracellular nitrate (NO_3^-) requires reduction to nitrite (NO_2^-) and subsequently to NH_4^+ by the enzymes nitrate reductase and nitrite reductase, respectively.

Given the scarcity of NH_3/NH_4^+ and NO_3^- in many aquatic and terrestrial ecosystems, the ability to fix atmospheric N$_2$ (comprising 78% of the atmospheric gas) became one of the most advantageous physiological strategies developed by microorganisms. In cyanobacteria, N$_2$-fixation occurs via nitrogenase complexes, the central enzyme of which is denominated nitrogenase 1. Nitrogenase 1 is a molybdenum-dependent adenosine triphosphate (ATP)-hydrolyzing complex of two metalloproteins: a MoFe-protein (dinitrogenase α2β2 heterotetramer) that contains the active site for the reduction of N$_2$ and a Fe-protein (dinitrogenase reductase γ2 homodimer) that transfers high-energy electrons to MoFe-containing dinitrogenase. Nitrogenase reduces one molecule of N$_2$ into two molecules of NH$_3$ (Eqn (1)) at a high-energy cost (16 ATP molecules), due to the need to break the stable triple bond between the N atoms in N$_2$- and generating H$_2$ as a byproduct (Eqn (1)):

$$N_2 + 8H^+ + 8e^- + 16\,ATP \rightarrow 2NH_3 + H_2 + 16\,ADP + 16\,Pi \tag{1}$$

Nitrogenase 1 is inactivated upon oxygen binding. This suggests that nitrogenase originated before the Great Oxygenation Event, more than 2200 million years ago. According to Latysheva et al. (2012), cyanobacteria were the first organisms releasing oxygen to the atmosphere as a byproduct of photosynthesis. To overcome the incompatibility of N$_2$-fixation in an oxygenic atmosphere, cyanobacteria have developed a number of strategies to counteract the inactivation of nitrogenase by photosynthetic O$_2$, including the separation of photosynthesis and N$_2$-fixation in space (e.g., in the heterocystous cyanobacterium *Anabaena*) and in time (e.g., in the nonheterocystous *Lyngbya*), as well as the restriction of N$_2$-fixation to microaerobic or anaerobic conditions (e.g., in nonheterocystous *Leptolyngbya boryanum*; Table 1).

1.2 Separation of N$_2$-Fixation and Photosynthesis in Space: Heterocystous Cyanobacteria

Following the traditional classification of cyanobacteria into five orders, heterocystous cyanobacteria are located within orders Nostocales and Stigonematales (Table 1), equivalent to subsections IV and V in the bacteriological classification. To date, more than 100 heterocytous genera have been described, some of the most common are the nostocalean *Anabaena*, *Aphanizomenon*, *Nostoc*, *Calothrix*, and *Tolypothrix* and the true-branching stigonematalean *Fischerella* and *Mastigocladus*.

The heterocyst and vegetative cells of cyanobacteria are examples of cellular specialization normally associated with higher organisms (plants and animals). Heterocystous cyanobacteria are filamentous organisms capable of aerobic N$_2$-fixation due to the production of specialized, differentiated, nonphotosynthetic cells called heterocysts. Heterocysts possess only photosystem I providing ATP for N$_2$ fixation, but not the water-splitting and oxygen-producing photosystem II; in addition, they are surrounded by a thick wall to limit the entry of oxygen. The heterocyst wall is a thick double layer, with the

TABLE 1   Classification on N$_2$-Fixing Cyanobacteria

Group	N$_2$-fixation behavior	Separation of photosynthesis and N$_2$-fixation	Morphology	Orders	Some genera
Heterocystous	Aerobic	Spatial	Filamentous	Nostocales	*Anabaena* *Nostoc*
				Stigonematales	*Fischerella* *Mastigocladus*
Non-heterocystous	Aerobic	Temporal	Unicellular	Chroococcales	*Gloeothece* *Cyanothece*
				Pleurocapsales	*Chroococcidiopsis*
			Filamentous	Oscillatoriales	*Lyngbya* *Microcoleus*
	Aerobic	Spatial and temporal	Filamentous	Oscillatoriales	*Trichodesmium* *Katagyneme*
	Anaerobic	-	Filamentous	Oscillatoriales	*Leptolyngbya*

Bergman, B., Gallon, J., Rai, A., Stal, L., 1997. N$_2$ Fixation by non-heterocystous cyanobacteria. FEMS Microbiol. Rev. 19, 139–185.
Gallon, J., 2005. N$_2$ Fixation by Non-Heterocystous Cyanobacteria, Genetics and Regulation of Nitrogen Fixation in Free-Living Bacteria. Springer, pp. 111–139.

first polysaccharide layer being for mechanical support. The second "laminated" layer limits oxygen diffusion into the cell and is composed of unusual glycolipids (heterocyst glycolipids) (Wörmer et al., 2012). Heterocysts also lack ribulose-1,5-biphosphate carboxylase (RuBisCO), thus being incapable of fixing CO_2 and instead relying on adjacent photosynthetic vegetative cells for their carbon supply (Figure 1). Organic carbon is delivered to the heterocyst in the form of disaccharides (e.g., maltose), whose metabolism supplies the reducing power (NADPH) required for N_2 reduction via the oxidative pentose phosphate pathway enzymes (glucose-6-phosphate and 6-phosphogluconate dehydrogenases). Glutamine synthetase (GS) catalyzes the incorporation of ammonia, produced by nitrogen fixation, to form glutamine. The heterocystous glutamate pool is generated either by endogenous synthesis or by transport from the vegetative cell. In return, glutamine is exported to the vegetative cell, where it forms glutamate via glutamate synthase (GOGAT) for incorporation into vegetative cell metabolites.

The heterocysts store fixed N in cyanophycin polar bodies (Figure 1), which are clearly visible under the light microscope (Figure 2(a)).

1.3 Nonheterocystous Cyanobacteria

Until the 1960s, only heterocystous cyanobacteria were believed to fix N_2. Since then, N_2-fixation has been described in about 17 genera, including 70 strains of nonheterocystous cyanobacteria within orders Chroococcales, Pleurocapsales, and Oscillatoriales (Table 1), in most cases when incubated in micro-oxic or anoxic conditions (Bergman et al., 1997). Although less conspicuous than their heterocystous counterparts, nonheterocystous cyanobacteria are important contributors to global N_2-fixation. The highly important nonheterocystous *Trichodesmium* alone is estimated to contribute 42% (240 Tg N_2 year^{-1}) of newly fixed N_2 to the earth's nitrogen budget, with the total input through nonheterocystous cyanobacteria believed to exceed 50% of the annual global N_2-fixation rate (Berman-Frank et al., 2003; Gallon, 2001).

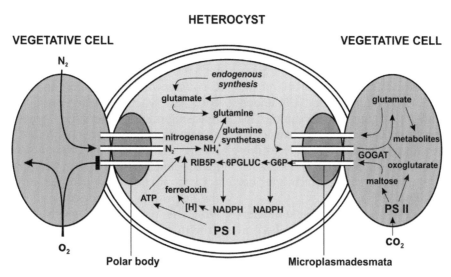

FIGURE 1   Schematic view of N_2-fixation in heterocysts and carbon–nitrogen exchanges with vegetative cells in diazotrophic filamentous cyanobacteria. PS I, photosystem I; PS II, photosystem II; RIB5P, ribulose-5-phosphate; 6PGLUC, 6-phosphogluconic acid; G6P, glucose-6-phosphate; GOGAT, glutamate synthase. *Modified from Lea (1997).*

FIGURE 2   Planktonic N$_2$-fixing cyanobacteria in temperate freshwaters. (a) *Anabaena crassa* (syn. *Dolichospermum crassum*), one of the most common species in temperate freshwaters. H, heterocyst (N$_2$-fixing cell); Ak, akinete (resting stage). (b) Mass proliferation (bloom) of *Anabaena flos-aquae* in a Mediterranean freshwater body (Cueva Foradada reservoir, northeast Spain).

The filamentous nonheterocystous *Leptolyngbya boryanum* (previously known as *Plectonema boryanum*) can fix N$_2$ under continuous illumination at low-light intensity if an O$_2$-free atmosphere is generated by sparging gas (Ar, Ar/CO$_2$, or N$_2$/CO$_2$) or by adding inhibitors of photosynthetic O$_2$ production (DCMU (3-(3,4-dichlorophenyl)-1,1-dimethylurea)) or sodium sulfide) (Bergman et al., 1997). When transferred to N$_2$-free medium, N$_2$ fixation is derepressed for 2–24 h in *L. boryanum*, after which temporal separation of O$_2$ production and N$_2$ fixation is effected under continuous illumination, but not when grown in dark/light cycles (Bergman et al., 1997; Gallon, 2005). Although the physiology of anaerobic O$_2$ fixation is not fully understood, experiments suggest that during the photosynthetic phase light energy is transferred to the O$_2$-evolving photosystem II, whereas during the N$_2$-fixing phase light energy is transferred preferentially to photosystem I, which does not generate O$_2$ (Gallon, 2005).

Although more restricted, aerobic nonheterocystous N$_2$-fixation has been described in several unicellular and filamentous species capable of circadian separation of photosynthesis and N$_2$ fixation (e.g., unicellular *Gloeothece* and filamentous *Lyngbya*) and/or containing certain cellular adaptations to create microaerobic environments (e.g., filamentous *Trichodesmium*) (Table 1). With the exception of *Trichodesmium*, *Katagyneme*, and *Symploca*, which confine nitrogenase to specialized but not differentiated cells (see Section 2.3 for details on *Trichodesmium* N$_2$ fixation), nonheterocystous cyanobacteria fix N$_2$ during the dark phase of an alternating light/dark cycle. Circadian control of N$_2$ fixation is achieved through differential expression of the nitrogenase-encoding genes (*nif* genes, see Section 1.3), with transcription being restricted to the night phase and destruction of the enzyme during the day phase in the unicellular *Gloeothece* and *Cyanothece*. In contrast, genes encoding reaction center proteins of the O$_2$-releasing photosystem II (*psb* genes) and photosystem I (*psa* genes) are only transcribed during the light phase (Gallon, 2005). In *Lyngbya*, nitrogenase levels persist during the light period and activity is controlled through modification of the Fe-protein of nitrogenase to a larger form, which is presumably inactive under aerobic growth conditions (Gallon, 2005). Thus, a reciprocal rhythm of photosynthetic O$_2$ production and O$_2$ consumption to support N$_2$-fixation is created, with increased respiratory requirements for N$_2$ fixation supported by enhanced carbohydrate breakdown (Gallon, 2005).

1.4 *Nif* Genes: Phylogeny and Molecular Methods

As detailed in Section 1.1, N_2 fixation in cyanobacteria is carried out by the central enzyme nitrogenase 1, a molybdenum-dependent complex of two metalloproteins: a FeMo-dinitrogenase heterotetramer whose α and β subunits are encoded by the *nifD* and *nifK* genes, respectively, and a Fe-dinitrogenase reductase homodimer encoded by the gene *nifH*. Other than the *nifHDK* genes, at least 13 other genes are involved in N_2-fixation, 10 of which are also denoted as *nif* genes (*nifWX-NEUSBVZT*) arranged in a tight 15-kb cluster (Stucken et al., 2010). In addition to nitrogenase 1, some heterocystous cyanobacteria, such as *Anabaena variabilis* ATCC29413, contain a second alternative Mo-nitrogenase (encoded by the *nif*2 operon) that is transcribed in both vegetative cells and heterocysts under anaerobic conditions and/or an alternative vanadium-dependent nitrogenase (encoded by the *vnf* genes) that is transcribed in heterocysts under molybdenum deficiency (Thiel and Pratte, 2013).

Although the origin of *nif* genes has not been fully determined yet, phylogenetic studies using 49 cyanobacterial genomes suggest that they may have a very ancient precyanobacterial origin, arising approximately 3 billion years ago under the Early Earth atmosphere (Latysheva et al., 2012). Despite their long evolutionary history, the apparently low rates of horizontal gene transfer resulted in highly conserved sequences, thus making *nif* genes suitable as molecular markers. Among the *nif* operon, *nifH* is the most sequenced. For example, 23,847 *nifH* sequences are available in the database by Gaby and Buckley (2012), hence becoming the marker gene of choice for researchers studying the phylogeny, diversity, and abundance of nitrogen-fixing microorganisms, including cyanobacteria. Furthermore, a wide range of oligonucleotides for polymerase chain reaction (PCR) amplification of cyanobacterial *nifH* in cultures and environmental samples exist (Table 2). Generally, *nifH*-based phylogenetic trees show an overall separation of nonheterocystous and heterocystous

TABLE 2   Selection of PCR Primers Suitable for *nifH* Amplification in Cyanobacteria

Primer	Sequence (5′–3′)	Product size (bp)	Specificity	References
nifH3	ATRTTRTTNGCNGCRTA	473	Universal, external	Zehr and Turner (2001)[c]
nifH4	TTYTAYGGNAARGGNGG			
nifH1	TGYGAYCCNAARGCNGA	359	Universal, internal	
nifH2	ADNGCCATCATYTCNCC			
CNF	CGTAGGTTGCGACCCTAAGGCTGA	375	Cyanobacteria-specific	Olson et al. (1998)[b]
CNR	GCATACATCGCCATCATTTCACC			
cylnif-F	TAARGCTCAAACTACCGTAT	220	*Cylindrospermopsis*-specific	Dyble et al. (2002)[a]
cylnif-R	ATTTAGACTTCGTTTCCTAC			

Note: NifH3 and nifH4 are often used as external primers in combination with internal nifH1 and nifH2 in a nested PCR approach (Zehr and Turner, 2001).

[a]*Dyble, J., Paerl, H.W., Neilan, B.A., 2002. Genetic characterization of Cylindrospermopsis raciborskii (Cyanobacteria) isolates from diverse geographic origins based on nifH and cpcBA-IGS nucleotide sequence analysis. Appl. Environ. Microbiol. 68, 2567–2571.*

[b]*Olson, J., Steppe, T., Litaker, R., Paerl, H., 1998. N₂-fixing microbial consortia associated with the ice cover of Lake Bonney, Antarctica. Microbial Ecol. 36, 231–238.*

[c]*Zehr, J., Turner, P., 2001. Nitrogen fixation: Nitrogenase genes and gene expression. Methods Microbiol. 30, 271–286.*

cyanobacteria, with subgroups within heterocystous (branching Stigonematales vs. Nostocales) and nonheterocystous (filamentous Oscillatoriales vs. unicellular Chrococcales). However, *nifH* trees are often more convoluted than this, because some nonheterocystous members cluster close to heterocystous taxa. There are also many cyanobacterial species containing two to three nonidentical *nifH* copies, leading to clustering in different parts of the trees (Yeager et al., 2007).

Besides its generalized use in phylogeny, *nifH* has been widely used in studies of N₂-fixing cyanobacterial diversity in environmental samples, such as by density gradient gel electrophoresis and clone libraries of DNA (Ininbergs et al., 2011; Omoregie et al., 2004) as well as in reverse-transcriptase PCR approaches to evaluate gene expression (*nifH* transcription levels) in a range of aquatic and terrestrial ecosystems (Omoregie et al., 2004; Zani et al., 2000) or in response to environmental factors in cultures (Vintila and El-Shehawy, 2007).

2. N₂-FIXING CYANOBACTERIA IN THE ENVIRONMENT

2.1 Soils and Agroecosystems

N₂-fixing cyanobacteria can be considered one of the greatest natural biological fertilizers in soils with varied agricultural uses. In those ecosystems, N₂-fixing cyanobacteria thrive in benthic, epiphytic, and subaerophytic habitats, as free forms or in symbiotic association with plants such as *Azolla*.

Within the variety of crop-derived ecosystems, rice fields are particularly remarkable given their economic relevance and the ecological importance of N₂-fixing cyanobacteria for rice field sustainability. Approximately 90% of the rice production occurs in wetland fields, constituting the most dominant anthropogenic wetland ecosystem worldwide (Scott and Marcarelli,

2012) and a major habitat for N₂-fixing cyanobacterial communities. The diversity of those communities is illustrated by a study of 38 soil samples from rice fields in Bangladesh, where 84 species of cyanobacteria, 12 of which were heterocystous diazotrophic species, of 14 different genera were identified (Khan et al., 1994). Additional studies of 102 soils from four countries found that half of the cyanobacterial genera reported were heterocystous (*Anabaena, Aulosira, Calothrix, Cylindrospermum, Fischerella, Gloeotrichia, Nostoc, Scytonema, Tolypothrix, Wollea*); their presence was strongly correlated with pH and P concentrations (Whitton, 2002 and references therein). Within those genera, *Fischerella, Nostoc*, and *Calothrix* are considered widespread, often persisting as epiphytic communities, such as *Nostoc* sp. growing on the macrophyte *Chara vulgaris*. Additionally, the symbiosis between the freshwater fern *Azolla* and the N₂-fixing cyanobacterium *Anabaena azollae*—used for centuries as a natural biofertilizer for rice fields in China and Vietnam—is widely distributed in Asian paddy fields.

A number of in situ investigations have revealed the enormous magnitude of cyanobacterial N₂-fixation in rice fields. Roger and Ladha (1992) estimated the N contribution from cyanobacteria to ~80 kg N ha^{-1} crop, clearly outperforming N input from heterotrophic N-fixing bacteria in the rice rhizosphere (31 kg N ha^{-1} crop). In Spanish rice fields, indigenous cyanobacterial N₂-fixation seems to follow a crop-cycle seasonal pattern, with low values at the beginning of the crop (May), maximum values at the end of the tillering stage (June), and declining again with the end of the cultivation cycle (September) (Quesada et al., 1998). Over the annual cycle, N₂-fixation by indigenous cyanobacteria has been estimated to provide 0.23–75.5 kg N ha^{-1} year^{-1}, representing a remarkable natural N input that could potentially reduce the need for urea fertilizer by 25–35% (Whitton, 2002 and references therein). In a pilot study in Chile, Pereira et al. (2009)

demonstrated that the use of biofertilizer from indigenous *Anabaena iyenganii* and *Nostoc* spp. reduced the amount of synthetic nitrogen fertilizer (50 kg N ha^{-1} year^{-1}) by 50% while providing the same yield of 7.4 t rice ha^{-1}. The agricultural tradition in Asia (e.g., in Bangladesh) includes growing cyanobacteria in open cultures settled on ground parcels with nutrients and afterward draining the parcels and drying the biomass under the sun or directly applying it as fertilizer to the rice field. This potential of cyanobacterial biofertilizers has been overlooked in rice fields elsewhere in favor of using synthetic fertilizers, which cause serious eutrophication of water systems and gravely compromise the environmental sustainability of rice production.

2.2 Freshwaters

N$_2$-fixing cyanobacteria are widely distributed in lentic freshwater ecosystems (lakes, dammed rivers, lagoons, wetlands) and running waters of all latitudes. Investigations since the early 1980s have traditionally linked the dominance of N$_2$-fixing cyanobacteria to low nitrogen:phosphorus ratios in water. Nevertheless, extensive monitoring data seem to contradict this commonly accepted dogma, confirming a lack of relationship between N:P ratios and the abundance of N$_2$-fixing cyanobacteria in 102 German lakes (Dolman et al., 2012). Diazotrophic cyanobacteria are of crucial importance for the N cycle in freshwaters, where N$_2$-fixation can provide up to 1.5 g N m^{-2} year^{-1} (Scott and Grantz, 2013), compensating for up to 35% of N-deficiency in freshwater ecosystems (Wiedner et al., 2014). Other than nutrient availability, temperature seems to be directly correlated with the efficiency of freshwater cyanobacterial N$_2$-fixation (Scott and Marcarelli, 2012).

Deep thermally stratified lakes and reservoirs in temperate regions host important populations of planktonic N$_2$-fixing cyanobacteria, such as the bloom-forming nostocalean genera *Anabaena*, *Anabaenopsis, Aphanizomenon,* and *Cylindrospermopsis*, with benthic members developing mostly in shallow waters. N$_2$-fixing planktonic nostocalean contain intracellular gas vesicles (aerotopes), enabling them to regulate their position in the water column in response to environmental gradients. During certain periods of high temperatures and calm winds, these taxa generate mass proliferations (blooms), which visibly accumulate in the water surface due to overbuoyancy (Figure 2(b)) and on occasion produce cyanotoxins (microcystins, anatoxin-a, saxitoxins, cylindrospermopsin). Most N$_2$-fixing planktonic nostocalean differentiate akinetes (Figure 2(a)), resting stages that are able to survive desiccation and low temperatures, which germinate when water conditions improve (Cirés et al., 2013). Akinetes remain viable after transport by winds or migratory birds, allowing for the invasion of new habitats (Sukenik et al., 2012).

In running freshwaters, N$_2$-fixing cyanobacteria develop as benthic forms in epilithic habitats, forming mats together with other filamentous and unicellular cyanobacteria. Some of the most common N$_2$-fixing taxa are the heterocystous *Calothrix, Nostoc, Rivularia,* and *Tolypothrix*, and nonheterocystous *Schizothrix* (Mateo et al., 2010). N$_2$ fixation rates by benthic diazotrophic cyanobacteria can reach 8.4 mg m^{-2} h^{-1}, with spatial and temporal differences driven by nutrient availability (Scott and Marcarelli, 2012). Additionally, some of these N$_2$-fixers (e.g., *Rivularia, Tolypothrix, Nostoc, Schizothrix*) are important contributors to the recycling of organic P in oligotrophic streams due to the production of extracellular phosphatase enzymes (Mateo et al., 2010).

2.3 Oceans: *Trichodesmium*

Trichodesmium is an important tropical, filamentous, nonheterocystous cyanobacterium, providing new nitrogen to the oligotrophic seas. Filaments can cluster into puff or tuff colonies; the reasons for these filament bundlings are poorly understood and not species specific. *Trichodesmiun* can occur at a depth of 200 m and can represent up to 60% of the chlorophyll *a* in the top 50 m of the water column. It also contributes more than 20% of the primary productivity of the area. *Trichodesmium* forms extensive annual blooms (Figure 3(a)), stretching for 1600 km along the Queensland coast and covering an area of up to 52,000 km^2, as observed by Joseph Banks in 1770 on the *Endeavor* with James Cook "Vast quantities of little substances ... floating upon the water in large lines a mile or so long ... either immediately upon the surface or not many inches under it. The seamen ... began to call it Sea Sawdust" (Walsby, 1978).

Blooms can be terminated through cyanophage infection, grazing by the harpacticoid copepod *Macrosetella gracilis*, oxidative stress, ultraviolet (UV) damage, or nutrient limitation (N and P). Nutrient limitation activates programmed cell death via the proteolytic caspase pathway in *Trichodesmium*, a pathway typically thought to be confined to higher order plants and animals (Berman-Frank et al., 2004). As the blooms decay, they also provide an important source of organic carbon (Figure 3(b)) and phosphate (P). The end of a *Trichodesmium* bloom

closes the nutrient recycling cycle and ensures proliferation of the food web; thus, *Trichodesmium* blooms make a direct contribution to high tropical ecosystem biodiversity.

nifH gene sequences are unique for the genus *Trichodesmium*, containing roughly seven recognized species identified by filament width and cell length:width ratios (Janson et al., 1995). The 16S sequences place them close to species in the order Oscillatoriales (Capone et al., 1997). *Trichodesmium* contains the strongest gas vesicles discovered in cyanobacteria, withstanding pressures of ~3.5 MPa in *T. contortum* and *T. thiebautii*, 14 times stronger than those of freshwater cyanobacteria, which presumably represents and adapts to occurrence at great depths (200 m = 2 MPa) (Walsby, 1978). Gas vesicle collapse studies showed that large colonies (larger filament diameter species of *Trichodesmium*) had greater floating velocities and sank faster after gas vesicle collapse induction at 6 MPa, with initial floatation speed to sinking speeds after pressurization ratios of 1.17, 0.61, and 0.54 for *T. erythraeum*, *T. thiebautii*, and *T. contortum* (Walsby, 1978).

Trichodesmium shows Diel vertical migration, retreating to nutrient-rich lower water

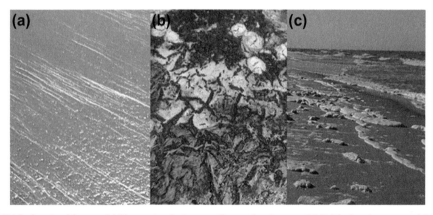

FIGURE 3   *Trichodesmium* blooms. (a) Bloom streaks in open Queensland water. (b) *Trichodesmium* crust of dried washed-up material (*Courtesy of John Webster, EPA*). (c) *Trichodesmium* bloom decay (foam on the beach) at Saunders Beach, Townsville, Queensland Australia. *Courtesy of John Webster, EPA.*

during the night and migrating to the surface at the onset of day. This is achieved through carbohydrate and polyphosphate ballasting. Cyanophycin content does not appear to have a functional role (Romas et al., 1994).

Trichodesmium is a nonheterocystous cyanobacterium where N$_2$-fixation is also under circadian control. However, in contrast to N$_2$-fixation by other nonheterocystous diazotrophic cyanobacteria, nitrogenase activity is downregulated during the night and is established 1—3 h after the transition to the light phase, reaching peak activity at midday (Ohki and Fujita, 1988). The oxygen sensitivity of the nitrogenase complex is also solved through cell differentiation, with the production of diazocytes instead of heterocysts. Unlike heterocysts, diazocytes are indistinguishable from vegetative cells at the light microscopical level but are ultrastructurally different; they are characterized by dense thylakoid networks partitioning the vacuole-like space, less extensive gas vacuoles, and fewer and smaller cyanophycin granules (Fredericksson and Bergman, 1997). In addition, *Trichodesmium* is capable of both temporal and spatial segregation of oxygenic photosynthesis and N$_2$-fixation; photosynthetic activity is highest in the early morning and lowest at midday, when nitrogenase activity is highest and/or diazocytes may be aggregated in clusters, allowing for simultaneous N$_2$ fixation and oxygenic photosynthesis (Berman-Frank et al., 2001). Large blooms of *Trichodesmium* are supported by an organism's capability to simultaneously utilize additional nitrogen sources, such as ammonium generated by grazers and other mat-associated organisms with cell surface amino acid oxidases (Mulholland and Capone, 2000).

Dissolved inorganic P is typically low in tropical waters, but alternative pathways exist for phosphate acquisition, such as the use of dissolved organic P pools (DOP), which are a significant portion of the total dissolved P pool (TDP) or alkaline-phosphate liberated P, as realized in many microalgal species. DOP exists as monophosphate esters, making up 75% of the TDP, with the rest being phosphonates (25%; many herbicides are phosphonates). *Trichodesmium* is supposed to have acquired the genes necessary for the phosphonate-lyase pathway, typically present in bacteria, through lateral gene transfer before speciation occurred (Dyhrman et al., 2006). Harboring the phosphonate-lyase pathway is a unique feature of *Trichodesmium*, as it has not been found thus far in other cyanobacteria, and it is nutritionally activated.

3. BIOTECHNOLOGICAL APPLICATIONS OF N$_2$-FIXING CYANOBACTERIA

3.1 Bioremediation of Wastewaters

N$_2$-fixing cyanobacteria have been the subject of broad research for wastewater bioremediation, given their natural presence in wastewaters (e.g., *Nostoc* sp. found in municipal wastewater from Brazil; Furtado et al., 2009), their well-known metabolic flexibility, and their tolerance to harsh conditions. Studies have used cultures and consortia, including diazotrophic Nostocales (*Anabaena*, *Calothrix*, *Cylindrospermum*, *Nostoc*, *Rivularia*, and *Tolypothrix*), Stigonematales (*Hapalosiphon*, *Mastigocladus*, and *Stigonema*), and less frequently, Chroococcales (*Gloeocapsa* and *Cyanothece*), to remove contaminants from a variety of domestic, industrial, and synthetic wastewaters (Table 3). Contaminants remediated included N (NO$_2^-$, NO$_3^-$, NH$_4^+$) and P (PO$_4^{3-}$) sources responsible for eutrophication of aquatic systems, organic matter (chemical oxygen demand), and persistent and highly toxic pesticides (e.g., lindane) and heavy metals (e.g., Cd, Cr, Hg, Pb). Removal efficiencies obtained were generally high ($>80\%$), although with wide variations (13—100%) depending on the target contaminant, water characteristics, and the species used (Table 3).

TABLE 3   Wastewater Bioremediation by N_2-Fixing Cyanobacteria

Type of wastewater	Species	Target contaminants	Removal efficiency (%)
Open fish pond effluent	*Aulosira fertilissima*	NO_2^-, NH_4^+	100
Primary treated effluent	*Anabaena* sp., *Westelliopsis* sp, *Fischerella* sp.[a]	NO_3-N, NH_4-N	90–100
		PO_4-P	98
		COD	87
Mixed domestic–industrial	*Anabaena oryzae*	COD	74
	Tolypothrix ceytonica	Cu, Zn	86–94
Industrial (plating industry)	*Nostoc* sp. PCC7936	Cr(III), Cr(VI)	60–99
Industrial (paper production)	*Anabaena subcylindrica*	Cu, Co, Pb, Mn	33–86
	Nostoc muscorum	Cu, Co, Pb, Mn	22–85
Synthetic (pesticide-containing)	*Nostoc* sp.	Lindane[b]	87–94
	Nodularia sp.	Lindane	76–98
Synthetic metal mixtures	*Anabaena* spp.	Ag, Au, Cd, Cu, Hg, Pb, Zn	29–85
	Calothrix spp.	Ag, Au, Cd, Cu, Hg, Pb, Zn	13–88
	Cylindrospermum sp.	Cd, Pb	52–65
	Gloeocapsa sp.	Cd, Pb	96
	Hapalosiphon spp.	Cd, Hg, Pb	13–90
	Mastigocladus spp.	Cd, Hg, Pb, Zn	29–89
	Nostoc spp.	Cd, Cr, Cu, Hg, Ni, Pb, Zn	22–94
	Rivularia sp.	Cd, Hg, Pb	76–88
	Scytonema schmidlei	Cd	98
	Stigonema sp.	Cd, Hg, Pb	80–89
	Tolypothrix tenuis	Cd, Cu, Hg, Pb, Zn	53–94

Removal efficiency is expressed as a percentage of contaminant removed compared to the initial concentration (growth periods varying between 2 and 15 days). COD, chemical oxygen demand.
[a]*Consortium includes N_2-fixing and non-fixing strains (Renuka et al., 2013).*
[b]*Lindane [γ-hexachlorocyclohexane (γ-HCH)].*
Information from De Philippis and Micheletti (2009), El-Bestawy et al. (2007), Brayner et al. (2007), Colica et al. (2010), El-Sheekh et al. (2005).

The capacity of N_2-fixing species to remove metals and metalloids from synthetic metal mixtures is particularly remarkable (Table 3). Although mechanisms behind this phenomenon are not fully understood, a big part of the metal retention is attributed to the exopolysaccharides (EPS) usually produced in high amounts by many N_2-fixing strains. Cyanobacterial EPS are exocellular polysaccharidic layers classified according to their features, such as sheaths (thick layers with high mechanical and physicochemical stability), capsules (a gelatinous layer associated with the cell surface), and slime (an amorphous mucilaginous material loosely dispersed around the microorganism) (De Philippis and Micheletti, 2009). The complex chemical composition of EPS seems to be strain dependent. Metal sorption occurs by binding of cationic metals to the negatively charged surface of EPS, mainly via carboxylic groups (in particular at low pH), with a minor role for other functional groups (e.g., sulfonate and amino groups, particularly at high pH). Besides EPS, the intracellular polyphosphate granules and the metal-chelating proteins metalothionines are also involved in the sequestration of metals by cyanobacteria (Turner and Robinson, 1995).

3.2 Bioproducts and Bioenergy

Cyanobacteria are a prolific source of bioproducts, derived from their very active and versatile metabolisms. Within cyanobacteria, diazotrophic species share most applications with their non-nitrogen-fixing counterparts but with the added advantage of potentially lower costs (and/or carbon and water footprints) for mass production. N$_2$-fixing cyanobacteria offer potentially reduced culturing costs (no nitrogen fertilizer necessary) and harvesting—dewatering costs. Many species are self-settling, thus reducing centrifugation/filtration costs, and/or can be grown in low-water containing biofilms. However, these applications have not been fully explored on a commercial scale.

Cyanobacterial bioproducts include high-value compounds with an ongoing worldwide market, such as phycocyanin and fatty acids (Table 4); these are currently produced commercially from non-nitrogen-fixing microorganisms, such as the cyanobacterium *Spirulina* (phycocyanin), or eukaryotic microalgae, such as *Nannochloropsis* (fatty acids) (Borowitzka, 2013). In addition, N$_2$-fixing cyanobacteria

produce uncountable products that may reach high economic value but whose markets are still to be developed, including cyanotoxin standards and cyanotoxin-derived pharmaceuticals, UV sunscreens, anticancer compounds, EPS-derived cosmetics, and metal nanoparticles. Furthermore, N$_2$-fixing cyanobacteria represent an environmentally friendly source for bioplastics and biofertilizers, which may reach an enormous market as substitutes for traditional highly energy-consuming and contaminant petrol-derived plastics and chemical fertilizers (Table 4).

Furthermore, N$_2$-fixing cyanobacteria may also play a role in the increasing field of biofuel and bioenergy. The production of biohydrogen, with applications for transport and electricity, has been described in at least 14 cyanobacterial genera, including the N$_2$-fixing *Anabaena*, *Calothrix*, *Chroococcidiopsis*, *Cyanothece*, *Gloeobacter*, and *Nostoc*. Cyanobacteria generate hydrogen (H$_2$) either as a byproduct of N$_2$ fixation using nitrogenase (see Eqn (1) in Section 1.1) or by a reversible NADPH-dependent [NiFe]-hydrogenase (Peters et al., 2013); however, more research is still needed to understand the metabolic pathways that influence H$_2$ rates and yields in these organisms. In addition, N$_2$-fixing cyanobacteria are often rich in carbohydrates (starch), therefore representing a good feedstock for bioethanol generation by yeast fermentation (John et al., 2011). Studies also demonstrate the natural production of hydrocarbons (alkanes and alkenes) by at least 13 N$_2$-fixing cyanobacterial genera (Coates et al., 2014). Some of these hydrocarbons (e.g., pentadecane) could be directly used as biofuel without the need for transesterification, as required for fatty-acid-based microalgal biodiesel. Additionally, the feedstock-agnostic process of hydrothermal liquefaction is opening promising avenues for the use of wet cyanobacterial biomass to generate cheaper jet biofuel, which remains to be explored in N$_2$-fixing cyanobacteria.

TABLE 4   Bioproducts and Bioenergy from N₂-Fixing Cyanobacteria

Product	Applications	Main N₂-fixing producers	Price ($/kg)	Global market (million US$/annum)[c]
Phycobiliproteins (phycocyanins, phycoerythrin)	Biomedicine (fluorescent markers) Food coloring Pharmaceuticals Cosmetics	Nostoc, Anabaena[a]	50,000	60
Fatty acids (omega-3 and omega-6)	Neutraceuticals Animal feed (aquaculture)	Anabaena, Anabaenopsis, Aphanizomenon, Calothrix, Nodularia, Nostoc	0.88–3.8	700
Bioplastics (Polyhydroxyalkanoates)	Substitute for nonbiodegradable petrochemical-based plastics	Anabaena, Aulosira, Chlorogloea Gloeocapsa. Gloethece, Nostoc, Scytonema, Trichodesmium	1.5	—
Metal nanoparticles (Ag, Au)	Chemical industry (catalysis) Environmental remediation Biomedicine (gene therapy, biomarkers)	Anabaena, Calothrix	—	—
Ultraviolet sunscreens (Scytonemin, mycosporine-like aminoacids)	New-generation sunscreens Pharmaceuticals (anti-inflammatory, antiproliferative,) and cosmetics	Anabaena, Aphanizomenon, Aphanothece, Calothrix, Chlorogloeopsis, Gloeapsa, Gloethece, Scytonema	—	—
Cytotoxic (anticancer) compounds	Alternative anticancer therapies (leukemia cell-apoptogens, protection against chemoresistance)	Anabaena, Calothrix, Nostoc, Nodularia	—	—
Biofertilizers and phytohormones (gibereline-like compounds, indole-3 acetic acid)	Agriculture (N, P, and trace element sources; soil conditioners; enhancement of plant growth)	Anabaena spp, Aulosira, Chroococcidiopsis, Nostoc, Tolypothrix., Scytonema sp.	—	5 × 10⁹

Exopolysaccharides	Cosmetics Biomedicine (antioxidants, blood-clotting agents) Chemical industry (thickening agents)	Anabaena, Calothrix, Cylindrospermum, Gloeocapsa, Nostoc, Tolypothrix	—
Cyanotoxins (microcystins, anatoxins, saxitoxins, cylindrospermopsin)	Analytical standards Biomedicine (potential pharmaceutical applications)	Anabaena, Aphanizomenon, Cylindrospermopsis, Raphidiopsis, Umezakia	1–3
Hydrogen	Bioenergy	Anabaena, Calothrix, Chroococcidiopsis, Cyanothece, Gloeobacter, Nostoc	—
Bioethanol (from yeast-fermented cyanobacteria biomass)	Bioenergy (liquid biofuel)	Nostoc[b]	—

Borowitzka, M., 2013. High-value products from microalgae—their development and commercialisation. J. Appl. Phycol. 25, 743–756.

Brayner, R., Barberousse, H., Hemadi, M., Djedjat, C., Yéprémian, C., Coradin, T., Livage, J., Fiévet, F., Couté, A., 2007. Cyanobacteria as bioreactors for the synthesis of Au, Ag, Pd, and Pt nanoparticles via an enzyme-mediated route. J. Nanosci. Nanotech. 7, 2696–2708.

Sharma, N.K., Rai, A.K., Stal, L.J., 2014. Cyanobacteria: An Economic Perspective. John Wiley & Sons, UK.

Liu, L., Herfindal, L., Jokela, J., Shishido, T.K., Wahlsten, M., Døskeland, S.O., Sivonen, K., 2014. Cyanobacteria from terrestrial and marine sources contain apoptogens able to overcome chemoresistance in acute myeloid leukemia cells. Mar. Drugs 12, 2036–2053.

[a]Phycobiliproteins ubiquitous in cyanobacteria, only high yield (8–17% dry weight) phycocyanin-containing N_2-fixers specified (Moreno, J., Rodríguez, H., Vargas, M.A., Rivas, J., Guerrero, M.G., 1995. Nitrogen-fixing cyanobacteria as source of phycobiliprotein pigments. Composition and growth performance of 10 filamentous heterocystous strains. J. Appl. Phycol. 7, 17–23.).

[b]Potentially from all high carbohydrate N_2-fixers, but only genera with high-starch content after oil extraction are included (John, R.P., Anisha, G., Nampoothiri, K.M., Pandey, A., 2011. Micro and macroalgal biomass: a renewable source for bioethanol. Biores. Technol. 102, 186–193).

[c]Actual market values for phycobiliproteins and fatty acids in 2013 (Markou, G., Nerantzis, E., 2013. Microalgae for high-value compounds and biofuels production: A review with focus on cultivation under stress conditions. Biotechnol. Adv. 31, 1532–1542.) and projected market estimates for biofertilizers (soil conditioner from N_2-fixers) and toxins (Sharma, N.K., Rai, A.K., Stal, L.J., 2014. Cyanobacteria: An Economic Perspective. John Wiley & Sons, UK.).

References

Bergman, B., Gallon, J., Rai, A., Stal, L., 1997. N$_2$ fixation by non-heterocystous cyanobacteria. FEMS Microbiol. Rev. 19, 139–185.

Berman-Frank, I., Bidle, K.D., Haramaty, L., Falkowski, P.G., 2004. The demise of the marine cyanobacterium, *Trichodesmium* spp., via an autocatalyzed cell death pathway. Limnol. Oceanogr 49, 997–1005.

Berman-Frank, I., Lundgren, P., Falkowski, P., 2003. Nitrogen fixation and photosynthetic oxygen evolution in cyanobacteria. Res. Microbiol. 154, 157–164.

Berman-Frank, I., Pernilla Lundgren, P., Chen, Y.-B., Küpper, H., Kolber, Z., Bergman, B., Falkowski, P., 2001. Segregation of nitrogen fixation and oxygenic photosynthesis in the marine cyanobacterium *Trichodesmium*. Science 294, 1534–1537.

Borowitzka, M., 2013. High-value products from microalgae— their development and commercialisation. J. Appl. Phycol. 25, 743–756.

Brayner, R., Barberousse, H., Hemadi, M., Djedjat, C., Yéprémian, C., Coradin, T., Livage, J., Fiévet, F., Couté, A., 2007. Cyanobacteria as bioreactors for the synthesis of Au, Ag, Pd, and Pt nanoparticles via an enzyme-mediated route. J. Nanosci. Nanotech. 7, 2696–2708.

Capone, D.G., Zehr, J.P., Paerl, H.W., Bergman, B., Carpenter, E.J., 1997. *Trichodesmium*, a globally significant marine cyanobacterium. Science 276, 1221–1229.

Cirés, S., Wörmer, L., Agha, R., Quesada, A., 2013. Overwintering populations of *Anabaena*, *Aphanizomenon* and *Microcystis* as potential inocula for summer blooms. J. Plankton Res. 35, 1254–1266.

Coates, R.C., Podell, S., Korobeynikov, A., Lapidus, A., Pevzner, P., Sherman, D.H., Allen, E.E., Gerwick, L., Gerwick, W.H., 2014. Characterization of cyanobacterial hydrocarbon composition and distribution of biosynthetic pathways. PloS One 9, e85140.

Colica, G., Mecarozzi, P., De Philippis, R., 2010. Treatment of Cr(VI)-containing wastewaters with exopolysaccharide-producing cyanobacteria in pilot flow through and batch systems. Appl. Microbiol. Biotechnol. 87, 1953–1961.

De Philippis, R., Micheletti, E., 2009. Heavy metal removal with exopolysaccharide-producing cyanobacteria. In: Wang, L.K., Chen, J.P., Hung, Y.-T., Shammas, N.K. (Eds.), Heavy Metals in the Environment. CRC Press, Boca Raton, USA, pp. 89–122.

Dolman, A.M., Rücker, J., Pick, F.R., Fastner, J., Rohrlack, T., Mischke, U., Wiedner, C., 2012. Cyanobacteria and cyanotoxins: the influence of nitrogen versus phosphorus. PLoS One 7, e38757.

Dyhrman, S.T., Chappell, P.D., Haley, S.T., Moffett, J.W., Orchard, E.D., Waterbury, J.B., Webb, E.A., 2006. Phosphonate utilization by the globally important marine diazotroph *Trichodesmium*. Nature 439, 68–71.

El-Bestawy, E.A., El-Salam, A.Z.A., Mansy, A.E.-R.H., 2007. Potential use of environmental cyanobacterial species in bioremediation of lindane-contaminated effluents. Int. Biodeterior. Biodegrad. 59, 180–192.

El-Sheekh, M.M., El-Shouny, W.A., Osman, M.E., El-Gammal, E.W., 2005. Growth and heavy metals removal efficiency of *Nostoc muscorum* and *Anabaena subcylindrica* in sewage and industrial wastewater effluents. Environ. Toxicol. Pharmacol. 19, 357–365.

Fredericksson, C., Bergman, B., 1997. Ultrastructural characterisation of cells specialised for nitrogen fixation in a non-heterocystous cyanobacterium, *Trichodesmium* spp. Protoplasma 197, 76–85.

Furtado, A., Calijuri, M., Lorenzi, A., Honda, R., Genuário, D., Fiore, M., 2009. Morphological and molecular characterization of cyanobacteria from a Brazilian facultative wastewater stabilization pond and evaluation of microcystin production. Hydrobiologia 627, 195–209.

Gaby, J.C., Buckley, D.H., 2012. A comprehensive evaluation of PCR primers to amplify the *nifH* gene of nitrogenase. PloS One 7, e42149.

Gallon, J., 2005. N$_2$ Fixation by Non-Heterocystous Cyanobacteria, Genetics and Regulation of Nitrogen Fixation in Free-living Bacteria. Springer, pp. 111–139.

Gallon, J.R., 2001. N$_2$ fixation in phototrophs: adaptation to a specialized way of life. Plant Soil 230, 39–48.

Ininbergs, K., Bay, G., Rasmussen, U., Wardle, D.A., Nilsson, M.C., 2011. Composition and diversity of *nifH* genes of nitrogen-fixing cyanobacteria associated with boreal forest feather mosses. New Phytol. 192, 507–517.

Janson, S., Siddiqui, P.J.A., Walsby, A.E., Romas, K.M., Carpenter, E.J., Bergman, B., 1995. Cytomorphological characterization of the planktonic, diazotrophic cyanobacteria *Trichodesmium* spp. from the Indian Ocean, the Carribean and the Sargasso Seas. J. Phycol. 31, 463–477.

John, R.P., Anisha, G., Nampoothiri, K.M., Pandey, A., 2011. Micro and macroalgal biomass: a renewable source for bioethanol. Biores. Technol. 102, 186–193.

Khan, Z., Begum, Z.T., Mandal, R., Hossain, M., 1994. Cyanobacteria in rice soils. World J. Microbiol. Biotechnol. 10, 296–298.

Latysheva, N., Junker, V.L., Palmer, W.J., Codd, G.A., Barker, D., 2012. The evolution of nitrogen fixation in cyanobacteria. Bioinformatics 28, 603–606.

Lea, P.J., 1997. Primary nitrogen metabolism. In: Dey, P.M., Harborne, J.B. (Eds.), Plant Biochemistry. Academic Press, pp. 273–314.

Mateo, P., Berrendero, E., Perona, E., Loza, V., Whitton, B.A., 2010. Phosphatase activities of cyanobacteria as indicators of nutrient status in a Pyrenees river. Hydrobiologia 652, 255–268.

Mulholland, M.R., Capone, D.G., 2000. The nitrogen physiology of the marine N$_2$-fixing cyanobacteria *Trichodesmium* spp. Trends Plant Sci. 5, 148−153.

Ohki, K., Fujita, Y., 1988. Aerobic nitrogen acitivity measured acetylene reduction in the marine non-heterocystous cyanobacterium *Trichodesmium* spp. grown under artificial conditions. Mar. Biol. 98, 111−114.

Omoregie, E.O., Crumbliss, L.L., Bebout, B.M., Zehr, J.P., 2004. Determination of nitrogen-fixing phylotypes in *Lyngbya* sp. and *Microcoleus* chthonoplastes cyanobacterial mats from Guerrero Negro, Baja California, Mexico. Appl. Environ. Microbiol. 70, 2119−2128.

Pereira, I., Ortega, R., Barrientos, L., Moya, M., Reyes, G., Kramm, V., 2009. Development of a biofertilizer based on filamentous nitrogen-fixing cyanobacteria for rice crops in Chile. J. Appl. Phycol. 21, 135−144.

Peters, J.W., Boyd, E.S., D'Adamo, S., Mulder, D.W., Therien, J., Posewitz, M.C., 2013. Hydrogenases, Nitrogenases, Anoxia, and H$_2$ Production in Water-oxidizing Phototrophs, Algae for Biofuels and Energy. Springer, pp. 37−75.

Quesada, A., Nieva, M., Leganés, F., Ucha, A., Martín, M., Prosperi, C., Fernández-Valiente, E., 1998. Acclimation of cyanobacterial communities in rice fields and response of nitrogenase activity to light regime. Microb. Ecol. 35, 147−155.

Renuka, N., Sood, A., Ratha, S.K., Prasanna, R., Ahluwalia, A.S., 2013. Evaluation of microalgal consortia for treatment of primary treated sewage effluent and biomass production. J. Appl. Phycol. 25, 1529−1537.

Roger, P.-A., Ladha, J., 1992. Biological N$_2$ Fixation in Wetland Rice Fields: Estimation and Contribution to Nitrogen Balance, Biological Nitrogen Fixation for Sustainable Agriculture. Springer, pp. 41−55.

Romas, K.M., Carpenter, E.J., Bergman, B., 1994. Buoyancy regulation in the colonial diazotrophic cyanobacterium *Trichodesmium tenue*: ultrastructure and storage of carbohydrate, polyphosphate and nitrogen. J. Phycol. 30, 935−942.

Scott, J.T., Grantz, E.M., 2013. N$_2$ fixation exceeds internal nitrogen loading as a phytoplankton nutrient source in perpetually nitrogen-limited reservoirs. Freshwater Sci. 32, 849−861.

Scott, J.T., Marcarelli, A.M., 2012. Cyanobacteria in freshwater benthic environments. In: Ecology of Cyanobacteria II. Springer, pp. 271−289.

Stucken, K., John, U., Cembella, A., Murillo, A.A., Soto-Liebe, K., Fuentes-Valdés, J.J., Friedel, M., Plominsky, A.M., Vásquez, M., Glöckner, G., 2010. The smallest known genomes of multicellular and toxic cyanobacteria: comparison, minimal gene sets for linked traits and the evolutionary implications. PLoS One 5, e9235.

Sukenik, A., Hadas, O., Kaplan, A., Quesada, A., 2012. Invasion of Nostocales (cyanobacteria) to subtropical and temperate freshwater lakes—physiological, regional, and global driving forces. Front. Microbiol. 3, 86.

Thiel, T., Pratte, B.S., 2013. Alternative nitrogenases in *Anabaena variabilis*: the role of molybdate and vanadate in nitrogenase gene. Adv. Microbiol. 3, 87−95.

Turner, J.S., Robinson, N.J., 1995. Cyanobacterial metallothioneins: biochemistry and molecular genetics. J. Ind. Microbiol. 14, 119−125.

Vintila, S., El-Shehawy, R., 2007. Ammonium ions inhibit nitrogen fixation but do not affect heterocyst frequency in the bloom-forming cyanobacterium *Nodularia spumigena* strain AV1. Microbiol. 153, 3704−3712.

Walsby, A.E., 1978. The properties and buoyancy providing role of gas vacuoles in *Trichodesmium* Ehrenberg. Br. Phycol. J. 13, 103−116.

Whitton, B.A., 2002. Soils and Rice-fields, the Ecology of Cyanobacteria. Springer, pp. 233−255.

Wiedner, C., Dolman, A.M., Rücker, J., Knie, M., 2014. Does nitrogen fixation matter? In: Joint Aquatic Sciences Meeting 2014, Portland, Oregon.

Wörmer, L., Cirés, S., Velázquez, D., Quesada, A., Hinrichs, K.-U., 2012. Cyanobacterial heterocyst glycolipids in cultures and environmental samples: diversity and biomarker potential. Limnol. Oceanogr 57, 1775.

Yeager, C.M., Kornosky, J.L., Morgan, R.E., Cain, E.C., Garcia-Pichel, F., Housman, D.C., Belnap, J., Kuske, C.R., 2007. Three distinct clades of cultured heterocystous cyanobacteria constitute the dominant N$_2$-fixing members of biological soil crusts of the Colorado Plateau, USA. FEMS Microbiol. Ecol. 60, 85−97.

Zani, S., Mellon, M.T., Collier, J.L., Zehr, J.P., 2000. Expression of *nifH* genes in natural microbial assemblages in Lake George, New York, detected by reverse transcriptase PCR. Appl. Environ. Microbiol. 66, 3119−3124.

34

An Overview of Harmful Algal Blooms on Marine Organisms

Panchanathan Manivasagan, Se-Kwon Kim

Pukyong National University, Department of Marine-Bio Convergence Science and Marine Bioprocess Research Center, Busan, South Korea

1. INTRODUCTION

Harmful algal blooms (HABs) are a global threat to living marine resources and human health (Band-Schmidt et al., 2010). These events impact all coastal U.S. states and large portions of coastal Canada and Mexico (Mudie et al., 2002; Hernández-Becerril et al., 2007; Anderson et al., 2008; Band-Schmidt et al., 2010). HABs have had significant ecological and socioeconomic impacts on the Pacific coastal communities of North America for decades, and their prevalence and impacts on living resources in this region have increased markedly in frequency and geographical distribution in recent years (Anderson et al., 2008). The HABs that threaten the water quality, health of living resources, and economies of Pacific coast communities are diverse and often extend beyond jurisdictional boundaries. A comprehensive understanding of the causes and impacts of west-coast HABs will therefore require a regionally integrated approach. Effective HAB management will depend on interstate and international collaboration and coordination (Lewites et al., 2012).

Communities of microscopic organisms are ubiquitous and perform an enormous range of ecologically important functions. The organisms present within a community contribute to ecological functions by their individual metabolic capabilities and through interacting metabolic networks. Examples include marine systems, where autotrophy make a significant contribution to global carbon cycling. In contrast to any positive impact they might have on atmospheric carbon, periodic blooms of dinoflagellates can have devastating ecological and economic consequences through toxin production and localized oxygen depletion (Morey et al., 2011). The processes of bloom initiation, maintenance, and termination are poorly understood (Cloern et al., 2005; Anderson and Rengefors, 2006; Adolf et al., 2008).

Large accumulations of phytoplankton, macroalgae, and, occasionally, colorless heterotrophic protists are increasingly reported throughout the coastal areas of all continents. Aggregations of these organisms can discolor the water, giving rise to red, mahogany, brown, or green tides; they can float on the surface in scums, cover beaches with biomass or exudates (foam), and deplete oxygen levels through excessive respiration or decomposition. Alternatively, certain species in HABs can exert their effects through the synthesis of compounds (e.g., toxins) that can alter the cellular process of other organisms, from plankton to humans. The most severe—and therefore most memorable—effects of HABs include fish, bird, and mammal (including human) mortalities; respiratory or digestive tract problems; memory loss; seizures; lesions and skin irritation; as well as losses of coastal resources, such as submerged aquatic vegetation and benthic epifauna and infauna (Sellner et al., 2003). In this chapter, we evaluate HABs on marine organisms.

2. MARINE ALGAE

Algae are eukaryotic organisms. The presence of chlorophyll and other pigments helps in carrying out photosynthesis. The true roots, stems, or leaves are absent. Algae can be multicellular or unicellular. Mostly, they are *photoautotrophic* and carry on photosynthesis; some of these are *chemoheterotrophic* and obtain energy from chemical reactions, as well as nutrients from preformed organic matter. Microalgae can fix CO_2 using solar energy, with an efficiency that is 10 times greater than terrestrial plants (Wang et al., 2011). Many species of algae are present, such as green, red, and brown algae, which belong to the groups of *Chlorophyta*, *Rhodophyta*, and *Phaeophyta*, respectively. Algae belong to a wide range of habitats, such as fresh water and marine water, in deep oceans and in rocky shores. The planktonic and benthic algae can become important constituents of soil flora and can exist even in extreme conditions, such as in snow, sands/desert, or hot springs (temperatures above 80 °C).

3. HARMFUL ALGAL BLOOMS

HABs are often linked to significant economic losses through massive fish killings, shellfish harvest closures, and the potential threat to humans from shellfish poisonings. To manage

and mitigate the adverse impact of HABs, various strategies have been applied to control their outbreak and persistence, involving treatment with chemical agents such as copper sulfate (Anderson, 1997), flocculation of microalgae with clay (Sengco and Anderson, 2003), and other physical techniques.

Although effective in controlling blooms, chemical and physical approaches are considered to be potentially dangerous because chemical agents could cause serious secondary pollution, and they could indiscriminately kill multiple organisms in the aquatic ecosystem, which may alter marine food webs and eventually impact natural fish communities (Jeong et al., 2008). Biological agents, including bacteria (Mayali and Azam, 2004), viruses (Nagasaki et al., 2004), protozoa (Jeong et al., 2008), and macrophytes (Nakai et al., 1999; Jin and Dong, 2003), are considered to be potential suppressors in controlling the outbreak and maintenance of algal blooms.

Bacteria play an important role in nutrient regeneration and energy transformation in aquatic ecosystems (Azam et al., 1983). Therefore, algal—bacterial interactions are of particular interest; they have been considered to be potentially important regulators of algal growth and toxin production (Doucette et al., 1998). Research into the relationships between algae and bacteria have resulted in the isolation of strains of algicidal bacteria, which mainly belong to the *Cytophaga/Flavobacterium/Bacteroidetes* group or to the γ-proteobacteria group, and to the genera *Cytophaga*, *Saprospira*, *Alteromonas*, and *Pseudoalteromonas*. These bacteria show algicidal activity through either direct or indirect attack on the target algal cells (Mayali and Azam, 2004; Su et al., 2011).

4. HARMFUL CYANOBACTERIA BLOOMS

Cyanobacteria (blue-green algae) are the earth's oldest known oxygen-producing organisms, with fossil remains dating back ~3.5 billion years (Schopf, 2000). Cyanobacterial proliferation during the Precambrian period is largely responsible for the modern-day, oxygen-enriched atmosphere and subsequent evolution of higher plant and animal life (Schopf, 2000; Whitton and Potts, 2000). This long evolutionary history has served cyanobacteria well, enabling them to develop diverse and highly effective ecophysiological adaptations and strategies for ensuring survival and dominance in aquatic environments undergoing natural and human-induced environmental change (Hallock, 2005). Today, they enjoy a remarkably broad geographic distribution, ranging from polar to tropical regions in northern and southern hemispheres, where they are capable of dominating planktonic and benthic primary production in diverse habitats.

As a "microalgal" group, the cyanobacteria exhibit highly efficient nutrient (N, P, Fe, and trace metal) uptake and storage capabilities. They are the only oxygenic phototrophs capable of using atmospheric dinitrogen (N_2) as a nitrogen source to support growth via N_2 fixation (Gallon, 1992). Furthermore, many planktonic genera are capable of rapid vertical migration by altering their buoyancy, allowing them to exploit deeper, nutrient-rich waters while also taking advantage of radiant-rich conditions near the surface (Ibelings et al., 1991; Walsby et al., 1997). Lastly, some genera have formed symbioses (as endosymbionts) in diatoms, sponges, corals, lichens, ferns, and mutualistic associations with a variety of other organisms, which provide protection and enhance nutrient cycling and availability in nutrient-depleted waters (Paerl and Pinckney, 1996; Carpenter et al., 2001).

Over the past several centuries, human nutrient over-enrichment (particularly nitrogen and phosphorus) associated with urban, agricultural, and industrial development has promoted accelerated rates of primary production—that is, eutrophication. Eutrophication favors periodic proliferation and dominance of harmful blooms of cyanobacteria (cyanoHABs), both in planktonic (Steinberg and Hartmann, 1988) and benthic (Baker et al., 2001) environments.

Mass development of cyanoHABs increases turbidity and hence restricts light penetration in affected ecosystems. This, in turn, suppresses the establishment and growth of aquatic macrophytes and benthic microalgae, thereby negatively affecting the underwater habitat for benthic flora and fauna (Scheffer et al., 1997; Jeppesen et al., 2007). CyanoHABs also cause nighttime oxygen depletion through respiration and bacterial decomposition of dense blooms, which can result in fish kills and loss of benthic infauna and flora (Watkinson et al., 2005; García and Johnstone, 2006). Persistence of cyanoHABs can lead to long-term loss of benthic habitat. Lastly, numerous planktonic and benthic cyanobacterial bloom genera produce toxic peptides and alkaloids (Berry et al., 2008; Liu and Rein, 2010). Ingestion of these cyanotoxins has been linked to liver, digestive, and skin diseases; neurological impairment; and even death (Carmichael, 2001). Hence, toxic cyanoHABs are a major threat to the use of freshwater ecosystems and reservoirs for drinking water and irrigation, and freshwater and marine fishing and recreation (Carmichael, 2001).

5. THE HARMFUL PROPERTIES OF HABs

HABs are harmful in two fundamental ways (although not all HABs have both properties). The first is the production of toxins that may kill fish and shellfish or harm human consumers. The range of toxins produced by HABs is quite extensive, including brevetoxins, which cause neurotoxic shellfish poisoning; saxitoxins, which cause paralytic shellfish poisoning; okadaic acid, which causes diarrheic shellfish poisoning; domoic acid, which causes amnesic shellfish poisoning; azaspiracid, which causes azaspiracid shellfish poisoning; and numerous others

(Landsberg, 2002). Toxins may kill shellfish or fish directly. Alternatively, they may have little effect on them, but may cause illness or death when shellfish that have accumulated the algal toxins are eaten by humans or other consumers in the food web (Landsberg, 2002). There are no known antidotes for poisonings caused by HAB toxins. Some HABs, while not directly toxic, have physical structures, such as spines, that can lodge in gills, causing irritation and eventual suffocation.

The other fundamental way in which HABs are harmful is through high biomass accumulation, which may lead to environmental damage, including hypoxia, anoxia, and shading of submerged vegetation. Each of these, in turn, can lead to a multitude of negative environmental consequences (Landsberg, 2002).

6. HARMFUL ALGAE BLOOM FORMATION

6.1 Formation of HABs

The primary factor causing the occurrence of HABs is the presence of high nutrient concentrations in water bodies (Cloern, 1999; Houser and Richardson, 2010). Several research studies have demonstrated that the effect of these nutrients on HAB formation is intensified by wind monsoons, water temperature, and cold eddies. The following sections present the multiple sources of nutrient (natural or anthropogenic) discerned over the region.

6.2 Dust Deposition

Dust deposition can be a main source of nutrients (iron). These aerosols are emitted naturally from the surrounding areas by wind, then transported to precipitate later over the seas when the wind speed slows down, or they are washed out during rainfall events. Aerosol concentration decreases in the winter and increases in the summer, when aerosol optical depth values range between 0.4 and 0.7 (Vinoj and Satheesh, 2003; Gherboudj and Ghedira, 2014). At high concentrations, the deposition of aerosols can increase the activities of toxic phytoplankton, as was observed during the dry seasons of 2000 and 2008 over the Arabian Gulf (Richlen et al., 2010; Al-Shehhi et al., 2012).

6.3 Seabed

The seabed is also considered to be an important source of nutrients for phytoplankton blooms. In fact, a common hypothesis in the region of the Arabian Sea is that the formation and distribution of blooms are related to the cold eddy, which in turn is associated with the poor oxygenation of water and the presence of nutrients that are brought up from the seabed to the surface. If the cyclonic eddy is affected by higher vertical velocity and more heat loss, the growth of phytoplankton increases (Tang et al., 2002). For instance, a blooming period of a month was observed in the Arabian Sea in November 1996 over 100 km, up to the Gulf of Oman, where a cold eddy had occurred (Richlen et al., 2010).

6.4 Human Activities

The increasing population worldwide has resulted in increased demands for food, medicine, goods, and habitats. All of these issues are directly or indirectly associated with marine problems, such as overfishing, increasing emissions of pharmaceuticals and other emerging contaminants, increasing activity of Ship breaking and recycle industries (SBRIs), and increases in plastic waste, oil exploration, transportation, and algal blooms (Figure 1). In particular, algal blooms are closely connected to increases in organic matter (DOM and particulate organic matter) inputs and the effects of global warming (Mostofa et al., 2013b).

The fast depletion of fish stocks by overfishing and environmental deterioration is both an economic and an ecological problem, ruining

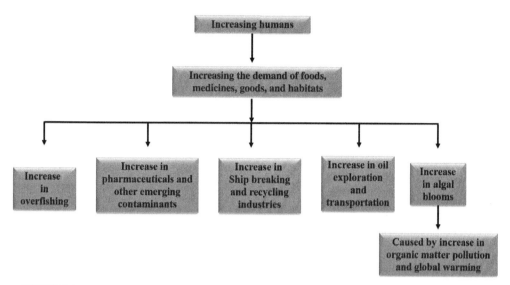

FIGURE 1   Relationship between increasing human population and problems in marine ecosystems.

fishing communities and seriously damaging the whole fishing-based supply chain. An example is demonstrated by the difficulties encountered by fisheries in the Mediterranean Sea, which has undergone overfishing for decades (Grigorakis and Rigos, 2011). The release of pollutants to marine environments is a serious threat to human health, because food consumption (including, most notably, seafood) is a major route of transmission of Emerging contaminants (ECs) to both humans and other organisms (Figure 2).

To have an idea of the pollution load, one can consider that the world's population was 3 billion in 1960, 7 billion in 2012, and will be approximately 10.6 billion in 2050 (UNFPA, 2011). The present pollution of marine waters by human activities can be roughly assessed by considering that each person can pollute $20\,L\,day^{-1}$, which equals approximately 5.1×10^4 billion $L\,year^{-1}$ worldwide (Mostofa et al., 2013a). This volume might seem small when compared to the total volume of waters in oceans, which is approximately 1.37×10^{12} billion L (Garrison, 2006). However, one should consider that a considerable fraction of the pollution is concentrated in coastal or

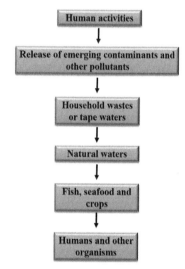

FIGURE 2   Transmission of contaminants to humans and other organisms through food consumption.

estuarine zones, which can be key breeding areas for some marine species. Considering the demands of the world's population, marine ecosystems could be polluted approximately three times more in the next 50 years compared with the last 50 years (Mostofa et al., 2013a). At equal technology, there seems to be little doubt that some control of the world's population could be important to solve problems in marine ecosystems.

7. ALGAL TOXINS

Algal toxins, or red-tide toxins, are naturally derived and toxic emerging contaminants produced during HABs in surface waters (Imai et al., 2006; Prince et al., 2008; Castle and Rodgers, 2009; Yates and Rogers, 2011). The occurrence, abundance, and geographical distribution of toxin-producing algae or cyanobacterial blooms have substantially increased during the last few decades because of increased anthropogenic input of organic matter pollution and nutrients and because of global warming (Phlips et al., 2004; Yan and Zhou, 2004; Luckas et al., 2005; McCarthy et al., 2007; Mostofa and Sakugawa, 2009).

8. IMPACTS OF ALGAL TOXINS

Algal toxins produced during algal blooms in surface waters are responsible for physiological, ecological, and environmental adverse effects, including the following:

- Deterioration of water quality with high eutrophication (Howarth, 2008; Castle and Rodgers, 2009)
- Depletion of dissolved oxygen below the pycnocline (Jeong et al., 2008)
- Loss of seagrasses and benthos (Bricelj and Lonsdale, 1997)
- Loss of phytoplankton competitor motility (Prince et al., 2008)
- Inhibition of enzymes and photosynthesis (Prince et al., 2008)
- Cell and membrane damage (Prince et al., 2008)
- Mortality of fish, coral reefs, livestock, and wildlife (Bricelj and Lonsdale, 1997; Imai and

Kimura, 2008; Southard et al., 2010; Yates and Rogers, 2011)
- Shellfish or finfish poisoning caused by neurotoxic compounds (brevetoxins), produced by blooms of red tide dinoflagellates such as *Karenia brevis* or other algae (Backer et al., 2005; Moore et al., 2008)
- Illness or even death of higher organisms or humans, associated with consumption of contaminated fish, seafood, and water; inhalation of contaminated aerosol; and contact with contaminated water during outdoor recreational or occupational activities (Backer et al., 2005; Fleming et al., 2005; Moore et al., 2008)
- Adverse health effects (e.g., eczema or acute respiratory illness) from direct contact with, ingestion, or inhalation of cyanobacteria or various toxins during recreational or occupational activities (e.g., water skiing, water craft riding, swimming, fishing) (Fleming et al., 2005; Jeong et al., 2008; Moore et al., 2008); such effects can be observed when algal scum appears on the water surface, in coastal sea beaches, or in freshwater ecosystems.

9. POSSIBLE SOLUTIONS TO CONTROLLING HABs

Remedial measures are needed for controlling algal blooms, particularly in coastal seawaters (McCarthy et al., 2007; Prince et al., 2008; Castle and Rodgers, 2009; Yates and Rogers, 2011). Prevention measures are basically centered on avoiding eutrophication (Ollikainen and Honkatukia, 2001; Imai et al., 2006; Elofsson, 2010). In fact, control of organic matter inputs, including both dissolved organic matter (DOM) and particulate organic matter, can reduce the regeneration of photoproducts, microbial products, and nutrients (NH^+_4, NO_3^-, and PO_4^{3-}). Such measures would reduce photosynthesis and, as a consequence, primary production in natural waters, while also limiting the positive feedback processes described above. Unfortunately, such measures may be less effective in already-eutrophic environments because of the nutrient regeneration phenomenon. In such cases, removal of algae or phytoplankton during algal blooms using fine, small-mesh nets and the removal of sediments (when feasible) could reduce the further photo-induced and microbial release of DOM and nutrients from primary production (Mostofa et al., 2013c).

10. CLIMATE CHANGE

Lyngbya majuscula occurs in tropical and subtropical environments and grows at maximal rates between 24 and 30 °C (Watkinson et al., 2005). The high-temperature requirements for *L. majuscula* are similar to those reported for other cyanobacteria (Robarts and Zohary, 1987; Paul et al., 2005; Paerl and Huisman, 2009). Projected temperature increases this century may increase the bloom persistence and duration of *L. majuscula* where they already occur, as well as increase its geographic range. This has already been observed in Moreton Bay, where small populations of *L. majuscula* now persist through winter months. Its northern range on the east coast of the United States may be expanding, with extensive blooms observed during the summer months in Provincetown, Massachusetts, and Penobscot Bay, Maine.

Beyond affecting growth, temperature, and physical factors in the environments where blooms occur may also influence secondary metabolite accumulation in cyanobacteria (Watanabe and Oishi, 1985; Sivonen, 1990). Although temperature has not been directly linked to increased toxin production in *L. majuscula*, maximum toxin concentrations typically occur at the peak of bloom abundances, which often coincide with temperature and growth maxima (Osborne et al., 2007). Given that bloom initiation of *L. majuscula* is in the benthos,

periods of higher temperatures and water column stability coincide with higher benthic light penetration and thus increase growth and productivity for *L. majuscula* (Watkinson et al., 2005). These changes may also directly or indirectly change other physiological features, such as toxin production. For instance, it has been demonstrated that concentrations of the bioactive compound pitipeptolide A in *L. majuscula* increase under high-light levels.

11. CONCLUSION

In conclusion, the HAB problem is significant and growing worldwide. HABs pose a major threat to public health, ecosystem health, fisheries, and economic development. The HAB problem and its impacts are diverse, as are the causes and underlying mechanisms controlling the blooms. HABs may be caused by the explosive growth of a single species that rapidly dominates the water column, but they may also be the result of highly toxic cells that do not accumulate in high numbers. Winds, tides, currents, fronts, and other features can create discrete patches or streaks of cells at all scales.

A full understanding of the many biological, chemical, and physical processes that underlie HABs will continue to be a challenge, given the many different species and hydrographic systems involved. HABs are a serious and growing problem in the global coastal ocean—one that requires the interplay of all oceanographic disciplines, as well as other fields, such as public health and resource management. Only through a recognition of the diversity of these interactions will progress be made toward the goal of scientifically based management of HAB-threatened resources.

Acknowledgments

This research was supported by a grant from the Marine Bioprocess Research Center of the Marine Biotechnology Program, funded by the Ministry of Oceans and Fisheries, R and D/2004-6002, Republic of Korea.

References

Adolf, J.E., Bachvaroff, T., Place, A.R., 2008. Can cryptophyte abundance trigger toxic *Karlodinium veneficum* blooms in eutrophic estuaries. Harmful Algae 8 (1), 119–128.

Al-Shehhi, M.R., Gherboudj, I., Ghedira, H., 2012. Temporal-spatial analysis of chlorophyll concentration associated with dust and wind characteristics in the Arabian Gulf. In: OCEANS, 2012-Yeosu: IEEE, pp. 1–6.

Anderson, D., Rengefors, K., 2006. Community assembly and seasonal succession of marine dinoflagellates in a temperate estuary: the importance of life cycle events. Limnol. Oceanogr. 51 (2), 860–873.

Anderson, D.M., 1997. Turning back the harmful red tide. Nature 388 (6642), 513–514.

Anderson, D.M., Burkholder, J.M., Cochlan, W.P., Glibert, P.M., Gobler, C.J., Heil, C.A., Kudela, R.M., Parsons, M.L., Rensel, J., Townsend, D.W., 2008. Harmful algal blooms and eutrophication: examining linkages from selected coastal regions of the United States. Harmful Algae 8 (1), 39–53.

Azam, F., Fenchel, T., Field, J., Gray, J., Meyer-Reil, L., Thingstad, F., 1983. The ecological role of water-column microbes in the sea. Mar. Ecol. Prog. Ser. 10 (3), 257–263.

Backer, L.C., Kirkpatrick, B., Fleming, L.E., Cheng, Y.S., Pierce, R., Bean, J.A., Clark, R., Johnson, D., Wanner, A., Tamer, R., 2005. Occupational exposure to aerosolized brevetoxins during Florida red tide events: effects on a healthy worker population. Environ. Health Perspect. 1135, 644–649.

Baker, P.D., Steffensen, D.A., Humpage, A.R., Nicholson, B.C., Falconer, I.R., Lanthois, B., Fergusson, K.M., Saint, C.P., 2001. Preliminary evidence of toxicity associated with the benthic cyanobacterium *Phormidium* in South Australia. Environ. Toxicol. 16 (6), 506–511.

Band-Schmidt, C.J., Bustillos-Guzmán, J.J., López-Cortés, D.J., Gárate-Lizárraga, I., Núñez-Vázquez, E.J., Hernández-Sandoval, F.E., 2010. Ecological and physiological studies of *Gymnodinium catenatum* in the Mexican Pacific: a review. Mar. Drugs 8 (6), 1935–1961.

Berry, J.P., Gantar, M., Perez, M.H., Berry, G., Noriega, F.G., 2008. Cyanobacterial toxins as allelochemicals with potential applications as algaecides, herbicides and insecticides. Mar. Drugs 6 (2), 117–146.

Bricelj, V.M., Lonsdale, D.J., 1997. *Aureococcus anophagefferens*: causes and ecological consequences of brown tides in US mid-Atlantic coastal waters. Limnol. Oceanogr. 42 (5), 1023–1038.

Carmichael, W.W., 2001. Health effects of toxin-producing cyanobacteria: "The CyanoHABs". Hum. Ecol. Risk Assess. 7 (5), 1393–1407.

Carpenter, S.R., Cole, J.J., Hodgson, J.R., Kitchell, J.F., Pace, M.L., Bade, D., Cottingham, K.L., Essington, T.E., Houser, J.N., Schindler, D.E., 2001. Trophic cascades, nutrients, and lake productivity: whole-lake experiments. Ecol. Monogr. 71 (2), 163–186.

Castle, J.W., Rodgers Jr., J.H., 2009. Hypothesis for the role of toxin-producing algae in Phanerozoic mass extinctions based on evidence from the geologic record and modern environments. Environ. Geosci. 16 (1), 1–23.

Cloern, J.E., 1999. The relative importance of light and nutrient limitation of phytoplankton growth: a simple index of coastal ecosystem sensitivity to nutrient enrichment. Aquat. Ecol. 33 (1), 3–15.

Cloern, J.E., Schraga, T.S., Lopez, C.B., Knowles, N., Grover Labiosa, R., Dugdale, R., 2005. Climate anomalies generate an exceptional dinoflagellate bloom in San Francisco Bay. Geophys. Res. Lett. 32 (14).

Doucette, G., Kodama, M., Franca, S., Gallacher, S., 1998. Bacterial interactions with harmful algal bloom species: bloom ecology, toxigenesis, and cytology. In: Anderson, D.M., Cembella, A.D., Hallegraeff, G.M. (Eds.), Physiological Ecology of Harmful Algal Bloom, 41. Springer-Verlag, Heidelberg, Berlin, pp. 619–648.

Elofsson, K., 2010. The costs of meeting the environmental objectives for the Baltic Sea: a review of the literature. AMBIO 39 (1), 49–58.

Fleming, L.E., Kirkpatrick, B., Backer, L.C., Bean, J.A., Wanner, A., Dalpra, D., Tamer, R., Zaias, J., Cheng, Y.S., Pierce, R., 2005. Initial evaluation of the effects of aerosolized Florida red tide toxins (brevetoxins) in persons with asthma. Environ. Health Perspect. 1135, 650–657.

Gallon, J., 1992. Tansley Review No. 44. Reconciling the incompatible: N_2 fixation and O_2. New Phytol. 122, 571–609.

García, R., Johnstone, R.W., 2006. Effects of *Lyngbya majuscula* (Cyanophycea) blooms on sediment nutrients and meiofaunal assemblages in seagrass beds in Moreton Bay, Australia. Mar. Freshwater Res. 57 (2), 155–165.

Garrison, T., 2006. Oceanography: An Invitation to Marine Science, sixth ed. Thomson Brooks/Cole Pub Co., Belmont.

Gherboudj, I., Ghedira, H., 2014. Spatiotemporal assessment of dust loading over the United Arab Emirates. Int. J. Climatol. 34, 3321–3335. http://dx.doi.org/10.1002/joc.3909.

Grigorakis, K., Rigos, G., 2011. Aquaculture effects on environmental and public welfare—the case of Mediterranean mariculture. Chemosphere 85 (6), 899–919.

Hallock, P., 2005. Global change and modern coral reefs: new opportunities to understand shallow-water carbonate depositional processes. Sediment. Geol. 175 (1), 19–33.

Hernández-Becerril, D.U., Alonso-Rodríguez, R., Alvarez-Góngora, C., Barón-Campis, S.A., Ceballos-Corona, G., Herrera-Silveira, J., Meave del Castillo, M.E., Juárez-Ruiz, N., Merino-Virgilio, F., Morales-Blake, A.,

Ochoa, J.L., Orellana-Cepeda, E., Ramirez-Camarena, C., Rodríguez-Salvador, R., 2007. Toxic and harmful marine phytoplankton and microalgae (HABs) in Mexican Coasts. Journal of Environmental Science and Health Part A 42, 1349–1363.

Houser, J.N., Richardson, W.B., 2010. Nitrogen and phosphorus in the upper Mississippi river: transport, processing, and effects on the river ecosystem. Hydrobiologia 640 (1), 71–88.

Howarth, R.W., 2008. Coastal nitrogen pollution: a review of sources and trends globally and regionally. Harmful Algae 8 (1), 14–20.

Ibelings, B.W., Mur, L.R., Walsby, A.E., 1991. Diurnal changes in buoyancy and vertical distribution in populations of Microcystisin two shallow lakes. J. Plankton Res. 13 (2), 419–436.

Imai, I., Kimura, S., 2008. Resistance of the fish-killing dinoflagellate *Cochlodinium polykrikoides* against algicidal bacteria isolated from the coastal sea of Japan. Harmful Algae 7 (3), 360–367.

Imai, I., Yamaguchi, M., Hori, Y., 2006. Eutrophication and occurrences of harmful algal blooms in the Seto Inland Sea, Japan. Plankton Benthos Res. 1 (2), 71–84.

Jeong, H.J., Kim, J.S., Yoo, Y.D., Kim, S.T., Song, J.Y., Kim, T.H., Seong, K.A., Kang, N.S., Kim, M.S., Kim, J.H., 2008. Control of the harmful alga *Cochlodinium polykrikoides* by the naked ciliate *Strombidinopsis jeokjo* in mesocosm enclosures. Harmful Algae 7 (3), 368–377.

Jeppesen, E., Søndergaard, M., Meerhoff, M., Lauridsen, T.L., Jensen, J.P., 2007. Shallow lake restoration by nutrient loading reduction—some recent findings and challenges ahead. Hydrobiologia 584 (1), 239–252.

Jin, Q., Dong, S., 2003. Comparative studies on the allelopathic effects of two different strains of *Ulva pertusa* on *Heterosigma akashiwo* and *Alexandrium tamarense*. J. Exp. Mar. Biol. Ecol. 293 (1), 41–55.

Landsberg, J.H., 2002. The effects of harmful algal blooms on aquatic organisms. Rev. Fish. Sci. 10 (2), 113–390.

Lewitus, A.J., Horner, R.A., Caron, D.A., Garcia-Mendoza, E., Hickey, B.M., Hunter, M., Huppert, D.D., Kudela, R.M., Langlois, G.W., Largier, J.L., 2012. Harmful algal blooms along the North American west coast region: history, trends, causes, and impacts. Harmful Algae 19, 133–159.

Liu, L., Rein, K.S., 2010. New peptides isolated from *Lyngbya* species: a review. Mar. Drugs 8 (6), 1817–1837.

Luckas, B., Dahlmann, J., Erler, K., Gerdts, G., Wasmund, N., Hummert, C., Hansen, P., 2005. Overview of key phytoplankton toxins and their recent occurrence in the North and Baltic Seas. Environ. Toxicol. 20 (1), 1–17.

Mayali, X., Azam, F., 2004. Algicidal bacteria in the sea and their impact on algal blooms. J. Eukaryotic Microbiol. 51 (2), 139–144.

McCarthy, M.J., Lavrentyev, P.J., Yang, L., Zhang, L., Chen, Y., Qin, B., Gardner, W.S., 2007. Nitrogen dynamics and microbial food web structure during a summer cyanobacterial bloom in a subtropical, shallow, well-mixed, eutrophic lake (Lake Taihu, China). Hydrobiologia 581, 195–207.

Moore, S.K., Trainer, V.L., Mantua, N.J., Parker, M.S., Laws, E.A., Backer, L.C., Fleming, L.E., 2008. Impacts of climate variability and future climate change on harmful algal blooms and human health. Environ. Health 7 (2), 1–12.

Morey, J.S., Monroe, E.A., Kinney, A.L., Beal, M., Johnson, J.G., Hitchcock, G.L., Van Dolah, F.M., 2011. Transcriptomic response of the red tide dinoflagellate, *Karenia brevis*, to nitrogen and phosphorus depletion and addition. BMC Genomics 12 (1), 346–364.

Mostofa, K.M., Sakugawa, H., 2009. Spatial and temporal variations and factors controlling the concentrations of hydrogen peroxide and organic peroxides in rivers. Environ. Chem. 6 (6), 524–534.

Mostofa, K.M., Liu, C.-Q., Gao, K., Vione, D., Ogawa, H., 2013a. Challenges and solutions to marine ecosystems (Invited Speaker). In: Proceedings of BIT's Second Annual World Congress of Marine Biotechnology WCMB-2012, September 19–23, Dalian, China.

Mostofa, K.M., Liu, C.-Q., Vione, D., Gao, K., Ogawa, H., 2013b. Sources, factors, mechanisms and possible solutions to pollutants in marine ecosystems. Environ. Pollut. 182, 461–478.

Mostofa, K.M., Liu, C.-Q., Gao, K., Li, S., Vione, D., Mottaleb, M.A., 2013c. Impacts of global warming on biogeochemical cycles in natural waters. In: Mostofa, K.M.G., Yoshioka, T., Mottaleb, A., Vione, D. (Eds.), Photobiogeochemistry of Organic Matter: Principles and Practices in Water Environment. Springer, New York, pp. 851–914.

Mudie, P.J., Rochon, A., Levac, E., 2002. Palynological records of red tide-producing species in Canada: pasttrends and implications for the future. Palaeogeography, Palaeoclimatology, Palaeoecology 180, 159–186.

Nagasaki, K., Tomaru, Y., Katanozaka, N., Shirai, Y., Nishida, K., Itakura, S., Yamaguchi, M., 2004. Isolation and characterization of a novel single-stranded RNA virus infecting the bloom-forming diatom *Rhizosolenia setigera*. Appl. Environ. Microbiol. 70 (2), 704–711.

Nakai, S., Inoue, Y., Hosomi, M., Murakami, A., 1999. Growth inhibition of blue-green algae by allelopathic effects of macrophytes. Water Sci. Technol. 39 (8), 47–53.

Ollikainen, M., Honkatukia, J., 2001. Towards efficient pollution control in the Baltic Sea: an anatomy of current failure with suggestions for change. AMBIO 30 (4), 245–253.

Osborne, N., Shaw, G.R., Webb, P., 2007. Health effects of recreational exposure to Moreton Bay, Australia waters during a *Lyngbya majuscula* bloom. Environ. Int. 33 (3), 309–314.

Paerl, H., Pinckney, J., 1996. A mini-review of microbial consortia: their roles in aquatic production and biogeochemical cycling. Microb. Ecol. 31 (3), 225–247.

Paerl, H.W., Huisman, J., 2009. Climate change: a catalyst for global expansion of harmful cyanobacterial blooms. Environ. Microbiol. Rep. 1 (1), 27–37.

Paul, V.J., Thacker, R.W., Banks, K., Golubic, S., 2005. Benthic cyanobacterial bloom impacts the reefs of South Florida (Broward County, USA). Coral Reefs 24 (4), 693–697.

Phlips, E., Badylak, S., Youn, S., Kelley, K., 2004. The occurrence of potentially toxic dinoflagellates and diatoms in a subtropical lagoon, the Indian River Lagoon, FL, USA. Harmful Algae 3 (1), 39–49.

Prince, E.K., Myers, T.L., Kubanek, J., 2008. Effects of harmful algal blooms on competitors: allelopathic mechanisms of the red tide dinoflagellate *Karenia brevis*. Limnol. Oceanogr. 53 (2), 531.

Richlen, M.L., Morton, S.L., Jamali, E.A., Rajan, A., Anderson, D.M., 2010. The catastrophic 2008–2009 red tide in the Arabian gulf region, with observations on the identification and phylogeny of the fish-killing dinoflagellate *Cochlodinium polykrikoides*. Harmful Algae 9 (2), 163–172.

Robarts, R.D., Zohary, T., 1987. Temperature effects on photosynthetic capacity, respiration, and growth rates of bloom-forming cyanobacteria. N. Z. J. Mar. Freshwater Res. 21 (3), 391–399.

Scheffer, M., Rinaldi, S., Gragnani, A., Mur, L.R., van Nes, E.H., 1997. On the dominance of filamentous cyanobacteria in shallow, turbid lakes. Ecology 78 (1), 272–282.

Schopf, J.W., 2000. The fossil record: tracing the roots of the cyanobacterial lineage. In: Whitton, B.A., Potts, M. (Eds.), The Ecology of Cyanobacteria. Kluwer Academic Publishers, Dordrecht, pp. 13–35.

Sellner, K.G., Doucette, G.J., Kirkpatrick, G.J., 2003. Harmful algal blooms: causes, impacts and detection. J. Ind. Microbiol. Biotechnol. 30 (7), 383–406.

Sengco, M., Anderson, D., 2003. Controlling harmful algal blooms through clay flocculation. J. Eukaryotic Microbiol. 51 (2), 169–172.

Sivonen, K., 1990. Effects of light, temperature, nitrate, orthophosphate, and bacteria on growth of and hepatotoxin production by *Oscillatoria agardhii* strains. Appl. Environ. Microbiol. 56 (9), 2658–2666.

Southard, G.M., Fries, L.T., Barkoh, A., 2010. *Prymnesium parvum*: the Texas experience. J. Am. Water Resour. Assoc. 46, 14–23.

Steinberg, C.E., Hartmann, H.M., 1988. Planktonic bloom-forming cyanobacteria and the eutrophication of lakes and rivers. Freshwater Biol. 20 (2), 279–287.

Su, J., Yang, X., Zhou, T., Zheng, T., 2011. Marine bacteria antagonistic to the harmful algal bloom species *Alexandrium tamarense* (Dinophyceae). Biol. Control 56 (2), 132–138.

Tang, D., Kawamura, H., Luis, A.J., 2002. Short-term variability of phytoplankton blooms associated with a cold eddy in the Northwestern Arabian Sea. Remote Sens. Environ. 81 (1), 82–89.

UNFPA, 2011. The State of World Population 2011. United Nations Population Fund, New York, USA. www.unfpa.org.

Vinoj, V., Satheesh, S., 2003. Measurements of aerosol optical depth over Arabian Sea during summer monsoon season. Geophys. Res. Lett. 30 (5), 1263.

Walsby, A.E., Hayes, P.K., Boje, R., Stal, L.J., 1997. The selective advantage of buoyancy provided by gas vesicles for planktonic cyanobacteria in the Baltic Sea. New Phytol. 136 (3), 407–417.

Wang, X., Hao, C., Zhang, F., Feng, C., Yang, Y., 2011. Inhibition of the growth of two blue-green algae species (*Microsystis aruginosa* and *Anabaena spiroides*) by acidification treatments using carbon dioxide. Bioresour. Technol. 102 (10), 5742–5748.

Watanabe, M.F., Oishi, S., 1985. Effects of environmental factors on toxicity of a cyanobacterium (*Microcystis aeruginosa*) under culture conditions. Appl. Environ. Microbiol. 49 (5), 1342–1344.

Watkinson, A., O'Neil, J., Dennison, W., 2005. Ecophysiology of the marine cyanobacterium, *Lyngbya majuscula* (*Oscillatoriaceae*) in Moreton Bay, Australia. Harmful Algae 4 (4), 697–715.

Whitton, B.A., Potts, M., 2000. The Ecology of Cyanobacteria: Their Diversity in Time and Space. Springer, ISBN 0792347358.

Yan, T., Zhou, M.-J., 2004. Environmental and health effects associated with harmful algal bloom and marine algal toxins in China. Biomed. Environ. Sci. 17 (2), 165–176.

Yates, B.S., Rogers, W.J., 2011. Atrazine selects for ichthyotoxic *Prymnesium parvum*, a possible explanation for golden algae blooms in lakes of Texas, USA. Ecotoxicology 20 (8), 2003–2010.

Microalgae-Derived Toxic Compounds

Zhong-Ji Qian[1], Kyong-Hwa Kang[2], BoMi Ryu[3]

[1]College of Food Science and Technology, Guangdong Ocean University, Zhanjiang, PR China;
[2]Department of Marine-Bio. Convergence Science and Marine Bioprocess Research Center, Pukyong National University, Busan, Republic of Korea; [3]School of Pharmacy, The University of Queensland, Brisbane, QLD, Australia

1. INTRODUCTION

Microalgae are eukaryotic photosynthetic microorganisms that can produce hydrocarbons, proteins, polyunsaturated fatty acids, vitamins, carotenoids, and other chemicals (Borowitzka and Borowitzka, 1988). A review by Raposo et al. (2013) outlined the great potential of marine microalgae for applications in a variety of areas, including for human nutrition and feed in aquaculture; as biofertilizers and in the treatment of effluents; and as anti-inflammatory, antiallergic, and analgesic agents. These researchers also highlighted the fact that some marine unicellular algae, such as *Porphyridium* and *Rhodella* (rhodophytes), and cyanobacteria, such as *Arthrospira*, can produce sulfated polysaccharides, which have already found applications as antiviral agents, either in vivo or in vitro (Huleihel et al., 2002), as well as nutraceuticals (Dvir et al., 2009), agents to prevent tumor cell growth (Gardeva et al., 2009), therapeuticals (Arad and Atar, 2007; Guzman et al., 2003), and even ion exchangers (Lupescu et al., 1991) and drag reducers (Gasljevic et al., 2008).

The toxins produced by some microalgae demonstrate the highly effective bioactivity of the compounds they may contain. This is mostly evidenced in shellfish, which are depurators and can accumulate high concentrations of toxins. At some times of the year, blooms of such microalgae can become very dangerous because of paralytic shellfish poisoning (neurotoxic compounds produced by the dinoflagellate *Alexandrium lusitanicum*), diarrheic shellfish poisoning (DSP; substances like okadaic acid (OA) and dynophysotoxin, produced by the dinoflagellate *Dinophysis*), and amnesic shellfish poisoning compounds (ASPs; such as the amino acid domoic acid (DA), produced by the diatom *Nitzschia pungens*; Pulz and Gross, 2004). On the other hand, the cytotoxic activity of these compounds has been applied in anticancer treatments (Sirenko et al., 1999). Furthermore, gambieric acids from the dinoflagellate *Gambierdiscus toxicus* have antifungal activity, as reported by Donia and Hamann (2003). The gambieric acids were 2000 times more potent than amphotericin B, with only moderate toxic side effects, both in vivo and in vitro (Nagai et al., 1992a, 1992b).

The therapeutic value of all toxins has not yet been completely investigated. Several studies have reported the following:

1. Cytotoxic activity has been found to be important for anticancer drugs (Sirenko et al., 1999).
2. Antiviral activities were found—mainly in cyanobacteria, but also in apochlorotic diatoms and the conjugaphyte *Spirogyra*, where certain sulfolipids are active (e.g., against the herpes simplex virus; Muller-Feuga et al., 2003).
3. Antimicrobial activity has been investigated as part of an effort to find new antibiotics. Although the success rate is about 1% (Muller-Feuga et al., 2003), there seems to be some promising substances from microalgae, such as the cyanobacterium *Scytonema*.
4. Antifungal activity has been found in different extracts of cyanobacteria (Nagai et al., 1992a, 1992b).
5. Antihelminthic effects were found for *Spirogyra* and *Oedogonium* (Muller-Feuga et al., 2003).

Many species of marine microalgae are capable of producing or accumulating biotoxins, which can accumulate in filter-feeding shellfish. This chapter reviews the most important features of microalgae-derived toxic compound applications.

2. TOXIC COMPOUNDS FROM MARINE MICROALGAE AND THEIR APPLICATIONS

2.1 Okadaic Acid

OA and its close chemical dinophysistoxin analogs are major DSP toxins found worldwide (Figure 1). OA was first isolated in the black sponge *Holichondria okadai* (Takai et al., 1987) and later in the dinoflagellate *Prorocentrum lima* and *Dinophysis acuta* (Daranas et al., 2001; Dickey et al., 1990; Vale and Botana, 2008).

Microalgae represent an important part of the aquatic ecosystem, being responsible for more than half of the global net primary biomass production (Field et al., 1998). However, several algal species are very sensitive to environmental stress, which can decrease their capacity for biomass production. For example, *Dunaliella tertiolecta*, a model marine alga, shows a high sensitivity to the effects of various pollutants, such as metals, solvents, pesticides, and petrochemicals (Samson and Popovic, 1988; Okumura et al., 2001; DeLorenzo and Serrano, 2003; Carrera-Martínez et al., 2010). In microalgae, OA was previously found to decrease the growth rate of *D. tertiolecta* and of other non-OA producing algae (*Dunaliella salina*, *Thalassiosira weissflogii*, *G. toxicus*, *Coolia monotis*) (Windust et al., 1996; Sugg and VanDolah, 1999). Conversely, species producing OA were found to be very resistant to OA's effects (Windust et al., 1996; Sugg and VanDolah, 1999). Although the role of OA as a major allelopathic compound responsible for the dominance of harmful algae in blooms was found to be unlikely (Sugg and VanDolah, 1999; Jonsson et al., 2009), the sensitivity of non-OA producing algae to the effects of OA indicates that the presence of this toxin in the aquatic ecosystem during harmful algae blooms may induce deleterious effects in the algal community, which are still unknown and should be investigated.

OA intoxication occurs rapidly after the ingestion of contaminated seafood, with symptoms such as vomiting and diarrhea. Although it is not fatal, it represents an increasing global concern regarding both health and economic burdens (Van Dolah, 2000). The main molecular targets of OA are the serine/threonine phosphoprotein phosphatase classes PP1A and PP2A, which have half maximal inhibitory concentration values in the nanomolar range; other classes of protein phosphatases are less sensitive or insensitive to the toxin (Bialojan and Takai, 1988; Cohen et al., 1990). Acting in opposition to protein kinases and phosphorylases, protein phosphatases are an important group of enzymes involved in the

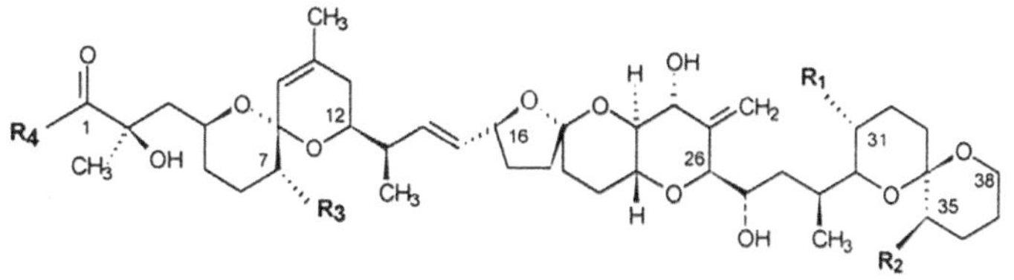

		R1	R2	R3	R4	R5
	OA	CH₃	H	OH	OH	-
I	DTX1	CH₃	CH₃	OH	OH	-
	DTX2	H	CH₃	OH	OH	-
II	"DTX3"	(H or CH₃)	(H or CH₃)	Acyl	OH	-
	OA methyl-ester	CH₃	H	OH	OMe	-
III	OA ethyl-ester	CH₃	H	OH	OEt	-
	OA diol-ester	CH₃	H	OH	Diol	-

FIGURE 1   Chemical structure of okadaic acid and dinophysistoxin-1, -2, and -3.

regulation of numerous signaling pathways; consequently, they help to controlling a number of cell functions, including cell cycle progression, metabolism, ion balance, gene expression, cytoskeletal rearrangements, and cell movement (Vale and Botana, 2008; Van Dolah, 2000). This broad spectrum of potential targets probably explains why OA has also been termed a tumor promoter, due to its ability to induce skin carcinogenesis in the mouse (Suganuma et al., 1988). Previous works demonstrated that OA induces severe genotoxic and cytotoxic effects in a cell type–dependent manner. These effects include DNA strand breaks and micronuclei induction, alterations in cell cycle and DNA repair, oxidative DNA damage, decrease of viability, and apoptosis (Le Hégarat et al., 2004; Lago et al., 2005; Souid-Mensi et al., 2008; Valdiglesias et al., 2010, 2011a,b,c).

2.2 Dinophysistoxin

The main toxins responsible for DSP are OA and dinophysistoxin (DTX)-1, -2, and -3 (Yasumoto et al., 1985; Figure 1). OA and the DTXs are polyketide compounds containing furan and pyran-type ether rings and an alpha-hydroxycarboxylic function; the only difference between them is the number or position of the methyl groups (Dominguez et al., 2010). DTX-1 is the dominant toxin in Japan, Canada, and Norway (Lee et al., 1989; Quilliam et al., 1993). Whereas OA is the predominant toxin in Europe; high amounts of DTX-2 have been detected in Spain and Portugal (Blanco et al., 1995; Vale and Sampayo, 2000), and it is also the predominant species in Irish mussels (Carmody et al., 1996).

OA and the DTXs are free polyether acids that inhibit serine/threonine phosphatase; they affect the secretion and gene transcription of nerve growth factor (Pshenichkin and Wise, 1995; Garcia et al., 2003). OA and its derivatives, DTX-1 and DTX-2, were found to be associated with dinoflagellates of the genus *Dinophysis*; they are considered to be diarrhetic (Hamano et al., 1986) and tumorogenic phycotoxins (Fujiki and Suganuma, 1993).

DTX-1 is the methyl derivative of OA, whereas DTX-3 is the DTX-1 ester (7-O-acyl-derivatives of dinophysistoxin-1). The 7-OH in DTX-1 can be esterified with fatty acids ranging from tetradecanoic acid (C14:0) to docosahexaenoic acid (C22:6w3); palmitic acid is the most common fatty acid found in DTX-3 (Yasumoto et al., 1985). DTX-1 and DTX-3 have been only isolated from shellfish samples, and DTX-3 is absent in wild and cultivated plankton samples;

for this reason, it has been proposed that the acylation of the 7-OH in DTX-1 only occurs in shellfish. The underlying mechanism of action associated with these toxin activities is explained mainly by their potent inhibitory action against ser/thre protein phosphatase (PP) 2A, 1, and 2B; the latter is inhibited only at high concentrations of phycotoxin (Bialojan and Takai, 1988; Rivas et al., 2000). DTX-3 does not inhibit the enzymes (Takai et al., 1992) but is easily hydrolyzed to DTX-1 by digestive enzymes, such as lipase.

DTX-2 is the primary DTX produced by *Dinophysis* spp., but DTX can be found in a variety of forms, with at least 15 different derivatives presently identified (Miles, 2007; Anonymous, 2009). Most of these are believed to occur only as metabolites in shellfish, however (Suzuki et al., 1999). *Dinophysis acuta* normally contains OA and PTX-2, as well as either DTX-1 or DTX-2, but the cellular content of each toxin can vary a lot (Pizarro et al., 2008, 2009; Fux et al., 2010). Aune et al. determined that DTX-2 is approximately 40% less toxic than OA by intraperitoneal injection in mice (Aune et al., 2007), and it is also less toxic in primary cultures of cerebellar neurons (Perez-Gomez et al., 2004).

2.3 Domoic Acid

To date, production of the neurotoxin DA has been found in 12 *Pseudo-nitzschia* species from different parts of the world (Lundholm et al., 2012; Trainer et al., 2012); most research on toxic diatoms has focused on these species (Figure 2). In addition to *Pseudo-nitzschia* spp., DA has been reported in the benthic diatom *Halamphora coffeaeformis* (Agardh) Levkov (previously *Amphora coffeaeformis*) (Maranda et al., 1990) and in several strains of a single *Nitzschia* species, *Nitzschia navis-varingica* Lundholm and Moestrup (Kotaki et al., 2004, 2005). DA has been reported worldwide and has caused ASP in sea birds, mammals (sea otters, sea lions, and whales), and humans. ASP was first discovered

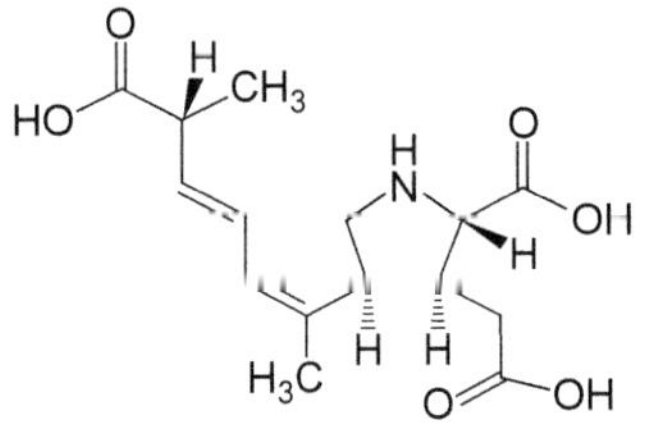

FIGURE 2   Structure of domoic acid.

in 1987, when blue mussels grown off Prince Edward Island, Canada became toxic (Wright et al., 1989) after a long-lasting bloom of *Pseudonitzschia multiseries* (Hasle) Hasle, intoxicating humans (Bates et al., 1989). Apart from DA, other less toxic isomers of DA exist, of which two—isodomoic acid A and isodomoic acid B—have been found in *Nitzschia navis-varingica* in different combinations and relative amounts (Kotaki et al., 2005; Romero et al., 2011); isodomoic acid C has been found in *Pseudo-nitzschia australis* (Holland et al., 2005).

DA functions as a glutamate agonist and exerts excitotoxicity via overstimulation of glutamate receptors (AMPA, NMDA) in the central nervous system (CNS) (Berman and Murray, 1997). DA-mediated increases in glutamatergic activation may alter cellular transcriptional responses; if gene product formation follows transcriptional changes, this could lead to changes in synapse structure and function, dendrite and nerve terminal integrity, and neuronal injury. Changes in synaptic function and loss of select glutamatergic neurons, or their synapses, occurring during neurodevelopment or in the aged may be particularly deleterious for normal CNS function (Morrison and Hof, 1997). Toxicological studies with animals show clinical signs and brain lesions, which are mostly consistent with those in naturally exposed humans and sea lions. Those clinical signs included seizures, periods of marked lethargy and inappetence, vomiting, muscular twitching, central blindness, blepharospasm (often unilateral), and abnormal behavior (Goldstein et al., 2008). Based on the data for sea lions, the authors suggested that the conversion from acute to chronic exposure is possible.

In a model system with rats, seizures with increasing magnitude over a period of 6 months were induced in animals by the application of low-dose, repetitive DA (Muha and Ramsdell, 2011). Low levels of DA (0.20–0.75 ppm) caused no toxic symptoms in humans and nonhuman primates, but clinical effects were apparent at a concentration of 1.0 ppm. The tolerable daily intake (TDI) of DA for humans was calculated to be 0.075 ppm. For razor clams and crabs, the TDIs were found to be 19.4 and 31.5 ppm, respectively (Costa et al., 2010).

2.4 Ciguatoxin

Ciguatera fish poisoning (CFP) is a seafood intoxication caused in humans by the consumption of tropical coral reef fishes that have accumulated ciguatoxins (CTXs) in their tissues (Lewis, 2001). Produced by *Gambierdiscus* dinoflagellates, CTXs are potent polyether neurotoxins

that bioaccumulate and biotransform in fish tissues up the trophic chain (Yasumoto, 2005; Figure 3). Dinoflagellates of the genera *Prorocentrum*, *Ostreopsis*, *Coolia*, and *Amphidinium* coexist with *Gambierdiscus* spp. in ciguatera endemic regions and have sometimes been associated with ciguatera (Anderson and Lobel, 1987). However, their involvement in CFP is debatable because these genera do not produce CTXs but produce other types of toxins (Holmes et al., 1995; Kobayashi et al., 1991; Murakami et al., 1982; Taniyama et al., 2003).

The hazard of ciguatera in a specific area may be difficult to assess. The processes of transmission, bioaccumulation, and biotransformation of CFP toxins through the food web are poorly understood and involve long-term toxin transfer. The CFP toxin transvectors (e.g., fish) are mobile and toxins accumulated in the fish could result from long-term processes, so their toxin content may not represent the cell density of microalgae at the time and place when the fish have been caught. Rarely, ciguatera intoxication

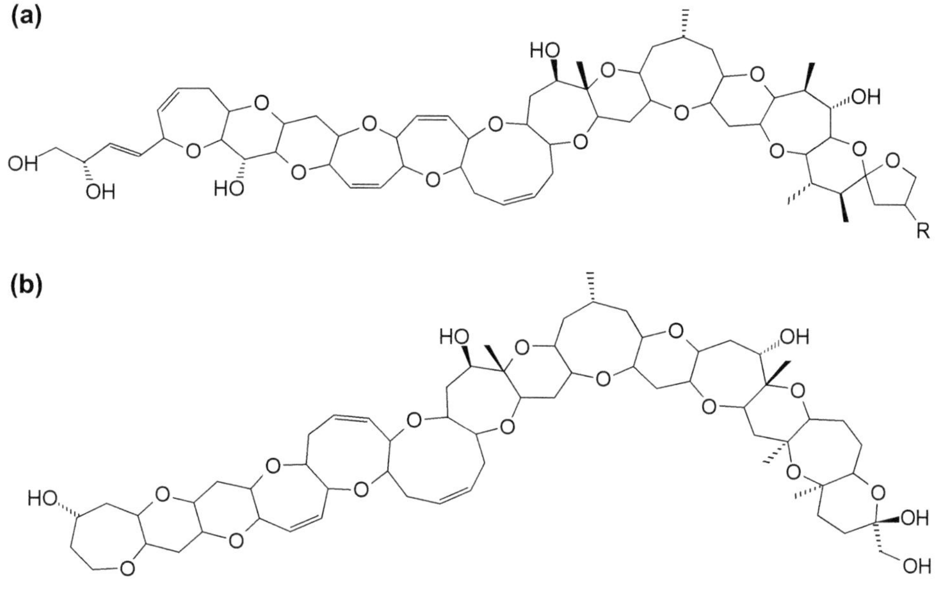

FIGURE 3 Structure of ciguatoxins. (a) The major Pacific ciguatoxins: P-CTX-1 R=OH, P-CTX-2, and P-CTX-3 R=H. (b) The major Caribbean ciguatoxin, C-CTX-1.

or accumulation of toxins in fish has been related to observable punctual events of *Gambierdiscus* blooms. The cell abundance and toxicity of natural populations of *Gambierdiscus* in fishing areas may be used as an indicator of local ciguatera risk (Chinain et al., 2010; Darius et al., 2007). However, the cell density of *Gambierdiscus* in natural populations does not always correlate with its CTX production. The existence of genetically determined CTX "superproducing" and nonproducing strains within the same natural population of *Gambierdiscus* was proposed by Holmes et al. (1991).

Clinically, ciguatera is associated with gastrointestinal disturbances of limited duration, particularly nausea, diarrhea, and abdominal pain, with neurological disturbances being the predominant presentation. The neurological symptoms of ciguatera include distressing, often persistent, sensory disturbances such as perioral and distal paraesthesias, dysesthesias, pruritus, headache, and asthenia (Pearn et al., 2001; Schnorf et al., 2002). Of these neurological disturbances, temperature dysesthesia, or cold allodynia, is considered pathognomonic and occurs in up to 95% of those with ciguatera (Bagnis et al., 1979; Schnorf et al., 2002). At the molecular level, CTX is the most potent known activator of voltage-gated sodium channels (Nav) (Strachan et al., 1999). CTX also inhibits neuronal potassium channels (Birinyi-Strachan et al., 2005), resulting in further increased neuronal excitability. The pharmacological action of CTXs on Nav in excitable cells results in a range of pathophysiological effects, including spontaneous action potential discharge, release of neurotransmitters, increase of intracellular Ca^{2+}, and axonal Schwann cell edema (Katharina et al., 2013).

2.5 Gambieric Acids

CTXs accumulate in fish through the food chain, starting from the dinoflagellate *G. toxicus* (Adachi and Fukuyo, 1979). Some strains produce not only CTXs but also other polycyclic ethers, such as gambierol (Morohashi et al., 1998) and gambieric acids (Morohashi et al., 2000). Interestingly, gambierol exhibits toxicity toward mice, whereas gambieric acids are nontoxic but potent antifungals. Gambieric acids A–D (1–4; Figure 4) are marine polycyclic ether natural products isolated from the culture medium of the ciguatera causative dinoflagellate *G. toxicus* (Nagai et al., 1992a, 1992b).

2.6 Karatungiol A

Karlotoxins (KmTxs) are a group of potent amphipathic ichthyotoxins produced by the

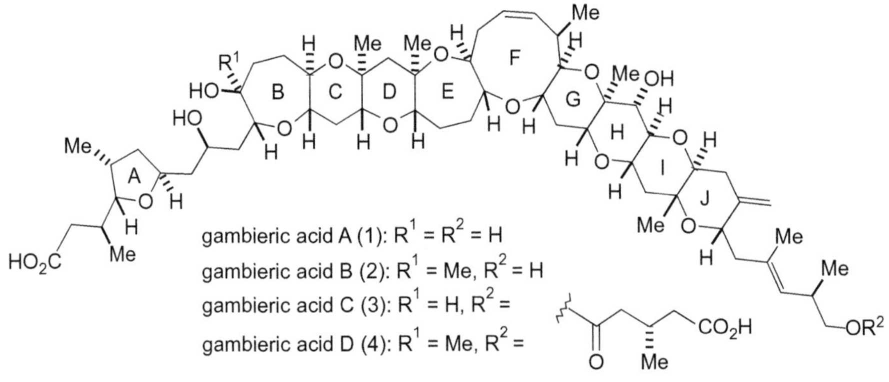

gambieric acid A (1): $R^1 = R^2 = H$
gambieric acid B (2): $R^1 = Me, R^2 = H$
gambieric acid C (3): $R^1 = H, R^2 =$
gambieric acid D (4): $R^1 = Me, R^2 =$

FIGURE 4   Structure of gambieric acid A–D (1–4).

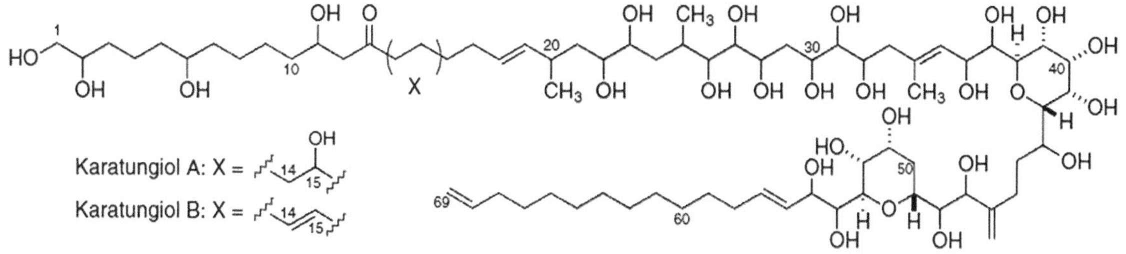

FIGURE 5   Structure of karatungiols A and B (1−4).

dinoflagellate *Karlodinium veneficum*. This organism, first described as *Gymnodinium galatheanum*, was collected from Walvis Bay, Namibia, in 1950 during the second famed Danish *Galathea* expedition (Braarud, 1957). *Gymnodinium veneficum* was collected from the English Channel in the same year. The organism has been associated with fish kills worldwide ever since (Kempton et al., 2002).

The taxonomic identity of the organism has changed several times, with synonyms now including *Gymnodinium/Gyrodinium galatheanum*, *Gymnodinium micrum*, *Gymnodinium veneficum*, and *Karlodinium micrum* (Bergholtz et al., 2006) Several hemolytic, cytotoxic, and ichthyotoxic compounds were first described from then *K. micrum* following an investigation of a large mortality event at HyRock fish farm in Maryland, USA, in 1996.

Karlotoxins appear to function by nonspecifically increasing the ionic permeability of biological membranes, resulting in osmotic cell lysis. They kill fish through damage to sensitive gill epithelial tissues (Deeds et al., 2006). The physiological effects of the karlotoxins suggest similarities with the amphidinols, a series of amphipathic linear polyketides isolated from various species of the dinoflagellate *Amphidinium*.

Karatungiols A and B (Figure 5) were isolated from the cultured marine dinoflagellate *Amphidinium* sp. Using spectroscopic analysis and degradation reactions, the structures of A and B were determined to be novel polyol compounds (Kazuto et al., 2006). Karatungiols were

structurally similar to lingshuiol (Huang et al., 2004) and were amphidinol analogs with a ketone moiety and a terminal saturated alkyl chain moiety, as in lingshuiol. Karatungiol A exhibited potent antifungal activity against NBRC4407 *Aspergillus niger* and antiprotozoan activity against *Tritrichomonas foetus*. Karatungiol A showed antiprotozoan activity against *Trichomonas* sp. as an amphidinol derivative.

3. CONCLUSIONS

Marine microalgae and cyanobacteria are very rich in several chemical compounds. Therefore, they may be used in several biological applications with related health benefits. This chapter reviewed the bioactive compounds produced by marine unicellular algae, as directly or indirectly related to human health. Many species of marine microalgae are capable of producing or accumulating biotoxins, which can accumulate in filter-feeding shellfish. *Dinophysis* spp. are among the very few toxic microalgae that rely on chloroplasts sequestered from their prey. The main microalgae-derived toxic compounds (OA, dinophysistoxin, DA, CTX, gambieric acids, karatungiol A) and bioactives are potentially useful in many applications. These special toxic properties increase the applicability of marine microalgae-derived compounds in biological and biomedical fields, especially in the protection of the marine environments.

References

Adachi, R., Fukuyo, Y., 1979. The thecal structure of a marine toxic dinoflagellate *Gambierdiscus toxicus* gen. et sp. nov. collected in a ciguatera-endemic area. Bull. Jpn. Soc. Sci. Fish 45, 67–71.

Anderson, D., Lobel, P., 1987. The continuing enigma of ciguatera. Biol. Bull. 172 (1), 89.

Anonymous, 2009. Scientific opinion of the panel on contaminants in the food chain on a request from the European Commission on marine biotoxins in shellfish-pectenotoxin group. EFSA J. 1109, 1–47.

Arad, S.(M.), Atar, D., 2007. Viscosupplementation with Algal Polysaccharides in the Treatment of Arthritis. Patent no WO/2007/066340 (Ben Gurion University of the Negev Research and Development Authority), Minneapolis, MN.

Aune, T., Larsen, S., Aasen, J.A., Rehmann, N., Satake, M., Hess, P., 2007. Relative toxicity of dinophysistoxin-2 (DTX-2) compared with okadaic acid, based on acute intraperitoneal toxicity in mice. Toxicon 49 (1), 1–7.

Bagnis, R., Kuberski, T., Laugier, S., 1979. Clinical observations on 3,009 cases of ciguatera (fish poisoning) in the South Pacific. Am. J. Trop. Med. Hyg. 28, 1067–1073.

Blanco, J., Fernández, M., Marino, J., Reguera, B., Míguez, A., Maneiro, J., et al., 1995. From *Dinophysis* spp. toxicity to DSP outbrakes: a preliminary model of toxin accumulation in mussels. In: Lassus, P.A.G., Erard, E., Gentien, P., Marcaillou, C. (Eds.), Harmful Marine Algal Blooms. Lavoisier Science Publishers, Paris, pp. 777–782.

Bates, S.S., Bird, C.J., De Freitas, A.S.W., Foxall, R., Gilgan, M., Hanic, L.A., et al., 1989. Pennate diatom *Nitzschia pungens* as the primary source of domoic acid, a toxin in shellfish from eastern Prince Edward Island, Canada. Can. J. Fish. Aquat. Sci. 46, 1203–1215.

Bergholtz, T., Daugbjerg, N., Moestrup, Ø., Fernandez-Tejedor, M., 2006. On the identity of *Karlodinium veneficum* and description of *Karlodinium armiger* sp. nov. (Dinophyceae), based on light and electron microscopy, nuclear-encoded LSU rDNA, and pigment composition. J. Phycol. 42, 170–193.

Berman, F.W., Murray, T.F., 1997. Domoic acid neurotoxicity in cultured cerebellar granule neurons is mediated predominantly by NMDA receptors that are activated as a consequence of excitatory amino acid release. J. Neurochem. 69 (2), 693–703.

Bialojan, C., Takai, A., 1988. Inhibitory effect of a marine-sponge toxin, okadaic acid, on protein phosphatases. Specificity and kinetics. Biochem. J. 256, 283–290.

Birinyi-Strachan, L.C., Gunning, S.J., Lewis, R.J., Nicholson, G.M., 2005. Block of voltagegated potassium channels by Pacific ciguatoxin-1 contributes to increased neuronal excitability in rat sensory neurons. Toxicol. Appl. Pharmacol. 204, 175–186.

Borowitzka, M.A., Borowitzka, L.J., 1988. Microalgae Biotechnology. Cambridge University Press, Cambridge.

Braarud, T., 1957. A red water organism from Walvis Bay (*Gymnodinium galatheanum n. sp.*). Galathea Rep. 1, 137–138.

Carmody, E.P., James, K.J., Kelly, S.S., 1996. Dinophysistoxin-2: the predominant diarrhoetic shellfish toxin in Ireland. Toxicon 34 (3), 351–359.

Carrera-Martínez, D., Mateos-Sanz, A., López-Rodas, V., Costas, E., 2010. Microalgae response to petroleum spill: an experimental model analysing physiological and genetic response of *Dunaliella tertiolecta* (Chlorophyceae) to oil samples from the tanker Prestige. Aquat. Toxicol. 97, 151–159.

Chinain, M., Darius, H., Ung, A., Fouc, M., Revel, T., Cruchet, P., et al., 2010. Ciguatera risk management in French Polynesia: the case study of Raivavae Island (Australes Archipelago). Toxicon 56 (5), 674–690.

Cohen, P., Holmes, C.F., Tsukitani, Y., 1990. Okadaic acid: a new probe for the study of cellular regulation. Trends Biochem. Sci. 15, 98–102.

Costa, L.G., Giordano, G., Faustman, E.M., 2010. Domoic acid as a developmental neurotoxin. Neurotoxicology 31 (5), 409–423.

Daranas, A.H., Norte, M., Fernandez, J.J., 2001. Toxic marine microalgae. Toxicon 39, 1101–1132.

Darius, H.T., Ponton, D., Revel, T., Cruchet, P., Ung, A., Fouc, M.T., Chinain, M., 2007. Ciguatera risk assessment in two toxic sites of French Polynesia using the receptor-binding assay. Toxicon 50, 612–626.

Deeds, J.R., Reimschuessel, R., Place, A.R., 2006. Histopathological effects in fish exposed to the toxins from *Karlodinium micrum* (Dinophyceae). J. Aquat. Anim. Health 18, 136–148.

DeLorenzo, M.E., Serrano, L., 2003. Individual and mixture toxicity of three pesticides; atrazine, chlorpyrifos, and chlorothalonil to the marine phytoplankton species *Dunaliella tertiolecta*. J. Environ. Sci. Health B 38, 529–538.

Dickey, R.W., Bobzin, S.C., Faulkner, D.J., Bencsath, F.A., Andrzejewski, D., 1990. Identification of okadaic acid from a Caribbean dinoflagellate, *Prorocentrum concavum*. Toxicon 28, 371–377.

Dominguez, H.J., Paz, B., Daranas, A.H., Norte, M., Franco, J.M., Fernandez, J.J., 2010. Dinoflagellate polyether within the yessotoxin, pectenotoxin and okadaic acid toxin groups: characterization, analysis and human health implications. Toxicon 56 (2), 191–217.

Donia, M., Hamann, M.T., 2003. Marine natural products and their potential applications as anti-infective agents (review). Lancet Infect. Dis. 3, 338–348.

Dvir, I., Stark, A.H., Chayoth, R., Madar, Z., Arad, S.(M.), 2009. Hypocholesterolemic effects of nutraceuticals produced from the red microalga *Porphyridium* sp. in rats. Nutrients 1, 15667.

Field, C.B., Behrenfeld, M.J., Randenon, J.T., Falkowski, P.G., 1998. Primary production of the biosphere: integrating terrestrial and oceanic components. Science 281, 237−240.

Fujiki, H., Suganuma, M., 1993. Tumor promotion by inhibitors of protein phosphatases 1 and 2A: the okadaic acid class of compounds. Adv. Cancer Res. 61, 143−194.

Fux, E., Gonzalez-Gil, S., Lunven, M., Gentien, P., Hess, P., 2010. Production of diarrhetic shellfish poisoning toxins and pectenotoxins at depths within and below the euphotic zone. Toxicon 56, 1487−1496.

Goldstein, T., Mazet, J.A., Zabka, T.S., Langlois, G., Colegrove, K.M., Silver, M., et al., 2008. Novel symptomatology and changing epidemiology of domoic acid toxicosis in California sea lions (Zalophus californianus): an increasing risk to marine mammal health. Proc. R. Soc. B 275, 267−276.

Garcia, A., Cayla, X., Guergnon, J., Dessauge, F., Hospital, V., Rebollo, M.P., Fleischer, A., Rebollo, A., 2003. Serine/threonine protein phosphatases PP1 and PP2A are key players in apoptosis. Biochimie 85, 721−726.

Gardeva, E., Toshkova, R., Minkova, K., Gigova, L., 2009. Cancer protective action of polysaccharide derived from microalga Porphyridium cruentum a biological background. Biotechnol. Equip. 23 (2), 783−787.

Gasljevic, K., Hall, K., Chapman, D., Matthys, E.F., 2008. Drag-reducing polysaccharides from marine microalgae: species productivity and drag reduction effectiveness. J. Appl. Phycol. 20, 299−310.

Guzman, S., Gato, A., Lamela, M., Freire-Garabal, M., Calleja, J.M., 2003. Anti-inflammatory and immunomodulatory activities of polysaccharide from Chlorella stigmatophora and Phaeodactylum tricornutum. Phytother. Res. 17, 665−670.

Hamano, Y., Kinoshita, Y., Yasumoto, T., 1986. Enteropathogenicity of diarrhetic shellfish toxins in intestinal models. J. Food Hyg. Soc. Jpn. 27, 375−379.

Holland, P.T., Selwood, A.I., Mountfort, D.O., Wilkins, A.L., McNabb, P., Rhodes, L., et al., 2005. Isodomoic acid C, an unusual amnesic shellfish poisoning toxin from Pseudonitzschia australis. Chem. Res. Toxicol. 18, 814−816.

Holmes, M.J., Lewis, R.J., Jones, A., Wong Hoy, A.W., 1995. Cooliatoxin, the first toxin from Coolia monotis (dinophyceae). Nat. Toxin. 3 (5), 355−362.

Holmes, M., Lewis, R., Poli, M., Gillespie, N., 1991. Strain dependent production of ciguatoxin precursors (gambiertoxins) by Gambierdiscus toxicus (Dinophyceae) in culture. Toxicon 29 (6), 761−775.

Huleihel, M., Ishanu, V., Tal, J., Arad, S.(M.), 2002. Activity of Porphyridium sp. polysaccharide against herpes simplex viruses in vitro and in vivo. J. Biochem. Biophys. Methods 50, 189−200.

Huang, X.C., Zhao, D., Guo, Y.W., Wu, H.M., Lin, L.P., Wang, Z.H., Ding, J., Lin, S., 2004. Lingshuiol, a novel polyhydroxyl compound with strongly cytotoxic activity from the marine dinoflagellate Amphidinium sp. Bioorg. Med. Chem. Lett. 14, 3117−3120.

Jonsson, P.R., Pavia, H., Toth, G., 2009. Formation of harmful algal blooms cannot be explained by allelopathic interactions. Proc. Natl. Acad. Sci. USA 106, 11177−11182.

Katharina, Z., Jennifer, R.D., Marco, C.I., Lindon, S.C., Barbara, N., Peter, J.C., et al., 2013. Analgesic treatment of ciguatoxin-induced cold allodynia. Pain 154, 1999−2006.

Kazuto, W., Tomoyuki, K., Kaoru, Y., Masaki, K., Daisuke, U., 2006. Karatungiols A and B, two novel antimicrobial polyol compounds, from the symbiotic marine dinoflagellate Amphidinium sp. Tetra Lett. 47, 2521−2525.

Kempton, J.W., Lewitus, A.J., Deeds, J.R., Law, J.M., Place, A.R., 2002. Toxicity of Karlodinium mierum (Dinophyceae) associated with a fish kill in a South Carolina brackish retention pond. Harmful Algae 1, 233−241.

Kobayashi, J., Shigemori, H., Ishibashi, M., Yamasu, T., Hirota, H., Sasaki, T., 1991. Amphidinolides G and H: new potent cytotoxic macrolides from the cultured symbiotic dinoflagellate Amphidinium sp. J. Org. Chem. 56 (17), 5221−5224.

Kotaki, Y., Furio, F.F., Lundholm, N., Katayama, T., Koike, K., Fulgueras, V.P., Bajarias, F.A., Takata, Y., Kobayashi, K., Sato, S., Fukuyo, Y., Kodama, M., 2005. Production of isodomoic acids A & B as major toxin components of a pennate diatom Nitzschia navis-varingica. Toxicon 46, 946−953.

Kotaki, Y., Lundholm, N., Onodera, H., Kobayashi, K., Bajarias, F.A., Furio, E.F., Iwataki, M., Fukuyo, Y., Kodama, M., 2004. Wide distribution of Nitzschia navis-varingica, a new domoic acid-producing benthic diatom found in Vietnam. Fish. Sci. 70, 28−32.

Lago, J., Santaclara, F., Vieites, J.M., Cabado, A.G., 2005. Collapse of mitochondrial membrane potential and caspases activation are early events in okadaic acid-treated Caco-2 cells. Toxicon 46, 579−586.

Lee, J.S., Igarashi, T., Fraga, S., Dahl, E., Hovgaard, P., Yasumoto, T., 1989. Determination of diarrhetic shellfish toxins in various dinoflagellate species. J. Appl. Phycol. 1 (2), 147−152.

Le Hégarat, L., Fessard, V., Poul, J.M., Dragacci, S., Sanders, P., 2004. Marine toxin okadaic acid induces aneuploidy in CHO-K1 cells in presence of rat liver postmitochondrial fraction, revealed by cytokinesis-block micronucleus assay coupled to FISH. Environ.Toxicol. 19, 123−128.

Lewis, R.J., 2001. The changing face of ciguatera. Toxicon 39, 97−106.

Lundholm, N., Bates, S.S., Baugh, K.A., Brian, D., Bill, B.D., Connell, L.B., Léger, C., Trainer, V.L., 2012. Cryptic and pseudo-cryptic diversity in diatoms—with descriptions of Pseudo-nitzschia hasleana sp. nov. and P. fryxelliana sp. nov. J. Phycol. 48, 436−454.

Lupescu, N., Geresh, S., Arad, S.(M.), Bernstein, M., Glaser, R., 1991. Structure of some sulfated sugars isolated after acid hydrolysis of the extracellular polysaccharide of *Porphyridium* sp. unicellular red alga. Carbohydr. Res. 210, 349–352.

Maranda, L., Wang, R., Masuda, K., Shimizu, Y., 1990. Investigation of the source of domoic acid in mussels. In: Granéli, E., Sundström, B., Edler, L., Anderson, D.M. (Eds.), Toxic Marine Phytoplankton. Elsevier Science Publishing Co., Inc., New York, pp. 300–304.

Miles, C.O., 2007. Pectenotoxins. Chapter 9. In: Botana, L. (Ed.), Phycotoxins—Chemistry and Biochemistry. Blackwell Publishing, IA, pp. 159–186.

Morohashi, A., Satake, M., Nagai, H., Oshima, Y., Yasumoto, T., 2000. The absolute configuration of gambieric acid A-D, potent antifungal polyethers, isolated from the marine dinoflagellate *Gambierdiscus toxicus*. Tetrahedron 56, 8995–9001.

Morohashi, A., Satake, M., Yasumoto, T., 1998. The absolute configuration of gambierol, a toxic marine polyether from the dinoflagellate, gambierdiscus toxicus. Tetrahedron Lett. 39, 97–100.

Morrison, J.H., Hof, P.R., 1997. Life and death of neurons in the aging brain. Science 278 (412), 412–419.

Muha, N., Ramsdell, J.S., 2011. Domoic acid induced seizures progress to a chronic state of epilepsy in rats. Toxicon 57, 168–171.

Muller-Feuga, A., Moal, J., Kaas, R., 2003. The microalgae for aquaculture. In: Stottrup, J.G., McEvoy, L.A. (Eds.), Life Feeds in Marine Aquaculture. Blackwell, Oxford.

Murakami, Y., Oshima, Y., Yasumoto, T., 1982. Identification of okadaic acid as a toxic component of a marine dinoflagellate *Prorocentrum lima*. Bull. Jpn. Soc. Sci. Fish 48, 69–72.

Nagai, H., Murata, M., Torigoe, K., Satake, M., Yasumoto, T., 1992a. Gambieric acids: new potent antifungal substances with unprecedented polyether structures from a marine dinoflagellate *Gambierdiscus toxicus*. J. Org. Chem. 57, 5448–5453.

Nagai, H., Torigue, K., Satake, M., Murata, M., Yasumoto, T., Hirota, H., 1992b. Gambieric acids: unprecedented potent antifungal substances isolated from cultures of a marine dino-agellate *Gambierdiscus toxicus*. J. Am. Chem. Soc. 114, 1102–1103.

Okumura, Y., Koyama, J., Takaku, H., Satoh, H., 2001. Influence of organic solvents on the growth of marine microalgae. Arch. Environ. Contam. Toxicol. 41, 123–128.

Pearn, J., 2001. Neurology of ciguatera. J. Neurol. Neurosurg. Psychiatry 70, 4–8.

Perez-Gomez, A., Garcia-Rodriguez, A., James, K.J., Ferrero-Gutierrez, A., Novelli, A., Fernandez-Sanchez, M.T., 2004. The marine toxin dinophysistoxin-2 induces differential apoptotic death of rat cerebellar neurons and astrocytes. Toxicol. Sci. 80 (1), 74–82.

Pizarro, G., Escalera, L., Gonzalez-Gil, S., Franco, J.M., Reguera, B., 2008. Growth, behaviour and cell toxin quota of *Dinophysis acuta* during a daily cycle. Mar. Ecol.-Prog. Ser. 353, 89–105.

Pizarro, G., Paz, B., Gonzalez-Gil, S., Franco, J.M., Reguera, B., 2009. Seasonal variability of lipophilic toxins during a *Dinophysis acuta* bloom in Western Iberia: differences between picked cells and plankton concentratos. Harmful Algae 8, 926–937.

Pshenichkin, S.P., Wise, B.C., 1995. Okadaic acid increases nerve growth-factor secretion, messenger-RNA stability, and gene-transcription in primary cultures of cortical astrocytes. J. Biol. Chem. 270, 5994–5999.

Pulz, O., Gross, W., 2004. Valuable products from biotechnology of microalgae. Appl. Microbiol. Biotechnol. 65, 635–648.

Quilliam, M.A., Gilgan, M.W., Pleasance, S., deFreitas, A.S.W., Douglas, D., Fritz, L., et al., 1993. Confirmation of an incident of diarrhetic shellfish poisoning in eastern Canada. In: Smayda, T., Shimuzu, Y. (Eds.), Toxic Phytoplankton Blooms in the Sea. Elsevier, Amsterdam, pp. 547–552.

Raposo, M.F.J., Morais, R.M.S.C., Morais, A.M.M.B., 2013. Bioactivity and applications of sulphated polysaccharides from marine microalgae (review). Mar. Drugs 11 (1), 233–252.

Rivas, M., García, C., Liberona, J.L., Lagos, N., 2000. Biochemical characterization and inhibitory effects of dinophysistoxin-1, okadaic acid and microcystine L-R on protein phosphatase 2A purified from the mussel *Mytilus chilensis*. Biol. Res. 33, 197–206.

Romero, M.L.J., Kotaki, Y., Lundholm, N., Thoha, H., Ogawa, H., Relox, J.R., et al., 2011. Unique amnesic shellfish toxin composition found in the South East Asian diatom *Nitzschia navis-varingica*. Harmful Algae 10, 456–462.

Samson, G., Popovic, R., 1988. Use of algal fluorescence for determination of phytotoxicity of heavy metals and pesticides as environmental pollutants. Ecotoxicol. Environ. Saf. 16, 272–278.

Schnorf, H., Taurarii, M., Cundy, T., 2002. Ciguatera fish poisoning: a double-blind randomized trial of mannitol therapy. Neurology 58, 873–880.

Sirenko, L.A., Kirpenko, Y.A., Kirpenko, N.I., 1999. Influence of metabolites of certain algae on human and animal cell cultures. Int. J. Algae 1, 122–126.

Souid-Mensi, G., Moukha, S., Mobio, T.A., Maaroufi, K., Creppy, E.E., 2008. The cytotoxicity and genotoxicity of okadaic acid are cell-line dependent. Toxicon 51, 1338–1344.

Strachan, L.C., Lewis, R.J., Nicholson, G.M., 1999. Differential actions of pacific ciguatoxin-1 on sodium channel subtypes in mammalian sensory neurons. J. Pharmacol. Exp. Ther. 288, 379–388.

Suganuma, M., Fujiki, H., Suguri, H., Yoshizawa, S., Hirota, M., Nakayasu, M., Ojika, M., Wakamatsu, K., Yamada, K., Sugimura, T., 1988. Okadaic acid: an additional non-phorbol-12-tetradecanoate-13-acetate-type tumor promoter. Proc. Natl Acad. Sci. USA 85, 1768—1771.

Sugg, L.M., VanDolah, F.M., 1999. No evidence for an allelopathic role of okadaic acid among ciguatera-associated dinoflagellates. J. Phycol. 35, 93—103.

Suzuki, T., Ota, H., Yamasaki, M., 1999. Direct evidence of transformation of dinophysistoxin-1 to 7-O-acyl-dinophysistoxin-1 (dinophysistoxin-3) in the scallop *Patinopecten yessoensis*. Toxicon 37, 187—198.

Takai, A., Bialojan, C., Troschka, M., Ruegg, J.C., 1987. Smooth muscle phosphatase inhibition and force enhancement by block sponge toxin. FEBS Lett. 217, 81—84.

Takai, A., Murata, M., Torigoe, K., Isobe, M., Mieskes, G., 1992. Inhibitory effects of okadaic acid derivatives on protein phosphatases: a study on structure—affinity relationship. Biochem. J. 284, 539—544.

Taniyama, S., Arakawa, O., Terada, M., Nishio, S., Takatani, T., Mahmud, Y., Noguchi, T., 2003. *Ostreopsis* sp., a possible origin of palytoxin (PTX) in parrotfish *Scarus ovifrons*. Toxicon 42 (1), 29—33.

Trainer, V.L., Bates, S.S., Lundholm, N., Thessen, A.E., Cochlan, W.P., Adams, N.G., Trick, C.G., 2012. *Pseudonitzschia* physiological ecology, phylogeny, toxicity, monitoring and impacts on ecosystem health. Harmful Algae 14, 271—300.

Vale, C., Botana, L.M., 2008. Marine toxins and the cytoskeleton: okadaic acid and dinophysistoxins. FEBS J. 275, 6060—6066.

Valdiglesias, V., Laffon, B., Pásaro, E., Méndez, J., 2011a. Okadaic acid induces morphological changes, apoptosis and cell cycle alterations in different human cell types. J. Environ. Monit. 13, 1831—1840.

Valdiglesias, V., Laffon, B., Pásaro, E., Méndez, J., 2011b. Okadaic acid-induced genotoxicity in human cells evaluated by micronucleus test and γH2AX analysis. J. Toxicol. Environ. Health 74, 980—992.

Valdiglesias, V., Méndez, J., Pásaro, E., Cemeli, E., Anderson, D., Laffon, B., 2011c. Induction of oxidative DNA damage by the marine toxin okadaic acid depends on human cell type. Toxicon 57, 882—888.

Valdiglesias, V., Méndez, J., Pásaro, E., Cemeli, E., Anderson, D., Laffon, B., 2010. Assessment of okadaic acid effects on cytotoxicity, DNA damage and DNA repair in human cells. Mutat. Res. 689, 74—79.

Vale, P., Sampayo, M.A., 2000. Dinophysistoxin-2: a rare diarrhoeic toxin associated with *Dinophysis acuta*. Toxicon 38 (11), 1599—1606.

Van Dolah, F.M., 2000. Marine algal toxins: origins, health effects, and their increased occurrence. Environ. Health Perspect. 108, 133—141.

Windust, A.J., Wright, J.L.C., McLachlan, J.L., 1996. The effects of the diarrhetic shellfish poisoning toxins, okadaic acid and dinophysistoxin-1, on the growth of microalgae. Mar. Biol. 126, 19—25.

Wright, J.L.C., Boyd, R.K., De Freitas, A.S.W., Falk, M., Foxall, R.A., Jamieson, W.D., et al., 1989. Identification of domoic acid, a neuroexcitatory amino acid, in toxic mussels from eastern Prince Edward Island. Can. J. Chem. 67, 481—490.

Yasumoto, T., 2005. Chemistry, etiology, and food chain dynamics of marine toxins. Proc. Jpn. Acad. Ser. B 81, 43—51.

Yasumoto, T., Murata, M., Oshima, Y., Sano, M., Matsumoto, G.K., Clardy, J., 1985. Diarrhetic shellfish toxins. Tetrahedron 41, 1019—1025.

36

Toxicity Bioassays on Benthic Diatoms

Cristiano V.M. Araújo[1,2], Ignacio Moreno-Garrido[1]

[1]Institute of Marine Sciences of Andalusia (CSIC), Research group of Ecotoxicology, Ecophysiology and Biodiversity of Aquatic Systems, Excellence International Campus of the Seas (CEIMAR), Puerto Real, Cádiz, Spain; [2]Universidad Laica Eloy Alfaro de Manabí (ULEAM), Central Department of Research (DCI), Manta, Ecuador

1. INTRODUCTION

At the end of World War II, the old stated dilution paradigm (the solution for pollution is dilution) began to be replaced by the boomerang paradigm (everything you throw away can come back and beat you) (Newman and Unger, 2003). Concern about ecotoxicology slowly took place in governments and populations, and regulatory guidelines began to be legislated. In marine environments, sediments can act as sink but also as a future source of pollutants (Chapman and Wang, 2001; Radakovitch et al., 2008). In spite of this, little attention has been paid to the effects of pollutants on the microflora that inhabits the sediments (Langston et al., 2010).

Microalgae are responsible for more than a half of total carbon primary production in our planet (Falkowski, 1980). Any process able to alter microalgal populations will affect trophic nets in aquatic systems. Microphytobenthos (microscopic algae that inhabit the first millimeters of the submerged or intertidal sediments) (MacIntyre et al., 1996; Miller et al., 1996) play a key role in the coastal trophic aquatic nets (Araújo et al., 2010a) as well as in the geological coastal dynamics, as exopolysaccharides extruded by these organisms confer

stability to fine sediments (de Brower et al., 2000). In freshwater, the importance of epipelic microflora (the term *microphytobenthos* is related more often to estuarine and marine environments) as a key compartment in nutrient recycling and pollutant dynamics is also recognized (Admiraal et al., 1999).

The need for the establishment of normalized and reproducible methods to measure toxicity of sediments is a keystone in ecotoxicology (Simpson et al., 2005). This chapter provides a brief survey of the normalized planktonic microalgal toxicity test; a description of a revision on the rationale, previous attempts, and potential problems for sediment toxicity tests on sediments to microalgae; and a proposal for a guideline for a sediment bioassay on benthic diatoms. Finally, a survey on the new perspectives in benthic diatom bioassays, including new endpoints and methodologies, is presented.

2. TOXICITY BIOASSAYS ON MICROALGAE

First attempts to measure the effects of pollution on microalgae were done in the first decade

Handbook of Marine Microalgae
http://dx.doi.org/10.1016/B978-0-12-800776-1.00036-4

of the twentieth century, but it was not until the mid-1960s that microalgal bioassays methods were validated and published. From the pioneering Algal Assay Procedure Bottle Test (US EPA, 1971), designed in order to detect aquatic eutrophication, many guidelines for microalgal toxicity tests have been developed (Rand, 1995).

Basically, the current standard microalgal bioassays for substances and effluents involve batch cultures of 50 mL in borosilicate flasks with an initial cellular density of 10^4 cells mL^{-1}, continuous light, temperatures of $20 \pm 1\,°C$ and 72 h incubation, with four total cellular density counts (day 0, 1, 2, and 3). A good example of these tests can be found in ISO (1995) for seawater (involving *Phaeodactylum tricornutum* or *Skeletonema costatum*) and in OECD (1998) for freshwater environments (involving *Raphidocelis subcapitata*, formerly known as *Selenastrum capricornutum* and *Pseudokirchneriella subcapitata*, *Scenedesmus subspicatus*, or *Chlorella vulgaris*). Culture media and certain other parameters could vary slightly, but some validation criteria stands in almost all the guidelines: cell density in controls should increase at least 16 times in 72 h and pH value should not vary more than one unit. Added nutrients must be carefully watched as some chelators such as EDTA can interfere with toxicity of some pollutants (metals, for instance). Bioassay data treatments for microalgae are clearly and substantially discussed in the reference work of Nyholm (1990).

Batch cultures for toxicity tests have often been criticized—for good reasons. Toxicity of substances is often underestimated when those protocols are used, as part of the toxicants can be removed from the media by adsorption to vial walls or the surfaces of living or dead cells. In the case of metals, for instance, adsorption to the cellular surface seems to be a very quick process (Garnham et al., 1992) and real metal concentration in culture media can dramatically decrease in a few hours or even minutes. Cells can also accumulate toxicants in special structures in their cytoplasm (Maeda and Sakaguchi,

1990), and a limited quantity of atoms or molecules of the toxicant in a closed environment removes the results from environmentally relevant data. Pollutants can also be inactivated by cellular exudates or metabolized by microalgae (Ozturk et al., 2010). Associated bacterial flora in non-axenic cultures can also interfere with toxicity results (Levy et al., 2009), with the same result. Some of those problems could be solved with the design of continuous-flow toxicity tests (Chen et al., 1997), although other problems could appear, such as the dilution of the cellular density. This can be avoided by cellular immobilization, but new adsorption surfaces for pollutants in the immobilization matrix are then provided and immobilization can protect cells from toxicity (Moreno-Garrido, 2008). Thus, in spite of their limitations, microalgal bioassays in batch cultures can provide useful and comparable information about the toxicity of substances and effluents.

3. SEDIMENT TOXICITY TESTING ON MICROALGAE: RATIONALE, PREVIOUS ATTEMPTS, AND POTENTIAL PROBLEMS

Even though batch toxicity bioassays present clear limitations (discussed above), environment mimetic conditions in bioassays should be maintained as far as possible, and bioassays on extracts, elutriates, or interstitial water could not be as complete as direct toxicity tests on whole sediments. Thus, the design of bioassays involving the photosynthetic microorganisms that inhabit sediments (microphytobenthos) and low altered sediments was lacking and needed to be developed. The use of laboratory strains instead of local natural microalgal assemblages is strongly recommended as adaptation to toxicity can occur in local populations submitted to pollutant pressure (Erickson et al., 1984; García Balboa et al., 2013).

Some benthic microalgae have already been used in toxicity bioassays. Stauber and Florence

(1985) determined the influence of iron in copper toxicity to the benthic diatom *Cylindrotheca closterium*. Other research by the same authors determined the toxic mechanisms of copper on this species (Florence and Stauber, 1986; Stauber and Florence, 1987). In later years, attention was also paid to other benthic diatom species such as *Entomoneis* cf. *punctulata* or *Nitzschia* cf. *paleacea*, as they showed better results when flow-cytometry and esterase inhibition techniques were used (Franklin et al., 2001). All of those studies on benthic diatoms were not performed in sediments but in aqueous matrixes. First attempts at designing bioassays on whole sediments were carried out on artificial sediments spiked with metals or surfactants by Moreno-Garrido et al. (2003a,b, 2006) and Mauffret et al. (2010) as well as on and natural polluted sediments (Moreno-Garrido et al., 2006). In these cases, *C. closterium* was selected as the target organism. Those incipient bioassays determined that particle-size distribution could affect population growth if cultures were shaken daily, but this effect did not appear if cells were gently disposed over the sediments and a single cell count was performed at the end of the test (72 h). Normalized bioassays on phytoplankton stated an initial cellular density of 10,000 cells mL^{-1}, as this is the lower threshold of counting for Neubauer counting chambers. In bioassays involving planktonic species this density is not completely realistic as it is quite elevated, but in sediment bioassays, assuming 3 cm of radius in the borosilicate flasks used, this cellular density will rend a cellular surface concentration of 1.8×10^3 cells cm^{-2}, which is within the range of the cellular concentrations found in natural locations (Delgado, 1989).

However, the good results obtained on artificial sediments were not expected on natural sediments from real locations, as some problems can arise when this material is used. First, redox conditions of the sediment can alter the pH of the used media. Therefore, only surface sediment (first 10–50 mm) (Lohse et al., 1995) should be used for these bioassays, as this is the natural habitat of microphytobenthos. Second, in natural sediments local flora and, generally local fauna, could greatly influence the laboratory microalgal population growth by nutrient competition or directly by predation. Thus, those populations must be eliminated from the sediments altering the sediment as little as possible. Finally, benthic microalgae are similar in size and Stokes numbers (sedimentation velocities) to a large part of the particles present in fine sediment, and thus it is not easy to separate the microalgal populations from sediment particles. Generally, in fine sediments counts at bright-field microscopy of the microalgal population can present difficulties as discrimination of cells is not easy for nontrained eyes. All of these problems need to be solved when toxicity tests on natural sediments are performed. This will be discussed in the following section. The benthic diatoms (Bacillariophyceae, order Pennales) *C. closterium* (for freshwater environments) and *Nitzschia palea* (for freshwater environments) have demonstrated to be good target organisms for whole-sediment bioassays. For estuarine environments, the limit of salinity between these two species seems to be around 12, salinity that does not imply an inhibition of growth higher than 20% with respect to the assumed optimal salinity for each species (36 for the marine species and 0 for the freshwater species) (Figure 1).

4. SEDIMENT TOXICITY BIOASSAY ON BENTHIC DIATOMS: A GUIDELINE

4.1 Sediment Sampling and Preparation

As mentioned, sediment bioassays involving microphytobenthos are restricted to surface sediments. Thus, if specific devices such as those described in Araújo et al. (2009) are not available, attention must be paid when sediments are sampled and care must be taken in order

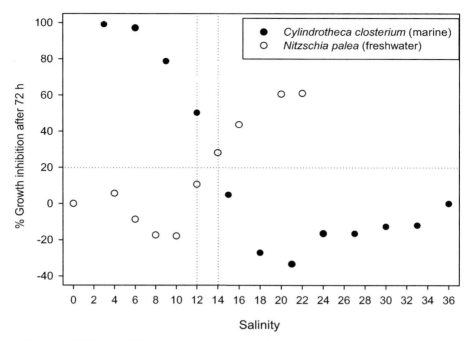

FIGURE 1   Growth inhibition for *Cylindrotheca closterium* (●) and *Nitzschia palea* (○) after 72 h when cultured at different salinities. Control (assumed 0% inhibition) for *C. closterium* is at salinity of 36; control for *N. palea* is at salinity of 0. Salinity of around 12–14 does not imply higher inhibitions from 20% (dotted horizontal line) (figure is original from the authors).

to collect only the first sediment layer. Samples should be stored in a dark, cold place (a refrigerator) until treated at the laboratory. Once in the laboratory, sediment samples should be mechanically homogenized with a nonmetallic spatula and submitted to a frozen pulse by immersion in liquid nitrogen in order to eliminate microfauna and microflora (Araújo et al., 2009, 2010b,c). It should be noted that this is not a method for sediment storage, as sediment conditions will change with time (Malueg et al., 1986; Araújo et al., 2008). This pulse can be performed dropping liquid nitrogen on a tray containing the sediment or, more properly, submerging plastic vials containing the sediment in nitrogen for 5 min. After this, sediments must be immediately thawed. Subsamples should be taken in order to measure, as soon as possible, water content of the sediment, as equivalence of dry sediment will be used in

the bioassays. For control flasks, acid-washed sand from a nonpolluted area should be used.

4.2 Bioassay

Bioassays will be performed by triplicate. Properly washed borosilicate flasks of 125-mL capacity, topped with artificial cotton (Perlon) will be used. Wet sediment weight equivalent to 2 g of dry sediment will be disposed in each flask, and 45 mL of synthetic marine or freshwater media will be added. The use of synthetic media is recommended: ASTM Substitute Ocean Water (ASTM, 1975), enriched with NO_3^- (150 mg L^{-1}), PO_4^{3-} (10 mg L^{-1}) and SiO_2 (50 mg L^{-1}) can be used for marine microalgae. Media similar to those described in Fábregas et al. (2000) can be used for freshwater species, but metals and EDTA must not be used in all the cases for the fabrication of experimental media in order to avoid potential chelation of metals present in sediments. Thus, aquatic experimental media and sediments must be mixed, pH values adjusted to 8.2 in marine media and to 7.0 in freshwater media, and then sedimentation of the particles must be permitted for at least 1 h.

Three-day-old (exponential-growing) cultures of the diatoms should be used for the bioassays: appropriate volumes should be gently centrifuged and resuspended two times in the experimental media in order to remove the stock culture media. Care should be taken when centrifuging some microalgal cells: high speeds or long centrifuging times can damage cells that would be apparently unharmed after centrifugation. For the two species recommended, (*C. closterium* and *N. palea*) 0.6 rfc and 4 min is enough to ensure good cell recovery and cellular integrity.

After sediment decantation, 5 mL of concentrated culture with a fixed cellular density of 0.1×10^6 cells mL^{-1} should be gently dropped on the flasks without disturbing the deposited sediment. Flasks must be placed for 72 h in culture chambers under continuous light (around $40 \, \mu mol \, m^{-2} \, s^{-1}$) and $20 \pm 1 \, °C$ and not shacked until the end of the bioassay.

Cellular counts, after vigorously shaking the flasks, will be performed in Neubauer-type counting chambers, after 72-h incubation, by the use of fluorescence microscopy (Moreno-Garrido et al., 2007b). The use of excitation blue light and a barrier filter of 530 nm permits to discriminate the living microalgae by the intense red fluorescence of the chloroplasts. Assuming 100% growth of the cellular concentration of controls after incubation, percentage of growth inhibition can be calculated for a determined sample.

The present bioassay used to assess whole-sediment toxicity on benthic diatoms was applied in a ring test in which six laboratories in Spain and Portugal participated (Araújo et al., 2010d). Data indicated that the whole-sediment assay on *C. closterium* was considered

sufficiently successful for possible use as a standard toxicity test due to ecological relevance, feasibility, and good interlaboratory reproducibility.

5. NEW PERSPECTIVES IN BENTHIC DIATOM BIOASSAYS: IN SITU BIOASSAYS AND AVOIDANCE BIOASSAYS

In the 1990s, a new approach for toxicity testing was adopted in order to increase the ecological relevance of the toxicity bioassays. The idea was to carry the target organisms to the field instead of taking samples to the laboratory and there expose the organisms to the sampled water or sediment (Pereira et al., 2000). The development of these so-called *in situ* bioassays has a clear restriction due to the small size of the involved organisms. Munawar and Munawar (1987) tried to use dialysis-membrane chambers containing algae but some problems arose from the difficulty of water exchange in low-energy aquatic systems and the material in suspension in high-energy aquatic systems, which could cover the pores of the membrane. A possible solution is the use of calcium-alginate immobilized microalgal cells (Moreira dos Santos et al., 2002). Many microalgal species (including diatoms) have been shown to be able to live and grow inside this matrix (Moreno-Garrido et al., 2005). In spite of this, many problems, such as the covering of the calcium-alginate beads by the tide-moving sediment, need to be solved in order to translate the good results from the laboratory to the field (Moreno-Garrido et al., 2007a).

Avoidance assays are a recent approach used in ecotoxicology to assess the ability of organisms to detect contamination and move toward a less-contaminated environment. Recently, assays in which avoidance has been used as an endpoint have employed a multicompartmental system where a contamination gradient is formed and organisms can freely move and choose less-contaminated compartments (Araújo et al., 2014a). Avoidance has been studied on freshwater organisms with considerable active displacement (e.g., cladocera, tadpoles, fish), and results have shown the high sensitivity and reliability of this response to different contaminants (Rosa et al., 2012; Araújo et al., 2014a,b). For organisms that are not expected to move quickly, drift, characterized by the displacement of the organisms by means of the water flow (Brittain and Eikeland, 1988), may be an alternative mechanism to move sufficiently long distances. Benthic diatoms are organisms that present different modalities of movement (Apoya-Horton et al., 2006), but they are not distinguished by high displacement velocity. Therefore, drift could be an alternative to run much more quickly if environmental conditions are disturbed. The first approach employing drift as response to stressful conditions developed on the estuarine benthic microalgae *C. closterium* was studied by Araújo et al. (2013). The ability of *C. closterium* to drift when exposed to stressful decreasing salinity was compared with standard population growth inhibition assays; based on this, both growth and drift responses were combined to estimate how population could decline (Araújo et al., 2013). These authors showed that the species responded by drifting when exposed to stress by salinity; nonetheless, population growth was slightly more sensitive. Although effects on population growth have previously been the most frequent endpoint used in toxicity testing, drift may also have important effects on the population dynamics of benthic diatoms, as a mechanism for avoiding a less favorable area. Therefore, given that spatial distribution of benthic microalgae could be partially determined by disturbance level that could trigger displacement by drifting to different environments, the use of drift response of benthic diatoms to assess effects of contamination should be encouraged.

References

Admiraal, W., Blanck, H., Buckert-de Jong, M., Guasch, H., Ivorra, N., Lehmann, V., Nyström, B.A.H., Paulsson, M., Sabater, S., 1999. Short-term toxicity of zinc to microbenthic algae and bacteria in a metal polluted stream. Wat. Res. 33 (9), 1989—1996.

Apoya-Horton, M.D., Yin, L., Underwood, G.J.C., Gretz, M.R., 2006. Movement modalities and response to environmental changes of the mudflat diatom *Cylindrotheca closterium* (Bacillariophyceae). J. Phycol. 42, 379—390.

Araújo, C., Blasco, J., Moreno-Garrido, I., 2010a. Microphytobenthos in ecotoxicology: a review of the use of marine benthic diatoms in bioassays. Environ. Int. 36, 637—646.

Araújo, C.V.M., Shinn, C., Mendes, L.B., Delello-Schneider, D., Sanchez, A.L., Espíndola, E.L.G., 2014a. Avoidance response of *Danio rerio* to a fungicide in a linear contamination gradient. Sci. Tot. Environ. 484, 35—42.

Araújo, C.V.M., Shinn, C., Moreira-Santos, M., Lopes, I., Espíndola, E.L.G., Ribeiro, R., 2014b. Copper-driven avoidance and mortality by temperate and tropical tadpoles. Aquat. Tox. 146, 70—75.

Araújo, C.V.M., Diz, F.R., Laiz, L., Lubián, L.M., Blasco, J., Moreno-Garrido, I., 2009. Sediment integrative assessment of the Bay of Cádiz (Spain): an ecotoxicological and chemical approach. Environ. Int. 35, 831—841.

Araújo, C.V.M., Diz, F.R., Lubián, L.M., Blasco, J., Moreno Garrido, I., 2010b. Sensitivity of *Cylindrotheca closterium* to copper: Influence of three endpoints and two test methods. Sci. Tot. Environ. 408, 3696—3703.

Araújo, C.V.M., Diz, F.R., Tornero, V., Lubián, L.M., Blasco, J., Moreno-Garrido, I., 2010c. Ranking sediment samples from three Spanish estuaries in relation to its toxicity for two benthic species: the microalga *Cylindrotheca closterium* and the copepod *Tisbe battagliai*. Environ. Tox. Chem. 29 (2), 393—400.

Araújo, C.V.M., Moreno-Garrido, I., Diz, F.R., Lubián, L.M., Blasco, J., 2008. Effects of cold-dark storage on growth of *Cylindrotheca closterium* and its sensitivity to copper. Chemosphere 72, 1366—1372.

Araújo, C.V.M., Romero-Romero, S., Lourençato, L.F., Moreno-Garrido, I., Blasco, J., Gretz, M.R., Moreira-Santos, M., Ribeiro, R., 2013. Going with the flow: detection of drift in response to hypo-saline stress by the estuarine benthic diatom *Cylindrotheca closterium*. Plos One 8 (1), e81073, 1—7.

Araújo, C.V.M., Tornero, V., Lubián, L.M., Blasco, J., van Bergeijk, S.A., Cañavate, P., Cid, A., Franco, D., Prado, R., Bartual, A., Gil-López, M., Ribeiro, R., Moreira dos Santos, M., Torreblanca, A., Jurado, B., Moreno-Garrido, I., 2010d. Ring test for whole-sediment toxicity assay with a benthic marine diatom. Sci. Tot. Environ. 408, 822—828.

ASTM (American Standard for Testing and Materials), 1975. Standard Specification for Substitute Ocean Water. Designation D 1141−75.

Brittain, J.E., Eikeland, T.J., 1988. Invertebrate drift—a review. Hydrobiologia 166, 77−93.

Chapman, P.M., Wang, F., 2001. Assessing sediment contamination in estuaries. Environ. Tox. Chem. 20 (1), 3−22.

Chen, Ch-Y., Lin, K.-C., Yang, D.T., 1997. Comparison of the relative toxicity relationships based on batch and continuous algal toxicity tests. Chemosphere 35 (9), 1959−1965.

de Brower, J.F.C., Bjelic, S., Deckere, E.M.G.T., Stal, L.J., 2000. Interplay between biology and sedimentology in a mudflat (Biezelingse Ham, Westerchelde, The Netherlands). Cont. Shelf. Res. 20, 1159−1177.

Delgado, M., 1989. Abundance and distribution of microphytobentos in the bays of Ebro delta (Spain). Estuarine, Coastal Shelf Sci. 29, 183−194.

Erickson, J.M., Rahire, M., Bennoun, P., Delepelaire, P., Diner, B., Rochaix, J.-D., 1984. Herbicide resistance in *Chlamydomonas reindhartii* results from a mutation in the chloroplast gene for the 32-kiloDalton protein of photosystem II. PNAS 81, 3617−3621.

Fábregas, J., Domínguez, A., García-Álvarez, D., Lamela, T., Otero, A., 2000. Optimization of cultura médium for the continuous cultivation of the microalga *Haematococcus pluvialis*. Appl. Microbiol. Biotechnol. 53, 530−535.

Falkowsky, P.G., 1980. Primary productivity in the sea. Environ. Sci. Res., 19. In: Falkowski (Ed.), Plenum press, New York and London.

Florence, T.M., Stauber, J.L., 1986. Toxicity of copper complexes to the marine diatom *Nitzschia closterium*. Aquat. Tox. 8, 223−229.

Franklin, N.M., Stauber, J.L., Apte, S.C., Lim, R.P., 2001. Development of flow cytometry-based algal bioassays for assessing toxicity of copper in natural waters. Environ. Tox. Chem. 20 (1), 160−170.

García-Balboa, C., Baselga-Cervera, B., García-Sánchez, A., Igual, J.M., Lopez-Rodas, V., Costas, E., 2013. Rapid adaptation of microalgae to bodies of water with extreme pollution from uranium mining: an explanation of how mesophilic organisms can rapidly colonise extremely toxic environments. Aquat. Tox. 144−145, 116−123.

Garnham, G.W., Codd, G.A., Gadd, G.M., 1992. Kinetics of uptake and intracellular location of cobalt, manganese and zinc in the estuarine green alga *Chlorella salina*. Appl. Microbiol. Biotechnol. 37, 270−276.

ISO, 1995. ISO 10253:1995(E). Water Quality—Marine Algal Growth Inhibition Test with *Skeletonema Costatum* and *Phaeodactylum Tricornutum*, pp. 1−8.

Langston, W.J., Pope, N.D., Jonas, P.J.C., Nikitic, C., Field, M.D.R., Dowell, B., Shillabeer, N., Swarbrick, R.H., Brown, A.R., 2010. Contaminants in fine sediments and their consequences for biota of the Severn Estuary. Mar. Pollut. Bull. 61, 68−82.

Levy, J., Stauber, J.L., Wakelin, S.A., Jolley, D., 2009. The effect of bacteria on the sensitivity of microalgae to copper in laboratory bioassays. Chemosphere 74, 1266−1274.

Lohse, L., Epping, E.H.G., Helder, W., van Raaphorst, W., 1995. Oxygen pore water profiles in continental shelf sediments of the North Sea: turbulent versus molecular diffusion. Mar. Ecol. Prog. Ser. 145, 63−75.

MacIntyre, H.L., Geider, R.J., Miller, D.C., 1996. Microphytobenthos: the ecological role of the "secret garden" of unvegetated, shallow-water marine habitats. I. Distribution, abundance and primary production. Estuaries 19 (2A), 186−201.

Maeda, S., Sakaguchi, T., 1990. Accumulation and detoxification of toxic elements by algae. In: Akatsuka, I. (Ed.), Introduction to Applied Phycology. SPB Academic Publishing bv, The Hage, The Netherlands.

Malueg, K.W., Schuytema, G.S., Krawczyk, D.F., 1986. Effects of sample storage on a copper-spiked freshwater sediment. Environ. Tox. Chem. 5, 245−253.

Mauffret, A., Moreno-Garrido, I., Blasco, J., 2010. The use of marine benthic diatom in a growth inhibition test with spiked whole-sediment. Ecotox. Environ. Safe 73, 262−269.

Miller, D.C., Geider, R.J., MacIntyre, H.L., 1996. Microphytobenthos: the ecological role of the "secret garden" of unvegetated, shallow-water marine habitats. II. Role in sediment stability and shallow-water food webs. Estuaries 19 (2A), 202−212.

Moreira dos Santos, M., Moreno-Garrido, I., Gonçalves, F., Soares, A.M.V.M., Ribeiro, R., 2002. An in situ bioassay with microalgae for estuarine environments. Environ. Tox. Chem. 21 (3), 567−574.

Moreno-Garrido, I., 2008. Immobilized microalgae: current techniques and uses (review). Bioresour. Technol. 99, 3949−3964.

Moreno-Garrido, I., Campana, O., Lubián, L.M., Blasco, J., 2005. Calcium-alginate immobilised marine microalgae: experiments on growth and short-term heavy metal accumulation. Mar. Pollut. Bull 51, 823−829.

Moreno-Garrido, I., Lubián, L.M., Blasco, J., 2007a. Sediment toxicity tests involving immobilized microalgae (*Phaeodactylum tricornutum* Bohlin). Environ. Int. 33, 481−495.

Moreno-Garrido, I., Lubián, L.M., Jiménez, B., Soares, A.M.V.M., Blasco, J., 2007b. Estuarine toxicity tests on diatoms: sensitivity comparison for three species. Estuarine Coastal Shelf Sci. 71, 278−286.

Moreno-Garrido, I., Robveille, N., Riba, I., DelValls, T.A., 2006. Toxicity of sediment from a mining spill to *Cylindrotheca closterium* (Ehremberg) Lewin and Reimann (Bacillariophyceae). Bull. Environ. Contam. Toxicol. 76 (1), 66−72.

Moreno-Garrido, I., Hampel, M., Lubián, L.M., Blasco, J., 2003a. Marine benthic microalgae *Cylindrotheca closterium* (Ehremberg) Lewin and Reimann (Bacillariophyceae) as a tool for measuring toxicity of linear alkylbenzene sulfonate (LAS) in sediments. Bull. Environ. Contam. Toxicol. 70 (2), 242−247.

Moreno-Garrido, I., Hampel, M., Lubián, L.M., Blasco, J., 2003b. Sediment toxicity tests using benthic marine microalgae *Cylindrotheca closterium* (Ehremberg) Lewin and Reimann (Bacillariophyceae). Ecotox. Environ. Safe 54, 290–295.

Munawar, M., Munawar, I.F., 1987. Phytoplankton bioassays for evaluating toxicity of *in situ* sediment contaminants. Hydrobiologia 149, 87–105.

Newman, M.C., Unger, M.A. (Eds.), 2003. Fundamentals of Ecotoxicology, second ed. CRC Press, Boca Ratón (Lewis Publishers).

Nyholm, N., 1990. Expresion of results from growth inhibition toxicity tests with algae. Arch. Environ. Contam. Toxicol. 19, 518–522.

OECD (Organization for the Economic Cooperation and Development), 1998. OECD Guideline for Testing of Chemicals: Alga, Growth Inhibition Test. OECD, Paris, France.

Ozturk, S., Aslim, B., Suludere, Z., 2010. Cadmium (II) sequestration characteristics by two isolates of *Synechocystis sp.* in terms of exopolysaccharide (EPS) production and monomer composition. Bioresour. Technol. 101, 9742–9748.

Pereira, A.M.M., Soares, A.M.V.M., Gonçalvez, F., Ribeiro, R., 2000. Water-column, sediment and in situ chronic bioassays with cladocerans. Ecotox. Environ. Safe. 47, 27–38.

Radakovitch, O., Roussiez, V., Ollivier, P., Ludwig, W., Grenz, C., Probst, J.L., 2008. Input of particulate heavy metals from rivers and associated sedimentary deposits on the Gulf of Lion continental shelf. Estuarine Coastal Shelf Sci. 77, 285–295.

Rand, G.M., 1995. Fundamentals of Aquatic Ecotoxicology, second ed. Ecological Services Inc., FL.

Rosa, R., Materatski, P., Moreira-Santos, M., Sousa, J.P., Ribeiro, R., 2012. A scaled-up system to evaluate zooplankton spatial avoidance and population immediate decline concentration. Environ. Tox. Chem. 31, 1301–1305.

Simpson, S.L., Batley, G.E., Chariton, A.A., Stauber, J.L., King, C.K., Chapman, J.C., Hyne, R.V., Gale, S.A., Roach, A.C., Maher, W.A., 2005. Handbook for Sediment Quality Assessment. CSIRO, Bangor, NSW.

Stauber, J.L., Florence, T.M., 1985. The influence of iron on copper toxicity to the marine diatom, *Nitzschia closterium* (Ehrenberg) W. Smith. Aquat. Tox. 6, 297–305.

Stauber, J.L., Florence, T.M., 1987. Mechanism of toxicity of ionic copper and copper complexes to algae. Mar. Biol. 94, 511–519.

U.S. EPA (US Environmental Protection Agency), 1971. Algal Assay Procedure: Bottle Test. National Environmental Research Centre, Corvallis, OR.

Ciguatera: Tropical Reef Fish Poisoning

Kirsten Heimann[1,2], Leanne Sparrow[1,2]

[1]James Cook University, College of Marine and Environmental Sciences, Townsville, QLD, Australia;
[2]James Cook University, Centre for Sustainable Fisheries and Aquaculture, Townsville, QLD, Australia

1. INTRODUCTION

Ciguatera (tropical reef fish poisoning) is the most reported seafood-associated illness in the world occurring regularly in tropical and subtropical regions (Donati, 2006). Despite the origin being tropical/subtropical, it is now well recognized that tropical reef fish exports make ciguatera a global, yet neglected disease. It is estimated to annually affect 50,000–500,000 people globally (Anderson and Lobel, 1987; Arena et al., 2004; Chinain et al., 2010). This 10-fold difference in estimates can be attributed to several reasons, including the geographic isolation of many tropical island nations, lack of medical and technical infrastructure, preference for local bush remedies, and not seeking medical assistance in mild cases. It does not help that no quick diagnostic ciguatera-specific test kits exist for either unequivocally diagnosing the illness or to enable testing for ciguateric fish either at the markets (prior to distribution to shops) or on the boat for commercial/recreational fishers.

Economic losses (hundreds of thousands of dollars) induced by ciguatera can be grouped into losses in (1) the primary industry through loss of faith in the product or market/export restriction for certain fish species from certain areas, (2) gastronomy/tourism in endemic ciguatera regions, and (3) the income of affected people (Bagnis et al., 1990; Botana, 2000; Lewis, 1986a,b). In addition, public health budgets are also being penalized for the diagnosis and treatment of ciguatera, as treatment costs are estimated to range from US$1850 to US$8950 per person (Arena et al., 2004). Less severe cases of ciguatera are often misdiagnosed as people only experience gastrointestinal symptoms that are often related to flu, viral, or general food poisoning illnesses—this is further impacted by travelers who are misdiagnosed after returning to nonendemic ciguatera regions. These factors have contributed to an estimated less than 10–20% of ciguatera cases being reported in endemic regions (Arena et al., 2004; Lewis, 2006). Reported cases of ciguatera have expanded into nonendemic regions throughout the world, particularly over the last 10 years, which has been linked not only to the increased affordability of travel but also to the increased demand for the import of frozen tropical fish supported with modern technological advances. Currently, preventative solutions such as an affordable and reliable clinical test kit for detection of ciguatoxins in fish and monitoring systems for ciguatera-causing dinoflagellates have not yet been developed. Prevention remains limited to restrictions and bans on fish species for commercial purchase—this method has been in use by the Sydney Fish Markets since the major outbreak of ciguatera in 1987 from fish purchased at this market. Although no other ciguatera outbreaks have occurred since the implication of these restrictions and bans, this method has negative implications for commercial fisheries in the tropical regions. The development of local marine fisheries in the Caribbean and Pacific regions is constrained by ciguatera with many dependent on export trade in addition to local consumption (Anderson and Lobel, 1987; Olsen et al., 1984). The local economy of the Pacific island nation of Kiribati was severely impacted after the loss of their export trade with Hong Kong due to an outbreak of ciguatera from their exported fish. The loss of this export trade closed the local fishery trade for Kiribati as well as caused a loss of AU$250,000 annual income to their local economy (Laurent et al., 2005). Yet, to truly reflect ciguatera-associated economic losses and health costs, a much more in-depth understanding of distribution and seasonality patterns, prevalence, and incident rates is required.

Fish become ciguateric (capable of inducing ciguatera in consumers) via trophic accumulation and bioconversion of toxins produced by dinoflagellates of the genus *Gambierdiscus* (Figure 1(a)), yet the trophic transfer routes are poorly understood or unambiguously documented. Likewise, it is presently still debated whether the severity of the illness is solely influenced by the dosage of toxin ingested and/or the toxin type or whether toxins produced by co-occurring dinoflagellates, such as species of *Prorocentrum*, *Coolia*, and *Ostreopsis* (Figure 1(b)–(d)) (Nishimura et al., 2014) influence severity and symptomology (Lewis, 2006; Tosteson et al., 1988). Regardless, ciguatera induced more intense research in areas with endemic ciguatera, and this has revealed that the genus *Gambierdiscus* is more complex than previously assumed. There are 11 additional species of *Gambierdiscus* (*Gambierdiscus australes*,

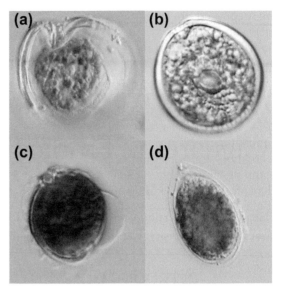

FIGURE 1　Toxic benthic dinoflagellates co-occurring with *Gambierdiscus* spp. Light micrographs of *Gambierdiscus* sp. (a) *Prorocentrum* (b), *Coolia* sp. (c) and *Ostreopsis* (d).

Gambierdiscus belizeanus, *Gambierdiscus carolinianus*, *Gambierdiscus carpenteri*, *Gambierdiscus caribaeus*, *Gambierdiscus exentricus*, *Gambierdiscus polynesiensis*, *Gambierdiscus pacificus*, *Gambierdiscus ruetzleri*, *Gambierdiscus scabrosus* sp. nov., and *Gambierdiscus yasumotoi*; Nishimura et al., 2014 and references therein) currently being recognized in addition to *Gambierdiscus toxicus*, the latter believed to be the main causative organism for ciguatera-poisoning events (Lewis, 2006). The genus, however, contains cryptic species, and it can be expected that future morphological and molecular analyses of field samples will add to the taxonomic diversity of this genus (Murray et al., 2014).

2. MICROALGAE CAUSING CIGUATERA

2.1 *Gambierdiscus* spp.: Gambiertoxin-Producing Dinoflagellates

Species of the genus *Gambierdiscus* are armored dinoflagellates, where cellulose plates are deposited in thecal vesicles between the inner and outer continuous membranes (the *Amphiesma*). The regions where the thecal plates overlap (the sutures) gives rise to a complex pattern of plate arrangements called tabulation, which is used to distinguish genera and species of armored dinoflagellates (Nishimura et al., 2014 and references therein). For species of the genus *Gambierdiscus*, however, differences in tabulation are very subtle, so that light- and scanning-electron microscopy alone is insufficient for unequivocal identification (Murray et al., 2014; Nishimura et al., 2014).

Cells of *Gambierdiscus* are typically photosynthetic, anteriorly/posteriorly compressed and lentil to ellipsoid in shape (except for *G. ruetzleri* and *G. yasumotoi*, where the cells are globular (Litaker et al., 2009)) containing a dorsally located u-shaped nucleus (Nishimura et al., 2014). The $2''''$ (1p) plate of the hypotheca and the $2'$ plate of the epitheca are useful morphological descriptors for differentiating between morphospecies of this genus (Litaker et al., 2009), in addition to differences in the apical pore complex (APC) (MacKenzie, 2008). The morphology of the $2''''$ plate allows for differentiation between two morpho groups of *Gambierdiscus*, *G. caribaeus*, *G. carolinianus*, *G. carpenteri*, *G. polynesiensis*, and *G. toxicus* having a broad $2''''$ plate with an average length-to-width ratio of 1.48, while *G. australes*, *G. belizeanus*, *G. pacificus*, and *G. scabrosus* have a narrow $2''''$ plate (mean L: W ratio 2.15) (Nishimura et al., 2014 and references therein). Morphology of the $2'$ plate reveals another grouping where *G. belizeanus*, *G. carolinianus*, *G. pacificus*, and *G. toxicus* are characterized by a hatchet-shaped plate, while *G. australes*, *G. caribaeus*, *G. carpenteri*, and *G. scabrosus* have a rectangular-shaped plate (Nishimura et al., 2014 and references therein). Combining these two plate morphologies allows for four groupings within the genus *Gambierdiscus*: *G. belizeanus*, *G. pacificus*, *G. ruetzleri*, and *G. yasumotoi* having a narrow $2''''$ plate and a hatchet-shaped $2'$ plate, *G. australes*, *G. scabrosus*, and

TABLE 1 Morphological Differentiation between Eight Species of *Gambierdiscus* Based on $2''''$ and $2'$ Plate Morphology

Gambierdiscus species	$2''''$ plate morphology	$2'$ plate morphology
G. belizeanus[a]	Narrow	Hatchet
G. pacificus[a]	Narrow	Hatchet
G. ruetzleri[b]	Narrow	Hatchet
G. yasumotoi[c]	Narrow	Hatchet
G. australes[a]	Narrow	Rectangular
G. scabrosus[a]	Narrow	Rectangular
G. excentricus[d]	Narrow	Rectangular
G. carolinianus[a]	Broad	Hatchet
G. toxicus[a]	Broad	Hatchet
G. polynesiensis[e]	Broad	Hatchet
G. caribaeus[a]	Broad	Rectangular
G. carpenteri[a]	Broad	Rectangular

[a]*Nishimura et al., 2014.*
[b]*Chinain et al., 1999.*
[c]*Holmes, M.J., 1998. Gambierdiscus yasumotoi sp. nov. (Dinophyceae), a toxic benthic dinoflagellate from southeastern Asia. J. Phycol. 34, 661–668.*
[d]*Fraga, S., Rodríguez, F., Caillaud, A., Diogéne, J., Raho, N., Zapata, M., 2011. Gambierdiscus excentricus sp. nov. (Dinophyceae), a benthic toxic dinoflagellate from the Canary Islands (NE Atlantic Ocean). Harmful Algae 11, 10–22.*
[e]*Litaker et al., 2009.*

Gambierdiscus excentricus also having a narrow $2''''$ plate, but a rectangular $2'$ plate, and *G. carolinianus*, *G. polynesiensis*, and *G. toxicus* having a broad $2''''$ and hatchet-shaped $2'$ plate, and *G. caribaeus* and *G. carpenteri* also having a broad $2''''$ but rectangular $2'$ plate (Table 1). As such, these two morphological characteristics allow forming four distinct groupings within the genus *Gambierdiscus*.

Further differentiation based on morphology between the species is difficult; however, the $3''$ plate (equal to the $4''$ plate in (Chinain et al., 1999; Faust, 1995; Litaker et al., 2009)) of the

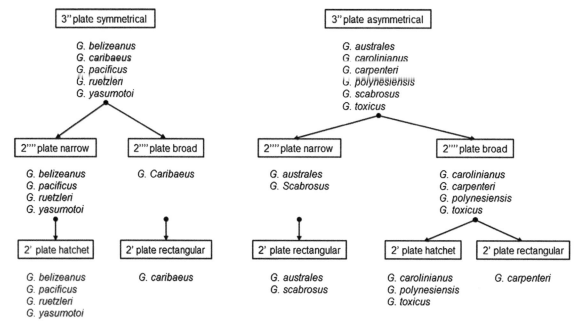

FIGURE 2   Plate morphological characteristics used to differentiate between species of *Gambierdiscus*.

epitheca is valuable to differentiate between *G. belizeanus*, *G. caribaeus*, *G. carpenteri*, and *G. scabrosus* being symmetrical in *G. caribaeus* and *G. belizeanus* but asymmetrical in *G. carpenteri* and *G. scabrosus* (Nishimura et al., 2014). Except for *G. exentricus*, for which no plate symmetry information for plate 3″ could be obtained, only two of the 12 described *Gambierdiscus* species can be unambiguously identified: *G. caribaeus* and *G. carpenteri* (Figure 2). Further differentiating characteristics are cell shape and size and additional details of the tabulation, but the APC details appear to be of little value (Litaker et al., 2009).

2.2 Occurrence of *Gambierdiscus* spp.

Gambierdiscus spp. are marine benthic species occurring on a variety of substrates ranging from life corals, dead corals, and sand grains to macroalgae. They are distributed throughout tropical and subtropical regions across the world

(Aligizaki and Nikolaidis, 2008; Litaker et al., 2009; Villareal, 2006), yet occurrences in temperate regions have been reported for Australia and Japan (Heimann et al., 2011; Nishimura et al., 2014). For example, *Gambierdiscus* sp. is now regularly found in the cold waters of Merimbula, New South Wales, Australia (Heimann et al., 2011), and *G. scabrosus* occurs in Itsumo, Kishimoto, Wakayama, Japan (Nishimura et al., 2014).

Whether or not these temperate distributions are temporary, for example, occurring only during the warmer weather and distributed there passively on warm ocean currents or whether these populations are permanent, for example, adapted to the cooler/colder winter periods is yet to be established, but the global warming of sea surface temperatures and the strengthening of, for example, the East Australian current can replenish these populations and assist their range expansions (Heimann et al., 2011). Although it is unknown whether species occurring in temperate regions produce toxins and which factors regulate toxin production, the wider range distribution of some species bears the risk of broadening the endemic range of occurrences of ciguatera.

Based on reported ciguatera cases, potential distribution hot spots can be identified (Donati, 2006). For Australia, high seasonal ciguatera risk exists at Bremer Island, East Woody Island, Cape Arnhem, and North East Island, and Connexion Island, northeast and northwest off Groote Eylandt, respectively (Donati, 2006). In contrast, although ciguateric fish are most frequently obtained from Queensland waters, only two high-risk areas have been identified for Spanish mackerel: Hervey Bay and Platypus Bay on the Hervey Bay side of Fraser Island (Heimann et al., 2008). Other hot spots for ciguatera are the Virgin Islands (UK), New Caledonia, French Polynesia, and Tuvalu, while other small islands have lower ciguatera incident rates (Heimann et al., 2008).

The validity of basing the distribution of *Gambierdiscus* spp. on the occurrences of ciguateric high-order predatory fish is questionable, as fish are typically not affected by the toxins, and buildup of the toxins, which persists for years in the fish flesh, could be a result of repeated ingestion of small amounts via their diet rather than accumulation originating from a *Gambierdiscus* bloom (high cell numbers per unit volume) event. Likewise, it remains to be resolved whether high abundances of *Gambierdiscus* spp. are a sound indication for ciguatera-risk areas, as these could simply be the result of reduced predation, which would in itself limit the trophic transfer of the toxins. In addition to these issues, it is known that the same species can be toxic or nontoxic, with the drivers that induce toxicity yet to be determined. Furthermore, not all species may be toxic, but in the absence of identified toxin genes, the potential for toxicity is hard to establish, especially since toxin production is typically not observed in cultured material.

2.3 Coinhabiting Dinoflagellates and Their Characteristics

Several species of *Gambierdiscus* typically occur together (McCormack, 2007), and the community also harbors other toxic, benthic dinoflagellates, such as species of *Prorocentrum, Coolia, Ostreopsis* (Figure 1), and *Amphidiniium* (Nishimura et al., 2014). *Coolia* spp. produce cooliatoxins, *Osteropsis* spp. ostreotoxin and palytoxin analogs, *Prorocentrum* spp. okadaic acid, and *Amphidinium* spp. hemolytic monoacyl-galactolipids (Botana, 2000; Tosteson et al., 1988).

Prorocentrum species are characterized by desmokont flagellation, where the transverse and longitudinal flagella arise apically (Heimann et al., 1995; Roberts et al., 1995), while all other species show dinokont flagellation, where the transverse and longitudinal flagella arise ventrally. In the athecate (unarmored) *Amphidinium*, the cingulum is apically displaced resulting in a very small epitheca with a nose-like appearance (Patterson and Burford, 2001). Unlike the thecate dinoflagellates, species identification of *Amphidinium* using plate tabulation is not possible, as the *Amphiesma* vesicles do not contain cellulose plates. In species of *Ostreopsis*, the thecal plates are very thin, but the epitheca and hypotheca are not noticeably different in size and cells are anteriorly/posteriorly compressed (Faust and Gulledge, 2002). In contrast, species of *Coolia* are more or less spherical with a median cingulum dividing the cell into an epitheca and hypotheca of roughly equal size (Momigliano et al., 2013). The genus *Coolia* contains many cryptic species, and morphological identification of species must be accompanied by molecular data for unambiguous identification (Momigliano et al., 2013). For example, the species *Coolia monotis* was described in 1919 and remained monotypic for 90 years before molecular data shed light on the true biodiversity of the genus (Momigliano et al., 2013).

3. CIGUATOXINS

Ciguatoxins are highly oxygenated lipid soluble cyclic (13 or 14 rings) polyether compounds and are heat stable (not destroyed through cooking or freezing), and different strains of *Gambierdiscus* produce different toxins that have varying levels of potency (Lewis, 2006). Ciguatoxins are bioconverted within the food web from the precursor toxin, gambiertoxin, the latter of which is produced by species of the genus *Gambierdiscus*. The organism and toxin were named after the Gambier Islands in French Polynesia where ciguatera is endemic with high incident rates (Heimann et al., 2008). The trophic conversion of gambiertoxin results in an array of ciguatoxins that differ in potency and induced symptomologies of ciguatera. Moreover, toxin levels vary between herbivorous and carnivorous fish in certain regions of French Polynesia (Bagnis et al., 1990; Lewis, 2006), indicating that toxin dose and storage play a fundamental role in ciguatera severity. The variation in toxin levels and profiles within different fishes is not well established and may depend on ratios within the diet, the fishes' digestive properties (e.g., adsorption, metabolism, and excretion of the toxins), and the oxidation state of the toxins (Lehane and Lewis, 2000).

3.1 How Many Different Ciguatoxins Exist?

Twelve different ciguatoxins have been identified to date showing geographic specificity (Lewis, 2006). C-CTX 1 and 2 are characteristic for the Caribbean with reported potencies of 3.6 and 1 µg kg^{-1}, respectively, while I-CTX 1 and 2 (0.5 µg kg^{-1}) are characteristic for the Indian Ocean, both being more toxic than their Caribbean counter parts. While there appears to be only two dominant ciguatoxins for the Caribbean and Indian Ocean, eight ciguatoxins are characteristic for the Pacific Ocean (P-CTX 1

and 2, P-CTX 3, P-CTX 3C, 2,3 dihydroxy P-CTX 3C, 51-hydroxy P-CTX 3C, P-CTX 4A and B with levels of toxicity being 0.25, 2.3, 0.9, 2, 1.8, 0.27, 2, and 4 µg kg^{-1}, respectively (Lewis, 2006)). Thus among the ciguatoxin derivatives, P-CTX 1, P-CTX 3, I-CTX 1 and 2, P-CTX 3, and C-CTX 2 being the most toxic ($\leq$1 µg kg^{-1}) and P-CTX 4B, C-CTX 1, P-CTX 1, P-CTX 3C, and P-CTX 4A being the least toxic ($\geq$2 µg kg^{-1}). Different toxins show strain and predator characteristic distributions with P-CTX 4A and B being derived from *G. toxicus* and found in herbivorous fish, while P-CTX 3C, PTX 4A and B are found within the dinoflagellate itself (Lehane and Lewis, 2000; Lewis, 2006), providing evidence that bioconversion of toxins is possible without trophic transfer. All other ciguatoxins have been found in carnivorous fish (Lewis, 2006).

3.2 Mechanism of Toxicity

Like the dinoflagellate brevetoxins (Heimann et al., 2012 and references therein), gambiertoxins and gambiertoxins exert their toxicity through interaction with voltage-gated sodium channels (Lewis et al., 2000). Sodium channels are membrane integral transmembrane proteins forming a sodium-selective pore within the plasma membrane for the passive diffusion of Na$^+$ along the concentration gradient. The sodium channel consists of approximately 2000 amino acids, which are organized into four homologous repeats (I–IV) where domain IV interacts with domain I to form the channel. Each repeat contains six α-helical segments (S1–S6) (Lewis et al., 2000). Binding of ciguatoxins to segment 5 via interaction with a receptor near the S5–S6 loop of domain IV, which is the same receptor used by brevetoxins, opens the sodium channel (Strachan et al., 1999). Competition experiments with CTX 1 and later with CTX 2 and 3 established that the receptor for ciguatoxin is the same as for brevetoxin, but the receptor had a 30 times greater affinity for CTX 1 over brevetoxin (Lehane and Lewis, 2000). In contrast, competition experiments with tetrodotoxin, the toxin of the puffer fish, abolished binding of ciguatoxins (Lewis et al., 2000). Other polyethers produced by *Gambierdiscus* spp. include gambierol and gamberic acid, which competitively interact with the voltage-gated sodium channel in the same way as brevenal, preventing ciguatoxin binding (MacKenzie, 2008). All *Gambierdiscus* species have been shown to produce potent water-soluble maitotoxins, which have no structural similarity to ciguatoxins but cause similar symptoms (MacKenzie, 2008), yet their poor accumulation in fish flesh and oral uptake efficiency predict that these toxins are less likely to contribute or induce ciguatera (Lewis, 2006), although they accumulate in large quantities in the gut of herbivorous fish (MacKenzie, 2008). The discovery that maitotoxins can be bioconverted to ciguatoxins (Hambright et al., 2014) and the production of maitotoxin precursors in other toxic benthic microalgae that are part of *Gambierdiscus* communities (see below), however, indicates that the above traditional view of the contribution of maitotoxins to ciguatera must be taken with caution. As explained below, it will be essential to conduct controlled feeding experiments to resolve the contribution of maitotoxins, as well as bioconversion routes and trophic vectors, with any degree of certainty.

Sodium influx upon ciguatoxin binding leads to membrane depolarization, which at a threshold level generates an action potential, which in severe cases can lead to severe nerve depolarization, which can explain the tingling and the sensation of temperature reversal (Lehane and Lewis, 2000). It has been shown though that ciguatoxins are also stored within fat tissues, but it is yet to be determined which mechanisms govern interactions with voltage-gated sodium channels versus storage. A possible route for storage would be that less oxidized and hence less potent ciguatoxins are diverted to storage while those with a high level

of oxidation and potentially a higher affinity for binding to the receptor of the voltage-gated sodium channel exert immediate toxicity effects with the level of severity depending on the concentrations ingested at an incident event.

4. BIOCONVERSION OF TOXINS AND TOXIN ACCUMULATION

4.1 Bioconversion of Ciguatoxins

It is thought that the interaction of environmental and physiochemical parameters in association with predation results in the sporadic and patchy distribution of *Gambierdiscus* populations. Ciguatoxin produced by *Gambierdiscus* cells also retains uncertainty in key parameters that drive toxin production. Ciguatoxin is hypothesized to accumulate along coral reef fish food chains resulting in higher order predatory fish, such as Spanish mackerel and coral trout, accumulating measurable concentrations of ciguatoxin (Lewis and Sellin, 1992). While trace metals have been shown to effect toxin production by benthic toxic dinoflagellates, *Prorocentrum lima* and *Ostreopsis siamensis* (Rhodes et al., 2006), this has not been explored with *Gambierdiscus* species. The influx of some trace metals (Figure 4), through either natural or anthropogenic events, may be influential in the production or inhibition of ciguatoxin thereby contributing to its sporadic occurrence within fish sourced from ciguateric-endemic tropical regions.

The bioconversion of gambiertoxin to ciguatoxins is not yet resolved, but it has been suggested that cytochromes catalyze the oxidation reactions in the liver of fish (Lehane and Lewis, 2000). It has been proposed that P-CTX 2 and CTX 4 are derived through oxidation from CTX 4A, while CTX 4B is the possible precursor toxin for CTX 3 and CTX 1 (Guzmán-Pérez and Park, 2000). The oxidative metabolism in the predators yields toxins within increasing polarity, and it appears that this is positively correlated with toxicity (Guzmán-Pérez and Park, 2000). In contrast, CTX 4B, P-CTX 3, and P-CTX 1 can be derived through a reduction of CTX 4A, P-CTX 2, and P-CTX 4, respectively (Guzmán-Pérez and Park, 2000).

While maitotoxins are traditionally not implicated in ciguatera (see below), it has been shown that these can be bioconverted to ciguatoxins by herbivorous fish and invertebrates (Hambright et al., 2014). This is disturbing as maitotoxin precursors are also produced by the co-occurring benthic, toxic dinoflagellates *Prorocentrum* spp., *Ostreopsis*, spp., *C. monotis*, and *Amphidinium carterae* (Hambright et al., 2014). It could thus be possible that maitotoxins are an additional inducer of ciguatera albeit causing different symptoms; they interact with voltage-gated calcium channels, thereby causing depolarization of the membrane and calcium influx. The Ca-ions in turn interact with smooth and skeletal muscle in vitro (Hambright et al., 2014). Activation of the voltage-gated calcium channels also leads to the secretion of hormones and neurotransmitters and the breakdown of phosphoinositides, which play an important role in regulating the function of integral membrane proteins (Hambright et al., 2014).

4.2 Potential Bioaccumulation Routes

There is a paucity of information on vectors for gambiertoxin and ciguatoxin biotransfer, their distribution, and seasonal abundances, as well as their mobility for direct uptake, conversion, and bioaccumulation (Heimann et al., 2008). The resolution of these processes is important for developing biomonitoring processes and to more comprehensively establish risk areas, risk diets, and risk groups in people, specifically in remote island locations. Most information comes from gut analyses of fish, ciguateric or otherwise, which can provide misleading information due to the potential for toxin storage and accumulation in fatty tissues. As *Gambierdiscus* spp. occur primarily on macroalgal

substrates, herbivorous fish have been traditionally implicated in the acquisition of toxin precursors and their partial oxidation to the more potent ciguatoxins (Cruz-Rivera and Villareal, 2006; Haloc et al., 1999; Heil et al., 2004). Macroalgae, however, display different levels of palatability to herbivorous fish, and *Gambierdiscus* loads can vary between patches and with macroalgal strain (Cruz-Rivera and Villareal, 2006), that is, unpalatable species of *Halimeda* spp., *Caulerpa* spp., *Chaetomorpha* spp., and *Laurencia* spp. can carry high loads of *Gambierdiscus* spp. (Heimann et al., 2008). As such, the contribution of unpalatable macroalgae with high loads of *Gambierdiscus* to trophic transfer requires further investigation. Another aspect is that present research focuses on high abundances of *Gambierdiscus* on their substrates, yet high abundances would suggest low grazing pressures (i.e., low ingestion rates). It will therefore be essential to correlate fish migration and the seasonality of *Gambierdiscus* abundance with potential vector (prey, food chain organisms) abundances and distributions to unambiguously resolve trophic transfer routes.

In the context of the above, the potential role and contribution of invertebrates have received little attention to date (Heimann et al., 2008), although herbivorous invertebrates could play a significant role in the trophic transfer of ciguatoxins (Lewis, 2006). Coral trout, grouper, mackerel, and snapper are sought-after table fish with a diet consisting of benthic crustaceans and bony fish (Heimann et al., 2008). For example, the trophic transfer of ciguatoxins to Spanish mackerel (*Scomberomerus commerson*) has been suggested based on gut content analysis to occur via the Alpheid shrimps and Saddle Grunt (*Pomodasys maculatus*) in Platypus Bay, Queensland, Australia (Lewis, 2006). In contrast, *Sphoeroides maculatus* (another Spanish mackerel) predates round herring (*Etrumerus teres*) and anchovy (*Anchoa hepsetus*) (both bony fish), which feed on planktonic crustaceans (Heimann et al., 2008). The latter case illustrates that gut content—based analyses are insufficient for the resolution of trophic transfer routes for ciguatoxins and call for direct feeding experiments. The latter are hampered by the fact that the level of toxicity of wild populations and the taxonomic mix of such communities are difficult to establish, while toxicity is typically not present in cultured communities. In addition, maintenance requirements for many invertebrate vectors are yet to be established, a research area in which aquaculture can make a significant contribution.

5. CIGUATERA SYMPTOMOLOGY

This illness remains diagnostic and is represented by an array of gastrointestinal, neurological, and cardiovascular symptoms. Diagnosis is further complicated as the severity is person and toxin dose dependent as well as variability in the time delay to onset of symptoms (generally 1—48 h after consuming fish), absence in sequence of symptoms experienced, and regional variation in symptoms. The toxins affect the digestive system, the cardiovascular and the neurological systems (Botana, 2000) can persist for months or even years and can be retriggered by the consumption of nonseafood-based foods,

such as chicken, peanuts, alcohol etc. (Donati, 2006). Digestive system symptoms are nausea, vomiting, diarrhea, and abdominal cramps, while cardiovascular symptoms are low blood pressure and an erratic pulse (Donati, 2006). Neurological symptoms are headache, sweating, convulsions, fatigue, fainting, intense itching and numbness, tingling, joint and muscle pain, audiovisual hallucinations, sensation of temperature cold-hot reversal, and muscular paralysis (breathing difficulties) (Donati, 2006).

In the Caribbean, gastrointestinal symptoms appear to be more frequent in contrast to the Pacific region where neurological symptoms predominate (Lewis, 2001). The average percentage of symptoms in reported ciguatera cases also indicates the dominance of gastrointestinal symptoms in the Caribbean (Figure 3(a)). The most frequently reported neurological symptoms include paresthesia of extremities, joint pain, muscle pain, temperature reversal, circumoral paresthesia, weakness, and fatigue (Figure 3(b)). With the exception of weakness and fatigue, the average percentage of these neurological symptoms in reported ciguatera cases supports the dominance within the Indo-Pacific regions including Australia (Figure 3(b)). These data support that toxin-dose but also the ciguatoxin type (see above) affects ciguatera symptomology.

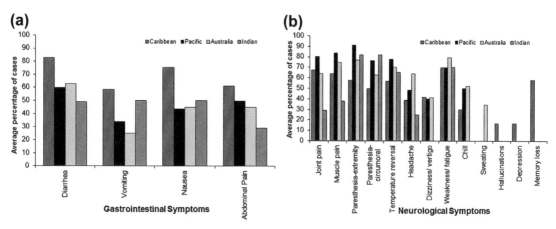

FIGURE 3   Averaged percentage of gastrointestinal symptoms (a) and neurological symptoms (b) in reported ciguatera cases.

6. GEOGRAPHIC RANGE OF CIGUATERA

Ciguatera-causing dinoflagellates have a distribution from 35°N to 35°S and typically do not grow to large populations (bloom, high cell numbers per volume) (Lucas et al., 1997). The geographic range of ciguatera is influenced by several factors, including population growth of *Gambierdiscus* spp. and other benthic toxic dinoflagellates, movement of ciguateric fish, as well as previously mentioned affordable travel to popular exotic locations and export of frozen tropical fish (see the chapter introduction).

The expansion and migration of *Gambierdiscus* populations can result in a temporary or permanent increase of ciguatera risk within endemic geographic boundaries as well as expanding into new geographic regions. In ciguatera-endemic regions, coral reefs are dynamic ecosystems with a number of parameters that can

be attributed to the temporary or permanent reversal of dominance from coral to macroalgal substrate. These parameters include natural disturbances such as cyclones, storms, crown-of-thorn plagues, and coral bleaching as well as man-made disturbance events including dredging, anchorage, overfishing, and pollutants (Figure 4). It is difficult to identify key parameters that influence the expansion of macroalgal substrates as quite often a reef ecosystem experiences more than one parameter within a short time frame.

It is thought that the dominance or expansion of macroalgal substrates is followed by the expansion of *Gambierdiscus* populations that in turn leads to increases in the incidence of ciguatera. An increase in incident rate has been observed 3 months following a population boom of *Gambierdiscus* cells (Chateau-Degat et al., 2005). An increase in ciguateric fish has also been observed following disturbance events, particularly herbivorous fish species.

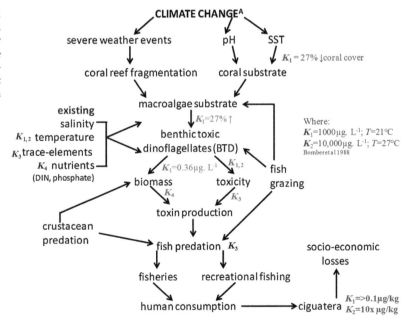

FIGURE 4 A hypothetical model of ciguatoxin bioaccumulation within a coral reef ecosystem under existing seasonal and climate change conditions. K_x: scenario under applied parameters to predict ciguatoxin accumulation in fish food chains.

There is insufficient information to support any correlation between *Gambierdiscus* population peaks and increases in ciguateric predatory fish.

Concurrent with the expansion of macroalgal substrates, it is known that *Gambierdiscus* populations require warm sea surface temperatures (SST) and adequate nutrients; however, the influence of other parameters such as the predation by fish, crustaceans, and other marine invertebrates as well as dinoflagellate community structure challenge understanding in the growth and toxicity of *Gambierdiscus* populations.

Migratory predatory fish species such as mackerel can extend the range of ciguatera incidence to outside endemic regions. In Queensland, Australia, Spanish mackerel migrate annually from northern Queensland down the coastline, through the Great Barrier Reef, to subtropical waters off the highly populated southeastern Queensland coastline. Spanish mackerel is a very popular table fish in Queensland, and the prevalence of this fish in the southeastern Queensland waters coincides with an increase in reported ciguatera cases in adjoining subtropical and temperate regions of Australia. In addition, transportation via ocean currents, ship ballast water, drifting macroalgal and other structures has thought to facilitate migration and expansion of *Gambierdiscus* populations into nonendemic geographic regions.

Warmer sea surface temperatures experienced through climate change fuel opportunities for dinoflagellate populations to expand their endemic range. Warm sea surface temperatures, protected bodies of water, and suitable benthic surfaces are required to optimize population growth, which can result in ciguatera hotspots due to an increased occurrence of ciguateric fish. The expansion of *Gambierdiscus* populations into nonendemic marine habitats may be associated with warmer sea surface temperatures, and populations have been documented in temperate regions over at least the last 10 years (Heimann et al., 2011).

References

Aligizaki, K., Nikolaidis, G., 2008. Morphological identification of two tropical dinoflagellates of the genera *Gambierdiscus* and *Sinophysis* in the Mediterranean Sea. J. Biol. Res.—Thessaloniki 9, 75–82.

Anderson, D.M., Lobel, P.S., 1987. The continuing enigma of ciguatera. Biol. Bull. 172, 89–107.

Arena, P., Levin, B., Fleming, L.E., Friedman, M.A., Blythe, D., 2004. A pilot study of the cognitive and psychological correlates of chronic ciguatera poisoning. Harmful Algae 3, 51–60.

Bagnis, R., Spiegel, A., N'Guyen, L., Plichart, R., 1990. Public health, epidemiological and socioeconomic patterns of ciguatera in Tahiti. In: Tosteson, T.R. (Ed.), Proceedings of the 3rd International Conference on Ciguatera Fish Poisoning. Polyscience, Quebec, pp. 157–168.

Botana, L.M., 2000. Seafood and Freshwater Toxins: Pharmacology, Physiology, and Detection. CRC Press.

Chateau-Degat, M.L., Chinain, M., Cerf, N., Gingras, S., Hubert, B., Dewailly, E., 2005. Seawater temperature, *Gambierdiscus* spp. variability and incidence of ciguatera poisoning in French Polynesia. Harmful Algae 4, 1053–1062.

Chinain, M., Darius, H.T., Ung, A., Fouc, M.T., Revel, T., Cruchet, P., Pauillac, S., Laurent, D., 2010. Ciguatera risk management in French Polynesia: the case study of Raivavae island (Australes Archipelago). Toxicon 56, 674–690.

Chinain, M., Faust, M.A., Pauillac, S., 1999. Morphology and molecular analyses of three toxic species of *Gambierdiscus* (Dinphyceae): *G. pacificus* sp. nov., *G. australes* sp. nov., and *G. polynesiensis* sp. nov. J. Phycol. 35, 1282–1296.

Cruz-Rivera, E., Villareal, T.A., 2006. Macroalgal palatability and the flux of ciguatera toxins through marine food webs. Harmful Algae 5, 497–525.

Donati, A.C., 2006. Ciguatera poisoning. In: Fishnote. Nothern Territory Government.

Faust, M.A., 1995. Observation of sand-dwelling toxic dinoflagellates (Dinophyceae) from widely differing sites, including two new species. J. Phycol. 31, 996–1003.

Faust, M.A., Gulledge, R.A., 2002. Identifying Harmful Marine Dinoflagellates. Department of Systematic Biology—Botany. National Museum of Natural History, Washington, DC.

Guzmán-Pérez, S.E., Park, D.L., 2000. Ciguatera toxins: chemistry and detection. In: Botana, L.M. (Ed.), Seafood and Freshwater Toxins Pharmacology, Physiology and Detection. Marcel Dekker Inc., USA.

Hales, S., Weinstein, P., Woodward, A., 1999. Ciguatera (fish poisoning), El Nino, and Pacific sea surface temperatures. Ecosys. Health 5, 20–25.

Hambright, K.D., Zamor, R.M., Easton, J.D., Allison, B., 2014. Algae. In: Wexler, P. (Ed.), Encyclopedia of Toxicology, third ed. Elsevier Inc., Academic Press, pp. 130–141.

Heil, C.A., Chaston, K., Jones, A., Bird, P., Longstaff, B., Costanzo, S., Dennison, W.C., 2004. Benthic microalgae in coral reef sediments of the southern Great Barrier Reef, Australia. Coral Reefs 23, 336—343.

Heimann, K., Capper, A., Sparrow, L., 2011. Ocean Surface Warming: impact on Toxic Benthic Dinoflagellates Causing Ciguatera. The Encyclopaedia of Life Sciences A23373. Wiley Publishing.

Heimann, K., Hansen, G., Roberts, K.R., 2012. Gymnodinium. ELS. John Wiley & Sons, Ltd.

Heimann, K., Roberts, K.R., Wetherbee, R., 1995. Flagellar apparatus transformation and development in *Prorocentrum micans* and *P. minimum* (Dinophyceae). Phycologia 34, 323—335.

Heimann, K., Sparrow, L., Blair, D., 2008. Report on reported ciguatera poisoning incidents, distribution, and seasonality based on analysis of database records in fishbase and compiled by the OzFoodNet Working Group. In: Heimann, K., Sparrow, L., Blair, D. (Eds.), Marine and Tropical Science Research Facility. Reef and Rainforest Research Centre Limited, Cairns, p. 73.

Laurent, D., Yeeting, B., Labrosse, P., Gaudechoux, J.P., 2005. Ciguatera: A Field Reference Guide. Secretariat of the Pacific Community, Noumea, New Caledonia.

Lehane, L., Lewis, R.J., 2000. Ciguatera: recent advances but the risk remains. Int. J. Food Microbiol. 61, 91—125.

Lewis, J., 2006. Ciguatera: Australian perspectives on a global problem. Toxicon 48, 799—809.

Lewis, N.D., 1986a. Disease and development: ciguatera fish poisoning. Soc. Sci. Med. 23, 983—993.

Lewis, N.D., 1986b. Epidemiology and impact of ciguatera in the Pacific: a review. Mar. Fish. Rev. 48, 6—13.

Lewis, R.J., 2001. The changing face of ciguatera. Toxicon 39, 97—106.

Lewis, R.J., Molgó, J., Adams, D.J., 2000. Ciguatera toxins: pharmacology of toxins involved in ciguatera and related fish poisoning. In: Botana, L.M. (Ed.), Seafood and Freshwater Toxins Pharmacology, Physiology and Detection. Marcel Dekker Inc., USA.

Lewis, R.J., Sellin, M., 1992. Short communications: multiple ciguatoxins in the flesh of fish. Toxicon 30, 915—919.

Litaker, R.W., Vandersea, M.W., Faust, M.A., Kibler, S.R., Chinain, M., Holmes, M.J., Holland, W.C., Tester, P.A., 2009. Taxonomy of *Gambierdiscus* including four new species, *Gambierdiscus caribaeus, Gambierdiscus carolinianus, Gambierdiscus carpenteri* and *Gambierdiscus ruetzleri* (Gonyaulacales, Dinophyceae). Phycologia 48, 344—390.

Lucas, R.E., Lewis, R.J., Taylor, J.M., 1997. Pacific ciguatoxin-1 associated with a large common-source outbreak of ciguatera in East Arnhem Land, Australia. Nat. Toxin. 5, 136—140.

MacKenzie, L., 2008. Ecobiology of the brevetoxin, ciguatoxin and cyclic imine producers. In: Botana, L.M. (Ed.), Sea food and Freshwater Toxins: Pharmacology, Physiology and Detection, second ed. CRC Press, Boca Raton, Florida, pp. 433—469.

McCormack, G., 2007. In: Trust, N.H. (Ed.), Cook Islands Biodiversity Database.

Momigliano, P., Sparrow, S., Blair, D., Heimann, K., 2013. The diversity of *Coolia* spp. (Dinophyceae, Ostreopsidaceae) in the Central Great Barrier Reef region. PLoS ONE 8, e79278.

Murray, S., Momigliano, P., Heimann, K., Blair, D., 2014. Molecular phylogenetics and morphology of *Gambierdiscus yasumotoi* from tropical eastern Australia. Harmful Algae 39, 242—252.

Nishimura, T., Sato, S., Tawong, W., Sakanari, H., Yamaguchi, H., Adachi, M., 2014. Morphology of *Gambierdiscus scabrosus* sp. nov. (Gonyaulacales): a new epiphytic toxic dinoflagellate from coastal areas of Japan. J. Phycol. 50, 506—514.

Olsen, D.A., Nellis, D.W., Wood, R.S., 1984. Ciguatera in the eastern Caribbean. Mar. Fish. Rev. 46, 13—18.

Patterson, D.J., Burford, M.A., 2001. A Guide to the Protozoa of Marine Aquaculture Ponds. CSIRO Publishing, Melbourne.

Rhodes, L., Selwood, A., McNabb, P., Briggs, L., Adamson, J., Van Ginkel, R., Laczka, O., 2006. Trace metal effects on the production of biotoxins by microalgae. Afr. J. Mar. Sci. 28, 393—397.

Roberts, K.R., Heimann, K., Wetherbee, R., 1995. The flagellar apparatus and canal structure in *Prorocentrum micans* (Dinophyceae). Phycologia 34, 313—322.

Strachan, L.C., Lewis, R.J., Nicholson, G.M., 1999. Differential actions if pacific ciguatoxin-1 on sodium channel subtypes in mammalian sensory neurons. J. Pharmacol. Exp. Ther. 288, 379—388.

Tosteson, T.R., Ballantine, D.L., Durst, H.D., 1988. Seasonal frequency of ciguatoxic barracuda in southwest Puerto Rico. Toxicon 26, 795—801.

Villareal, T.A., 2006. Macroalgal palatability and the flux of ciguatera toxins through marine food webs. Harmful Algae 5, 497—525.

Dunaliella Identification Using DNA Fingerprinting Intron-Sizing Method and Species-Specific Oligonucleotides

New Insights on *Dunaliella* Molecular Identification

Jorge Olmos Soto

Molecular Microbiology Laboratory, Department of Marine Biotechnology, Centro de Investigación Científica y de Educación Superior de Ensenada (CICESE), Ensenada, B.C., México

1. INTRODUCTION

Biflagellate and halotolerant species from the *Dunaliella* genus are the microalgae most ubiquitous in hypersaline environments (Massyuk, 1973). Taxonomic studies among *Dunaliella* have identified several new species since the initial identification by Teodorescos in 1905. However, the high plasticity in the green stage and the almost indistinguishable differences in the red phase make identification and differentiation of species very difficult and time-consuming (Olmos et al., 2000). The *Dunaliella* genus includes a reduced number of species that when exposed to elevated luminous intensities, high saline concentrations, and limited conditions of oxygen and nitrogen accumulate great amounts of β-carotene and glycerol (Ben-Amotz et al., 1982; Borowitzka, 1981). *Dunaliella salina* and *Dunaliella bardawil* are the most important species of the genus for β-carotene

production. Several investigations have demonstrated that both species produce more than 10% of the dry weight (Olmos-Soto et al., 2002). Lutein, chlorophyll, and other pigments and carotenoids are also produced by species of *Dunaliella*, under the same stressful environmental conditions (Guedes, 2011; Lordan et al., 2011; Skjanes et al., 2013). Lipids for aquaculture, human nutrition, and biodiesel production have also been investigated in *Dunaliella* species (Breuer et al., 2012; Da Silva et al., 2013; Fu et al., 2012).

There are 28 reported species in the *Dunaliella* genus, but only 23 have been accepted taxonomically. *D. tertiolecta*, *D. salina*, *D. bardawil*, and *D. parva* are the most studied and cultivated on a large scale for commercial and scientific applications. However, their identification is time-consuming and nonspecific, making the production of commercially important compounds difficult. For this reason, the application of molecular methodologies such as the intron-sizing method described here is a useful tool to simplify the specific identification of *Dunaliella* species (Olmos et al., 2009; Olmos-Soto et al., 2012).

2. GENERAL CHARACTERISTICS OF THE GENUS *DUNALIELLA*

The first description of a unicellular biflagellate red-colored alga living in concentrated brines was made in 1838 by Dunal, who reported occurrence of the organism we know today as *D. salina* in the salterns of Montpellier, on the Mediterranean coast of France (Dunal, 1838; Oren, 2005). Descriptions of *Dunaliella* (*Chlorophyta, Chlorophyceae, Chlamydomonadales, Dunaliellaceae*) as a new genus were presented in 1905 by Teodoresco from Bucharest using Romanian salt lake samples and by Clara Hamburger from Heidelberg using samples from Cagliari, Sardinia (Hamburger, 1905). The *Dunaliella* genus mainly includes halophilic species adapted to hypersaline environments (1–5 M NaCl), high luminosity (200–1000 µmol photons $m^{-2} s^{-1}$), high pH (8–9), and low concentrations of oxygen and nitrogen, with sizes varying 5–25 µm in length and 3–13 µm in width (Ben-Amotz et al., 1982; Borowitzka and Borowitzka, 1988; Chen and Jiang, 2009; Hejazi and Wijffels, 2003; Ramos et al., 2011). Due to the extreme environmental conditions in which these algae grow and because *Dunaliella* do not have cell walls but are only enclosed by a thin elastic plasma membrane, species of the genus present a vast morphological variability with respect to environmental conditions (Butcher, 1959; Lerche, 1937; Massyuk, 1973). The cells of *Dunaliella* can be ellipsoid, ovoid to almost spherical, pyriform, or fusiform. *Dunaliella* cells are radially symmetrical, bilateral, or slightly asymmetrical (Figure 1). Motile cells are biflagellate, with the flagella inserted at the anterior end of the cell and with flagella length varying between

species. There is a single, large posterior chloroplast occupying most of the cell volume. It is either cup-, dish-, or bell-shaped and contains a pyrenoid in the thickened basal part in all species except some of the freshwater species (Borowitzka and Siva, 2007).

Vegetative cell division in *D. salina* initiates with nuclear division that is followed almost immediately by an in-furrowing of the cell, usually first observed at the flagellar (anterior) end of the cell between the flagella and then, soon after, at the opposite (posterior) end of the cell (Borowitzka and Siva, 2007). Isogamic sexual reproduction in *Dunaliella* initiates with the fusion of two equally sized gametes to form a zygote (Hamburger, 1905; Massyuk, 1973; Teodoresco, 1905). Lerche in 1937 reported sexual zygote formation in five of the six species studied (*D. salina, D. parva, D. peircei, D. euchlora,* and *D. minuta*).

3. COMMERCIALLY IMPORTANT COMPOUNDS PRODUCED BY *DUNALIELLA*

Microalgae include a very diverse group of eukaryotic organisms that play important ecological–environmental roles and constitute a natural source of a variety of bioactive compounds and molecules for pharmaceutical, food, and cosmetic applications (Guedes, 2011; Paniagua-Michel et al., 2009). With fast growth rates and low production costs, microalgae provide useful cell factories for the production of valuable compounds and recombinant products like lipids, biofuels, vaccines, antibodies, and carotenoids (Fletcher et al., 2007; Greenwell et al., 2010; Mayfield et al., 2007; Skjanes et al., 2013; Vilchez et al., 2011).

The commercial production of β-carotene that comes from *Dunaliella* species has been the third most important microalgae industry since 1986. The companies involved in its production are mainly in Australia, Israel, India, Spain, China,

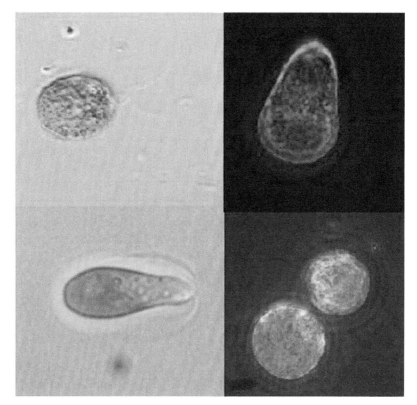

FIGURE 1   *Dunaliella* species isolated from environmental samples.

and the United States. The annual worldwide production of *Dunaliella* reported in 2006 was 1200 tons of dry weight (Spolaore et al., 2006). β-Carotene products derived from *Dunaliella* include extracts of pure β-carotene for medical and pharmaceutical use, *Dunaliella* powder for human food, and *Dunaliella* powder for animal feed. The prices of these *Dunaliella* products are between US$3000 and US$300/kg, respectively (Ben-Amotz, 2004).

Microalgae are also used as a source of biofuel such as biodiesel or bioethanol, and significant research has been done over the last several years in order to make conversion of microalgae biomass to fuel a viable process. A combined multidisciplinary process for using solar energy to capture CO_2 while producing biofuels and different high-value products has been presented (Skjanes et al., 2007, 2013). Production of biofuel from algae is dependent on the microalgal biomass production rate and lipid content (Mata et al., 2010). Both biomass production and lipid accumulation are limited by several factors, of which nutrients play a key role (Chen & Jiang, 2009). Microalgae-derived lipids are an alternative to vegetable and fossil oils, but lipid content and quality vary among microalgae strains (Da Silva et al., 2013). Selection and identification of a suitable strain for lipid production are therefore of paramount importance (Breuer et al., 2012). Additionally, palmitic, linolenic, and oleic acids account for more than 85% of the total fatty acid content of *D. salina* (Lordan et al., 2011).

4. THE *18S RDNA* GENE AS A MOLECULAR MARKER OF *DUNALIELLA* SPECIES IDENTIFICATION

The 16S/18S molecules have been utilized extensively for facilitating the identification of microorganisms and elucidating the confusion between species of different genera, and even from the same genus, especially when they look very similar (Amann et al., 2001; De Long et al., 1989; García and Olmos, 2007; Hernández and Olmos, 2006; Olmos-Soto et al., 2012; Olsen et al., 1986).

In more than a century that has passed since its formal description (Teodoresco, 1905), *Dunaliella* has become a convenient model organism for the study of salt adaptation in algae. The establishment of the concept of organic compatible solutes to provide osmotic balance was largely based on the study of *Dunaliella* species. Moreover, the massive accumulation of β-carotene by some strains under suitable growth conditions has led to interesting biotechnological applications (Oren, 2005). However, several *Dunaliella* strains in culture collections are misidentified, and unfortunately these names have been perpetuated in the literature, leading to difficulty in reconciling published information, resulting in potential confusion and loss of information, especially if no information is given on the strain being studied (Borowitzka and Borowitzka, 1988; Loeblich, 1982). In this sense, the utilization of molecular markers like the *18S rDNA* gene since the year 2000 has led to easy, fast, and precise identification and differentiation of *Dunaliella* species from environmental samples and culture collections (Olmos et al., 2000). Molecular characterization of the 18S between *Dunaliella* showed different sizes and sequence variability mainly on introns of β-carotene producer strains (Olmos-Soto et al., 2002). Therefore, the intron-sizing method is based on the size and also the sequence variability

presented by the 18S introns of *Dunaliella* species (Figure 2). These characteristics offered the possibility to develop a consistent and very powerful *18S rDNA*-fingerprinting technique to make an easy, fast, and precise identification of *D. tertiolecta*, *D. salina*, *D. bardawil*, and *D. parva*, obtained from natural samples or culture collections. With this methodology, neither cultivation nor exhausting taxonomic studies of the species are needed, avoiding large periods of purification and microscopic identification (Olmos et al., 2009). However, we do agree that a first correlation must be done appropriately between taxonomic and molecular characterization, to assign a unique and precise name to *Dunaliella* species (Borowitzka and Siva, 2007).

5. *DUNALIELLA* MA1-MA2 CONSERVED AND DSs-, DBs-, DPs- SPECIFIC OLIGONUCLEOTIDES

The classification of organisms based on conserved and specific sequences from the *16S/18S rDNA* is a common procedure in phylogenetic studies (Amann et al., 2001; DeLong et al., 1989; Olsen et al., 1986). In this sense, the identification of conserved sequences preferentially from the 5′ and 3′ regions of the *18S rDNA* opened the possibility to amplify by PCR the complete sequence of this gene. MA1 (forward) and MA2 (reverse) conserved oligonucleotides were designed from the beginning and the end, allowing the amplification of the 18S complete sequence from *Dunaliella* species (Olmos et al., 2000). MA1-MA2 18S amplification generated PCR products of ~1700 bp to *D. tertiolecta* (no introns), ~2100 bp to *D. salina* (one intron) and ~2500 bp to *D. bardawil* (two introns) and *D. parva* (two introns) (Table 1). The obtained results offered the opportunity to distinguish between *Dunaliella* strains with and without introns inside the *18S rDNA* gene (Figure 2). Introns from the analyzed strains presented an

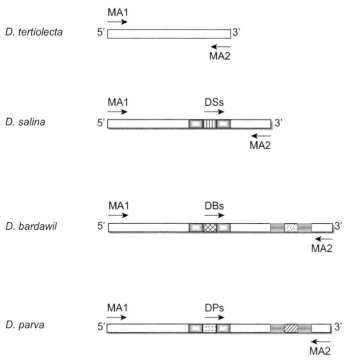

FIGURE 2   Graphical representation of *18S rDNA* gene in different species of *Dunaliella*. The arrows indicate the binding sites of conserved and specific primers. ▢ Intron I conserved sequence. ▦ Intron II conserved sequence. ▨▨▨ Intron different specific sequences. ▢ Exons conserved sequence.

approximate size of 400 bp and contain the most variable sequences of the *18S rDNA* genes, opening the possibility to design specific oligonucleotides to differentiate between *Dunaliella* species (Olmos et al., 2009; Olmos-Soto et al., 2012; Wilcox et al., 1992).

   D. salina and *D. bardawil* are β-carotene overproducer strains that in the green and red stages do not present enough taxonomic differences to be classified as two *Dunaliella* species (Borowitzka and Borowitzka, 1988). This lack of differences induced a disagreement between two of the most prestigious researchers in the field, which still remains. Additionally, in the gene bank of the NCBI a great confusion remains with respect to *Dunaliella* classification, due noncertification is being applied when sequences are submitted. Similarly, some *Dunaliella* strains from culture collections are misclassified and unnecessary names have emerged, increasing the confusion (Olmos et al., 2009). In this sense, the

finding of species-specific sequences contained exclusively on the *18S rDNA* introns created the opportunity to design species-specific oligonucleotides and to make an easy, fast, and precise identification of *Dunaliella* species, avoiding confusion and misclassification (Olmos-Soto et al., 2012). Thus, *D. salina*-specific (DSs), *Dunaliella bardawill*-specific (DBs) and *D. parva*-specific (DPs) oligonucleotides were designed selecting sequence variability localized precisely on their introns (Table 1). The utilization of each one of these specific oligonucleotides in combination with the MA2 conserved primer (DSs-MA2, DBs-MA2, DPs-MA2) produced PCR products with a specific fingerprinting profile for each of the analyzed species (Figure 2). Knowing the fingerprinting pattern for the species of *Dunaliella*, their identification in the green or red stage is not difficult because with the intron-sizing method we do not depend on the taxonomic characteristics but only need to know that the

TABLE 1　Sequence of the Primers Used to Amplify the *18S rDNA* Gene of *Dunaliella* and Size of the PCR-Amplified Products

	Primers	Sequence[a]	Strain	PCRp[b]
Conserved	MA1	5′ CTACTCATATCCTTCTCTC 3′		
	MA2	5′ CTTCTGCAGGTTCACC 3′		
	MA1-MA2		*D. tertiolecta*	~1700
	MA1-MA2		*D. salina*	~2100
	MA1-MA2		*D. bardawil*	~2500
	MA1-MA2		*D. parva*	~2500
Specific	DSs	5′ GCAGGAGAGCTAATAGGA 3′		
	DBs	5′ GGGAGTCTTTTTCCACCT 3′		
	DPs	5′ GTAGAGGGTAGGAGAAGT 3′		
	DSs-MA2		*D. salina*	~750
	DBs-MA2		*D. bardawil*	~1000
	DPs-MA2		*D. parva*	~1000

[a]*Does not include restriction sites.*
[b]*PCRp = PCR products.*

strain belongs to the *Dunaliella* genus. Furthermore, β-carotene producer species (*D. salina, D. bardawil,* and *D. parva*) coming from culture collection or natural samples are also identified comparing their fingerprinting profiles, and no cultivation nor taxonomic characterizations are needed (Olmos et al., 2009; Paniagua-Michel et al., 2009). Additionally, glycerols, polyunsaturated fatty acids, luteins, and other important pigments and carotenoids are mostly produced by the same *Dunaliella* overproducer species (Breuer et al., 2012; Lordan et al., 2011; Mata et al., 2010; Paniagua-Michel et al, 2012; Skjanes et al., 2013). Moreover, genome sequencing, development of molecular tools, and the construction of recombinant strains are being principally applied on *D. salina* (Smith et al., 2010). In this sense, the utilization of the intron-sizing method assures an easy, fast, and precise identification of the most commercially important *Dunaliella* species and avoids wasting time and a great deal of money.

6. IDENTIFICATION OF *DUNALIELLA* SPECIES FROM CERTIFIED CULTURE COLLECTIONS AND ENVIRONMENTAL SAMPLES

At the end of the 1990s, *D. bardawil* LB 2538 was obtained from the University of Texas (UTEX) the Culture Collection of Algae, the chromosomal DNA was purified, and a PCR amplification of the complete *18S rDNA* gene was done using the MA1-MA2 conserved primers (Olmos et al., 2000). The PCR product was evaluated to corroborate the presence of introns inside the 18S and to elucidate for the first time its complete sequence (AF150905). Two introns were localized to provide an 18S full size of ~2500 bp using the MA1-MA2 conserved primers. Additionally, DBs species-specific oligonucleotide designed from the first intron of *D. bardawil* presented a PCR product of ~1000 bp in combination with MA2 conserved primer (Olmos-Soto et al., 2002). In those times, we isolated some β-carotene hyperproducer strains of *Dunaliella* from places around Baja California, Mexico. One of these strains matched perfectly with the fingerprinting profile obtained from the *D. bardawil* strain isolated from Israel and deposited by Dr. Ben-Amotz (LB 2538). However, 10 years later we acquired the same strain from UTEX and, surprisingly, this amplified a PCR product of ~2100 bp utilizing the MA1–MA2 conserved primers and ~700 bp using the DSs-specific primer, indicating that we were working with *"Dunaliella salina var Teod"* instead of *D. bardawil* (Olmos et al., 2009). The amount of β-carotene produced by these two species is practically the same (~10%), and also their taxonomic characteristics; however, the growth conditions and environmental parameters required for carotenoids production are quite different between both species (in preparation). In this sense, β-carotene production in *D. bardawil* is more difficult to induce than in *D. salina*, therefore some isolates of *D. bardawil* have been identified and classified as non-β-carotene producers (Jayappriyan et al., 2011).

Microscopic analysis of *D. salina* 19/18 sample obtained from the Culture Collection of Algae and Protozoa (CCAP) showed two different strains of *Dunaliella*: one red and one green. Molecular fingerprinting determination using MA1–MA2 *18S rDNA* conserved primers from the 19/18 DNA sample presented two PCR products: one band of ~2100 bp that belongs to the β-carotene hyperproducer species of *D. salina*, and a second product of ~1700 bp from a *Dunaliella* species that never turns red-colored, indicating that the original sample contains other nonidentified *Dunaliella* strains (Olmos et al., 2009). In this sense, the intron-sizing method provides an easy, fast, and precise identification that is based on the *18S rDNA*-fingerprinting profile of *Dunaliella* species. Thus, this methodology could be recommended for its utilization by the algae culture collections mentioned above (UTEX, CCAP) to improve the certification of *Dunaliella* species that they are providing to the scientific community.

The precise identification of *Dunaliella* species from environmental samples has increased exponentially since the first publication of the MA1–MA2 conserved and DSs-, DBs-, DPs-specific primers (Olmos et al., 2000; Olmos-Soto et al., 2002). In this sense, we are going to mention some of the researchers from around the world who have been utilizing our conserved and specific oligonucleotides to identify its isolated strains (Elsaied, 2013; Raja et al., 2007; Sathasivam et al., 2012; Sharma et al., 2012). All of them concluded that the intron-sizing method facilitated the precise identification of *Dunaliella* species that came from environmental samples. Additionally, other researchers combine molecular identification (intron-sizing method) with taxonomical characterization of *Dunaliella* species, achieving excellent results (Jayappriyan et al., 2011; Mishra et al., 2008; Preetha et al., 2012; Tempesta et al., 2011; Wang et al., 2014). Furthermore, research such as that by Dr. Hejazi and coworkers when using the intron-sizing method has served to identify new species of *Dunaliella* with a different intron arrangement (Hejazi et al., 2010). Moreover, our conserved and specific oligonucleotide methods that have been done for the identification of microalgae strains have great potential for biofuel production (Davey et al., 2012; Pereira et al., 2011, 2013; Ramos et al., 2011; Yilancioglu et al., 2014).

7. CONCLUDING REMARKS

The production of commercially important compounds like pigments, carotenoids, lipids, glycerols, and other bioactive metabolites by *Dunaliella* and other microalgae genera represents a great opportunity to solve several needs of human consumption. However, the isolation and growth of the strains and the production and purification of their compounds are done at huge cost. Therefore, we need to assure the precise identification of the microalgal-isolated strains being used in order to avoid wasting both time and money. Molecular identification represents a common practice in microbiology where taxonomic characterization is complicated due to the size of the microbes. Additionally, molecular identification is used in forensic applications such as paternity tests. The intron-sizing method that is based on the *18S rDNA* fingerprinting profiles is a tool that was developed to allow an easy, fast, and precise *Dunaliella* species identification due to high phenotypic similarities between the species of the genus.

Acknowledgments

The authors wish to thank Rosalia Contreras Flores for figures and tables development.

References

Amann, R., Fuchs, B., Behrens, S., 2001. The identification of microorganisms by fluorescence in situ hybridization. Curr. Opin. Biotech. 12, 231–236.

Ben-Amotz, A., 2004. Industrial production of microalgal cell mass and secondary products—major industrial species: *Dunaliella*. In: Richmond, A. (Ed.), Handbook of Microalgal Culture. Blackwell Publishing, pp. 57–82.

Ben-Amotz, A., Katz, A., Avron, M., 1982. Accumulation of β-carotene in halotolerant algae: purification and characterization of β-carotene-rich globules from *Dunaliella bardawil* (Chlorophyceae). J. Phycol. 18, 529–537.

Borowitzka, L.J., 1981. The microflora: adaptations to life in extremely saline lakes. Hydrobiologia 81, 33–46.

Borowitzka, M.A., Borowitzka, L.J., 1988. Micro-Algal Biotechnology. Cambridge University Press, Cambridge.

Borowitzka, M.A., Siva, C.J., 2007. The taxonomy of the genus *Dunaliella* (Chlorophyta, Dunaliellales) with emphasis on the marine and halophilic species. J. Appl. Phycol 19, 567–590.

Breuer, G., Lamers, P., Martens, E., Draaisma, H., Wijffels, H., 2012. The impact of nitrogen starvation on the dynamics of triacylglycerol accumulation in nine microalgae strains. Bioresour. Technol. 124, 217–226.

Butcher, R.W., 1959. An Introductory Account of the Smaller Algae of British Coastal Waters. Part I: Introduction and *Chlorophyceae*. London: Ministry for Agriculture, Fisheries and Food, Fishery Investigations, Series IV. Her Majesty's Stationery Office.

Chen, H., Jiang, J.G., 2009. Osmotic responses of *Dunaliella* to the changes of salinity. J. Cell Physiol. 219, 251–258.

Da Silva, C.M., Gomez, A.D.A., Couri, S., 2013. Morphological and chemical aspect of Chlorella pyrenoidosa, *Dunaliella tertiolecta, Isochrysis galbana* and *Tetraselmis gracilis* microalgae. Nat. Sci. 5, 783–791.

Davey, P., Hiscox, W., Lucker, B., O'Fallon, J., Chen, S., Helms, G., 2012. Rapid triacylglyceride detection and quantification in live micro-algal cultures via liquid state 1H NMR. Algal Res. 1, 166–175.

DeLong, E.F., Wickham, G.S., Pace, N.R., 1989. Phylogenetic stains: ribosomal RNA-based probes for the identification of single cells. Science 243, 1360–1363.

Dunal, F., 1838. Extrait d'un mémoire sur les algues qui colorent en rouge certains eaux des marais salants méditerranéens. Ann. Sc. Nat. Bot. 9, 172.

Elsaied, H.E., 2013. Monitoring of uncultured *Dunaliella sp.* in an Egyptian solar saltern field based on RuBisCO-encoding gene. Afr. J. Biotechnol. 12, 5361–5369.

Fletcher, S.P., Muto, M., Mayfield, S.P., 2007. Optimization of recombinant protein expression in the chloroplasts of green algae. Adv. Exp. Med. Biol. 616, 90–98.

Fu, W., Magnúsdóttir, M., Brynjólfson, S., Palsson, B., Paglia, G., 2012. UPLC-UV-MSE analysis for quantification and identification of major carotenoid and chlorophyll species in algae. Anal. Bioanal. Chem. 404, 3145–3154.

García, T.A., Olmos, S.J., 2007. Quantification by fluorescent in situ hybridization of bacteria associated with *Litopenaeus vannamei* larvae in Mexican shrimp hatchery. Aquaculture 262, 211–218.

Greenwell, H.C., Laurens, L.M.L., Shields, R.J., Lovitt, R.W., Flynn, K.J., 2010. Placing microalgae on the biofuels priority list: a review of the technological challenges. J. R. Soc. Interface 7, 703–726.

Guedes, A.C., 2011. Microalgae as source of high added-value compounds: a brief review of recent work. Biotechnol. Progr. 27, 597–613.

Hamburger, C., 1905. Zur Kenntnis der *Dunaliella salina* und einer Amöbe aus Salinenwasser von Cagliari. Arch. F. Protistenkd. 6, 111–131.

Hejazi, M.A., Barzegari, A., Gharajeh, N.H., Hejazi, M.H., 2010. Introduction of a novel *18S rDNA* gene arrangement along with distinct ITS region in the saline water microalga *Dunaliella*. Sal. Syst. 6, 4.

Hejazi, M.A., Wijffels, R.H., 2003. Effect of light intensity on b-carotene production and extraction by *Dunaliella salina* in two-phase bioreactors. Biomol. Eng. 20, 171–175.

Hernández, Z.G., Olmos, S.J., 2006. Identification of bacterial diversity in the oyster *Crassostrea gigas* by fluorescent in situ hybridization and polymerase chain reaction. J. Appl. Microbiol. 100, 664–672.

Jayappriyan, K.R., Rajkumar, R., Rengasamy, R., 2011. Unusual occurrence of non carotenogenic strain of *Dunaliella bardawil* and *Dunaliella parva* in India. J. Basic Microb. 51, 473–483.

Lerche, W., 1937. Untersuchungen über Entwicklung und Fortpflanzung in der Gattung Dunaliella. Arch. F. Protistenkd. 88, 236–268.

Loeblich, L., 1982. Photosynthesis and pigments influenced by light intensity and salinity in the halophile *Dunaliella Salina* (Chlorophyta). J. Mar. Biol. Asso. UK. 62, 493–508.

Lordan, S., Ross, P., Stanton, C., 2011. Marine bioactives as functional food ingredients: potential to reduce the incidence of chronic diseases. Mar. Drugs 9, 1056–1100.

Massyuk, N.P., 1973. Morphology, Taxonomy, Ecology and Geographic Distribution of the Genus *Dunaliella Teod* and Prospects for its Potential Utilization Kiev. Naukova Dumka.

Mata, T.M., Martins, A.A., Caetano, N.S., 2010. Microalgae for biodiesel production and other applications: a review. Renew. Sust. Energ. Rev. 14, 217–232.

Mayfield, S.P., Manuell, A.L., Chen, S., Wu, J., Tran, M., Siefker, D., Muto, M., Marin-Navarro, J., 2007. *Chlamydomonas reinhardtii* chloroplasts as protein factories. Curr. Opin. Biotechnol. 18, 126–133.

Mishra, A., Mandoli, A., Jha, B., 2008. Physiological characterization and stress-induced metabolic responses of *Dunaliella salina* isolated from salt pan. J. Ind. Microbiol. Biotechnol. 35, 1093–1101.

Olmos, J., Paniagua, J., Contreras, R., 2000. Molecular identification of *Dunaliella sp.* utilizing the *18S rDNA* gene. Lett. Appl. Microbiol. 30, 80–84.

Olmos-Soto, J., Paniagua-Michel, J., Contreras, R., Trujillo, L., 2002. Molecular identification of β-carotene hyperproducing strain of *Dunaliella* from saline environment using species-specific oligonucleotides. Biotechnol. Lett. 24, 365–369.

Olmos, J., Ochoa, L., Paniagua-Michel, J., Contreras, R., 2009. DNA fingerprinting differentiation between β-carotene hyperproducer strain of *Dunaliella* from around the world. Sal. Syst. 5, 5.

Olmos-Soto, J., Paniagua-Michel, J., Contreras, R., Ochoa, L., 2012. DNA fingerprinting intron-sizing, and sensitive identification of carotenogenic *Dunaliella* species. In: Barredo, J.L. (Ed.), Microbial Carotenoids from Bacteria and Microalgae. Humana Press, New York, pp. 269–281. c/o Springer Science + Bussines Media, LLC.

Olsen, G.J., Lane, D.J., Giovannoni, S.J., Pace, N.R., Stahl, D.A., 1986. Microbial ecology and evolution: a ribosomal RNA approach. Annu. Rev. Microbiol. 40, 337–365.

Oren, A., 2005. A hundred years of *Dunaliella* research: 1905–2005. Sal. Syst. 1, 2.

Paniagua-Michel, J., Capa-Robles, W., Olmos-Soto, J., Gutierrez-Millan, L.E., 2009. The carotenogenic pathway via the isoprenoid β-carotene interference approach in a new strain of *Dunaliella salina* isolated from baja California Mexico. Mar. Drugs 7, 45–56.

Paniagua-Michel, J., Olmos-Soto, J., Acosta, R.M., 2012. Pathway of carotenoids biosynthesis in bacteria and microalgae. In: Barredo, J.L. (Ed.), Microbial Carotenoids from Bacteria and Microalgae. Humana Press, New York, pp. 1–12. c/o Springer Science + Bussines Media, LLC.

Pereira, H., Barreira, L., Mozes, A., Florindo, C., Polo, C., Duarte, C., Custódio, L., Varela, J., 2011. Microplate-based high throughput screening procedure for the isolation of lipid-rich marine microalgae. Biotechnol. Biofuels 4, 61–72.

Pereira, H., Barreira, L., Custódio, L., Alrokayan, S., Mouffouk, F., Varela, J., Abu-Salah, K., Ben-Hamadou, R., 2013. Isolation and fatty acid profile of selected microalgae strains from the red sea for biofuel production. Energies 6, 2773–2783.

Preetha, K., John, L., Subin, S.S., Vijayan, K.K., 2012. Phenotypic and genetic characterization of *Dunaliella (Chlorophyta)* from Indian salinas and their diversity. Aquat. Biosyst. 8, 27.

Raja, R., Hema, I.S., Balasubramanyam, D., Rengasamy, R., 2007. PCR-identification of *Dunaliella salina (Volvocales, Chlorophyta)* and its growth characteristics. Microbiol. Res. 162, 168–176.

Ramos, A., Polle, J., Tran, D., Cushman, J.C., Jin, E., Varela, J., 2011. The unicellular green alga *Dunaliella salina Teod.* as a model for abiotic stress tolerance: genetic advances and future perspectives. Algae 26, 3–20.

Sathasivam, R., Kermanee, P., Roytrakul, S., Juntawong, N., 2012. Isolation and molecular identification of β-carotene producing strains of *Dunaliella salina* and *Dunaliella bardawil* from salt soil samples by using species-specific primers and internal transcribed spacer (ITS) primers. Afr. J. Biotechnol. 11, 16677–16687.

Sharma, P., Agarwal, V., Mohan, M., Kachhwaha, S., Kothari, S., 2012. Isolation and characterization of *Dunaliella* species from sambhar lake (India) and its phylogenetic position in the genus *Dunaliella* using 18S rDNA. Natl. Acad. Sci. Lett. 35, 207–213.

Skjanes, K., Lindblad, P., Muller, J., 2007. BioCO2 – a multidisciplinary, biological approach using solar energy to capture CO_2 while producing H_2 and high value products. Biomol. Eng. 24, 405–413.

Skjanes, K., Rebours, C., Lindblad, P., 2013. Potencial for green microalgae to producer hydrogen, pharmaceuticals and other high value products in a combined process. Crit. Rev. Biotechnol. 33, 172–215.

Smith, D.R., Lee, R.W., Cushman, J.C., Magnuson, J.K., Tran, D., Polle, J., 2010. The *Dunaliella saline* organelle: large sequences, inflated with intronic and intergenic DNA. BMC Plant Biol. 10, 83.

Spolaore, P., Joannis-Cassan, C., Duran, E., Isambert, A., 2006. Commercial applications of microalgae. J. Biosci. Bioeng. 101, 87–96.

Tempesta, S., Paoletti, M., Pasqualetti, M., 2011. Morphological and molecular identification of a strain of the unicellular green alga *Dunaliella sp.* isolated from Tarquinia Salterns. Transit. Waters. Bull. 4, 60–70.

Teodoresco, E.C., 1905. Organisation et développement du *Dunaliella*, nouveau genre de Volvocacee-Polyblepharidée. Beih. Z. Bot. Centralbl. Bd XVIII, 215–232.

Vílchez, V., Forján, E., Cuaresma, M., Bédmar, F., Garbayo, I., Vega, J., 2011. Marine carotenoids: biological functions and commercial applications. Mar. Drugs 9, 319–333.

Wang, F., Feng, J., Xie, S., 2014. Phylogenetic and morphological investigation of a *Dunaliella* strain isolated from Yuncheng Salt Lake, China. Plant 2, 20–26.

Wilcox, L.W., Lewis, L.A., Fuerst, P.A., Floyd, G.L., 1992. Group I Introns within the nuclear-encoded small-subunit rRNA gene of three green algae. Mol. Biol. Evol. 9, 1103–1118.

Yilancioglu, K., Cokol, M., Pastirmaci, I., Erman, B., Cetiner, S., 2014. Oxidative stress is a mediator for increased lipid accumulation in a newly isolated *Dunaliella salina* strain. PLoS ONE 9, e91957.

Index

Note: Page numbers followed by "b", "f" and "t" indicate boxes, figures and tables respectively.

Printed and bound by CPI Group (UK) Ltd, Croydon, CR0 4YY

08/05/2025

01864996-0001